Electrical Machine Fundamentals with Numerical Simulation using MATLAB/SIMULINK

Electrical Machine Fundamentals with Numerical Simulation using MATLAB/SIMULINK

Atif Iqbal
Qatar University
Doha
Qatar

Shaikh Moinoddin
Aligarh Muslim University
Aligarh
India

Bhimireddy Prathap Reddy
Qatar University
Doha
Qatar

WILEY

This edition first published 2021

Registered Office(s)
John Wiley & Sons, Inc., 111 River Street, Hoboken, NJ 07030, USA
John Wiley & Sons Ltd, The Atrium, Southern Gate, Chichester, West Sussex, PO19 8SQ, UK

Editorial Office
The Atrium, Southern Gate, Chichester, West Sussex, PO19 8SQ, UK

For details of our global editorial offices, customer services, and more information about Wiley products visit us at www.wiley.com.

Wiley also publishes its books in a variety of electronic formats and by print-on-demand. Some content that appears in standard print versions of this book may not be available in other formats.

Library of Congress Cataloging-in-Publication Data

Names: Iqbal, Atif, author. | Moinoddin, Shaikh, 1954- author. | Reddy, Bhimireddy Prathap, 1993- author.
Title: Electrical machine fundamentals with numerical simulation using MATLAB/SIMULINK / Atif Iqbal, Shaikh Moinoddin, Bhimireddy Prathap Reddy.
Description: Hoboken, NJ : Wiley, 2021. | Includes bibliographical references and index.
Identifiers: LCCN 2020051131 (print) | LCCN 2020051132 (ebook) | ISBN 9781119682639 (cloth) | ISBN 9781119682653 (adobe pdf) | ISBN 9781119682660 (epub)
Subjects: LCSH: MATLAB. | SIMULINK. | Electric machinery.
Classification: LCC TK2000 .I76 2021 (print) | LCC TK2000 (ebook) | DDC 621.31/042028553–dc23
LC record available at https://lccn.loc.gov/2020051131
LC ebook record available at https://lccn.loc.gov/2020051132

Cover Design: Wiley
Cover Image: © nadla/Getty Images

Set in 9.5/12.5pt STIXTwoText by SPi Global, Chennai, India

10 9 8 7 6 5 4 3 2 1

Dedicate this book to my teacher of Electric Machine Late Prof. Salman Beg, Late Parents (Syed M. Naseer Alam and Akbari Begum), Late Parent in-laws (Syed Jafer Imam and Zainab Jafer), my Elder Sisters Nishat Fatma and Arfa Tabbasum, Elder Brother Asif Iqbal, Wife Shadma Jafer, Sons S.M. Abuzar Iqbal and Abu Baker Iqbal Daughter Noorin Iqbal and Niece Erina Iqbal—Prof Atif Iqbal

Dedicate this book to my wife Meher Fatema, son Fakhruddin, and daughters Sayeeda Fatima, Umme Salma, daughter-in-law Saba Haris and son-in-laws Sabir Ali, and Adnan Ahmad.—Dr Shaikh Moinoddin

Dedicate this book to my parents (Smt. Anusuyamma, Shri Venkatarami Reddy) and gurus (especially Prof. Atif Iqbal and Prof. Sivakumar and all gurus), my brothers (Ashok, Nagaraj and Madhu) and sisters-in-law (Latha, Shilpa and Swapna)—Dr. Bhimireddy Prathap Reddy

Contents

Preface

Electric machines are taught in all universities and colleges in Electrical/Electronics/Computer/Mechanical/Chemical Engineering degree programmes. This book elaborates on the fundamental concepts of electrical machines along with several numerical problems and simulation models using Matlab/Simulink. We have been teaching this course for more than 25 years and developed some easy ways to understand the concept. This book contains a great deal of analytics with adequate physical explanations supported with many numerical examples. We have conveyed the electrical machine concepts using simple language that students will find easy to follow, which is the main aim of our book. The major highlights of the book include:

- Simple and easy to understand the language
- Clear elaboration of Fundamental concepts
- Adequate analytical treatment
- Extensive numerical problems
- Matlab/Simulink simulation models for all machine types discussed
- Wide coverage of the topics taking into consideration the syllabus of most of international universities
- Concept of Multiphase (more than three-phase)
- Introduction to Finite Element Method of Analysis of Electric Machines.

Electrical Motors are bulk energy consumers and almost 70% of the electricity generated worldwide is used by them. Electrical machines is a general term used for both electrical generators and electrical motors. Broadly classified, electrical machines can be either AC or DC. A special machine called the 'universal machine' can be operated by both AC and DC. Electrical energy is invariably generated using electrical generators called synchronous generators which are found in conventional electrical power plants. Synchronous generators are available in different ratings ranging from kWs to MWs. Electrical motors are used in almost every industry and household application, in transportation such as electrical or diesel traction, electric and hybrid electric vehicles, ship propulsion and 'more electric aircrafts' etc. Modern-day society relies upon electrical machines. Due to the importance of electrical machines, this subject is taught as a core course in the discipline of Electrical Engineering. In addition, it is also taught Mechanical/Chemical/Petrochemical/Computer and Electronic Engineering students across the globe. In some places, it takes the form of a one-semester course while in other institutions, it is a two-semester course. The machines are taught at different levels; fundamental and advanced level courses. This book is aimed at covering the fundamental concepts of machines; however, it will also be of use at the advanced level.

Advancement in technology has led to automation of processes and control. Nowadays, digital computer control is invariably used for model and design verification and also for real-time system control. The most popular software package that is used and taught is Matlab/Simulink. The software is highly user friendly and offers extensive simulation tools. The book provides easy to understand models for all types of machines. The models will provide an in-depth understanding of the concepts and students can also design and analyse the machines on their own. This book provides clarity on the concepts of such machines. The basic electrical engineering laws that required for understanding the concepts are shown and discussed with adequate examples in the first chapter. The book is built following a logical sequence with discussion of transformers (static electric machine), followed by the basic concepts of rotating electrical machines and the winding design. Three-phase induction machine is elaborated on followed by synchronous machines and then single-phase induction machines. DC machines are discussed at length where first-generator action is explained followed by a discussion on motoring action. Each machine is accompanied by its Matlab/Simulink model. Some basic numerical problems are given in each chapter which will aid in the understanding of the underlying principles. One chapter is dedicated to advanced electrical machines such as the switched reluctance machine, brushless DC machine, permanent magnet machines and their variations. Moreover, this book describes the importance of the multiphase machines (more than three phases of the machine); and simulation of electrical machines with Finite Element Method (FEM) analysis. This book is unique in the sense that it covers almost all types of machines within one single volume encompassing theoretical concepts, analytical analysis, extensive numerical problems and simulation models.

To motivate graduating students towards the research in electric machines, in this textbook, in addition to the traditional three-phase machines, advanced electrical machines such as multiphase machines are also presented. In this chapter, authors are presenting the necessity for and advantages of multiphase phases, associated winding structures, control, modelling etc., which are not included in other textbooks.

Moreover, the basics related to simulation of electrical machines with FEM are also presented. FEM analysis mainly helps to analyse flux distribution, forces acting on the machine, mmf, flux density, etc., which are not possible with Matlab/Simulink.

Most importantly, the book describes the modelling procedure to implement electrical machines in a Matlab/Simulink environment. This is a major distinctive feature of the book. Matlab/Simulink Models for Chapters 3, 5, 6, 7, 8, and 9 are provided as additional resources, available via the Instructor Companion Site on the book's home page on www.wiley.com\go\iqbal\electricalmachine

Readers can modify the given models (as it is open source) and study and analyse the performance of electric machines. This is a highly beneficial additional learning resource.

Thus, the single volume of this book fulfils the requirements of the basic and advanced courses of electrical machines in most universities. We hope this book will provide a handy tool for students in both undergraduate and graduate courses.

Prof. Atif Iqbal
Dr. Shaikh Moinoddin
Dr. Bhimireddy Prathap Reddy

Acknowledgements

We acknowledge the help and support of our graduate students and our research assistants who assisted in preparing this book. We wish to especially thank Dr. Pandav Kiran Maroti, Dr. Salman Ahmed and Mr. Marif Duala for their contributions. Mr. Marif has mostly contributed in Chapters 5 and 8. Most of the diagrams have been drawn by Dr. Kiran as he possesses excellent skills and has provided nice figures. Dr. Salman helped in Chapter 7 on the section of voltage regulation for synchronous machine. We would like to thank Prof. Adel Gastli and Prof Lazhar Ben-Brahim, Dept. of Electrical Engineering, Qatar University, some of the figures and texts are used in the book from their lecture notes. We would like to thank Prof. Imtiaz Ashraf, Dept. of Electrical Engineering, Aligarh Muslim University, Aligarh, India for his encouragement in writing this book. Our thanks also go to Prof. Haitham Abu-Rub, Texas A&M University at Qatar, Doha, Qatar for his constant support in all academic endeavors. We also wish to thank our family members for their help and support.

1

Fundamentals of Electrical Machines

1.1 Preliminary Remarks

The study of the construction of electrical machines, along with the principles and laws on which they operate, gives the knowledge to predict behaviour and characteristics of the machines. Therefore, the study of electric machines requires an understanding of basic electromagnetic fields. The purpose of this chapter is to give some familiarity with the principles and laws that help understand the principle of Electric Machinery.

1.2 Basic Laws of Electrical Engineering

1.2.1 Ohm's Law

The current through an ideal *resistance* element is directly proportional to the voltage drop across it, as shown in Figure 1.1.

Ideal resistance element here refers to the element without any parasitic inductance or capacitance.

The relationship between voltage and current in an ideal resistor is given in Eq. (1.1) in which v is in volts, i is in amps, and the constant of proportionality is resistance R measured in ohms (Ω).This simple formula is known as Ohm's law in honour of the German physicist, Georg Ohm, whose original experiments led to this incredibly useful and important relationship.

$$v = Ri \tag{1.1}$$

Notice that voltage v is measured across the resistor. That is, it is the voltage at point A with respect to the voltage at point B. When the current is in the direction shown, the voltage at A with respect to B is positive, so it is quite common to say that there is a *voltage drop* across the resistor.

An equivalent relationship for a resistor is given in Eq. (1.2), where the current is given in terms of voltage and the proportionality constant is conductance G with units of Siemens (S). In older literature, the unit of conductance was mhos.

$$i = Gv \tag{1.2}$$

Electrical Machine Fundamentals with Numerical Simulation using MATLAB/SIMULINK, First Edition.
Atif Iqbal, Shaikh Moinoddin, and Bhimireddy Prathap Reddy.

Companion website: www.wiley.com/go/iqbal/electricalmachine

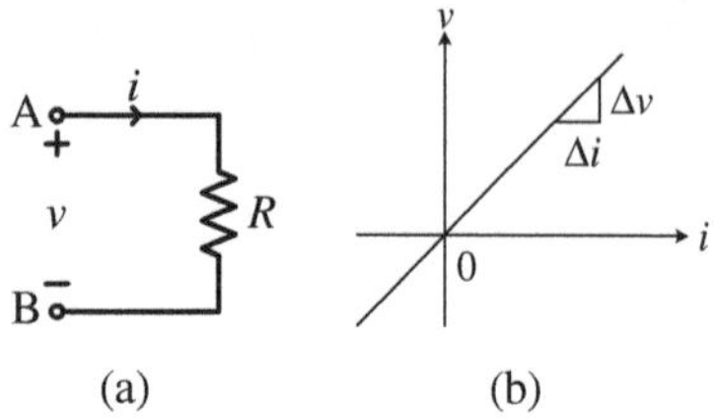

Figure 1.1 (a) An ideal resistor symbol. (b) voltage–current relationship.

Magnetic field problems involving components such as current coils, ferromagnetic cores, and air gaps can be solved as magnetic circuits according to the analogous behaviour of the magnetic quantities to the corresponding electric quantities in an electric circuit.

1.2.2 Generalization of Ohm's Law

Generalization of Ohm's law means expressing the law in terms of **E** (electric field in V/m) and **J** (current density in Amp/m^2), rather than V and I. Consider a length l of a conductor of uniform cross-sectional area A through which a current I flows. In general, the electrical resistance of the conductor is proportional to its length, l, and inversely proportional to its cross-sectional area, A. Thus, resistance can be expressed as

$$R = \rho \frac{l}{A} = \frac{1}{\sigma} \frac{l}{A} \tag{1.3}$$

Here, the constants ρ and σ are called the *resistivity* and *conductivity* of the conducting medium, respectively, and are measured in units of ohm-meters and (ohm-meter)$^{-1}$. Therefore, Ohm's law is stated as

$$V = \rho \frac{l}{A} I \tag{1.4}$$

But, $I/A = J_z$ (assuming that the conductor is aligned along the z-axis) and $V/l = E_z$, hence, Eq. (1.4) modifies to

$$E_z = \rho J_z \tag{1.5}$$

Equation (1.5) can be generalized as

$$\mathbf{E} = \rho \mathbf{J} \tag{1.6a}$$

or

$$\mathbf{J} = \sigma \mathbf{E} \tag{1.6b}$$

where σ is the conductivity of the medium. Equations (1.6a) and (1.6b) are the vector form of Ohm's law.

1.2.2.1 Derivation of Eq. (1.6)

Let a unit volume of conductor contains n number of free electrons. In the presence of electric field **E**, the free electrons accelerate with drift velocity

$$\mathbf{v}_d = -\frac{q\tau}{m_e} \mathbf{E} \tag{1.7}$$

where q, τ, and m_e are charged in coulomb, relaxation time in seconds and mass of electron respectively. The total charge of electrons $(-nq)$ is moving with velocity $(\mathbf{v}_d)$. Therefore, the current density

$$\mathbf{J} = (-nq)\,\mathbf{v}_d = \frac{nq^2\tau}{m_e}\mathbf{E} = \sigma\mathbf{E} \tag{1.8}$$

Thus comparing Eqs. (1.6b) and (1.8), the conductivity can be defined as

$$\sigma = \frac{1}{\rho} = \frac{ne^2\tau}{m_e} \tag{1.9}$$

Let P be the power dissipated per unit volume inside the conducting medium. P is given as

$$\mathrm{P} = \mathbf{j}\cdot\mathbf{E} = \rho \mathrm{j}^2 \left(\left(\mathrm{Amp/m^2}\right).\left(\mathrm{V/m}\right) = \mathrm{Volt.Amp/m^3} = \mathrm{Watt/m^3}\right) \tag{1.10}$$

1.2.3 Ohm's Law for Magnetic Circuits

This law is also called as Hopkinson's law, after *John Hopkinson*, but was formulated earlier by *Henry Augustus Rowland* in 1873. It states that

$$\mathfrak{F} = \phi\mathfrak{R} \tag{1.11}$$

where $\mathfrak{F}$, ϕ, and $\mathfrak{R}$ are magneto motive force (MMF) in Amp-turn (A.t), flux in Wb and reluctance in A.t/Wb, respectively.

1.2.4 Kirchhoff's Laws for Magnetic Circuits

As in electric circuits, Kirchhoff's laws can be applied in magnetic circuits because Kirchhoff's voltage and Kirchhoff's current laws are analogous to Ampere's law and Gauss Law, respectively.

Magnetic circuits obey other laws of electric circuits such as reluctances in series $\left(\mathfrak{R}_{total} = \mathfrak{R}_1 + \mathfrak{R}_2 + \mathfrak{R}_3 + \ldots\right)$ and parallel $\left(\frac{1}{\mathfrak{R}_{total}} = \frac{1}{\mathfrak{R}_1} + \frac{1}{\mathfrak{R}_2} + \frac{1}{\mathfrak{R}_3} + \ldots\right)$

A space containing a quasi-stationary electric or magnetic field may be partitioned into *flux tubes:* geometrical figures in which all lines of flux are perpendicular to their bases and no lines of flux cut their sides; see Figure 1.2.

Lines of the equal magnetic scalar potential, u, are perpendicular to lines of flux, ϕ; therefore, the bases of a flux tube are equipotential planes. The magnetic scalar potential difference between the bases equals the MMF drop. Generally, the ratio of potential difference at the ends of the flux tube that contains no current and the flux through it, is a function of flux tube geometry and the characteristics of the medium. Mathematically, this ratio is equal to:

$$R = \int_0^l \frac{dx}{c(x)A(x)} \tag{1.12}$$

Figure 1.2 A flux tube in a field [1].

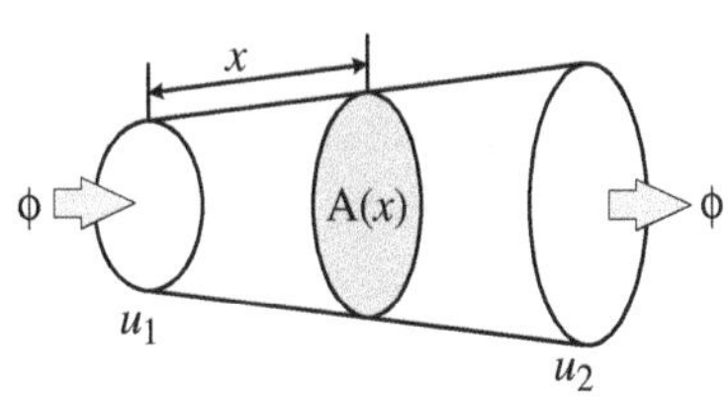

where l is the total flux tube length, A is its cross-sectional area, and c is a function of material properties. The quantity R is defined as reluctance in magnetic fields and is analogous to resistance in electric fields and the inverse of capacitance in electrostatic fields. R is a function of field quantities and also a function of geometry. The quantity c is equal to the flux tube material permeability, μ, for magnetic fields. For electric fields, c is equal to the flux tube conductivity, σ, while for electrostatic fields c is the permittivity, ϵ. For the three field types, one can write:

$$R_{magnetic} = \int_0^l \frac{dx}{\mu(x) A(x)} = Reluctance \tag{1.13}$$

$$R_{electrostatic} = \int_0^l \frac{dx}{\varepsilon(x) A(x)} = Inverse\ of\ capacitance \tag{1.14}$$

$$R_{electric} = \int_0^l \frac{dx}{\sigma(x) A(x)} = Resistance \tag{1.15}$$

The analogy between the three types of fields is almost complete, except in one detail: to maintain a certain energy level in a magnetic and an electrostatic field, no support from outside the field is needed. On the contrary, in an electric field, there must be a current from a source to cover losses. Electrostatic and magnetic fields can store energy, whereas in an electric field, the complete energy is irreversibly converted into heat. If all quantities in a field are constant, then stored energy in a magnetic or an electrostatic field is constant. Under the same circumstances, the energy lost in an electric field increases in proportion to time. A comparison of the field is tabulated in Table 1.1.

By comparing expressions for stored energy in electrostatic and magnetic fields, one can answer the question of why the magnetic field is chosen as an electromechanical energy conversion medium. Assuming the same level of field strength in both types of fields in free space, one can store μ_0/ϵ_0, or approximately 1.42×10^5, more energy per volume in a magnetic than in an electrostatic field. Stored energy per volume is representative of electro-mechanical energy conversion capability and, therefore, the choice of the magnetic field as a medium is obvious.

Table 1.1 Analogies between different types of fields [1].

	Type of field		
	Magnetic	**Electrostatic**	**Electric**
Material Constant	μ	ϵ	Σ
Flux or current density	$\overline{B}$	$\overline{D}$	$\overline{J}$
Field strength	$\overline{H}$	$\overline{E}$	$\overline{E}$
Reluctance, capacitance, resistance	$R_m = \int_0^l \frac{dx}{\mu(x)A(x)} = \frac{N^2}{L}$	$R_{es} = \int_0^l \frac{dx}{\varepsilon(x)A(x)} = \frac{1}{C}$	$R_{el} = \int_0^l \frac{dx}{\sigma(x)A(x)}$
Potential difference	$\mathfrak{F} = \int_0^l \overline{H} \cdot d\overline{l}$	$V = \int_0^l \overline{E} \cdot d\overline{l}$	$V = \int_0^l \overline{E} \cdot d\overline{l}$
Flux, charge, current	$\phi = \int \overline{B} \cdot d\overline{A} = \frac{\mathfrak{F}}{R_m}$	$Q = \int \overline{D} \cdot d\overline{A} = \frac{V}{R_{es}} = VC$	$I = \int \overline{J} \cdot d\overline{A} = \frac{V}{R_{el}}$
Energy	$\int \mathfrak{F} d\phi$	$\int VdQ$	$\int IVdt$

Further comparison between electric and magnetic fields shows that magnetic circuits almost always operate in a saturated, i.e. nonlinear mode, while most elements in electric circuits are linear. There is also a large difference in the ratio of material constants between non-conducting and conducting media in magnetic and electric fields. The electrical conductivity of air is typically more than six orders of magnitude less than the conductivity of a current-carrying medium while, in a magnetic field, the permeability of air is hardly ever less than four orders of magnitude smaller than that of the flux-carrying medium. The ratio of air to medium permeability increases as the magnetic medium becomes more saturated. Therefore, in a magnetic field, one must take into account the non-negligible part of the flux that goes through the air, a phenomenon that is unknown in an electric field.

1.2.5 Lorentz Force Law

Electric and magnetic fields are set up by moving charged particles or current flow. This is described by Maxwell's equations, but the force acting on the moving charge q in the presence of magnetic fields can be found using Lorentz force law. It can be stated as

$$\mathbf{F} = q\mathbf{v} \times \mathbf{B} \tag{1.16}$$

or

$$F = qvB \sin\theta \tag{1.17}$$

where q is the charge of particles moving with a velocity of $\mathbf{v}$ in a magnetic field whose density is $\mathbf{B}$. The magnitude of the force is equal to the product of velocity, flux density, and sine of the angle between $\mathbf{v}$ and $\mathbf{B}$. The direction of the force can be determined by right corkscrew rule as shown in Figure 1.3. It can be seen that the direction of force is perpendicular to the v-B plane. One Newton of force will be exerted when a charge of 1 coulomb is moving with a velocity of 1 m/s in a magnetic field having a flux density of 1 T with an angle between velocity and flux density of 90°.

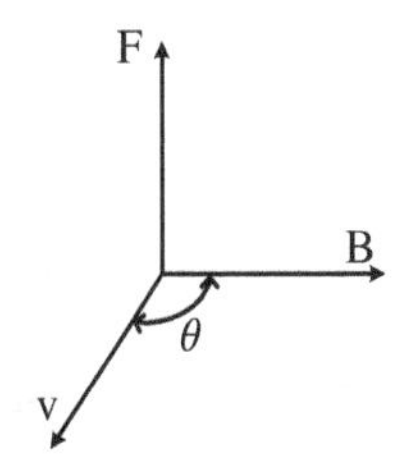

Figure 1.3 Force F, Flux density B, and velocity v vectors 90° displaced.

Equation (1.16) can be written in an incremental form as

$$d\mathbf{F} = dq\mathbf{v} \times \mathbf{B} = dq\frac{d\mathbf{l}}{dt} \times \mathbf{B} = \frac{dq}{dt}\mathbf{l} \times \mathbf{B} = id\mathbf{l} \times \mathbf{B} \tag{1.18}$$

Equation (1.18) shows that the moving charges constitute a line current and only have mathematical significance because, practically, $id\mathbf{l}$ itself does not exist. So, the force can be found by integrating Eq. (1.18).

$$\mathbf{F} = \oint id\mathbf{l} \times \mathbf{B} \tag{1.19}$$

Equations (1.18) and (1.19) are very useful in the analysis and design of electric motors. If the angle between $d\mathbf{l}$ and $\mathbf{B}$ is 90°, then Eq. (1.19) becomes

$$F = ilB \tag{1.20}$$

One can observe the following points when charged particles of various types are moving in a magnetic field:

(i) The magnetic force F is proportional to charge q and velocity v of the particle.
(ii) No force will be experienced by the conductor when it moves parallel to the field.
(iii) The direction of force is perpendicular to both **v** and **B** assuming that the angle between them is not equal to zero.
(iv) The direction of the magnetic force will reverse if the nature of the charge (positive or negative) is reversed.
(v) The magnetic force F is also proportional to $\sin\theta$ where θ is the angle between **v** and **B**.

The differences between Electric ($F_e = qE$) and magnetic forces are:

(i) The direction of the electric force is in the direction of the electric field but the direction of the magnetic force is perpendicular to the magnetic field (Lorentz Force law).
(ii) The electric force acts on a stationary or moving charged particle whereas the magnetic force acts when the charge is moving in a magnetic field but not in parallel to the magnetic field.
(iii) There is work done by the electric force by displacing the charged particle whereas there is no work done by the magnetic force by displacing the charged particle because the angle between the direction of movement of the particle and the magnetic force is 90°.

1.2.6 Biot-Savart Law

Biot and Savart (pronounced 'Bee-oh' and 'Suh-var') law states that magnetic field intensity (dH) at any point P in free space is proportional to the product of differential current segment idl and sine of the angle between the line segment and the line joining the segment and the point; and inversely proportional to the squire of the distance between the line segment and the point as shown in Figure 1.4. The direction of dH can be found using the right-handed corkscrew rule (Figure 1.5a) or right-hand rule (Figure 1.5b). The directions of dH at different points are shown in Figure 1.5c.

$$dH \propto \frac{idl \sin \alpha}{R^2}$$

where R is the distance between the point and the segment and α is the angle between dl and R.

$$dH = k\frac{idl \sin \alpha}{R^2} \tag{1.21}$$

Here k is constant of proportionality which equals to $^1/_{4\pi}$ in SI units.

$$k = \frac{1}{4\pi}$$

$$dH = \frac{1}{4\pi}\frac{idl \sin \alpha}{R^2} \tag{1.22}$$

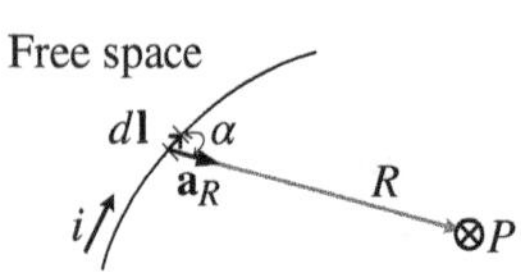

Figure 1.4 Magnetic field intensity $d\mathbf{H}$, directed into the page, due to $id\mathbf{L}$.

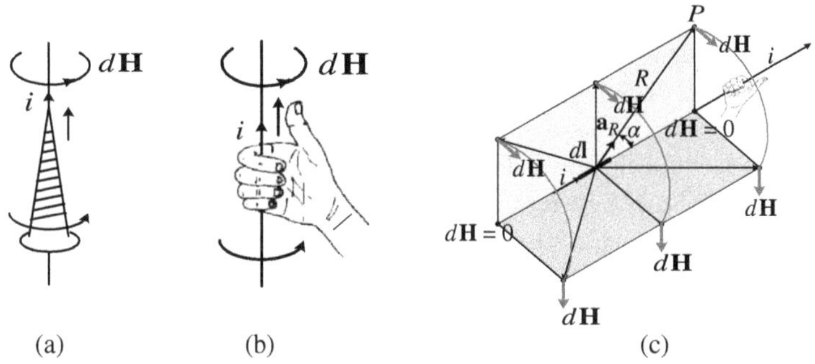

Figure 1.5 (a) Right-handed corkscrew rule (b) Right-hand rule (c) Directions of *dH* at various points.

The equation above can be represented in vector cross-product form

$$d\mathbf{H} = \frac{id\mathbf{l} \times \mathbf{a}_R}{4\pi R^2} = \frac{id\mathbf{l} \times \mathbf{R}}{4\pi R^3} \tag{1.23}$$

where $d\mathbf{l}$ and $\mathbf{R}$ are differential line vector and vector from $d\mathbf{l}$ to point P respectively; $R = |\mathbf{R}|$ and the unit vector $\mathbf{a}_R = \frac{\mathbf{R}}{|\mathbf{R}|}$.

Equation (1.23) can also be written for magnetic flux density as

$$d\mathbf{B} = \mu_o \frac{id\mathbf{l} \times \mathbf{a}_R}{4\pi R^2} = \mu_o \frac{id\mathbf{l} \times \mathbf{R}}{4\pi R^3} \tag{1.24}$$

Integrating Eq. (1.24), one gets the magnitude of **B** which is constant around the circle, with a radius of *R*, centring the conductor when the current flowing through a long conductor as

$$B = \frac{\mu_o i}{2\pi R} \tag{1.25}$$

Example 1.1 ***Magnetic Field Surrounding a Thin, Straight Conductor***
Consider a thin, straight wire of finite length carrying a constant current *I* and placed along the *x*-axis as shown in Figure E1.1 [2]. Determine the magnitude and direction of the magnetic field at point *P* due to this current.

Figure E1.1 (Example 1.1) (a) A thin, straight wire carrying a current *I*. (b) The angles θ_1 and θ_2 used for determining the net field.

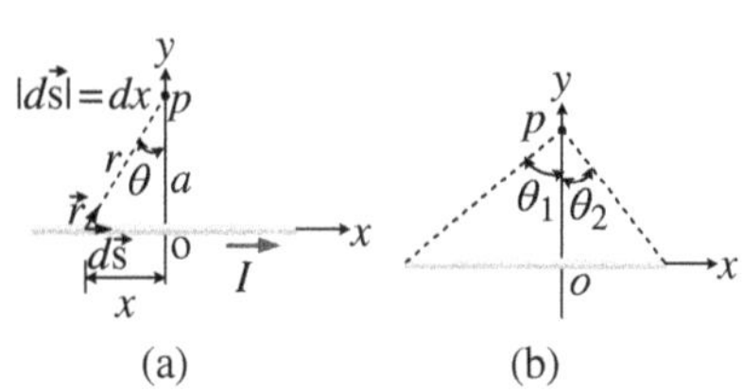

Solution
From the Biot–Savart law, we expect that the magnitude of the field is proportional to the current in the wire and decreases as the distance '*a*' from the wire to point *P* increases. We also expect the field to depend on the angles θ_1 and θ_2 in Figure E1.1b. We place the origin at *O* and let point *P* be along the positive *y*-axis, with $\hat{y}$ being a unitvector pointing out of the page.

We are asked to find the magnetic field due to a simple current distribution, so this example is a typical problem for which the Biot–Savart law is appropriate. We must find the field contribution from a small element of current and then integrate over the current distribution.

Let's start by considering a length element $d\vec{s}$l located a distance r from P. The direction of the magnetic field at point P due to the current in this element is out of the page because of $d\vec{s} \times \vec{r}$ is out of the page. In fact, because *all* current elements $Id\vec{s}$ lie in the plane of the page, they all produce a magnetic field directed out of the page at point P. Therefore, the direction of the magnetic field at point P is out of the page and we need only find the magnitude of the field.

Evaluate the cross product in the Biot–Savart law:

$$d\vec{s} \times \vec{r} = |d\vec{s} \times \vec{r}|\,\hat{y} = \left[dx \sin\left(\frac{\pi}{2} - \theta\right)\right]\hat{y} = (dx \cos\theta)\,\hat{y}$$

Substitute for Eq. (1.24):

$$d\vec{B} = (dB)\,\hat{y} = \frac{\mu_o I}{4\pi}\frac{dx \cos\theta}{r^2}\hat{y} \tag{1.26}$$

From the geometry in Figure E1.1a, express r in terms of θ:

$$r = \frac{a}{\cos\theta} \tag{1.27}$$

Notice that $\tan\theta = -\frac{x}{a}$ from the right triangle in Figure E1.1a (the negative sign is necessary because $d\vec{s}$ is located at a negative value of x) and solve for x:$x = -a\tan\theta$

Find the differential dx:

$$dx = -a\sec^2\theta d\theta = -\frac{ad\theta}{\cos^2\theta} \tag{1.28}$$

Substitute Eqs. (1.27) and (1.28) into the expression for the z component of the field from Eq. (1.26):

$$dB = -\frac{\mu_o I}{4\pi}\left(-\frac{ad\theta}{\cos^2\theta}\right)\left(\frac{1}{\left(\frac{a}{\cos\theta}\right)^2}\right)\cos\theta = -\frac{\mu_o I}{4\pi a}\cos\theta d\theta \tag{1.29}$$

Integrate Eq. (1.29) overall length elements on the wire, where the subtending angles range from θ_1 to θ_2 as defined in Figure E1.1b:

$$B = -\frac{\mu_o I}{4\pi a}\int_{\theta_1}^{\theta_2}\cos\theta d\theta = \frac{\mu_o I}{4\pi a}\left(\sin\theta_1 - \sin\theta_2\right) \tag{1.30}$$

We can use this result to find the magnitude of the magnetic field of *any* straight current-carrying wire if we know the geometry and hence the angles θ_1 and θ_2. Consider the special case of an infinitely long, straight wire. If the wire in Figure E1.1b becomes infinitely long, we see that $\theta_1 = \frac{\pi}{2}$ and $\theta_2 = -\frac{\pi}{2}$ for length elements ranging between positions $x = -\infty$ and $x = +\infty$. Because $\left(\sin\theta_1 - \sin\theta_2\right) = \left(\sin\frac{\pi}{2} - \sin\left(-\frac{\pi}{2}\right)\right) = 2$,

Equation (1.30) becomes

$$B = \frac{\mu_o I}{2\pi a} \tag{1.31}$$

Equations (1.30) and (1.31) both show that the magnitude of the magnetic field is proportional to the current, and decreases with increasing distance from the wire, as expected. Equation (1.31) has the same mathematical form as the expression for the magnitude of the electric field due to a long-charged wire.

Example 1.2 A long thin conductor is carrying a current of 1.5 A. Calculate the magnetic field strength at a distance of 20 cm from the wire.

Solution

The magnetic field is given by

$$B = \frac{\mu_0 I}{2\pi r} = \frac{(4\pi \times 10^{-7}\ T.m/A)\,(1.5\,A)}{2\pi\,(0.20\,m)} = 1.5 \times 10^{-6}\ T$$

Example 1.3 The magnitude of magnetic field of a large tornado was measured as 14 Nano T, pointing north when the Tornado was 10 km east of the measurement observatory. Calculate the current carried up or down the funnel of the Tornado. The vortex can be modelled as a long, straight conductor carrying a current.

Solution

Model the tornado as a long, straight, vertical conductor and imagine grasping it with the right hand so the fingers point northward on the western side of the tornado (that is, at the observatory's location). The thumb is directed downward, meaning that the conventional current is downward. The magnitude of the current is found from $B = \mu_0 I/2\pi r$ as

$$I = \frac{2\pi r B}{\mu_0} = \frac{2\pi\,(10 \times 10^3\ m)\,(1.40 \times 10^{-8}\ T)}{4\pi \times 10^{-7}\ T.m/A} = 700\,A$$

Thus, the current is 700 A, downward.

Example 1.4 **A square conducting loop is formed by a wire of sides** $l = 0.4$ m that carries a current $I = 5$ A as shown in Figure E1.4a.

(a) Compute the magnitude and direction of the magnetic field at the centre of the conducting loop.
(b) Compute magnitude and direction of the magnetic field at the centre, if the loop is formed in a circular shape with same current.

Figure E1.4a Square-conducting loop for Example 1.4.

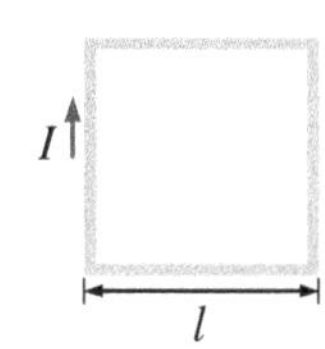

Solution

(a) Use Eq. (1.30) for the field produced by each side of the square.

$$B = \frac{\mu_0 I}{4\pi a}\,(\sin\theta_1 - \sin\theta_2)$$

Where $\theta_1 = 45^o$, $\theta_2 = -45^o$, and $a = \frac{l}{2}$

Figure E1.4b Square-conducting loop for Example 1.4.

Each side produces a field into the page. The four sides altogether produce

$$B_{centre} = 4B = 4\frac{\mu_o I}{4\pi a}\left(\sin\theta_1 - \sin\theta_2\right) = \frac{\mu_o I}{\pi\frac{l}{2}}\left[\sin 45^o - \sin\left(-45^o\right)\right] = \frac{2\mu_o I}{\pi l}\left[\frac{2}{\sqrt{2}}\right]$$

$$= \frac{2\sqrt{2}\mu_o I}{\pi l} = \frac{2\sqrt{2}\left(4\pi\times10^{-7}\ T.m/A\right)(5.0\,A)}{\pi\,(0.4\,m)} = \sqrt{2}\times10^{-5}\ T$$

$$= 14.15\ \mu T \text{ into the page}$$

(b) For a single circular turn with $4l = 2\pi R$, using equation $B = \frac{\mu_o I}{2R}$ for B at centre of the turn having radius of R,

$$B = \frac{\mu_o I}{2R} = \frac{\mu_o \pi I}{4l} = \frac{\pi\left(4\pi\times10^{-7}\ T.m/A\right)(5.0\,A)}{4\,(0.4\,m)} = 12.33\ \mu T \text{ into the page}$$

Example 1.5 In Niels Bohr's hydrogen atom model, an electron circles the proton at a distance of 5.29×10^{-11} m with a speed of $2.19\times10^{6}\ m/s$. Calculate the strength of the magnetic field this motion produces at the proton.

Solution

Treat the magnetic field as that produced in the centre of a ring of radius R carrying current I: from equation for the field is $B = \frac{\mu_o I}{2R}$.

The current due to the electron is

$$I = \frac{\Delta q}{\Delta t} = \frac{e}{2\pi\frac{R}{v}} = \frac{ev}{2\pi R}$$

so, the magnetic field is

$$B = \frac{\mu_o I}{2R} = \frac{\mu_o}{2R}\left(\frac{ev}{2\pi R}\right) = \frac{\mu_o}{4\pi}\frac{ev}{R^2} = \left(\frac{\left(4\pi\times10^{-7}\ T.m/A\right)}{4\pi}\right)$$

$$\times\frac{\left(1.6\times10^{-19}\ C\right)\left(2.19\times10^{6}\ m/s\right)}{\left(5.29\times10^{-11}\ m\right)^2} = 12.5\ T$$

Example 1.6 A conducting loop is formed in circular form and two straight conductors as shown in Figure E1.6. The radius of circular loop $R = 10.0$ cm. The wire lies in the plane of the paper and carries a current $I = 2.00$ A. Find the magnetic field at the centre of the loop.

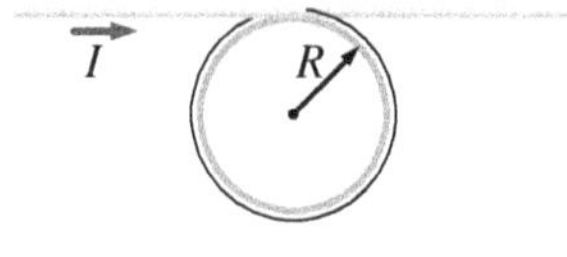

Figure E1.6 Circular-conducting wire for Example 1.6.

Solution

We can think of the total magnetic field as the superposition of the field due to the long, straight wire, having a magnitude $\frac{\mu_o I}{2\pi R}$ and directed into the page, and the field due to the circular loop, having a magnitude $\frac{\mu_o I}{2R}$ and directed to the page. The resultant magnetic field is:

$$\vec{B} = \left(1 + \frac{1}{\pi}\right)\frac{\mu_o I}{2R} = \left(1 + \frac{1}{\pi}\right)\frac{\left(4\pi \times 10^{-7}\, T.\frac{m}{A}\right)(2.0\, A)}{2\,(0.10\, m)} = 5.52 \times 10^{-6}$$
$$= 4\,\pi T \textit{ into the page}$$

Example 1.7 The conducting loop of Figure E1.6 is of radius R with two long, straight sections. The wire lies in the plane of the paper and carries a current I. (a) What is the direction of the magnetic field at the centre of the loop? (b) Find an expression for the magnitude of the magnetic field at the centre of the loop.

Solution

We can think of the total magnetic field as the superposition of the field due to the long, straight wire, having a magnitude $\frac{\mu_o I}{2\pi R}$ and directed into the page, and the field due to the circular loop, having a magnitude $\frac{\mu_o I}{2R}$ and directed to the page. The resultant magnetic field is:

$$\vec{B} = \left(1 + \frac{1}{\pi}\right)\frac{\mu_o I}{2R} \text{ directed into the page}$$

Example 1.8 Consider two long, straight, parallel conductors that carry currents directed perpendicular to the page as shown in Figure E1.8. Conductor 1 carries a current I_1 into the page (in the negative z-direction) and passes through the x-axis at $x = +a$. Conductor 2 passes through the x-axis at $x = -a$ and carries an unknown current I_2. The total magnetic field at the origin, due to the current-carrying conductors, has a magnitude of $2\mu_o I_1/(2\pi a)$. The current I_2 can have either of two possible values. (a) Determine the value of I_2 with the smaller magnitude, stating it in terms of I_1 and also find its direction. (b) Determine the other possible value of I_2. [2]

Figure E1.8 Current-carrying conductors for Example 1.8.

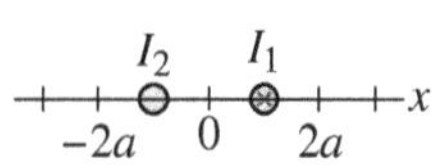

Solution

Wire 1 creates at the origin magnetic field:

$$\vec{B}_1 = \frac{\mu_o I}{2\pi r}\text{right-hand rule} = \frac{\mu_o I_1}{2\pi a} = \frac{\mu_o I_1}{2\pi a}\hat{y}$$

(a) If the total field at the origin is $\frac{2\mu_o I_1}{2\pi a}\hat{y} = \frac{\mu_o I_1}{2\pi a}\hat{y} + \vec{B}_2$ then the second wire must create a field according to $\vec{B}_2 = \frac{\mu_o I_1}{2\pi a}\hat{y} = \frac{\mu_o I_2}{2\pi(2a)}$ Then $I_2 = 2I_1$ out of paper

(b) The other possibility is $\vec{B}_1 + \vec{B}_2 = \frac{2\mu_0 I_1}{2\pi a}(-\hat{y}) = \frac{\mu_0 I_1}{2\pi a}\hat{y} + \vec{B}_2$. Then

$$\vec{B}_2 = \frac{3\mu_0 I_1}{2\pi a}(-\hat{y}) = \frac{\mu_0 I_2}{2\pi (2a)}$$ and $I_2 = 6I_1$ into the paper.

Example 1.9 An infinitely long wire carrying a current I is in the form of a right angle, as shown in Figure E1.9 Calculate the magnetic field at point P, located a distance x from the bend of the wire. [2]

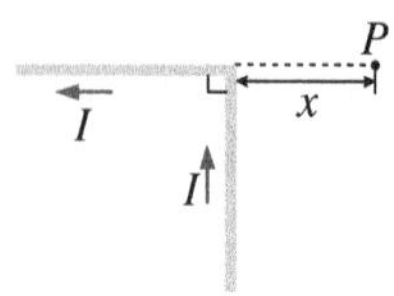

Figure E1.9 Current-carrying conductors for Example 1.9.

Solution

The vertical section of wire constitutes one half of an infinitely long, straight wire at distance x from P, so it creates a field equal to

$$\vec{B} = \frac{1}{2}\left(\frac{\mu_0 I}{2\pi x}\right)$$

Hold your right hand with extended thumb in the direction of the current; the field is away from you, into the paper. For each bit of the horizontal section of wire $d\vec{s}$ is to the left and $\hat{r}$ is to the right, so $d\vec{s} \times \hat{r} = 0$. The horizontal current produces zero fields at P. Thus,

$$\vec{B} = \frac{\mu_0 I}{4\pi x} \text{ into the paper}$$

Example 1.10 A right-angle bend made at the middle in a long, straight conductor carrying a current I, is shown in Figure E1.10. The bend forms an arc of a circle of radius r as shown in the Figure E1.10 Compute the magnetic field at point P, at the centre of the arc.

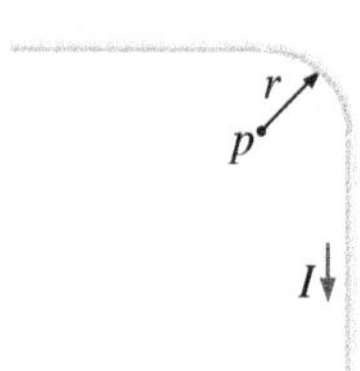

Figure E1.10 Current-carrying conductors for Example 1.8.

Solution

Every element of current creates a magnetic field in the same direction, into the page, at the centre of the arc. The upper straight portion creates one-half of the field that an infinitely long, straight wire would create. The curved portion creates one-quarter of the field that a circular loop produces at its centre. The lower straight segment also creates a field $\frac{1}{2}\left(\frac{\mu_0 I}{2\pi r}\right)$

The total field is

$$\vec{B} = \left(\frac{1}{2}\frac{\mu_0 I}{2\pi r} + \frac{1}{4}\frac{\mu_0 I}{2r} + \frac{1}{2}\frac{\mu_0 I}{2\pi r}\right) = \frac{\mu_0 I}{2r}\left(\frac{1}{\pi} + \frac{1}{4}\right) = \left(\frac{0.28415\mu_0 I}{r}\right) \textit{ into the page}$$

Example 1.11 Choose a flat, circular current loop of radius R carrying a current I. Take the x-axis to be along the axis of the loop, with the origin at the loop's centre. Plot a graph of the ratio of the magnitude of the magnetic field at coordinate x to that at the origin for $x = 0$ to $x = 5R$. *Use Matlab code to solve this problem*

Solution

Along the axis of a circular loop of radius R,

$$B = \frac{\mu_o I R^2}{\left(R^2 + x^2\right)^{\frac{3}{2}}}$$

Or

$$\frac{B}{B_o} = \frac{1}{\left(1 + \left(\frac{x}{R}\right)^2\right)^{\frac{3}{2}}}$$

Where

$$B_o = \frac{\mu_o I}{2R}$$

$\frac{x}{R}$	$\frac{B}{B_o}$
0.0	1
1.0	0.3535
2.0	0.0894
3.0	0.0316
4.0	0.0142
5.0	0.00754

MATLAB code

```
x=[0.0 1.0 2.0 3.0 4.0 5.0];
y=[1 0.3535 0.0894 0.0316 0.0142 0.00754];
xi = linspace(min(x), max(x), 150);    % Evenly-Spaced Interpolation Vector
yi = interp1(x, y, xi, 'spline', 'extrap');
figure(1)
plot(x, y, 'bp')
hold on
plot(xi, yi, '-r')
hold off
grid
xlabel('x/R')
ylabel('B/B_o')
legend('Original Data', 'Interpolation', 'Location', 'NE')
```

The plot is shown in Figure E1.11:

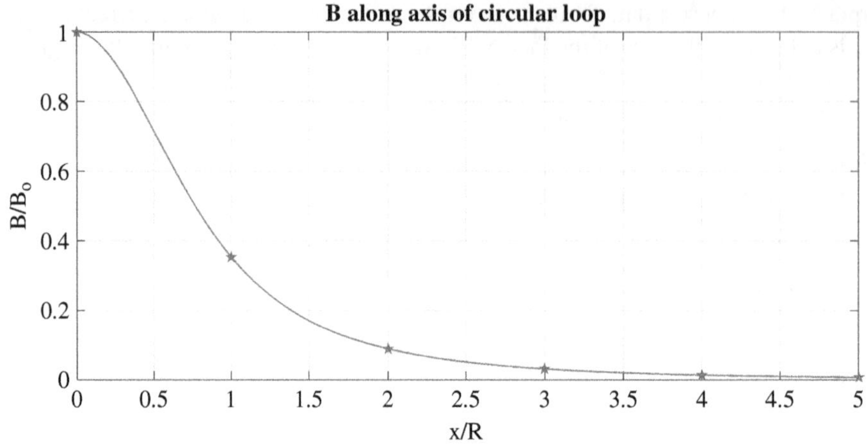

Figure E1.11 Output of Matlab program.

Example 1.12 A current loop is formed as shown in Figure E1.12a that produces a magnetic field at point P, which is at the centre of the arc. If the arc made an angle of θ = 45° and the radius of the arc is 0.5 m, calculate the magnitude and direction of the field produced at P if the current is 2.0 A?

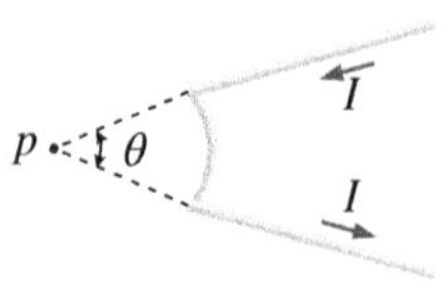

Figure E1.12a Current loop for Example 1.12.

Solution

We use the Biot-Savart law. For bits of wire along the straight-line sections, $d\vec{s}$ is at 0° or 180° to $\hat{r}$, so $d\vec{s} \times \hat{r} = 0$. Thus, only the curved section of wire contributes to $\vec{B}$ at P. Hence, $d\vec{s}$ is tangent to the arc and $\hat{r}$ is radially inward; so $d\vec{s} \times \hat{r} = \left|d\vec{s}\right| 1 \sin 90^o \otimes = \left|d\vec{s}\right| \otimes$. All points along the curve are the same distance $r = 0.600$ m from the field point, so

$$B = \int_{all\ curves} \left|d\vec{B}\right| = \int \frac{\mu_o I}{4\pi} \frac{\left|d\vec{s} \times \hat{r}\right|}{r^2} = \frac{\mu_o I}{4\pi r^2} \int \left|d\vec{s}\right| = \frac{\mu_o I}{4\pi r^2} s$$

Where s is the arc length of the curved wire

$$s = r\theta = (0.5\,m)\left(45^o\right)\left(\frac{2\pi}{360}\right) = 0.392\,m$$

Then

$$B = \frac{\mu_o I}{4\pi r^2} s = \frac{\left(4\pi \times 10^{-7}\,T.m/A\right)(2.0\,A)(0.392\,m)}{4\pi(0.5\,m)^2} = 313.6\,nT \text{ into the page}$$

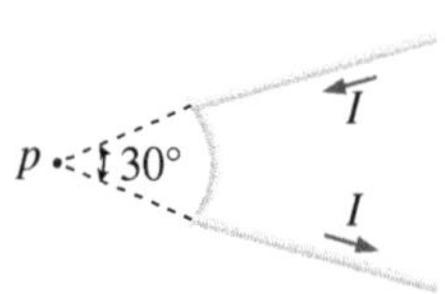

Figure E1.12b Solution for the Figure E1.12a.

Example 1.13 Current of 30 A passes through a long, straight conductor that lies in the left of the x-axis as shown in Figure E1.13. Current of 50 A passes through the second, long wire to the right along the line ($y = 0.280$ m, $z = 0$). (a) Compute the location where the total magnetic field is equal to zero (b) Consider a charge of -2.00 μC moving at a velocity of $150\hat{x}$ Mm/s along the line ($y = 0.100$ m, $z = 0$). Determine the vector magnetic force acting on the charge particle.

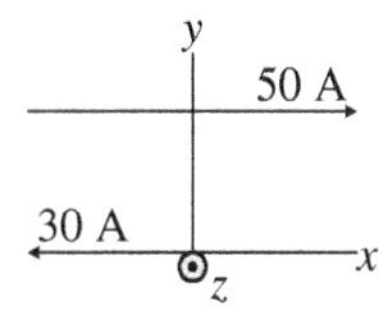

Figure E1.13 Current-carrying conductors for Example 1.13.

Solution

(a) Above the pair of wires, the field out of the page of the 50.0-A current will be stronger than the $(-\hat{z})$ field of the 30.0-A current, so they cannot add to zero. Between the wires, both produce fields into the page. They can only add to zero below the wires, at coordinate $y = -|y|$. Here the total field is

$$\vec{B} = \frac{\mu_o I}{2\pi r} ☞ + \frac{\mu_o I}{2\pi r} ☜$$

$$0 = \frac{\mu_o}{2\pi}\left[\frac{50.0\,A}{|y| + 0.28\,m}(-\hat{z}) + \frac{30.0\,A}{|y|}(\hat{z})\right]$$

$$50.0\,|y| = 30.0\,(|y| + 0.28\,m)$$

$$50.0\,(-y) = 30.0\,(-y + 0.28\,m)$$

$$-20.0y = 30.0\,(0.28\,m)$$

$$y = -0.420\,m$$

(b) At $y = 0.100$ m the total field is

$$\vec{B} = \frac{\mu_o I}{2\pi r} ☞ + \frac{\mu_o I}{2\pi r} ☜$$

$$\vec{B} = \frac{\mu_o I_1}{2\pi r_1}(-\hat{z}) + \frac{\mu_o I_2}{2\pi r_2}(-\hat{z}) = \frac{\mu_o}{2\pi}\left[\frac{I_1}{r_1}(-\hat{z}) + \frac{I_2}{r_2}(-\hat{z})\right] = \frac{\left(4\pi \times 10^{-7}\,T.m/A\right)}{2\pi}$$

$$\times\left[\frac{50.0\,A}{(0.28 - 0.10)\,m}(-\hat{z}) + \frac{30.0\,A}{0.10\,m}(-\hat{z})\right] = 1.16 \times 10^{-4}\,(-\hat{z})\;T$$

The force on the particle is

$$\vec{F} = q\vec{v} \times \vec{B} = \left(-2 \times 10^{-6}\,C\right)\left(150 \times 10^{-6}\,m/s\right)(\hat{x}) \times \left(1.16 \times 10^{-4}\frac{N.s}{C.m}\right)(-\hat{z})$$

$$= 3.47 \times 10^{-2}\,(-\hat{y})\;N$$

Note: Force (F) = Newton (N) = kg.m/s^2 and Magnetic Field (B) = T = kg/A.s^2, hence

$$T = \frac{N}{kg.m/s^2}\,\frac{kg}{\frac{C}{s}s^2} = \frac{N.s}{C.m}$$

Example 1.14 Three long, straight conductors are placed as shown in Figure E1.14a and they carry a current of 3 A each. The currents are coming out of the page.

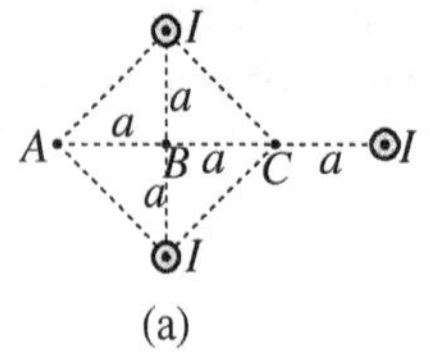

Figure E1.14a Current-carrying conductors for Example 1.14.

Given $a = 1.0$ cm, Compute the magnitude and direction of the magnetic field at (a) point A, (b) point B, and (c) point C.

Solution

Label the wires 1, 2, and 3 as shown in Sol. Figure E1.14b and let the magnetic field created by the currents in these wires be $\vec{B}_1$, $\vec{B}_2$, $\vec{B}_3$ respectively.

(a) At point A: $\left|\vec{B}_1\right| = \left|\vec{B}_2\right| = \frac{\mu_o I}{2\pi\left(a\sqrt{2}\right)}$, and $\left|\vec{B}_3\right| = \frac{\mu_o I}{2\pi(3a)}$

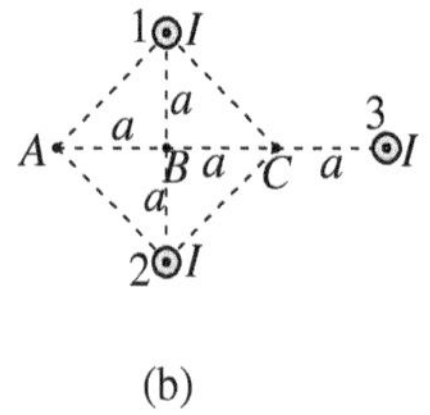

Figure E1.14b Solution for Example E1.14a.

The directions of these fields are shown in Sol. Figure E1.14c.

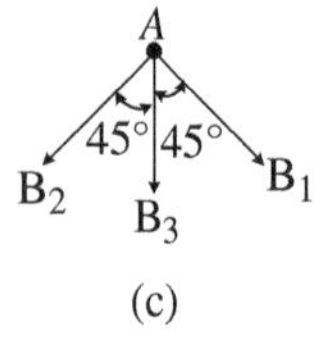

Figure E1.14c Solution for Example E1.14b.

Observe that the horizontal components of $\vec{B}_1$ and $\vec{B}_2$ cancel while their vertical components both add onto $\vec{B}_3$. Therefore, the net field at point A is

$$B_A = B_1\cos 45^o + B_2\cos 45^o + B_3 = \frac{\mu_o I}{2\pi a}\left[\frac{2}{\sqrt{2}}\cos 45^o + \frac{1}{3}\right]$$

$$= \frac{\left(4\pi\times 10^{-7}\,T.m/A\right)(3.0\,A)}{2\pi\left(1.0\times 10^{-2}\,m\right)}\left[\frac{2}{\sqrt{2}}\cos 45^o + \frac{1}{3}\right]$$

$$= 79.95\,\mu T \text{ towards bottom of the page}$$

(b) At point B:$\vec{B}_1$ and $\vec{B}_2$ cancel, leaving

$$B_B = B_3 = \frac{\mu_o I}{2\pi\,(2a)} = \frac{\left(4\pi \times 10^{-7}\,T.m/A\right)(3.0\,A)}{4\pi\left(1.0 \times 10^{-2}\,m\right)}$$

$$= 30.0\,\mu T \text{ towards bottom of the page}$$

(c) At point C:$\left|\vec{B}_1\right| = \left|\vec{B}_2\right| = \frac{\mu_o I}{2\pi\left(a\sqrt{2}\right)}$, and $\left|\vec{B}_3\right| = \frac{\mu_o I}{2\pi(a)}$ with the directions shown in Sol. Figure E1.14d. Again, the horizontal components of $\vec{B}_1$ and $\vec{B}_2$ cancel. The vertical components both oppose $\vec{B}_3$ giving

$$B_C = B_1 \cos 45^o + B_2 \cos 45^o - B_3 = \frac{\mu_o I}{2\pi a}\left[\frac{2}{\sqrt{2}}\cos 45^o - 1\right] = 0$$

Figure E1.14d Solution for Example E1.14c.

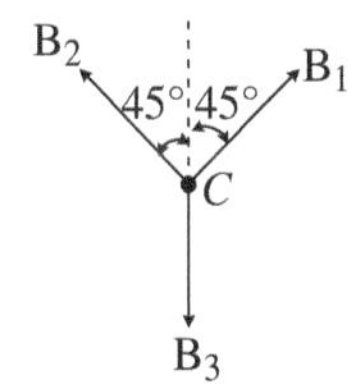

1.2.7 Ampere Circuital Law

Consider a long conductor carrying current i and, according to Biot-Sarvat law, sets up a magnetic field **H** or **B** around it at a radius of R. Ampere's law is derived using Biot-Savart law. The *line integral* of **B** around a closed path is denoted by

$$\oint \mathbf{B} \cdot d\mathbf{l} \tag{1.32}$$

To evaluate the product $\mathbf{B} \cdot d\mathbf{l}$ consider a small length element $d\mathbf{l}$ on the circular path and sum the products for all elements over the closed circular path. Along this path, the vectors $d\mathbf{l}$ and $\mathbf{B}$ are parallel at each point (Figure 1.6c), so $\mathbf{B} \cdot d\mathbf{l} = Bdl$. And also, the magnitude of $\mathbf{B}$ is constant on this circle and is given by Eq. (1.25). Therefore, the sum of the products Bdl over the closed path, which is equivalent to the line integral of $\mathbf{B} \cdot d\mathbf{l}$ is

$$\oint \mathbf{B} \cdot d\mathbf{l} = \oint B_{\parallel} \cdot dl = B \oint dl = \frac{\mu_o i}{2\pi R}(2\pi R) = \mu_o i$$

where $B_{\parallel}$ is the component of **B** which is parallel to $d\mathbf{l}$

or

$$\oint \mathbf{B} \cdot d\mathbf{l} = \mu_o i \tag{1.33}$$

In free space $\mathbf{B} = \mu_o \mathbf{H}$

$$\oint \mathbf{H} \cdot d\mathbf{l} = i \tag{1.34}$$

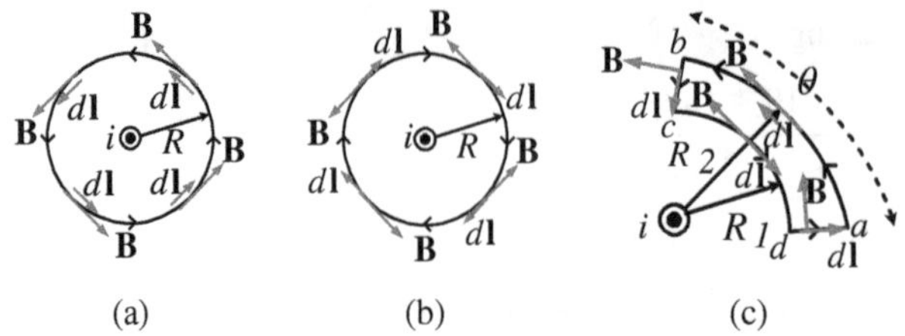

Figure 1.6 The current coming out of page (a) counter clockwise $\oint \mathbf{B} \cdot d\mathbf{l} = \mu_o i$. Here, the angle between **B** and $d\mathbf{l}$ is 0° (b) clockwise $\oint \mathbf{B} \cdot d\mathbf{l} = -\mu_o i$. Here, the angle is 180° (c) Contour *abcda* not enclosing the current $\oint \mathbf{B} \cdot d\mathbf{l} = 0$ because fields along curves get cancelled and field at *da* and *bc* are zero since the angle is 90°.

Concerning Figure 1.6, one can show that the *enclosed current 'i'* in Eq. (1.34) is equal to the integral of the current density **J** over any surface bounded by the closed path *C*.

$$\oint_C \mathrm{H} \cdot dl = \int_S \mathrm{J} \cdot d\mathrm{S} \tag{1.35}$$

where **J** and **S** are the current density (A/m^2) and surface area (m^2) vectors respectively.

The path need not always be circular. It can take any shape but it should enclose the current line. Equations (1.34) and (1.35) are applicable for static magnetic fields. It can be said that *the line integral of* **H** *around a closed path is equal to the total current flowing through the surface bounded by that path.*

Applying Maxwell's equation for the time-varying electric field, Ampere's circuital law can be modified that the MMF around a closed path C is equal to the sum of the current enclosed by that path due to actual flow of charges and the displacement current due to the time rate of increase of the electric flux (or displacement flux) enclosed by that path; that is

$$\oint_C \mathrm{H} \cdot dl = \int_S \mathrm{J} \cdot d\mathrm{S} + \frac{d}{dt}\int_S \mathrm{D} \cdot d\mathrm{S} \tag{1.36}$$

where **D** is electric flux density (C/m^2), as shown, for example, in Figure 1.7

The first part of Eq. (1.36) is called conduction or convection current which flows through wires or leads of a capacitor and the second part is called displacement current which flows through a dielectric between palates of a capacitor under time-varying conditions.

In Figure 1.7a, we see that the total current entering the hemisphere through the base plate is equal to the total current leaving the hemisphere through the spherical surface. In Figure 1.7b, the base plate from the hemisphere is removed. The closed path C is formed by

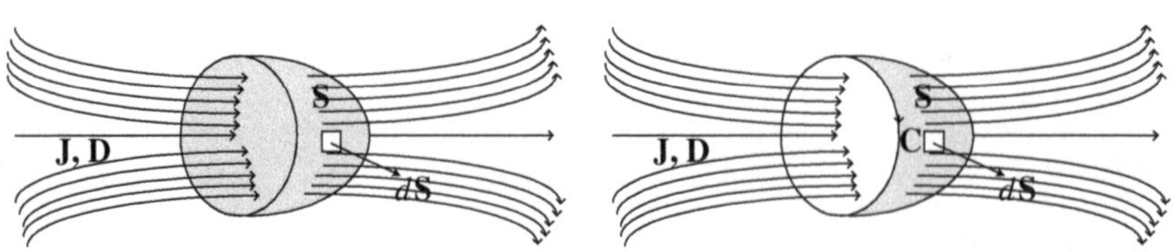

Figure 1.7 Illustration of Ampere's circuital law. (a) Closed hemisphere shell (b) Closed hemisphere shell but base plate removed; edge rim forms a closed path C.

the circular rim of the hemisphere. It can be seen in Figure 1.7b, the current enclosed by C is equal to the current flowing through the surface S bounded by C.

Example 1.15 Niobium metal becomes a superconductor when cooled below 9 K. Its superconductivity is destroyed when the surface magnetic field exceeds 0.100 T. In the absence of any external magnetic field, determine the maximum current a 2.00-mm-diameter niobium wire can carry and remain superconducting [2].

Solution

From $\oint \overline{B} \cdot d\bar{l} = \mu_o I \rightarrow I = \frac{2\pi r B}{\mu_o} = \frac{2\pi(1.0\times10^{-3} m)(0.1\ T)}{(4\pi\times10^{-7} T\cdot m/s)} = 500\,A$

Example 1.16 A coaxial cable is shown in Figure E1.16 where the central conductor is surrounded by a rubber layer, an outer conductor, and another rubber layer. The current in the inner conductor is I_1 = 2.00 A coming out of the page and the current in the outer conductor is I_2 = 5.00 A going into the page. Assuming the distance d = 1.00 mm, determine the magnitude and direction of the magnetic field at (a) point a and (b) point b.

Figure E1.16 Coaxial cable of Example 1.16.

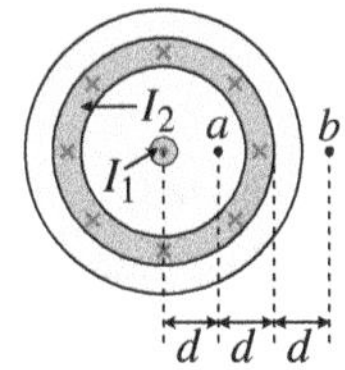

Solution

(a) From Ampère's law, the magnetic field at point a is given by $B_a = \frac{\mu_o I_a}{2\pi r_a}$, where I_a is the net current through the area of the circle of radius r_a. In this case, I_a = 1.00 A out of the page (the current in the inner conductor), so

$$B_a = \frac{\mu_o I_a}{2\pi r_a} = \frac{(4\pi \times 10^{-7} T \cdot m/s)\,(2.0\,A)}{2\pi\,(1.0 \times 10^{-3} m)} = 400\ \mu T \text{ towards the top of page}$$

(b) Similarly, at point b: $B_b = \frac{\mu_o I_b}{2\pi r_b}$, where I_b is the net current through the area of the circle having radius r_b. Taking out of the page as positive, I_b = 2.00 A – 5.00 A = –3.00 A, or I_b = 3.00 A into the page. Therefore

$$B_b = \frac{\mu_o I_b}{2\pi r_b} = \frac{(4\pi \times 10^{-7} T \cdot m/s)\,(3.0\,A)}{2\pi\,(3.0 \times 10^{-3} m)} = 199.5\ \mu T \text{ towards the bottom of page}$$

Example 1.17 A toroidal shape magnetic coils is used inside a fusion reactor having an inner radius of 0.700 m and an outer radius of 1.30 m. The magnetic coil has 1000 turns that carries a current of 15.0 kA. Determine the magnitude of the magnetic field inside the toroid along (a) the inner radius and (b) the outer radius.

Solution

(a) $B_{inner} = \frac{\mu_o NI}{2\pi r} = \frac{(4\pi\times10^{-7} T\cdot m/s)(900)(15\times10^3 A)}{2\pi(0.7\ m)} = 3.86\ T$

(b) $B_{outer} = \frac{\mu_o NI}{2\pi r} = \frac{(4\pi\times10^{-7} T\cdot m/s)(900)(15\times10^3 A)}{2\pi(1.3\ m)} = 2.07\ T$

Example 1.18 A packed bundle of 100 long, straight, insulated wires forms a cylinder of radius $R = 0.500$ cm (Figure E1.18). If each wire carries 2.00 A, what are (a) the magnitude and (b) the direction of the magnetic force per unit length acting on a wire located 0.200 cm from the centre of the bundle? (c) Would a wire on the outer edge of the bundle experience a force greater or smaller than the value calculated in parts (a) and (b)? Give a qualitative argument for your answer [2].

Solution

By Ampère's law, the field at the position of the wire at distance r from the centre is due to the fraction of the other 99 wires that lie within the radius r.

$$\oint \overline{B} \cdot d\overline{s} = \mu_o I \to B(2\pi r) = \mu_o \left[99I \left(\frac{\pi r^2}{\pi R^2} \right) \right] \to B = \frac{\mu_o (99I)}{2\pi r} \left(\frac{r^2}{R^2} \right) = \frac{\mu_o (99I)}{2\pi R} \left(\frac{r}{R} \right)$$

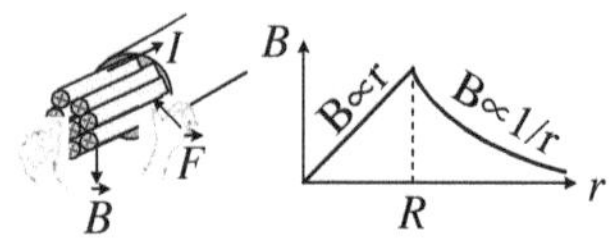

Figure E1.18 Cable and flux density vs radius of Example 1.18.

The field is proportional to r, as shown in Sol. Figure E1.18. This field points tangent to a circle of radius r and exerts a force $\overline{F} = I\overline{l} \times \overline{B}$ on the wire towards the centre of the bundle. The magnitude of the force is

$$\frac{F}{l} = IB \sin\theta = I \left[\frac{\mu_o (99I)}{2\pi R} \left(\frac{r}{R} \right) \right] \sin 90^o = \frac{\mu_o (99) I^2}{2\pi R} \left(\frac{r}{R} \right)$$

$$= \frac{(4\pi \times 10^{-7} T \cdot m/A)(99)(2.00\,A)^2}{2\pi \left(0.5 \times 10^{-2} m\right)} \left(\frac{0.2}{0.5} \right) = 6.34 \times 10^{-2} N/m$$

(a) 6.34×10^{-3} N/m

(b) Referring to the figure, the field is clockwise, so at the position of the wire, the field is downwards, and the force is inward towards the centre of the bundle.

(c) $B \propto r$, so B is greatest at the outside of the bundle. Since each wire carries the same current, F is greatest at the outer surface.

1.2.8 Faraday's Law

Faraday's law states that voltage will be induced across a loop when there is a rate of change of flux. If the circuit is closed, the induced current will flow. For N number of loops or turns, the law can be stated in mathematic form as

$$\varepsilon = -N\frac{d\phi}{dt} \tag{1.37a}$$

The negative sign is because of Lenz's law which states that the induced current must be directed so that the magnetic field it produces opposes the change in the external magnetic flux. There will be opposition not only to the change of flux but also opposition to the movement of coil or magnets. Figure 1.8 explains Lenz's law.

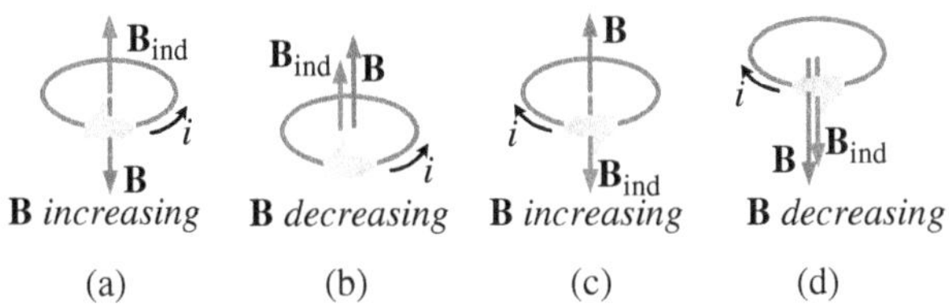

Figure 1.8 Induced current (i) induces field ($\mathbf{B}_{ind}$) which opposes the change. Fingers point induced current (i) direction while thumb points the induced field ($\mathbf{B}_{ind}$)(a) increasing **B** induces i with $\mathbf{B}_{ind}$ *opposes the change*, (b) decreasing **B** induces i with $\mathbf{B}_{ind}$ *opposes the change* (c) increasing **B** induces i with $\mathbf{B}_{ind}$ *opposes the change* (d) decreasing **B** induces i with $\mathbf{B}_{ind}$ *opposes the change*.

Faraday's law can be re-written as (considering only the magnitude, without direction);

$$\varepsilon = \frac{d\,(N\phi)}{dt}, \quad \text{and } L = \frac{N\phi}{i}$$

Therefore, $N\phi = Li$

$$\varepsilon = \frac{d\,(N\phi)}{dt} = \frac{d\,(Li)}{dt} = L\frac{di}{dt}$$

$$\varepsilon = L\frac{di}{dt} \tag{1.37b}$$

Generally, one can change the magnetic flux through a coil by the following means:

a. Change the magnitude **B** of the magnetic field with time within the coil.
b. Change the total area of the coil with time (for example by expanding the coil).
c. Change the portion of that area that lies within the magnetic field (for example, by sliding it into or out of the field).
d. Change the angle between the direction of the magnetic field **B** and normal to the plane of the coil. This can be achieved by rotating the coil in the magnetic field.
e. Any combination of the above will affect the change in the flux.

Emf can also be induced by moving the coil or conductor in a stationary magnetic field as well as by moving the conductor in a changing magnitude of the field with time. This induced emf is called *motional* emf.

Example 1.19 A long, straight wire carries a current I. A rectangular loop with two sides parallel to the straight wire has sides a and b, with its near side a distance d from the straight wire, as shown in Figure E1.19. (*a*) Compute the magnetic flux through the rectangular loop. (Hint: Calculate the flux through a strip of area $dA = b\,dx$ and integrate from $x = d$ to $x = d + a$.) (b) Evaluate your answer for $a = 5$ cm, $b = 10$ cm, $d = 2$ cm, and $I = 20$ A [3].

Figure E1.19 Rectangular loop of Example 1.19.

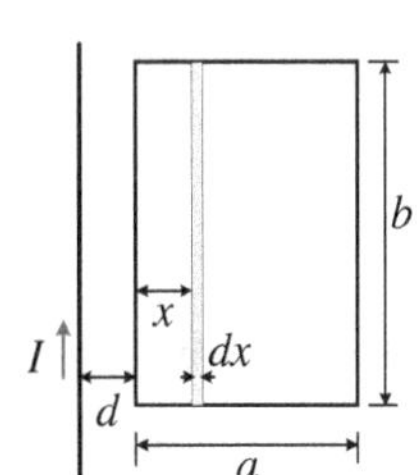

Solution

We can use the hint to set up the element of area dA and express the flux $d\phi_m$ through it and then carry out the details of the integration to express ϕ_m.

(a) Express the flux through the strip of area dA:	$d\phi_m = BdA$ where $dA = bdx$.
Express B at a distance x from a long, straight wire:	$B = \frac{\mu_o}{2\pi}\frac{I}{r} = \frac{\mu_o}{4\pi}\frac{2I}{x} = \frac{\mu_o}{2\pi}\frac{I}{x}$
Substitute to obtain:	$d\phi_m = \frac{\mu_o}{2\pi}\frac{I}{x}bdx = \frac{\mu_o Ib}{2\pi}\frac{dx}{x}$
Integrate from x = d to x = d + a:	$\phi_m = \frac{\mu_o Ib}{2\pi}\int_d^{d+a}\frac{dx}{x} = \frac{\mu_o Ib}{2\pi}\ln\left(\frac{d+a}{d}\right)$
(b) Substitute numerical values and evaluate ϕ_m:	$\phi_m = \frac{\mu_o Ib}{2\pi}\ln\left(\frac{d+a}{d}\right)$ $= \frac{(4\pi\times10^{-7}\,N/A^2)\,(20\,A)\,(0.1\,m)}{2\pi}\ln\left(\frac{(2+5)\,cm}{2\,cm}\right)$ $= 5.01\times10^{-7}\,Wb$

Example 1.20 A rectangular coil in the plane of the page has dimensions a and b. A long wire that carries a current I is placed directly above the coil (Figure E1.20). (a) Obtain an expression for the magnetic flux through the coil as a function of x for $0 \le x \le 2b$. (b) For what value of x, is the flux through the coil a maximum? For what value of x, is the flux a minimum? [3]

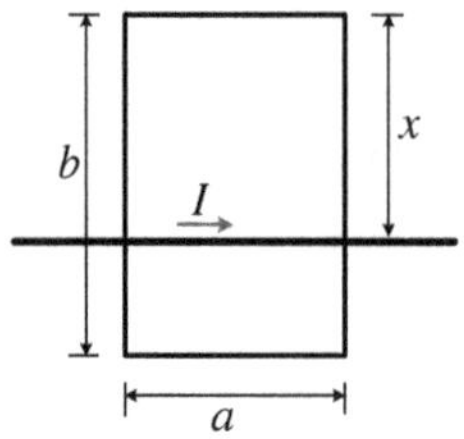

Figure E1.20 Rectangular loop of Example 1.19.

Solution

We can use its definition to express the flux through the rectangular region and Ampere's law to relate the magnetic field to the current in the wire and the position of the long, straight wire.

(a) Note that for $0 \le x \le b$, B is symmetric about the wire, into the paper for the region below the wire and out of the paper for the region above the wire. Thus, for area $2(b - x)a$:	$\phi_{m\_net} = 0$
To find the flux through the remaining area of the rectangle, express the flux through a strip of area dA:	$d\phi_m = BdA$ where $dA = adx$.
Using Ampere's law, express B at a distance x from a long, straight wire:	$B = \frac{\mu_o}{2\pi}\frac{I}{r} = \frac{\mu_o}{4\pi}\frac{2I}{x} = \frac{\mu_o}{2\pi}\frac{I}{x}$

Substitute to obtain:	$d\phi_m = \frac{\mu_0}{2\pi}\frac{I}{x}adx = \frac{\mu_0 Ia}{2\pi}\frac{dx}{x}$
For $0 \le x \le b$, integrate from $x' = b - x$ to $x' = x$:	$\phi_m\,(0 \le x \le b) = \frac{\mu_0 Ia}{2\pi}\int_{b-x}^{x}\frac{dx'}{x'} = \frac{\mu_0 Ia}{2\pi}\ln\left(\frac{x}{b-x}\right)$
For $x \ge b$, integrate from $x' = x$ to $x' = x + b$:	$\phi_m\,(x \le b) = \frac{\mu_0 Ia}{2\pi}\int_{x}^{x+b}\frac{dx'}{x'} = \frac{\mu_0 Ia}{2\pi}\ln\left(\frac{x+b}{x}\right)$
(*b*) From the expressions derived in (*a*), we see that $\phi_m \to \infty$ as:	$x \to 0$
The flux is a minimum ($\phi_m = 0$) for:	$x = \frac{1}{2}b$ as expected from symmetry.

Example 1.21 A uniform magnetic field B is established perpendicular to the plane of a loop of radius 5 cm, resistance 0.4 Ω, and negligible self-inductance. The magnitude of $\vec{B}$ is increasing at a rate of 40 mT/s. Find (a) the induced emf $\mathcal{E}$ in the loop, (b) the induced current in the loop, and (c) the rate of joule heating in the loop. [3]

Solution

We can find the induced emf by applying Faraday's law to the loop. The application of Ohm's law will yield the induced current in the loop and we can find the rate of joule heating using $P = I^2R$.

(*a*) Apply Faraday's law to express the induced emf in the loop in terms of the rate of change of the magnetic field:	$\lvert\mathcal{E}\rvert = \frac{d\phi_m}{dt} = \frac{d}{dt}(AB) = A\frac{dB}{dt} = \pi r^2\frac{dB}{dt}$
Substitute numerical values and evaluate ε:	$\lvert\mathcal{E}\rvert = \pi(0.05\,m)^2\,(40\,\text{m}T/s) = 0.314\,\text{m}V$
(*b*) Using Ohm's law, relate the induced current to the induced voltage and the resistance of the loop and evaluate *I*:	$I = \frac{\mathcal{E}}{R} = \frac{0.314\,mV}{0.4\,\Omega} = 0.785\,\text{m}A$
(*c*) Express the rate at which power is dissipated in a conductor in terms of the induced current and the resistance of the loop and evaluate *P*:	$P = I^2R = (0.785\,\text{m}A)^2(0.4\,\Omega) = 0.247\mu\text{W}$

Example 1.22 The flux through a loop is given by $\phi_m = (t^2 - 4\,t) \times 0^{-1}$ Wb, where t is in seconds. (a) Find the induced emf $\mathcal{E}$ as a function of time. (b) Find both ϕ_m and ε at $t = 0$, $t = 2$ s, $t = 4$ s, and $t = 6$ s [3].

Solution

Given ϕ_m as a function of time, we can use Faraday's law to express $\mathcal{E}$ as a function of time.

(*a*) Apply Faraday's law to express the induced emf in the loop in terms of the rate of change of the magnetic field:	$\mathcal{E} = -\frac{d\phi_m}{dt} = -\frac{d}{dt}\left[(t^2 - 4t) \times 10^{-1}\text{Wb}\right]$ $= -(2t - 4) \times \frac{10^{-1}\text{Wb}}{\text{s}} = -(0.2t - 0.4)\,\text{V}$
(*b*) Evaluate ϕ_m at $t = 0$:	$\phi_m(0s) = [0^2 - 4(0)] \times 10^{-1}\text{Wb} = 0$
Evaluate $\mathcal{E}$ at $t = 0$:	$\mathcal{E}\,(0s) = -[0.2\,(0) - 0.4]\,V = 0.4\,V$

Proceed as above to complete the table to the right:

t	ϕ_m	$\mathcal{E}$
(s)	(Wb)	(V)
0	0	0.4
2	−0.4	0
4	0	−0.4
6	1.2	−0.8

Example 1.23 (a) For the flux given in Example 1.22, sketch graphs of ϕ_m and ε versus *t*. (b) At what time is the flux at its minimum? What is the emf at this time? (c) At what times is the flux zero? What is the emf at these times? [3]

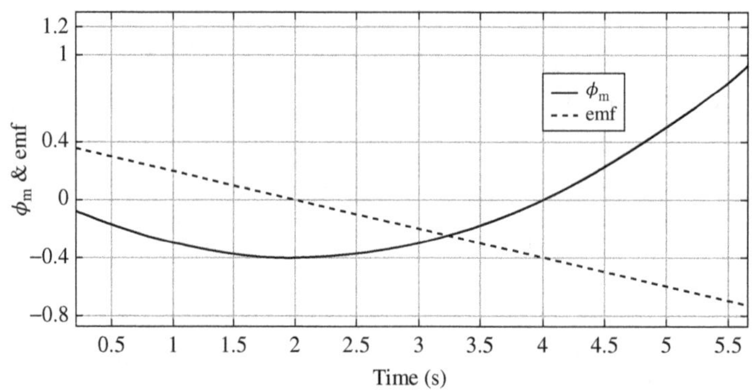

Figure E1.23 The flux and the induced emf as functions of time, t.

Solution

We can find the time at which the flux is at its minimum by looking for the lowest point on the graph of $\mathcal{E}$ versus *t* and the emf at this time by determining the value of *V* at this time from the graph. We can interpret the graphs to find the times at which the flux is zero and the corresponding values of the emf.

(a) The flux, ϕ_m, and the induced emf, $\mathcal{E}$, are shown in Figure E1.23 as functions of *t* in the following graph. The solid curve represents ϕ_m, the dashed curve represents $\mathcal{E}$.
(b) Referring to the graph, we see that the flux is a minimum at t = 2 s and that $\mathcal{E}$ = 0 V at this instant.
(c) The flux is zero at t = 0 and t = 4 s. At these times, $\mathcal{E}$=0.4 V and – 0.4 V, respectively.

1.2.8.1 Motional emf

A conductor of length *l* as shown in Figure 1.9 moves through a uniform magnetic field **B** directed into the page. It is assumed that the conductor moves in a direction perpendicular to the field **B** with constant velocity using some external agent. A magnetic force $\mathbf{F}_B$ acts

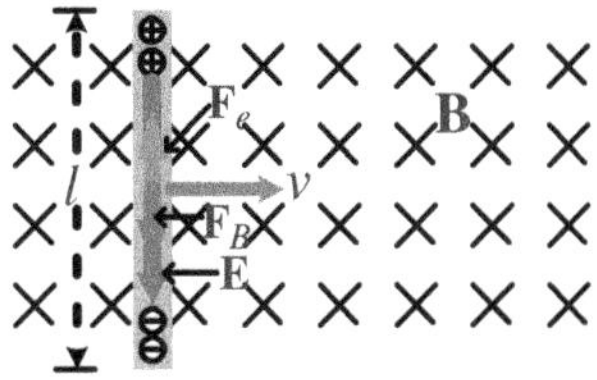

Figure 1.9 A conductor of length l moves with a velocity v through a uniform magnetic field B whose direction is perpendicular to v. Electrons are accumulated at the lower end of the conductor and equal and opposite charges become accumulated at the upper end of the conductor because of magnetic force F_B. This establishes an electric field E in the conductor. In steady-state, the electric and magnetic forces on the electrons in the wire are balanced.

upon the electrons in the conductor which is given as $\mathbf{F}_B = q\mathbf{v}\times\mathbf{B}$ whose direction is along the length l, perpendicular to both $\mathbf{v}$ and $\mathbf{B}$ (Eq. (1.16)). Electrons are accumulated at the lower end of the conductor and equal and opposite charges are accumulated at the upper end of the conductor because of magnetic force $\mathbf{F}_B$. The accumulation of charges continues until the downward magnetic field $\mathbf{F}_B$ $(= qvB)$ and upward electric force $\mathbf{F}_e(= qE)$ become balanced. This establishes an electric field $\mathbf{E}$ in the conductor. At equilibrium, the forces on the electrons become balanced.

$$qE = qvB \quad or \quad E = vB \tag{1.38}$$

The potential difference between any two points in the electric field is given as

$$V_{AB} = \Delta V = V_B - V_A = -\int_A^B E \cdot dl = -El$$

or

$$|\Delta V| = V_B - V_A = El \tag{1.39}$$

Therefore, for the equilibrium condition

$$\Delta V = El = Blv \tag{1.40}$$

In Figure 1.9, the upper end of the conductor is at a higher potential than the lower end. Hence, the potential difference between two ends of a conductor is established when the conductor moves in a uniform magnetic field. If the movement of the conductor is reversed, then the potential difference polarity is also reversed.

A conducting bar of length l is forced to move along two conducting rails with a velocity $\mathbf{v}$ by applying a force $\mathbf{F}_{app}$ in a uniform magnetic field $\mathbf{B}$, whose direction is into the page, as shown in Figure 1.10a. Assume that the bar and the rails have negligible resistance. A resistance R is connected between the rails to complete the circuit. Free charges in the bar move towards the ends of the bar which sets up an electric potential difference between the ends. The potential difference makes the current i flow in the resistance. The bar moves a distance of x at that instant. The area covered during that time is xl. The total flux (ϕ) equals to Blx Webers.

According to Faraday's law

$$\mathcal{E} = -\frac{d\phi_B}{dt} = -\frac{d}{dt}(Blx) = -Bl\frac{dx}{dt}$$

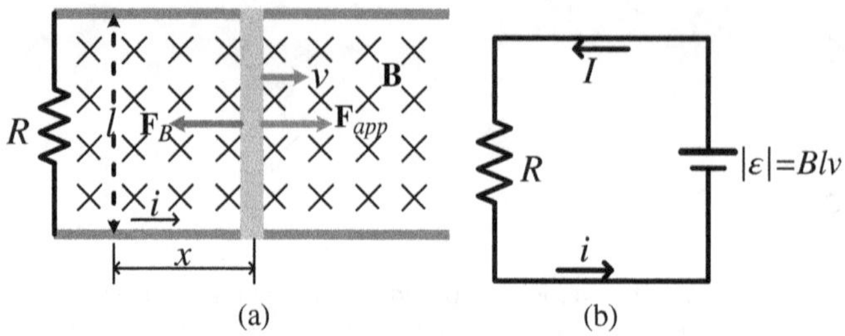

Figure 1.10 (a) A conducting bar has negligible resistance. Due to an applied force F_{app} the bar slides with a velocity v along two conducting rails. The magnetic force F_B opposes the motion, and the induced current *i* flows through the resistance R from the upper end to the lower end. (b) The equivalent circuit diagram.

Since

$$\frac{dx}{dt} = v$$

Therefore, motional emf is

$$\mathcal{E} = -Blv \tag{1.41}$$

Because the resistance of the circuit is R, the magnitude of the induced current is

$$i = \frac{|\mathcal{E}|}{R} = \frac{Blv}{R} \tag{1.42}$$

The equivalent circuit diagram is shown in Figure 1.10b.

Equation (1.41) can also be derived using the principle of conservation of energy. When the bar is moved by the force F_{app} with a velocity v, then at equilibrium, F_{app} experiences the equal and opposite magnetic force F_B.

Using Eq. (1.20)

$$F_B = -F_{app} = ilB$$

The power delivered by the magnetic force is

$$P = \left(F_B\right)(v) = (ilB)(v) = \left[\frac{Blv}{R}(lB)\right](v) = \frac{B^2l^2v^2}{R} = \frac{\mathcal{E}^2}{R} \tag{1.43}$$

Equation (1.38) shows that power is the rate of energy delivered to the resistance R.

Example 1.24 In Figure E1.24, $R = 6.00\,\Omega$, $l = 1.20\,\text{m}$, and a uniform 2.50-T magnetic field is directed into the page. (a) Find the speed of the bar to induce a current of 0.5 A in the resistor. (b) Find applied force to obtain a speed of 2 m/s to induce 0.5 A of current. (c) Find the rate of energy delivered to the resistor.

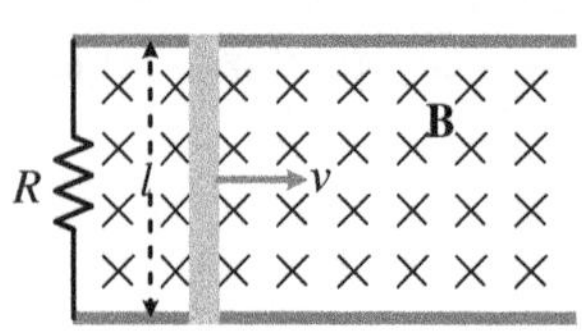

Figure E1.24 Conductor moving in the presence of magnetic field.

Solution

(a) From Eq. (1.40),

$$i = \frac{Blv}{R}$$

or speed is

$$v = \frac{iR}{Bl} = \frac{(0.5\,A)\,(6.0\,\Omega)}{(2.5\,T)\,(1.2\,m)} = 1\,m/s$$

(b)

$$F_B = |F_{app}| = (i)\,(lB) = \left(\frac{Blv}{R}\right)(lB) = \frac{B^2l^2v}{R} = \frac{(2.5)^2(1.2)^2\times 2}{6} = 3.0\text{ N}$$

(c) Rate of energy means power delivered to the resistor

$$P = \frac{B^2l^2v^2}{R} = \frac{(2.5)^2(1.2)^2\times(2)^2}{6} = 6.0\,Watts$$

Alternatively

$$P = F_Bv = 3.0\text{ N}\times 2.0\frac{\text{m}}{\text{s}} = 6.0\ \ Watts$$

Example 1.25 An electrical conductor (or rod) of l m in length and mass of m kg is made to move by a force F on two frictionless, horizontally placed rails at a speed of v m/s in a magnetic field B which is directed into the page as shown in Figure E1.24. A resistance of $R\ \Omega$ is connected between the conducting rails. Establish a relationship for the distance x m it travels in terms of R, B, l, m, and v.

Solution

Let the initial speed of the rod be v m/s and at a later time of t seconds the speed is u m/s.

The motional emf is given by

$$\mathcal{E} = Blv$$

Induced current is

$$i = \frac{\mathcal{E}}{R} = \frac{Blv}{R}$$

The magnetic force is opposite to applied mechanical force (= ma)

$$F = -ilB = -\frac{B^2l^2v}{R} = ma = m\frac{dv}{dt}$$

Separating the variables

$$-\frac{B^2l^2}{mR}dt = \frac{dv}{v}$$

Integrating both sides

$$\int_0^t\left(-\frac{B^2l^2}{mR}\right)dt = \int_v^u\frac{dv}{v}$$

Gives

$$-\frac{B^2l^2}{mR}(t-0) = ln\,(u) - ln\,(v) = ln\left(\frac{u}{v}\right)$$

Or

$$\frac{u}{v} = e^{-\frac{B^2l^2}{mR}t}$$

Or

$$u = ve^{-\frac{B^2l^2}{mR}t} = \frac{dx}{dt}$$

The distance travelled is given by

$$dx = ve^{-\frac{B^2l^2}{mR}t}dt$$

Integrating both sides with limits of x is 0 to $x_{\max}$ and that of time t is 0 to ∞.

$$\int_0^{x_{max}} dx = \int_0^{\infty} ve^{-\frac{B^2l^2}{mR}t}dt$$

$$[x]_0^{x_{max}} = v\int_0^{\infty} e^{-\frac{B^2l^2}{mR}t}dt = v\left[\frac{1}{\left(-\frac{B^2l^2}{mR}\right)}e^{-\frac{B^2l^2}{mR}t}\right]_0^{\infty}$$

Or

$$x_{max} - 0 = v\left(-\frac{mR}{B^2l^2}\right)\left(e^{-\infty} - e^0\right)$$

Or

$$x_{max} = \left(-\frac{vmR}{B^2l^2}\right)(0-1) = \frac{vmR}{B^2l^2}$$

Alternative Method: applying Newton's second law

$$-\frac{B^2l^2v}{R} = \left(-\frac{B^2l^2}{R}\right)\frac{dx}{dt} = m\frac{dv}{dt}$$

or

$$\left(-\frac{B^2l^2}{R}\right)dx = mdv$$

Integrating both sides from initial to the stopping point gives

$$\int_0^{x_{max}}\left(-\frac{B^2l^2}{R}\right)dx = \int_v^0 mdv$$

Or

$$\left(-\frac{B^2l^2}{R}\right)[x]_0^{x_{max}} = m\,[v]_v^0$$

Or

$$\left(-\frac{B^2l^2}{R}\right)(x_{max} - 0) = m\,(0 - v)$$

Therefore

$$x_{max} = \frac{vmR}{B^2l^2}$$

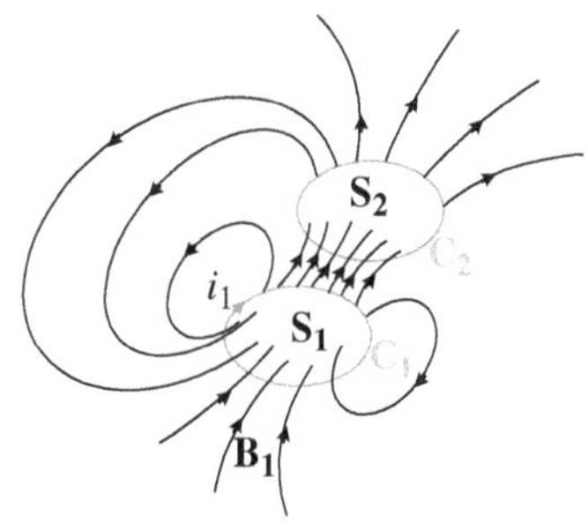

Figure 1.11 Magnetic coupling between two conducting loops.

1.2.9 Flux Linkages and Induced Voltages

Flux linkage is an extension of magnetic flux. Flux linkage is equivalent to the total flux passing through the surface formed by a coil and it is determined by knowing the value of flux (ϕ) and the number of turns as the following relation expresses.

$$\lambda = \int \mathbf{B} \cdot d\mathbf{S} = \int \mathcal{E} dt = N\phi \tag{1.44}$$

The closed loops C_1 and C_2 are placed near to each other whose surface area S_1 and S_2 respectively. C_1 carries current i_1 as shown in Figure 1.11. C_1 produces flux $\mathbf{B}_1$ because of the flow of current i_1. Part of the total flux passes through the surface S_2 bounded by C_2. Let that flux be ϕ_{12}. The magnetic flux ϕ_{12} is produced because of current i_1 flowing in C_1 is given by

$$\phi_{12} = \int_{S_2} \mathrm{B} \cdot d\mathrm{S} \tag{1.45}$$

Consider Figure 1.12 which has two turns. The magnetic flux ϕ_{12} links surface S_{21} and S_{22}. The total surface area will be $S_2 = S_{21} + S_{22}$. We see that the same magnetic flux passes through S_2 twice in the same direction. So, the magnetic flux linking C_2 must be $2\phi_{12}$. Hence, it can be said that the magnetic flux linkage is the product of ϕ_{12} and the number of turns.

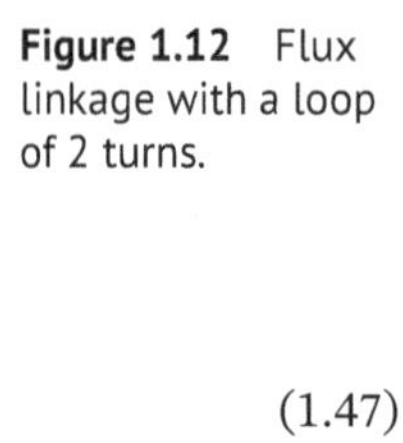

Figure 1.12 Flux linkage with a loop of 2 turns.

If C_2 has N_2 turns, then the magnetic flux linkage is given as

$$\lambda_{12} = N_2\phi_{12} \tag{1.46}$$

Here, we assumed the notation of subscript 12 as linking of flux due to current flow in C_1 to C_2. Hence, in general, the flux linkage is stated as

$$\lambda = N\phi \tag{1.47}$$

1.2.10 Induced Voltages

Induces voltage as per Faraday's law is given by

$$\mathcal{E} = \frac{d\lambda}{dt} = -N\frac{d\phi}{dt} \tag{1.48}$$

or by neglecting the negative sign, flux linkage is given by

$$\lambda = \int \left(N\frac{d\phi}{dt} \right) dt = \int N d\phi = N\phi \quad (1.49)$$

This states that Faraday's law i.e. rate of change of flux linkage will induce an electromotive force (emf).

1.2.11 Induced Electric Fields

Electric fields are induced by both static charges and changing magnetic fields. Both electric fields exert a force equal to q_0E. But there is a difference between them. When a positive test charge moved from point A to point B in an electric field **E**, electric force q_0E is exerted on the test charge, and then we say work is done by the conservative force. The work done is given by

$$W = -\int \mathbf{F}_e \cdot d\mathbf{l} = -q_0 \int \mathbf{E} \cdot d\mathbf{l} \quad (1.50)$$

The potential difference between point A and point B is

$$\Delta V = V_B - V_A = \frac{W}{q_0} = -\int_A^B \mathbf{E} \cdot d\mathbf{l} \quad (1.51)$$

But when the test charge moves from a particular point and comes back to the same point, then we say there is no work is done by Electric force (F_e). Therefore, the closed integral will be equal to zero.

$$\oint \mathbf{E} \cdot d\mathbf{l} = 0 \quad (1.52)$$

Now, consider a copper ring of radius r is immersed in a gradually increasing magnetic field of radius R whose direction is into the page as shown in Figure 1.13a, an emf (ε) is induced which makes current i to flow in the ring in a counter clockwise direction as per Lenz's law. The current set up an electric field in the same direction. As stated earlier, the

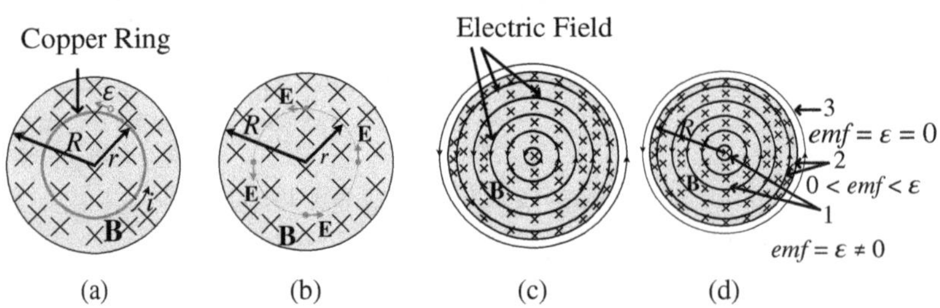

Figure 1.13 (a) When the magnetic field is changing at a uniform rate, then there will be inducement of current which flows in the copper ring of radius r. (b) Even if no ring is in the changing magnetic field, there is an induced electric field E. The four points are shown. (c) The electric fields are induced in every circular imaginary ring. (d) Three similar closed paths that enclose identical areas. Equal EMFs are induced around paths of region 1, which lie entirely within the region of changing the magnetic field. A smaller emf is induced around path 2, which only partially lies in that region. No net emf is induced around path 3, which lies entirely outside the magnetic field [4].

electric field is induced when there is a change of flux even if there is no current flowing. This has been shown in Figures 1.13b,c in that no copper ring is placed in the magnetic field which is changing with time. This induced electric field originates from a point and returns to the same point, or in other words, it has a closed path whereas the electric field, due to static charges, is radial i.e. it originates from a positive charge and ends up a negative charge. Hence, the close line integral of the Electric field due to static charges is zero (Eq. (1.50)). The electric field **E**, induced due to changing magnetic field **B** with time, induces EMFs. The imaginary ring outside B induces zero ems, rings near outer edge induces non zero emf but less than the inner rings. This is shown in Figure 1.13d.

1.2.12 Reformulation of Faraday's Law

In Figure 1.13b, a work W has done on a particle q_0 in moving around the circular path is equal to εq_0, where ε is the induced emf. Another way of expressing work W done is

$$W = \int \mathbf{F} \cdot \mathbf{dl} = \left(q_o E\right)(2\pi r) \tag{1.53}$$

where q_0E is the magnitude of the force acting on the charge and $2\pi r$ is the distance over which that force acts.

Or the Eq. (1.46) can be written as

$$W = \left(q_o E\right)(2\pi r) = q_o \mathcal{E}$$

or

$$\mathcal{E} = 2\pi r E \tag{1.54}$$

Rewriting Eq. (1.53)

$$W = \int \mathbf{F} \cdot \mathbf{dl} = q_0 \oint \mathbf{E} \cdot \mathbf{dl} = q_o \mathcal{E}$$

or

$$\mathcal{E} = \oint \mathbf{E} \cdot \mathbf{dl} \tag{1.55}$$

The integral in Eq. (1.54) becomes Eq. (1.53) when Figure 1.13b is considered.

According to Faraday's law or rewriting Eq. (1.37a)

$$\mathcal{E} = -\frac{d\phi}{dt}$$

Therefore, reformulated Faraday's law equation shall be

$$\oint_C \mathrm{E} \cdot \mathrm{dl} = -\frac{d\varphi}{dt} = -\frac{d}{dt}\int_S \mathrm{B} \cdot d\mathrm{S} \tag{1.56}$$

where S is the surface area enclosed by C as shown in Figure 1.14

Faraday's law can be stated as a time-varying magnetic field that gives rise to an electric field. Specifically, the electromotive force around a closed path C is equal to the negative of the time rate of increase of the magnetic flux enclosed by that path.

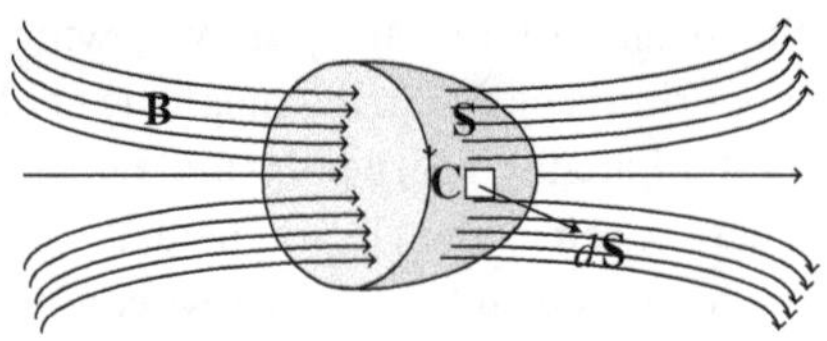

Figure 1.14 Illustration of Faraday's law.

Hence, Faraday's law can be applied to any closed path. In Figure 1.13d, it is shown that when there is a change in a magnetic field, then EMFs are induced: the inner most region '1' experiences inducement of emf = ε, outside magnetic field region '3' emf = 0, and region '2' near the outer edge of the changing magnetic field emf < ε.

Example 1.26 A conducting wire moves in a plane perpendicular to magnetic field of 40 mT. The length of the wire is 50 cm and the speed of the wire is 10 m/s. Determine (a) the force exerted on an electron in the wire, (b) the electrostatic field $\vec{E}$ in the wire, and (c) the potential difference produced between the ends of the wire.

Solution

We can apply the equation for the force on a charged particle moving in a magnetic field to find the magnetic force acting on an electron in the rod. We can use $\vec{E} = \vec{v} \times \vec{B}$ to find E and $V = El$, where l is the length of the rod, to find the potential difference between its ends.

(*a*) Relate the magnetic force on an electron in the rod to the speed of the rod, the electronic charge, and the magnetic field in which the rod is moving:	$\vec{F} = q\vec{v} \times \vec{B}$ and $F = qvB \sin\theta$
Substitute numerical values and evaluate F:	$F = (1.6 \times 10^{-19}\text{C})(10\text{ m/s})(0.04\text{ T}) \sin 90^o = 6.4 \times 10^{-20}$
(*b*) Express the electrostatic field $\vec{E}$in the rod in terms of the magnetic field $\vec{B}$	$\vec{E} = \vec{v} \times \vec{B}$ and $E = vB \sin\theta$
Substitute numerical values and evaluate B:	$E = (10\text{ m/s})(0.04\text{ T}) \sin 90^o = 0.4\text{ V/m}$
(*c*) Relate the potential difference between the ends of the rod to its length l and the electric field E:	$V = El$
Substitute numerical values and evaluate V:	$V = (0.4\text{ V/m})(0.5\text{ m}) = 0.20\text{ V}$

Example 1.27 Find the speed of the rod in Example 1.26 if the potential difference between the ends is 6 V.

Solution

We can use $\vec{E} = \vec{v} \times \vec{B}$ to relate the speed of the rod to the electric field in the rod and magnetic field in which it is moving and $V = El$ to relate the electric field in the rod to the potential difference between its ends.

Express the electrostatic field $\vec{E}$ in the rod in terms of the magnetic field $\vec{B}$ and solve for v:	$\vec{E} = \vec{v} \times \vec{B}$ and $v = \dfrac{E}{B \sin \theta}$
Relate the potential difference between the ends of the rod to its length l and the electric field E and solve for E:	$V = El \Rightarrow E = \frac{V}{l}$
Substitute for E to obtain:	$v = \dfrac{V}{Bl \sin \theta}$
Substitute numerical values and evaluate v:	$v = \dfrac{6\text{ V}}{(0.04\text{ T})(0.5\text{ m})} = 300\text{ m/s}$

Example 1.28 Given a conducting rod of 50 cm length moving at velocity of 10 m/s in the presence of a magnetic field of 1 T as shown in Figure E1.28. The resistive load R = 1 Ω Calculate (a) the induced emf in the circuit, (b) the current in the circuit, and (c) the force needed to move the conductor with constant velocity assuming negligible friction (d) the power input by the force found in Part (c), and (e) the rate of joule heat production I^2R.

Figure E1.28 A conductor moving in the presence of magnetic field.

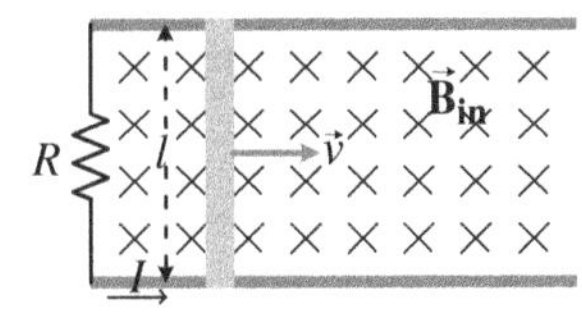

Solution

Because the speed of the rod is constant, an external force must action the rod to counter the magnetic force acting on the induced current. We can use the motional-emf equation $\mathcal{E} = vBl$ to evaluate the induced emf, Ohm's law to find the current in the circuit, Newton's 2nd law to find the force needed to move the rod with constant velocity, and $P = Fv$ to find the power input by the force.

(*a*) Relate the induced emf in the circuit to the speed of the rod, the magnetic field, and the length of the rod:	$\mathcal{E} = vBl = (10\text{ m/s})(1.0\text{T})(0.5\text{ m}) = 5.0\text{ V}$
(*b*) Using Ohm's law, relate the current in the circuit to the induced emf and the resistance of the circuit:	$I = \dfrac{\mathcal{E}}{R} = \dfrac{5.0\text{ V}}{1\,\Omega} = 5.0\text{ A}$ Note that, because the rod is moving to the right, the flux in the region defined by the rod, the rails, and the resistor is increasing. Hence, by Lenz's law, the current must be counter clockwise if its magnetic field is to counter this increase in flux.
(*c*) Because the rod is moving with constant velocity, the net force acting on it must be zero. Apply Newton's second law to relate F to the magnetic force F_m:	$\sum F_x = F - F_m = 0$ and $F = F_m = BIl = (1.0\text{ T})(5.0\text{ A})(0.5\text{ m}) = 2.5\text{ N}$
(*d*) Express the power input by the force in terms of the force and the velocity of the rod:	$P = Fv = (2.5\text{ N})(10\text{ m/s}) = 25\text{ W}$
(*e*) The rate of Joule heat production is given by:	$P = I^2R = (5\text{ A})^2(1\,\Omega) = 25\text{ W}$

Example 1.29 A 10 cm by 5 cm rectangular loop with resistance 2.5 Ω is pulled through a region of uniform magnetic field B = 1.7 T (Figure E1.29a) with constant speed $v = 2.4$ cm/s. The front of the loop enters the region of the magnetic field at time $t = 0$. (a) Find and graph the flux through the loop as a function of time. (b) Find and graph the induced emf and the current in the loop as a function of time. Neglect any self-inductance of the loop and extend your graphs from $t = 0$ to $t = 16$ s. [3]

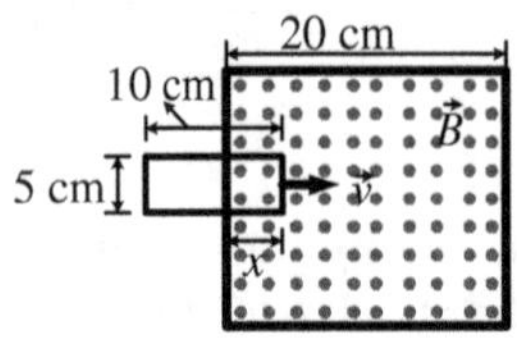

Figure E1.29a Rectangular Loop for Example 1.29.

Solution

We'll need to determine how long it takes for the loop to completely enter the region in which there is a magnetic field, how long it is in the region, and how long it takes to leave the region. Once we know these times, we can use its definition to express the magnetic flux as a function of time. We can use Faraday's law to find the induced emf as a function of time.

(a) Find the time required for the loop to enter the region where there is a uniform magnetic field:	$t = \frac{l_{\text{side of loop}}}{v} = \frac{10\text{ cm}}{2.4\text{ cm/s}} = 4.17\text{ s}$
Letting w represent the width of the loop, express and evaluate ϕ_m for $0 < t < 4.17$ s:	$\phi_m = NBA = NBwvt$ $= (1.7\text{ T})(0.05\text{ m})(0.024\text{ m/s})t$ $= (2.04\ m\text{Wb/s})t$
Find the time during which the loop is fully in the region where there is a uniform magnetic field:	$t = \frac{l_{\text{side of loop}}}{v} = \frac{20\text{ cm}}{2.4\text{ cm/s}} = 8.33\text{ s}$ i.e., the loop will begin to exit the region when $t = 8.33$ s.
Express ϕ_m for $4.17\text{ s} < t < 8.33$ s:	$\phi_m = NBA = NBlw$ $= (1.7\text{ T})(0.1\text{ m})(0.05\text{ m})$ $= (8.50\ m\text{Wb/s})$
The left-end of the loop will exit the field when $t = \frac{30\text{ cm}}{2.4\text{ cm/s}} = 12.5$ s. Express ϕ_m for $8.33\text{ s} < t < 12.5$ s:	$\phi_m = mt + b$ where m is the slope of the line and b is the ϕ_m-intercept.
For $t = 8.33$ s and $\phi_m = 8.50$ mWb:	$8.50\ m\text{W/s} = m\,(8.33\text{ s}) + b$
For $t = 12.5$ s and $\phi_m = 0$:	$0 = m\,(12.5\text{ s}) + b$
Solve Eqs. (1) and (2) simultaneously to obtain $m = -(2.04\ m\text{Wb/s})$ and $b = 25.5\ m$Wb:	$\phi_m = mt + b = -(2.04\ m\text{Wb/s})t + 25.5\ m\text{Wb}$
The loop will be completely out of the magnetic field when $t > 12.5$ s and:	$\phi_m = 0$

The following graph of (t) vs $\phi_m$was plotted using a Matlab program, shown in Figure E1.29b.

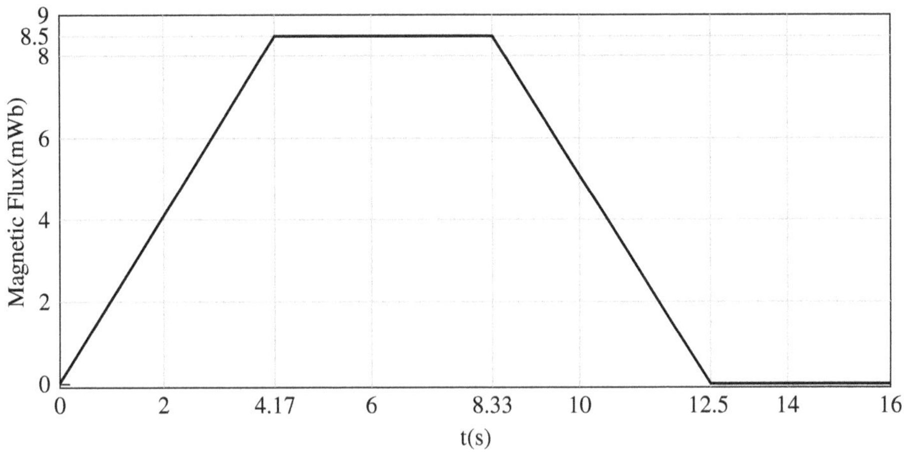

Figure E1.29b Graph of time (t) vs flux (ϕ_m).

(b) Using Faraday's law, relate the induced emf to the magnetic flux:	$\mathcal{E} = -\dfrac{d\phi_m}{dt}$
During the interval $0 < t < 4.17$ s:	$\mathcal{E} = -\dfrac{d}{dt}\left[(2.04\ m\text{Wb/s})\,t\right] = -2.04\ \text{mV}$
During the interval $4.17\ \text{s} < t < 8.33$ s:	$\mathcal{E} = -\dfrac{d}{dt}\left[8.50\ m\text{Wb/s}\right] = 0$
During the interval $8.33\ \text{s} < t < 12.5$ s:	$\mathcal{E} = -\dfrac{d}{dt}\left[-(2.04\ m\text{Wb/s})\,t + 25.5\ m\text{Wb}\right] = 2.04\ \text{mV}$
For $t > 12.5$ s:	$\mathcal{E} = 0$

The following graph of (t) vs ε was plotted using a Matlab program in Figure E1.29c.

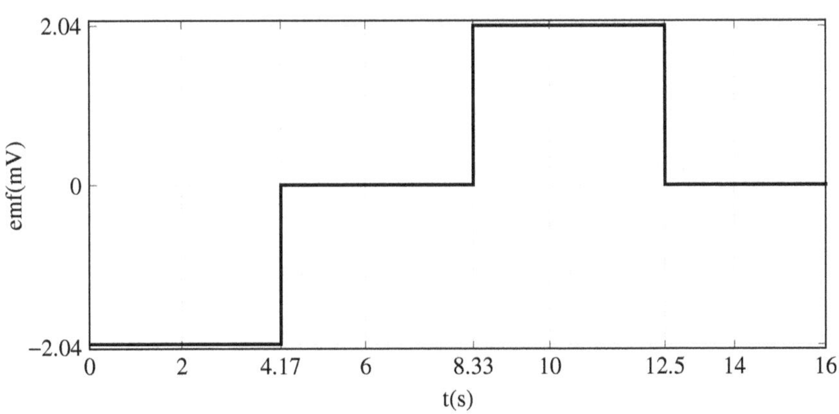

Figure E1.29c Graph of time (t) vs emf (ε).

Example 1.30 A conductor of length 15 cm lies parallel to the y-axis and oscillates in the x-direction with displacement given by $x = (2\text{ cm})\cos 120\pi t$. If a uniform magnetic field of magnitude 1.0 T is in the z-direction. Calculate induced emf in the conductor

Solution

The rod is executing simple harmonic motion in the xy plane, i.e. in a plane perpendicular to the magnetic field. The emf induced in the rod is a consequence of its motion in this magnetic field and is given by $|\mathcal{E}| = vBl$. Because we're given the position of the oscillator as a function of time, we can differentiate this expression to obtain v.

Express the motional emf in terms of v, B, and l:	$|\mathcal{E}| = vBl = Bl\dfrac{dx}{dt}$
Evaluate dx/dt:	$\dfrac{dx}{dt} = \dfrac{d}{dt}[(2\text{ cm})\ \cos 120\pi t]$
	$= -(2\text{ cm})\left(120\text{ s}^{-1}\right)\pi \sin 120\pi t$
	$= -(7.54\text{ m/s})\sin 120\pi t$
Substitute numerical values and evaluate $|\varepsilon|$:	$= -(1.0\text{ T})(0.15\text{ m})(7.54\text{ m/s})\sin 120\pi t$
	$= -(1.133\text{ V})\sin 120\pi t$

Example 1.31 A conducting rod with a resistance R moving across the horizontal rails which have negligible resistance as shown in Figure E1.31. A battery of emf $\mathcal{E}_b$ and negligible internal resistance is connected across the horizontal rail between points a and b so that the current in the rod is flowing downward. The rod is placed at rest at $t = 0$. (a) Find the force on the rod as a function of the speed v. (b) Show that the rod moves at a terminal speed and find an expression for it. (c) What is the current when the rod will approach its terminal speed?

Figure E1.31 Conductor moving in a magnetic field, Example 1.31.

Solution

Let m be the mass of the rod and F be the net force acting on it due to the current in it. We can obtain the equation of motion of the rod by applying Newton's second law to relate its acceleration to B, I, and l. The net emf that drives I in this circuit is the emf of the battery minus the emf induced in the rod as a result of its motion.

(a) Letting the direction of motion of the rod be the positive x-direction, apply $\sum F_x = ma_x$ to the rod:	$BIl = m\dfrac{dv}{dt}$ where $I = \dfrac{\mathcal{E}_b - Blv}{R}$
Substitute to obtain:	$\dfrac{dv}{dt} = \dfrac{Bl}{mR}\left(\mathcal{E}_b - Blv\right)$
(b) Express the condition on dv/dt when the rod has achieved its terminal speed:	$\dfrac{Bl}{mR}\left(\mathcal{E}_b - Blv_t\right) = 0$
Solve for v_t to obtain:	$v_t = \dfrac{\mathcal{E}_b}{Bl}$
(c) Substitute v_t for v in Eq. (2) to obtain:	$I = \dfrac{\mathcal{E}_b - Bl\frac{\mathcal{E}_b}{Bl}}{R} = 0$

Example 1.32 A conductor of mass m moves along frictionless conducting rails in a region of the static uniform magnetic field $\vec{B}$ directed into the page (Figure E1.32). An external agent is pushing the conductor, maintaining its motion to the right at constant speed v_0. At time $t = 0$, the external force acting on the conductor is removed and the conductor continues forward, being slowed by the magnetic force. Calculate the speed v of the rod as a function of time. Also, find the total distance travelled by the rod, find the total energy dissipated in the resistance and show that it is equal to mv_0^2.

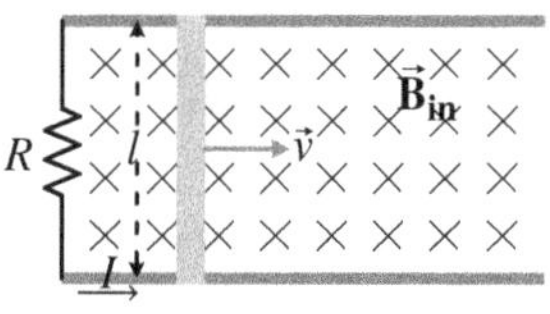

Figure E1.32 Conductor moving in a magnetic field, Example 1.32.

Solution

The speed of the rod changes because a magnetic force acts on the induced current. The motion of the rod through a magnetic field induces an emf $\mathcal{E} = Blv$ and, therefore, a current in the rod, $I = \frac{\mathcal{E}}{R}$. This causes a magnetic force to act on the rod, $F = IBl$. With the force known, we apply Newton's second law to find the speed as a function of time. Take the positive x-direction as being to the right.

1. Apply Newton's second law to the rod:	$F_x = ma_x = m\dfrac{dv}{dt}$
2. The force exerted on the rod is the magnetic force, which is proportional to the current and in the negative x-direction, as shown in Figure E1.32	$F_x = -IBl$
3. The current equals the motional emf divided by the resistance of the rod:	$I = \dfrac{\mathcal{E}}{R} = \dfrac{Blv}{R}$
4. Combining these results, we find the magnitude of the magnetic force exerted on the rod:	$F_x = -IBl = -\dfrac{Blv}{R}Bl = -\dfrac{B^2l^2v}{R}$
5. Newton's second law then gives:	$-\dfrac{B^2l^2v}{R} = m\dfrac{dv}{dt}$
6. Separate the variables, then integrate the velocity from v_0 to v_f and integrate the time from 0 to t_f:	$\dfrac{dv}{v} = -\dfrac{B^2l^2}{mR}dt$ $\displaystyle\int_{v_0}^{v_f}\frac{dv}{v} = -\frac{B^2l^2}{mR}\int_0^{t_f}dt$ $\ln\left(\frac{v_f}{v_0}\right) = -\dfrac{B^2l^2}{mR}t_f$
7. Let $v = v_f$ and $t = t_f$, then solve for v:	$v = v_0e^{-\frac{t}{\tau}}$, where $\tau = \dfrac{mR}{B^2l^2}$
8. In step 7 it is shown that the speed of the rod is given by $v = v_0e^{-(B^2l^2/mR)t}$. We can write v as dx/dt, separate the variables and integrate to find the total distance travelled by the rod.	
Apply the result from step 7 to obtain:	$\dfrac{dx}{dt} = v_0e^{-Ct}$ where $C = \dfrac{B^2l^2}{mR}$

Separate variables and integrate x' from 0 to x and t' from 0 to ∞:	$\int_0^x dx' = v_0 \int_0^\infty e^{-Ct} dt$
Evaluate the integrals to obtain:	$x = \frac{v_0}{C}$
Substitute for C and simplify:	$x = \frac{mRv_0}{B^2 l^2}$
9. In step 7, it is shown that the speed of the rod is given by $= v_0 e^{-(B^2l^2/mR)t}$. We can use the definition of power and the expression for a motional emf to express the power dissipated in the resistance in terms of B, l, v, and R. We can then separate the variables and integrate overall time to show that the total energy dissipated is equal to the initial kinetic energy of the rod.	
Express the power dissipated in terms of $\mathcal{E}$ and R:	$P = \frac{\mathcal{E}^2}{R}$
Express $\mathcal{E}$ as a function of B, l, and v:	$\mathcal{E} = Blv$ where $v = v_0 e^{-(B^2l^2/mR)t}$
Substitute to obtain:	$P = \frac{(Blv)^2}{R}$
The total energy dissipated as the rod comes to rest is obtained by integrating $dE = Pdt$:	$E = \int_0^\infty \frac{(Blv)^2}{R} dt = \int_0^\infty \frac{\left(Blv_0 e^{-(B^2l^2/mR)t}\right)^2}{R} dt = \frac{B^2l^2v_0^2}{R} \int_0^\infty e^{-2(B^2l^2/mR)t} dt$
Evaluate the integral (by changing variables to $u = -\frac{2B^2l^2}{mR}$) to obtain:	$E = \frac{B^2l^2v_0^2}{R}\left(\frac{mR}{2B^2l^2}\right) = \frac{1}{2}mv_0^2$

REMARKS If the force were constant, the rod's speed would decrease linearly with time. However, because the force is proportional to the rod's speed, as found in step 4, the force is large initially but the force decreases as the speed decreases. In principle, the rod never stops moving. Even so, the rod travels only a finite distance.

The general equation for motional emf is

$$\mathcal{E} = \oint_C \left(\vec{v} \times \vec{B}\right) \cdot \vec{dl} = -\frac{d\phi_m}{dt}$$

where v is the velocity of the wire at the element $\vec{dl}$. The integral is taken at an instant in time.

1.3 Inductance

An inductor (symbol:) is a solenoid wound on a core having N turns. When current i flows through, it produces a magnetic field. The inductance (L) of an inductor is defined as:

$$L = \frac{N\phi}{i} \tag{1.57}$$

The product $N\phi$ is called flux linkage. So, the inductance of an inductor is defined as the flux linkage per unit of current. SI unit of inductance is *henry* (H) which is equal to tesla-m^2/A.

1.3.1 Application of Ampere's Law to Find B in a Solenoid

Let us consider a solenoid of Figure 1.15 which carries current i. The direction of current and the magnetic flux is shown as per right-hand thumb rule. Select a rectangular Amperian path $abcda$. According to Ampere's law

$$\oint_C \mathrm{B} \cdot dl = \mu_o i_{encl}$$

$$\oint_C \mathbf{B} \cdot d\mathbf{l} = \int_a^b \mathbf{B} \cdot d\mathbf{l} + \int_b^c \mathbf{B} \cdot d\mathbf{l} + \int_c^d \mathbf{B} \cdot d\mathbf{l} + \int_d^a \mathbf{B} \cdot d\mathbf{l} + Bh + 0 + 0 + 0$$
$$= Bh = \mu_o i_{encl}$$

Here, the net current enclosed by the Amperian path i_{encl} is not the solenoid current i because the windings pass more than once through the enclosed path. Let l be the length of the solenoid having N number of turns. Let $n(=N/l)$ be the number of turns per unit length. Therefore, the number of turns in the enclosed path will be equal to nh. Enclosed current in terms of solenoid current will be

$$i_{encl} = \textit{number of turns in the enclosed path} \times i = nhi$$

From Ampere's circuital law

$$\oint_C \mathbf{B} \cdot d\mathbf{l} = \mu_o i_{encl} = \mu_o nhi = Bh$$

or

$$B = \mu_o ni \tag{1.58}$$

Equation (1.58) is derived for long solenoid but it is also useful for the short solenoid to find magnetic flux at the inner part of the solenoid. Flux does not depend upon the diameter of the solenoid.

Now flux in the solenoid is

$$\phi = BA$$

where A is the cross-sectional area of the solenoid.

Using Eq. (1.52) and taking $n = N/l$, Eq. (1.44) for inductance (in *henries*) becomes

$$L = \frac{N\phi}{i} = \frac{NBA}{i} = \frac{N(\mu_o ni)A}{i} = \frac{\mu_o NNA}{l} = \frac{\mu_o AN^2}{l} = \mu_o An^2 l \tag{1.59}$$

or

Inductance per unit length (H/m)

$$\frac{L}{l} = \frac{N(\mu_o ni)A}{i} = \mu_o An^2 \tag{1.60}$$

Figure 1.15 Application of Ampere's law to a section of a long ideal solenoid carrying a current i. The Amperian path is the rectangle $abcda$.

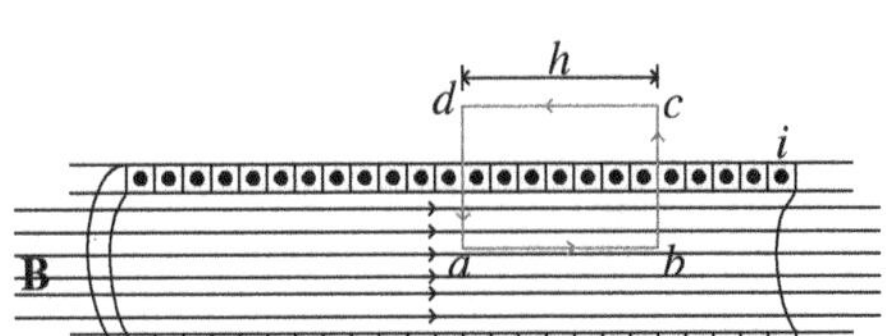

1.3.2 Magnetic Field of a Toroid

A toroid is shown in Figure 1.16a. Figure 1.16b shows the horizontal cross section of the toroid. The magnetic field inside the hollow portion of the toroid is found by applying Ampere's Circuital Law. From Eq. (1.39) one gets

$$(B)(2\pi r) = \mu_0 Ni$$

where i is the current in the winding and N is the total number of turns. The current i is positive for those windings enclosed by the Amperian path. From the above equation, one can have

$$B = \frac{\mu_0 Ni}{2\pi r} \text{ (toroid)}. \tag{1.61}$$

Unlike a solenoid, B is not constant over the cross section of the toroid. It is assumed that the toroid is an ideal one, i.e. $B = 0$ outside the toroid. The direction of the magnetic field can be found using the right-hand thumb rule i.e. when the toroid is grasped, the fingers point in the direction of current and the thumb points in the direction of the magnetic field.

1.3.3 The Inductance of Circular Air-Cored Toroid

Magnetic field density $\mathbf{B}$ is given by

$$\mathbf{B} = B_\phi \mathbf{a}_\phi = \frac{\mu_0 Ni}{2\pi r}\mathbf{a}_\phi$$

From Figure 1.17

$$r = r_o - \rho\cos\phi$$

Using a cylindrical coordinate system, flux is

$$\phi = \int_S \mathbf{B}\cdot\mathbf{A} = \int_0^b \int_0^{2\pi} \frac{\mu_0 Ni}{2\pi r}\mathbf{a}_\phi \cdot (\rho d\rho d\phi)\,\mathbf{a}_\phi = \frac{\mu_0 Ni}{2\pi}\int_0^b \int_0^{2\pi} \frac{\rho d\rho d\phi}{r_o - \rho\cos\phi}$$

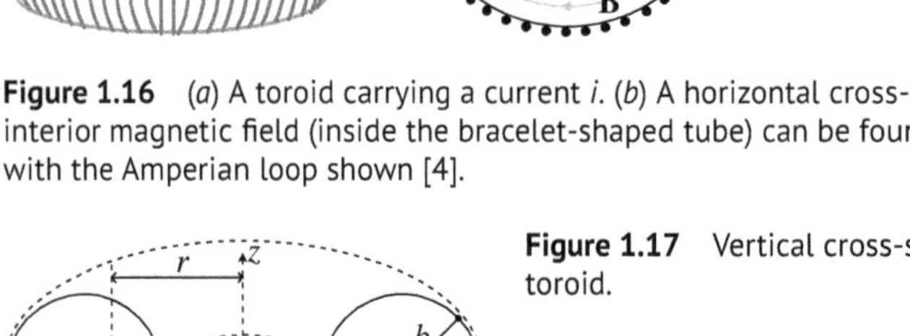

Figure 1.16 (a) A toroid carrying a current i. (b) A horizontal cross-section of the toroid. The interior magnetic field (inside the bracelet-shaped tube) can be found by applying Ampere's law with the Amperian loop shown [4].

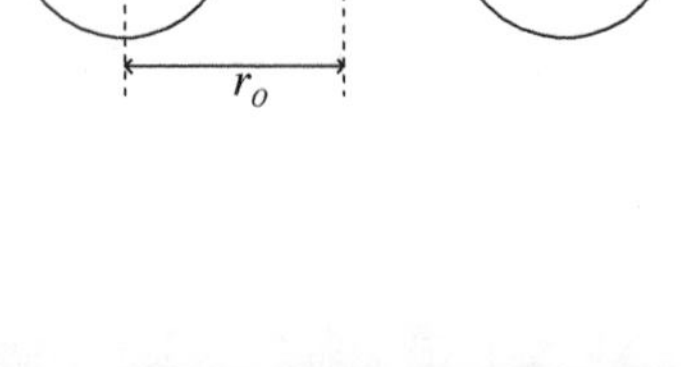

Figure 1.17 Vertical cross-section of the air-cored circular toroid.

or

$$\phi = \frac{\mu_o Ni}{2\pi}\int_0^b\int_0^{2\pi}\frac{\rho d\rho d\phi}{r_o - \rho\cos\phi} = \frac{\mu_o Ni}{2\pi}\int_0^b \rho d\rho \int_0^{2\pi}\frac{d\phi}{r_o - \rho\cos\phi}$$

$$= \frac{\mu_o Ni}{2\pi}\int_0^b \rho d\rho\,[I_1] \tag{1.62}$$

Where $I_1 = \int_0^{2\pi}\frac{d\phi}{r_o - \rho\cos\phi}$

Applying property of definite integration

$$\int_0^{2a} f(x)\,dx = 2\int_0^a f(x)\,dx$$

If

$$f(x) = f(2a - x)$$

Here $f(\phi) = \frac{1}{r_o - \rho\cos\phi}$ and $f(2\pi - \phi) = \frac{1}{r_o - \rho\cos(2\pi - \phi)} = \frac{1}{r_o - \rho\cos\phi} = f(\phi)$

$$I_1 = \int_0^{2\pi}\frac{d\phi}{r_o - \rho\cos\phi} = 2\int_0^{\pi}\frac{d\phi}{r_o - \rho\cos\phi}$$

Let $t = \tan\left(\frac{\phi}{2}\right)$ so that $dt = \frac{1}{2}\sec^2\left(\frac{\phi}{2}\right)d\phi = \frac{1}{2}\left(1 + \tan^2\left(\frac{\phi}{2}\right)\right)d\phi = \frac{1}{2}\left(1 + t^2\right)$ or $d\phi = \frac{2dt}{(1+t^2)}$

The limits of integration are converted from ϕ to t as:

ϕ	0	π
$t = \tan\left(\frac{\phi}{2}\right)$	$t = 0$	$t = \infty$

Also $\cos\phi = \frac{1-\tan^2\left(\frac{\phi}{2}\right)}{1+\tan^2\left(\frac{\phi}{2}\right)} = \frac{1-t^2}{1+t^2}$

$$I_1 = 2\int_0^{\pi}\frac{d\phi}{r_o - \rho\cos\phi} = 2\int_0^{\infty}\frac{\frac{2dt}{(1+t^2)}}{r_o - \rho\left(\frac{1-t^2}{1+t^2}\right)} = 4\int_0^{\infty}\frac{\frac{dt}{(1+t^2)}}{\frac{(1+t^2)r_o - \rho(1-t^2)}{(1+t^2)}}$$

$$= 4\int_0^{\infty}\frac{dt}{(r_o+\rho)\,t^2 + (r_o - \rho)} = \frac{4}{(r_o+\rho)}\int_0^{\infty}\frac{dt}{t^2 + \left(\sqrt{\frac{(r_o-\rho)}{(r_o+\rho)}}\right)^2}$$

$$= \frac{4}{(r_o+\rho)}\cdot\frac{1}{\sqrt{\frac{(r_o-\rho)}{(r_o+\rho)}}}\left[\tan^{-1}t\right]_0^{\infty} = \frac{4}{(r_o+\rho)}\frac{\sqrt{(r_o+\rho)}}{\sqrt{(r_o-\rho)}}\left[\tan^{-1}(\infty) - \tan^{-1}(0)\right]$$

$$= \frac{4}{\sqrt{(r_o+\rho)}}\cdot\frac{1}{\sqrt{(r_o-\rho)}}\cdot\left[\frac{\pi}{2} - 0\right] = \frac{2\pi}{\sqrt{r_o^2 - \rho^2}} \tag{1.63}$$

Now substituting the value of $I1$ from (1.63) in (1.62), we get

$$\phi = \frac{\mu_o Ni}{2\pi}\int_0^b \rho d\rho\,[I_1] = \frac{\mu_o Ni}{2\pi}\int_0^b \rho d\rho\left[\frac{2\pi}{\sqrt{r_o^2 - \rho^2}}\right] = \mu_o Ni\int_0^b\frac{\rho d\rho}{\sqrt{r_o^2 - \rho^2}}$$

Let $u = r_o^2 - \rho^2$ so that $du = -2\rho d\rho$ or $\rho d\rho = -\frac{du}{2}$
The limits of integration are converted from ρ to u as:

ρ	0	b
$u = r_o^2 - \rho^2$	$u = r_o^2$	$u = r_o^2 - b^2$

$$\phi = \mu_o Ni \int_0^b \frac{\rho d\rho}{\sqrt{r_o^2 - \rho^2}} = \mu_o Ni \int_{r_o^2}^{r_o^2 - b^2} \frac{-\frac{du}{2}}{\sqrt{u}} = -\frac{\mu_o Ni}{2} \int_{r_o^2}^{r_o^2 - b^2} \frac{du}{\sqrt{u}} = -\frac{\mu_o Ni}{2} \left[\frac{\sqrt{u}}{\frac{1}{2}} \right]_{r_o^2}^{r_o^2 - b^2}$$

$$= -\mu_o Ni \left[\sqrt{r_o^2 - b^2} - \sqrt{r_o^2} \right] = \mu_o Ni \left(r_o - \sqrt{r_o^2 - b^2} \right) \tag{1.64}$$

Therefore, the inductance of air-cored circular toroid will be

$$L = \frac{N\phi}{i} = \frac{N\left[\mu_o Ni \left(r_o - \sqrt{r_o^2 - b^2} \right) \right]}{i} = \mu_o N^2 \left(r_o - \sqrt{r_o^2 - b^2} \right) \tag{1.65}$$

If $r_o \gg b$, $B_\phi \cong \frac{\mu_o Ni}{2\pi r_o}$ (constant)

$$\phi = B_\phi . S = B_\phi \left(\pi b^2 \right) \cong \frac{\mu_o Ni}{2\pi r_o} \left(\pi b^2 \right) \cong \frac{\mu_o Nib^2}{2r_o}$$

The inductance of thin toroid will be

$$L = \frac{N\phi}{i} \cong \frac{N}{i} \left(\frac{\mu_o Nib^2}{2r_o} \right) \cong \frac{\mu_o N^2 b^2}{2r_o} \tag{1.66}$$

Equations (1.59), (1.60), (1.65), and (1.66) show that inductance depends upon the physical dimensions of the structure.

Example 1.33 A coil with a self-inductance of 10H carries a current of 5 A that is changing at a rate of 200 A/s. Determine (*a*) the magnetic flux through the coil and (*b*) the induced emf in the coil.

Solution

We can use $\phi_m = LI$ and the dependence of I on t to find the magnetic flux through the coil. We can apply Faraday's law to find the induced emf in the coil.

(*a*) Use the definition of self-inductance to express ϕ_m:	$\phi_m = LI$
Express I as a function of time:	$I = 5\text{A} + (200\ \text{A/s})t$
Substitute to obtain:	$\phi_m = L[5\text{A} + (200\ \text{A/s})t]$
Substitute numerical values and express ϕ_m:	$\phi_m = (10\text{H})[5\text{A} + (200\ \text{A/s})t]$ $= 50\text{Wb} + (2000\ \text{H. A/s})t$
(*b*) Use Faraday's law to relate $\mathcal{E}$, L, and dI/dt:	$\mathcal{E} = -L\frac{dI}{dt}$
Substitute numerical values and evaluate $\mathcal{E}$:	$\mathcal{E} = -(10\text{H})\left(200\frac{\text{A}}{\text{s}}\right) = -2.0\text{kV}$

Example 1.34 A coil with self-inductance L carries a current I, given by $I = I_0 \sin(2\pi ft)$. Determine the relationship between the flux ϕ_m and the self-induced emf as functions of time.

Solution

We can apply $\phi_m = LI$ to find ϕ_m and Faraday's law to find the self-induced emf as functions of time.

Use the definition of self-inductance to express ϕ_m:	$\phi_m = LI = LI_0 \sin 2\pi ft$

The graph of the flux ϕ_m as a function of time shown in figure E1.34a was plotted using a Matlab program. The maximum value of the flux is LI_0 and we have chosen $2\pi f = 1$ rad/s.

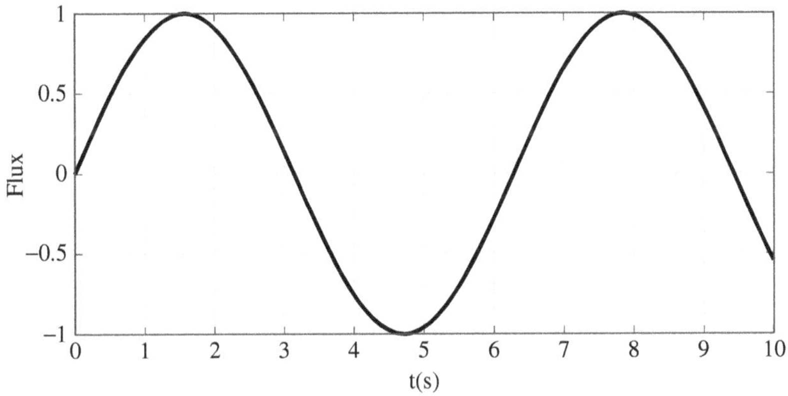

Figure E1.34a Graph of the flux ϕ_m as a function of time.

Apply Faraday's law to relate $\mathcal{E}$, L, and dI/dt:	$\mathcal{E} = -L\frac{dI}{dt} = -L\frac{d}{dt}\left[I_0 \sin 2\pi ft\right] = -2\pi fLI_0 \cos 2\pi ft$

The graph of the emf $\mathcal{E}$ as a function of time shown in figure E1.34b was plotted using a Matlab program. The maximum value of the induced emf is $2\pi fLI_0$ and we have chosen $2\pi f = 1$ rad/s.

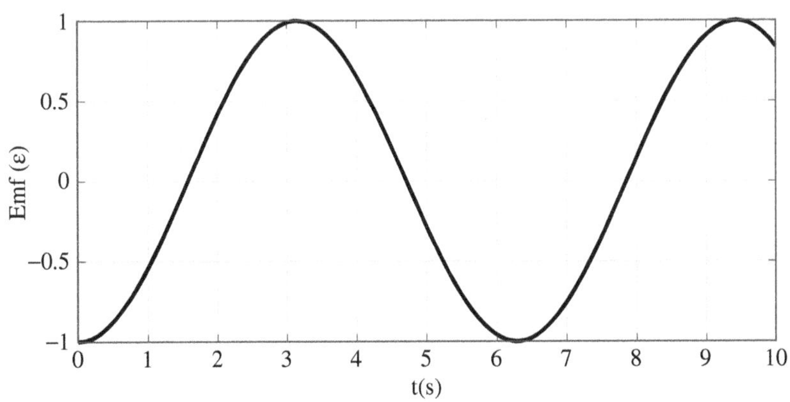

Figure E1.34b Graph of the emf $\mathcal{E}$ as a function of time.

Example 1.35 A solenoid has a length of 25 cm, a radius of 1 cm, 500 turns, and carries a current of 5-A. Determine (*a*) the flux density *B* on the axis at the centre of the solenoid; (b) the flux through the solenoid, assuming flux density *B* to be uniform; (*c*) the self-inductance of the solenoid; and (*d*) the induced emf in the solenoid when the current changes at 100 A/s.

Solution

We can use $B = \mu_o nI$ to find the magnetic field on the axis at the centre of the solenoid and the definition of magnetic flux to evaluate ϕ_m. We can use the definition of magnetic flux in terms of *L* and *I* to find the self-inductance of the solenoid. Finally, we can use Faraday's law to find the induced emf in the solenoid when the current changes at 100 A/s.

(*a*) Apply the expression for *B* inside a long solenoid to express and evaluate *B*:	$B = \mu_o nI = (4\pi \times 10^{-7}\mathrm{N/A^2})\left(\frac{400}{0.25\ \mathrm{m}}\right)(5\mathrm{A}) = 10.05\ mT$
(*b*) Apply the definition of magnetic flux to obtain:	$\phi_m = NBA = 500(6.03\ mT)\pi(0.01\ \mathrm{m})^2 = 9.475 \times 10^{-4}\mathrm{Wb}$
(c) Relate the self-inductance of the solenoid to the magnetic flux through it and it's current:	$L = \frac{\phi_m}{I} = \frac{7.58 \times 10^{-4}\mathrm{Wb}}{5\mathrm{A}} = 0.1516\mathrm{mH}$
(*d*) Apply Faraday's law to obtain:	$\mathcal{E} = -L\frac{dI}{dt} = -(0.253\ \mathrm{mH})\left(100\frac{\mathrm{A}}{\mathrm{s}}\right) = -25.3\mathrm{mV}$

Example 1.36 Two solenoids of radii 2 and 5 cm are coaxial. They are each 25 cm long and have 300 turns and 5000 turns, respectively. Calculate their mutual inductance

Solution

We can find the mutual inductance of the two coaxial solenoids using

$$M_{2,1} = \frac{\phi_{m2}}{I_1} = \mu_o n_2 n_1 l \pi r^2$$

Substitute numerical values and evaluate $M_{2,1}$:

$$M_{2,1} = (4\pi \times 10^{-7}\mathrm{N/A^2})\left(\frac{300}{0.25\ \mathrm{m}}\right)\left(\frac{1000}{0.25\ \mathrm{m}}\right)(0.25\ \mathrm{m})\ \pi(0.02\ \mathrm{m})^2 = 0.945\mathrm{mH}$$

1.3.4 Mutual Inductance

Consider, Figure 1.18 where part of or perhaps the whole of magnetic flux density $\mathbf{B}_1$ induced due to time-varying current i_1 flowing in coil C_1 links with coil C_2. Let the part of that magnetic flux be ϕ_{12}. The flux linkage shall be $\lambda_{12} = N_2\phi_{12}$. Here, we say the two loops are magnetically coupled through mutual flux. Hence, mutual inductance between two coils $C_1$and $C_2$is defined as

$$M_{12} = \frac{\lambda_{12}}{i_1} = \frac{N_2}{i_1}\int_{S_2} \mathbf{B}_1 \cdot d\mathbf{S} \tag{1.67}$$

Ampere's circuital law $\nabla \times \mathbf{H} = \mathbf{J}$ can be written for a medium of μ permeability as $\nabla \times \frac{\mathbf{B}}{\mu} = \mathbf{J}$. It shows that magnetic field density $\mathbf{B}_1$ is proportional to current i_1 flowing in coil C_1. The permeability μ is independent of i_1 and $\mathbf{B}_1$ and its direction. Hence, the flux

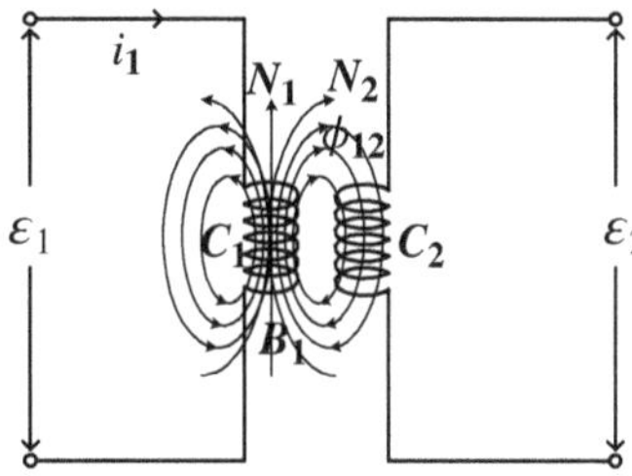

Figure 1.18 Mutually coupled coils.

linkage $\lambda_{12} = N_2\phi_{12}$ is proportional to i_1 where N_2 is the number of turns of coil C_2. The constant of proportionality is called mutual inductance; it can be seen in Eq. (1.67).

The magnetic flux passing through each turn of the coil C_2 is

$$\phi_{12} = \mu_o n_1 i_1 \pi r_2^2 \tag{1.68}$$

where r_2, and $n_1 \left(= \frac{N_1}{l_1}\right)$ are the mean radius, number of turns per unit length of coils C_2 and C_1 respectively.

Therefore, the total flux linking the coil C_2 is

$$N_2\phi_{12} = \mu_o \pi r_2^2 n_2 l_2 n_1 i_1 = M_{12} i_1$$

where $N_2 (= n_2 l_2 =$ turns per unit length × length of coil C_2) is the number of turns in the coil C_2.

$$M_{12} = \mu_o \pi r_2^2 n_1 n_2 l_2 \tag{1.69}$$

Equation (1.69) shows that the mutual inductance depends upon the physical data of the coils and the medium in which they are placed.

Similarly, when C_2 is excited with current i_2, the mutual inductance M_{21} is given by

$$M_{21} = \mu_o \pi r_1^2 n_1 n_2 l_1 \tag{1.70}$$

where r_1 and l_1 are mean radius and length of coil C_1.

If $r_1 = r_2 = r$ *and* $l_1 = l_2 = l$, then

$$M_{12} = M_{21} = M = \mu_o \pi r^2 n_1 n_2 l \tag{1.71}$$

Using Eq. (1.59), self-inductances of coils C1 and C2 can be written as

$$L_1 = \mu_o \pi r^2 n_1^2 l \tag{1.72}$$

$$L_2 = \mu_o \pi r^2 n_2^2 l \tag{1.73}$$

From Eqs. (1.72) and (1.73)

$$M = \sqrt{(\mu_o \pi r^2 n_1^2 l)\,(\mu_o \pi r^2 n_2^2 l)} = \sqrt{L_1 L_2} = \mu_o \pi r^2 n_1 n_2 l \tag{1.74}$$

It should be noted that Eq. (1.67) depends on the assumption that *all* of the flux produced by coil C_1 passes through the coil C_2. But practically, some of the flux leaks out, so that the mutual inductance is somewhat less than that given in Eq. (1.74). The mutual inductance can be written as

$$M = k\sqrt{L_1 L_2} \tag{1.75}$$

where the constant k is called the *coefficient of coupling* and lies in the range $0 \le k \le 1$.

Example 1.37 Derive an expression for the mutual force between two coaxial circular coils of radii R_1, R_2, turn N_1, N_2 and currents I_1, 1_2 respectively, if the distance between their centres is L, as shown in Figure E1.37.

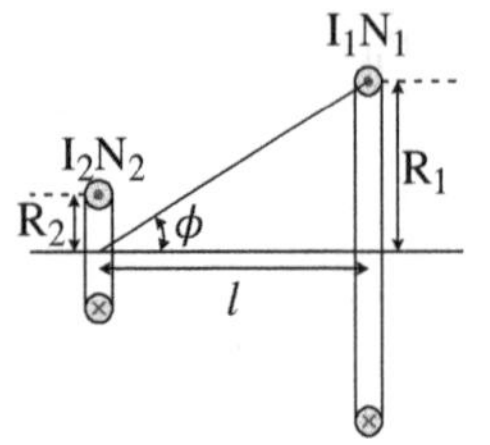

Figure E1.37 Coaxial circular coils of Example 1.37.

Assume the smaller coil is so situated that the flux, set up by the larger coil, completely envelops it and can be treated as uniform over the area of the smaller coil, together with the fact that the turns of each coil are co-planar. Applying the Biot-Savart law to the larger coil, to find the magnetic strength of the field on the co-axis, it can be seen that

$$H_1 = \frac{I_1 N_1 \sin \phi R_1}{2\left(R_1^2 + l^2\right)}$$

and since

$$\sin \phi = \frac{R_1}{\sqrt{\left(R_1^2 + l^2\right)}}$$

$$H_1 = \frac{I_1 N_1 R_1^2}{2\left(R_1^2 + l^2\right)^{\frac{3}{2}}}$$

thus, flux density B_1

$$B_1 = \mu_o H_1 = \frac{\mu_o I_1 N_1 R_1^2}{2\left(R_1^2 + l^2\right)^{\frac{3}{2}}}$$

Hence

flux linkage the small coil $\phi_2 = \pi R_2^2 B_1 = \frac{\pi R_2^2 \mu_o I_1 N_1 R_1^2}{2(R_1^2+l^2)^{\frac{3}{2}}}$

flux linkage with coil $2 = \phi_2 N_2 = \frac{\pi R_2^2 \mu_o I_1 N_1 R_1^2 N_2}{2(R_1^2+l^2)^{\frac{3}{2}}}$

The mutual inductance between the coils is

$$M_{12} = \frac{\text{flux linkage with coil 2}}{\text{current of coil 1}} = \frac{\pi R_2^2 \mu_o I_1 N_1 R_1^2 N_2}{2\left(R_1^2 + l^2\right)^{\frac{3}{2}} I_1} = \frac{\mu_o \pi N_1 N_2 R_1^2 R_2^2}{2\left(R_1^2 + l^2\right)^{\frac{3}{2}}}$$

Since the potential energy of one circuit in the field of another is given by $M_{12}I_1I_2$, the force between the coils $= I_1 I_2 \frac{dM_{12}}{dl}$

$$F = \frac{3\mu_o \pi N_1 N_2 R_1^2 R_2^2 I_1 I_2 l}{2\left(R_1^2 + l^2\right)^{\frac{5}{2}}}$$

1.4 Energy

Let a coil having an inductance of L Henry with the resistance of R Ω connected to a voltage source of V volts, resulting in a flow of current i amperes as shown in Figure 1.19. The KVL equation for the circuit is

$$V = L\frac{di}{dt} + Ri \tag{1.76}$$

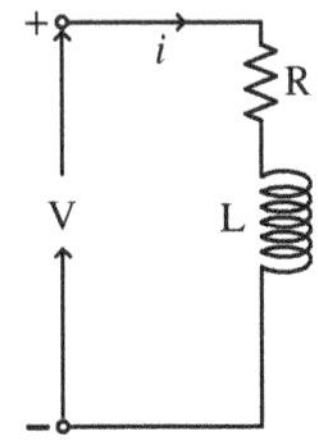

Figure 1.19
RL circuit.

To find the power supplied by the source to the circuit, multiply both sides of Eq. (1.76) by i, and the following is obtained

$$Vi = Li\frac{di}{dt} + Ri^2 \tag{1.77}$$

Equation (1.77) shows the work done per unit time by the source. The total work is done by the battery in raising the current in the circuit from zero at time $t = 0$ to i_T at time $t = T$ is

$$W = \int_0^T Vidt \tag{1.78}$$

Using Eq. (1.77), in Eq. (1.78), we get

$$W = \int_0^T \left(Li\frac{di}{dt} + Ri^2\right) dt$$

or

$$W = \frac{1}{2}Li_T^2 + R\int_0^T i^2dt \tag{1.79}$$

The first part of Eq. (1.79) shows the energy stored in the inductor which can be used once the battery is disconnected and the second part is the energy dissipated in the resistance of the coil but this energy is irreversible or it is the loss of energy.

$$W_{stored} = \frac{1}{2}Li^2 \tag{1.80}$$

where L is the self-inductance. We know from Eqs. (1.72), and (1.73) that

$$L = \mu_o n^2 \pi r^2 l \tag{1.81}$$

where n is the number of turns per unit length of the solenoid, r the radius, and l the length. The field inside the solenoid is uniform, with a magnitude

$$B = \mu_o ni \tag{1.82}$$

and is zero outside the solenoid. Equation (1.80) can be rewritten

$$W = \frac{B^2}{2\mu_o}V_{vol} \tag{1.83}$$

where $V_{vol} = \pi r^2 l$ is the volume of the solenoid. The above formula strongly suggests that a magnetic field possesses an energy density

$$U_B = \frac{B^2}{2\mu_o} \tag{1.84}$$

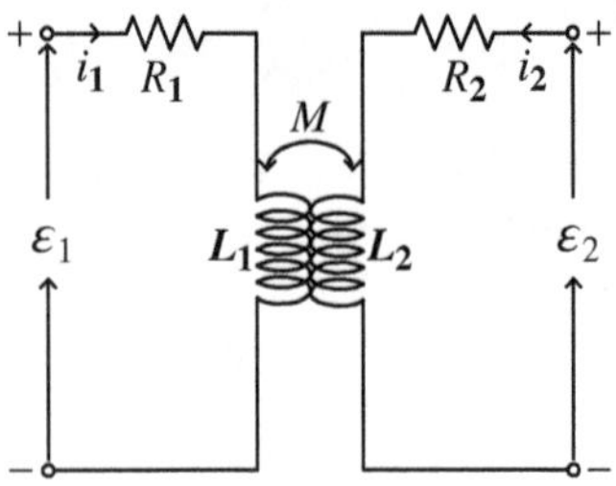

Figure 1.20 Two magnetically coupled coils.

Now, in Figure 1.20, two coils are wound one over the other and are connected to individual sources. The KVL equations are

$$\mathcal{E}_1 = R_1 i_1 + L_1 \frac{di_1}{dt} + M\frac{di_2}{dt} \tag{1.85a}$$

$$\mathcal{E}_2 = R_2 i_2 + L_2 \frac{di_2}{dt} + M\frac{di_1}{dt} \tag{1.85b}$$

The energy supplied by the two sources in increasing the currents in the two circuits, from zero at time 0, to i_1 and i_2 at time T, respectively, is

$$W = \int_0^T (\mathcal{E}_1 i_1 + \mathcal{E}_2 i_2)\,dt = \int_0^T (R_1 i_1^2 + R_2 i_2^2) + \frac{1}{2}L_1 i_1^2 + \frac{1}{2}L_2 i_2^2 + M\int_0^T \left(i_1\frac{di_2}{dt} + i_2\frac{di_1}{dt}\right)dt \tag{1.86}$$

Thus,

$$W = \int_0^T (R_1 i_1^2 + R_2 i_2^2) + \frac{1}{2}L_1 i_1^2 + \frac{1}{2}L_2 i_2^2 + M i_1 i_2 \tag{1.87}$$

Hence, the total magnetic energy stored in the two coils is

$$W_{stored} = \frac{1}{2}L_1 i_1^2 + \frac{1}{2}L_2 i_2^2 + M i_1 i_2 \tag{1.88}$$

The sign of mutual inductance M will be positive if the currents i_1 and i_2 flow in such a direction that the magnetic fields produced help each other. Then, the energy stored increases and whereas the sign of mutual inductance M will be negative if their directions are such that fields oppose each other, then, the stored energy decreases. The stored energy cannot be negative.

$$\frac{1}{2}L_1 i_1^2 + \frac{1}{2}L_2 i_2^2 + M i_1 i_2 \geq 0 \tag{1.89}$$

or

$$\frac{1}{2}\left(i_1\sqrt{L_1} + i_2\sqrt{L_2}\right)^2 - i_1 i_2\left(\sqrt{L_1 L_2} - M\right) \geq 0 \tag{1.90}$$

If we assume that $i_1 i_2 < 0$, then

$$M \leq \sqrt{L_1 L_2} \tag{1.91}$$

Equation (1.91) is the same as the Eq. (1.75).

1.5 Overview of Electric Machines

Electric Machines are the electromechanical energy conversion devices used to transform mechanical energy to electrical energy and vice-versa. A variety of electric machines are used in residential, commercial and industrial applications. Electric machines are classified into several types depending upon different parameters. Here some classifications are explained.

An electric Machine is a general term used for both motors and generators. Hence, the first classification is based on the fact that if the electrical power is converted to mechanical power (motors) or mechanical power is converted to electrical power (generator), as shown in Figure 1.21.

An electric machine is classified by the way it processes energy, stationary or rotating as shown in Figure 1.22. Stationary electric machines are 'Transformers'. Transformers are electric machines that transform electric energy from one circuit through to the magnetic field by changing the voltage and current levels without changing the frequency of the input and output signals. The most commonly used transformer has two windings coupled by a magnetic core. The source is connected to one winding called the 'primary winding' and a load is connected to the second winding, called the 'secondary winding'. The number of turns in each winding decides the voltage transfer ratio. Transformers are used in a huge number of applications from small power supplies (small appliances, mobile phone

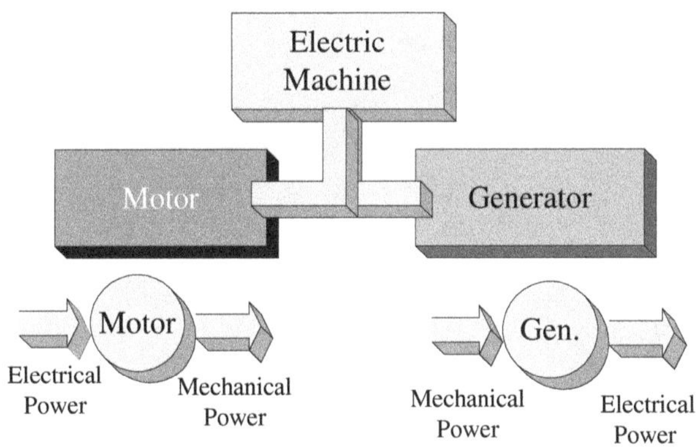

Figure 1.21 Electric machine classification based on their function.

Figure 1.22 Electric Machine classification, static or rotating.

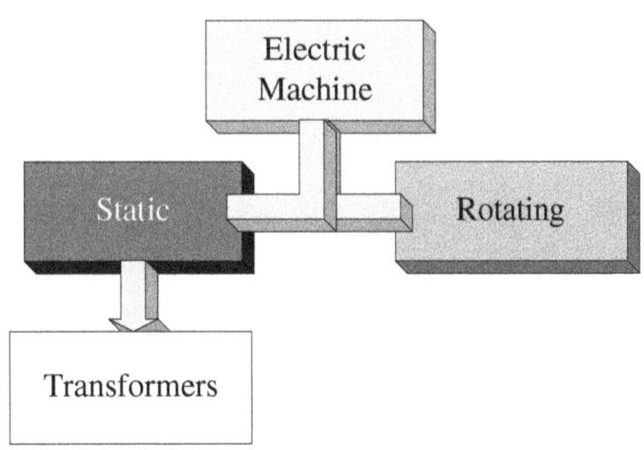

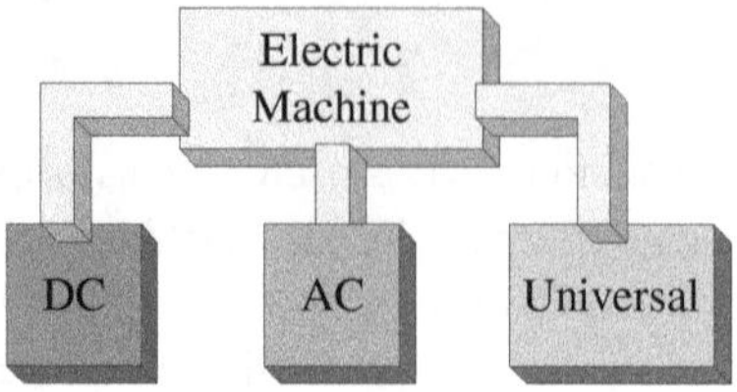

Figure 1.23 Electric Machine classification based on types of supply.

chargers, laptop chargers, home appliances) to big power stations and power transmission and distribution networks. A rotating machine has two parts; stator and rotor. One part is stationary, called the stator and other part is rotating called the 'rotor'. Stator and rotor are separated by a small air gap. The length of air-gap is kept as small as mechanically possible. The smaller the air-gap length, the smaller the magnetizing current drawn by the machine to establish the working flux will be.

Another classification of electric machine is based on the type of electric supplies it works on as shown in Figure 1.23. Hence a machine can be DC or AC. However, there exists a special machine that operates with both AC and DC, called a universal machine. Universal motors are most commonly used in Home Appliances for high-speed applications.

A classification of the DC machine is given in Figure 1.24, where all different possible types are listed. A PM machine has a permanent magnet for producing main flux. Thus, the field system is uncontrollable. However, there is space saving and no field losses, thus PM machines offer higher power density and greater efficiency. Major advantages of a DC machine include its simple and flexible control, especially shunt and separately excited machines where flux and torque can be independently controlled by the field current and

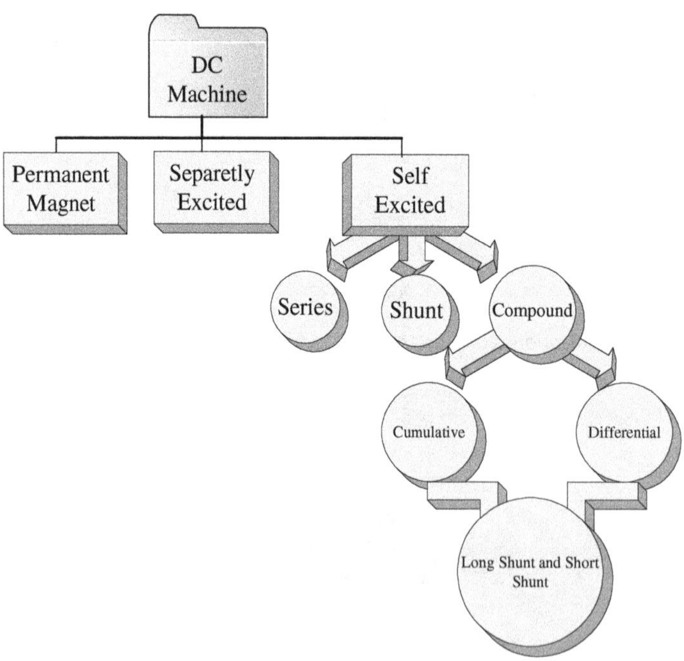

Figure 1.24 Classification of DC machines.

the armature current, respectively. The major problem with a DC machine is due to commutator and brushes. Due to commutator, a ripple in power occurs and it also limits the speed of the machine. Brushes cause friction and electromagnetic interference. Brushes wear out quickly hence, regular maintenance is required. Further, the size of the DC machine is comparatively bigger.

An AC machine classification is given in Figure 1.25. The major types are Induction and Synchronous. They are further classified according to the number of phases; single-phase, three-phase and multi-phase (more than three-phase), this classification is not shown in the figure. Machine refers to both the motor and generator. However, in conventional power stations (Thermal power plant, Hydroelectric power plant and Nuclear Power Plant) all the generators used are of the Synchronous type. This is because a synchronous machine

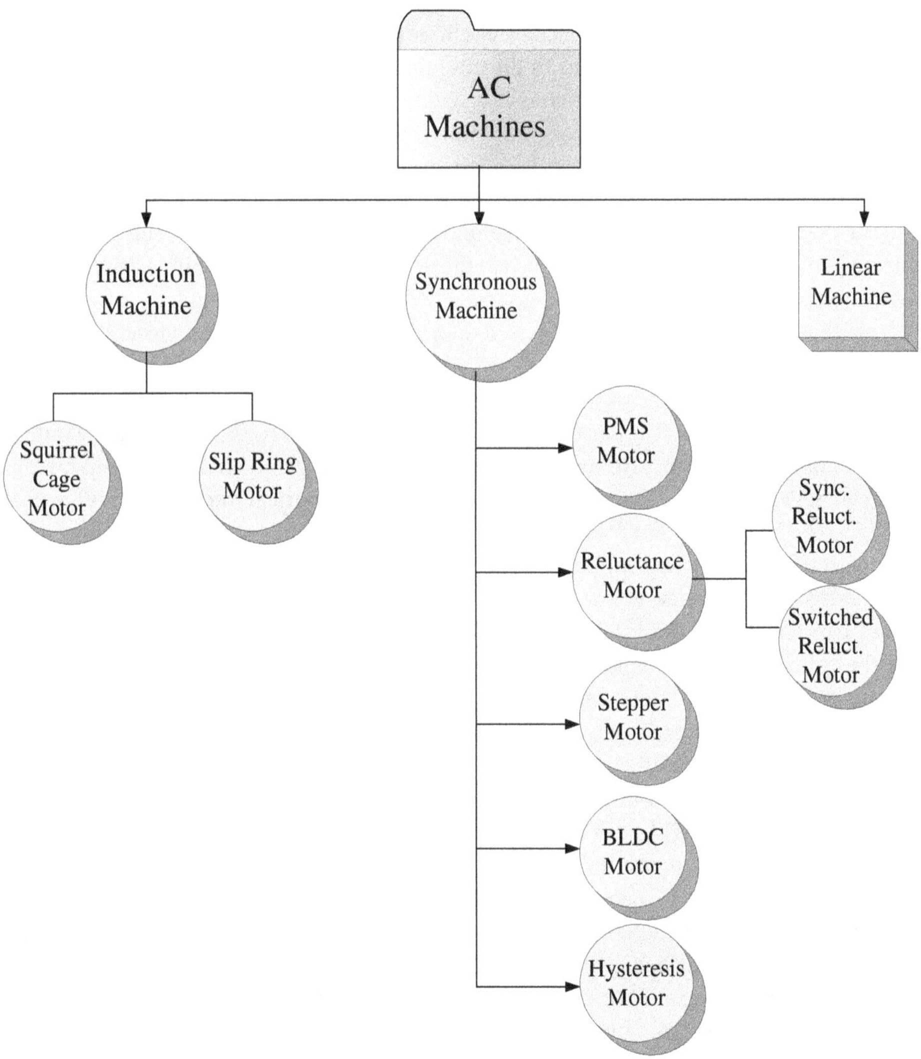

Figure 1.25 AC machine classification.

when operated at constant speed, gives a constant frequency voltage and current output. Induction generators are used mostly used in wind energy generation systems where the output frequency is variable and is controlled to be fixed using power electronic converters.

Three-phase induction motors are the most commonly used motors in industrial, commercial, and domestic applications. In domestic applications, mostly single-phase induction motors are used. Induction motors are called the workhorse of industry. Induction motors are reliable and rugged motors without any commutator or brushes. They are used for continuous operation in all types of environment. The principle of operation is based on Faraday's law of electromagnetic induction, hence the name. There are two types of induction machines based on their rotor constructions. If the rotor is a cage type where copper bars are embedded with short circuited end-rings without any winding, it is called a 'Squirrel cage' induction machine. The stator has three-phase balanced, either concentrated or distributed, windings. If the rotor has three-phase winding, it is called a 'wound rotor' or 'slip ring' type induction machine. In an induction machine, the field produced, due to stator current and rotor current, rotates at a synchronous speed (Ns = 120f/P, f-supply frequency and P is the number of poles of winding) while the rotor rotates at a speed lower than the synchronous and hence, this is called an asynchronous machine. Three-induction motors are self-starting while single-phase induction motors are not-self-starting. When the rotor of a wound rotor type induction machine is supplied by a power converter with variable frequency and the stator is connected to the grid, it is known as a 'Double Fed Induction Generator, DFIG'. It is mostly used in a Wind Power Generation system.

Synchronous machines are mostly used as a generator. Synchronous motors are only used for high-power applications in the MW range. As the name suggests, this is a constant speed machine that always rotates at a synchronous value (Ns = 120f/P). The stator carries AC armature winding (three-phase distributed) and the rotor has DC field winding. The rotor is supplied by an external DC source via the slip ring and brush arrangement. The rotor can have two shapes; cylindrical and salient pole. Cylindrical rotor machines are used for high-speed (thermal and nuclear power plants) and salient pole is used for low-speed (hydroelectric power plant) applications.

The cross-sectional view of the DC, induction and synchronous machines are given in Figure 1.26. The stator and rotor structures are clearly seen [5].

If the rotor winding is replaced by Permanent Magnets, the machine is called a Permanent Magnet Synchronous machine (PMSM) or sinusoidal PMSM. Use of PM offers several advantages, for example, no rotor losses (hence higher efficiency), faster response time as the electromechanical time constant is reduced. The PM material used for the rotor are; ferrite, aluminium-nickel-cobalt (Al-Ni-Co), samarium-cobalt (Sm-Co) and neodymium-iron-born (Nd-Fe-B). The most preferable PM material is Nd-Fe-B since it has high remanence which measures the strength of the magnetic field, high coercivity which denotes the resistance to becoming demagnetized and large energy product ($B.H_{max}$) which represents the density of magnetic energy. Permanent magnet, if used in the outer surface of the rotor, is called a surface mount PM machine, if PM is used inside the rotor surface it is called an inset PM machine and if PM is embedded inside the rotor, it is called an interior PM machine as shown in Figure 1.27.

The back emf of PMSM is sinusoidal in nature. The applied current is also sinusoidal. If the back emf is made trapezoidal, the machine becomes a Brushless DC Machine (BLDC).

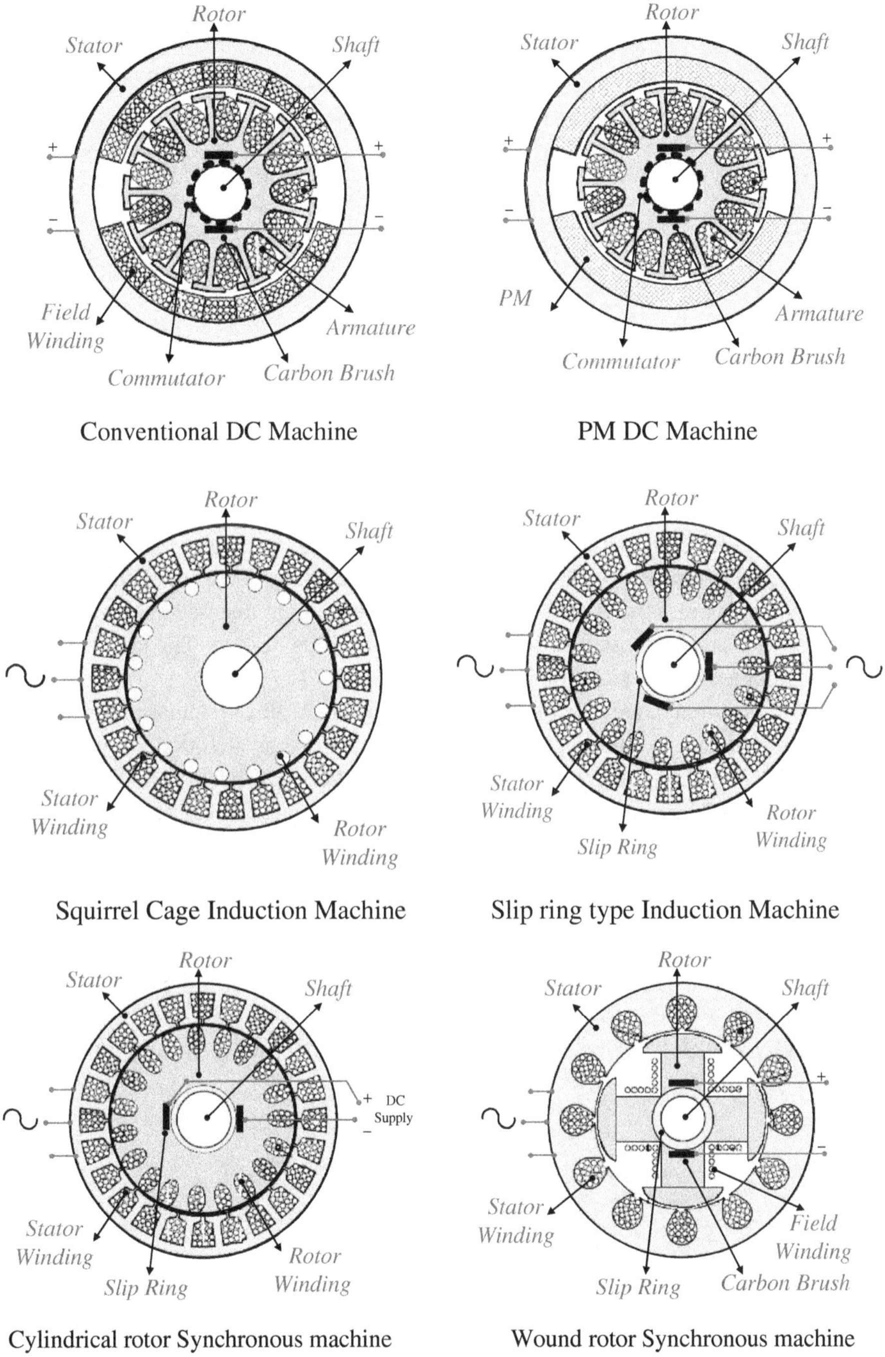

Figure 1.26 Cross-sectional view of different machines.

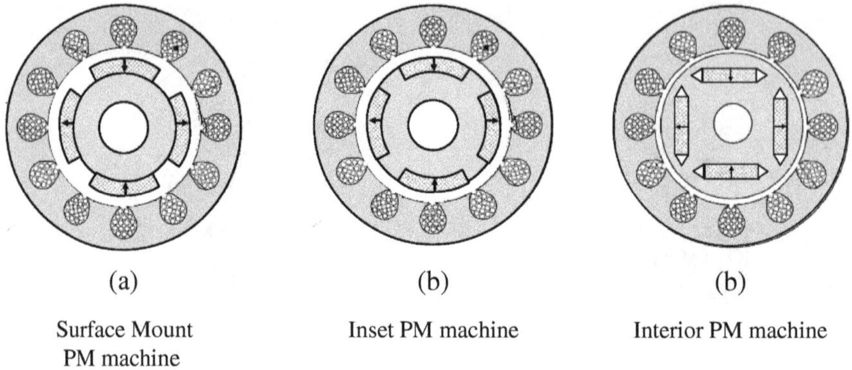

Figure 1.27 Sinusoidal PMSM machines.

The applied current is a quasi-rectangular waveform. For motoring operations, the supplied current is made positive for positive emf and negative for negative emf. Hence, their product is always positive. BLDC are ac synchronous machines with permanent magnets on rotor and trapezoidal back emf shape. The stator poles (projected type) are supplied by current to produce magnetic poles that attract/repel the rotor permanent magnets. The stator poles are energized in a proper sequence to produce a continuous motion. The cross-sectional view of a three-phase BLDC machine is shown in Figure 1.28.

The synchronous reluctance machine operates on the principle of reluctance torque. The torque is produced due to saliency in the rotor. The stator has a slotted structure with three-phase distributed winding placed inside the slots. The rotor has no windings and no permanent magnets. The number of poles on stator and rotors are same. The cross-sectional view of a synchronous reluctance machine is shown in Figure 1.29.

The switched reluctance machine has salient structure on both stator and rotor. The number of poles on stator and rotor are not integer multiples. This is done to avoid all stator and rotor poles aligning simultaneously. Under all aligned conditions, no torque is produced and the rotor is locked to the stator. Generally, the numbers of stator poles are higher than the rotor poles. The number of stator and rotor poles are 6/4, 8/6, 12/10, etc. The two windings on two opposite stator poles are connected in series or parallel to form one supply

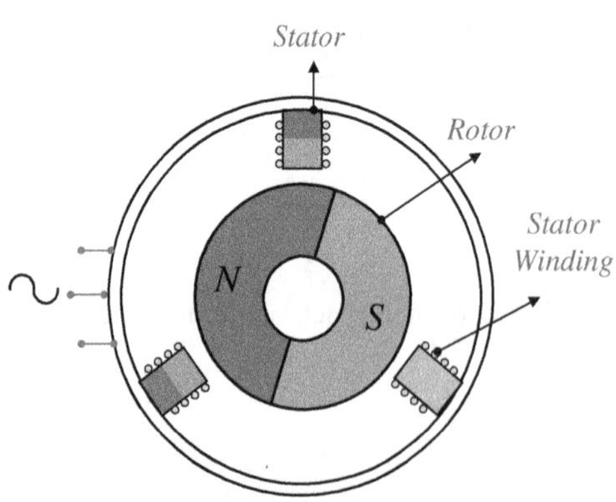

Figure 1.28 BLDC machine.

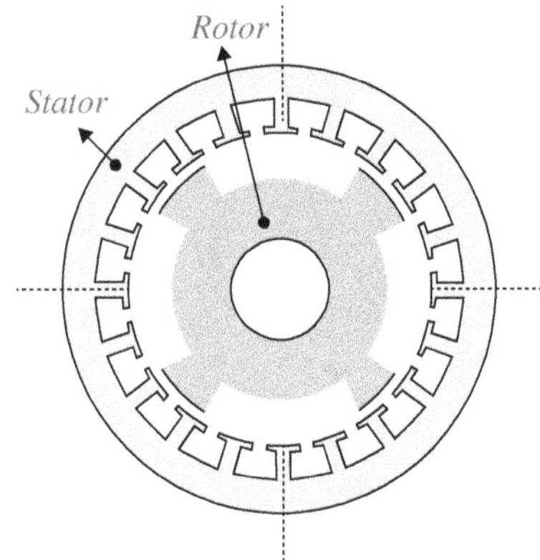

Figure 1.29 Synchronous reluctance machine.

phase. As such, all the supply phases are independent from one another. When a stator pole is equidistant from the two adjacent rotor poles, it is known as a 'fully unaligned position'. This is the position of maximum reluctance. When two (or more) stator poles are fully facing two (or more) rotor poles, it is called a 'fully aligned position' which is a position of minimum reluctance.

When a stator pole is excited, torque is produced in such a direction that the rotor moves toward minimum reluctance. Thus, the nearest rotor pole is pulled from the unaligned position into alignment position with the stator poles (lower reluctance position). Thus, by appropriately switching the phases, the machine rotates continuously in one direction. The major advantages of the SRM are their simple construction, high-power density and great efficiency. Since the rotor has no winding and no magnet, it is light in weight and there are no copper losses. Nowadays, SRM is mostly used in Electric Vehicle applications. Cross-sectional view of a 6/4 SRM machine is shown in Figure 1.30. The aligned and non-aligned positions in an 8/6 SRM machine is shown in Figure 1.31. SRM machines are rugged, reliable, have high-power density and require a simple unidirectional power converter. SRM requires rotor position sensor, has higher torque ripple with noisy machine operation. The current shape is non-sinusoidal. The machine construction is simple; however, the control is complex.

A stepper motor is not a continuously rotating machine but rather it rotates in steps. It rotates by a specific number of degrees. A train of input current pulses are supplied and

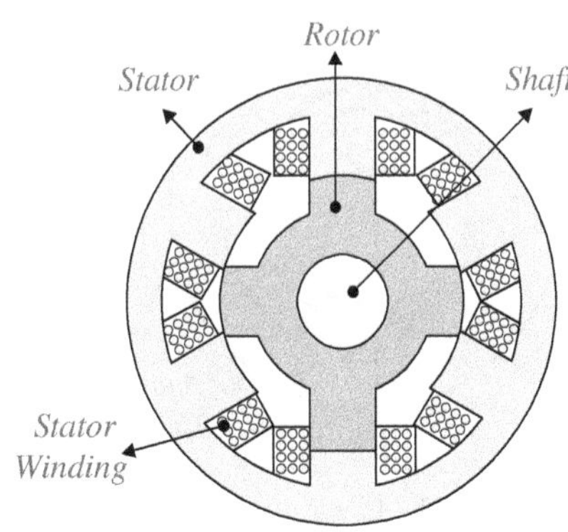

Figure 1.30 A 6/4 SRM motor.

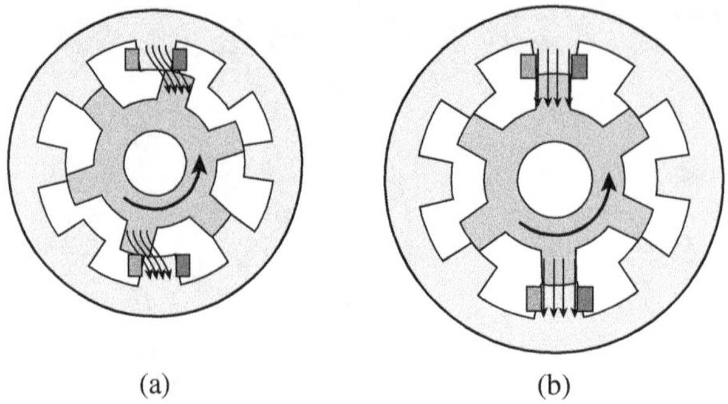

Figure 1.31 SRM, (a) unaligned, (b) aligned position.

the motor moves one step for each input current pulse. Typical steps sizes are 2°, 2.5°, 5°, 7.5°, and 15°. It is a digitally controlled control whose speed or steps are determined by the frequency of the current pulses. The machine does not need any position sensor or feedback system. There are two types of stepper motors namely; variable-reluctance type and permanent-magnet type.

Linear Motors gives linear or translational motion. Linear motor is obtained by simply cutting the round machine and unrolled onto a flat surface as shown in Figure 1.32. The stator and rotor in a linear motor is termed as primary and secondary, respectively. Any type of rotating machine can be converted into a linear motor version. In general, induction and synchronous linear motors are most commonly used. In a linear induction motor (LIM), the squirrel cage rotor is replaced by a cylinder of conductor (generally made of Aluminium) that encloses the rotor magnetic core. When three-phase supply is given to the primary, travelling flux wave is produced as shown in Figure 1.33. This flux links with the secondary, induces current in the secondary and produces a thrust (linear force). The air-gap length in an LIM is greater than rotatory IM. The typical value of air gap in an LIM is 15–30 mm. Larger air-gap length needs a larger magnetizing current and the power factor is poor. LIM operate at a greater slip value and hence losses are high. The efficacy of LIM is low. LIM can be both single-sided and double-sided. In double-sided, there are two primaries enclosing the secondary to enhance the efficiency of the machine as shown

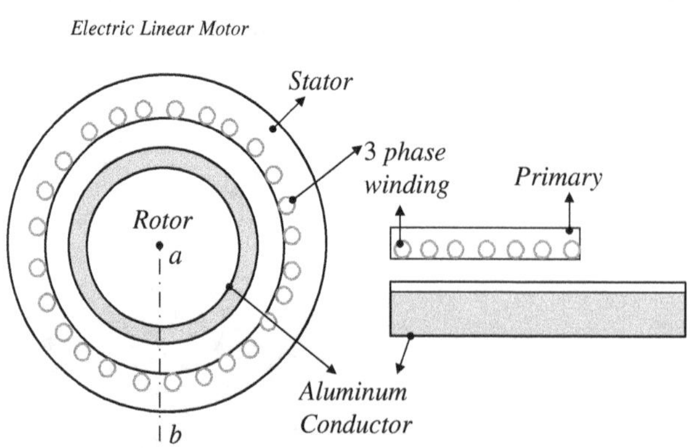

Figure 1.32 Linear Induction motor.

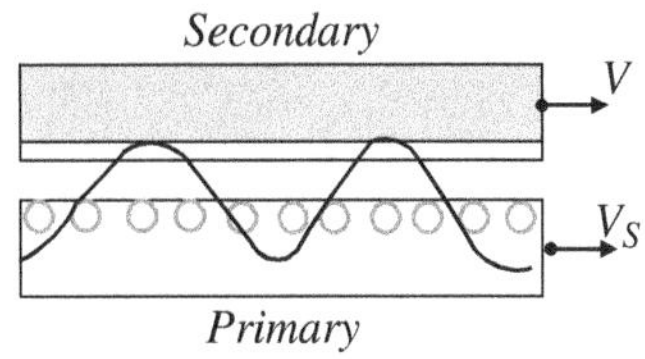

Figure 1.33 Linear Induction motor showing travelling wave.

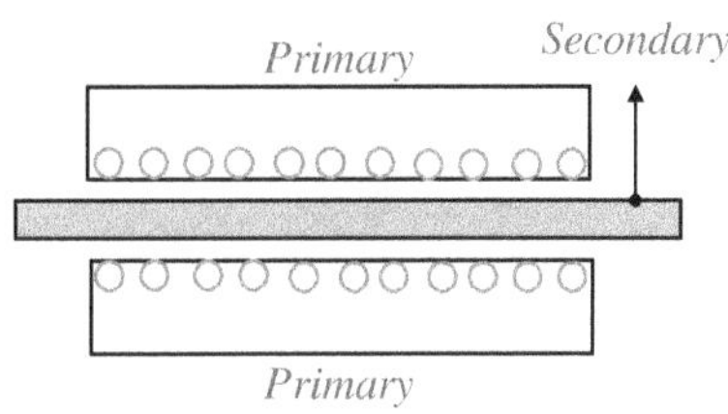

Figure 1.34 Double-sided Linear Induction motor.

in Figure 1.34. Major applications include the automatic sliding doors in trains, elevator etc., conveyor belts, high-speed trains, material handling, in automobile manufacturing industries etc. Conversion of a rotatory PM machine to Linear PM machine is shown in Figure 1.35.

Hysteresis motors are special synchronous motors that rely on hysteresis properties to produce torque. The stator is similar to induction and synchronous machines with either single- or three-phase winding. The rotor is smooth without any slots or any winding. It consists of hard magnetic material such as cobalt-steel wrapped around a non-magnetic material such as aluminium. The inside rotor material is chosen to reduce the rotor weight as shown in Figure 1.36.

When stator is supplied by current, a rotating magnetic field is produced that links with the rotor. The rotor is hard magnetic material with high retentivity. The rotor flux will lag the stator flux due to hysteresis property (retentivity). The angle between the stator and rotor flux causes torque production. Thus, the torque in a round rotor machine is given by

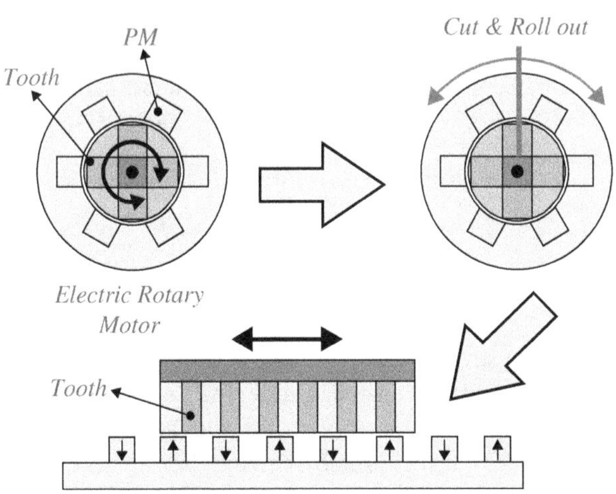

Figure 1.35 Linear PM synchronous machine.

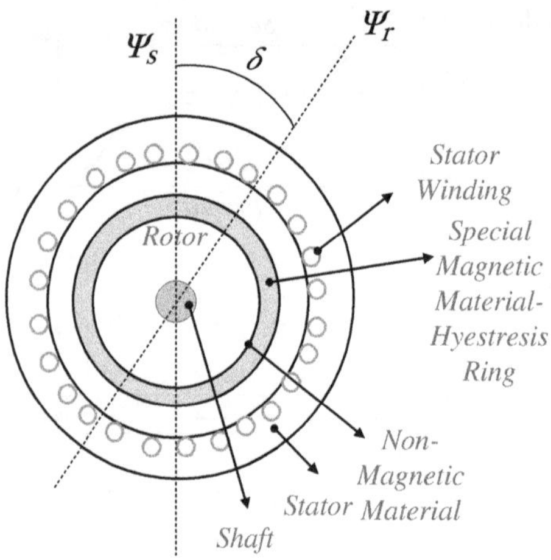

Figure 1.36 Hysteresis motor.

the relation;

$$T = K\psi_s\psi_r \sin(\delta) \tag{1.92}$$

The higher the value of δ, the higher the amount of torque will be. The angle δ depends upon the magnetic material property. The value of δ will be high if the hysteresis loop has a higher area. The rotor is smooth and hence is a noiseless operation. The major applications areas are timers, electric clocks and Magnetic tape recorders.

1.6 Summary

This chapter gives a foundation in electrical and magnetic systems covering the basic laws that govern their operation. All fundamental laws including Ohms law, Biot-Savart law, Ampere's law are discussed in detail with many solved numerical problems for clear understanding. The basics of Magnetic circuit are also included and which are further discussed in Chapter 2. Basic electrical components such as Resistance and Inductance are elaborated with numerical examples. Magnetically coupled circuits are also taken up for discussion. Principles and overview of electrical machines are given in the last section of the chapter.

Problems

Problems with The Biot-Savart Law

1.1 In a long, straight, vertical lightning stroke, electrons move downward and positive ions move upward and constitute a current of magnitude 20.0 kA. At a location 50.0 m east of the middle of the stroke, a free electron drifts through the air towards the west with a speed of 300 m/s. (a) Make a sketch showing the various vectors involved. Ignore the effect of the Earth's magnetic field. (b) Find the vector force the lightning stroke exerts on the electron. (c) Find the radius of the electron's path. (d) Is it a good approximation to model the electron as moving in a uniform field? Explain your answer. (e) If it does not collide with any obstacles, how many revolutions will the electron complete during the 60.0-ms duration of the lightning stroke? [2]

1.2 Determine the magnetic field (in terms of I, a, and d) at the origin due to the current loop in Figure P1.2. The loop extends to infinity above figure [2].

Figure P1.2

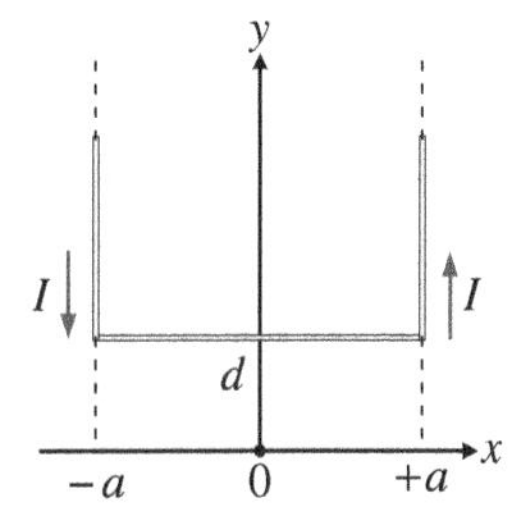

1.3 A wire carrying a current I is bent into the shape of an equilateral triangle of side L. (a) Find the magnitude of the magnetic field at the centre of the triangle. (b) At a point halfway between the centre and any vertex, is the field stronger or weaker than at the centre? Give a qualitative argument for your answer [2].

1.4 The two wires shown in Figure P1.4 are separated by $d = 10.0$ cm and carry currents of $I = 5.00$ A in opposite directions. Find the magnitude and direction of the net magnetic field (a) at a point midway between the wires; (b) at point P_1, 10.0 cm to the right of the wire on the right; and (c) at point P_2, $2d = 20.0$ cm to the left of the wire on the left. [2].

Figure P1.4

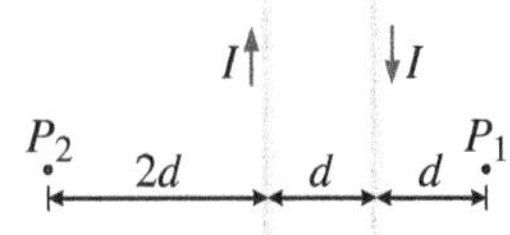

1.5 Two long, parallel wires carry currents of $I_1 = 3.00$ A and $I_2 = 5.00$ A in the directions indicated in Figure P1.5. (a) Find the magnitude and direction of the magnetic field at a point midway between the wires. (b) Find the magnitude and direction of the magnetic field at point P, located $d = 20.0$ cm above the wire carrying the 5.00-A current. [2].

Figure P1.5

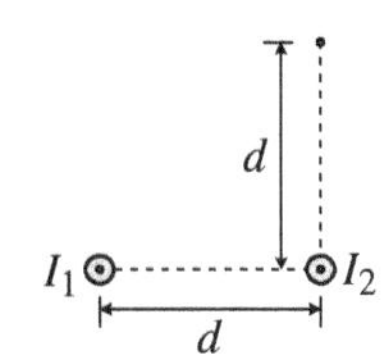

1.6 A current $I = 1.00$ A circulates in a round thin-wire loop of radius $R = 100$ mm. Find the magnetic induction
(a) at the centre of the loop;
(b) at the point lying on the axis of the loop at a distance $x = 100$ mm from its centre [6].

1.7 A current I flows along a thin wire, shaped as a regular polygon with n sides, which can be inscribed into a circle of radius R. Find the magnetic induction at the centre of the polygon. Analyse the obtained expression at $n \to \infty$ [6].

1.8 Find the magnetic induction at the centre of a rectangular wire frame whose diagonal is equal to $d = 16$ cm and the angle between the diagonals is equal to $\phi = 30°$; the current flowing in the frame equals $I = 5.0$ A [6].

1.9 A current $I = 5.0$ A flows along a thin wire shaped as shown in Figure P1.9. The radius of a curved part of the wire is equal to $R = 120$ mm, the angle $2\phi = 90°$. Find the magnetic induction of the field at point *0* [6].

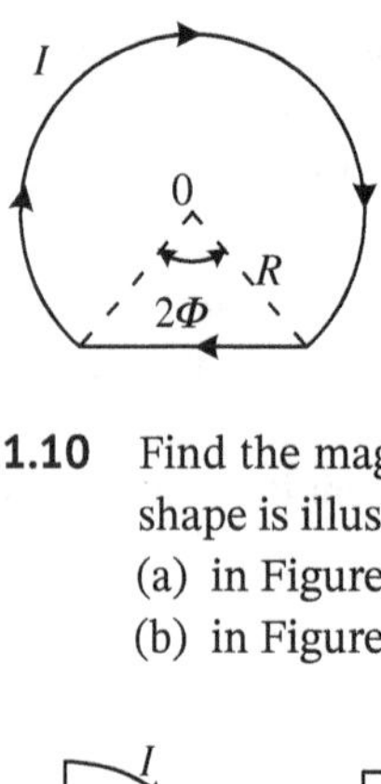

Figure P1.9

1.10 Find the magnetic induction of the field at point *0* of a loop with current I, whose shape is illustrated
(a) in Figure P1.10a, the radii a and b, as well as the angle ϕ are known;
(b) in Figure P1.10b, the radius a, and the side b are known [6].

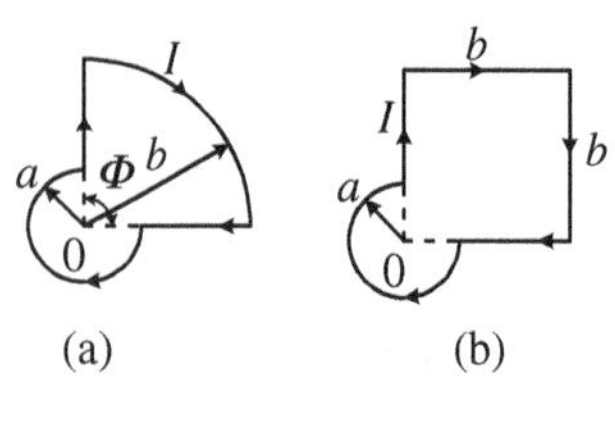

Figure P1.10

1.11 A current I flows along a lengthy thin-walled tube of radius R with the longitudinal slit of width h. Find the induction of the magnetic field inside the tube under the condition $h << R$ [6].

1.12 A current I flows in a long straight wire with a cross-section having the form of a thin half-ring of radius R (Figure P1.12). Find the induction of the magnetic field at point *0* [6].

Figure P1.12

1.13 Inside a long straight uniform wire of round cross-section, there is a long round cylindrical cavity whose axis is parallel to the axis of the wire and displaced from the latter by a distance ***l***. A direct current of density j flows along the wire. Find the magnetic induction inside the cavity. Consider, in particular, the case $l = 0$ [6].

Problems with Ampere's law

1.14 The magnetic field created by a large current passing through plasma (ionized gas) can force current-carrying particles together. This *pinch effect* has been used in designing fusion reactors. It can be demonstrated by making an empty aluminium can carry a large current parallel to its axis. Let R represent the radius of the can and I the current, uniformly distributed over the can's curved wall. Determine the magnetic field (a) just inside the wall and (b) just outside. (c) Determine the pressure on the wall [2].

1.15 A long, cylindrical conductor of radius R carries a current I as shown in Figure P1.15. The current density J, however, is not uniform over the cross-section of the conductor but rather is a function of the radius according to $J = br$, where b is a constant. Find an expression for the magnetic field magnitude B (a) at a distance $r_1 < R$ and (b) at a distance $r_2 > R$, measured from the centre of the conductor [2].

Figure P1.15

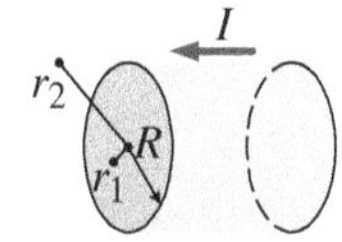

1.16 Four long, parallel conductors carry equal currents of $I = 5.00$ A. Figure P1.16 is an end view of the conductors. The current direction is into the page at points A and B and out of the page at points C and D. Calculate (a) the magnitude and (b) the direction of the magnetic field at point P, located at the centre of the square of edge length $l = 0.200$ m [2].

Figure P1.16

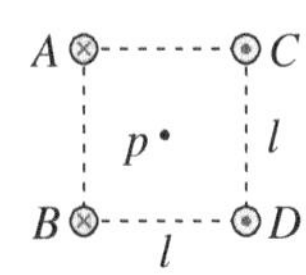

1.17 What is the line integral of $\overline{B}$ between points i and f in Figure P1.17? [7]

Figure P1.17

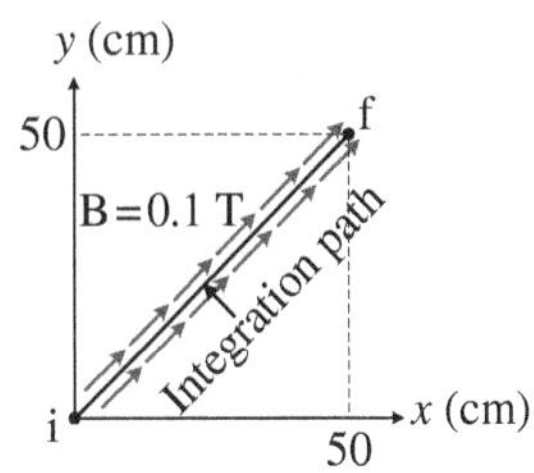

1.18 What is the line integral of $\overline{B}$ between points i and f in Figure P1.18? [7]

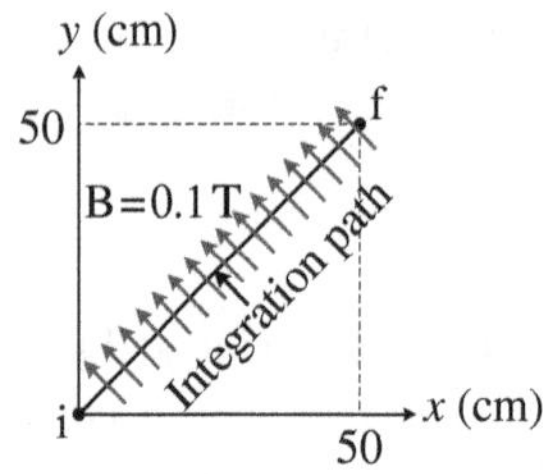

Figure P1.18 For problem 1.18

1.19 The value of the line integral of $\overline{B}$ around the closed path in Figure P1.19 is 3.77×10^{-6} Tm. What is I_3? [7]

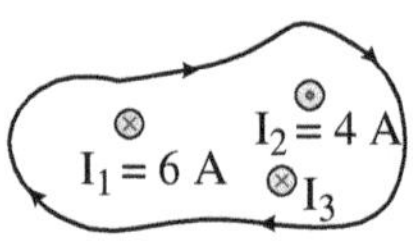

Figure P1.19

1.20 The value of the line integral around the closed path in Figure P1.20 is 1.38×10^{-5} Tm. What are the directions (in or out of the page) and the magnitude of I_3 [7].

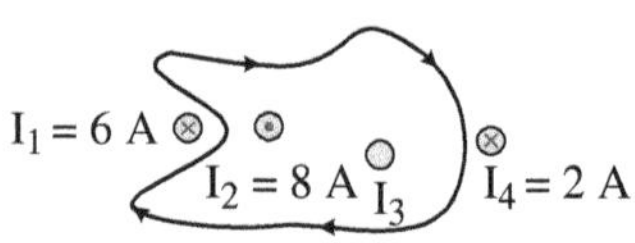

Figure P1.20

Induced EMF and Faraday's law

1.21 A solenoid of length 25 cm and radius 0.8 cm with 400 turns, is in an external magnetic field of 600 G that makes an angle of 50° with the axis of the solenoid. (a) Find the magnetic flux through the solenoid. (b) Find the magnitude of the emf induced in the solenoid if the external magnetic field is reduced to zero in 1.4 seconds [3].

1.22 A 100-turn circular coil has a diameter of 2 cm and a resistance of 50 Ω. The plane of the coil is perpendicular to a uniform magnetic field of magnitude 1 T. The direction of the field is suddenly reversed. (a) Find the total charge that passes through the coil. If the reversal takes 0.1 seconds, find (b) the average current in the coil and (c) the average emf in the coil. [3]

1.23 At the equator, a 1000-turn coil with a cross-sectional area of 300 cm^2 and a resistance of 15 Ω is aligned with its plane perpendicular to the earth's magnetic field of 0.7 G. If the coil is flipped over, how much charge flows through the coil? [3]

1.24 A circular coil of 300 turns and a radius of 5 cm is connected to a current integrator. The total resistance of the circuit is 20 Ω. The plane of the coil is originally aligned perpendicular to the earth's magnetic field at some point. When the coil is rotated through 90°, the charge that passes through the current integrator is measured to be 9.4 μC. Calculate the magnitude of the earth's magnetic field at that point [3].

1.25 A rectangular coil in the plane of the page has dimensions a and b. A long wire that carries a current I is placed directly above the coil (Figure P1.25). The wire is placed at $x = b/4$. (a) Obtain an expression for the emf induced in the coil if the current varies with time according to $I = 2t$. (b) If $a = 1.5$ m and $b = 2.5$ m, what should the resistance of the coil be so that the induced current is 0.1 A? What is the direction of this current? [3]

Figure P1.25

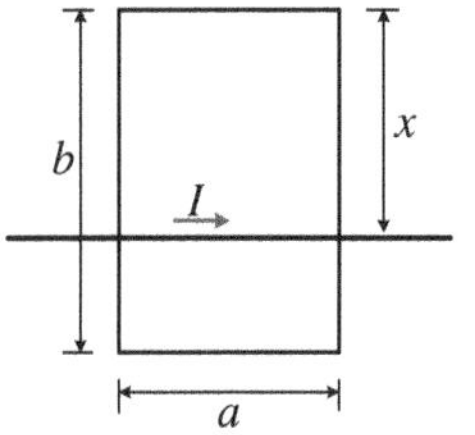

1.26 Repeat Problem 1.25 if the wire is placed at $x = b/3$. [3]

Problems on Motional EMF

1.27 In Figure P1.27, the rod has a resistance R and the rails have negligible resistance. A capacitor with charge Q_0 and capacitance C is connected between points a and b so that the current in the rod is downward. The rod is placed at rest at $t = 0$. (a) Write the equation of motion for the rod on the rails. (b) Show that the terminal speed of the rod down the rails is related to the final charge on the capacitor [3].

Figure P1.27

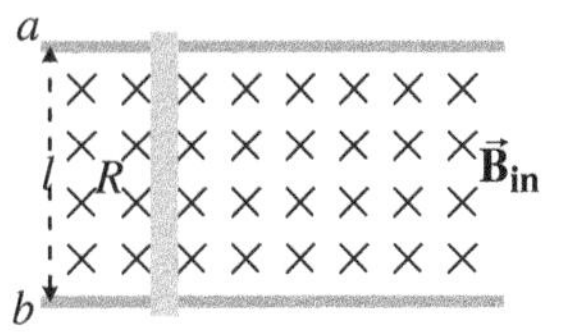

1.28 In Figure P1.28, a conducting rod of mass m and negligible resistance is free to slide without friction along two parallel rails of negligible resistance separated by a distance l and connected by a resistance R. The rails are attached to a long, inclined plane that makes an angle θ with the horizontal. There is a magnetic field B directed upward. (a) Show that there is a retarding force directed up the incline given by $F = \frac{B^2 l^2 v \cos^2\theta}{R}$. (b) Show that the terminal speed of the rod is $v_t = \frac{mgR \sin\theta}{B^2 l^2 \cos^2\theta}$ [3].

Figure P1.28

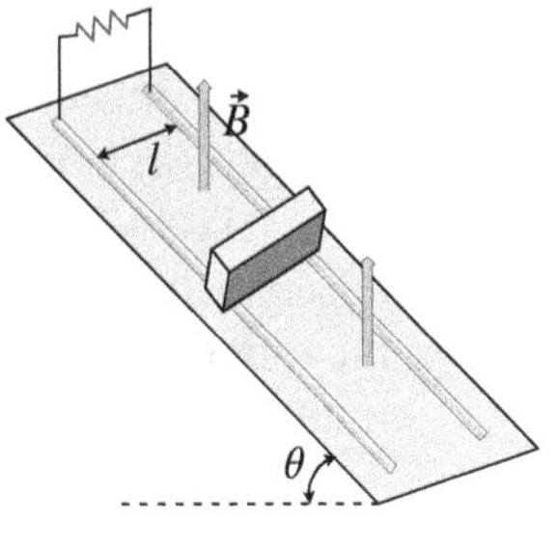

1.29 A square loop of conducting wire (area A) is pulled out of a region of constant, very high magnetic field B that is directed perpendicular to the plane of the wire. Half of the wire is in the field and half of the wire is out of the field when the wire is pulled out. A constant force F is exerted on the wire to pull the wire out. The wire is pulled out in time t. All else being equal, if the force were doubled, approximately how long would it take to pull the wire out? (*a*) t (*b*) $t/\sqrt{2}$ (c) $t/2$ (*d*) $t/4$ [3]

1.30 If, instead of doubling the force the resistance of the wire in Problem 1.29 was halved (all else being equal), what would the new time be? (*a*) t (*b*) $2t$ (*c*) $t/2$ (*d*) $t/\sqrt{2}$ [3]

1.31 A wire lies along the z-axis and carries current $I = 20$ A in the positive z-direction. A small conducting sphere of radius $r = 2$ cm is initially at rest on the y-axis at a distance $h = 45$ m above the wire. The sphere is dropped at time $t = 0$. (*a*) What is the electric field at the centre of the sphere at $t = 3$ s? Assume that the only magnetic field is the magnetic field produced by the wire. (b) What is the voltage across the sphere at $t = 3$ s? [3]

1.32 In Figure P1.32, let $\theta = 30°$; $m = 0.4$ kg, $l = 15$ m, and $R = 2\,\Omega$. The rod starts from rest at the top of the inclined plane at $t = 0$. The rails have negligible resistance. There is a constant, vertically directed magnetic field of magnitude $B = 1.2$ T. (*a*) Find the emf induced in the rod as a function of its velocity down the rails. (*b*) Write Newton's law of motion for the rod; show that the rod will approach a terminal speed and determine its value [3].

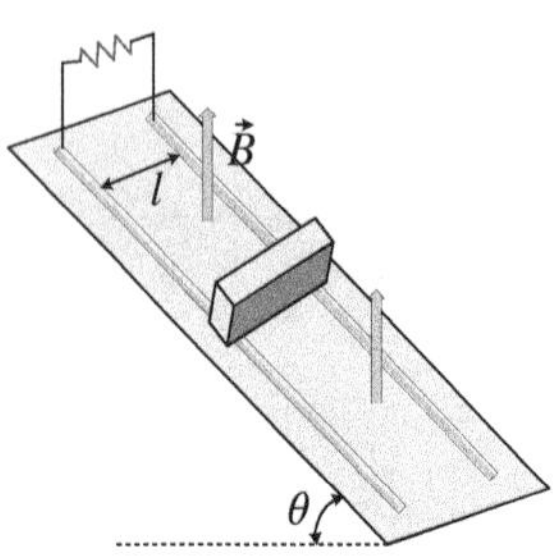

Figure P1.32

1.33 A solid conducting cylinder of radius 0.1 m and mass 4 kg rests on horizontal conducting rails (Figure P1.33). The rails, separated by a distance $a = 0.4$ m, have a rough surface, so the cylinder rolls rather than slides. A 12-V battery is connected to the rails as shown. The only significant resistance in the circuit is the contact resistance of 6 Ω between the cylinder and rails. The system is in a uniform vertical magnetic field. The cylinder is initially at rest next to the battery. (*a*) What must the magnitude be and the direction of $\vec{B}$ so that the cylinder has an initial acceleration of 0.1 m/s^2 to the right? (*b*) Find the force on the cylinder as a function of its speed v. (*c*) Find the terminal velocity of the cylinder. (*d*) What is the kinetic energy of the cylinder when it has reached its terminal velocity? (Neglect the magnetic field due to the current in the battery-rails-cylinder loop, and assume that the current density in the cylinder is uniform.) [3]

Figure P1.33

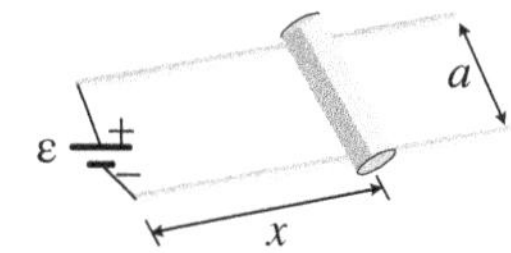

1.34 The loop in Figure P1.34 moves away from the wire with a constant speed v. At time $t = 0$, the left side of the loop is a distance d from the long straight wire. (a) Compute the emf in the loop by computing the motional emf in each segment of the loop that is parallel to the long wire. Explain why you can neglect the emf in the segments that are perpendicular to the wire. (b) Compute the emf in the loop by first computing the flux through the loop as a function of time and then using $\mathcal{E} = -\frac{d\phi_m}{dt}$. Compare your answer with that obtained in Part (a). Given: $a = 5\,\text{cm}$, $b = 10\,\text{cm}$, $d = 2\,\text{cm}$, and $I = 20\,\text{A}$ [3].

Figure P1.34

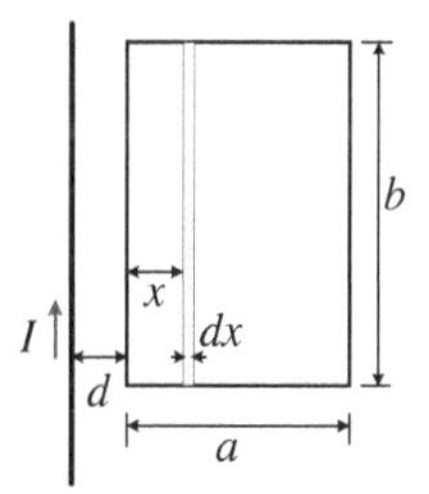

1.35 A conducting rod of length l rotates at a constant angular velocity about one end, in a plane perpendicular to a uniform magnetic field B (Figure P1.35). (a) Show that the magnetic force on a body whose charge is q at a distance r from the pivot is $Bqr\omega$. (b) Show that the potential difference between the ends of the rod is $V = \frac{1}{2}B\omega l^2$ (c) Draw any radial line in the plane from which to measure $\theta = \omega t$. Show that the area of the pie-shaped region between the reference line and the rod is $A = \frac{1}{2}l^2\theta$. Compute the flux through this area, and show that $\mathcal{E} = \frac{1}{2}B\omega l^2$ follows when Faraday's law is applied to this area [3].

Figure P1.35

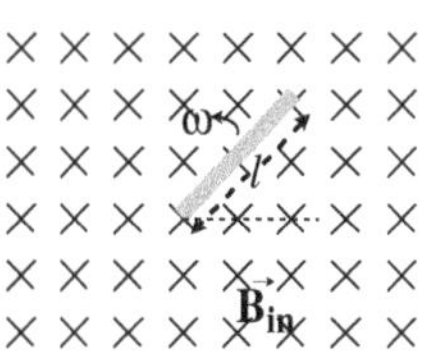

Problems with Inductance

1.36 A long insulated wire with a resistance of 18 Ω/m is to be used to construct a resistor. First, the wire is bent in half and then the doubled wire is wound in a cylindrical form, as shown in Figure P1.36. The diameter of the cylindrical form is 2 cm, its length is 25 cm, and the total length of wire is 9 m. Find the resistance and inductance of this wire-wound resistor. [3]

Figure P1.36

1.37 In Figure P1.37, circuit 2 has a total resistance of 300 Ω. A total charge of 2×10^{-4} C flows through the galvanometer in circuit 2 when switch S in circuit 1 is closed. After a long time, the current in circuit 1 is 5 A. What is the mutual inductance between the two coils? [3]

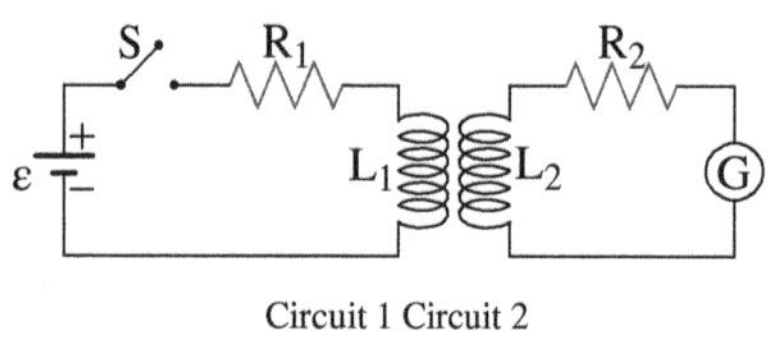

Figure P1.37

1.38 Show that the inductance of a toroid of rectangular cross-section, as shown in Figure P1.38, is given by $L = \frac{\mu_o NH \ln(b/a)}{2\pi}$, where N is the total number of turns, a is the inside radius, b is the outside radius, and H is the height of the toroid [3].

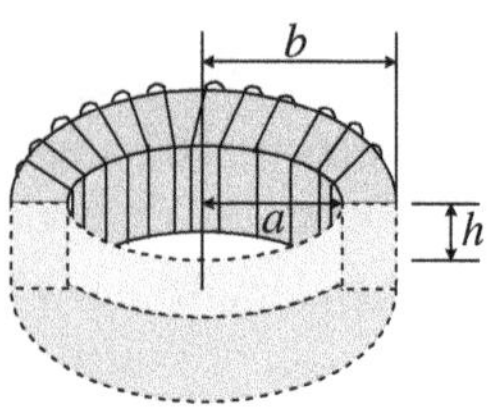

Figure P1.38

1.39 Using the result of Figure P1.38, calculate the self-inductance of an inductor wound from 10 cm of wire with a diameter of 1 mm into a coil with radius $R = 0.25$ cm [3].

1.40 A toroid of mean radius 25 cm and circular cross-section of radius 2 cm is wound with a superconducting wire of length 1000 m that carries a current of 400 A. (*a*) What is the number of turns on the coil? (*b*) What is the magnetic field at the mean radius? (c) Assuming that B is constant over the area of the coil, calculate the magnetic energy density and the total energy stored in the toroid [3].

References

1 Ostovic, V. (ed.) (1989). *Dynamics of Saturated Electric Machines*. Springer-Verlag.

2 Serway, R.A. and Jewett, J.W. Jr. (1996). *Physics for Scientists and Engineers with Modern Physics*, 9e. Boston, MA: Brooks/Cole.

3 Tipler, P. and Mosca, G. *Physics for Scientists and Engineers*, 5e. W. H. Freeman and Company.

4 Walker, J.(eds. J. Walker, D. Halliday and R. Resnick). *Fundamentals of Physics*, 10e. Wiley.

5 Chau, K.T., Li, W.L., and Lee, C.H.T. (2012). Challenges and Opportunities of Electric Machines for Renewable Energy (invited paper). In: *Progress In Electromagnetics Research B*, vol. 42, 45–74.

6 Irodov, I. (1979). *Problems in General Physics*. Mir Publishers of Moscow.

7 Knight, R.D. *Physics for Scientists and Engineers: A Strategic Approach*, 2e. Addison Wesley (book), Benjamin Cummings(solution manual).

2

Magnetic Circuits

2.1 Preliminary Remarks

There are many basic principles or laws related to electric machines that must be known and well understood before studying theory and analysis of electric machines. This chapter briefly reviews some of the fundamentals of magnetism. Properties of magnetic materials such as permeability, magnetic field intensity, hysteresis loop, core and eddy current losses and energy stored in magnetic fields are explained.

According to Oersted, we can have an electromagnet. Oersted stated that any current carrying conductor is always surrounded by a magnetic field. The property of such a current is called the magnetic effect of an electric current. A natural magnet or an electromagnet, both have a close relation with electromotive force (emf), mechanical force experienced by conductor, electric current etc. To understand this relationship, it is necessary to study the fundamental concepts of magnetic circuits. In this chapter, we shall study the laws of magnetism, magnetic field due to current-carrying conductor, magnetomotive force, simple series and parallel magnetic circuits.

Apart from the above stated topics, excitation and reluctance torques of electromechanical conversion systems, multiple excitation systems, and the concept of rotating magnetic field are also discussed.

2.2 Permeability

Magnetic Materials are those materials which can be magnetized and after magnetization, set up a magnetic field around its space. The motion of electrons in atoms gives rise to the phenomenon called magnetism. Basically, two types of motions are associated with the electrons:

- Motion of electrons in an orbit within the nucleus.
- Spin of electrons around its axis.

The spin and orbital moments impart magnetic moment on each electron, forcing them to behave as tiny magnets.

Electrical Machine Fundamentals with Numerical Simulation using MATLAB/SIMULINK, First Edition.
Atif Iqbal, Shaikh Moinoddin, and Bhimireddy Prathap Reddy.

Companion website: www.wiley.com/go/iqbal/electricalmachine

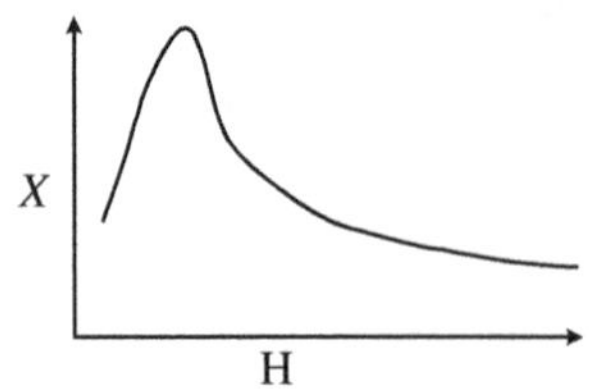

Figure 2.1 Susceptibility versus field strength. Source: Ferromagnetic materials.

These orbits can be regarded as circulating currents which develop a magneto motive force (mmf). In magnetized specimens, a number of dipoles may remain aligned in the direction of the external field and thus exhibit permanent magnetism. The permeability (μ) of the material is the measure of accepting magnetism. The permeability of a material is given as:

$$\mu = \mu_0 \mu_r \tag{2.1}$$

where $\mu_0[4\pi.10^{-7}]$ is the permeability of free space in Henry/m and μ_r is the unit less relative permeability of the material.

The magnetic flux density (B) is the actual magnetic field within a material regarded as a concentration of magnetic field lines, or flux, per unit cross-sectional area.

$$B = \frac{\phi}{A}\ \text{Wb/m}^2$$

The magnetic field strength (H, unit is Ampere-turn/m) is the magnetizing field produced by electric current flow in a coil. Mathematically, the magnetic permeability (μ) is defined as

$$\mu = B/H.$$

The permeability of most non-magnetic materials is equal to μ_0 H/m. The permeability of magnetic materials is μ H/m which is different for a variety of magnetic materials. The relative permeability of magnetic materials varies with the degree of magnetization (Figure 2.1). The dimension-less susceptibility ($\mathcal{X}$) is given by

$$\mathcal{X} = \mu_r - 1 \tag{2.2}$$

The magnetic susceptibility depends on the type of the magnetic material and on its physical state (temperature, etc.). When material is cold worked its susceptibility may change. Susceptibility of copper (for illustration) may change from negative to positive when cold worked and after cold work, on annealing it ($\mathcal{X}$) may go back to negative.

If one knows the force exerted on the magnetic material which is placed in magnetic field, the susceptibility can be identified. The susceptibility of ferromagnetic material differs with the variation of field strength.

2.3 Classification of Magnetic Materials

Magnetic materials, for which a linear relationship between M and H exists, are divided into classes depending upon the sign of $\mathcal{X}$. Materials which have a negative value of $\mathcal{X}$ of the order of 10^{-4} to 10^{-6} are called diamagnetic and those which have a positive value of $\mathcal{X}$

Figure 2.2 Different permeabilities.

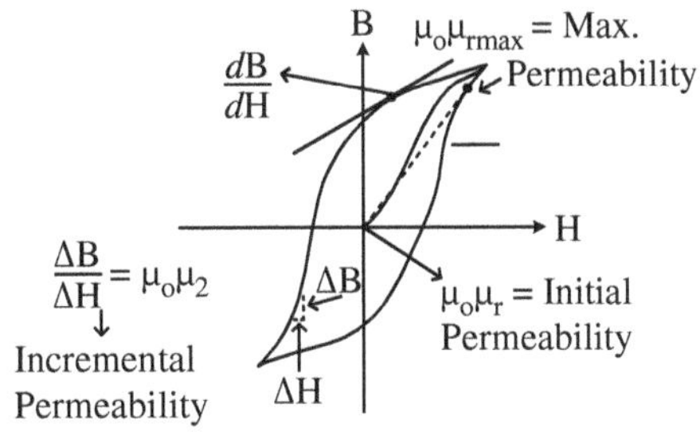

Figure 2.3 Arrangement of spins: (a) paramagnetic, (b) ferromagnetic, (c) anti-ferromagnetic, (d) Ferrimagnetic.

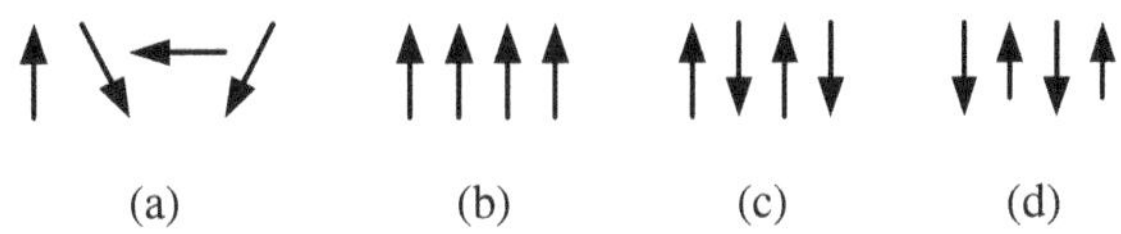

of about the same order of magnitude are called paramagnetic. The materials in which the resultant magnetization is one to several orders of magnitude greater than $\mu_0 H$ are called ferromagnetic. Figure 2.2 shows a number of permeabilities.

Another classification of magnetic materials is based on the presence or absence of permanent magnetic dipoles. Materials which lack permanent magnetic dipoles are called diamagnetic. If permanent magnetic dipoles are present in the atoms of a material, it may be paramagnetic, ferromagnetic, anti-ferromagnetic or ferromagnetic depending on the interaction between the individual dipoles. If the interaction between the atomic permanent dipole moments is zero or negligible and the individual dipole moments are oriented at random as shown in Figure 2.3a, the material will be paramagnetic. If the dipoles tend to line up in parallel, as shown in Figure 2.3b, the material will be ferromagnetic. When neighbouring moments are aligned in anti-parallel as in Figure 2.3c, the phenomenon is called anti-ferromagnetism. When the order of the magnetic moments is as shown in Figure 2.3d, the phenomenon is known as ferrimagnetism. The molecular moments in ferromagnetic materials will have values in between the values for ferromagnetic and anti-ferromagnetic materials.

The magnetic properties of materials are characterized by their relative permeabilities. Based on this, magnetic materials can be divided into three categories viz. (i) ferromagnetic materials, the relative permeabilities of which are far greater than unity and are dependent on the field strengths. Examples of paramagnetic materials are Fe, Co, Ni. (ii) Paramagnetic materials, which have relative permeabilities slightly greater than unity. Examples of paramagnetic materials are Al, Cr, Mo, Ti, Zr. (iii) diamagnetic materials, the relative permeabilities of which are slightly less than unity. Materials such as Cu, Al_2O_3, Si, Zn have stronger diamagnetic property.

When a paramagnetic or a diamagnetic material is placed in a magnetic field, the distortion of the field is negligible. On the other hand, when a ferromagnetic material is placed on the field, there is considerable distortion. Diamagnetism is a universal property of all materials. However, the diamagnetic properties are weaker than the paramagnetic ones and still weaker than the ferromagnetic properties. The peculiarities of ferromagnetic behaviour are due to a very specific property, viz. the formation within the ferromagnetic material of vast regions or domains within which the magnetic moments of a large number of atoms are arranged parallel to one another giving magnetic saturation in each domain.

2.3.1 Uniform Magnetic Field

The simplest magnetic field is one that is uniform – it doesn't change in magnitude or direction from one point to another [1]. A perfectly uniform field over a large area is not easy to produce. But the field between two flat parallel pole pieces of a magnet is almost uniform if the area the pole faces is large compared to their separation, as shown in Figure 2.4. At the edges, the field 'fringes' out somewhat: the magnetic field lines are no longer quite parallel and uniform. The parallel evenly spaced field lines in the central region of the gap indicate that the field is uniform at points not too near the edges.

2.3.2 Magnetic-Field Intensity

Any object of finite size contains a large number of atoms. In general, the magnetic moments (μ_m) of these atoms are randomly oriented and there is no net magnetic moment (μ_m) in any volume of the material that contains more than a several thousand atoms (Figure 2.5a). This volume is still quite small at microscopic scale. However, when the material is kept in an external magnetic field, torques act on the atomic dipoles and these torques try to align them parallel to the field (Figure 2.5b). The alignment is only partial because the thermal motion of the atoms frequently changes the orientation of the atoms and hence tries to randomize the magnetic moments (μ_m). The degree of alignment increases if the strength of the applied field is increased and also if the temperature is decreased. With sufficiently strong fields, the alignment is near perfect. We then say that the material is magnetically *saturated*.

When the atomic dipoles are aligned, partially or fully, there is a net magnetic moment (μ_m) in the direction of the field in any small volume of the material. We define the *magnetization vector* $\vec{M}$ as the magnetic moment $(\vec{\mu}_m)$ per unit volume. It is also called the *intensity of magnetization* or simply *magnetization*, Thus,

$$\vec{M} = \frac{\text{magnetic moment}}{\text{volume}} = \frac{\vec{\mu}_m}{V_{vol}} \tag{2.3}$$

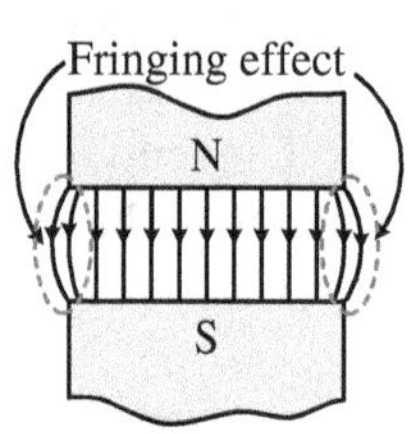

Figure 2.4 Magnetic field between two wide poles of a magnet is nearly uniform, except near the edges.

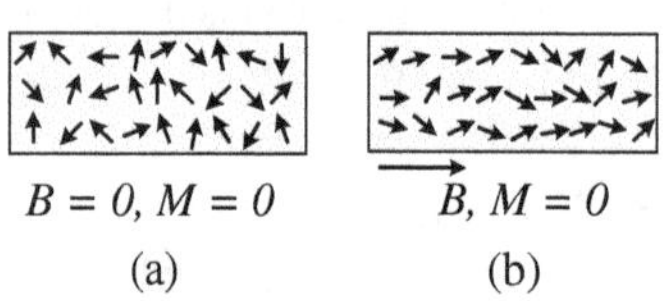

Figure 2.5 Magnetic diploes.

The unit of magnetic moment $(\vec{\mu}_m)$ is ampere-metre^2 so that from Eq. (2.4) the unit of $\vec{M}$ is ampere/metre.

Consider a bar magnet of pole strength m, length l and area of cross-section A. The magnetic moment (μ_m) of the bar magnet is $\mu_m = ml$. The intensity of magnetization is

$$M = \frac{\mu_m}{V_{vol}} = \frac{(m\ \text{Ampere.metre})\,(l\ \text{metre})}{(A\ \text{metre}^2)\,(l\ \text{metre})} = \frac{(m\ \text{Ampere.metre})}{(A\ \text{metre}^2)}$$

Thus, for a bar magnet, the intensity of magnetization may be defined as the *pole strength per unit face area.*

Example 2.1 A bar magnet made of steel has a magnetic moment of 2.5 A.m^2 and a mass of 6.6×10^{-3} kg. If the density of steel is 7.9×10^3 kg/m^3, find the intensity of magnetization of the magnet.

Solution

The volume of the bar magnet is

$$V_{vol} = \frac{\text{mass}}{\text{density}} = \frac{6.6 \times 10^{-3}\text{kg}}{7.9 \times 10^3\text{kg/m}^3} = 8.3 \times 10^{-7}\text{m}^3$$

The intensity of magnetization is

$$M = \frac{\mu_m}{V_{vol}} = \frac{2.5\ \text{A.m}^2}{8.3 \times 10^{-7}\text{m}^3} = 3.0 \times 10^8\text{A/m}$$

When a magnetic field is applied to a material, the material becomes magnetized. The actual magnetic field inside the material is the sum of the applied magnetic field and the magnetic field due to magnetization. It is convenient to define a new vector field

$$\vec{H} = \frac{\vec{B}}{\mu_o} - \vec{M} \tag{2.4}$$

where $\vec{B}$ is the resultant magnetic field and $\vec{M}$ is the intensity of magnetization. This quantity $\vec{H}$ is called *magnetic intensity* or *magnetizing field intensity*. The unit of H is the same as that of M, that is, ampere/metre. If no material is present (vacuum), $\vec{M} = 0$ and we have

$$\vec{H} = \frac{\vec{B}}{\mu_o} \tag{2.5}$$

Thus, the magnetic intensity due to a current element $id\vec{l}$ is, from Biot-Savart law,

$$d\vec{H} = \frac{1}{4\pi}\frac{id\vec{l} \times \vec{r}}{r^3} \tag{2.6}$$

The magnetic intensity due to a magnetic pole of pole strength m at a distance r from it is

$$H = \frac{m}{4\pi r^2} \tag{2.7}$$

Whenever the end effects of a magnetized material can be neglected, the magnetic intensity due to magnetization is zero. This may be the case with ring-shaped material or in the middle portion of a long rod. The magnetic intensity in a material is then determined by the external sources only, even if the material is magnetized.

Example 2.2 Find the magnetic intensity **H** at the centre of a long solenoid, having n turns per unit length, and carrying a current I (a) when no material is kept in it and (b) when a long copper rod is inserted in the solenoid.

Solution

(a) When there is no rod, the magnetic field at the centre of the solenoid is given by

$$B = \mu_0 ni$$

The magnetic intensity is

$$H = \frac{B}{\mu_0} = ni$$

(b) As the solenoid and the rod are long and we are interested in the magnetic intensity at the centre, the end effect may be neglected. There is no effect of the rod on the magnetic intensity at the centre. Its value in both cases are the same. Thus $H = ni$.

2.4 Hysteresis Loop

The magnetization in a ferromagnetic material not only depends on the magnetic intensity H but also on the previous history of the specimen. Suppose a ferromagnetic material is formed in the shape of a ring and placed inside a toroid having n turns per unit length. A current i can be passed through the toroid to produce a magnetic intensity H. The magnetic field produced by the current is

$$B_0 = \mu_0 ni$$

and hence

$$H = \frac{B_0}{\mu_0} = ni \tag{2.8}$$

Note that B_0 is the field produced by the toroid current only. The ring becomes magnetized and produces an extra field due to magnetization. The total field in the ring is

$$B = \mu_0 (H + M)$$

or

$$M = \frac{B}{\mu_0} - H = \frac{B}{\mu_0} - ni \tag{2.9}$$

One can measure the total field B inside the ring by using an apparatus known as Rowland's ring. The intensity of magnetization M can then be obtained from equation (ii). Thus, from (i) and (ii) one can obtain H and M for any current.

Figure 2.6 shows a typical magnetization curve when the current is changed. In the beginning, the current is zero and the sample has no magnetization. On the same graph, the **B-H** characteristic for non-magnetic material is shown to show the relative magnitudes involved.

Thus, $H = 0$ and $M = 0$. This corresponds to the point O. As the current is increased, H increases and the magnetization increases. As the current is increased to a maximum, H becomes H_{sat} and the magnetization M becomes almost saturated. In the whole process, the magnetization varies along the path Oa. Now, suppose the current is gradually

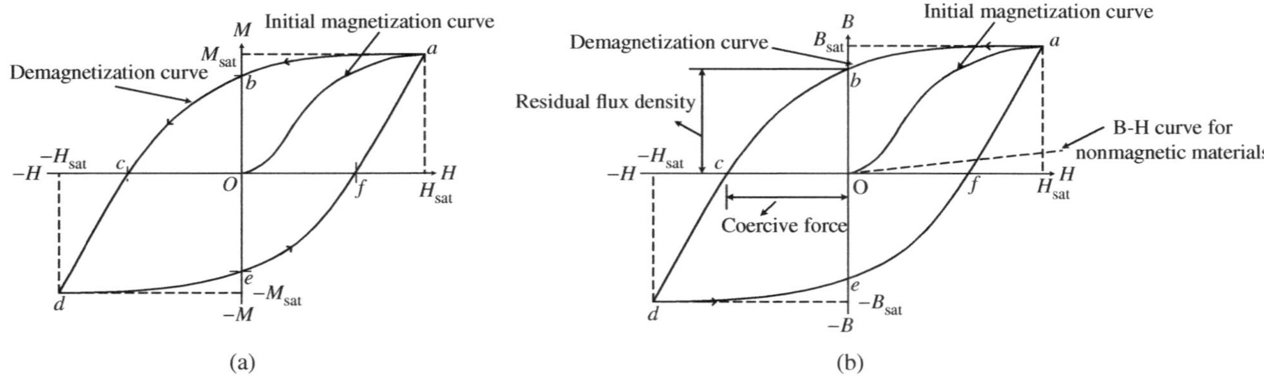

Figure 2.6 Hysteresis loop. (a) *H* Vs *M*. (b) *H* Vs *B*.

decreased, the magnetization decreases but the path *Oa* is not retraced. As the current reduces to zero, H becomes zero. However, there is still some magnetization left in the ring. Those domains which were aligned at the time of increasing H, are not completely randomized as the magnetic intensity H is reduced to zero. The remaining value of M at the point b is called the *retentivity* of the material. To reduce M to zero, a current must be passed in the opposite direction so as to dis-align the domains forcibly. The value of H needed to make $M = 0$ is called *coercive force*. In Figure 2.6, the coercive force is represented by the magnitude of H corresponding to *Oc*. As the current is increased further in the opposite direction, the magnetization M follows the path *cd* as the magnetic intensity becomes $-H_{sat}$. If the current is now reduced to zero, the magnetization M follows the path *de*. Finally, if the current is increased in the original direction, the point a is reached via *efa*. If we repeat the current cycle so that H changes from H_{sat} to $-H_{sat}$ to H_{sat}, the curve *abcdefa* is retraced.

As H is increased and then decreased to its original value, the magnetization M, in general, does not return to its original value. This fact is called *hysteresis*. The curve *abcdefa* is called the *hysteresis loop*. The area of the hysteresis loop is proportional to the thermal energy developed per unit volume of the material as it goes through the hysteresis cycle.

2.4.1 Hysteresis Loop for Soft Iron and Steel

Figure 2.7 shows hysteresis loops for soft iron and steel. The retentivity and the coercive force are larger for steel than for soft iron. The area of the hysteresis loop is also larger for steel than for soft iron. Soft iron is, therefore, easily magnetized by a magnetizing field but only a small amount of magnetization is retained when the field is removed. Also, the loss of energy, as the material is taken through periodic variations in the magnetizing fields, is small. Materials like soft iron are suitable for making electromagnets and cores inside current-carrying coils to increase the magnetic field. In transformers, moving coil galvanometers etc., soft iron core is used in the coils.

On the other hand, steel and similar materials are suitable for making permanent magnets. Large magnetizing fields are needed to appreciably magnetize the material. However, once magnetized, the magnetization is retained to a large extent even when the magnetizing field is removed (retentivity is large). The magnetization is not easily destroyed even if the material is exposed to stray reverse fields (coercive force is large).

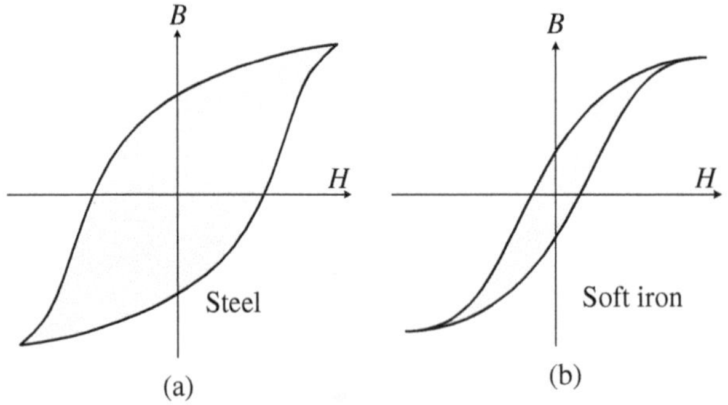

Figure 2.7 Hysteresis loop, (a) steel and (b) soft iron.

Materials with a broad hysteresis curve as in Figure 2.7a are said to be magnetically 'hard' and make good permanent magnets. On the other hand, a hysteresis curve such as that in Figure 2.7b occurs for 'soft' iron, which is preferred for electromagnets and transformers since the field can be more readily switched off, and the field can be reversed with less loss of energy.

A ferromagnetic material can be demagnetized – that is, made un-magnetized. This can be done by reversing the magnetizing current repeatedly while decreasing its magnitude. This results in the curve of Figure 2.8. The heads of a tape recorder are demagnetized in this way. The alternating magnetic field acting at the heads due to a handheld demagnetizer is strong when the demagnetizer is placed near the heads and decreases as it is moved slowly away. Video and audio tapes themselves can be erased and ruined by a magnetic field, as can computer hard disks, other magnetic storage devices, and the magnetic stripes on credit cards.

It should be noted that for each maximum value of the ac magnetic-field intensity cycle, there is a steady-state loop, as shown in Figure 2.9. In the figure, the dashed curve connecting the tips of the loops is the de-magnetization curve for the material.

The distinction between hard and soft magnetic material on the basis of their hysteresis loops is shown in Figure 2.10. Wide and narrow loops are for hard and soft magnetic materials, respectively. It is evident from the figure that the coercive force $\mathbf{H}_c$ for the soft magnetic material is much lower than that for a hard, magnetic material.

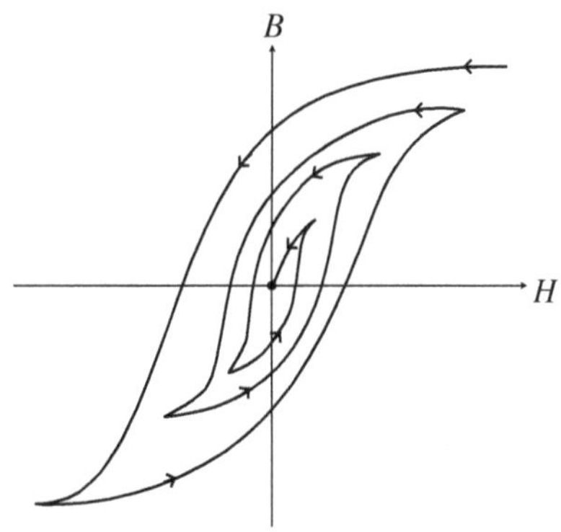

Figure 2.8 Successive hysteresis loops during demagnetization.

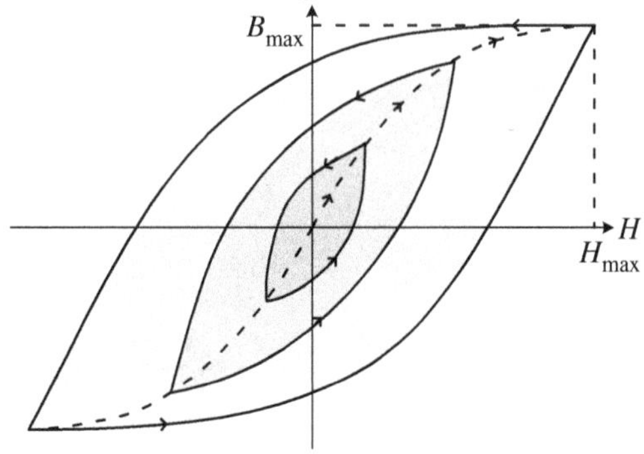

Figure 2.9 Hysteresis loops at different maximum values of AC magnetic field intensity.

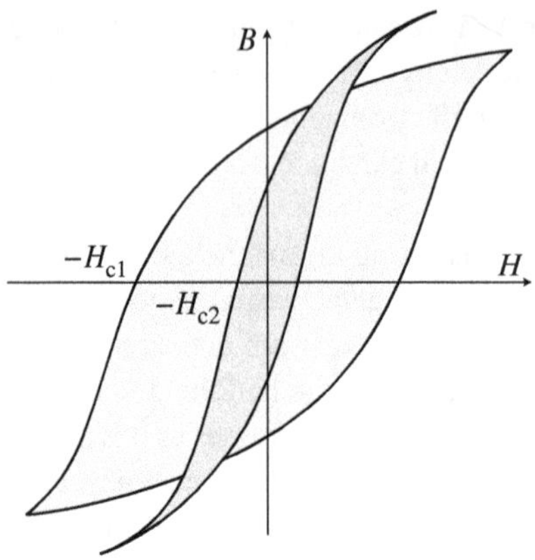

Figure 2.10 Hysteresis loop for hard and soft magnetic materials.

The magnetic field reverses direction every one-half cycle of the applied voltage and energy is expended in the core to accomplish the cyclic reversals of the field. This loss component is known as the hysteresis loss P_h. The hysteresis loss per second is approximated empirically by [1]:

$$P_h = K_h\, V_{vol}\, f\, B_m^{1.5\ to\ 2.5} \text{Watt} \tag{2.10}$$

where K_h is a proportionality constant dependent on the characteristics of iron, V_{vol} is the volume of the core in cubic meters, f is the frequency in Hz, and B is the maximum flux density in Tesla.

2.5 Eddy-Current and Core Losses

Induced currents are not always confined to well-defined paths such as those in wires. Consider, for example, the rotating metal wheel in Figure 2.11a. An external magnetic field is applied to a limited area of the wheel as shown and points into the page. The section of wheel in the magnetic field has an emf induced because the conductor is moving, carrying electrons. The flow of induced (conventional) current in the wheel is upward in the region of the magnetic field (Figure 2.11b), and the current follows a downward return path outside that region. Why? According to Lenz's law, the induced currents oppose the change that causes them. Consider the part of the wheel labelled c in Figure 2.11b, where the magnetic field is zero but is just about to enter a region where $\vec{B}$ points into the page. To oppose

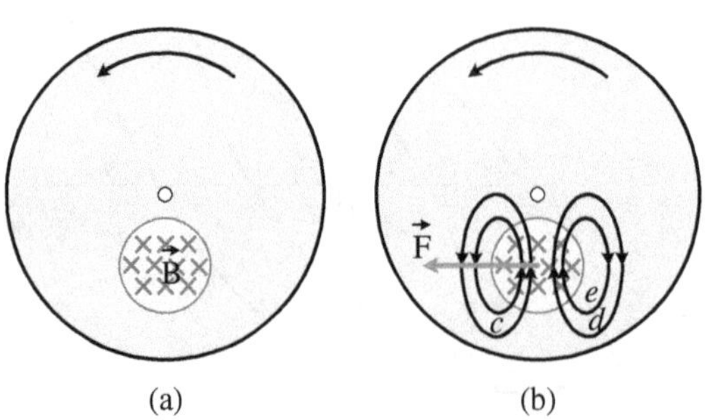

Figure 2.11 (a) Production of eddy currents in a rotating wheel. The grey lines in (b) indicate induced current.

this inward increase in magnetic field, the induced current is counter clockwise to produce a field pointing out of the page (right-hand-rule). Similarly, region d is about to move to e, where $\vec{B}$ is zero; hence the current is clockwise to produce an inward field opposed to this decreasing flux inward. These currents are referred to as eddy currents. They can be present in any conductor that is moving across a magnetic field or through which the magnetic flux is changing [1].

In Figure 2.11b, the magnetic field exerts a force $\vec{F}$ on the induced currents it has created, and that force opposes the rotational motion. Eddy currents can be used in this way as a smooth braking device on, say, a rapid-transit car. In order to stop the car, an electro-magnet can be turned on that applies its field either to the wheels or to the moving steel rail below. Eddy currents can also be used to dampen (reduce) the oscillation of a vibrating system. Eddy currents, however, can be a problem. For example, eddy currents induced in the armature of a motor or generator produce heat and waste energy. To reduce the eddy currents, the armatures are *laminated*; that is, they are made of very thin sheets of iron that are well insulated from one another. The total path length of the eddy currents is confined to each slab, which increases the total resistance; hence the current is less and there is less wasted energy [1].

Walk-through metal detectors at airports detect metal objects using electromagnetic induction and eddy currents. Several coils are situated in the walls of the walk-through at different heights. In a technique called 'pulse induction,' the coils are given repeated brief pulses of current (on the order of microseconds), hundreds or thousands of times a second. Each pulse in a coil produces a magnetic field for a very brief period of time. When a passenger passes through the walk-through, any metal object being carried will have eddy currents induced in it. The eddy currents persist briefly after each input pulse, and the small magnetic field produced by the persisting eddy current (before the next external pulse) can be detected, setting off an alert or alarm. Stores and libraries sometimes use similar systems to discourage theft.

In the examples we have discussed, currents were induced in thin wires or rods. However, a changing flux often induces circulating currents, which are called *eddy currents*, in a piece of bulk metal like the core of a transformer. The heat produced by such current constitutes a power loss in the transformer. Consider a conducting slab between the pole faces of an electromagnet (Figure 2.12). If the magnetic field $\vec{B}$ between the pole faces is changing with time (as it will if the current in the magnet windings is alternating current), the flux through any closed loop in the slab, such as through the curve C indicated in the figure, will change. Consequently, there will be an induced emf around C. Because path C is in a conductor, the emf will drive currents in the conductor [2].

The existence of eddy currents can be demonstrated by pulling a copper or aluminium sheet through the region between the poles of a strong permanent magnet (Figure 2.13). Part of the area enclosed by curve C in the figure is in the magnetic field, and part of

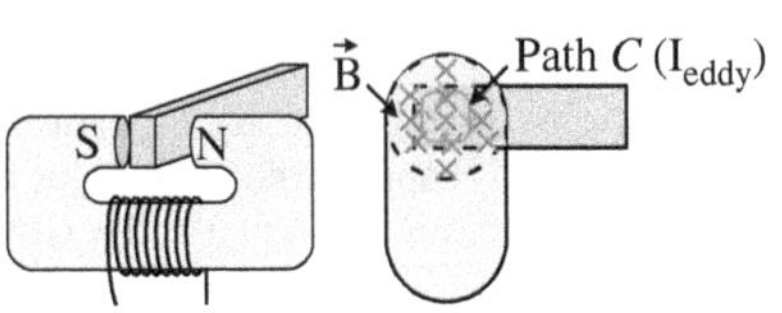

Figure 2.12 Eddy currents. When the magnetic field through a metal slab is changing, an emf is induced in any closed loop in the metal, such as loop C. The induced emfs drive currents, which are called eddy currents.

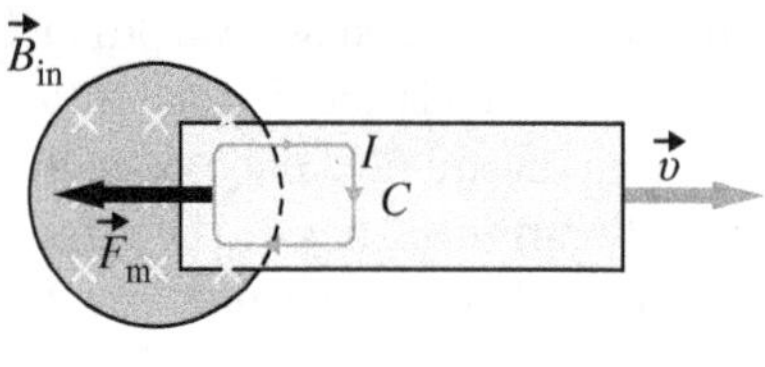

Figure 2.13 Demonstration of eddy currents. When the metal sheet is pulled to the right, there is a magnetic force to the left on the induced current opposing the motion.

the area enclosed by curve C is outside the magnetic field. As the sheet is pulled to the right, the flux through this curve decreases (assuming that into the paper is the positive normal direction). A clockwise emf is induced around this curve. This emf drives a current that is directed upward in the region between the pole faces, and the magnetic field exerts a force on this current to the left opposing motion of the sheet. You can feel this drag force on the sheet if you pull a conducting sheet rapidly through a region that has a strong magnetic field.

Eddy currents are frequently undesirable because power is lost due to Joule heating by the current, and this dissipated energy must be transferred to the environment. The power loss can be reduced by increasing the resistance of the possible paths for the eddy currents, as shown in Figure 2.14a. Here, the conducting slab is laminated; that is, the conducting slab is made up of small strips glued together. Because insulating glue separates the strips, the eddy currents are essentially confined to the individual strips. The large eddy-current loops are broken up, and the power loss is greatly reduced. Similarly, if there are cuts in the sheet, as shown in Figure 2.14b, the eddy currents are reduced and the magnetic force is greatly reduced.

Eddy currents are not always undesirable. For example, eddy currents are often used to damp unwanted oscillations. With no damping present, sensitive mechanical balance scales that are used to measure small masses might oscillate back and forth around their equilibrium reading many times. Such scales are usually designed so that a small sheet of aluminium (or some other metal) moves between the poles of a permanent magnet as the scales oscillate. The resulting eddy currents dampen the oscillations so that equilibrium is quickly reached. Eddy currents also play a role in the magnetic braking systems of some rapid transit cars. A large electromagnet is positioned in the vehicle over the rails. If the magnet is energized by a current in its windings, eddy currents are induced in the rails by the motion of the magnet and the magnetic forces provide a drag force on the magnet that slows the car.

We have seen earlier in this chapter that a current loop resists being pulled from or pushed into a magnetic field because the field induces currents that exert forces that act against any change in flux. Any conducting material, such as a sheet of copper, will resist being pushed into or pulled out of a magnetic field, because the changing field causes currents to loop

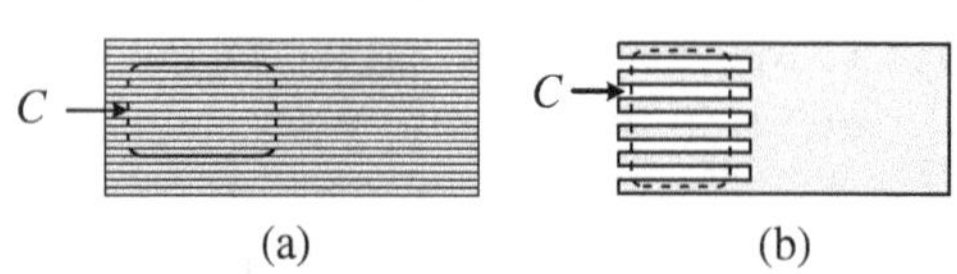

Figure 2.14 Disrupting the conduction paths in the metal slab can reduce the eddy current. (a) If the slab is constructed from strips of metal glued together, the insulating glue between the slabs increases the resistance of the closed loop C. (b) Slots cut into the metal slab also reduce the eddy current.

through the entire conductor as if it were many, many parallel wires. We call these currents 'eddy currents' [3].

Eddy currents are remarkably important, as they are a source of *energy loss* whenever we attempt to alter a magnetic field in the vicinity of *any conductor*. Eddy currents produce *Joule heating* of the conducting material very readily – one can actually cook food on stoves that use a rapidly varying magnetic field to directly heat metal pots placed in the field. Transformers (covered later) rely on rapidly varying, ferromagnetically enhanced magnetic fields to step up or step-down voltage, and unless care is taken to prevent eddy currents in the design of the magnetic cores, much of the energy being transmitted through the transformer will be lost to heating the cores. Eddy currents cancel electromagnetic radiation at the surfaces of conductors, both heating the conductors slightly and causing the electromagnetic field to *reflect* from the surface rather than be transmitted. It seems worthwhile to spend a moment trying to understand them.

In Figure 2.15, a sheet of copper being pulled rapidly out of a strong magnetic field is illustrated. It is moving at some speed v to the right. As it is pulled out, the magnetic flux through the *entire sheet* is reduced. This creates an induced field in the conductor and its associated induced voltage that (because it is a *good conductor*) can and does drive a large current in the copper. This current is not isolated or confined in the conductor – the conducting sheet is like an entire field of parallel resistance pathways and the current spreads out to use them.

We will return to the notion of eddy currents when we treat transformers because the iron cores of transformers are usually *laminated* – made of thin sheets or wires of iron coated with, and separated by, an insulating resin – precisely to prevent eddy currents from the rapidly changing magnetic fields they help support from heating the iron and hence was the energy in the time-varying magnetic field.

The eddy-current loss is empirically approximated as [4]

$$P_e = K_e V_{\text{vol}} f^2 \tau^2 B_m^2 \text{ Watts} \tag{2.11}$$

where K_e is a constant dependent on the characteristics of iron, V_{vol} is the volume of iron, f is the frequency in Hz, τ is the lamination thickness (usually a stack of thin laminations makes up the core), and B_m is the maximum core flux density. By choosing very thin laminations (making τ smaller), the eddy-current loss can be reduced.

Core or iron loss in an electrical machine is comprised of hysteresis losses and eddy current losses.

$$P_{\text{core}} = P_{\text{h}} + P_{\text{e}} \tag{2.12}$$

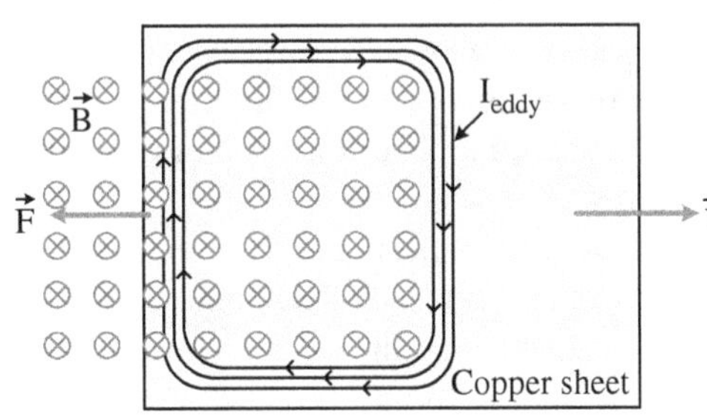

Figure 2.15 A sheet of copper being pulled rapidly out of a field has induced *eddy currents*. The forces from these currents, according to Lenz's Law, *resist* the motion, causing a magnetic 'drag force' similar to that observed in the rod on rails problem. The kinetic energy of the object is transformed into heat by these currents (resistive Joule heating) [3].

2.6 Magnetic Circuits

2.6.1 The Magnetic Circuit Concept

When a field of magnetic flux is desired in engineering practice, specially-shaped structures of ferromagnetic material and appropriately located current-carrying conductors, or appropriately located permanent magnets, are used. The ability to compute the magnetic field intensity H and the flux density B throughout such a structure is necessary for the solution of many problems. In general, H and B are functions of space and time, determined by the geometry of the conductors and the magnetic structure, by the properties and history of the magnetic material, and by the values and derivatives or frequencies of the currents or the strength of the magnets. A field problem involving materials which have nonlinear characteristics is thus encountered. Suitable simplifying assumptions, in many instances, make the reduction of the general field problem to a simpler one possible by involving the circuit concept [5].

In general, it is not possible to calculate magnetic fields accurately, when magnetic materials are involved. There are several reasons for this, (i) The relation between **H** and **B** for ferromagnetic materials is non-linear. It even depends on the previous history of the material. Also, ferromagnetic materials are often non-isotropic. (ii) The iron cores that are used to confine and guide the magnetic flux are often quite inefficient. As we shall see, a large part of the flux can be situated outside the core, (iii) Permanent magnets are not as simple as one would like them to be; their magnetization M is non-uniform and depends on the presence of neighbouring magnetic materials.

It is nonetheless necessary to be able to make approximate calculations. The calculation serves to design a model on which magnetic fields or magnetic forces can be measured, and which can then be modified to give the final design.

2.6.2 Magnetic Circuits Terminology

Terms used in Magnetic Circuit

1. **Magnetomotive force (mmf)** drives or tends to drive flux through a magnetic circuit and corresponds to electromotive force (emf) in an electric circuit.
   mmf is equal to the work done in joules in carrying a unit magnetic pole once through the entire magnetic circuit. It is measured in ampere-turns.
   In fact, as p.d. between any two points is measured by the work done in carrying a unit charge from one point to another, similarly, mmf between two points is measured by the work done in joules in carrying a unit magnetic pole from one point to another.
2. **Ampere-turns (AT)** is the unit of magnetomotive force (mmf) and is given by the product of number of turns of a magnetic circuit and the current in amperes in those turns.
3. **Reluctance** is the name given to that property of a material which opposes the creation of magnetic flux in it. Reluctance, in fact, measures the opposition offered to the passage of magnetic flux through a material and is analogous to resistance in an electric circuit even in form. Its unit is **AT/Wb.**

$$\text{reluctance} = \frac{l}{\mu A} = \frac{l}{\mu_0 \mu_r A}$$
$$\text{resistance} = \rho \frac{l}{A} = \frac{l}{\sigma A}$$

In other words, the reluctance of a magnetic circuit is the number of amp-turns required per Weber of magnetic flux in the circuit. Since 1 AT/Wb = 1/H, the unit of reluctance is 'reciprocal Henry.'

4. **Permeance** is reciprocal of reluctance and implies the case or readiness with which magnetic flux is developed. It is analogous to conductance in electric circuits. It is measured in terms of Wb/AT or Henry.
5. **Reluctivity** is specific reluctance and corresponds to resistivity which is 'specific resistance'.
6. **Air Gap**: is the distance between the north and south poles of a magnetic circuit. In conducting pull tests, this is the distance between the working surface of the magnet and the testing apparatus.
7. **Isotropic:** (non-oriented) A material with no preferred direction of orientation resulting in the same magnetic characteristics through any axis
8. **Anisotropic:** (oriented) A material that has a preferred direction of magnetic orientation which produces superior magnetic characteristics through a particular axis.

Figure 2.16 shows a ferromagnetic core around which is wound a short coil of N turns carrying a current I. We wish to calculate the magnetic flux Φ through the core.

In the absence of ferromagnetic material, the lines of **B** are as shown in the Figure 2.16. At first sight, one expects the **B** inside the core to be much larger close to the winding than on the opposite side. This is not the case, however, and **B** is of the same order of magnitude at all points within the ferromagnetic material [5].

This can be understood as follows. The magnetic induction of the current magnetizes the core in the region near the coil, and this magnetization gives equivalent currents that both increase **B** and extend it along the core. This increases further and extends the magnetization, and hence **B**, until the lines of **B** extend all around the core.

Of course, some of the lines of **B** escape into the air and then return to the core to pass again through the coil. This constitutes the *leakage flux* that may, or may not, be negligible. For example, if the toroid is made up of a long thin wire, most of the flux leaks across from one side of the ring to the other and the flux at P is negligible compared to that near the coil.

Let us assume that the cross-section of the toroid is large enough to render the leakage flux negligible. Then, applying Ampere's circuital law, to a circular path of radius r going all around inside the toroid,

$$\oint \vec{H} \cdot d\vec{l} = NI \tag{2.13}$$

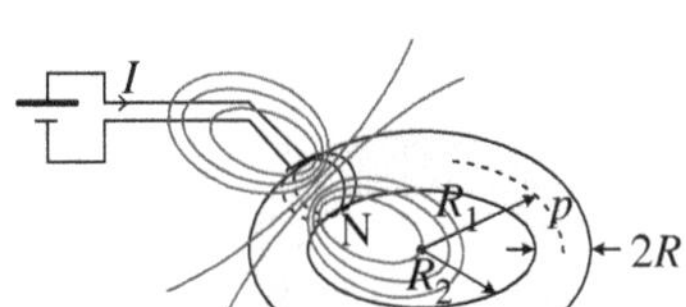

Figure 2.16 Ferromagnetic toroid with concentrated winding. The lines of force shown are in the plane of the toroid. They apply only when there is no iron is present [5].

$$\frac{2\pi rB}{\mu_r\mu_o} = NI, \tag{2.14}$$

$$B = \frac{\mu_r\mu_o NI}{2\pi r}. \tag{2.15}$$

Taking R_1 to be the radius corresponding to the average value of B, and R_2 to be the minor radius of the toroid.

$$\Phi = \frac{\mu_r\mu_o\pi R_2^2 NI}{2\pi R_1} \tag{2.16}$$

The flux through the core is therefore, the same as if we had a toroidal coil of the same size and if the number of ampere-turns were increased by a factor of μ_r. In other words, for each ampere-turn in the coil there are μ_r-1 ampere-turns in the core. The amplification can be as high as 10^5. This equation shows that the magnetic flux is given by the magnetomotive force NI multiplied by the factor

$$\frac{\mu_r\mu_o\pi R_2^2}{2\pi R_1}$$

which is called the *permeance* of the magnetic circuit. The inverse of the permeance is called the *reluctance*. Reluctance in a magnetic circuit is opposition to the flow of flux similar to resistance in electrical circuit which is opposition to the flow of current. Thus

$$\Phi = \frac{NI}{\mathcal{R}} = \frac{NI}{\left(\dfrac{2\pi R_1}{\mu_r\mu_o\pi R_2^2}\right)} \tag{2.17}$$

and the reluctance is

$$\mathcal{R} = \frac{2\pi R_1}{\mu_r\mu_o\pi R_2^2} \tag{2.18}$$

Thus, reluctance is given as;

$$\Re = \frac{l}{\mu A}$$

where l is the length of the flux path (flux is assumed to flow through the centre of the core) and A is the cross-sectional area of the core.

Reluctance is expressed in Henrys^{-1} or Amp-Turn/Wb.

The analogy with Ohm's law is obvious: if an electromotive force V were induced in the core, the current would be

$$I = \frac{V}{\dfrac{2\pi R_1}{\sigma\pi R_2^2}} \tag{2.19}$$

where $\frac{2\pi R_1}{\sigma\pi R_2^2}$ is the resistance of the core.

There is one important difference between electric and magnetic circuits: the magnetic flux cannot be made to follow a magnetic circuit in the manner that an electric current follows a conducting path. Indeed, a magnetic circuit behaves much as an electric circuit would if it were submerged in tap water: part of the current would flow through the components, and the rest would flow through the water. This means magnetic flux flows through components as well as leaks out of the components.

If a magnetic circuit is not properly designed, the leakage flux can easily be an order of magnitude *larger* than that flowing around the circuit.

Equivalent circuit of magnetic circuit is drawn by considering the mmf and reluctance as shown in Figure 2.17.

Ohm's law is not a correct analogy with Hopkinson's law in terms of modelling power and energy flow because there is the dissipation of power across resistances in electric circuits whereas in magnetic circuits there is no loss of energy in reluctances.

An analogy between electrical and magnetic circuit is illustrated in Table 2.1. The fundamental laws in electrical and magnetic circuits are given in Table 2.2.

Figure 2.17 Equivalent circuit of magnetic circuit.

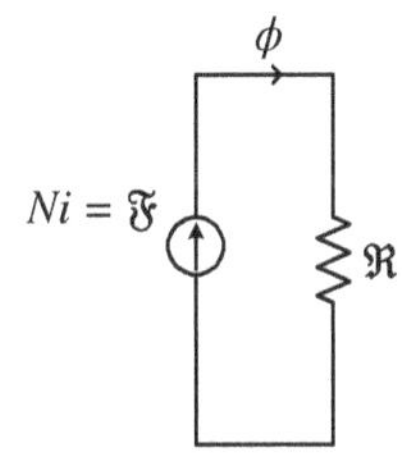

Table 2.1 Analogy between electrical and magnetic circuits.

Electric circuit	Magnetic circuit
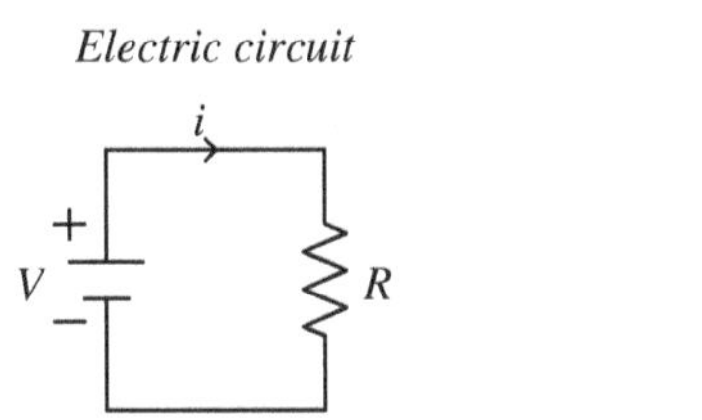	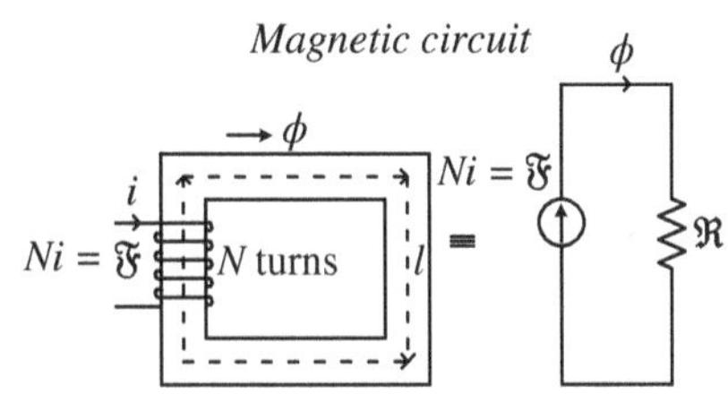
$V = \mathcal{E}$ = electromotive force (V) [emf]	$\mathfrak{F}$ = magnetomotive force (At) [mmf]
i = total current (A)	ϕ = total magnetic flux (Wb)
R = resistance (Ω)	$\mathfrak{R}$ = reluctance (H^{-1} or A.turn/Wb)
E = electric field (V/m or N/C)	H = magnetic field (A/m)
J = current density (A/m^2)	B = flux density (Wb/m^2 or Tesla [T])
$\mathbf{J} = \sigma\mathbf{E}$	$\mathbf{B} = \mu\mathbf{H}$
$V = \mathcal{E} = \oint E \cdot dl$	$\mathfrak{F} = \oint \mathbf{H} \cdot d\mathbf{l} = Ni$
$i = \iint_S \mathbf{J} \cdot d\mathbf{S}$	$\phi = \iint_S \mathbf{B} \cdot d\mathbf{S}$
$R = \frac{l}{\sigma A} = \frac{1}{G}$	$\mathfrak{R} = \frac{l}{\mu A} = \frac{1}{\mathfrak{p}}$
G = conductance (S or mho [℧])	$\mathfrak{p}$ = permeance (H or Wb/A.turn)
σ = electrical conductivity (S/m)	μ = permeability (H/m or Wb/A.m or T.m/A)

Table 2.2 Comparison of Fundamental laws for electric and magnetic circuits [9].

Law	Electric circuit	Magnetic circuit
Ohm's law	Resistance $R = \frac{V}{I}$	Reluctance $\mathfrak{R} = \frac{\mathfrak{F}}{\phi}$
	Conductance $G = \frac{I}{V}$	Permeance $\mathfrak{p} = \frac{\phi}{\mathfrak{F}}$
2nd Ohm's law	Resistance $R = \frac{l}{\sigma A}$	Reluctance $\mathfrak{R} = \frac{l}{\mu A}$
	Conductance $G = \frac{\sigma A}{l} = \frac{A}{\rho A}$	Permeance $\mathfrak{p} = \frac{\mu A}{l}$
Kirchhoff's current law	Sum of currents $\sum I = 0$	Sum of magnetic fluxes $\sum \phi = 0$
Kirchhoff's voltage law	Sum of voltage drops $\sum V - \sum RI = 0$	Sum of magnetic voltage drops $\sum \mathfrak{F} - \sum \mathfrak{R}\phi = 0$
Faraday's law/Ampere's law	EMF $\mathcal{E} = \sqrt{2}\pi f N\phi$	MMF $\mathfrak{F} = NI$

2.6.2.1 Limitations of the Analogy Between Electric and Magnetic Circuits

(i) In electric circuits, there is a flow of electrons whereas in the magnetic circuit there is no flow of any particle.

(ii) In magnetic circuits, there is leakage of flux because of the presence of magnetic permeability outside of the circuit, whereas in electric circuits the current remains in the circuit.

(iii) Resistance is constant in electric circuits whereas reluctance is not constant as it varies with the magnetic field. The magnetic circuits saturate which limits further increase of flux. At saturation, level reluctance increases rapidly.

(iv) In some magnetic materials, flux remains in the circuit even if mmf is turned off, whereas in electric circuits current becomes zero through the resistance as soon as emf is turned off.

2.6.3 Effect of Air Gaps

2.6.3.1 Magnetic Circuit with an Air Gap

Figure 2.18 shows a circuit with an air gap whose cross-section is different from that of the soft-iron yoke. Each winding provides $NI/2$ ampere-turns. We wish to calculate the magnetic induction in the air gap [5].

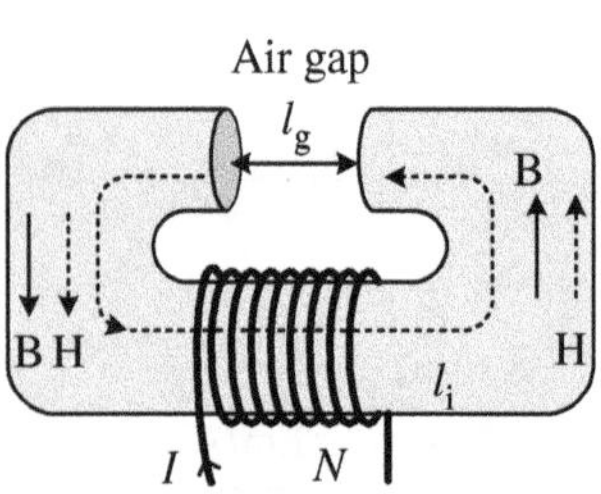

Figure 2.18 Electromagnet, the coils have been cut out to expose the iron core.

We assume that the leakage flux is negligible. As we shall see, this assumption will result in quite a large error.

Applying Ampere's circuital law to the circuit, the following is obtained

$$NI = H_i l_i + H_g l_g \tag{2.20}$$

where the subscript i refers to the iron core and g to the air gap; l_i, and l_g are the path lengths. The path length or length of flux path in the iron can be taken to be the length measured along the centre of the cross-section of the core.

If leakage flux is neglected, the flux of **B** must be the same over any cross-section of the magnetic circuit, so

$$B_i A_i = B_g A_g \tag{2.21}$$

where A_i and A_g are, respectively, the cross-sectional area of the iron core and of the air gap.

Combining these two equations.

$$B_g A_g \left[\frac{l_i}{\mu_r \mu_o A_i} + \frac{l_g}{\mu_o A_g} \right] = NI, \tag{2.22}$$

and the magnetic flux is

$$\Phi = B_g A_g = \frac{NI}{\left[\frac{l_i}{\mu_r \mu_o A_i} + \frac{l_g}{\mu_o A_g} \right]}; \tag{2.23}$$

The magnetic flux is, therefore, equal to mmf divided by the sum of the reluctances of the iron and of the air gap.

Since we have neglected leakage flux, the above equation can only provide an upper limit for Φ.

Now let $\mathcal{R}_i = \frac{l_i}{\mu_r \mu_o A_i}$ be the reluctance of the iron, and $\mathcal{R}_g = \frac{l_g}{\mu_o A_g}$ be the reluctance of the air gap. Note that

$$H_i l_i = \mathcal{R}_i \Phi. \tag{2.24}$$

$$H_g l_g = \mathcal{R}_g \Phi. \tag{2.25}$$

If $\mathcal{R}_i \ll \mathcal{R}_g$, as is usually the case, since $\mu_r \gg 1$, Eqs. (2.72) and (2.75) become

$$NI \approx H_g l_g, \tag{2.26}$$

$$\Phi \approx \mathcal{R}_g \Phi. \tag{2.27}$$

Equivalent circuit of magnetic circuit is drawn by considering the mmf and reluctances in series as shown in Figure 2.19.

Figure 2.19 Reluctances in series.

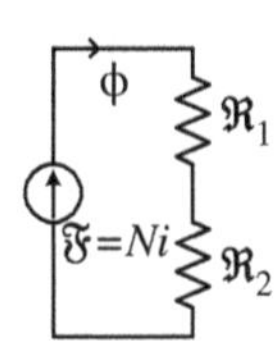

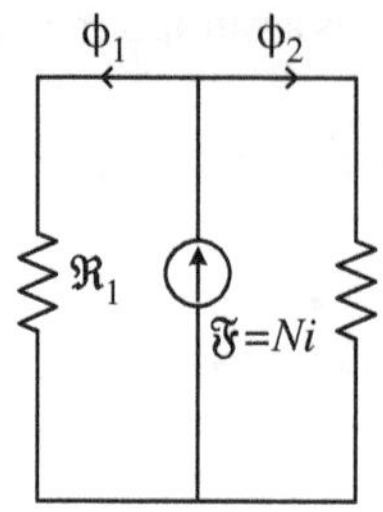

Figure 2.20 Reluctances in parallel.

Equivalent circuit of magnetic circuit is drawn by considering the mmf and reluctances in parallel as shown in Figure 2.20.

The reluctance of iron portion and air-gap are in series and can be combined by adding them.

This is a general law: reluctances in series in a magnetic circuit add in the same way as resistances in series in an electric circuit; permeances in parallel add like conductances in parallel.

Series equivalent of reluctances:

$$\mathfrak{R}_{eq} = \mathfrak{R}_1 + \mathfrak{R}_2 + \mathfrak{R}_3 + \ldots\ldots\ldots\ldots\ldots \mathfrak{R}_n$$

In terms of permeance for a series circuit;

$$\frac{1}{\wp_{eq}} = \frac{1}{\wp_1} + \frac{1}{\wp_2} + \frac{1}{\wp_3} + \ldots\ldots\ldots\ldots\ldots + \frac{1}{\wp_n}$$

For parallel circuit, equivalent reluctance is obtained as;

$$\frac{1}{\mathfrak{R}_{eq}} = \frac{1}{\mathfrak{R}_1} + \frac{1}{\mathfrak{R}_2} + \frac{1}{\mathfrak{R}_3} + \ldots\ldots\ldots\ldots\ldots\ldots + \frac{1}{\mathfrak{R}_n}$$

In terms of permeance for a parallel circuit;

$$\wp_{eq} = \wp_1 + \wp_2 + \wp_3 + \ldots\ldots\ldots\ldots + \wp_n$$

Example 2.3 ***Electromagnet***

Calculate B and the stored energy in the electromagnet of Figure 2.18. If there is a total of 10 000 turns in the winding, and setting $I = 1.00$ A. $A_i = 10^4$ square millimetres, $A_g = 5 \times 10^3$ square millimetres, $\mu_r = 1000$, $l_i = 900$ mm, and $l_g = 10$ mm, the leakage flux is 70% of the flux in the gap.

Solution

$$\Phi = \frac{NI}{\left[\dfrac{l_i}{\mu_r \mu_o A_i} + \dfrac{l_g}{\mu_o A_g}\right]} = \frac{10000 \times 1.0\,(\text{A.t})}{\dfrac{0.9\text{m}}{10^3 \times 4\pi \times 10^{-7} \times 10^{-2}} + \dfrac{10^{-2}\text{m}}{4\pi \times 10^{-7} \times 5 \times 10^{-3}}} \tag{2.28}$$

$$\Phi = 6.0 \times 10^{-3}\ \text{Wb}, \tag{2.29}$$

$$B_g = \frac{6.0 \times 10^{-3}}{5 \times 10^{-3}} = 1.2\ \text{T}. \tag{2.30}$$

The self-inductance is

$$L = \frac{N\Phi}{I} = \frac{10^4 \times 6.0 \times 10^{-3}}{1.0} = 60\text{ H}. \tag{2.31}$$

The length l_i is measured along the middle of the cross-section of the yoke.

The stored energy is

$$\frac{1}{2}LI^2 = \frac{1}{2} \times 60 \times (1.0)^2 = 30\text{ J}. \tag{2.32}$$

In this particular case, the leakage flux is 70% of the flux in the gap. In other words, the magnetic induction in the gap is not 1.2 T, but only 1.2/1.7 = 0.71 T. There exist empirical formulae for estimating leakage flux for simple geometries [5].

2.6.3.2 Magnetic Forces Exerted by Electromagnets

Electromagnets are commonly used to actuate various mechanisms such as switches, valves, and so forth [5]. A switch activated by an electromagnet is called a *relay* (Figure 2.21). These electromagnets comprise a coil and an iron core that is made in two parts, one fixed and one movable, called the *armature,* with an air gap between the two. The coil and its core together are usually termed a *solenoid*. The magnetic force is *attractive*.

In the simplest cases, one has a flat air gap, perpendicular to **B**. Then the magnetic force of attraction is $\frac{B^2}{2\mu_o}$ times the cross-section of the gap. If the solenoid is energized with direct current, then B is approximately proportional to the inverse of the gap length, and since the force is proportional to B^2, the force increases rapidly with decreasing gap length.

The air gap is often designed so that the attractive force will vary with the gap length in some prescribed way. The design of the magnetic circuit is largely empirical.

Example 2.4 ***Clip-On Ammeter***

It is often useful to be able to measure the current flowing in a wire without disturbing the circuit. This can be done with a clip-on ammeter. As a rule, clip-on ammeters are transformers, and can therefore be used only with alternating currents.

It is also possible to make a clip-on ammeter that will measure direct currents as in Figure E2.4. In this instrument, the magnetic flux through the yoke is measured with a Hall generator, or with a magneto-resistor situated in the gap of length l_g.

(a) Show that, if $l_g = \frac{l_i}{\mu_r}$ the magnetic induction in the gap is $\frac{\mu_o I}{l_g}$-What is the advantage of using an iron core?
(b) Is this magnetic induction affected by the position of the wire inside the ring?

Figure 2.21 The relay.

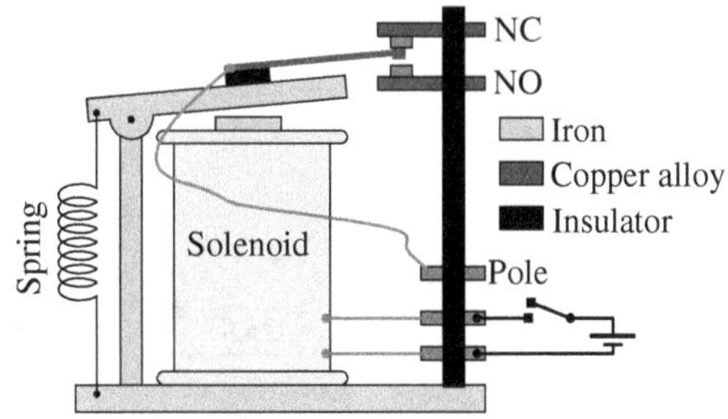

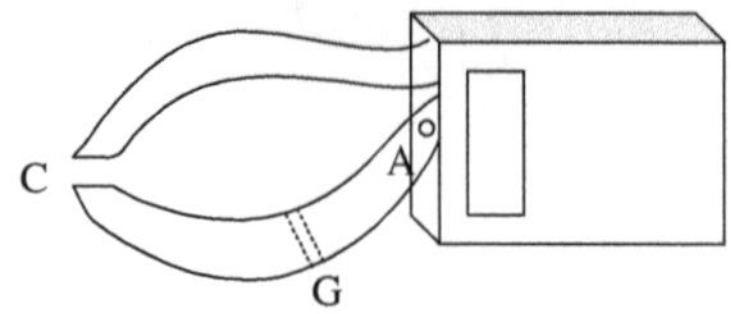

Figure E2.4 Clip-on ammeter for direct currents. A hinge '*A*' permits the iron yoke to be opened at *C* so that it can be clipped around the current-carrying wire. A gap *G* in the yoke contains either a Hall generator or a magneto-resistor. The magnetic induction in the gap is a measure of the current *I* [5].

Solution

Clip-On Ammeter

(a) $$\Phi = B_g A_g = \frac{NI}{\left[\dfrac{2\pi R - l_g}{\mu_r \mu_o A} + \dfrac{l_g}{\mu_o A}\right]} = \frac{\mu_o A I}{\dfrac{2\pi R - l_g}{\mu_r} + l_g} \approx \frac{\mu_o A I}{l_g},\ B = \frac{\mu_o I}{l_g}$$

Without the iron core, B is $\frac{\mu_o I}{2\pi R}$ and is much smaller.

(b) The position of the wire is unimportant.

Example 2.5 ***Magnetic Circuit***

The ductile cast-iron ring of Figure E2.5a is cut, leaving an air gap 1 mm long. The current in the toroidal coil is again 2.4 A.

Calculate *B* in the air gap using the relative permeability curve of Figure E2.5b. You will have to solve this problem by successive approximations. Start by assuming a reasonable value for μ_r, say 500. Find the corresponding *B*. This will probably not give the correct *B* on the curve. Then try another value of μ_r, etc. Draw a table of μ_r and of the calculated *B* as a help in selecting your next approximation. The calculated value of *B* need not agree to better than 10% with the *B* on the curve [5].

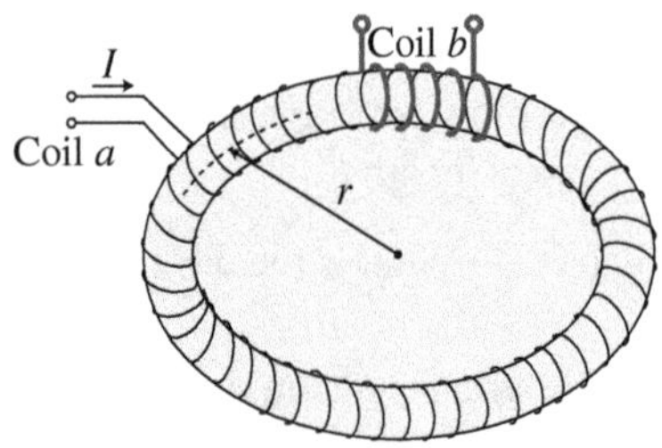

Figure E2.5a Rowland ring for the determination of *B* as a function of *H* in a ferromagnetic substance.

Solution

Magnetic Circuit

For $\mu_r = 500$

$$B = \frac{NI}{\left[\dfrac{2\pi R}{\mu_r \mu_o} + \dfrac{l_g}{\mu_o}\right]} = \frac{\mu_o NI}{\dfrac{2\pi R}{\mu_r} + l_g} = \frac{4\pi \times 10^{-7} \times 500 \times 2.4}{\dfrac{2\pi \times 0.2}{500} + 10^{-3}} = \frac{1.508 \times 10^{-3}}{\dfrac{1.257}{500} + 10^{-3}} = 0.43\ T$$

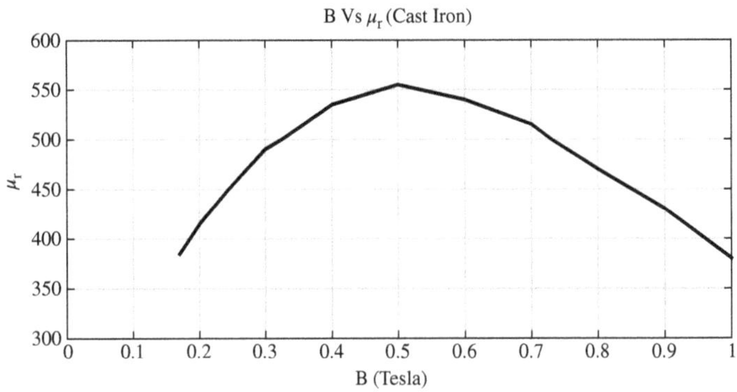

Figure E2.5b Relative permeability as a function of magnetic induction for ductile cast iron.

Table 2.3 Values of B calculated for various values of μ_r.

μ_r	501	502	503	504	505	506	507	508	509	510
B(T)	0.4298	0.4304	0.4311	0.4317	0.4323	0.4329	0.4335	0.4341	0.4347	0.4353
μ_r	511	512	513	514	515	516	517	518	519	520
B(T)	0.4359	0.4365	0.4371	0.4377	0.4384	0.439	0.4396	0.4402	0.4408	0.4414
μ_r	521	522	523	524	525	526	527	528	529	530
B(T)	0.442	0.4426	0.4432	0.4438	0.4444	0.445	0.4455	0.4461	0.4467	0.4473
μ_r	531	532	533	534	535	536	537	538	539	540
B(T)	0.4479	0.4485	0.4491	0.4497	0.4503	0.4509	0.4515	0.4521	0.4526	0.4532
μ_r	541	542	543	544	545	546	547	548	549	550
B(T)	0.4538	0.4544	0.455	0.4556	0.4562	0.4567	0.4573	0.4579	0.4585	0.4591

This B is too large: From curve: for $\mu_r = 500$, B = 0.32 T. Try $\mu_r = 550$, B = 0.459 T, instead of 0.5 on the graph (Figure E2.6a). This is satisfactory. At different μ_r, B is calculated using Matlab and is tabulated in Table 2.3.

The simple Matlab code is:

```
mur=linspace (500,550,51);
Num=pi*4*10^(-7)*500*2.4;
x=(pi*2*0.2)./mur;
Den=x+10^(-3);
B=Num./Den;
%plot(B,mur)
out=[mur;B];
```

Example 2.6 ***Rowland Ring***

A ring of ductile cast-iron (Figure E2.5a) has a major radius of 200 mm and a minor radius of 10 mm. A 500-turn toroidal coil is wound over it.

(a) What is the value of B inside the ring when the current through the coil is 2.4 A? Use Figure E2.6a.
(b) A 10-turn coil is wound over the first one. Calculate the voltage induced in it, if the current in the large coil suddenly increases by a small amount at the rate of 10 A per second. Assume that μ_r remains constant [5].

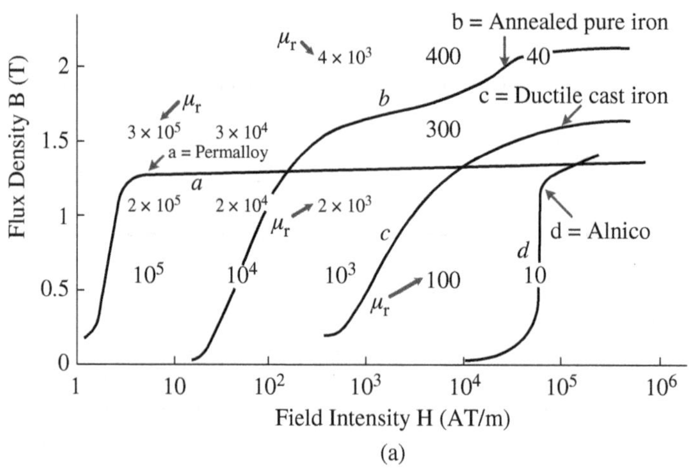

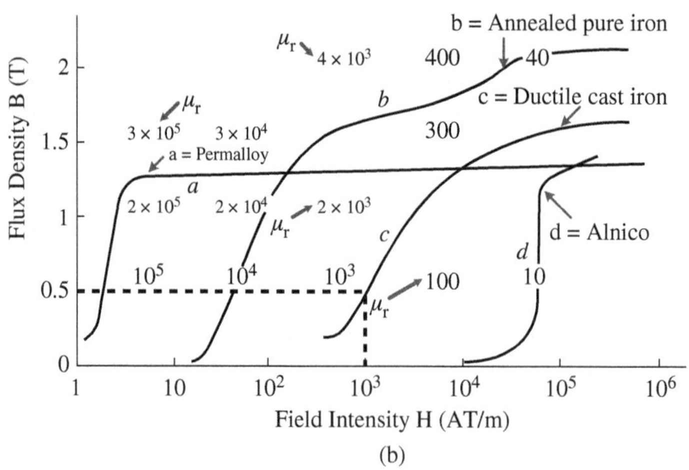

Figure E2.6 Magnetization curves for various magnetic materials: (*a*) Permalloy; (*b*) annealed pure iron; (*c*) ductile cast iron; (*d*) Alnico 5. The numbers shown are the relative permeabilities $\frac{B}{\mu_o H}$ [5].

Solution

Rowland Ring

(a) $H = \frac{NI}{2\pi R} = \frac{500\times 2.4}{2\pi\times 0.2} = 955 \approx 1000\,A/m$

From characteristic curve at H = 1000 A/m, B = 0.5 T (see Figure E2.6b).

(b) $V = N\frac{d\phi}{dt} = NA\frac{dB}{dt} = 10\times\pi\times\left(\frac{10}{1000}\right)^2\times 0.5\times\left(\frac{10}{2.4}\right) = 6.6\,mV$

Example 2.7 ***Magnetic Force on the Armature of an Electromagnet***

Figure E2.7 shows a small electromagnet. All dimensions are in millimetres. Each coil has 500 turns and carries a current of 2 A.

(a) Calculate $\boldsymbol{B}$ in the air gaps as a function of the gap length x. Set $\mu_r = 1000$, and neglect the leakage flux [5].

Figure E2.7 Electromagnet. All dimensions in mm.

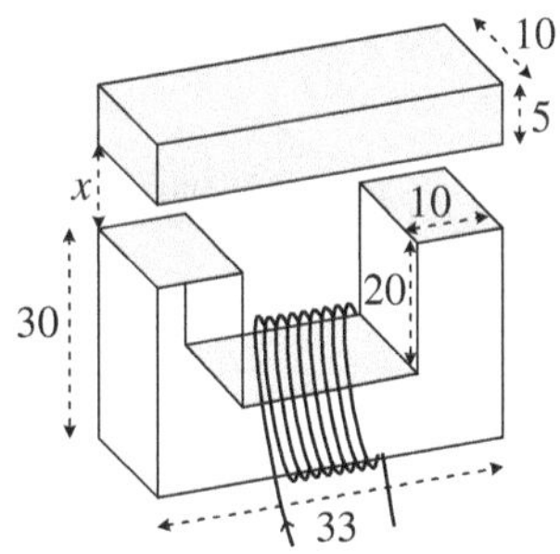

Solution

The mmf is

$$NI = 2\times 2\times 500 = 2000 \text{ ampere turns.} \tag{2.33}$$

The iron yoke has a cross-section of 10×10 mm², and a mean length of 2×25 mm for the vertical parts, plus 23 mm for the horizontal part. Hence its reluctance is

$$\mathcal{R}_y = \frac{73\times 10^{-3}}{\left(1000\times 4\pi\times 10^{-7}\times 10^{-4}\right)} = 5.809\times 10^5 \text{henry}^{-1}. \tag{2.34}$$

The armature has a cross-section of 50 mm². The mean path length of the flux is about 23 mm. Hence

$$\mathcal{R}_a = \frac{23\times 10^{-3}}{\left(1000\times 4\pi\times 10^{-7}\times 50\times 10^{-6}\right)} = 3.66\times 10^5 \text{henry}^{-1}. \tag{2.35}$$

Finally, the two air gaps have a reluctance

$$\mathcal{R}_g = \frac{2x}{\left(4\pi\times 10^{-7}\times 10^{-4}\right)} = 1.59\times 10^{10} x \text{ henry}^{-1}. \tag{2.36}$$

Thus, in the gaps

$$B = \frac{NI}{\mathcal{R}A} = \frac{2000}{\left(5.8 \times 10^5 + 3.7 \times 10^5 + 1.6 \times 10^{10}x\right) \times 10^{-4}}, \tag{2.37}$$

$$B = \frac{200}{9.5 + 1.6 \times 10^5 x}\,T. \tag{2.38}$$

b) Calculate the force of attraction exerted on the armature when $x = 5$ mm.

Solution

The force is equal to the product of the energy density in the gap, by its cross-section A:

$$B = \frac{200}{9.5 + 1.6 \times 10^5 x} = \frac{200}{9.5 + 1.6 \times 10^5 \times 5 \times 10^{-3}} = 0.247\,T$$

Energy density which is derived in energy section is given by

$$\frac{W}{V_{\text{vol}}} = \frac{B^2}{2\mu_0}$$

As we know that Work done = Force × distance = Force × Length

or

$$\text{Energy density} = \frac{\text{Work done}}{\text{volume}} = \frac{\text{Force} \times \text{Length}}{\text{volume}}$$

$$= \frac{\text{Force} \times \text{Length}}{\text{Area} \times \text{Length}} = \text{Force per unit area}$$

Hence

$$\text{Force per unit area} = \text{Energy density} = \frac{\text{F}}{\text{A}} = \frac{B^2}{2\mu_0}$$

$$F = \frac{2B^2 A}{2\mu_o}, \tag{2.39}$$

$$F = \frac{(0.247)^2 \times 10^{-4}}{4\pi \times 10^{-7}} = 4.855 \approx 5 = 4.855 \approx 5\text{N}. \tag{2.40}$$

c) Would it make sense to use Eq. (2.90) to calculate the force (i) at $x = 0.1$ mm? (ii) at $x = 20$ mm?

Solution

At $x = 0.1$ mm, Eq. (2.90) gives a magnetic induction of about 12 teslas. This is much more than the saturation field of one or two teslas for iron (Figure E2.6a). Eq. (2.90) is not valid at $x = 10^{-4}$.

At $x = 20$ mm, the reluctance of the air gap is large and the leakage flux is much greater than the flux in the gaps. The actual force is much smaller than that calculated from Eq. (2.91). Equation (2.90) again is not valid.

So, the magnetic induction found in Eq. (2.90) is valid only over a limited range of x.

Example 2.8 ***Relay***

Calculate the force of attraction on the armature of a relay such as the one shown in Figure E2.8 whose magnetic circuit has the following characteristics: Coil, 10 000 turns; resistance, 1000 Ω; rated voltage, 10 V. Gap length, 2 mm; gap cross-section, 1 cm^2.

For simplicity, let us assume that the gap has a uniform length, that the reluctance of the iron is negligible when the gap is open, and that there is zero leakage flux [5].

Figure E2.8 Relay. When the coil is energized by closing the switch, the armature falls, opens the upper contact, and closes the lower one. Tens of contacts can be actuated simultaneously in this way.

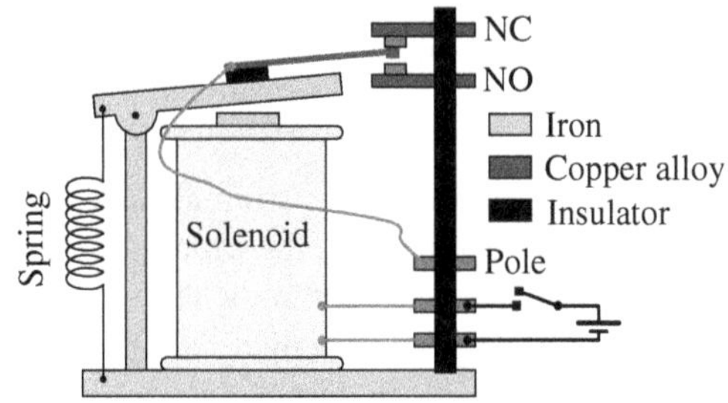

Solution
Relay

$$F = \left(\frac{B^2}{2\mu_o}\right)A = \left(\frac{A}{2\mu_o}\right)\left(\frac{\mu_o NI}{l_g}\right)^2 = \left(\frac{\mu_o A}{2}\right)\left(\frac{N\frac{V}{R}}{l_g}\right)^2$$

$$= \left(\frac{4\pi \times 10^{-7} \times 10^{-4}}{2}\right)\left(\frac{10^4 \times \frac{10}{10^3}}{2 \times 10^{-3}}\right)^2 = 1.571 \approx 1.6\,\text{N}$$

Lumped electromechanical elements [6].

Example 2.9 As an example of the calculation of lumped parameters, consider the magnetic field system of Figure E2.7. It consists of a fixed structure made of highly permeable magnetic material with an excitation winding of N turns. A movable plunger, also made of highly permeable magnetic material, is constrained by a non-magnetic sleeve to move in the x-direction. This is the basic configuration used for tripping circuit breakers, operating valves, and other applications in which a relatively large force is applied to a member that moves a relatively small distance.

We wish to calculate the flux linkage λ at the electrical terminal pair (as a function of current i and displacement x) and the terminal voltage v for a specified time variation of i and x.

Solution
To make the analysis of the system of Figure E2.7 more tractable, but still quite accurate, it is conventional to make the following assumptions:

1. The permeability of the magnetic material is high enough to be assumed infinite.
2. The air-gap lengths g and x are assumed to be small compared with transverse dimensions $g \ll w$, $x \ll 2w$, so that fringing at the gap edges can be ignored.
3. Leakage flux is assumed to be negligible; that is, the only appreciable flux passes through the magnetic material except for gaps g and x.

Needed to solve this problem are the quasi-static magnetic field Eqs. (2.41)–(2.44).

$$\oint_C \mathbf{H}\cdot d\mathbf{l} = \int_S \mathbf{J}_f \cdot \mathbf{n}da \tag{2.41}$$

$$\oint_S \mathbf{B}\cdot\mathbf{n}da = 0 \tag{2.42}$$

$$\oint_S \mathbf{J}_f \cdot \mathbf{n}da = 0 \tag{2.43}$$

$$\mathbf{B} = \mu_o (\mathbf{H} + \mathbf{M}) \tag{2.44}$$

We first assume that the terminal current is i. Then, by using (2.43) we establish that the current at each point along the winding is i. $\mathbf{M}\ (= \mathcal{X}_m \mathbf{H})$ is the magnetization density where $\mathcal{X}_m$ is the magnetic susceptibility. The magnetization density $\mathbf{M}$ is introduced to account for the effects of magnetizable materials. Next, we recognize that the specification of infinitely permeable magnetic material implies that we can write (2.44) as

$$\mathbf{B} = \mu \mathbf{H}$$

with $\mu \to \infty$. Thus, with finite flux density $\mathbf{B}$ the field intensity $\mathbf{H}$ is zero inside the magnetic material. Thus, the only nonzero $\mathbf{H}$ occurs in the air gaps g and x, where $\mathbf{M} = \mathbf{0}$, and (2.44) becomes

$$\mathbf{B} = \mu_o \mathbf{H}$$

The use of (2.41) with contour (2) in Figure E2.9 shows that the field intensities in the two gaps g are equal in magnitude and opposite in direction. This is expected from the symmetry of the system. Denoting the magnitude of the field intensity in the gaps g as H_1 and the field intensity in gap x by H_2, we can integrate (2.41) around contour (1) in Figure E2.9 to obtain

$$H_1 g + H_2 x = Ni, \tag{2.45a}$$

where H_2 is taken positive upward and H_1 is taken positive to the right. We now use (2.42) with a surface that encloses the plunger and passes through the gaps to obtain

$$\mu_o H_1 (2wd) - \mu_o H_2 (2wd) = 0. \tag{2.45b}$$

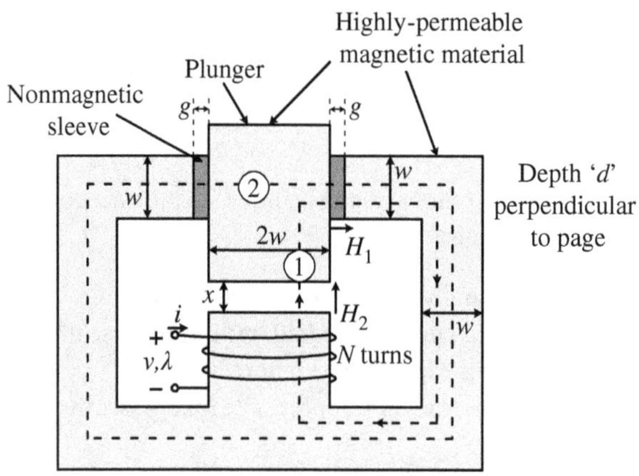

Figure E2.9 A magnetic field system.

Combine (2.45a) and (2.45b) to obtain

$$H_1 = H_2 = \frac{Ni}{g + x}$$

The flux through the centre leg of the core is simply the flux crossing the air gap $\boldsymbol{x}$ and is

$$\phi = \mu_o H_2 (2wd) = \frac{\mu_o (2wd) Ni}{g+x}$$

In the absence of leakage flux, this same flux links the N-turn winding $\boldsymbol{N}$ times; that is, when we evaluate

$$\int_S \mathbf{B} \cdot \boldsymbol{n} da$$

over a surface enclosed **by** the wire of the N-turn winding, we obtain the flux linkage λ as

$$\lambda = N\phi = \frac{2\mu_o wdN^2 i}{g+x} \tag{2.45c}$$

Note that because λ is a linear function of I, the system is electrically linear and we can write (2.45c) as

$$\lambda = L(x)\, i, \tag{2.45d}$$

where

$$L(x) = \frac{2\mu_o wdN^2}{g+x} \tag{2.45e}$$

When we assume that the current i and displacement x are specified functions of time, we can use (2.45d) with (2.45f) to evaluate the terminal voltage as

$$v = L(x)\frac{di}{dt} + i\frac{dL}{dx}\frac{dx}{dt} \tag{2.45f}$$

$$v = \frac{2\mu_o wdN^2}{g+x}\frac{di}{dt} - \frac{2\mu_o wdN^2 i}{(g+x)^2}\frac{dx}{dt}$$

The first term is the transformer voltage that will exist if x is fixed and i is varying. The second term is the speed voltage that will exist if i is constant and x is varying.

Example 2.10 As a second example, consider the system in Figure E2.10a which has two electrical terminal pairs and the mechanical displacement is rotational. This system consists of a fixed annular section of highly permeable magnetic material that is concentric with a cylindrical piece of the same material of the same axial length. Mounted in axial slots in the material are coils labelled with the numbers of turns and current directions. The angular position of the inner structure (rotor) relative to the outer structure (stator) is indicated by an angle θ which can vary with time. Current is fed to the coil on the rotor through sliding contacts (brushes that make contact with slip rings) [6].

The system in Figure E2.10a represents the basic method of construction of many rotating machines. In our solution, we discuss how this configuration is used with some variations to achieve the lumped parameters desired for rotating-machine operation.

We wish to calculate the two flux linkages λ_1 and λ_2 as functions of the currents i_1 and i_2 and the angular displacement θ. Voltages at the two terminal pairs are also to be found, assuming that i_1, i_2, and θ are specified functions of time.

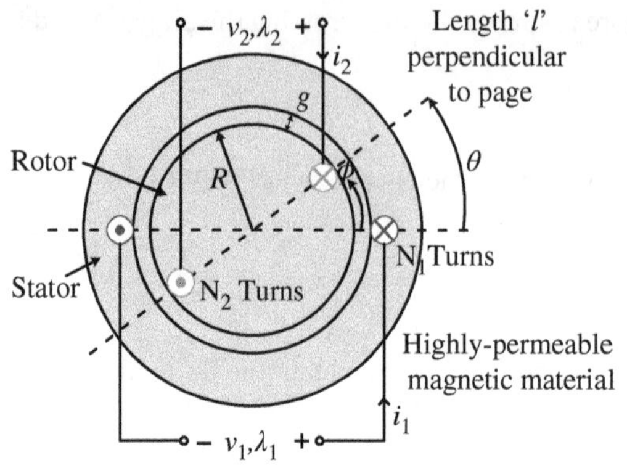

Figure E2.10a Doubly excited magnetic field system.

Solution

Electromechanical systems of the type illustrated in Figure E2.10a are normally constructed with relative dimensions and materials that allow reasonably accurate calculation of lumped parameters when the following assumptions are made:

1. The permeability of the magnetic material is high enough to be assumed infinite.
2. The radial air-gap length g is small enough compared with the radius R and axial length l to allow the neglect of fringing fields at the ends and radial variation of magnetic field intensity in the air gap.
3. The slots containing the windings are small enough, both radially and circumferentially, to perturb the fields a negligible amount; that is, the coils are considered to be infinitely thin.

Equations (2.41) and (2.42) are used to write the radial fields in the air gap in terms of the angular variable ϕ defined in Figure E2.10a. If we define the magnetic field as positive when directed radially outward and consider $0 < \phi < \pi$,

$$H_r = \frac{N_1 i_1 - N_2 i_2}{2g} \quad \text{for} \quad 0 < \phi < \theta,$$

$$H_r = \frac{N_1 i_1 + N_2 i_2}{2g} \quad \text{for} \quad \theta < \phi < \pi,$$

$$H_r = -\frac{N_1 i_1 - N_2 i_2}{2g} \quad \text{for} \quad \pi < \phi < \pi + \theta,$$

$$H_r = -\frac{N_1 i_1 + N_2 i_2}{2g} \quad \text{for} \quad \pi + \theta < \phi < 2\pi,$$

The flux linkages with the two windings can be found from the integrals

$$\lambda_1 = \int_0^{\pi} \mu_o N_1 H_r l R d\phi,$$

$$\lambda_2 = \int_0^{\pi+\theta} \mu_o N_2 H_r l R d\phi,$$

Evaluation of these integrals yields

$$\lambda_1 = L_1 i_1 + L_m i_2,$$

$$\lambda_2 = L_m i_1 + L_2 i_2,$$

where $L_1 = N_1^2 L_o$, $L_2 = N_2^2 L_o$, $L_m = L_o N_1 N_2 \left(1 - \frac{2\theta}{\pi}\right)$, for $0 < \theta < \pi$, $L_o = \frac{\mu_o l R \pi}{2g}$.

Similar arguments show that for $-\pi < \theta < 0$ the terminal relations have the same form except that

$$L_m = L_o N_1 N_2 \left(1 + \frac{2\theta}{\pi}\right)$$

Note that only the mutual inductance L_m is a function of angular displacement θ because the geometry seen by each coil individually does not change with θ; thus, the self-inductances are constants. The mutual inductance is sketched as a function of θ in Figure E2.10b.

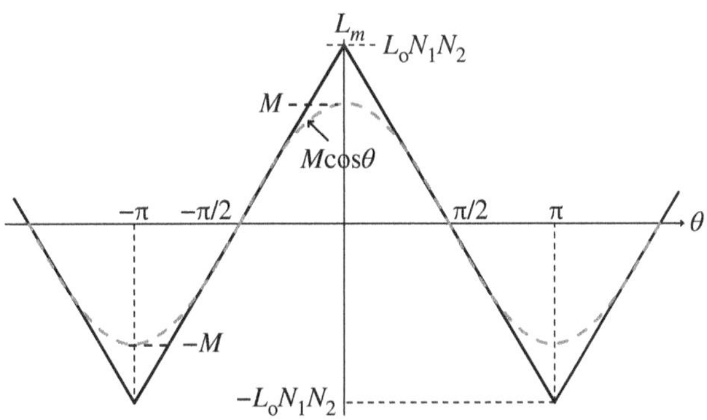

Figure E2.10b Mutual inductance *Lm* as a function of θ.

In the design of rotating machines, especially for operation on alternating currents, it is desirable to have a system similar to that in Figure E2.10a but to modify it in such a way that the mutual inductance varies cosinusoidally with $\theta(L_m = M\cos\theta)$. This is accomplished by putting additional slots and windings at different positions around the periphery of both members. By using a proper distribution of slots and numbers of turns, the dependence of L_m, can be made the cosinusoidal function shown by the dashed curve in Figure E2.10b. In many later examples, we assume that this design process has been followed.

When the two currents i_1 and i_2 and the angular position θ are functions of time and the mutual inductance is expressed as $L_m = M\cos\theta$, we can write the terminal voltages as

$$v_1 = \frac{d\lambda_1}{dt} = L_1\frac{di_1}{dt} + M\cos\theta\frac{di_2}{dt} - i_2 M\cos\theta\frac{d\theta}{dt}$$

$$v_2 = \frac{d\lambda_2}{dt} = L_2\frac{di_2}{dt} + M\cos\theta\frac{di_1}{dt} - i_1 M\cos\theta\frac{d\theta}{dt}$$

Note that the first term in each expression is a derivative with a constant coefficient, whereas the last two terms are derivatives with time-varying coefficients.

2.7 Field Energy

2.7.1 Energy Stored in a Magnetic Field

To calculate the work which must be done to establish a magnetic field, we shall calculate the energy supplied by a source to an isolated circuit when the current density increases from zero to some value **J**.

At a given point in the conducting medium (non-magnetic), there is a current density **J** and corresponding electric field intensity [7]

$$\mathbf{E} = \frac{\mathbf{J}}{\sigma} \tag{2.46}$$

where σ is the electrical conductivity in mhos/metre. This equation is always true, whatever the origin of **E** may be. It states that, under the action of an electric field intensity **E**, the electrons drift through the conductor at such a velocity that the results current density **J**/σ. This is Ohm's law.

The electric field intensity results from (*a*) the field $-\vec{\nabla}V$ produced by accumulations of charge on the terminals of the source, and on the surfaces of the conductor and from (*b*) the field $\left(-\frac{\partial\vec{A}}{\partial t}\right)$ induced by the vector potential $\vec{A}$ in the conductor, if it is time dependent. It should, however, be kept in mind that the net charge density ρ can be set equal to zero inside a conductor. In a wire, $\vec{\nabla}V$ adjusts itself such that the total field intensity **E** is along the axis.

Inside a source, we also have a third electric field which comes from the local generation of energy. This field can be written as $\mathbf{E}_e$; for example, inside a battery.

$$\frac{\mathbf{J}}{\sigma} = \mathbf{E} + \mathbf{E}_e \tag{2.47}$$

The work done per unit time and per unit volume on the moving charges at any point outside the source can be calculated as follows. Consider an element of volume having the form of a rectangular parallelepiped oriented such that one set of sides is parallel to the total current density **J**, as in Figure 2.22. In one second, a charge *Jds* goes in through the left-hand face and a similar charge comes out at the other end. The source which maintains a difference in potential of $\vec{\nabla}V\cdot\vec{dl}$ across these faces supplies to the element of volume *dv* an amount of power given by

$$\frac{dW}{dt} = -\vec{\nabla}V\cdot\vec{dl}\left(\vec{J}\vec{ds}\right) \tag{2.48}$$

$$\frac{dW}{dt} = \left(\vec{E} + \frac{\partial\vec{A}}{\partial t}\right)\cdot\vec{J}dv \tag{2.49}$$

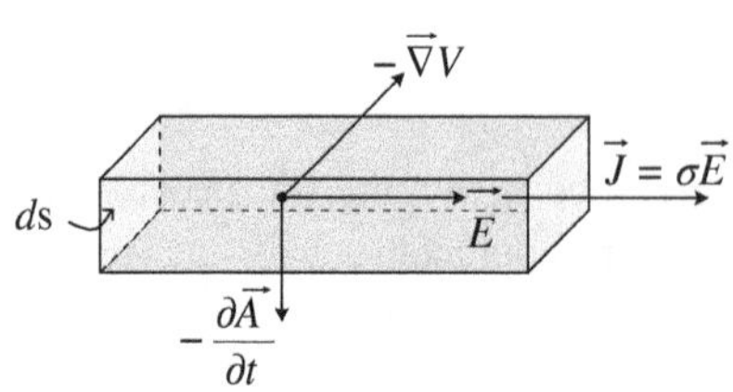

Figure 2.22 An element of volume having form of a rectangular parallelepiped oriented such that one set of sides is parallel to the total current density **J**.

and the integral of this expression over the volume v including all the current in the system, gives the total power spent by the source;

$$\frac{dW}{dt} = \int_v \frac{J^2}{\sigma} dv + \int_v \frac{\partial \vec{A}}{\partial t} \cdot \vec{J} dv \tag{2.50}$$

The first term of the right side gives the total power spent in Ohmic or Joule losses; the second gives the rate at which work is done by the source against the induced electromotance. This latter work is that which must be done to establish the magnetic field. We are mainly concerned with this and therefore shall disregard Joule losses and write W_m for the energy stored in the magnetic field. Then

$$\frac{dW}{dt} = \left(\int_v \frac{\partial \vec{A}}{\partial t} \cdot \vec{J} \right) dv \tag{2.51}$$

This expression is always true, even if there are magnetic materials such as iron present in the field. We shall now find other expressions For W_m, but those will not be as general as the one shown above.

2.7.1.1 The Magnetic Energy in Terms of the Magnetic Induction B

It is now desired to express the magnetic energy W_m in terms of the magnetic induction **B**, just as the electrostatic energy expressed in terms of the electrostatic field intensity **E**. We shall now restrict our discussion by assuming (*a*) there are no magnetic materials present and (*b*) the displacement current is negligible compared to the conduction current. Then

$$\mathbf{J} = \frac{1}{\mu_o} \left(\vec{\nabla} \times \vec{B} \right) \tag{2.52}$$

and

$$\frac{dW_m}{dt} = -\frac{1}{\mu_o} \int_v \left[\vec{E}_i \cdot \left(\vec{\nabla} \times \vec{B} \right) \right] dv \tag{2.53}$$

$$= \frac{1}{\mu_o} \int_v \left[\vec{\nabla} \cdot \left(\vec{E}_i \times \vec{B} \right) \right] dv - \frac{1}{\mu_o} \int_v \left[\vec{B} \cdot \left(\vec{\nabla} \times \vec{E}_i \right) \right] dv \tag{2.54}$$

where $\mathbf{E}_i$ is the induced electric field intensity. Using divergence theorem to transform the first term into a surface integral, and using $\vec{\nabla} \times \vec{E} = -\frac{\partial \vec{B}}{\partial t}$ in the second term, we obtain

$$\frac{dW_m}{dt} = \frac{1}{\mu_o} \int_s \left(\vec{E}_i \times \vec{B} \right) \cdot \vec{ds} + \frac{1}{\mu_o} \int_v \left(\vec{B} \cdot \frac{\partial \vec{B}}{\partial t} \right) dv \tag{2.55}$$

The volume v being any volume which includes all points where the current density **J** is not zero, and s being a corresponding surface.

Let us now choose v to include all space, in which case, the surface s is at infinity. The magnetic induction **B** falls off as $\frac{1}{r^3}$ at large distances. The induced electric field intensity $\vec{E}_i$ also falls off as $\frac{1}{r^2}$ (e.g. for a current loop). Since the surface area s increases only as r^2, the surface integral decreases as $\frac{1}{r^3}$ and vanishes as the surface of integration become infinite. Thus

$$\frac{dW_m}{dt} = \frac{1}{\mu_o} \int_\infty \left(\vec{B} \cdot \frac{\partial \vec{B}}{\partial t} \right) dv \tag{2.56}$$

Setting $W_m = 0$ when $\vec{B} = 0$, we have

$$dW_m = \frac{1}{\mu_0}\int_{\infty}\left(\vec{B}\cdot d\vec{B}\right)dv = \frac{1}{\mu_0}\int_{\infty}\left(\vec{B}\cdot d\vec{B}\right)dv$$

$$W_m = \frac{1}{2\mu_0}\int_{\infty} B^2 dv \tag{2.57}$$

The quantity W_m is the total work which must be done to establish a magnetic field in terms of the magnetic induction **B** either in free space or in non-magnetic matter.

It should be noted that the magnetic energy varies as the square of the magnetic induction B. Therefore, if several fields are superimposed, the total energy is not just the sum of the energies calculated for each separate field.

Just as in electrostatics, we may define an energy density

$$\frac{dW_m}{dv} = \frac{B^2}{2\mu_o} \tag{2.58}$$

associated with each point in space. This matter will be discussed further in the next section.

2.7.1.2 The Magnetic Energy in Terms of the Current Density J and the Vector Potential A

In an electrostatic field, the energy density is expressed either in the form $\frac{\varepsilon_o E^2}{2}$ or $\frac{\rho V}{2}$. In Eq. (2.58) the magnetic density has also been expressed as $\frac{B^2}{2\mu_o}$.

Let us now express it in terms of the Current Density **J** and the Vector Potential **A**.

We rewrite Eq. (2.58) as follows:

$$W_m = \frac{1}{2\mu_o}\int_{\infty}\left(\vec{B}\cdot\vec{\nabla}\times\vec{A}\right)dv \tag{2.59}$$

Using the vector identity

$$\vec{\nabla}\cdot\left(\vec{A}\times\vec{B}\right) = \vec{B}\cdot\left(\vec{\nabla}\times\vec{A}\right) - \vec{A}\cdot\left(\vec{\nabla}\times\vec{B}\right)$$

and

$$\int_{v}\left(\vec{\nabla}\times\vec{A}\right)dv = -\int_{s}\vec{A}\times d\vec{s}$$

we obtain

$$W_m = \frac{1}{2\mu_o}\int_{\infty}\left[\vec{A}\cdot\left(\vec{\nabla}\times\vec{B}\right)\right]dv - \frac{1}{2\mu_o}\int_{\infty}\left(\vec{A}\times\vec{B}\right)\cdot d\vec{s} \tag{2.60}$$

The surface integral vanishes as in Eq. (2.55), and when

$$\vec{\nabla}\times\vec{B} = \mu_o\vec{J}$$

$$W_m = \frac{1}{2\mu_o}\int_{\infty}\left[\vec{A}\cdot\mu_o\vec{J}\right]dv = \frac{1}{2}\int_{\infty}\left(\vec{J}\cdot\vec{A}\right)dv \tag{2.61}$$

where v is any volume which includes all regions where **J** is not zero.

It is convenient to design an energy density

$$\frac{dW_m}{dv} = \frac{1}{2}\left(\vec{J}\cdot\vec{A}\right) \tag{2.62}$$

to conductors carrying a current density **J**.

Again, as in electrostatics, the assignment of an energy density to a point in space is quite arbitrary and meaningless, except as a means of computing the overall magnetic energy W_m. Eqs. (2.58) and (2.62) are clearly contradictory in that the former assigns a finite energy density to all points where $B \neq 0$, whereas the latter makes the energy density zero wherever there is zero current density.

2.7.1.3 The Magnetic Energy in Terms of the Current I and of the Flux Ψ_m

For a filamentary circuit, we must also express the field energy in terms of the current and of the flux linking current. If we replace **J***dv* by I***dl*** in Eq. (2.61) ***dl*** being an element of the circuit which carries the current I, then

$$W_m = \frac{1}{2}I \int \vec{A} \cdot d\vec{l} \tag{2.63}$$

With real conductors, the current is not truly filamentary but is distributed over a small but finite area. We can then use the mean value of **A** over the cross-section.

Since the integral gives the flux Ψ_m linking the circuit

$$W_m = \frac{1}{2}I\Psi_m \tag{2.64}$$

The directions in which I and Ψ_m are taken to be positive are related as in the right-hand screw rule.

2.7.1.4 The Magnetic Energy in Terms of the Currents and Inductances

It is possible to express the energy stored in a magnetic field in yet another way, in terms of currents and inductances. In the above equation, the magnetic flux Ψ_m can be replaced by the product of the self-inductance L and the current I. Then

$$W_m = \frac{1}{2}I\Psi_m = \frac{1}{2}ILI = \frac{1}{2}LI^2 \tag{2.65}$$

For two circuits (Figure 2.23) carrying currents I_1 and I_2,

$$W_m = \frac{1}{2}I_1\Psi_1 + \frac{1}{2}I_2\Psi_2 \tag{2.66}$$

where $\Psi_1 (\equiv \Psi_{m1})$ and $\Psi_2 (\equiv \Psi_{m2})$ are the total fluxes linking circuits 1 and 2, respectively. The total fluxes consist of contributions from both circuits:

$$\Psi_1 = \Psi_{11} + \Psi_{12} \tag{2.67}$$

$$\Psi_2 = \Psi_{21} + \Psi_{22} \tag{2.68}$$

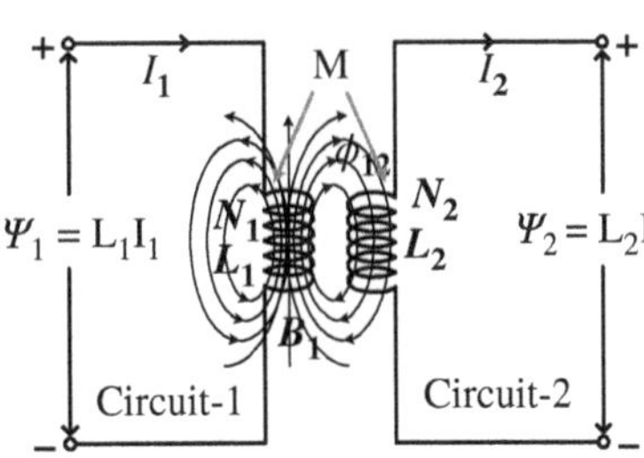

Figure 2.23 Mutually coupled two circuits.

Thus

$$W_m = \frac{1}{2}I_1\Psi_{11} + \frac{1}{2}I_2\Psi_{22} + \frac{1}{2}I_1\Psi_{12} + \frac{1}{2}I_2\Psi_{21} \tag{2.69}$$

$$W_m = \frac{1}{2}I_1\Psi_{11} + \frac{1}{2}I_2\Psi_{22} + \frac{1}{2}I_1\Psi_{12} \tag{2.70}$$

$$W_m = \frac{1}{2}I_1\Psi_{11} + \frac{1}{2}I_2\Psi_{22} + \frac{1}{2}I_2\Psi_{21} \tag{2.71}$$

or, from the definition of self and mutual inductance

$$\Psi_1 = L_1I_1 + MI_2 \tag{2.72}$$

$$\Psi_2 = L_2I_2 + MI_1 \tag{2.73}$$

and so

$$W_m = \frac{1}{2}I_1\Psi_1 + \frac{1}{2}I_2\Psi_2 = \frac{1}{2}I_1\left(L_1I_1 + MI_2\right) + \frac{1}{2}I_2\left(L_2I_2 + MI_1\right)$$

or

$$W_m = \frac{1}{2}L_1I_1^2 + \frac{1}{2}L_2I_2^2 + MI_1I_2 \tag{2.74}$$

The first two terms on the right are self-energies arising from the interaction of each current with its own field, whereas the third term is an interaction energy arising from the mutual inductance.

2.8 The Magnetic Energy for a Solenoid Carrying a Current I

Let us consider the relatively simple case of the long solenoid of length l and radius R with negligible end effects and calculate W_m by each of the four methods given above (Figure 2.24).

(1) We know that, neglecting end effects, the magnetic induction is uniform throughout the interior of the solenoid and is given by

$$B = \mu_0 nI \tag{2.75}$$

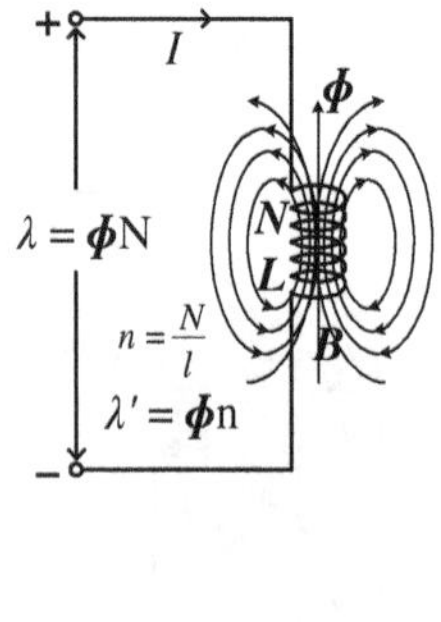

Figure 2.24 Solenoid carrying current *I*.

where n is the number of turns per metre. The magnetic induction is zero everywhere outside. Then

$$W_m = \frac{1}{2\mu_o}\int B^2 dv \tag{2.76}$$

$$W_m = \frac{B^2}{2\mu_o}\pi R^2 l = \frac{\pi}{2}\mu_o n^2 I^2 R^2 l \tag{2.77}$$

(2) Also

$$W_m = \frac{1}{2}\int \left(\vec{J}\cdot\vec{A}\right) dv \tag{2.78}$$

In this case, the current can be taken to be distributed, not over a volume, but over the surface of the solenoid with a density

$$\lambda' = \frac{\lambda}{l} = nI \tag{2.79}$$

The vector potential **A** is parallel to I and is obtained as follows:
To find **A** outside the solenoid, we will first determine **A** for a single turn and integrate it over the length of the solenoid. Thus, at any radius $r > R$, $|\mathbf{A}|$ is given by

$$A = \frac{\mu_o R^2}{2r}\frac{N}{l}I$$

Then at surface i.e. $r = R$

$$A = \frac{\mu_o R}{2}nI$$

$$W_m = \frac{1}{2}\int \lambda' A ds = \frac{1}{2}nI\frac{\mu_o R}{2}nI\int ds = \frac{1}{2}n^2 I^2\frac{\mu_o R}{2}(2\pi R l) = \frac{1}{2}\mu_o \pi n^2 I^2 R^2 l \tag{2.80}$$

As before

(3) Using still another method, we find that

$$W_m = \frac{1}{2}I\Psi_m = \frac{1}{2}IB\left(\pi R^2 nl\right) = \frac{1}{2}I\left(\mu_o nI\right)\left(\pi R^2 nl\right) = \frac{1}{2}\mu_o \pi n^2 I^2 R^2 l \tag{2.81}$$

(4) Finally,

$$W_m = \frac{1}{2}LI^2$$

And, from the value of L given as

$$L = \mu_o \pi n^2 R^2 l$$

We again find that

$$W_m = \frac{1}{2}\mu_o \pi I^2 n^2 R^2 l \tag{2.82}$$

2.9 Energy Flow Diagram

Figure 2.25 illustrates the conversion of electrical energy into mechanical energy, according to the following equation:

$$dW_e = dW_f + dW_{loss} + dW_{mech} \tag{2.83}$$

where dW_e is the electrical energy (input energy) dW_f is the energy stored in the magnetic field (coil), dW_{loss} are all the power losses and dW_{mech} is the mechanical energy (output energy) [8].

2.9.1 Power Flow Diagram of DC Generator and DC Motor

The Power Flow Diagram is used to determine the efficiency of a generator or motor. In Figure 2.26 below, it is initially shown that the mechanical power is given as an input which is converted into electrical power, and the output which is obtained is in the form of electrical power. There are various losses such as friction, windage, stray losses, and core losses.

The power flow diagram of DC Generator is shown in Figure 2.26.

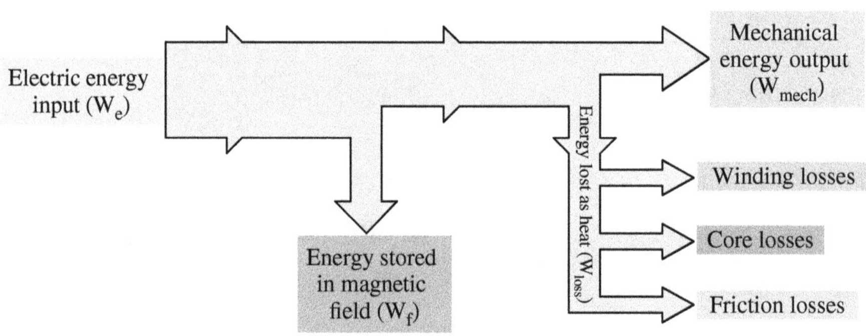

Figure 2.25 Energy flow in electromechanical energy conversion device.

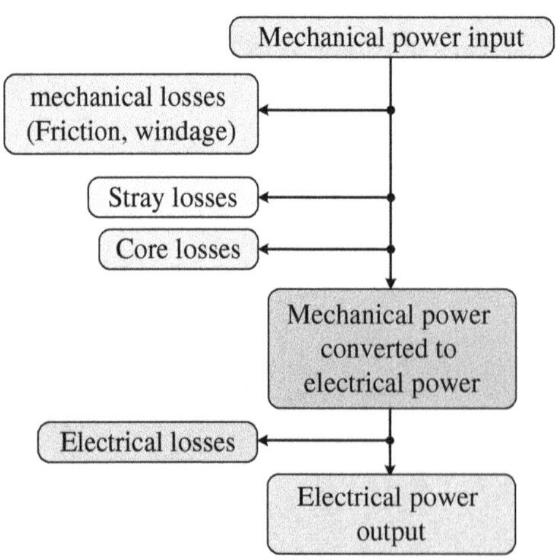

Figure 2.26 Power Flow Diagram of a DC Generator.

In a DC Generator, the input is the mechanical power. The power input is given by Eq. (2.84).

$$P_{\text{in}} = \omega_m T_{\text{app}} \tag{2.84}$$

where

ω_m is the angular speed of the armature in radian per second.
T_{app} is the applied torque in Newton-metre

The sum of stray, mechanical and core losses are subtracted from the input power, i.e. P_{in} to obtain the net mechanical power converted to electrical power by Electro Mechanical conversion.

$$P_{\text{conv}} = P_i - \text{stray loss} - \text{core losses}$$

$$P_{\text{conv}} = \omega_m T_{\text{app}} = \omega_m T_{\text{e}} \tag{2.85}$$

where T_e is the electromagnetic torque. The resulting electric power produced is given by the equation

$$P_{\text{conv}} = E_g I_{\text{a}} \tag{2.86}$$

The net electrical power output is obtained by subtracting electrical power I^2R losses and brush losses from P_{conv}

$$P_{out} = P_{\text{conv}} - \text{electrical } I^2R \text{ loss} - \text{brush losses} \tag{2.87}$$

$$P_{out} = V_T I_L \tag{2.88}$$

where

V_T is the terminal voltage, and I_L is the current delivered to the load.

The power flow diagram of DC Motor is shown in Figure 2.27.

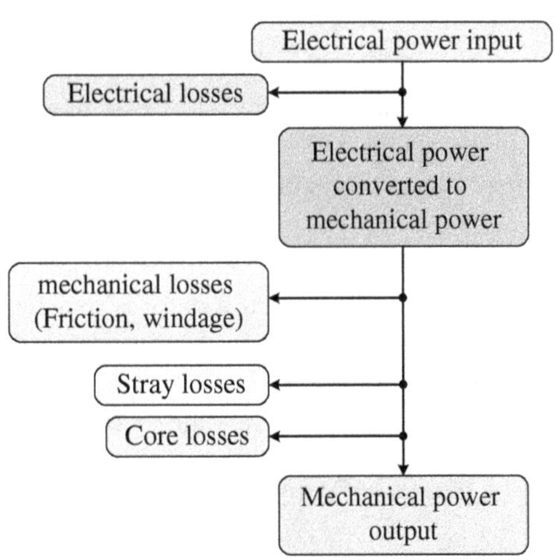

Figure 2.27 Power Flow Diagram of DC Motor.

From the power flow diagram of DC Motor, it is clear the input given to the motor is in electrical form which is converted into mechanical power in the second stage. The output is in the form of mechanical power.

In a DC motor, the input electrical power P_{in} is given by the equation shown below

$$P_{in} = V_T I_L \tag{2.89}$$

$$P_{conv} = P_i - \text{copper losses} \tag{2.90}$$

Power output is given by the equation shown below

$$P_{out} = \omega_m T_L \tag{2.91}$$

Also

$$P_{out} = P_{conv} - \text{core losses} - \text{mechnical losses} - \text{stray losses} \tag{2.92}$$

where

T_L is the load torque in Newton-metre

Thus, the power flow diagram gives an overview of how one form of energy is converted into another.

2.9.1.1 Power Flow Diagram and Losses of Induction Motor

The Power Flow Diagram of Induction Motor explains the input given to the motor, the losses occurring as well as the output of the motor. The input power given to an Induction motor is in the form of three-phase voltage and currents. The Power Flow Diagram of an Induction Motor is shown in Figure 2.28.

The power flow is given by the equation shown below

$$P_{in-stator} = \sqrt{3} V_L I_L \cos\phi_i = 3V_{ph} I_{ph} \cos\phi_i \tag{2.93}$$

Where $\cos\phi_i$ is the input power factor

The losses in the stator are I^2R losses in the stator-winding resistances. It is also known as Stator copper losses.

$$P_{stator\ copper\ loss} = 3I^2_{stator\ ph} R_{stator\ ph} \tag{2.94}$$

Hysteresis and eddy current losses in the stator core is $P_{stator(h+e)}$. These are known as Stator core losses.

The output power of the stator is given as

$$P_{out-stator} = P_{in-stator} - P_{stator\ copper\ loss} - P_{stator(h+e)} \tag{2.95}$$

This output power of the stator is transferred to the rotor of the machine across the air gap between the stator and the rotor. It is called the air gap Pg of the machine.

Thus

The Power output of the stator = air gap power = input power to the rotor

$$P_{out-stator} = P_{air\ gap} = P_{input\ rotor} \tag{2.96}$$

The losses in the rotor are as follows:

Figure 2.28 The power flow diagram of an induction motor.

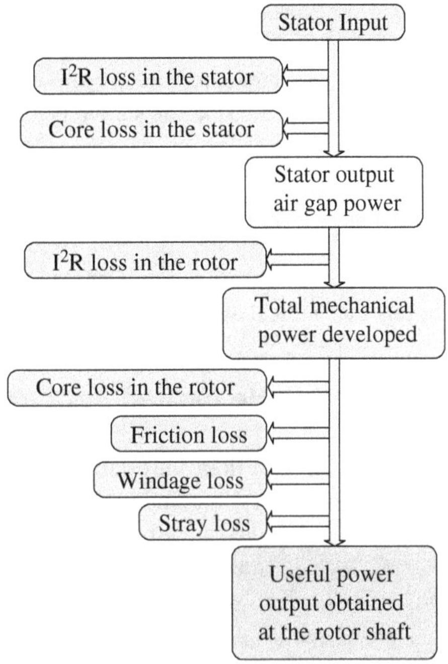

I^2R losses in the rotor resistance. They are also called Rotor copper losses and represented as

$$P_{rotor\ copper\ loss} = 3I^2_{rotor\ ph}R_{rotor\ ph} \tag{2.97}$$

Hysteresis and eddy current losses in the rotor core is $P_{rotor(h+e)}$. They are known as Rotor core losses.

Friction and Windage losses $P_{friction+wind}$ are due to friction and wind.

Stray-load losses P_{stray}, consisting of all losses not covered above, such as losses due to harmonic fields.

If the rotor copper losses are subtracted from rotor input power $P_{air\text{-}gap}$, the remaining power is converted from electrical to mechanical form. This is called Developed Mechanical Power $P_{devlop\ mech}$

Developed Mechanical power = Rotor input−Rotor copper loss

$$P_{devlop\ mech} = P_{input\ rotor} - P_{rotor\ copper\ loss} = P_{air\ gap} - P_{rotor\ copper\ loss} \tag{2.98}$$

$$P_{devlop\ mech} = P_{air\ gap} - 3I^2_{rotor\ ph}R_{rotor\ ph} \tag{2.99}$$

The output of the motor is given by the equation shown below

$$P_{out} = P_{devlop\ mech} - P_{\boldsymbol{friction+wind}} - P_{stray} \tag{2.100}$$

P_{out} is called the shaft power or the useful power.

2.9.1.2 Rotational Losses

At the beginning and during acceleration, the rotor core losses are high. With the increase in speed of the induction motor these losses decrease. The friction and windage losses are zero at the start. As the speed increases, the losses also start increasing. The sum of the friction, windage and core losses are almost constant with the change in speed. All these losses are added together and are known as Rotational Losses.

It is given by the equation shown below

$$P_{rotor} = P_{\boldsymbol{friction+wind}} + P_{rotor(h+e)} + P_{stray} \tag{2.101}$$

$$P_{out} = P_{devlop\ mech} - P_{rotor} = P_{devlop\ mech} - \left(P_{\boldsymbol{friction+wind}} + P_{rotor(h+e)} + P_{stray}\right) \tag{2.102}$$

Rotational losses are not represented by any element of the equivalent circuit as they are purely mechanical quantity.

2.10 Multiple Excited Systems

Systems with more than one exciting winding are referred to as multiple excited systems [9]. We encounter this situation in most rotating electromechanical energy conversion devices. The force (or torque) can be obtained by simple extension of the techniques discussed in the section. We will consider a system with three windings, as shown in Figure 2.29. Of course, our discussion can simply be extended to an arbitrary number of windings, *n*.

The differential electric-energy input is

$$dW_e = i_1 d\lambda_1 + i_2 d\lambda_2 + i_3 d\lambda_3 \tag{2.103}$$

The mechanical-energy increment is given by

$$dW_{mech} = T_{fld} d\theta$$

Thus, the field-energy increment is obtained as

$$dW_{fld} = dW_e - dW_{mech} = i_1 d\lambda_1 + i_2 d\lambda_2 + i_3 d\lambda_3 - T_{fld} d\theta \tag{2.104}$$

If we express W_{fld} in terms of λ_1, λ_2, λ_3, and θ, we have

$$dW_f = \frac{\partial W_f}{\partial \lambda_1} d\lambda_1 + \frac{\partial W_f}{\partial \lambda_2} d\lambda_2 + \frac{\partial W_f}{\partial \lambda_3} d\lambda_3 + \frac{\partial W_f}{\partial \theta} d\theta \tag{2.105}$$

As a result,

$$T_{fld} = \frac{\partial W_f\left(\lambda_1, \lambda_2, \lambda_3, \theta\right)}{\partial \theta} \tag{2.106}$$

$$i_1 = \frac{\partial W_f\left(\lambda_1, \lambda_2, \lambda_3, \theta\right)}{\partial \lambda_1} \tag{2.107}$$

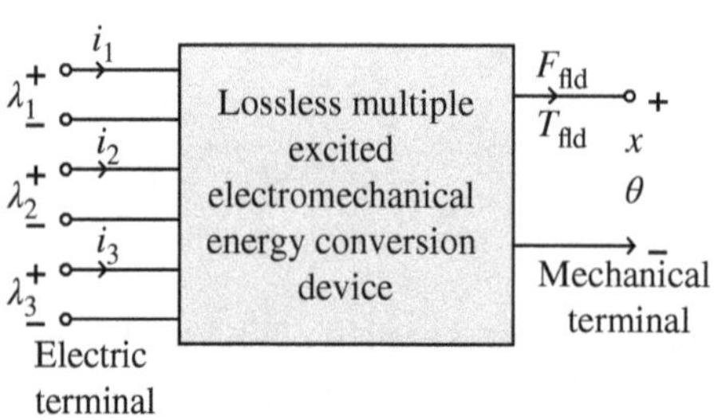

Figure 2.29 Multiple excited lossless electromechanical energy conversion device [9].

$$i_2 = \frac{\partial W_f\left(\lambda_1, \lambda_2, \lambda_3, \theta\right)}{\partial \lambda_2} \tag{2.108}$$

$$i_3 = \frac{\partial W_f\left(\lambda_1, \lambda_2, \lambda_3, \theta\right)}{\partial \lambda_3} \tag{2.109}$$

The singly excited case can be seen to be included as a special case of our present result. If we are dealing with a rotational system, we have

$$T_{fld} = \frac{\partial W_f\left(\lambda_1, \lambda_2, \lambda_3, \theta\right)}{\partial \theta} \tag{2.110}$$

We will need to find an expression for the field energy at a state corresponding to point *O*, where $\lambda_1 = \lambda_{1_0}, \lambda_2 = \lambda_{2_0}, \lambda_3 = \lambda_{3_0}$ and $\theta = \theta_0$. Again, as we are dealing with a conservative system (no loss), we can use a simple path of integration *ABCO* as follows (see Figure 2.30):

1. **A**$\lambda_1 = \lambda_2 = \lambda_3 = 0$and θvaries from 0 to θ_0. Thus $\lambda_1 = \lambda_2 = \lambda_3 = 0$and $T_{fld} = 0$, and we conclude that

$$\int_A^A dW_{fld} = \int_A^A \left(i_1 d\lambda_1 + i_2 d\lambda_2 + i_3 d\lambda_3 - T_{fld} d\theta\right) = 0 \tag{2.111}$$

2. AB$\lambda_1 = \lambda_2 = 0$, $\theta = \theta_0$and λ_3 is allowed to vary from 0 to λ_{3_0}.Thus$d\lambda_1 = d\lambda_2 = 0$, $d\theta = 0$, and we have

$$\int_A^B dW_{fld} = \int_A^B \left(i_1 d\lambda_1 + i_2 d\lambda_2 + i_3 d\lambda_3 - T_{fld} d\theta\right) = \int_{\lambda_3=0}^{\lambda_{3_0}} i_{3AB} d\lambda_3 \tag{2.112}$$

Here, we have

$$i_{3AB} = i_3\left(\lambda_1 = 0, \lambda_2 = 0, \lambda_3, \theta_0\right) \tag{2.113}$$

That is, λ_1 and λ_2 are held at zero, while θ is at θ_0 and the current is a function of $\lambda_3$only.

3. BC $\lambda_1 = 0$, $\lambda_3 = \lambda_{3_0}$, $\theta = \theta_0$, and $\lambda_2$is varied from zero to λ_{2_0}. Here we have $d\lambda_1 = d\lambda_3 = d\theta = 0$, and we obtain

$$\int_B^C dW_{fld} = \int_{\lambda_2=0}^{\lambda_{2_0}} i_{2BC} d\lambda_2 \tag{2.114}$$

where

$$i_{2BC} = i_2\left(\lambda_1 = 0, \lambda_2, \lambda_3 = \lambda_{3_0}, \theta_0\right) \tag{2.115}$$

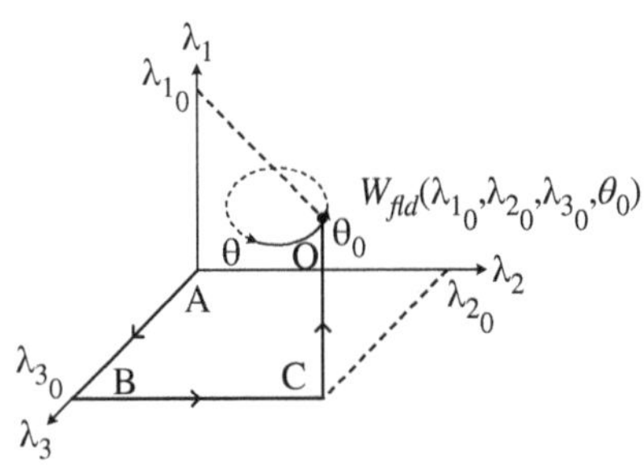

Figure 2.30 Integration path to obtain $W_{fld}\left(\lambda_{1_0}, \lambda_{2_0}, \lambda_{3_0}, \theta_{0.}\right)$.

4. CO$\lambda_2 = \lambda_{2_0}$, $\lambda_3 = \lambda_{3_0}$, $\theta = \theta_0$, and λ_1 is varied from zero to λ_{1_0}. We have

$$\int_C^O dW_{fld} = \int_{\lambda_1=0}^{\lambda_{1_0}} i_{1CO} d\lambda_1 \tag{2.116}$$

with

$$i_{1CO} = i_1\left(\lambda_1, \lambda_2 = \lambda_{2_0}, \lambda_3 = \lambda_{3_0}, \theta = \theta_0\right) \tag{2.117}$$

We can now conclude

$$W_{fld}\left(\lambda_{1_0}, \lambda_{2_0}, \lambda_{3_0}, \theta_0\right) = \int_A^A dW_{fld} + \int_A^B dW_{fld} + \int_B^C dW_{fld} + \int_C^O dW_{fld}$$

This now can be written as

$$W_{fld}\left(\lambda_{1_0}, \lambda_{2_0}, \lambda_{3_0}, \theta_0\right) = \int_{\lambda_3=0}^{\lambda_{3_0}} i_{3AB} d\lambda_3 + \int_{\lambda_2=0}^{\lambda_{2_0}} i_{2BC} d\lambda_2 + \int_{\lambda_1=0}^{\lambda_{1_0}} i_{1CO} d\lambda_1 \tag{2.118}$$

The energy stored in the field can be obtained using Eq. (2.118). The result is

$$W_{fld}\left(\lambda_{1_0}, \lambda_{2_0}, \lambda_{3_0}, \theta_0\right)$$

$$= \frac{1}{2}\left(\Gamma_{11}\lambda_{1_0}^2 + \Gamma_{22}\lambda_{2_0}^2 + \Gamma_{33}\lambda_{3_0}^2\right) + \Gamma_{12}\lambda_{1_0}\lambda_{2_0} + \Gamma_{23}\lambda_{2_0}\lambda_{3_0} + \Gamma_{13}\lambda_{1_0}\lambda_{3_0} \tag{2.119}$$

where

$$i_1 = \Gamma_{11}\lambda_1 + \Gamma_{12}\lambda_2 + \Gamma_{13}\lambda_3 \tag{2.120}$$

$$i_2 = \Gamma_{12}\lambda_1 + \Gamma_{22}\lambda_2 + \Gamma_{23}\lambda_3 \tag{2.121}$$

$$i_3 = \Gamma_{13}\lambda_1 + \Gamma_{23}\lambda_2 + \Gamma_{33}\lambda_3 \tag{2.122}$$

In compact form

$$\mathbf{I} = \mathbf{\Gamma}\boldsymbol{\lambda} \tag{2.123}$$

where the flux linkage vector λ is

$$\lambda = \begin{bmatrix} \lambda_1 \\ \lambda_2 \\ \lambda_3 \end{bmatrix} \tag{2.124}$$

The matrix Γ is the inverse of L: Here $L_{12} = L_{21}$, $L_{13} = L_{31}$ and $L_{32} = L_{31}$

$$\mathbf{\Gamma} = \mathbf{L}^{-1} = \begin{bmatrix} L_{11} & L_{12} & L_{13} \\ L_{12} & L_{22} & L_{23} \\ L_{13} & L_{23} & L_{33} \end{bmatrix}^{-1} \tag{2.125}$$

With elements Γll, ΓI2, …

$$\mathbf{\Gamma} = \begin{bmatrix} \Gamma_{11} & \Gamma_{12} & \Gamma_{13} \\ \Gamma_{12} & \Gamma_{22} & \Gamma_{23} \\ \Gamma_{13} & \Gamma_{23} & \Gamma_{33} \end{bmatrix} \tag{2.126}$$

The stored energy in the field is therefore

$$W_{fld}\left(\lambda_0, \theta_0\right) = \frac{1}{2}\lambda_0^t \Gamma \lambda_0 \tag{2.127}$$

We can express the energy stored in the field in terms of currents and inductances by substituting:

$$\lambda = \mathbf{LI} \tag{2.128}$$

$$W_{fld}\left(\lambda_0, \theta_0\right) = \frac{1}{2}\mathbf{I}_0^t \mathbf{\Gamma} \mathbf{L} \mathbf{I}_0 \tag{2.129}$$

Thus

$$W_{fld}\left(\lambda_0, \theta_0\right) = \frac{1}{2}\mathbf{I}_0^t \mathbf{L}^t \mathbf{I}_0 \tag{2.130}$$

2.11 Doubly Excited Systems

Multiple-Excited Magnetic Field Systems [10]

- Many electromechanical devices have multiple electrical terminals.
- Measurement systems: torque proportional to two electric signals; power as the product of voltage and current.
- Energy conversion devices: multiple-excited magnetic field system.
- A simple system with two electrical terminals and one mechanical terminal: Figure 2.31.

Three independent variables:$\{\theta, \lambda_1, \lambda_2\}$, $\{\theta, i_1, i_2\}$, $\{\theta, \lambda_1, i_2\}or\{\theta, i_1, \lambda_2\}$

$$dW_f\left(\lambda_1, \lambda_2, \theta\right) = i_1 d\lambda_1 + i_2 d\lambda_2 - T_f d\theta \tag{2.131}$$

$$i_1 = \left.\frac{\partial W_{fld}\left(\lambda_1, \lambda_2, \theta\right)}{\partial \lambda_1}\right|_{,\lambda_2,\theta} \tag{2.132}$$

$$i_2 = \left.\frac{\partial W_{fld}\left(\lambda_1, \lambda_2, \theta\right)}{\partial \lambda_2}\right|_{\lambda_1,\theta} \tag{2.133}$$

$$T_{fld} = -\left.\frac{\partial W_{fld}\left(\lambda_1, \lambda_2, \theta\right)}{\partial \theta}\right|_{\lambda_1,\lambda_2} \tag{2.134}$$

To find W_{fld}, use the path of integration in Figure 2.32.

Note that in each of these equations, *the partial derivative with respect to each independent variable must be taken, holding the other two independent variables constant.*

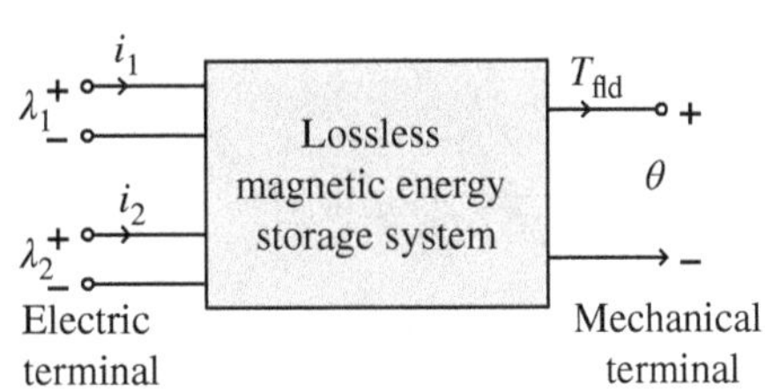

Figure 2.31 Multiple-excited magnetic energy storage system.

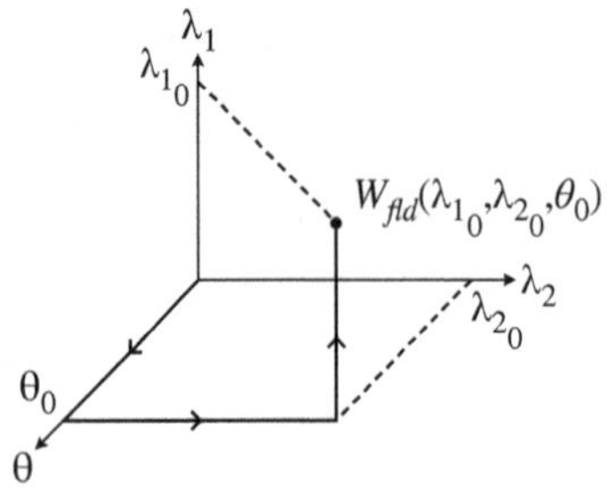

Figure 2.32 Integration path to obtain $W_{fld}\left(\lambda_{1_0}, \lambda_{2_0}, \theta_0\right)$.

The energy W_{fld} can be found by integrating Eq. (2.126). As in the singly-excited case, although the energy, at any given point, is independent of the integration path, this integration is most conveniently done by holding λ_1 and λ_2 fixed at zero and integrating first over θ; under these conditions, T_{fld} is zero, and thus this integral is zero. One can then integrate over λ_2 (while holding λ_1 zero) and finally over λ_1. Thus

$$W_{fld}\left(\lambda_{1_0}, \lambda_{2_0}, \theta_0\right) = \int_0^{\lambda_{20}} i_2\left(\lambda_1 = 0, \lambda_2, \theta = \theta_0\right) d\lambda_2 + \int_0^{\lambda_{10}} i_1\left(\lambda_1, \lambda_2 = \lambda_{2_0}, \theta = \theta_0\right) d\lambda_1 \tag{2.135}$$

In a magnetically-linear system

$$\lambda_1 = L_{11} i_1 + L_{12} i_2 \tag{2.136}$$

$$\lambda_2 = L_{21} i_1 + L_{22} i_2 \tag{2.137}$$

$$L_{12} = L_{21} \tag{2.138}$$

Note that $L_{ij} = L_{ji}(\theta)$

$$i_1 = \frac{L_{22}\lambda_1 - L_{12}\lambda_2}{D} \tag{2.139}$$

$$i_2 = \frac{-L_{21}\lambda_1 + L_{11}\lambda_2}{D} \tag{2.140}$$

$$D = L_{11}L_{22} - L_{12}L_{21} \tag{2.141}$$

The energy for this linear system is

$$\begin{aligned} W_{fld}\left(\lambda_{1_0}, \lambda_{2_0}, \theta_0\right) &= \int_0^{\lambda_{20}} \frac{L_{11}\left(\theta_0\right)\lambda_2}{D\left(\theta_0\right)} d\lambda_2 + \int_0^{\lambda_{10}} \frac{\left(L_{22}\left(\theta_0\right)\lambda_1 - L_{12}\left(\theta_0\right)\lambda_{2_0}\right)}{D\left(\theta_0\right)} d\lambda_1 \\ &= \frac{1}{2D\left(\theta_0\right)} L_{11}\left(\theta_0\right)\lambda_{2_0}^2 + \frac{1}{2D\left(\theta_0\right)} L_{22}\left(\theta_0\right)\lambda_{1_0}^2 - \frac{L_{12}\left(\theta_0\right)}{2D\left(\theta_0\right)}\lambda_{1_0}\lambda_{2_0} \end{aligned} \tag{2.142}$$

where the dependence of the inductances and the determinant $D(\theta)$ on the angular displacement θ has been explicitly indicated.

The co-energy function permits determination of force and torque directly in terms of the current for a single-winding system. A similar co-energy function can be defined in the case of systems with two windings as

$$W'_{fld}\left(i_1, i_2, \theta\right) = \lambda_1 i_1 + \lambda_2 i_2 - W_{fld} \tag{2.143}$$

It is a state function of the two terminal currents and the mechanical displacement. Its differential, following substitution of Eq. (2.126), is given by

$$W'_{fld}\left(i_1, i_2, \theta\right) = \lambda_1 i_1 + \lambda_2 i_2 + T_{fld} d\theta \tag{2.144}$$

$$\lambda_1 = \left.\frac{\partial W'_{fld}\left(i_1, i_2, \theta\right)}{\partial i_1}\right|_{i_2,\theta} \tag{2.145}$$

$$\lambda_2 = \left.\frac{\partial W'_{fld}\left(i_1, i_2, \theta\right)}{\partial i_2}\right|_{i_1,\theta} \tag{2.146}$$

Most significantly, the torque can now be determined directly in terms of the currents as

$$T_{fld} = \left.\frac{\partial W'_{fld}\left(i_1, i_2, \theta\right)}{\partial \theta}\right|_{i_1,i_2} \tag{2.147}$$

Analogous to Eq. (2.130), the co-energy can be found as

$$W'_{fld}\left(i_{1_0}, i_{2_0}, \theta_0\right) = \int_0^{i_{2_0}} \lambda_2\left(i_1 = 0, i_2, \theta = \theta_0\right) di_2 + \int_0^{i_{1_0}} \lambda_1\left(\lambda_1, i_2 = i_{2_0}, \theta = \theta_0\right) di_1 \tag{2.148}$$

For the linear system of Eqs. (2.131)–(2.133)

$$W'_{fld}\left(i_1, i_2, \theta\right) = \frac{1}{2} L_{11}(\theta) i_1^2 + \frac{1}{2} L_{22}(\theta) i_2^2 + L_{12}(\theta) i_1 i_2 \tag{2.149}$$

For such a linear system, the torque can be found either from the energy of Eq. (2.137) using Eq. (2.129) or from the co-energy of Eq. (2.144) using Eq. (2.142). It is at this point that the utility of the co-energy function becomes apparent. The energy expression of Eq. (2.144) is a complex function of displacement, and its derivative is even more so. Alternatively, the co-energy function is a relatively simple function of displacement and, from its derivative, a straight forward expression For torque can be determined as a function of the winding currents i_1 and i_2 as

$$T_{fld} = \left.\frac{\partial W'_{fld}\left(i_1, i_2, \theta\right)}{\partial \theta}\right|_{i_1,i_2} = \frac{i_1^2}{2}\left(\frac{dL_{11}(\theta)}{d\theta}\right) + \frac{i_2^2}{2}\left(\frac{dL_{22}(\theta)}{d\theta}\right) + i_1 i_2\left(\frac{dL_{12}(\theta)}{d\theta}\right) \tag{2.150}$$

Systems with more than two electrical terminals are handled in analogous fashion. As with the two-terminal-pair system above, the use of a co-energy function of the terminal currents greatly simplifies the determination of torque or force.

The derivation presented above for angular displacement can be repeated in an analogous fashion for the systems with linear displacement. If this is done, the expressions For energy and co-energy will be found to be

$$W_{fld}\left(\lambda_{1_0},\lambda_{2_0},x_0\right)$$
$$=\int_0^{\lambda_{20}} i_2\left(\lambda_1=0,\lambda_2,x=x_0\right)d\lambda_2+\int_0^{\lambda_{10}} i_1\left(\lambda_1,\lambda_2=\lambda_{2_0},x=x_0\right)d\lambda_1 \tag{2.151}$$

$$W'_{fld}\left(i_{1_0},i_{2_0},x_0\right)=\int_0^{i_{20}}\lambda_2\left(i_1=0,i_2,x=x_0\right)di_2+\int_0^{i_{10}}\lambda_1\left(\lambda_1,i_2=i_{2_0},x=x_0\right)di_1 \tag{2.152}$$

Similarly, the force can be found from

$$f_{fld}=\left.\frac{\partial W_{fld}\left(\lambda_1,\lambda_2,x\right)}{\partial x}\right|_{\lambda_1,\lambda_2} \tag{2.153}$$

or

$$f_{fld}=\left.\frac{\partial W'_{fld}\left(i_1,i_2,x\right)}{\partial x}\right|_{i_1,i_2} \tag{2.154}$$

For a magnetically linear system, the co-energy expression of Eq. (2.144) becomes

$$W'_{fld}\left(i_1,i_2,x\right)=\frac{1}{2}L_{11}(x)i_1^2+\frac{1}{2}L_{22}(x)i_2^2+L_{12}(x)i_1i_2 \tag{2.155}$$

and the force is thus given by

$$f_{fld}=\left.\frac{\partial W'_{fld}\left(i_1,i_2,x\right)}{\partial x}\right|_{i_1,i_2}=\frac{i_1^2}{2}\left(\frac{dL_{11}(x)}{dx}\right)+\frac{i_2^2}{2}\left(\frac{dL_{22}(x)}{dx}\right)+i_1i_2\left(\frac{dL_{12}(x)}{dx}\right) \tag{2.156}$$

2.11.1 Torque Developed

In the derivations for torque in simple machines, the following equations are used to find torque:

$$\tau=\vec{r}\times\vec{F} \tag{2.157}$$

or

$$\tau=\vec{r}\times i\left(\vec{l}\times\vec{B}\right) \tag{2.158}$$

The approach of finding torque from the Lorentz force is not practical in many cases and, in some cases (e.g. in the case when calculating the force between two permanent magnets) requires some fairly advanced approximations or assumptions. In the field of electric machines, there is usually a space between the moving surfaces. The principle of virtual work can be used to calculate the force in an air gap as the rate of change of stored energy in the air gap under constant flux conditions. Throughout the course, we use models that predict torque as functions of currents and fluxes, but it is important to recognize that we can always think of torque in terms of air gap magnetic fields density [11]. In rotating machines

$$\tau=\frac{d}{d\theta_R}W|_{constantflux} \tag{2.159}$$

where θ_R is the rotor position with respect to the stator and W is the stored energy. The energy density w of a magnetic field in air is defined as

$$w = \frac{1}{2}BH \tag{2.160}$$

or

$$w = \frac{1}{2\mu_o}B^2 \tag{2.161}$$

And the total stored energy in the air gap is given by integrating over the air-gap volume. Using an approximation that the flux density vector has only radial components (not circumferential) the energy in the air gap at an instant in time can be approximated as

$$w = rl\int_0^{2\pi} w\left(\theta_m\right) g\left(\theta_m\right) d\theta_m \tag{2.162}$$

or

$$w = \frac{rl}{2\mu_o}\int_0^{2\pi} B^2\left(\theta_m\right) g\left(\theta_m\right) d\theta_m \tag{2.163}$$

Considering the rotating mmf produced by three-phase AC coils on a stator

$$F_s = \frac{3IN}{\pi}\cos\left[\omega_e t - \frac{P}{2}\theta_m\right] \tag{2.164}$$

the resulting flux density can be written as

$$B_s\left(\theta_m, t\right) = \frac{\mu_o}{g}F_s\left(\theta_m, t\right) \tag{2.165}$$

or

$$B_s\left(\theta_m, t\right) = \frac{\mu_o}{g\left(\theta_m, t\right)}\hat{F}_s\cos\left[\omega_e t - \frac{P}{2}\theta_m\right] \tag{2.166}$$

Note that in the above equation, the air-gap length, g, is given as a function of position and time; it is not necessarily uniform. To simplify things, we can set the electrical frequency to zero, and consider forces in a static field

$$B_s\left(\theta_m\right) = \frac{\mu_o}{g\left(\theta_m\right)}\hat{F}_s\cos\left[\frac{P}{2}\theta_m\right] \tag{2.167}$$

In the sections below, the different components of torque are investigated using a simple two-pole case.

2.11.1.1 Excitation Torque

Excitation torque is the torque that occurs when two magnetic fields interact. Most of the machines covered in this material exploit excitation torque. This is similar to the force between two magnets. The animation below shows how the air-gap flux density changes with rotor position, and plots the normalized torque at each rotor position. A derivation of the torque function is shown below.

In the figures of Figure 2.33, the angle of a rotor magnetic field relative to a stationary stator magnetic field is varied in steps of 45°. The air gap between rotor and stator is constant, so the stator magnetic field can be written as

$$B_s\left(\theta_m\right) = \hat{B}_s\cos\left(\theta_m\right) \tag{2.168}$$

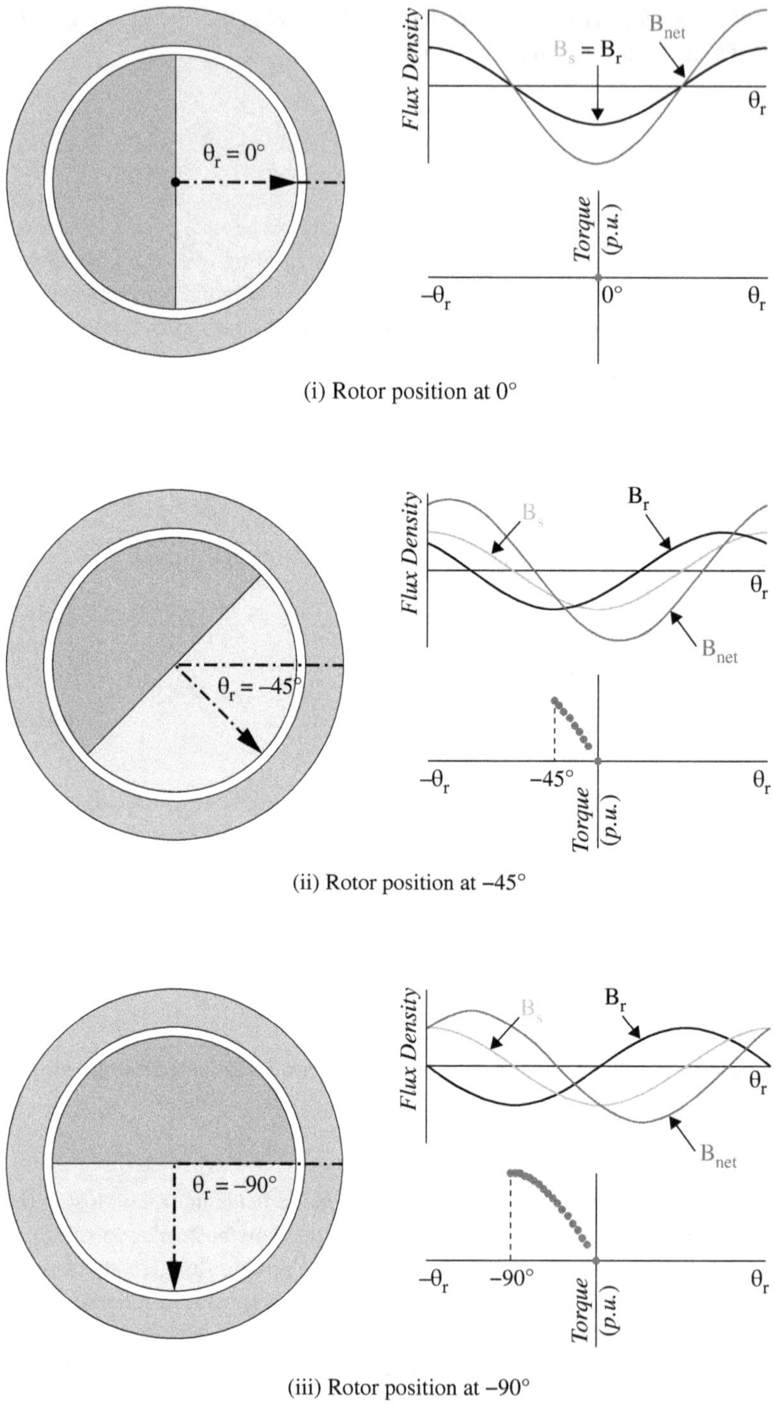

Figure 2.33 Graphical representation of B_s, B_r, B_{net} and normalized torque with respect to the rotor position (θ_r). (i) Rotor position at 0°, (ii) Rotor position at −45°, (iii) Rotor position at −90°, (iv) Rotor position at 45°, (v) Rotor position at 90°, (vi) Rotor position at 135°, (vii) Rotor position at 180°, (viii) Rotor position at 225°, (ix) Rotor position at 270°, (x) Rotor position at 315°, (xi) Rotor position at 360°.

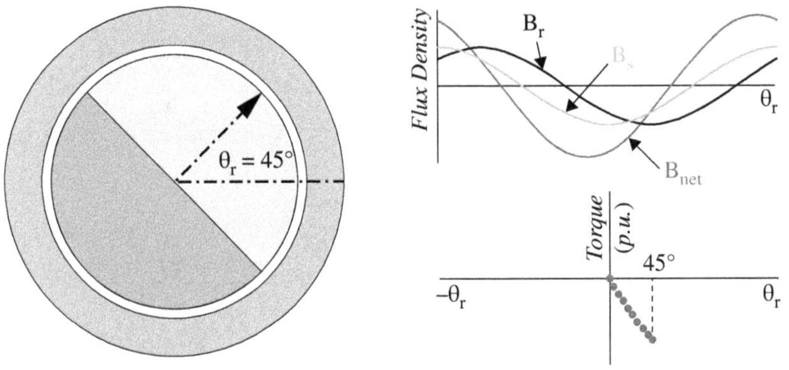

(iv) Rotor position at 45°

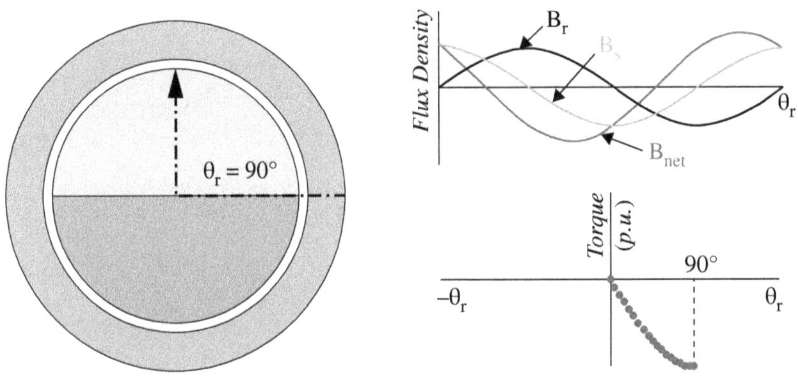

(v) Rotor position at 90°

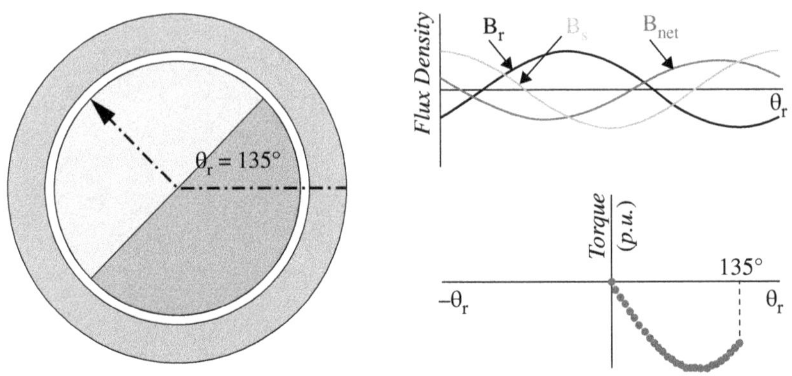

(vi) Rotor position at 135°

Figure 2.33 (*Continued*)

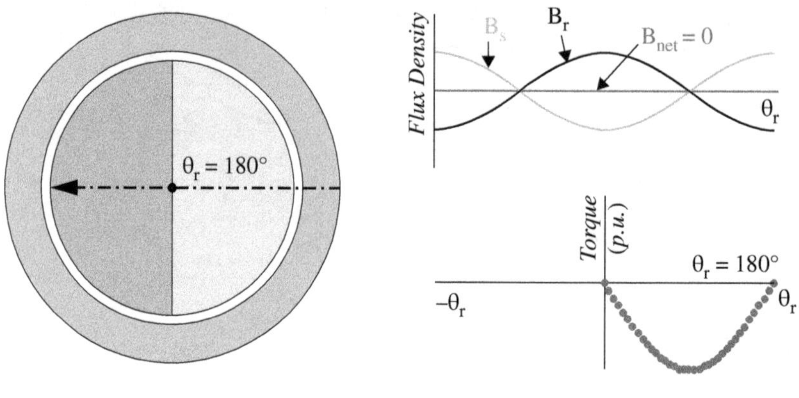

(vii) Rotor position at 180°

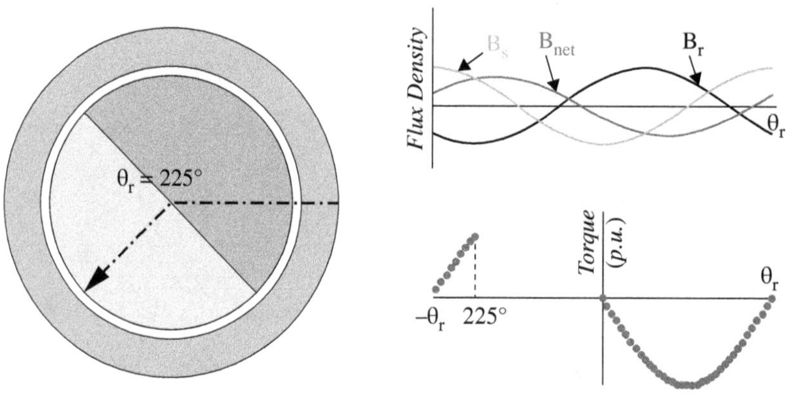

(viii) Rotor position at 225°

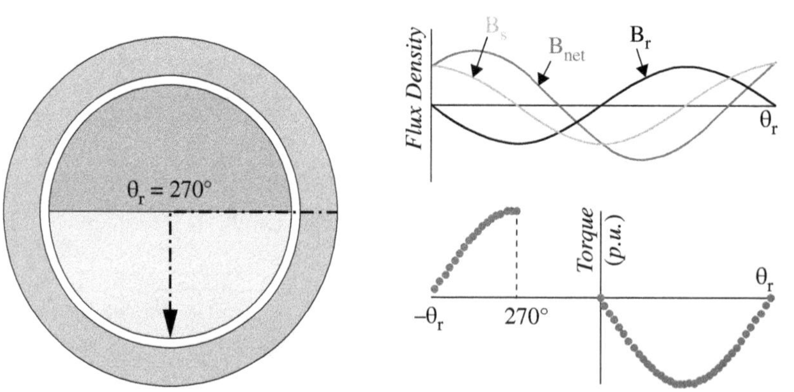

(ix) Rotor position at 270°

Figure 2.33 (*Continued*)

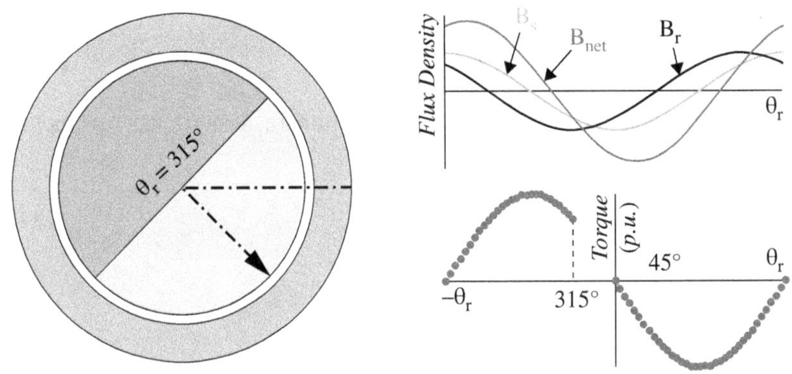

(x) Rotor position at 315°

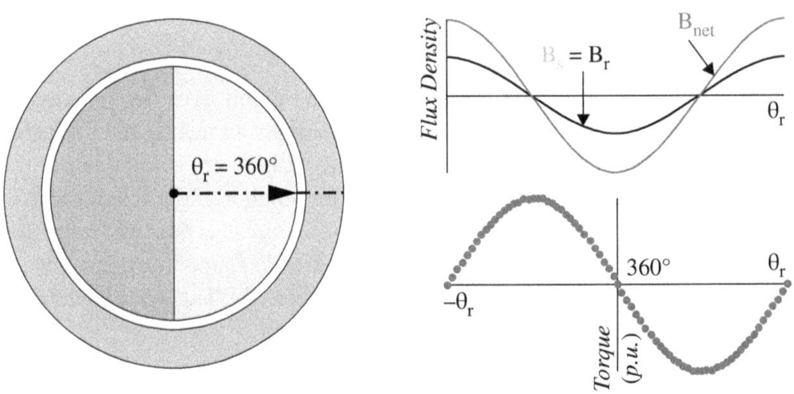

(xi) Rotor position at 360°

Figure 2.33 (*Continued*)

The rotor produces a magnetic field that is sinusoidal with respect to the rotor

$$B_r(\alpha) = \hat{B}_r \cos(\alpha) \tag{2.169}$$

And, therefore, when the rotor is at position θ_R, the rotor magnetic field seen from the stator is given by

$$B_r(\theta_m) = \hat{B}_r \cos(\theta_m - \theta_R) \tag{2.170}$$

Using the virtual work approach, the torque is given by

$$\tau = \frac{dW}{d\theta_R} \tag{2.171}$$

or

$$\tau = \frac{d}{d\theta_R} \frac{grl}{2\mu_o} \int_0^{2\pi} B^2(\theta_m)\, d\theta_m \tag{2.172}$$

or

$$\tau = \frac{d}{d\theta_R}\frac{grl}{2\mu_o}\int_0^{2\pi} \left(\hat{B}_s \cos\left(\theta_m\right) + \hat{B}_r \cos\left(\theta_m - \theta_R\right)\right) d\theta_m \tag{2.173}$$

Since the torque is the derivative with respect to θ_R, the integral terms that do not include θ_R can be neglected.

$$\tau = \frac{d}{d\theta_R}\frac{grl}{2\mu_o}\int_0^{2\pi} 2\hat{B}_s\hat{B}_r \cos\left(\theta_m\right)\cos\left(2\theta_m - \theta_R\right) d\theta_m \tag{2.174}$$

or

$$\tau = \frac{d}{d\theta_R}\frac{grl}{2\mu_o}\int_0^{2\pi} \hat{B}_s\hat{B}_r \left(\cos\left(\theta_R\right) + \cos\left(2\theta_m - \theta_R\right)\right) d\theta_m \tag{2.175}$$

or

$$\tau = \frac{d}{d\theta_R}\frac{grl\pi}{\mu_o}\hat{B}_s\hat{B}_r \cos\left(\theta_R\right) \tag{2.176}$$

or

$$\tau = -\frac{grl\pi}{\mu_o}\hat{B}_s\hat{B}_r \sin\left(\theta_R\right) \tag{2.177}$$

The above result gives the fact that the torque is zero when the flux densities are aligned, is maximum when the fields are perpendicular, and acts to apply a force on the rotor that opposes the movement. The sinusoidal variation of torque with angle will be seen in the section on synchronous machines. Excitation torque is the product of the magnitudes of the two phasors and the sine of the angle between them. If the angle is positive (rotor-flux density leads to stator-flux density) then the torque is negative. If stator-flux density leads to rotor-flux density then the angle is negative and torque is positive. This relationship can also be written as the cross-product of the two phasors:

$$\tau \propto \vec{B}_r \times \vec{B}_s \tag{2.178}$$

and

$$\vec{B}_{net} = \vec{B}_r + \vec{B}_s \tag{2.179}$$

hence

$$\tau \propto \vec{B}_r \times \vec{B}_{net} \tag{2.180}$$

2.11.1.2 Reluctance Torque

Reluctance torque, also known as alignment torque, is due to the forces that occur when a magnetic material interacts with a magnetic field. This is similar to the force between a magnet and a steel bar. The figures of Figure 2.34 show how the air gap flux density changes with rotor position and plots the normalized torque at each rotor position. A derivation of the torque function is shown below.

In the figures of Figure 2.34, the angle of a rotor made from magnetic material, relative to a stationary stator magnetic field, is varied in step of 45°. The air gap between rotor and stator is now variable, so the stator magnetic must now be written as

$$B_s\left(\theta_m\right) = \frac{\mu_o}{g\left(\theta_m\right)}\hat{F}_s \cos\left(\theta_m\right) \tag{2.181}$$

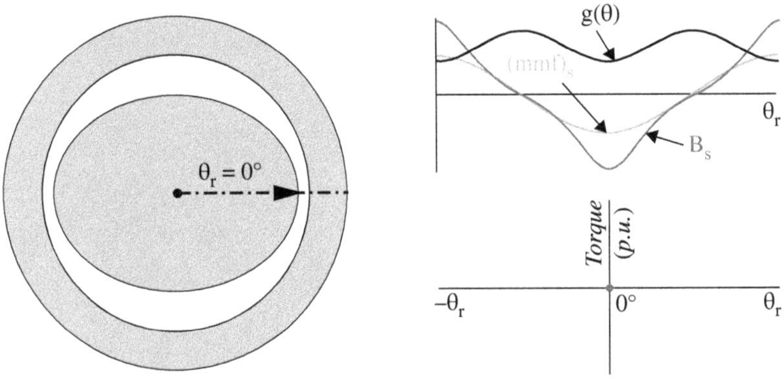

(i) Rotor position at 0°

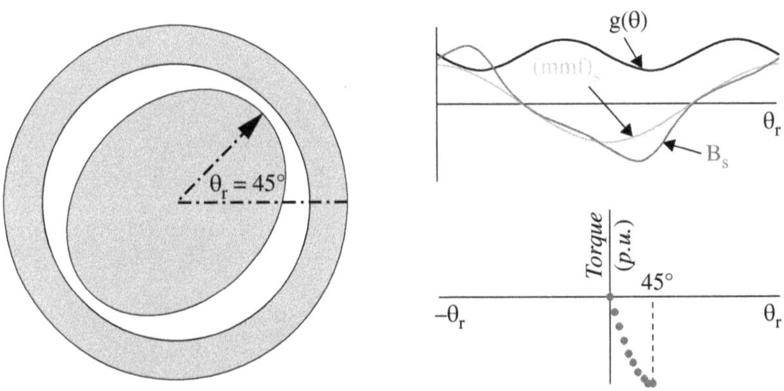

(ii) Rotor position at 45°

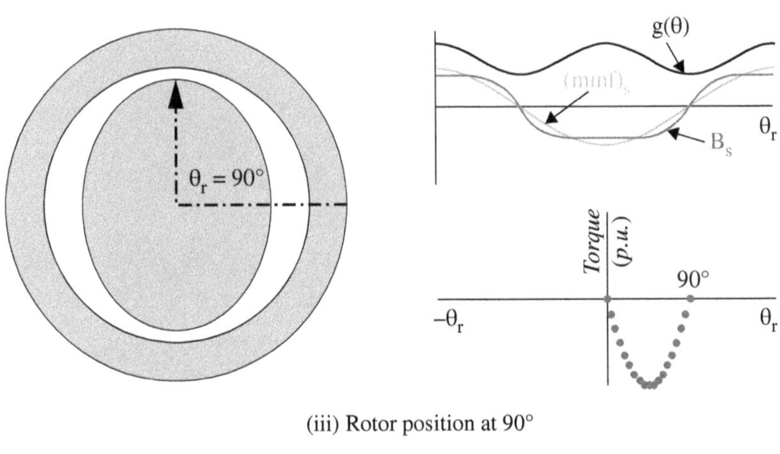

(iii) Rotor position at 90°

Figure 2.34 Graphical representation of B_s, B_r, B_{net} and normalized torque with respect to the rotor position of a reluctance motor (θ_r). (i) Rotor position at 0°, (ii) Rotor position at 45°, (iii) Rotor position at 90°, (iv) Rotor position at 135°, (v) Rotor position at 180°, (vi) Rotor position at 225°, (vii) Rotor position at 270°, (viii) Rotor position at 315°, (ix) Rotor position at 360°.

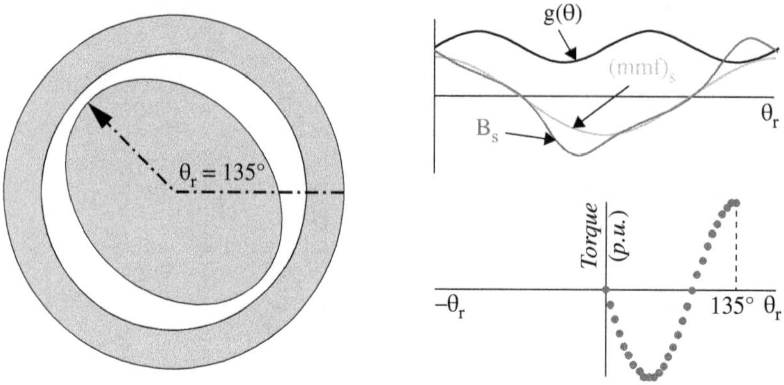

(iv) Rotor position at 135°

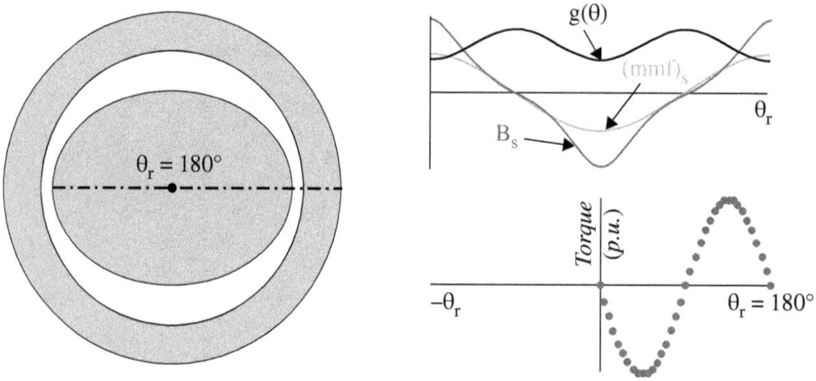

(v) Rotor position at 180°

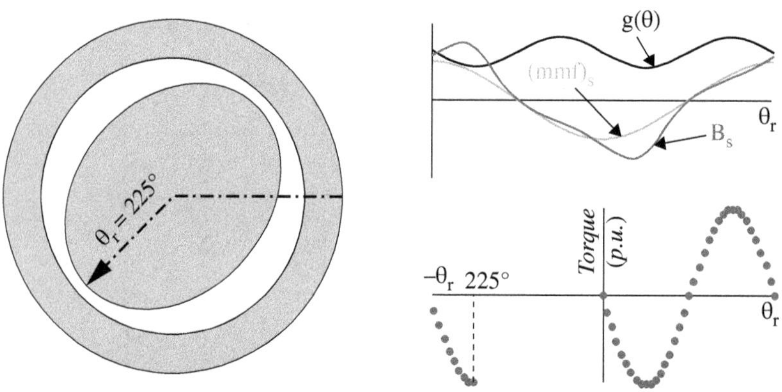

(vi) Rotor position at 225°

Figure 2.34 (*Continued*)

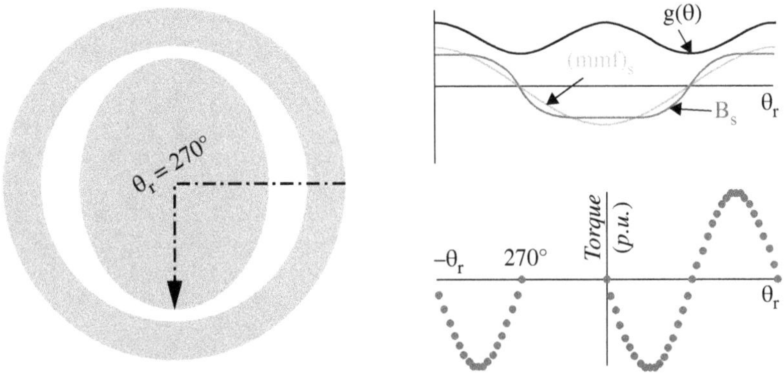

(vii) Rotor position at 270°

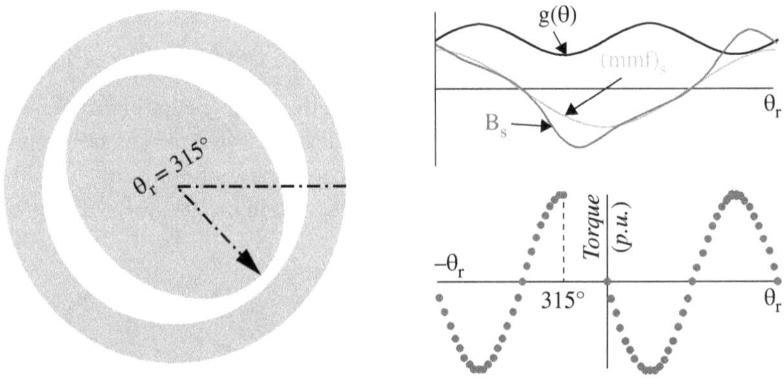

(viii) Rotor position at 315°

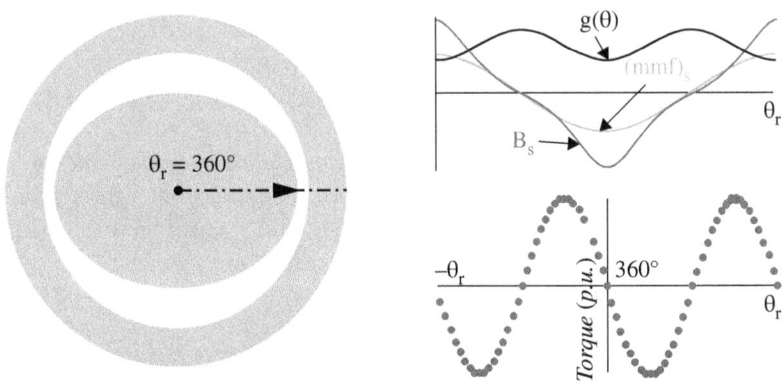

(xi) Rotor position at 360°

Figure 2.34 (*Continued*)

In the case of the elliptical rotor, a good approximation for the air gap length is given by

$$g(\alpha) = g_{ave} - \hat{g}\cos(2\alpha) \tag{2.182}$$

or

$$g(\theta_m) = g_{ave} - \hat{g}\cos(2\theta_m - 2\theta_R) \tag{2.183}$$

Using the virtual work approach, the torque is given by

$$\tau = \frac{dW}{d\theta_R}$$

$$\tau = \frac{d}{d\theta_R}\frac{rl}{2}\int_0^{2\pi} \frac{\mu_o}{g_{ave} - \hat{g}\cos(2\theta_m - 2\theta_R)}\hat{F}_s^2\cos^2\theta_m d\theta_m \tag{2.184}$$

Using the identity that $\frac{1}{(1-x)} = 1 + x^2 + x^3 + x^4 \ldots x < 1$, it is possible with some time, to show that the air gap energy density is proportional to $\cos(2\theta_m - 2\theta_R)\cos^2\theta_m$. As a result, using a similar approach to the one used to find excitation torque, it is possible to find the result that reluctance torque is proportional to $\sin(2\theta_R)$.

$$\tau_{rel} = -\hat{\tau} sin(2\theta_R) \tag{2.185}$$

Reluctance torque has traditionally been exploited, together with excitation torque, in large hydro-generators. Reluctance torque is exclusively used in synchronous reluctance and switched reluctance machines, and is a significant component of the torque in interior permanent magnet machines, which are commonly found in electric drive trains for vehicles such as the Toyota Prius.

2.12 Concept of Rotating Magnetic Field

2.12.1 Rotating Magnetic Field due to Three-Phase Currents

When a three-phase winding is energized from a three-phase supply, a rotating magnetic field is produced [12]. This field is such that its poles do not remain in a fixed position on the stator but go on shifting their positions around the stator. For this reason, it is called a rotating field. It can be shown that magnitude of this rotating field is constant and is equal to $1.5\phi_m$ where ϕ_m is the maximum flux of any phase.

To see how a rotating field is produced, consider a two-pole, three-phase winding as shown in Figure 2.36 (i) The three phases X, Y, and Z are energized from a three-phase source and currents in these phases are indicated as I_x, I_y, and I_z (see Figure 2.36 (ii)). Referring to Figure 2.36 (ii), the fluxes produced by these currents are given by:

$$\begin{aligned}\phi_x &= \phi_m \sin\omega t\\ \phi_y &= \phi_m \sin(\omega t - 120^o)\\ \phi_z &= \phi_m \sin(\omega t - 240^o)\end{aligned} \tag{2.186}$$

Here ϕ_m is the maximum flux due to any phase. Figure 2.35 shows the phasor diagram of the three fluxes. We shall now prove that this three-phase supply produces a rotating field of constant magnitude equal to $1.5\phi_m$.

Figure 2.35 The phasor diagram of the three fluxes.

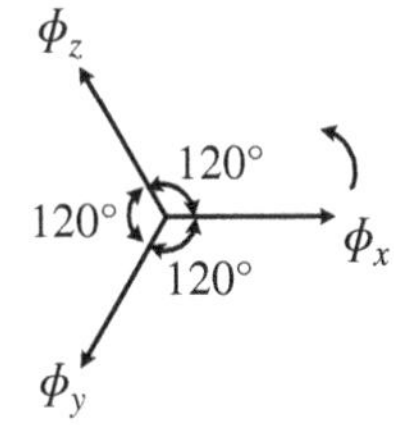

(i) At instant 1 (see Figure 2.36 (ii) and Figure 2.36 (iii)), the current in phase X is zero and currents in phases Y and Z are equal and opposite. The currents are flowing outward in the top conductors and inward in the bottom conductors. This establishes a resultant flux towards the right. The magnitude of the resultant flux is constant and is equal to $1.5\phi_m$ as proved under:
At instant 1, $\omega t = 0°$. Therefore, the three fluxes are given by;

$$\phi_x = \phi_m \sin \omega t = \phi_m \sin 0 = 0$$

$$\phi_y = \phi_m \sin\left(\omega t - 120^o\right) = \phi_m \sin\left(-120^o\right) = -\frac{\sqrt{3}}{2}\phi_m$$

$$\phi_z = \phi_m \sin\left(\omega t - 240^o\right) = \phi_m \sin\left(-240^o\right) = \frac{\sqrt{3}}{2}\phi_m$$

The phasor sum of $-\phi_y$ and ϕ_z is the resultant flux ϕ_r (see Figure 2.37). It is clear that:
Resultant flux

$$\phi_r = 2 \times \frac{\sqrt{3}}{2}\phi_m \cos\frac{60^o}{2} = 2 \times \frac{\sqrt{3}}{2}\phi_m \times \frac{\sqrt{3}}{2} = 1.5\phi_m$$

(ii) At instant 2, the current is maximum (negative) in ϕ_y phase Y and 0.5 maximum (positive) in phases X and Z. The magnitude of resultant flux is $1.5\phi_m$ as proved under:
At instant 2, $\omega t = 30°$. Therefore, the three fluxes are given by;

$$\phi_x = \phi_m \sin \omega t = \phi_m \sin 30^o = \frac{\phi_m}{2}$$

$$\phi_y = \phi_m \sin\left(\omega t - 120^o\right) = \phi_m \sin\left(30^o - 120^o\right) = \phi_m \sin\left(-90^o\right) = -\phi_m$$

$$\phi_z = \phi_m \sin\left(\omega t - 240^o\right) = \phi_m \sin\left(30^o - 240^o\right) = \phi_m \sin\left(-210^o\right) = \frac{\phi_m}{2}$$

The phasor sum of ϕ_x, $-\phi_y$ and ϕ_z is the resultant flux ϕ_r
Phasor sum of ϕ_x and ϕ_z, $\phi'_r = 2 \times \frac{\phi_m}{2} \cos\frac{120^o}{2} = \frac{\phi_m}{2}$
Phasor sum of ϕ'_r and $-\phi_y$, $\phi_r = \phi'_r - \phi_y = \frac{\phi_m}{2} + \phi_m = 1.5\phi_m$
Note that resultant flux is displaced 30° clockwise from position 1 (Figure 2.38).

(iii) At instant 3, the current in phase Z is zero and the currents in phases X and Y are equal and opposite (currents in phases X and Y are 0.866× max. value). The magnitude of resultant flux is $1.5\phi_m$ as proved under:

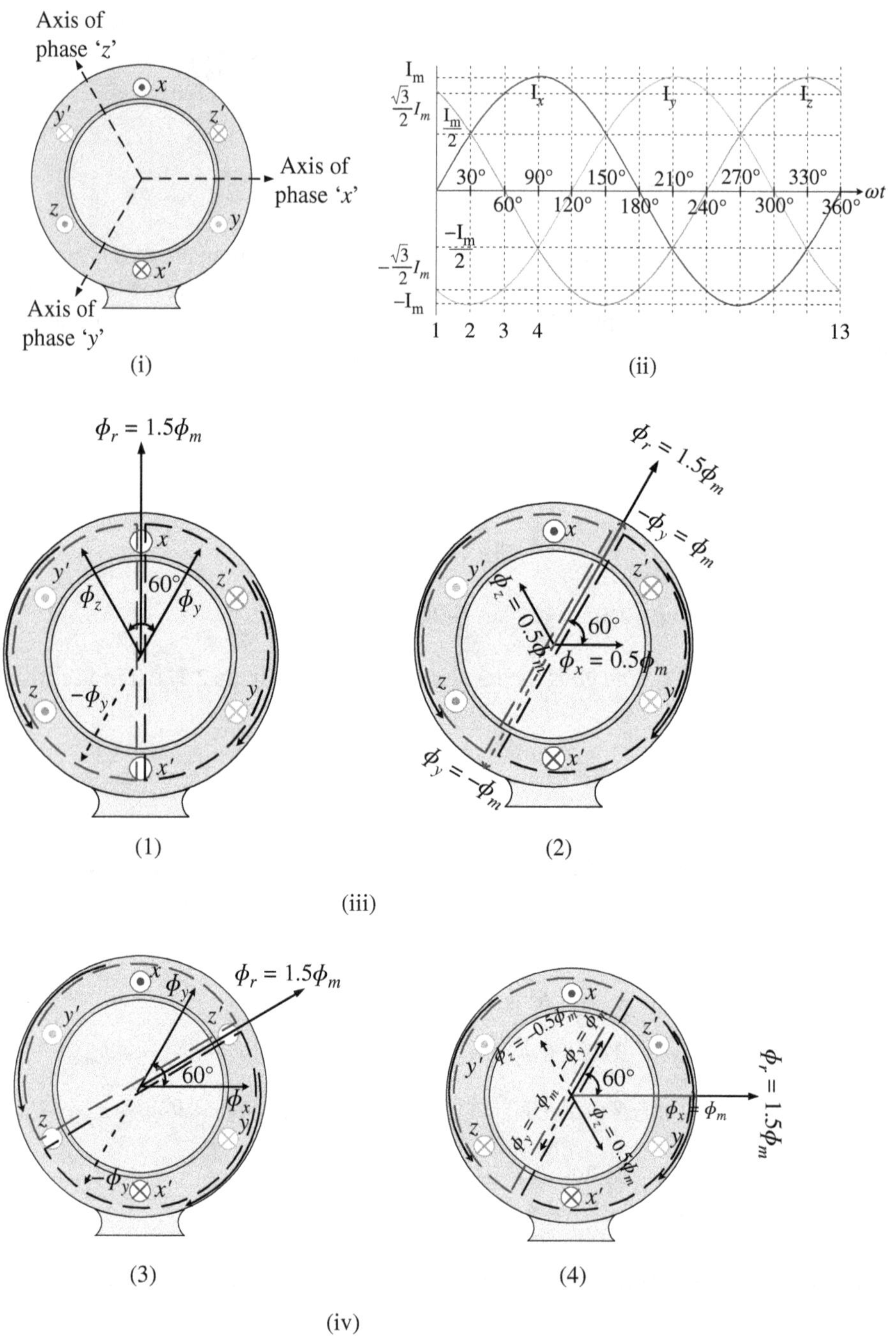

Figure 2.36 Production of rotating magnetic field

Figure 2.37 Phasors of the three fluxes at instant 1.

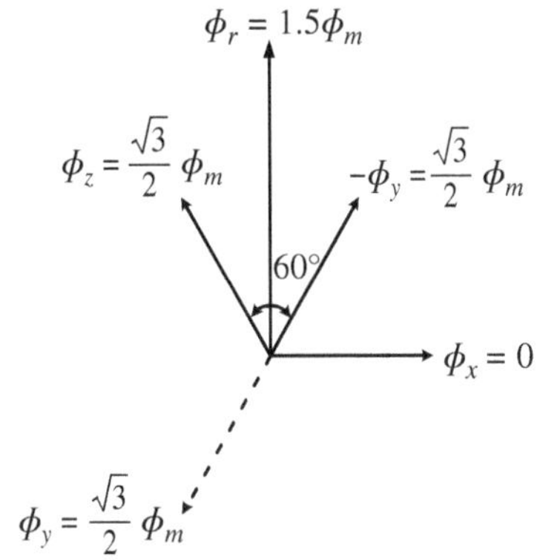

Figure 2.38 Phasors of the three fluxes at instant 2.

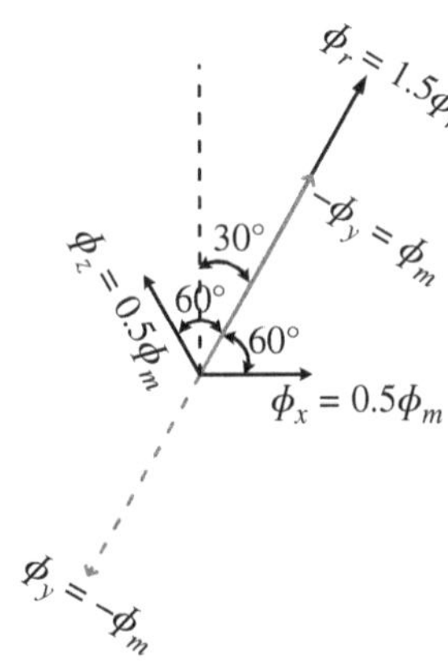

At instant 3, ωt = 60°. Therefore, the three fluxes are given by;

$$\phi_x = \phi_m \sin \omega t = \phi_m \sin 60^o = \frac{\sqrt{3}}{2}\phi_m$$

$$\phi_y = \phi_m \sin\left(\omega t - 120^o\right) = \phi_m \sin\left(60^o - 120^o\right) = \phi_m \sin\left(-60^o\right) = -\frac{\sqrt{3}}{2}\phi_m$$

$$\phi_z = \phi_m \sin\left(\omega t - 240^o\right) = \phi_m \sin\left(60^o - 240^o\right) = \phi_m \sin\left(-180^o\right) = 0$$

The resultant flux ϕ_r is the phasor sum of ϕ_x and $-\phi_y$ ($\because \phi_z = 0$).

$$\phi_r = 2\times\frac{\sqrt{3}}{2}\phi_m \cos\frac{60^o}{2} = 2\times\frac{\sqrt{3}}{2}\phi_m\times\frac{\sqrt{3}}{2} = 1.5\phi_m$$

Note that resultant flux is displaced 60° clockwise from position 1 (Figure 2.39).

(iv) At instant 4, the current in phase X is maximum (positive) and the currents in phases Y and Z are equal and negative (currents in phases Y and Z are 0.5× max. value). This establishes a resultant flux downward as shown under:
At instant 4, ωt = 90°. Therefore, the three fluxes are given by

$$\phi_x = \phi_m \sin \omega t = \phi_m \sin 90^o = \phi_m$$

$$\phi_y = \phi_m \sin\left(\omega t - 120^o\right) = \phi_m \sin\left(90^o - 120^o\right) = \phi_m \sin\left(-30^o\right) = -\frac{\phi_m}{2}$$

$$\phi_z = \phi_m \sin\left(\omega t - 240^o\right) = \phi_m \sin\left(90^o - 240^o\right) = \phi_m \sin\left(-150^o\right) = -\frac{\phi_m}{2}$$

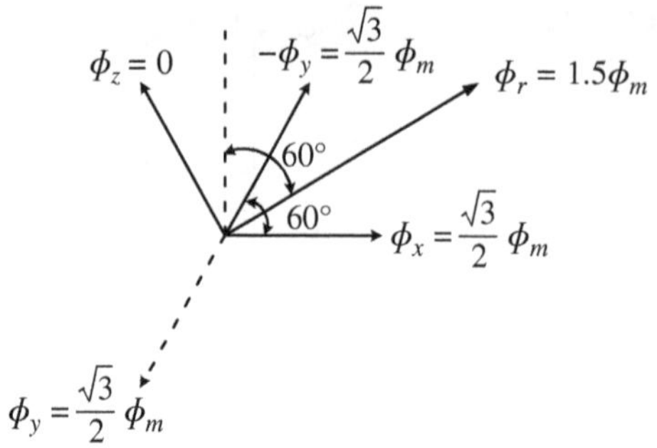

Figure 2.39 Phasors of the three fluxes at instant 3.

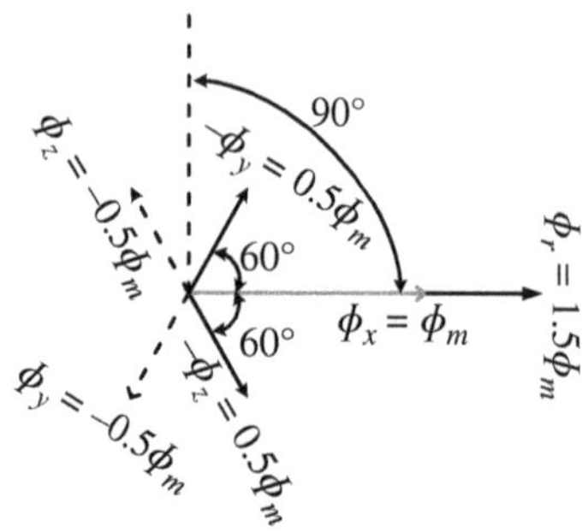

Figure 2.40 Phasors of the three fluxes at instant 4.

The phasor sum of ϕ_x, $-\phi_y$and $-\phi_z$ is the resultant flux ϕ_r
Phasor sum of $-\phi_z$and $-\phi_y$, $\phi_r' = 2 \times \frac{\phi_m}{2} \cos \frac{120^\circ}{2} = \frac{\phi_m}{2}$
Phasor sum of ϕ_r' and ϕ_x, $\phi_r = \phi_r' + \phi_x = \frac{\phi_m}{2} + \phi_m = 1.5\phi_m$
Note that the resultant flux is downward i.e., it is displaced 90° clockwise from position 1 (Figure 2.40).
It follows from the above discussion that a three-phase supply produces a rotating field of constant value (= $1.5\phi_m$, where ϕ_m is the maximum flux due to anyphase).

2.12.1.1 Speed of Rotating Magnetic Field

The speed at which the rotating magnetic field revolves is called the synchronous speed (N_s). Referring to Figure 2.36 (ii), the time instant 4 represents the completion of one-quarter cycle of alternating current I_x from the time instant 1. During this one-quarter cycle, the field has rotated through 90°. At a time instant, represented by 13 or one complete cycle of current I_x from the origin, the field has completed one revolution. Therefore, for a two-pole stator winding, the field makes one revolution in one cycle of current. In a four-pole stator winding, it can be shown that the rotating field makes one revolution in two cycles of current. In general, for P poles, the rotating field makes one revolution in P/2 cycles of current.

$$\therefore \text{Cycles of current} = \frac{P}{2} \times \text{revolutions of field}$$

or

$$\text{Cycles of current per second} = \frac{P}{2} \times \text{revolutions of field per second}$$

Since revolutions per second is equal to the revolutions per minute (N_s) divided by 60 and the number of cycles per second is the frequency f,

$$\therefore f = \frac{P}{2} \times \frac{N_s}{60} = \frac{PN_s}{120}$$

or

$$N_s = \frac{120f}{P} \tag{2.187}$$

The speed of the rotating magnetic field is the same as the speed of the alternator that is supplying power to the motor if the two have the same number of poles. Hence, the magnetic flux is said to rotate at synchronous speed.

2.12.1.2 Direction of Rotating Magnetic Field

The phase sequence of the three-phase voltage applied to the stator winding in Figure 2.36 (ii) is X-Y-Z. If this sequence is changed to X-Z-Y, it is observed that the direction of rotation of the field is reversed i.e., the field rotates counter clockwise rather than clockwise. However, the number of poles and the speed at which the magnetic field rotates remain unchanged. Thus, it is only necessary to change the phase sequence to change the direction of rotation of the magnetic field. For a three-phase supply, this can be done by interchanging any two of the three lines. As we shall see, the rotor in a three-phase induction motor runs in the same direction as the rotating magnetic field. Therefore, the direction of rotation of a three-phase induction motor can be reversed by interchanging any two of the three motor supply lines.

2.12.2 Alternate Mathematical Analysis for Rotating Magnetic Field

When a machine has more than two poles, only one single pair of poles needs to be considered because the electric, magnetic, and mechanical conditions associated with every other pole pair are repetitions of those for the pole pair under consideration. The angle subtended by one pair of poles in a P-pole machine (or one cycle of flux distribution) is defined as 360 *electrical degrees*, or 2π *electrical radians*. So, the relationship between the mechanical angle θ_m and the angle θ in electrical units is given by

$$\theta = \frac{P}{2}\theta_m \tag{2.188}$$

because one complete revolution has $P/2$ complete wavelengths (or cycles). In view of this relationship, for a two-pole machine, electrical degrees (or radians) will be the same as mechanical degrees (or radians).

In this section, we set out to show that a rotating field of constant amplitude and sinusoidal space distribution of mmf around a periphery of the stator is produced by a three-phase winding located on the stator and excited by balanced three-phase currents when the respective phase windings are wound $2\pi/3$ electrical radians (or 120 electrical degrees) apart in space [4]. Let us consider the two-pole, three-phase winding arrangement on the stator shown in Figure 2.41.The windings of the individual phases are displaced by 120 electrical degrees from each other in space around the air-gap periphery. The reference directions are given for positive phase currents.

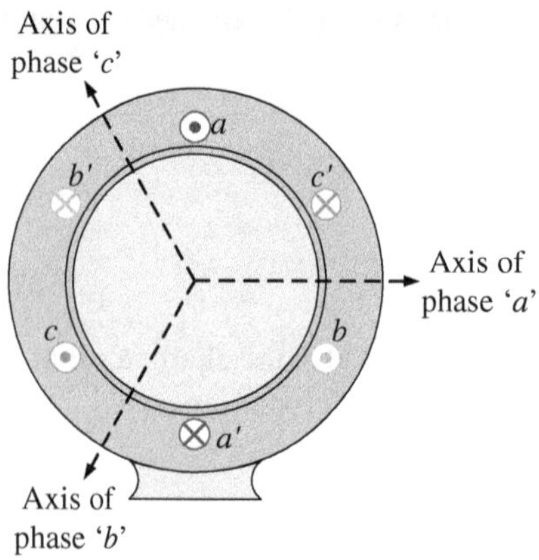

Figure 2.41 Simple two-pole, three-phase winding arrangement on a stator.

The concentrated full-pitch coils, shown here for simplicity and convenience, do in fact represent the actual distributed windings producing sinusoidal mmf waves centred on the magnetic axes of the respective phases. Thus, these three sinusoidal mmf waves are displaced by 120 electrical degrees from each other in space. Let a balanced three-phase excitation be applied with phase sequence *a*–*b*–*c*

$$i_a = I_m \cos \omega t; i_b = I_m \cos\left(\omega t - 120^o\right); i_b = I_m \cos\left(\omega t - 240^o\right) \tag{2.189}$$

where I_m is the maximum value of the current, and the time $t = 0$ is chosen arbitrarily when the *a*-phase current is a positive maximum. Each phase current is an ac wave varying in magnitude sinusoidally with time. Hence, the corresponding component mmf waves vary sinusoidally with time. The sum of these components yields the resultant mmf.

Analytically, the resultant mmf at any point at an angle θ from the axis of phase *a* is given by

$$F(\theta) = F_a \cos\theta + F_b \cos\left(\theta - 120^o\right) + F_c \cos\left(\theta - 240^o\right) \tag{2.190}$$

But the mmf amplitudes vary with time according to the current variations

$$F_a = F_m \cos \omega t; F_b = F_m \cos\left(\omega t - 120^o\right); F_b = F_m \cos\left(\omega t - 240^o\right) \tag{2.191}$$

Then, on substitution, it follows that

$$\begin{aligned} F(\theta, t) = {} & F_m \cos\theta \cos\omega t + F_m \cos\left(\theta - 120^o\right) \cos\left(\omega t - 120^o\right) \\ & + F_m \cos\left(\theta - 240^o\right) \cos\left(\omega t - 240^o\right) \end{aligned} \tag{2.192}$$

By the use of the trigonometric identity

$$\cos\alpha \cos\beta = \frac{1}{2}\left[\cos(\alpha - \beta) + \cos(\alpha + \beta)\right]$$

and noting that the sum of three equal sinusoids displaced in phase by 120° is equal to zero, Eq. (2.161) can be simplified as

$$F(\theta, t) = \frac{3}{2} F_m \cos(\theta - \omega t) \tag{2.193}$$

which is the expression For the resultant mmf wave. It has a constant amplitude 3/2 F_m, is a sinusoidal function of the angle θ, and rotates in synchronism with the supply frequency;

Figure 2.42 The individual flux directions are fixed but their magnitudes vary sinusoidally as does the current that produces them.

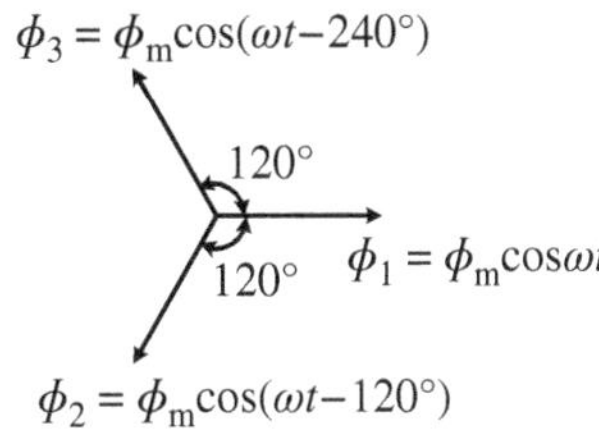

hence it is called a *rotating field*. The constant amplitude is 3/2 times the maximum contribution F_m of any one phase. The angular velocity of the wave is $\omega = 2\pi f$ electrical radians per second, where f is the frequency of the electric supply in hertz. For a P-pole machine, the rotational speed is given by

$$\omega_m = \frac{2}{P}\omega, \text{rad/s } or\ n = \frac{120f}{P}\text{rev/ min} \tag{2.194}$$

which is the synchronous speed.

The same result may be obtained graphically, as shown in Figure 2.42, which shows the spatial distribution of the mmf of each phase and that of the resultant mmf (given by the algebraic sum of the three components at any given instant of time).

Although the analysis here is carried out only for a three-phase case, it holds good for any q-phase ($q > 1$; i.e. polyphase) winding excited by balanced q-phase currents when the respective phases are wound $2\pi/q$ electrical radians apart in space. However, in a balanced two-phase case, note that the two-phase windings are displaced 90 electrical degrees in space, and the phase currents in the two windings are phase-displaced by 90 electrical degrees in time. The constant amplitude of the resultant rotating mmf can be shown to be $q/2$ times the maximum contribution of any one phase. Neglecting the reluctance of the magnetic circuit, the corresponding flux density in the air gap of the machine is then given by

$$B_g = \frac{\mu_0 F}{l_g} \tag{2.195}$$

where l_g is the length of the air gap.

We shall now use another useful method to find the magnitude and speed of the resultant flux due to three-phase currents. The three-phase sinusoidal currents produce fluxes ϕ_1, ϕ_2, and ϕ_3 which vary sinusoidally. The resultant flux at any instant will be the vector sum of all three at that instant. The fluxes are represented by three variable magnitude vectors (see Figure 2.42). In Figure 2.42, the individual flux directions are fixed but their magnitudes vary sinusoidally as does the current that produces them. To find the magnitude of the resultant flux, resolve each flux into horizontal and vertical components and then find their vector sum.

$$\begin{aligned}\phi_h = \phi_m \cos\omega t - \phi_m \cos\left(\omega t - 120^o\right)\cos 60^o \\ -\phi_m \cos\left(\omega t - 240^o\right)\cos 60^o = \frac{3}{2}\phi_m \cos\omega t\end{aligned}$$

$$\phi_v = 0 - \phi_m \cos\left(\omega t - 120^o\right)\cos 60^o - \phi_m \cos\left(\omega t - 240^o\right)\cos 60^o = \frac{3}{2}\phi_m \sin\omega t$$

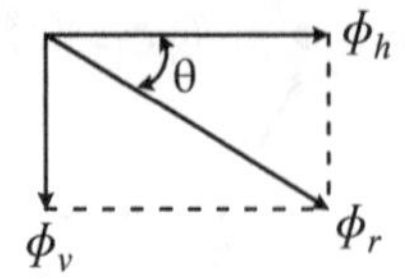

Figure 2.43 The phasor sum of the vertical and horizontal components of ϕ.

The resultant flux is given by

$$\phi_r = \sqrt{\phi_h^2 + \phi_v^2} = \frac{3}{2}\phi_m\sqrt{\cos\omega t^2 + \sin\omega t^2} = \frac{3}{2}\phi_m = 1.5\phi_m = \text{constant}$$

Thus, the resultant flux has constant magnitude (=$1.5\phi_m$) and does not change with time (Figure 2.43). The angular displacement of ϕ_r relative to the OX axis is

$$\tan\theta = \frac{\phi_v}{\phi_h} = \frac{\frac{3}{2}\phi_m \sin\omega t}{\frac{3}{2}\phi_m \cos\omega t} = \tan\omega t \rightarrow \therefore \theta = \omega t$$

Thus, the resultant magnetic field rotates at constant angular velocity $\omega(=2\pi f)$ rad/sec. For a P-pole machine, the rotation speed (ω_m) is

$$\omega_m = \frac{2}{P}\omega \sec/\text{rad}$$

or

$$\frac{2\pi N_s}{60} = \frac{2}{P} \times 2\pi f \ldots\ldots N_s \text{ is in r.p.m.}$$

$$\therefore N_s = \frac{120f}{P}$$

Thus, the resultant flux due to three-phase currents is of constant value (=$1.5\phi_m$where ϕ_m is the maximum flux in any phase) and this flux rotates around the stator winding at a synchronous speed of $\frac{120f}{P}$ rpm.

For example, for a six-pole, 50 Hz, three-phase induction motor, $N_s = \frac{120f}{P} = \frac{120\times50}{6} =$ 1000 rpm. It means that flux rotates around the stator at a speed of 1000 rpm.

2.13 Summary

This chapter is dedicated to magnetic circuit details. The electric machine is an electromechanical system that can be represented by electric and magnetic circuits. Both circuits have similarities and analogues to each other. Magnetic circuit material properties are elaborated. Many solved examples are given on magnetic circuit in order to understand the concepts clearly. Energy and co-energy are elaborated and the principle of operation of the rotating machine is explained. Production of the rotating magnetic circuit is elaborated. This is important to understand the operation of rotating electrical machines.

Problems

2.1 A toroidal solenoid with 500 turns and mean radius of 20 cm, carries a current of 2.0 A. The toroidal solenoid is filled with a magnetic material. The magnetic field inside the windings is 1.5 T. Determine (a) the relative permeability and (b) the magnetic susceptibility of the material that fills the toroid.

2.2 A paramagnetic material is placed in an external magnetic field of 1.8000 T, the field inside the material is measured to be 1.85 T. Calculate (a) the relative permeability and (b) the magnetic permeability of this material.

2.3 Given a toroidal coil as shown in Figure P2.3 made of magnetic material of relative permeability of the core material is 2000 and the outside radius of the toroid is 0.2 cm. The area of cross section is circular, with a diameter 0.05 cm. Calculate the magnetomotive force required to set up a flux density of 1.5 mT in the core. Calculate the flux in the core.

Figure P2.3 Toroid for Problem 2.3.

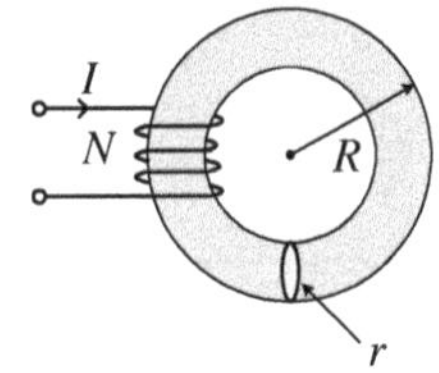

2.4 For the toroidal coil of Problem 2.3, find the flux density in the core for an mmf of 250 At.

2.5 Consider the toroid shown in Figure P2.5, which is wound with a coil of 200 turns. Assume that the current is 40 A and that the core is non-magnetic. The inner radius $a = 10$ cm, outer radius $b = 15$ cm, and $h = 8$ cm. It is required to calculate:
(a) The reluctance of the circuit.
(b) The flux and flux density inside the core.

Figure P2.5 Toroid for Problem 2.5.

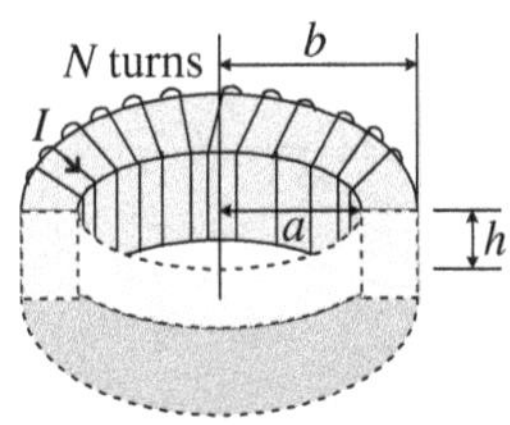

2.6 Consider the magnetic structure shown in Figure P2.6. Calculate the current in the 100-turns winding to set up a flux density of 1.0 T in the core, assuming that the relative permeability of the core material is 2000.

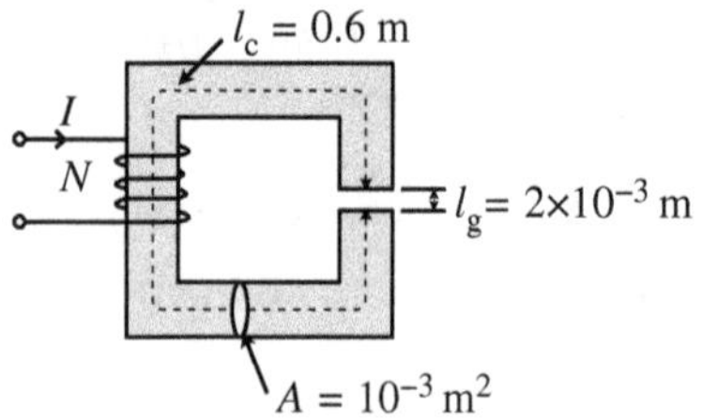

Figure P2.6 Magnetic structure for Problem 2.6.

2.7 For the magnetic structure shown in Figure P2.7, calculate the current I required to set up a flux density of 1.5 T in the air gap. Assume that the relative permeability of the core material is 1800.

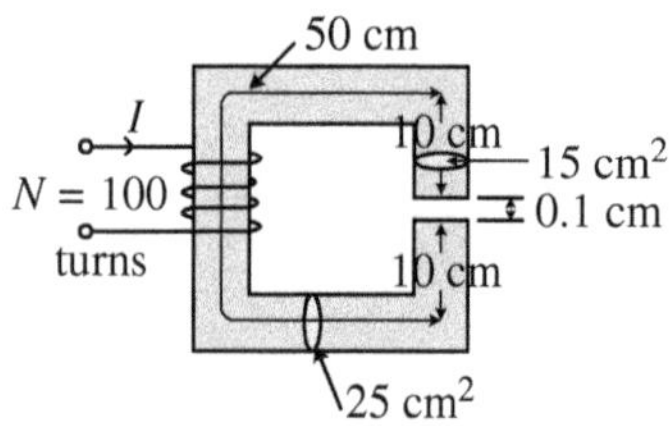

Figure P2.7 Magnetic structure for Problem 2.7.

2.8 For the magnetic structure of Problem 2.6, find the flux density in the air gap for a current of 50 A.

2.9 For the magnetic core of Problem 2.7, calculate the flux density in the air gap for a current of 10 A.

2.10 The core of the magnetic circuit of Figure P2.10 has a relative permeability of 2500. The coil current is 1.0 A and the corresponding flux is 10 mWb. Compute the depth d of the magnetic circuit.

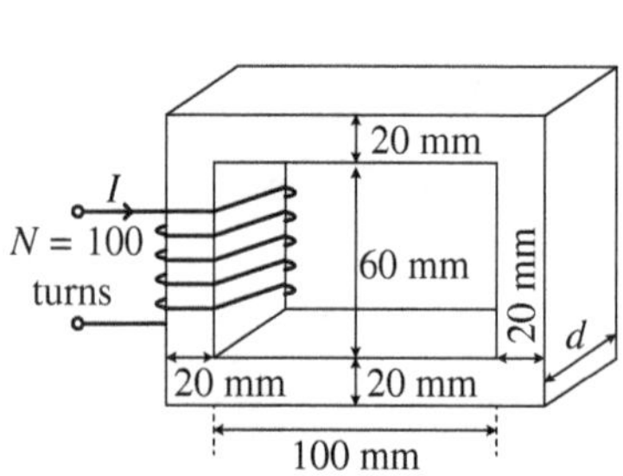

Figure P2.10 Magnetic structure for Problem 2.10.

2.11 A magnetic circuit is shown in Figure P2.11 and has mmf of 200 At. The two rotors are made of a magnetic material with relative permeability μ_r while the core material has a permeability of μ_i. If the rotors are absent, the flux is found to be 0.5236×10^{-5} Wb. With the rotors inside the circuit, the flux is found to be 3.122×10^{-5} Wb. Find the relative permeabilities μ_r and μ_i given that the cross section of the structure is uniform with area 4×10^{-4} m^2.

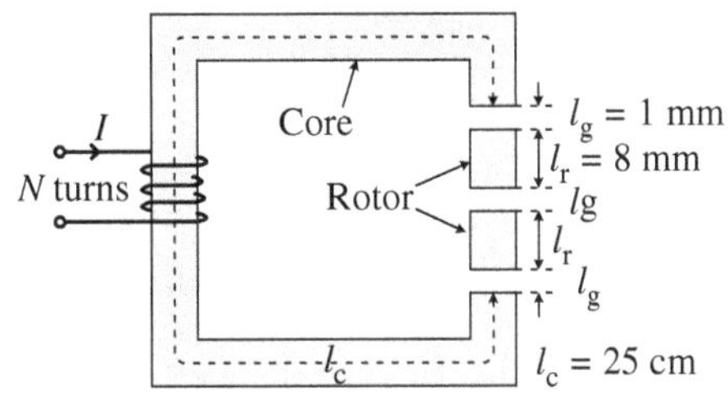

Figure P2.11 Magnetic structure for Problem 2.11.

2.12 The relative permeability of the core of the magnetic structure of Figure P2.12 is 2000. The mmf is 500 At and the flux is $7\ 10^{-4}$ Wb. Find the length l_o.

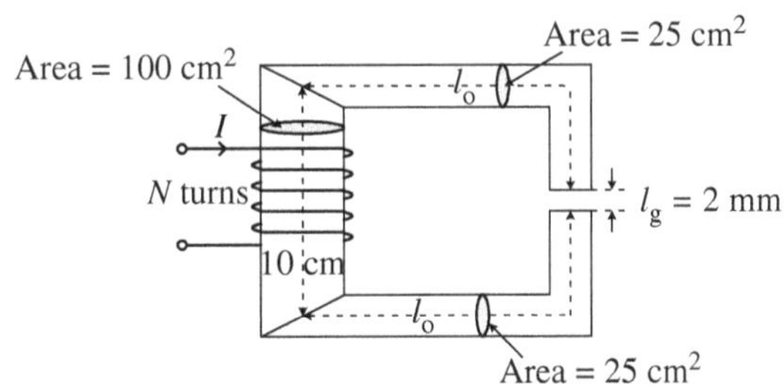

Figure P2.12 Magnetic structure for Problem 2.12.

2.13 Given the magnetic circuit shown in Figure P2.13. Draw an equivalent magnetic circuit for the system. Assume that the relative permeability of the core is 2500. Calculate the flux in the right-hand air gap for an mmf in the coil of 1000 At.

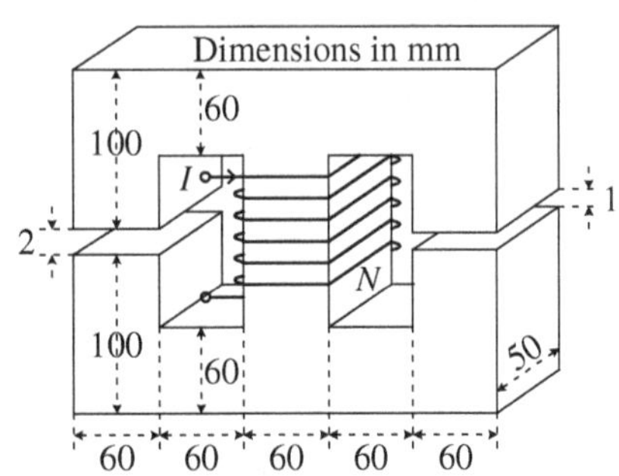

Figure P2.13 Magnetic structure for Problem 2.13.

2.14 Given the magnetic circuit shown in Figure P2.14. The flux density in the cast-iron portion is 0.6 T and the flux density in the sheet-steel portion is 1 T. This corresponds to an mmf of 410 At and a flux of 0.35 mWb. Assume that the permeability of cast iron is 4×10^{-4} and that the permeability of sheet steel is 4.348×10^{-3}. Determine the dimensions w_1, w_2, and d for this magnetic circuit.

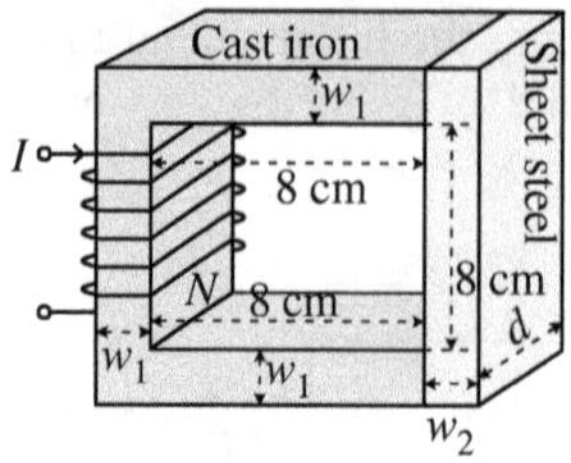

Figure P2.14 Magnetic structure for Problem 2.14.

2.15 Consider the magnetic circuit of Figure P2.15, made up of magnetic material with a relative permeability $\mu_r = 3000$. The magnetic core dimensions are: $l_a = 0.9$ m; $l_b = 0.8$ m; $l_c = 0.3$ m; $l_g = 1$ mm. The magnetic core has a cross-sectional area A = 6 x 10^{-3} m^2. Calculate the flux in the air gap for an MMF of 240 At.

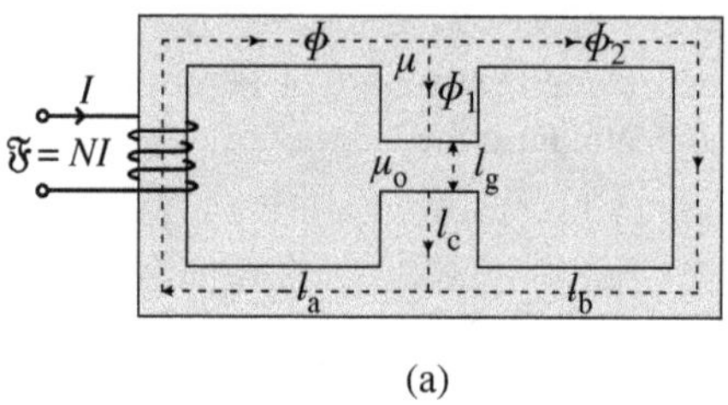

Figure P2.15 (a) Multiloop magnetic structure; (b) its equivalent circuit. Source: For Problem 2.15.

2.16 The magnetic core of a synchronous machine is shown in Figure P2.16. Assume that the rotor and stator core have relative permeabilities of 3500 and 1800, respectively. The outside and inside diameters of the stator yoke are 1 and 0.72 m, respectively. The rotor length is 0.7 and is 0.1 m wide. The air gaps are each 10 mm long. The axial length of the machine is 100 cm. Derive an equivalent magnetic circuit of the machine and calculate the flux density in the air gap for an MMF of 5000 At.

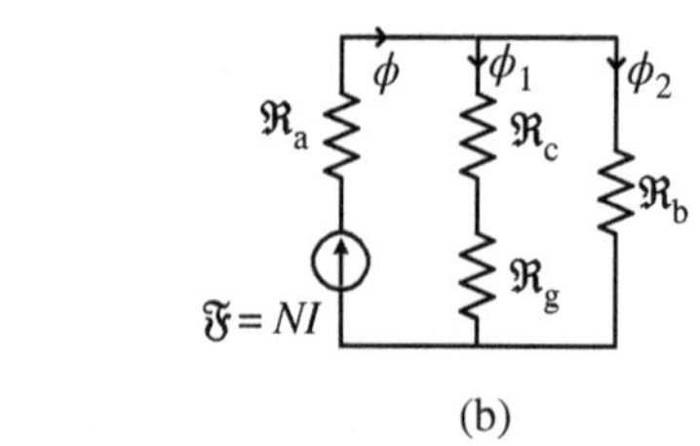

Figure P2.16 Magnetic structure for Problem 2.16.

2.17 Given the magnetic core of Problem 2.15 (Figure P2.17) and assuming that the air-gap length is 1.5 mm with all other data unchanged. Calculate the required mmf to set up a flux of 0.5×10^{-3} Wb in the centre section.

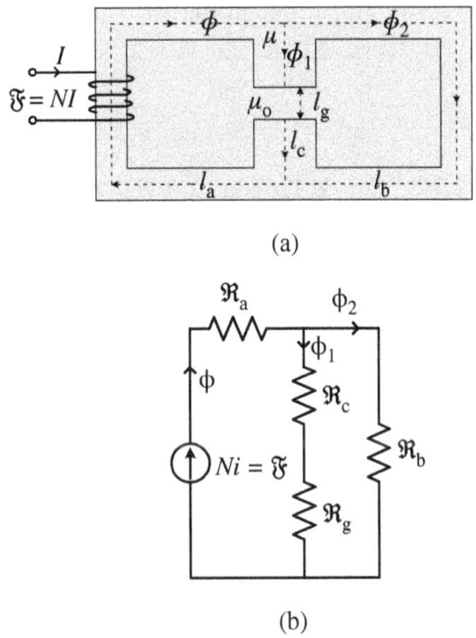

Figure P2.17 (a) Multiloop magnetic structure; (b) its equivalent circuit. Source: For Problem 2.17.

2.18 The magnetic structure of Figure P2.18 has relative permeability 2000. The air-gap length is 2 mm. Find the flux in the air gap given that $\mathfrak{F}_1 = \mathfrak{F}_2 = 4000$ At

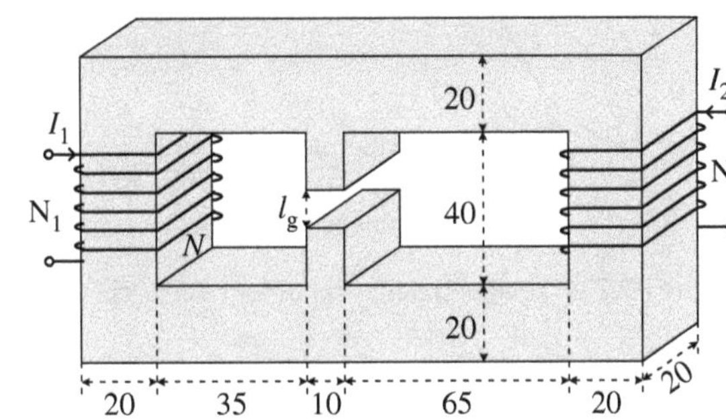

Figure P2.18 Magnetic structure for Problem 2.18.

2.19 For the magnetic structure of Problem 2.18, find $\mathfrak{F}_2$ given that $\mathfrak{F}_1 = 4000$ At, and that the flux in the air gap is 0.5×10^{-3} Wb.

2.20 Repeat Problem 2.6, accounting for fringing of flux in the air gap using 1.1 as a fringe factor: Fringe Factor $= \frac{\text{Cross-sectional area of flux lines}}{\text{Cross-sectional area of air gap}}$

2.21 Repeat Problem 2.7, (Figure P2.21), accounting for fringing effects assuming that the air gap is square in cross section.

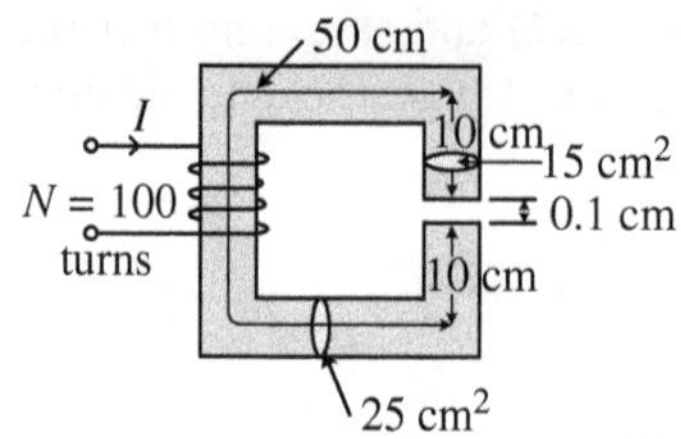

Figure P2.21 Magnetic structure for Problem 2.21.

2.22 Repeat Problem 2.6, (Figure P2.22b) assuming that the iron path is cast steel with a magnetization curve as shown in Figure P2.22a.

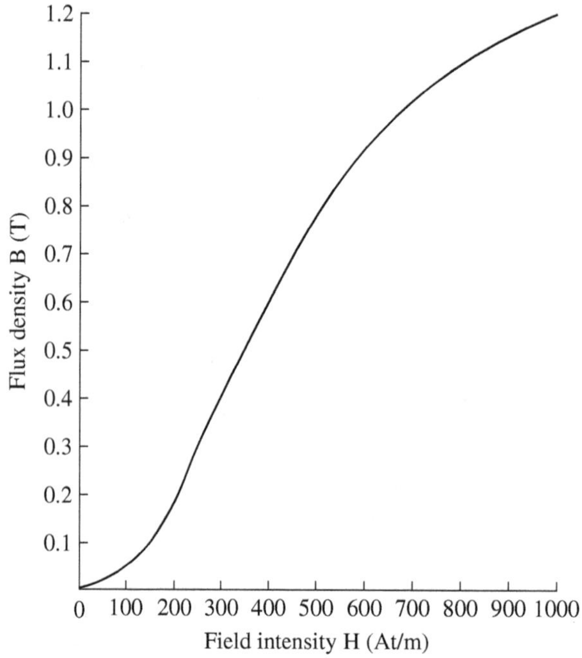

Figure P2.22a Magnetization curve for cast steel.

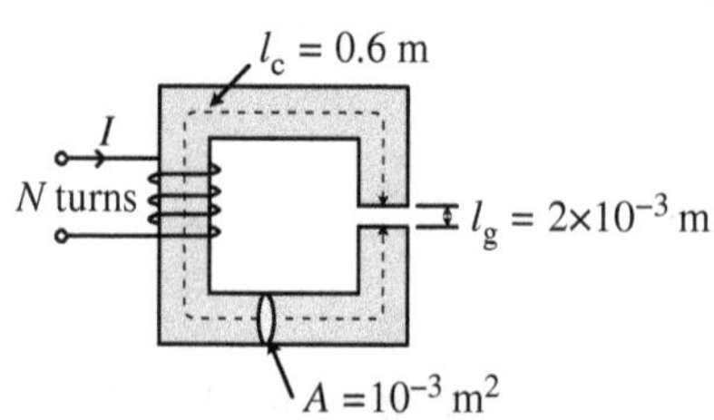

Figure P2.22b Magnetic structure for Problem 2.22.

2.23 Repeat Problem 2.19 assuming that the magnetization curve of the core is given in Figure P2.22a. Assume that the air-gap flux is 2×10^{-4} Wb and that $\mathfrak{F}_1$ is 1800 At.

2.24 Calculate the inductance of the coil of Problem 2.3 (Figure P2.24) assuming that $N = 10$ turns.

Figure P2.24 Toroid for Problem 2.24.

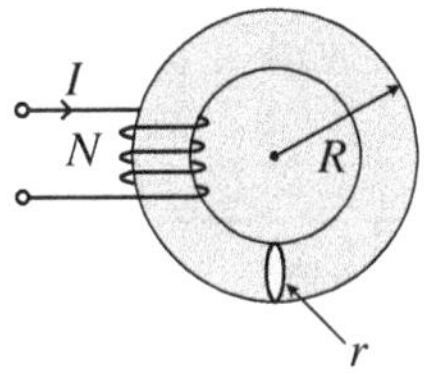

2.25 Calculate the inductance of the coil of Problem 2.6.

2.26 The flux linkages in a coil are related to the current by the relation $\phi = 1 - 0.5e^{-1.25i}$. Find the energy stored for a current of 0.8 A.

2.27 Consider the coil of Problem 2.6. Find the energy stored in the air gap and the magnetic core.

2.28 The eddy-current and hysteresis losses in a transformer are 500 and 600 W, respectively, when operating from a 60-Hz supply. Determine the eddy-current and hysteresis losses when the transformer is operated from a 50-Hz supply with an increase of 10% in flux density. Find the change in core losses.

2.29 The relationship between current, displacement, and flux linkages in a conservative electromechanical device is given by $i = \lambda[0.8\lambda + 2.2(x-1)^2]$. Find expressions For the stored energy and the magnetic-field force in terms of λ and x. Find the force for $x = 0.75$.

2.30 Repeat Problem 2.29 for the relationship: $i = \lambda^3 + \lambda(0.1\lambda + x)$.

2.31 Find an expression For the force exerted by the magnetic field in an electromechanical device with the following nonlinear characteristic: $i = I_0 \frac{a\lambda + b\lambda^3}{1+cx}$ where I_0, a, b, and c are given constants.

2.32 A plunger-type solenoid is characterized by the relation: $\lambda = \dfrac{10i}{1+\dfrac{10^4 x}{2.54}}$

Find the force exerted by the field for $x = 2.54 \times 10^{-3}$ and $i = 11$ A.

2.33 An electromagnet used to lift steel slabs is shown in Figure P2.33. Show that the minimum current required to lift a slab of mass M is given by $i = \dfrac{\Re_c + \left(\dfrac{2x}{\mu_0 A}\right)}{N}\sqrt{\mu_0 M A g}$. In the expression, $\Re_c$ is the reluctance of the magnetic core, A is the cross-sectional area of one air gap, and g is the acceleration of gravity.

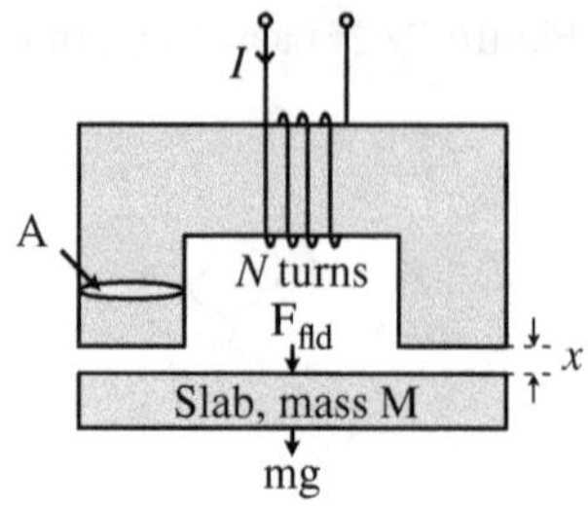

Figure P2.33 Electromagnet for Problem 2.33.

2.34 Find the minimum value of the current i required to lift a slab with $M = 100$ kg using the electromagnet of Problem 2.33, given that $A = 16 \times 10^{-4}$ m^2; $x = 0.15 \times 10^{-4}$ m; $N = 400$ turns. Neglect the reluctance of the magnetic core.

2.35 Consider the magnetic structure shown in Figure P2.35. Show that the dynamics of the system are described by

$$\frac{d\lambda(t)}{dt} = \frac{-(g_1 l_1 + x l_2)R}{\mu_o N^2 h l_1 l_2}\lambda(t) + RI(t)$$

$$\frac{d^2x}{dt^2} = \frac{-\lambda^2}{2\mu_o M N^2 h l_1} + g$$

where g is the acceleration of gravity. Neglect core reluctance.

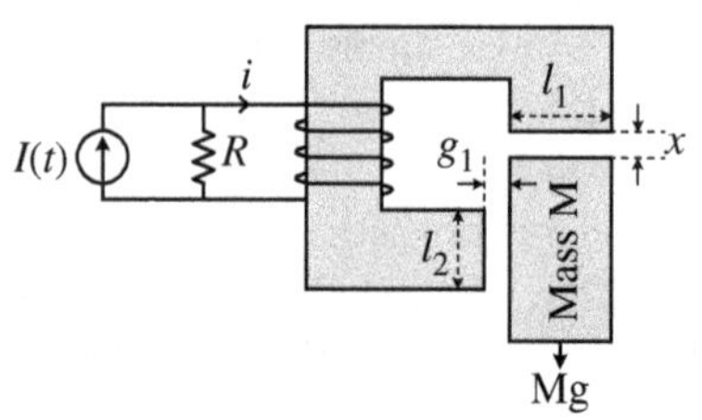

Figure P2.35 Structure for Problem 2.35.

2.36 Assume that the voltage applied to the coil of the magnetic structure shown in Figure P2.36 is given by $e(t) = E_m \cos \omega t$. Assume that $\lambda(0) = 0$, and find the force exerted on the movable part in terms of the structure geometry, E_m, ω, and time.

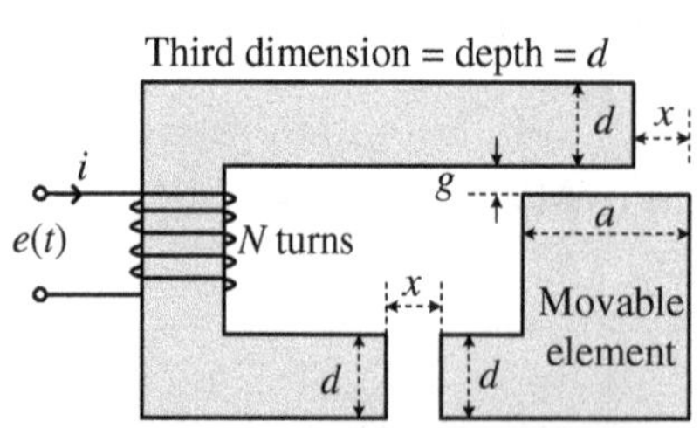

Figure P2.36 Magnetic structure for Problem 2.36.

2.37 For the structure of Problem 2.36, assume that $E_m = 10$ V and that the supply frequency is 60 Hz. The coil has 500 turns. The gap length $g = 0.1$ cm, $d = 10$ cm, and $a = 15$ cm. Find the force in terms of x and time. Calculate the average force in terms of x.

2.38 The inductance of a coil used with a plunger-type electromechanical device is given by

$$L = \frac{1.6 \times 10^{-5}}{x}$$

where x is the plunger displacement. Assume that the current in the coil is given by

$$i(t) = 5 \sin \omega t$$

where $\omega = 2\pi(60)$. Find the force exerted by the field for $x = 10^{-2}$ m. Assume that x is fixed and find the necessary voltage applied to the coil terminals, given that its resistance is 1 Ω.

2.39 The reluctance of the magnetic structure for a plunger-type relay is given by

$$\mathfrak{R} = \frac{200}{x(0.02 - x)}$$

Assume that the exciting coil has 1200 turns and that the exciting current is

$$i = 2 \cos$$

Find the magnetic-field force as a function of time and its average value if the displacement is given by

$$x(t) = 0.01 \cos t$$

2.40 The inductance of the exciting coil of a certain reluctance motor is given by

$$L = \frac{L_0}{1 + a \cos 4\theta}$$

Find an expression For the energy stored in the magnetic field, and hence find an expression For the torque developed in terms of flux linkages and the angular displacement θ and the system parameters. Find the torque developed for $\theta = 15°$ and $i = 10$ A. Assume that $L_0 = 0.05$ H and $a = -0.3$.

2.41 Repeat Problem 2.40 for the inductance expressed as

$$L = \frac{L_0}{1 + a(\cos 4\theta + \cos 8\theta)}$$

The value of a in this case is −0.25.

2.42 The reluctance of the magnetic circuit of a reluctance motor is given by

$$\mathfrak{R} = \mathfrak{R}_0 (1 + a \cos 4\theta)$$

Let $\mathfrak{R}_0 = 4000$ and assume that the number of turns on the exciting coil is 20 turns. Find the developed torque for $i = 5$ A, $\theta = 10°$, assuming that $a = -0.35$.

2.43 The developed torque for the motor of Problem 2.40 is given by

$$T = -15\lambda^2 \sin 4\theta$$

Assume that

$$\lambda(t) = 0.5 \cos \omega t$$

and that

$$\theta(t) = \omega_r t + a$$

Find the instantaneous torque developed and the condition For nonzero average torque.

2.44 A rotating electromechanical conversion device has a stator and rotor, each with a single coil. The inductances of the device are: $L_{11} = 0.5$ H; $L_{22} = 2.5$ H; $L_{12} = 1.5\cos\theta$ H, where the subscript 1 refers to stator and the subscript 2 refers to rotor. The angle θ is the rotor angular displacement from the stator coil axis. Express the torque as a function of currents i_1, i_2, and θ, and compute the torque for $i_1 = 3$ A and $i_2 = 1$ A.

2.45 Assume for the device of Problem 2.44 that $L_{11} = 0.4 + 0.1 \cos 2\theta$. All other parameters are unchanged. Find the torque in terms of θ for $i_1 = 3$ A and (a) $i_2 = 0$; (b) $i_2 = 1$ A.

2.46 Assume for the device of Problem 2.44 that the stator and rotor coils are connected in series, with the current being $i(t) = I_m \sin \omega t$. Find the instantaneous torque and its average value over one cycle of the supply current in terms of I_m and ω.

2.47 A rotating electromechanical energy conversion device has the following inductances in terms of θ in radians (angle between rotor and stator axes): $L_{11} = 0.764\theta$; $L_{22} = -0.25 + 1.719\theta$; $L_{12} = -0.75 + 1.337\theta$. Find the torque developed for the following excitations.

(a) $i_1 = 15$ A, $i_2 = 0$ A.
(b) $i_1 = 0$ A, $i_2 = 15$ A.
(c) $i_1 = 15$ A, $i_2 = 15$ A.
(d) $i_1 = 15$ A, $i_2 = -15$ A.

2.48 For the machine of Problem 2.44, assume that the rotor-coil terminals are shorted ($e_2 = 0$) and that the stator current is given by $i_1(t) = I \sin \omega t$. Find the torque developed as a function of I, θ, and time.

2.49 For the device of Problem 2.48, the rotor-coil terminals are connected to a 10-Ω resistor. Find the rotor current in the steady state and the torque developed.

References

1 Giancoli, D.C. *Physics for Scientists and Engineers with Modern Physics*, 4e. Upper Saddle River, NJ: Pearson Education, Inc. Pearson Prentice Hall, Pearson Education, Inc.

2 Tipler, P.A. and Mosca, G. *Physics for Scientists and Engineers with Modern Physics*, 6e. New York: W. H. Freeman and Company.

3 Brown, R.G. (2020). Introductory Physics II. http://webhome.phy.duke.edu/~rgb/Class/intro_physics_2.php (accessed 29 September 2020).
4 Sarma, M.S. (2001). *Introduction to Electrical Engineering*. Oxford University Press.
5 Lorrain, P. and Corson, D. (1979). *Electromagnetism: Principles and Applications*. W.H. Freeman & Co Ltd.
6 Woodson, H.H. and Melcher, J.R. (1968). *Electromechanical Dynamics*. New York, NY: Wiley.
7 Tewari, J.P. (1999). *Engineering Electromagnetics (Theory, Problems and Application)*, 2e. Khanna Publishers.
8 Gieras, J.F. (2017). *Electrical Machines: Fundamentals of Electromechanical Energy Conversion*. CRC Press, Taylor & Francis Group.
9 El-Hawary, M.E. *Principles of Electric Machines with Power Electronic Applications*, 2e. IEEE Press.
10 Umans, S.D. *Fitzgerald & Kingsley's Electric Machinery*, 7e. McGraw-Hill.
11 Knight, A. (2020). Electric machines: torque. https://people.ucalgary.ca/~aknigh/electrical_machines/fundamentals/f_ac_torque.html (accessed 28 October 2020).
12 Mehta, V.K. and Mehta, R. (2008). *Principles of Electrical Machines*. New Delhi: S Chand & Co Ltd.

3

Single-Phase and Three-Phase Transformers

3.1 Preliminary Remarks

A Transformer is an electrical device that converts electrical energy from one circuit to the other via a magnetic circuit with a change in the voltage and current levels without changing the frequency. Transformers are used in ac circuit only (DC circuit utilizes power electronic dc-dc converter such as buck/boost/buck-boost converters for changing voltage and or current level). Transformers are invariably used in every application requiring a change in voltage/current levels. It is the invention of transformer that has led to the adoption of AC for generation, transmission, distribution and utilization of electrical power. Transformers have made it possible to transmit electrical power to large distances without significant loss of power/energy. Transformers are used at electric power stations to increase (step-up) the voltage level, appropriate to match the transmission requirement. They are used at grid substations to reduce (step-down) the voltage level, appropriate to the distribution level. Moreover, transformers are used in a variety of electrical devices to reduce/increase the voltage level to match the requirement e.g. in TV sets, tape recorders, radios, PCs etc. Additionally, they are used for isolating high and low voltage circuits for protection, called isolation transformers. Instrument transformers are another class used for measurement purposes; current transformers for measuring currents and potential transformers for measuring voltages. Widespread use of transformers in the electrical power system is illustrated in Figure 3.1.

In its basic form, a transformer has two windings designated as 'Primary winding' and 'Secondary Winding', core on which windings are wound, enclosure, insulation system and cooling system. A winding in which supply is connected is called 'Primary' and the winding which supplies load is called 'Secondary'. An important feature of a transformer is that the frequency of the input quantities is same as that of the output quantities. Moreover, power in the primary and secondary winding is same.

The electrical energy fed to the primary winding is first converted to the magnetic form that is then reconverted to the electrical energy at the secondary winding. Primary and secondary windings are electrically isolated but magnetically coupled. A transformer is a static device without any moving parts and hence efficiency is high while the maintenance cost is low when compared with a rotating machine.

Electrical Machine Fundamentals with Numerical Simulation using MATLAB/SIMULINK, First Edition.
Atif Iqbal, Shaikh Moinoddin, and Bhimireddy Prathap Reddy.

Companion website: www.wiley.com/go/iqbal/electricalmachine

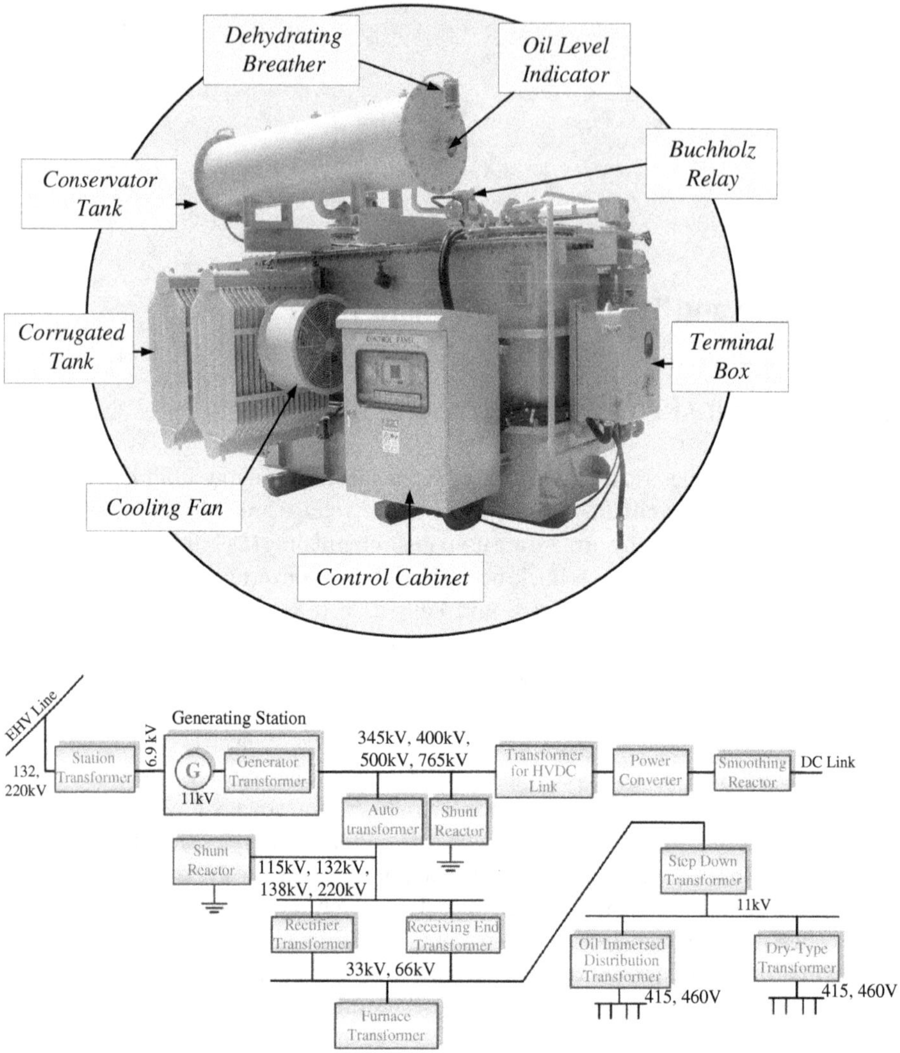

Figure 3.1 Use of transformers in a power system.

The core of the transformer is a magnetic material such as silicon steel, hot rolled grain-oriented silicon steel, and cold rolled grain oriented (CRGO) silicon steel. The main characteristics of the material used in a transformer is to obtain higher flux density. The research trend is in developing better core material to achieve higher flux density and to lower the core losses. For further core loss reduction, amorphous steel is used. However, this material is brittle and difficult to handle [1].

3.2 Classification of Transformers

Transformers are classified according to various parameters such a

a) According to the number of phases
   i) Single-phase
   ii) Three-phase
   iii) Multiphase
b) According to the operation
   i) Step-up
   ii) Step-down
c) According to the construction
   i) Core type
   ii) Shell type
d) According to the number of windings
   i) Single-winding (Auto transformer)
   ii) Two-winding
   iii) Three-winding (Tertiary winding)
e) According to the use
   i) Power Transformer
   ii) Distribution Transformer

3.2.1 Classification Based on Number of Phases

3.2.1.1 Single-Phase Transformers

Single-phase transformers are used for single-phase ac input and single-phase ac output. There is one primary winding with N1 number of turns and one secondary winding with N2 number of turns as shown in Figure 3.2.

3.2.1.2 Three-Phase Transformers

Three-phase transformers have three-phase input and three-phase output. There are three primary and three secondary windings. The primary and secondary windings are connected

Figure 3.2 Single-phase transformer.

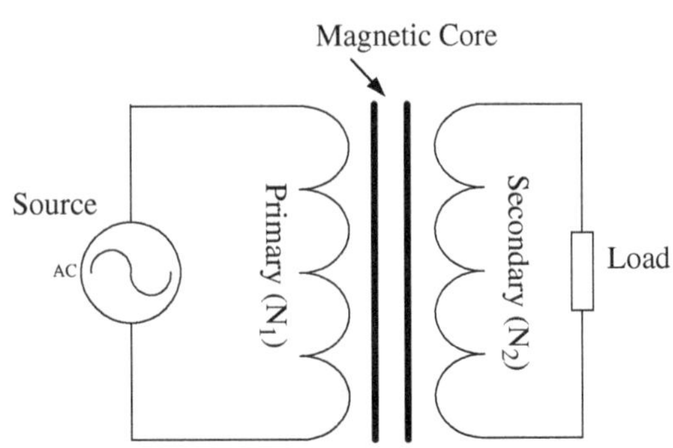

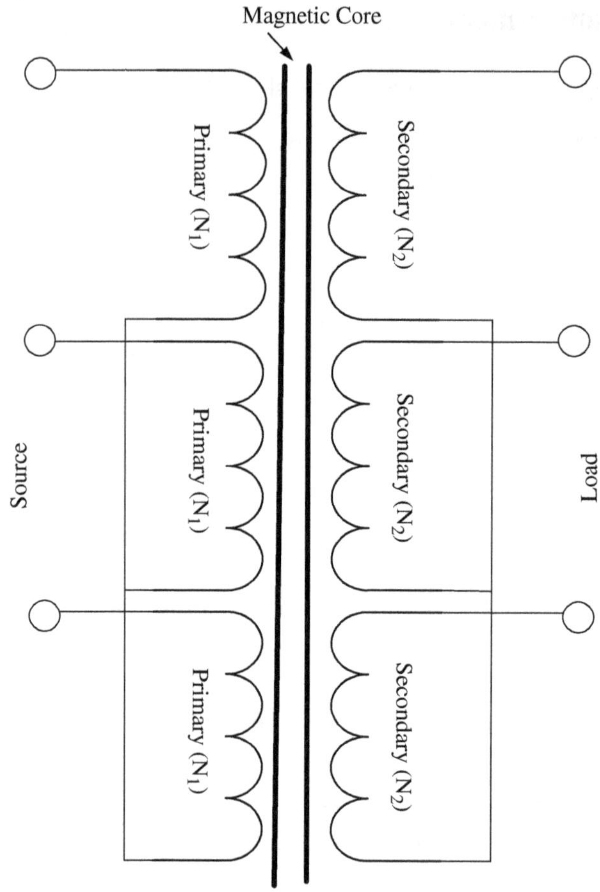

Figure 3.3 A three-phase transformer with star-star connection.

in either star and/or delta connection (shown in Figure 3.3 with star connected primary and star connected secondary). These transformers are used throughout the power system in generation, transmission, and distribution.

3.2.1.3 Multi-Phase Transformers

Multi-phase transformers can be three-phase to five-phase, three-phase to six-phase, three-phase to seven-phase, and three-phase to *n*-phase. These transformers can be used in special applications such as testing of multiphase motors and multiphase transmission. As an example, a three-phase to five-phase transformer is shown in Figure 3.4 [2]. Three-phase inputs are V_A, V_B, and V_C. The outputs are V_a, V_b, V_c, V_d, and V_e.

3.2.2 Classification Based on Operation

3.2.2.1 Step-Up Transformers

In this transformer, the output voltage is more than the input voltage while the output current is less than the input current.

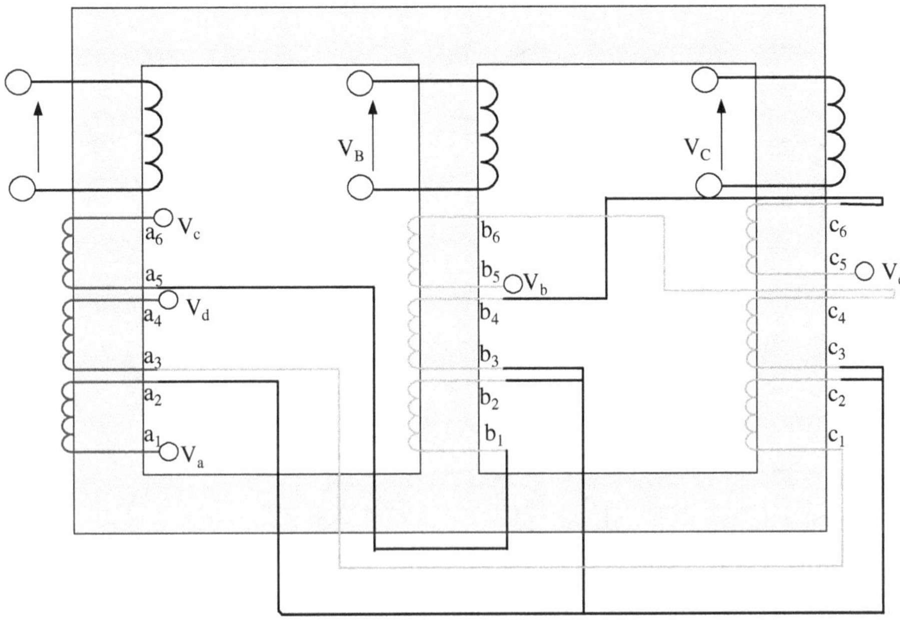

Figure 3.4 A three-to-five phase transformer connections.

3.2.2.2 Step-Down Transformers

In this transformer, the output voltage is less than the input voltage while the output current is more than the input current.

3.2.3 Classification Based on Construction

3.2.3.1 Core-Type Transformers

In this type of transformer, windings are put on the outer limb of the core and thus the windings are exposed and maintenance is easier. A core-type single-phase transformer is shown in Figure 3.5a where primary winding has N1 number of turns and secondary winding has N2 number of turns. A three-phase core-type transformer is shown in Figure 3.5b. Six windings are places on the transformer limbs. Both primary and secondary windings are placed on the same limb for each phase.

3.2.3.2 Shell-Type Transformers

In this type of transformer, windings are put on the central limb of the core and thus the windings are not easily accessible and maintenance is difficult, but the leakage reactance is lower than the core type and consequently, the voltage regulation is better. A single-phase shell type of transformer is shown in Figure 3.6. Both primary (N_1) and secondary winding (N_2) are placed one over the other. A three-phase shell type of transformer is shown in Figure 3.7.

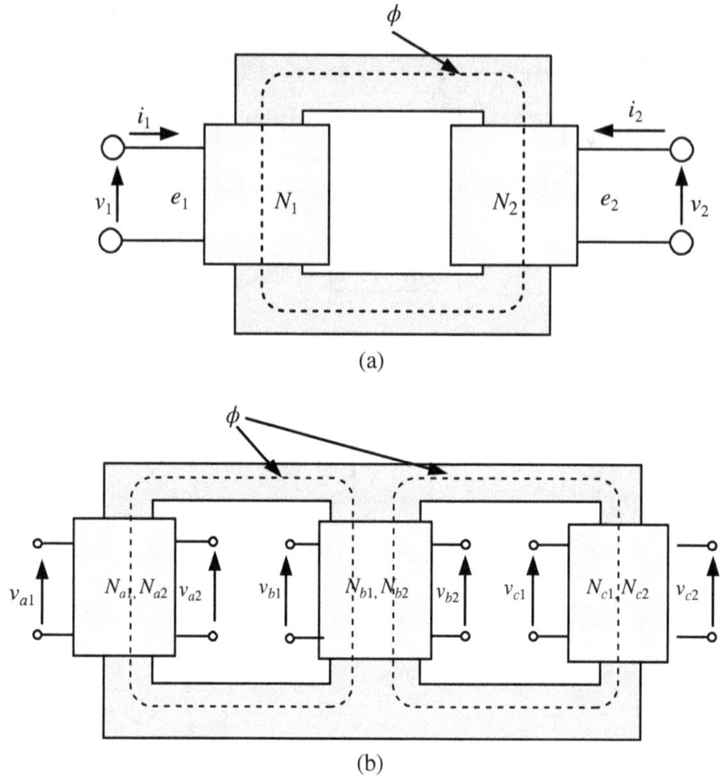

Figure 3.5 (a) Single-phase core type of transformer construction. (b) Three-phase core-type transformer construction.

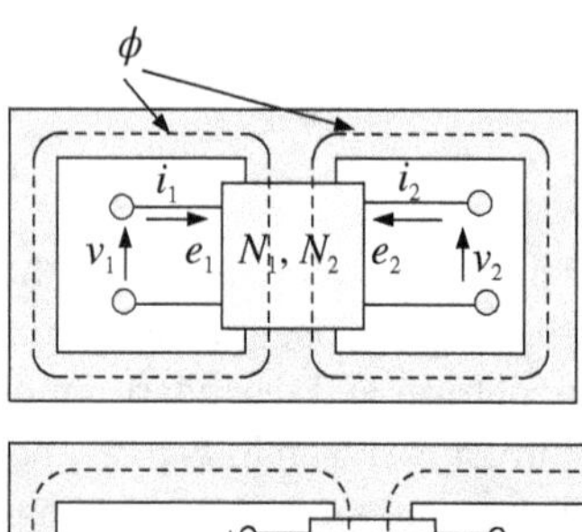

Figure 3.6 Single-phase shell type of transformer construction.

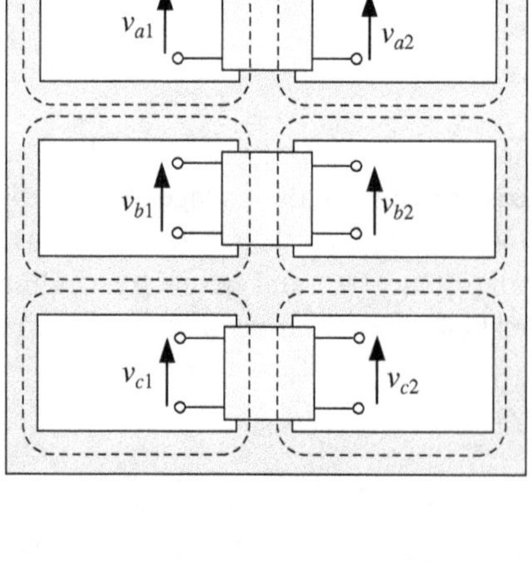

Figure 3.7 Three-phase shell-type transformer construction.

3.2.4 Classification Based on Number of Windings

3.2.4.1 Single-Winding Transformer

Transformers having only one winding are termed as 'Autotransformers' as shown in Figure 3.8. It has a slider on one end to provide variable output voltage. This transformer is used as a variable voltage supply. It transforms energy from one circuit to the other via induction as well as conduction since there is no electrical isolation between primary and secondary.

3.2.4.2 Two-Winding Transformer

This type of transformer has two separate windings called primary and secondary as shown in Figure 3.9.

3.2.4.3 Three-Winding Transformer

This is a special type of transformer having three windings namely primary, secondary and tertiary as shown in Figure 3.10. This transformer is used for special applications where two different voltage levels are required such as in substations for auxiliary loads.

3.2.5 Classification Based on Use

3.2.5.1 Power Transformer

The transformer used in power plant and grid substations for high voltage step up and step down are called power transformers. The generated voltage in the power plant (11–33 kV)

Figure 3.8 Basic autotransformer.

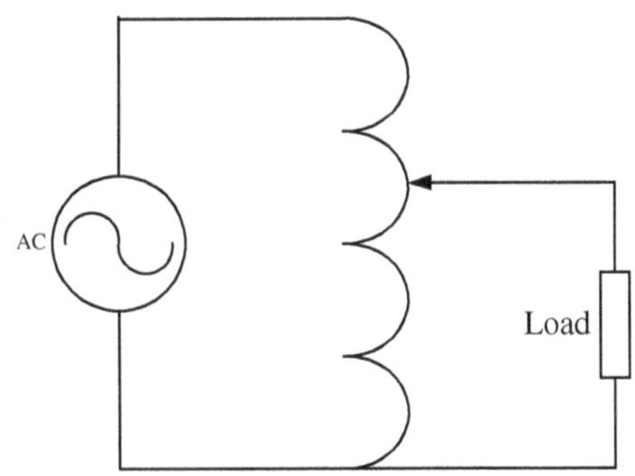

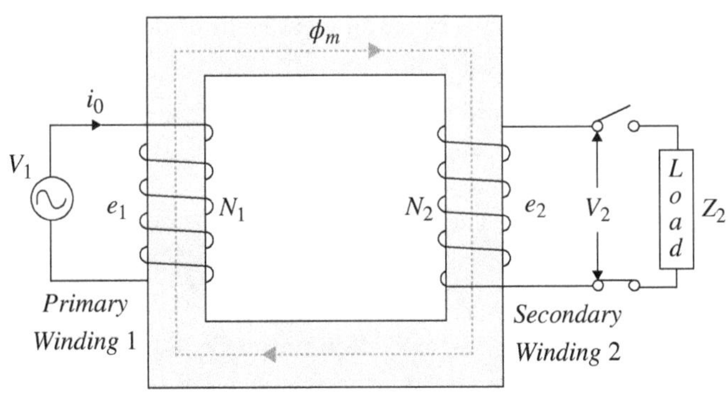

Figure 3.9 Two winding transformer.

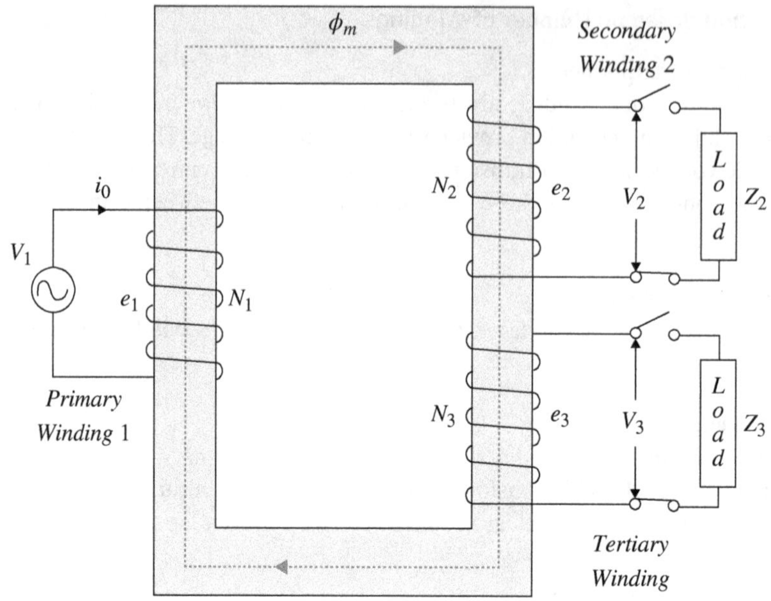

Figure 3.10 Three winding transformer.

is stepped-up to transmission voltage level (132, 220, and 400 kV). The same transmission voltage is then stepped down by power transformers to 66 and 33 kV in the grid substations. The power transformer is used to further step down 33–11 kV in the substation.

3.2.5.2 Distribution Transformer

This type of transformer is used for low-voltage distribution network to connect to the users. The 11 kV is stepped down to 440 V three-phase using distribution transformer to supply to the residential users.

3.3 Principle of Operation of the Transformer

3.3.1 Ideal Transformer

A transformer with the following characteristics is called an Ideal Transformer.

- The primary and secondary windings have no resistance.
- There is no leakage flux. All the flux produced by primary winding links with the secondary winding.
- There is no core/iron loss.
- Permeability of the core material is infinite. Therefore, the exciting current required to establish flux in the core is negligible.

An ideal transformer is first explained followed by a real transformer.

The principle of operation of a transformer is based on electromagnetic induction which states that a voltage is induced in a coil if time varying flux links the coil. The primary winding is connected to the source and the secondary winding is open circuited (under no-load conditions). The applied voltage to the primary windings is v_1. The applied voltage is AC that causes an AC current i_0 to flow in the winding. The current causes mmf (magneto motive force) to produce in the primary winding ($\Im_1 = i_0 N_1$). The mmf thus generated will cause a flux to produce. The flux will flow through the core linking the primary and secondary windings called mutual flux ϕ_m. The time varying flux linking the primary winding causes an emf e_1 to induce in the primary winding. The polarity of the induced emf induced is such as to oppose the applied voltage. The mutual flux also links the secondary winding causing an emf e_2 to induce. As this is an ideal transformer, there is no voltage drop in the winding and thus the applied voltage is equal to the induced emf.

$$v_1 = e_1 = N_1 \frac{d\phi_m}{dt} \tag{3.1}$$

$$v_2 = e_2 = N_2 \frac{d\phi_m}{dt} \tag{3.2}$$

The flux is sinusoidal and is given as;

$$\phi_m = \phi_{mp} \sin(\omega t) \tag{3.3}$$

Where ϕ_{mp} is peak value of the flux and the frequency is $\omega = 2\pi f$ rad/sec. The induced voltage is given as

$$e_1 = N_1 \frac{d}{dt}\left(\phi_{mp} \sin(\omega t)\right) = N_1 \omega \phi_m \cos(\omega t) = N_1 2\pi f \phi_m \cos(\omega t) = \sqrt{2} E_1 \phi_m \cos(\omega t) \tag{3.4}$$

The rms value of the induced voltage is obtained as

$$E_1 = \frac{N_1 2\pi f}{\sqrt{2}} = 4.44 N_1 f \phi_{mp} \tag{3.5}$$

Similarly, the rms value of the induced emf in the secondary winding is obtained as

$$E_2 = 4.44 N_2 f \phi_{mp} \tag{3.6a}$$

Dividing Eq. (3.1) by Eq. (3.2), the following is obtained

$$\frac{v_1}{v_2} = \frac{e_1}{e_2} = \frac{N_1}{N_2} = a = turn\ ratio \tag{3.6b}$$

The no-load current drawn by the primary winding has only one magnetizing component i_m assuming linear BH curve of the magnetic core. The magnetizing component is in phase with the mutual flux ϕ_m.

Now consider load on the secondary winding, current i_2 will flow in the secondary. Due to this current, secondary will provide an mmf $N_2 i_2$.This mmf will produce a flux ϕ_2. This flux will flow in the opposite direction to the mutual flux and will try to reduce/cancel the mutual flux. The mutual flux should not reduce since the applied voltage v_1 is constant (the mutual flux was produced due to the applied voltage on the primary winding and no-load current). In order to cancel the secondary flux ϕ_2, a new counter flux is produced from the primary winding ϕ_1. This flux is produced from the additional current i_1' (called load component) drawn by primary winding from the source such that the total primary current is sum of the load component and magnetizing component

$$i_1 = i_1' + i_o \tag{3.7}$$

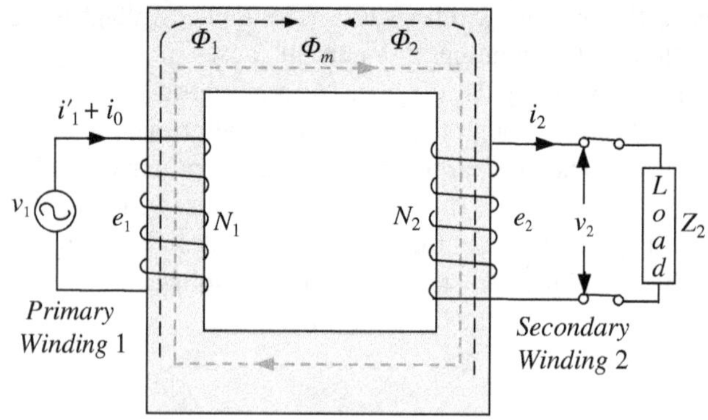

Figure 3.11 Loading effect in an ideal transformer.

The mmf produced from the load current of the primary current and the mmf produced due to secondary current are equal

$$N_1 i_1' = N_2 i_2 \tag{3.8}$$

The flux produced from the primary ϕ_1 due to additional load component of primary current cancels out the secondary flux ϕ_2 and thus the flux remain in the core is always the mutual flux ϕ_m as shown in Figure 3.11. In order to balance the mmf, the higher the current in the secondary, the higher the current will be drawn from the source in the primary.

The magnetizing component is ideally zero because of assumption of infinite permeability of the core and thus

$$i_1 = i_1' \tag{3.9}$$

The net mmf is zero (since it is an ideal transformer) because no mmf is required to produce flux and one car write the following relation

$$N_1 i_1 = N_2 i_2 \tag{3.10}$$

From this equation the ratio of currents is obtained as

$$\frac{i_1}{i_2} = \frac{N_2}{N_1} = \frac{1}{a} \tag{3.11}$$

The instantaneous power input to the transformer is equal to the instantaneous power output. This is valid because an ideal transformer is lossless and hence the following relation can be written

$$v_1 i_1 = v_2 i_2 \tag{3.12}$$

In terms of RMS quantities, the following relations can be written

$$V_1 I_1 = V_2 I_2$$
$$\frac{V_1}{V_2} = \frac{I_2}{I_1} = \frac{N_1}{N_2} = a \tag{3.13}$$

The turn ratio of the ideal transformer affects the magnitudes of the voltages and currents but not their angles.

Active or real power input in a transformer is given as

$$P_{in} = V_1 I_1 \cos(\theta_1) \tag{3.14}$$

Active or real power output in a transformer is given as

$$P_{out} = V_2 I_2 \cos(\theta_2) \tag{3.15}$$

Since voltage and current angles are un-affected by an ideal transformer, $\theta_1 = \theta_2$. Using Eq. (3.13), the following relation can be written

$$V_2 = \frac{V_1}{a}, \text{and } I_2 = aI_1 \tag{3.16}$$

The output Power can be written as

$$P_{out} = \frac{V_1}{a}(aI_1)\cos(\theta)$$

$$P_{out} = V_1 I_1 \cos(\theta) = P_{in} \tag{3.17}$$

The output real power of an ideal transformer is equal to its input power.

The reactive power input to the transformer is

$$Q_{in} = V_1 I_1 \sin(\theta_1) \tag{3.18}$$

The reactive power output from the transformer is

$$Q_{out} = V_2 I_2 \sin(\theta_2) \tag{3.19}$$

Using Eq. (3.16) and putting $\theta_1 = \theta_2$, the reactive power can be written as

$$Q_{in} = V_1 I_1 \sin(\theta) = V_2 I_2 \sin(\theta) = Q_{out} \tag{3.20}$$

The output reactive power of an ideal transformer is equal to its input reactive power.

3.4 Impedance Transformation

The impedance of the secondary side can be transferred or referred to the primary side and vice-versa. Consider an impedance connected to the secondary Z_2 given as

$$Z_2 = \frac{V_2}{I_2}$$

The primary side impedance can be obtained as

$$Z_1 = \frac{V_1}{I_1} = \frac{aV_2}{I_2/a} = a^2\frac{V_2}{I_2} = a^2 Z_2$$

$$Z_1 = a^2 Z_2 = Z_2' \tag{3.21}$$

An impedance Z_2 connected in the secondary will appear as an impedance Z_2' looking from the primary side. This is called secondary impedance referred to the primary side. This is illustrated in Figure 3.12. Thus, impedance can be transferred from secondary to primary if its value is multiplied by the square of the turn ratio.

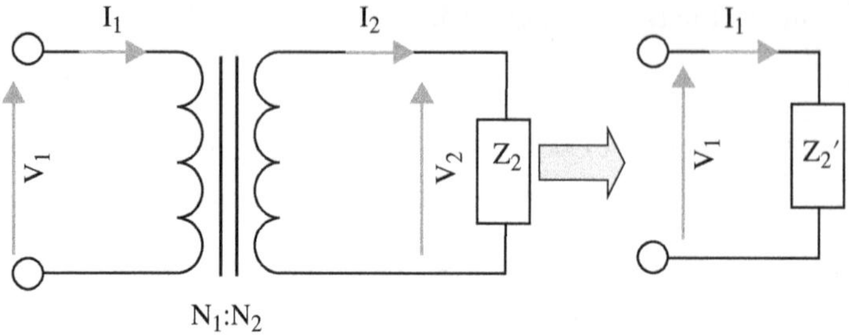

Figure 3.12 Secondary side impedance referred to the primary side.

An impedance can be transferred to the secondary side in the similar way (this is called primary impedance referred to the secondary side)

$$Z_1' = \frac{1}{a^2} Z_1 \tag{3.22}$$

Note:

The summary of transformations: From Eqs. (3.13), (3.16), (3.21), and (3.22), it can said that for an ideal transformer, voltage are transformed in the direct ratio of turns while current in inverse ratio of turns. The impedances in square of the turns and the powers remains unchanged.

3.5 DOT Convention

The voltage polarity and current directions in a transformer depend upon the way in which windings are wound. Since sense of windings turns are not seen from outside, it is common practice to put a dot at the terminals of the transformer as shown in Figure 3.13. The dotted terminals are positive at the same instant of time. If the primary voltage is ***positive*** at the dotted end of the winding, then the secondary voltage will also be ***positive*** at the dotted end. Voltage polarities are the same with respect to the dots on each side of the core. The currents entering dotted terminals produces flux in the same direction.

3.6 Real/Practical Transformer

The ideal transformer, discussed in the previous section, is merely to illustrate the fundamental concepts. However, no such transformer exists in reality. The real transformer has the following characteristics

- The primary and secondary windings are wound of copper wires and hence have resistances that can be shown as a lumped parameter in equivalent circuit.

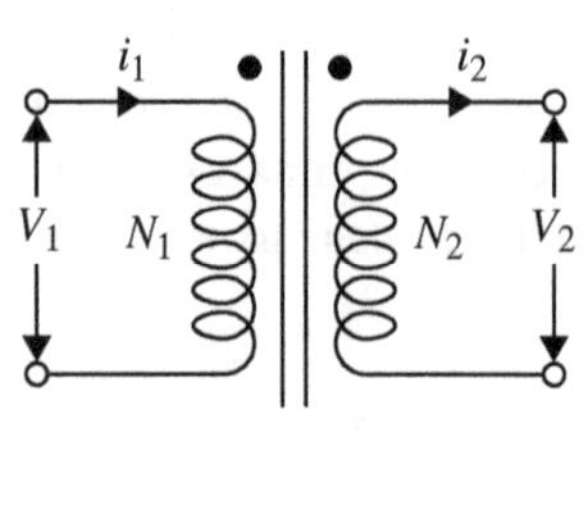

Figure 3.13 Transformer with dots shown on the terminals.

- The flux produced in the primary and secondary windings are not only confined to the core, but some the flux leaks and link with the same windings, called leakage flux.
- The leakage flux path is mostly in the air and hence it varies linearly with current.
- The effects of leakage flux is accounted by an inductance and hence called leakage reactance on both primary and secondary sides.

Transformer with leakage fluxes is shown in Figure 3.14.

In the core of a real transformer, two types of losses take place namely (i) Hysteresis loss, and (ii) Eddy current loss. These losses are due to cyclic magnetization of the core. These losses cause heating of the core and reduces the efficiency of the transformer. The hysteresis loss is minimized by using high-grade magnetic material while the eddy current loss is minimized by using laminated core.

The no-load or excitation current (I_0) has two components, (i) Magnetizing component (I_m), and (ii) Core loss component (I_c). The magnetizing component is responsible for production of mutual flux ϕ_m. The core loss component of current is responsible for production of core loss.

The core loss component of no-load current (I_c) is in phase with the induced emf. The magnetizing of current (I_m) is in phase with the mutual flux and lead I_c by 90°. The phasor sum of magnetizing current and core loss current is no-load current

$$\bar{I}_0 = \bar{I}_m + \bar{I}_c \tag{3.23}$$

The phasor diagram of a transformer under no-load condition is shown in Figure 3.15. Considering sinusoidal flux Eq. (3.3), the induced emf is co-sinusoidal Eq. (3.4). The flux is considered as reference then the induced emf leads by 90° as shown in Figure 3.15. The no-load current components are shown in the phasor diagram and they can be written as

$$I_c = I_0 \cos(\theta_0)$$
$$I_m = I_0 \sin(\theta_0)$$
$$I_0 = \sqrt{I_c^2 + I_m^2} \tag{3.24}$$

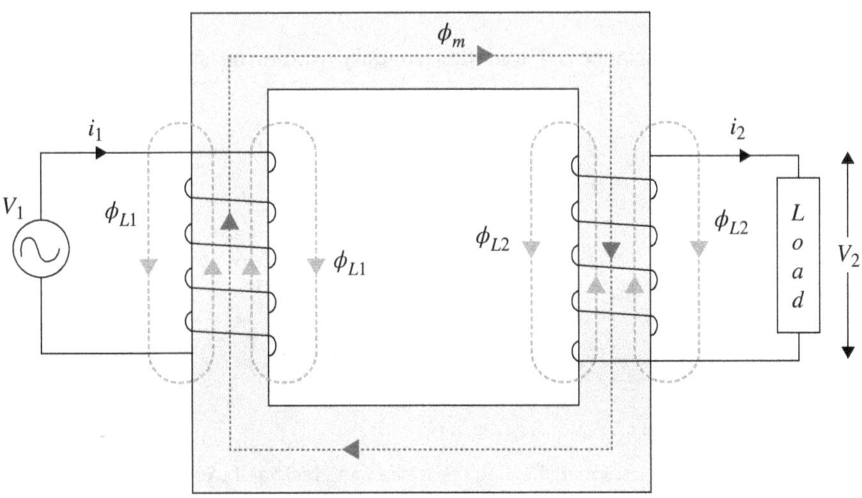

Figure 3.14 A real transformer showing leakage fluxes.

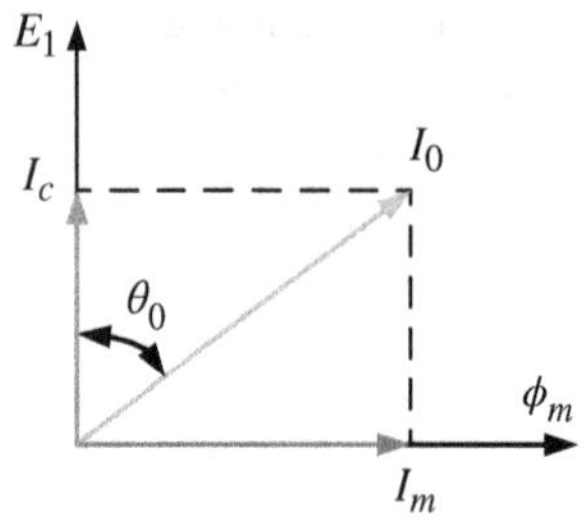

Figure 3.15 No-load phasor diagram of a single-phase transformer.

Example 3.1 A single-phase transformer has core loss component of no-load current I_c of 3 A and a magnetizing component of no-load current I_m of 6 A. Compute the no-load current, I_o and its power factor.

Solution

$$I_0 = \sqrt{I_c^2 + I_m^2} = \sqrt{(3)^2 + (6)^2} = 6.708A$$

$$I_c = I_0 \cos(\theta_0)$$

$$p.f. = \cos(\theta_0) = \frac{I_c}{I_0} = \frac{3}{6.708} = 0.447$$

3.7 Equivalent Circuit of a Single-Phase Transformer

The analysis of a transformer can be carried out by using an equivalent circuit. Equivalent circuit is derived considering the following

- The primary and secondary windings have finite resistances considered as lumped parameters.
- The leakage fluxes are modelled as leakage reactances in the equivalent circuit.
- The core-loss component of current is modelled using a shunt resistance.
- The magnetization of the core is modelled using a magnetizing reactance as a shunt branch.

The primary winding resistance is R_1 and the secondary winding resistance is R_2 as shown in Figure 3.16.

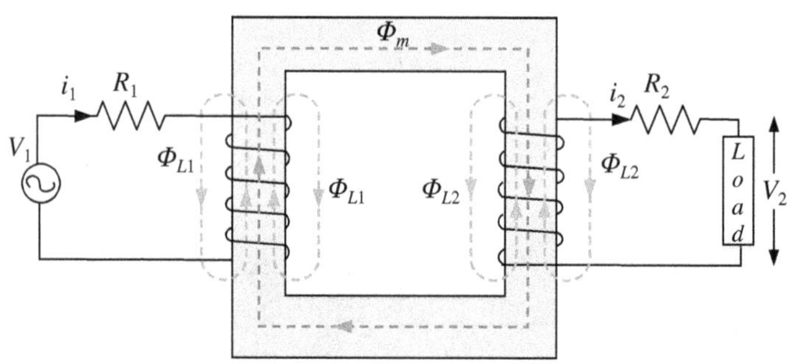

Figure 3.16 Single-phase transformer showing resistances and leakage fluxes.

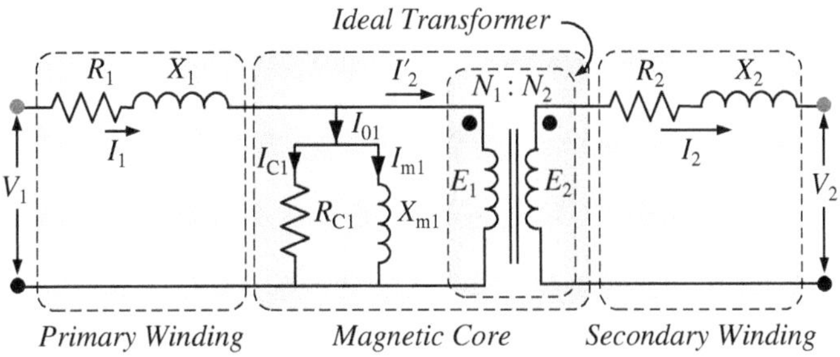

Figure 3.17 Equivalent circuit showing resistance and leakage reactance.

The first step in developing equivalent circuit is shown in Figure 3.17. The effect of leakage flux is shown as leakage reactance

$$X_{l1} = X_1 = 2\pi f L_{l1} \tag{3.25}$$

$$X_{l2} = X_2 = 2\pi f L_{l2}$$

The suffix 1 refers to the primary side parameters and suffix 2 represents secondary side quantities. In the equivalent circuit, the shunt branch resistance R_{c1} represent core losses, X_{m1} represent core magnetization. The currents and voltages are

I_1 – Primary side current
I_0- No-load/Excitation current component of primary current
I_{c1} – Core-loss component of no-load current
I_{m1} – Magnetizing component of no-load current
I_2' – Load component of primary current
I_2 – Secondary side current (load current)
V_1 – Applied primary side voltage
E_1 – Primary side induced emf
E_2 – Secondary side induced emf
V_2 – Secondary side terminal voltage

Applying KVL in the primary side of equivalent circuit of Figure 3.17.

$$\overline{V}_1 = \overline{E}_1 + \overline{I}_1 \ (R_1 + jX_1) \tag{3.26}$$

Applying KVL in the secondary side of equivalent circuit of Figure 3.17.

$$\overline{E}_2 = \left[\overline{V}_2 + \overline{I}_2 \left(R_2 + jX_2\right)\right] \tag{3.27}$$

$$\text{Since } \frac{V_1}{V_2} = \frac{E_1}{E_2} = \frac{I_2}{I_1} = a$$

Replacing E_1 with aE_2 in Eq. (3.26), the following can be written

$$\overline{V}_1 = \overline{E}_2 a + \overline{I}_1 \ (R_1 + jX_1) \tag{3.28}$$

Now substituting Eq. (3.27) in Eq. (3.28), the following is obtained

$$\overline{V}_1 = \left[\overline{V}_2 + \overline{I}_2 \left(R_2 + jX_2\right)\right] a + \overline{I}_1 \left(R_1 + jX_1\right) \tag{3.29}$$

$$\textit{but } I_2 = I_2' a$$

Hence Eq. (3.29) can be modified as

$$\overline{V}_1 = \left[\overline{V}_2 a + I_2'(a)^2 \left(R_2 + jX_2\right)\right] + \overline{I}_1 \left(R_1 + jX_1\right) \tag{3.30}$$

Let

$$R_2' = R_2(a)^2 \qquad \text{and} \qquad X_2' = X_2(a)^2 \tag{3.31}$$

Where R_2' is called secondary resistance referred to the primary side and X_2' is called secondary leakage reactance referred to the primary side.

$$\overline{V}_1 = \left[\overline{V}_2' + I_2' \left(R_2' + jX_2'\right)\right] + \overline{I}_1 \left(R_1 + jX_1\right) \tag{3.32}$$

Where

$V_2' = aV_2$ is called secondary voltage referred to the primary side
$I_2' = I_2/a$ is called secondary current referred to the primary side

Equation (3.32) can be written as;

$$\overline{V}_1 = \overline{I}_1 \left(R_1 + jX_1\right) + I_2' \left(R_2' + jX_2'\right) + \overline{V}_2' \tag{3.33}$$

The first term in the right side of equation is voltage drop in the primary side, the second term is the voltage drop on the secondary side while the third term is the secondary voltage referred to the primary side.

Replacing the secondary side quantities with their referred quantities to the primary side (Table 3.1), the Figure 3.17 can be drawn as Figure 3.18. The transformation relationship is summarized in Table 3.1.

Table 3.1 Referred quantities.

Referred to the primary	Referred to the secondary
$E_1 = E_2' = aE_2$	$E_1' = E_2 = E_1/a$
$V_2' = aV_2$	$V_1' = V_1/a$
$I_2' = I_2/a$	$I_1' = aI_1$
$X_2' = a^2X_2$	$X_1' = X_1/a^2$
$R_2' = a^2R_2$	$R_1' = R_1/a^2$

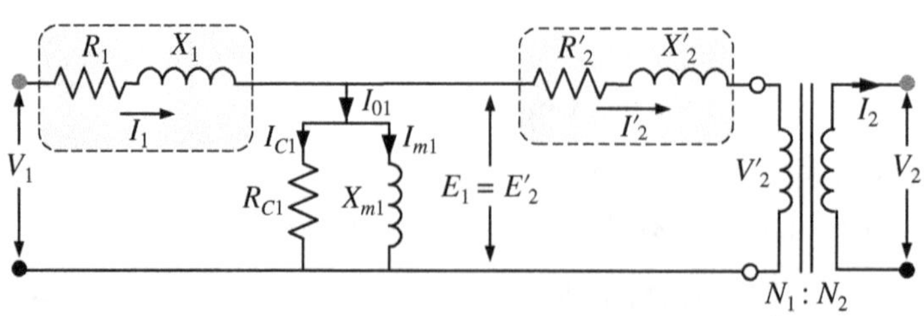

Figure 3.18 Modified equivalent circuit of a single-phase transformer.

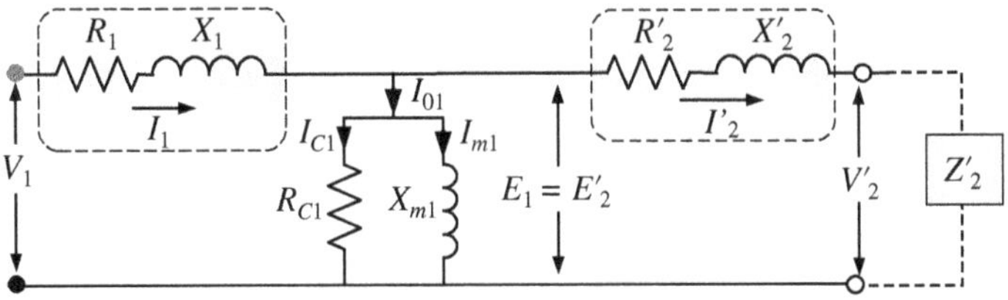

Figure 3.19 Exact equivalent circuit referred to the primary side.

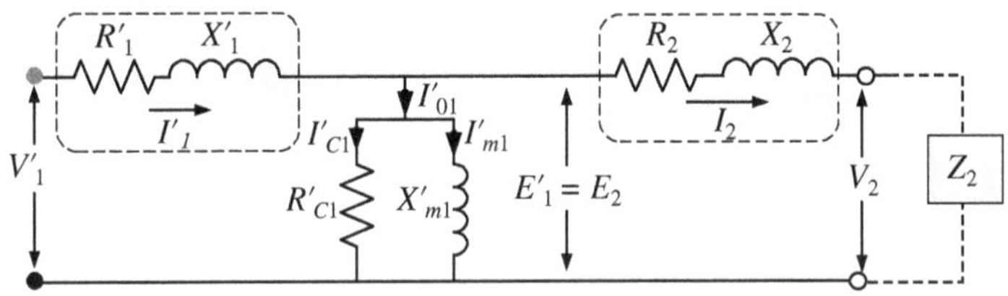

Figure 3.20 Exact equivalent circuit referred to the secondary side.

The ideal transformer turns can be removed and the equivalent circuit can be drawn as shown in Figure 3.19, called exact equivalent circuit referred to the primary side.

Similarly, equivalent circuit can be obtained referred to the secondary side as shown in Figure 3.20.

The equivalent circuit can be simplified by assuming small voltage drop across the primary impedance and $V_1 = E_1$. If the applied voltage and the induced emf are the same then the shunt branch can be moved across the source voltage and the approximate equivalent circuit is drawn as shown in Figure 3.21.

From the equivalent circuit the following relations are obtained

$$R_{eq1} = R_1 + R_2' = R_1 + a^2 R_2 \tag{3.34}$$

$$X_{eq1} = X_1 + X_2' = X_1 + a^2 X_2 \tag{3.35}$$

Where R_{eq1} is called equivalent resistance referred to the primary side and X_{eq1} is equivalent leakage reactance referred to the primary side.

Similarly, the following relations are obtained

$$R_{eq2} = R_2 + R_1' = R_2 + \frac{1}{a^2} R_1 \tag{3.36}$$

$$X_{eq2} = X_2 + X_1' = X_2 + \frac{1}{a^2} X_1 \tag{3.37}$$

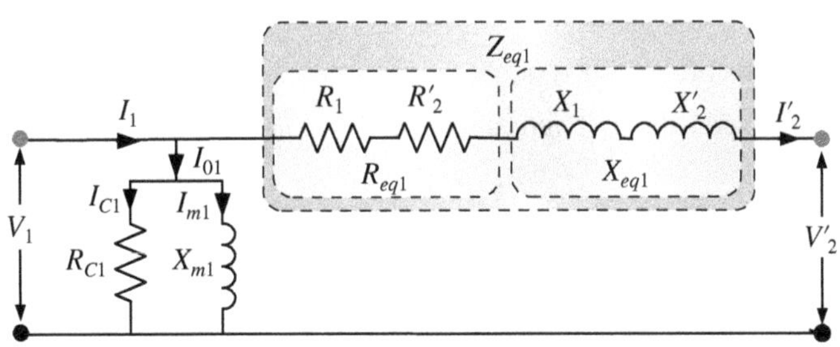

Figure 3.21 Approximate equivalent circuit referred to the primary side.

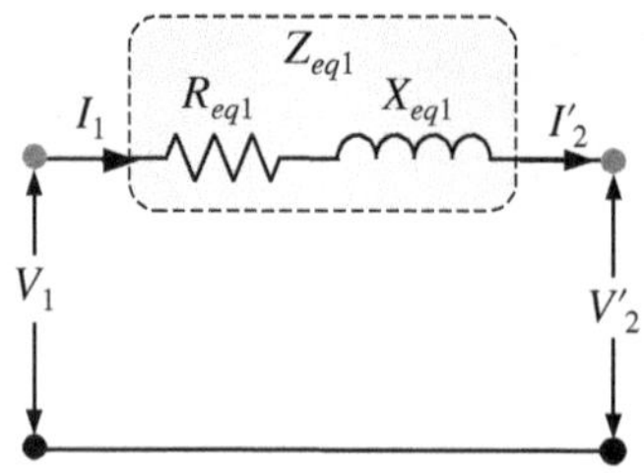

Figure 3.22 Approximate equivalent circuit of a single-phase transformer referred to the primary side.

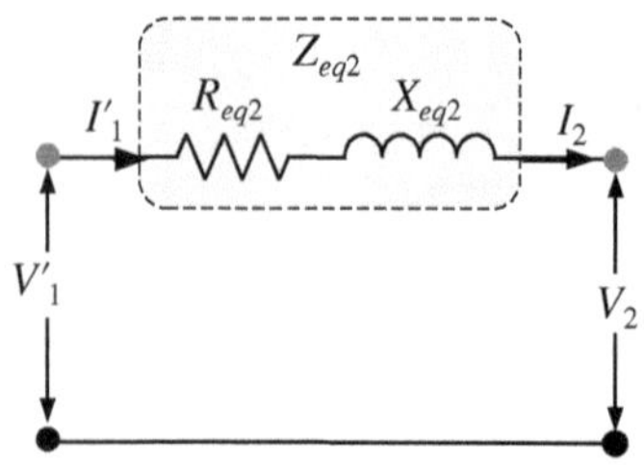

Figure 3.23 Approximate equivalent circuit of a single-phase transformer referred to the secondary side.

Where R_{eq2} is called equivalent resistance referred to the secondary side and X_{eq2} is equivalent leakage reactance referred to the secondary side.

Further approximation in the equivalent circuit is done. The no load current in a transformer is very small of the order of 2–5% of the rated current. Hence can be neglected and thus the shunt branch can be removed from the equivalent circuit of Figure 3.21. The final approximate equivalent circuit of a single-phase transformer is given as shown in Figures 3.22 and 3.23.

Example 3.2 Derive an expression for total resistive loss in a two-winding transformer in terms of equivalent resistances (a) referred to the primary side, and (b) referred to the secondary side.

Solution

Total Resistive loss in a two winding transformer is given as:

$$P_{resistive} = P_{cu} = I_1^2 R_1 + I_2^2 R_2$$

Resistive Loss is also called Copper Loss

Replacing I_2 with I_1, the experssion is given as;

$$\frac{I_2}{I_1} = a; I_2 = aI_a$$

$$P_{cu} = I_1^2 R_1 + (a)^2 I_1^2 R_2 = I_1^2 \left(R_1 + (a)^2 R_2\right) = I_1^2 R_{eq1}$$

Resistive loss in terms of secondary current;

Replacing I_1 with I_2, the experssion is given as;

$$\frac{I_2}{I_1} = a; I_1 = \frac{I_2}{a}$$

$$P_{cu} = \left(\frac{I_2}{a}\right)^2 R_1 + I_2^2 R_2 = I_2^2 \left(\left(\frac{1}{a^2}\right) R_1 + R_2\right) = I_2^2 R_{eq2}$$

Example 3.3 A 50 kVA, 11 kV/440 V, 50 Hz, single-phase transformer has the following data

Primary side: $R_1 = 2.8\,\Omega$, $X_1 = 6.9\,\Omega$
Secondary side: $R_2 = 0.005\,\Omega$, $X_2 = 0.010\,\Omega$

Calculate the following:

(a) Equivalent resistance and leakage reactance referred to the primary side
(b) Equivalent resistance and leakage reactance referred to the primary side
(c) Total resistive or copper loss
(d) Given the short-circuit current of the secondary side as 200 A, compute the voltage applied from the primary side (HV side).

Solution

(a) The equivalent resistance referred to the primary side when secondary is short circuited:

$$a = \frac{11000}{440} = 25$$
$$R_1 = 2.8\Omega, R_2 = 0.005\Omega$$
$$R_{eq1} = R_1 + (a)^2 R_2 = 2.8 + (25)^2 0.005 = 5.925\Omega$$

Equivalent leakage reactance referred to the primary side:

$$X_1 = 6.9\Omega, X_2 = 0.01\Omega$$
$$X_{eq1} = X_1 + (a)^2 X_2 = 6.9 + (25)^2 0.01 = 13.15\Omega$$

(b) The equivalent resistance referred to the secondary side:

$$R_1 = 2.8\Omega, R_2 = 0.005\Omega$$
$$R_{eq2} = R_2 + \left(\frac{1}{a}\right)^2 R_1 = 0.005 + \left(\frac{1}{25}\right)^2 2.8 = 0.00948\Omega$$

Equivalent leakage reactance referred to the primary side:

$$X_1 = 6.9\Omega, X_2 = 0.01\Omega$$
$$X_{eq2} = X_2 + \left(\frac{1}{a}\right)^2 X_1 = 0.01 + \left(\frac{1}{25}\right)^2 6.9 = 0.02104\Omega$$

(c) Total resistive loss

$$I_1 = \frac{S}{V_1} = \frac{50000}{11000} = 4.545A$$
$$I_2 = \frac{S}{V_2} = \frac{50000}{440} = 113.636A$$
$$P_{cu} = I_1^2 R_{eq1} = I_2^2 R_{eq2} = (4.545)^2 5.925 = 122.417W$$

or

$$(113.636)^2 0.00948 = 122.417W$$

(d) The equivalent circuit referred to the HV (primary) side is shown (Figure E3.3)

Figure E3.3 Equivalent circuit referred to the primary side when the secondary is short circuited.

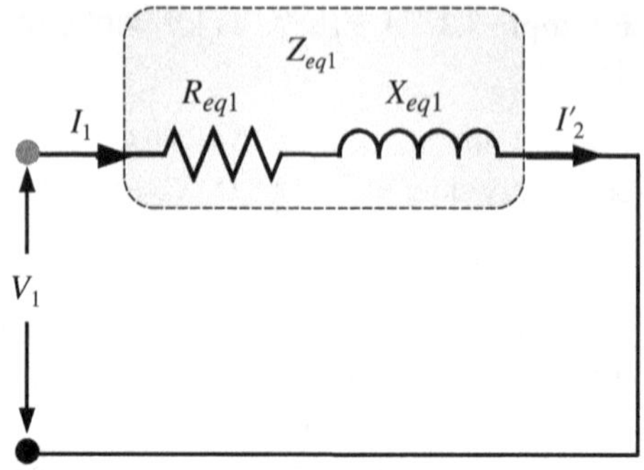

Given the secondary short circuit current = 200 A
The corresponding primary current $I_1 = \frac{I_2}{a} = \frac{200}{25} = 8A$
Voltage applied on primary side

$$V_1 = I_1 Z_{eq1} = I_1 \left(R_{eq1} + jX_{eq1}\right) = 4.545\,(5.925 + j13.15)$$
$$V_1 = 4.545\,(14.423) = 65.55V$$

3.8 Phasor Diagrams Under Load Condition

Considering the equivalent circuit referred to the primary side supplying load impedance Z_2 in Figure 3.24. Applying KVL in the circuit, the following voltage relation is written

$$\overline{V}_1 = \overline{V}_2' + \overline{I}_2' R_{eq1} + j\overline{I}_2' X_{eq1} \tag{3.38}$$

Phasor diagrams are drawn from Eq. (3.38). Assuming V_2' as references, the position of current depends upon the load power factor. For lagging power factor load the load current (I_2') lags the load terminal voltage (V_2'). The resistive voltage drop $I_2' R_{eq1}$ is in same phase as that of the load current I_2' and the reactive voltage drop $I_2' X_{eq2}'$ leads the voltage by 90°. The sum of these voltages is the applied voltage V_1 as shown in Figure 3.25. The power factor angle is θ.

The expression for V_1 is derived from the phasor diagram of Figure 3.26.

$$ab = I_2' R_{eq1} \cos(\theta)$$

$$bd = cf = I_2' R_{eq1} \sin(\theta)$$

$$bc = df = I_2' X_{eq1} \sin(\theta)$$

$$ef = I_2' X_{eq1} \cos(\theta)$$

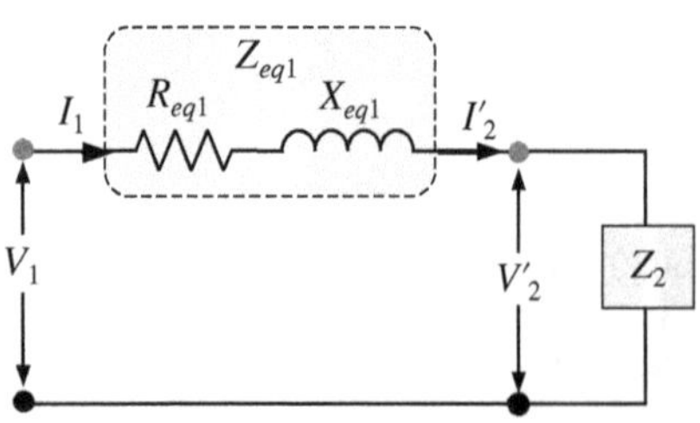

Figure 3.24 Equivalent circuit referred to the primary side when secondary is connected to load.

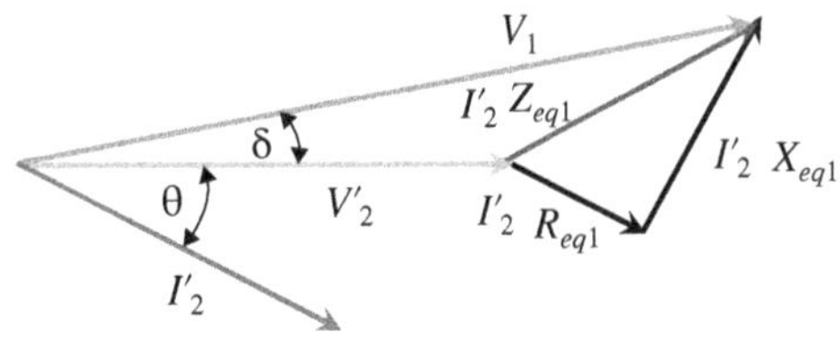

Figure 3.25 Phasor diagram under loaded condition with lagging power factor current.

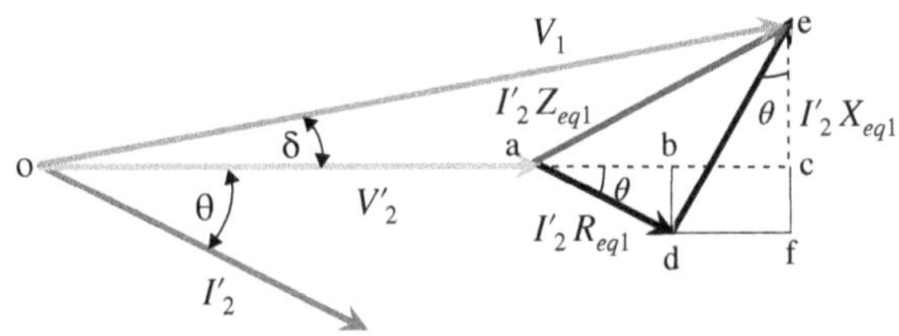

Figure 3.26 Derivation of equation for V1 using phasor diagram under lagging condition.

$$ec = ef - cf = I_2' X_{eq1} \cos(\theta) - I_2' R_{eq1} \sin(\theta)$$

$$oc = oa + ab + bc = V_2' + I_2' R_{eq1} \cos(\theta) + I_2' X_{eq1} \sin(\theta)$$

From triangle oec

$$oe = \sqrt{oc^2 + ec^2}$$

$$V_1 = \sqrt{\left(V_2' + I_2' R_{eq1} \cos(\theta) + I_2' X_{eq1} \sin(\theta)\right)^2 + \left(I_2' X_{eq1} \sin(\theta) - I_2' R_{eq1} \cos(\theta)\right)^2} \tag{3.39}$$

Since power factor angle θ is small about 5°. Thus, the approximate value of the primary side voltage is given as;

$$oe \cong oc = oa + ab + bc = V_1 \cong V_2' + I_2' R_{eq1} \cos(\theta) + I_2' X_{eq1} \sin(\theta)$$

$$V_1 = \sqrt{V_2' + I_2' R_{eq1} \cos(\theta) + I_2' X_{eq1} \sin(\theta)}$$

Considering leading power factor load, the phasor diagram is given as shown in Figure 3.27. Considering V_2' as reference, locate the position of load current I_2'. The resistive voltage drop has the same phase as that of load current and hence drawn in parallel to the load current. The reactive voltage drop is 90° leading the current. The sum of the load voltage, resistive voltage drop and reactive voltage drop gives the primary side applied voltage V_1.

The expression for V_1 can be obtained from the phasor diagram

$$\overline{V}_1 = \overline{V}_2' + \overline{I}_2' \overline{Z}_2' = \overline{V}_2' + \overline{I}_2' \left(R_{eq2} + jX_{eq2}\right)$$

$$\overline{V}_1 = V_2 + j0 + \left(I_2' \cos(\theta) + jI_2' \sin(\theta)\right)\left(R_{eq2} + jX_{eq2}\right)$$

$$= V_2 + I_2' \cos(\theta) R_{eq2} + jI_2' \cos(\theta) X_{eq2} + jI_2' \sin(\theta) R_{eq2} + j^2 I_2' \sin(\theta) X_{eq2}$$

$$= V_2 + \left(I_2' \cos(\theta) R_{eq2} - I_2' \sin(\theta) X_{eq2}\right) + j\left(I_2' \cos(\theta) X_{eq2} + I_2' \sin(\theta) R_{eq2}\right)$$

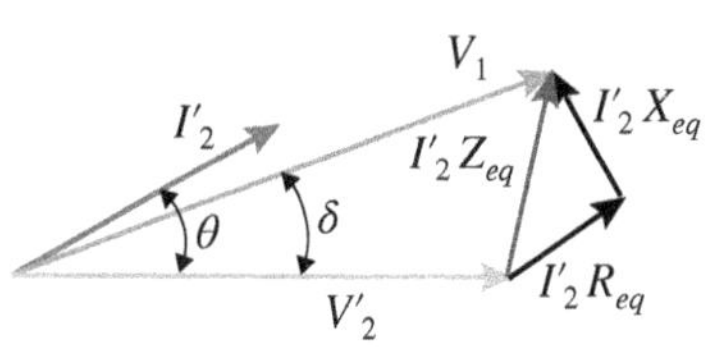

Figure 3.27 Phasor diagram for leading power factor load current.

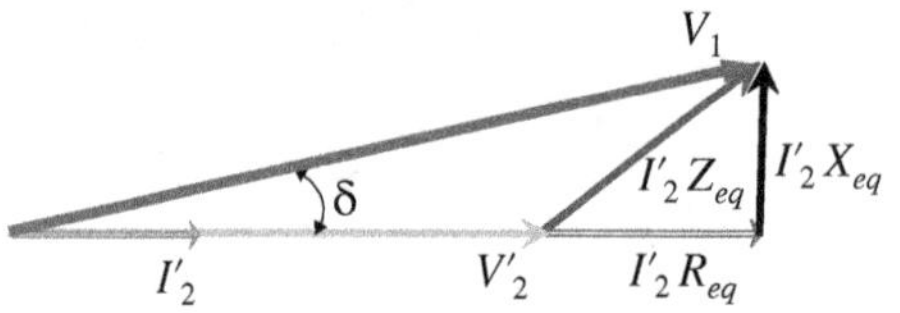

Figure 3.28 Phasor diagram for unity power factor load.

The imaginary terms are negligible compared to the real part and can be neglected, and the voltage can be written as

$$|V_1| \approx V_2 + \left(I_2' \cos(\theta) R_{eq2} - I_2' \sin(\theta) X_{eq2}\right) \tag{3.40}$$

The voltage expression can be derived by putting $\cos(\theta) = 1; \sin(\theta) = 0$ for unity power factor load as shown in Figure 3.28.

$$\overline{V}_1 = V_2 + \left(I_2' \cos(\theta) R_{eq2} - I_2' \sin(\theta) X_{eq2}\right) + j\left(I_2' \cos(\theta) X_{eq2} + I_2' \sin(\theta) R_{eq2}\right)$$

$$\overline{V}_1 = V_2 + I_2' R_{eq2} + jI_2' X_{eq2}$$

$$|V_1| = \sqrt{\left(V_2 + I_2' R_{eq2}\right)^2 + \left(I_2' X_{eq2}\right)^2} \tag{3.41}$$

Example 3.4 A 50 kVA, 11 kV/440 V, 50 Hz, single-phase transformer has the following data

Primary side: $R_1 = 2.8\,\Omega, X_1 = 6.9\,\Omega$
Secondary side: $R_2 = 0.005\,\Omega, X_2 = 0.010\,\Omega$

Calculate the secondary terminal voltage when the

(a) The low voltage winding is connected to a load impedance of $8 + j3\,\Omega$
(b) The transformer supplies its rated current at 0.85 pf lagging on the low voltage side.

In both cases, the primary voltage is kept constant at 11 kV.

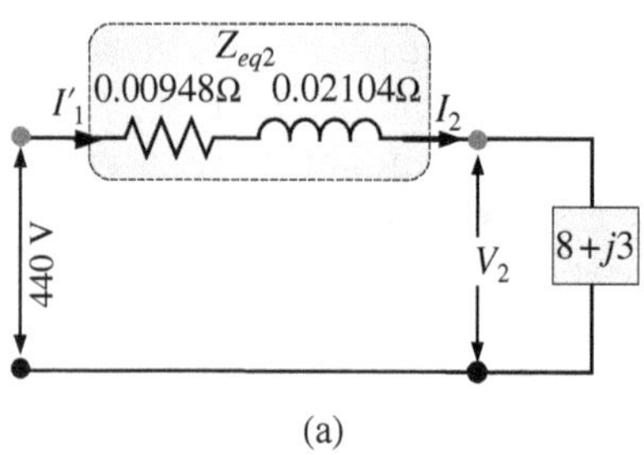

Figure E3.4a Equivalent circuit with known load impedance.

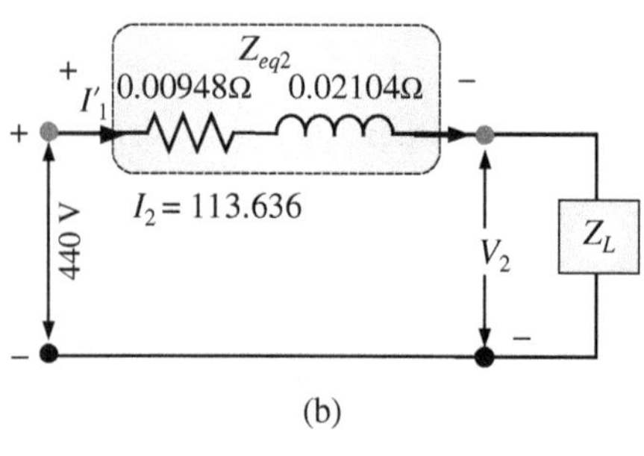

Figure E3.4b Equivalent circuit with unknown load impedance.

Solution

(a) The transformer equivalent circuit referred to the secondary side is shown in Figures E3.4a and E3.4b.

$$V_2' = \frac{V_1}{a} = \frac{11000}{25} = 440V$$

Total Impedance of the circuit:

$$\overline{Z} = (8 + 0.00948) + j\,(0.02104 + 3) = 8.00948 + j3.02104$$

$$\overline{Z} = 8.56\angle 20.66$$

$$\overline{I_2} = \frac{V_2'}{\overline{Z}} = \frac{440\angle 0}{8.56\angle 20.66} = 51.4\angle -20.66$$

The secondary terminal voltage is;

$$\overline{V_2} = \overline{I_2 Z_L} = 51.4\angle -20.66\,(8 + j3) = 51.4\angle -20.66\,(8.544\angle 20.55)$$

$$\overline{V_2} = 439.1616\angle -0.11$$

$$|V_2| = 439.1616V$$

(b) The rated current of the transformer is $I_2 = \frac{50000}{440} = 113.636A$

Applying KVL in the transformer circuit

$$pf = 0.85$$

$$\theta = \cos^{-1}(0.85) = 31.78$$

$$\overline{V_2} = \overline{V_1} - \overline{I_2 Z_{eq2}} = 440\angle 0 - 113.636\angle -31.78$$

$$= 440 + j0 - (113.636\cos(31.78) + j113.636\sin(-31.78))$$

$$= 440 - 96.6 + j59.85 = 343.4 + j59.85$$

$$= 348.57\angle 9.88$$

Example 3.5 The equivalent circuit parameters of a single-phase 240/2400, 50 Hz, transformer are $R_c = 600\,\Omega$, $X_m = 300\,\Omega$, $R_{eq1} = 0.25\,\Omega$, $X_{eq1} = 0.75\,\Omega$. The transformer is supplying a load of $400 + j200\,\Omega$. Keeping the primary voltage of 240 V, calculate the

(a) The secondary terminal voltage
(b) Current in the primary winding
(c) Power factor of the primary side
(d) Power output
(e) Power Input

Solution

The equivalent circuit with the parameter values are shown in Figure E3.5.

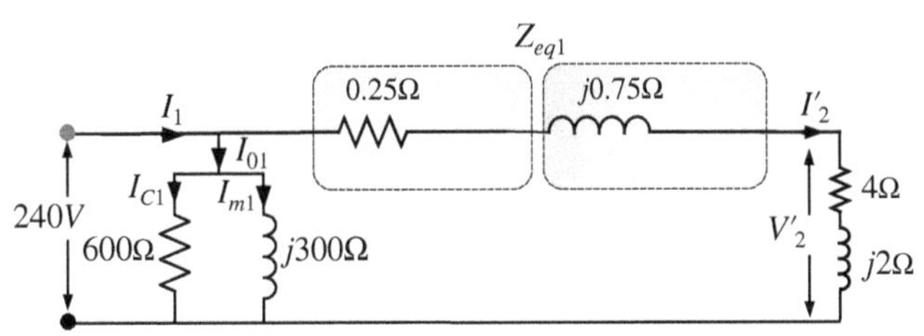

Figure E3.5

Since the equivalent circuit is referred to the low voltage (primary side), the load impedance is also transformed to the lv side.

$$a = \frac{V_1}{V_2} = \frac{240}{2400} = 0.1$$

$$\overline{Z}_L' = Z_L(a)^2 = (400 + j200)\,(0.1)^2 = 4 + j2$$

Total impedance in series

$$\overline{Z} = 0.25 + j0.75 + 4 + j2 = 4.25 + j2.75 = 5.062\angle 32.9$$

Assuming the primary voltage as reference

$$\overline{V}_1 = 240\angle 0$$

The current in the primary side

$$\overline{I_2'} = \frac{\overline{V_1}}{\overline{Z}} = \frac{240\angle 0}{5.062\angle 32.9} = 47.412\angle -32.9 = 39.8 - j25.753$$

(a) Secondary terminal voltage

$$V_2' = I_2' Z_L = 47.412\left(\sqrt{(4)^2 + (2)^2}\right) = 212.03V$$

(b) Primary current:

The core loss component of current $\overline{I}_{c1} = \frac{\overline{V}_1}{R_{c1}} = \frac{240\angle 0}{600} = 0.4\angle 0$

The magnetizing component of current $\overline{I}_{m1} = \frac{\overline{V}_1}{jX_{m1}} = \frac{240\angle 0}{j300} = 0.8\angle -90$

The no-load current $\overline{I}_{o1} = \overline{I}_{c1} + \overline{I}_{m1} = 0.4 - j0.8$

The primary current

$$\overline{I}_1 = \overline{I_2'} + \overline{I}_{o1} = 39.8 - j25.753 + 0.4 - j0.8 = 40.2 - j26.553 = 48.178\angle -33.44$$

(c) Power factor of the primary current $pf = \cos(33.44) = 0.834$ lagging

(d) Power output

$$P_{out} = V_2' I_2' \cos\left(\theta_2\right)$$
$$\cos\left(\theta_2\right) = \cos(32.9) = 0.839$$
$$P_{out} = 212.03\,(47.412)\,0.839 = 8.434kW$$

(e) Power Input

3.9 Testing of Transformer

Transformers are tested to determine the equivalent circuit parameters. Two tests are carried out are:

- Open-circuit Test, and
- Short-Circuit Test.

3.9.1 Open-Circuit Test

Open-Circuit Test is done to determine Shunt Branch parameters of the equivalent circuit:

- Secondary is open circuited
- Rated Voltage is applied to the Primary
- Voltage, Current, and Power are recorded.

Since the secondary is open circuited, the current through the primary winding is the 'no-load current'. Neglecting the small voltage drop across the series branch of equivalent circuit, the no-load current provides the magnetization of the core and accounts for the core/iron loss in the core of the transformer. Neglecting the small copper loss in the primary winding (since the current is very small only 2–5% of the rated value), the entire loss is the core loss in the transformer. The wattmeter measures the core loss.

The circuit diagram for conducting an open-circuit test is shown in Figure 3.29.

The equivalent circuit under open-circuit condition is only a shunt branch as shown in Figure 3.30. Since the test is conducted on the primary side, the shunt-branch parameters obtained are referred to the primary side.

To compute the parameters R_{c1} and X_{m1}, the following relations are used

$$Y_1 = \frac{1}{R_{c1}} + \frac{1}{jX_{m1}} = \frac{1}{R_{c1}} - j\frac{1}{X_{m1}}$$

$$Y_1 = \frac{I_{01}}{V_{oc}} \angle -\theta_0$$

$$\cos(\theta_0) = \frac{P_{oc}}{V_1 I_{01}} \tag{3.42}$$

The current components are computed as

$$I_{c1} = I_{01} \cos(\theta_0)$$
$$I_{m1} = I_{01} \sin(\theta_0) \tag{3.43}$$

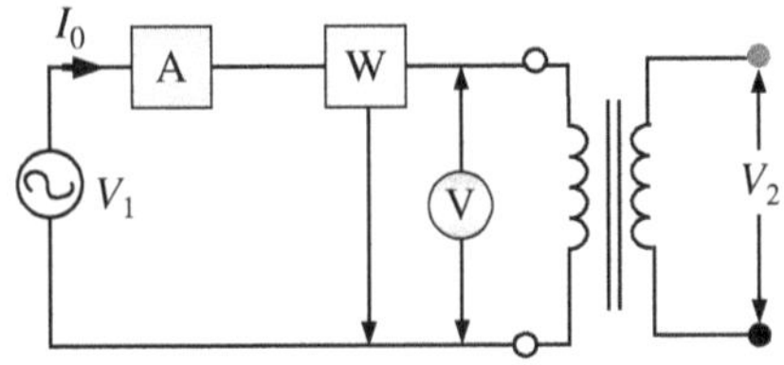

Figure 3.29 Circuit diagram for open-circuit test.

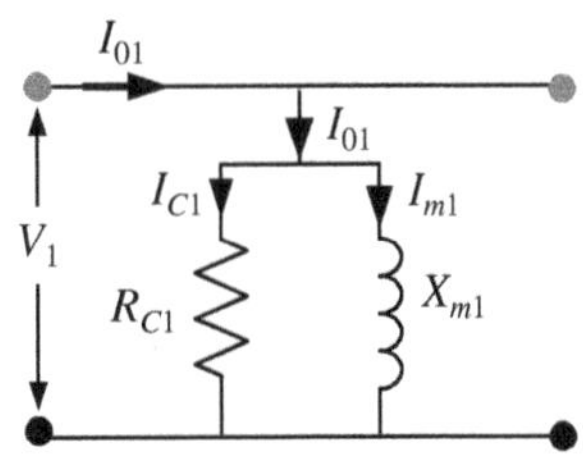

Figure 3.30 Equivalent circuit under open-circuit condition.

- The power angle is lagging (current lags behind voltage) because the circuit is inductive.
- The real part inverse of admittance is core-loss and an imaginary part is magnetizing reactance.

The alternative expressions to compute the shunt-branch parameters are given as

Step-1: Calculate the no-load power factor

$$P_{oc} = I_{oc}V_{oc}\cos(\theta_0)$$

$$\cos(\theta_0) = \frac{P_{oc}}{I_{oc}V_{oc}}$$

Step-2: Calculate the core-loss and magnetizing components of current

$$I_{c1} = I_{oc}\cos(\theta_0)$$

$$I_{m1} = I_{oc}\sin(\theta_0)$$

Step-3: Calculate the core-loss resistance and the magnetizing reactance

$$R_{c1} = \frac{V_1}{I_{C1}} \tag{3.44}$$

$$X_{m1} = \frac{V_1}{I_{m1}} \tag{3.45}$$

Note: The OCT is generally conducted on low-voltage side i.e. high-voltage side is open circuited and voltage is applied to the low voltage and metres are connected also on the low-voltage side. This is because low-voltage side is easily available in the lab for testing purposes.

3.9.2 Short-Circuit Test

Short-circuit test is done to determine the series branch parameters of the equivalent circuit

- Secondary is short circuited
- Reduced voltage is applied so that rated current flows
- Voltage, current, and Power are recorded.

In a short circuit test, reduced voltage is applied to the primary side, keeping the secondary short circuited. Since the secondary is short circuit, small reduced voltage is enough to flow rated current.

The applied voltage is about 30–40% of the rated voltage to obtain rated current. The core loss and the magnetization depend upon the applied voltage and since the applied voltage is small, the shunt branch is neglected. The equivalent circuit only contain the series branch parameters.

The circuit diagram for conducting a short-circuit test is shown in Figure 3.31.

The equivalent circuit under short-circuit condition is only series branch as shown in Figure 3.32. Since the test is conducted on the primary side the series branch parameters obtained is referred to the primary side.

The series parameters are computed as

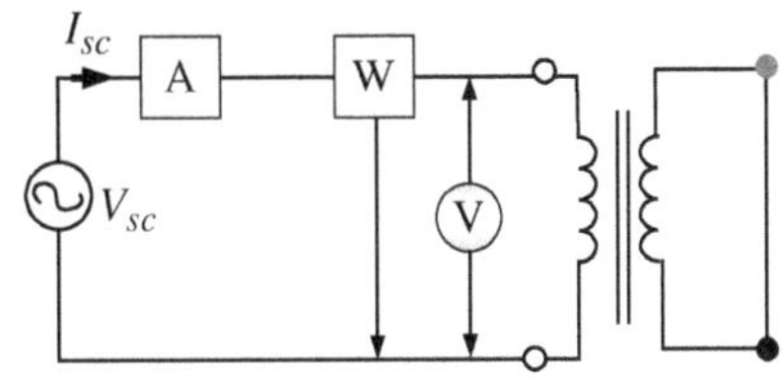

Figure 3.31 Circuit diagram for short-circuit test.

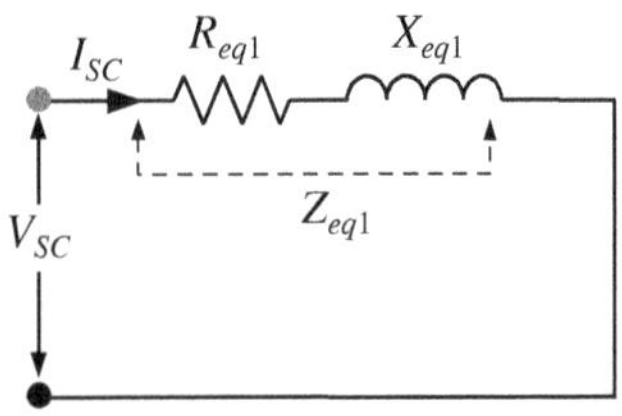

Figure 3.32 Equivalent circuit under short circuit condition.

Step 1: Compute power factor under short-circuit condition and the power factor angle

$$P_{sc} = I_{sc}V_{sc}\cos\left(\theta_{sc}\right)$$

$$\cos\left(\theta_{sc}\right) = \frac{P_{sc}}{I_{sc}V_{sc}}$$

$$\theta_{sc} = \cos^{-1}\left(\frac{P_{sc}}{I_{sc}V_{sc}}\right)$$

Step-2: Calculate the impedance

$$Z_{eq1} = \frac{V_{sc}\angle 0^0}{I_{sc}\angle - \theta_{sc}} = \left|\frac{V_{sc}}{I_{sc}}\right|\angle\theta_{sc}$$

Step-3: Calculate the equivalent resistance as the real part and equivalent leakage reactance as the imaginary part of the impedance

$$Z_{eq1} = R_{eq1} + jX_{eq1} \tag{3.46}$$

Alternative formulae to compute the series branch parameters is given as

$$R_{eq} = \frac{P_{sc}}{I_{sc}^2} \tag{3.47}$$

$$Z_{eq} = \frac{V_{sc}}{I_{sc}}$$

$$X_{eq} = \sqrt{Z_{eq}^2 - R_{eq}^2} \tag{3.48}$$

Note: The SCT is generally conducted on high-voltage side i.e. low-voltage side is short circuited and a reduced voltage is applied to the high voltage and metres are also connected on the high-voltage side. This is because on the high-voltage side the current is low and such a source is easily available in the lab for the testing purposes.

Example 3.6 Tests are conducted on a 50 kVA, 2400/240 V, single-phase 50 Hz transformer. The test results are given as

Open circuit test-240 V, 2.2 A, 340 W
Short circuit test-110 V, 20.8 A, 950 W

Calculate the approximate equivalent parameters of the transformer referred to the high voltage and low voltage sides.

Solution

Open-circuit test is conducted on lv side. All the quantities are thus referred to the lv side. The shunt-branch parameters are calculated as

$$\cos(\theta_0) = \frac{P_{oc}}{V_1 I_{o1}} = \frac{340}{240\,(2.2)} = 0.644$$

$$\theta_0 = \cos^{-1}(0.644) = 49.91^0$$

$$Y_2 = \frac{I_{o1}}{V_{oc}}\angle -\theta_0 = \frac{2.2}{240}\angle -49.91 = 0.00916\cos(49.91) - j0.00916\sin(49.91)$$

$$= 0.005899 - j0.007 = \frac{1}{R_{c2}} - j\frac{1}{X_{m2}}$$

$$R_{c2} = \frac{1}{0.005899} = 169.52\Omega; X_{m2} = \frac{1}{0.007} = 142.85\Omega$$

Shunt-branch parameters referred to the hv side are obtained by multiplying the quantities by square of the turn ratio

$$R_{c1} = (a)^2 R_{c2} = (10)^2 169.52 = 16.952k\Omega$$

$$X_{m1} = (a)^2 X_{m2} = (10)^2 142.85 = 14.285k\Omega$$

A short-circuit test is conducted on hv side and hence all quantities are referred to the hv side.

$$P_{sc} = I_{sc} V_{sc} \cos(\theta_{sc})$$

$$\cos(\theta_{sc}) = \frac{P_{sc}}{I_{sc} V_{sc}}$$

$$\theta_{sc} = \cos^{-1}\left(\frac{P_{sc}}{I_{sc} V_{sc}}\right) = \theta_{sc} = \cos^{-1}\left(\frac{950}{110\,(20.8)}\right) = \cos^{-1}(0.415) = 65.46^0$$

$$Z_{eq1} = \frac{V_{sc}\angle 0^0}{I_{sc}\angle -\theta_{sc}} = \left|\frac{V_{sc}}{I_{sc}}\right|\angle\theta_{sc} = \frac{110}{20.8}\angle 6546 = 5.28\angle 65.46 = 2.193 + j4.803$$

$$R_{eq1} = 2.193\Omega$$

$$X_{eq1} = 4.803\Omega$$

Series branch parameters referred to the lv side

$$R_{eq2} = \frac{R_{eq1}}{(a)^2} = \frac{2.193}{100} = 0.02193\Omega$$

$$X_{eq1} = \frac{X_{eq1}}{(a)^2} = \frac{4.803}{100} = 0.04803\Omega$$

Equivalent circuit referred to the hv side (Figures E3.6a and 6b):

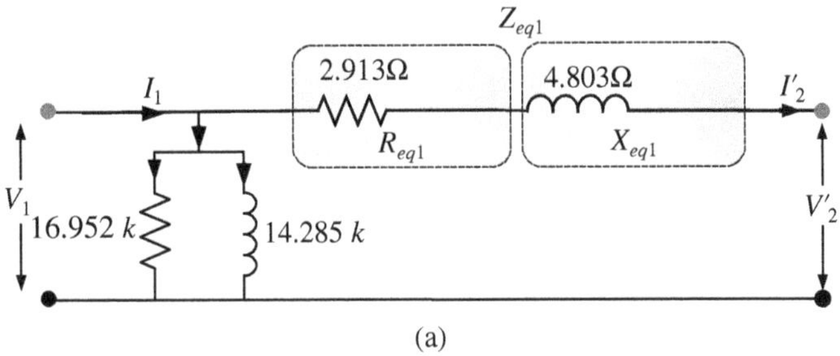

Figure E3.6a Obtained equivalent circuit referred to HV side.

Equivalent circuit referred to the lv side:

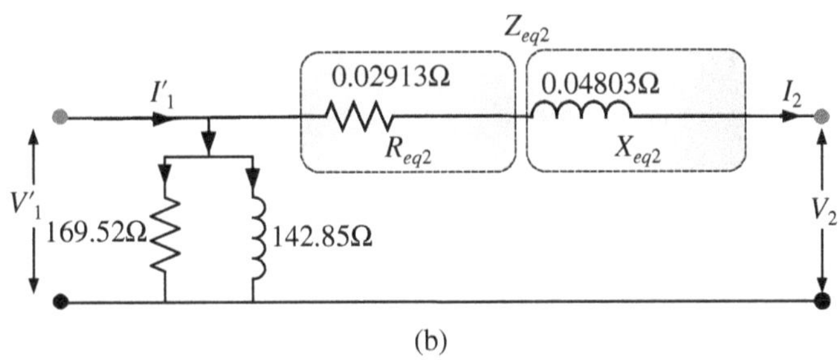

Figure E3.6b Obtained equivalent circuit referred to LV side.

3.10 Performance Measures of a Transformer

The major performance measures of a single-phase transformer are

- Voltage Regulation, and
- Efficiency.

The choice of a transformer for a specific application depends upon the above two performance indices.

3.10.1 Voltage Regulation

The loads connected to the secondary of the transformers are designed to operate at constant voltage and hence it is desired to keep the secondary terminal voltage constant.

However, the transformer secondary terminal voltage varies with the change in the load due to internal voltage drop in the transformer's impedance. This measure of the change in the secondary terminal voltage is an important parameter that defines the quality of the transformer.

Voltage Regulation is defined as the percentage change in the secondary terminal voltage when full-load is thrown off at a specified power factor, keeping the primary voltage constant.

The allowed voltage regulation in the transmission network is ± 12.5% while in distribution network is ± 5%.

The voltage regulation is defined mathematically as

$$\text{Regulation} = \frac{V_{s,no-load} - V_{s,full-load}}{V_{s,full-load}} * 100\% \tag{3.49}$$

Consider equivalent circuit referred to the primary side to determine the expression for voltage regulation (Figure 3.33).

$$\text{Voltage regulation} = \frac{|V|_{NL} - |V|_L}{|V|_L} \tag{3.50}$$

Under no-load condition $I_1 = I_2' = 0$, and hence $V_1 = V_2'$

$$\text{Voltage regulation} = \frac{|V_1| - |V_2'|_{load}}{|V_2'|_{load}} \tag{3.51}$$

Furthermore, the secondary terminal voltage under rated load condition is

$$|V_2'|_{load} = |V_2'| \tag{3.52}$$

Given the load voltage V_2', determine V_1 (the applied voltage), to determine the voltage regulation.

From the Figure 3.33, the following relation is written, to determine V_1 if the load terminal voltage and load current is known and then using Eq. (3.51).

$$V_1 = V_2' + I_2' Z_{eq1} \tag{3.53}$$

For equivalent circuit referred to the secondary side (Figure 3.34), the voltage regulation is given as

$$\text{Voltage regulation} = \frac{|V'_1| - |V_2|}{|V_2|} \tag{3.54}$$

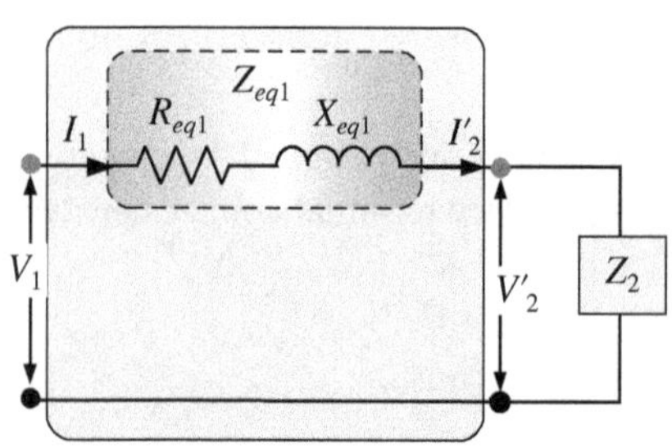

Figure 3.33 Equivalent circuit referred to the primary side.

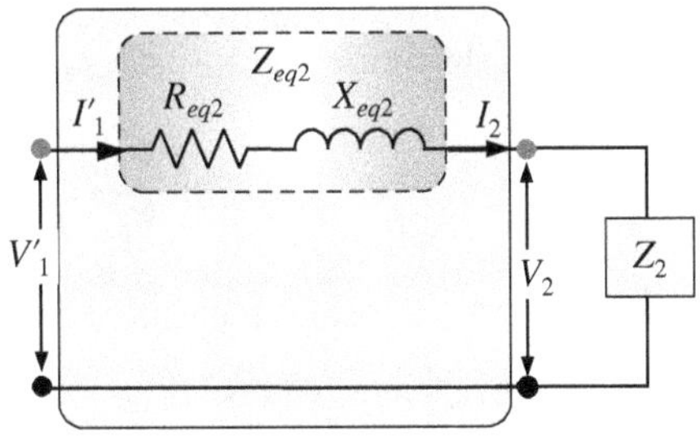

Figure 3.34 Equivalent circuit referred to the secondary side.

In this Eq. (3.54), V_2 is the secondary terminal voltage and is known. The applied voltage to the primary side is required to be calculated.

$$V_1' = V_2 + I_2 Z_{eq2}$$

$$\frac{V_1}{a} = V_2 + I_2 Z_{eq2} \tag{3.55}$$

Voltage regulation is given as

$$\frac{|V'_1| - |V_2|}{|V_2|} = \frac{\frac{V_1}{a} - V_2}{V_2} = \frac{I_2 Z_{eq2}}{V_2} = \frac{I_2 R_{eq2} \cos(\theta) + I_2 X_{eq2} \sin(\theta)}{V_2} \tag{3.56}$$

3.10.1.1 Condition for Maximum Voltage Regulation

The condition for maximum voltage regulation is obtained by differentiating the equation with respect to the power factor angle and putting it to zero.

$$V.R. = \frac{I_2 R_{eq2} \cos(\theta) + I_2 X_{eq2} \sin(\theta)}{V_2}$$

$$\frac{d}{d\theta}\left[\frac{I_2 R_{eq2} \cos(\theta) + I_2 X_{eq2} \sin(\theta)}{V_2}\right] = 0$$

$$-I_2 R_{eq2} \sin(\theta) + I_2 X_{eq2} \cos(\theta) = 0$$

$$I_2 R_{eq2} \sin(\theta) = I_2 X_{eq2} \cos(\theta)$$

$$\frac{\sin(\theta)}{\cos(\theta)} = \frac{X_{eq2}}{R_{eq2}}$$

$$\tan(\theta) = \frac{X_{eq2}}{R_{eq2}}$$

$$\theta = \tan^{-1}\left(\frac{X_{eq2}}{R_{eq2}}\right) \tag{3.57}$$

3.10.1.2 Condition for Zero Voltage Regulation

$$V.R. = \frac{I_2 R_{eq2} \cos(\theta) + I_2 X_{eq2} \sin(\theta)}{V_2}$$

$$V.R. = 0 = I_2 R_{eq2} \cos(\theta) + I_2 X_{eq2} \sin(\theta)$$

$$I_2 X_{eq2} \sin(\theta) = -I_2 R_{eq2} \cos(\theta)$$

$$\tan(\theta) = -\frac{R_{eq2}}{X_{eq2}}$$

$$\theta = -\tan^{-1}\left(\frac{R_{eq2}}{X_{eq2}}\right) \tag{3.58}$$

Example 3.7 Consider the data of Example 3.6, calculate the

(a) Voltage regulation for full load 0.8 pf lagging, unity pf and 0.6 leading pf
(b) Voltage regulation for half load 0.6 pf lagging and unity power factor.

Solution

Consider the equivalent circuit of the transformer referred to the high-voltage side (neglecting the shunt branch). The load voltage is assumed to be rated value of 240 V which is referred to the high-voltage side as 2400 V (Figure E3.7).

$$V_2' = aV_2 = 10(240) = 2400V.$$

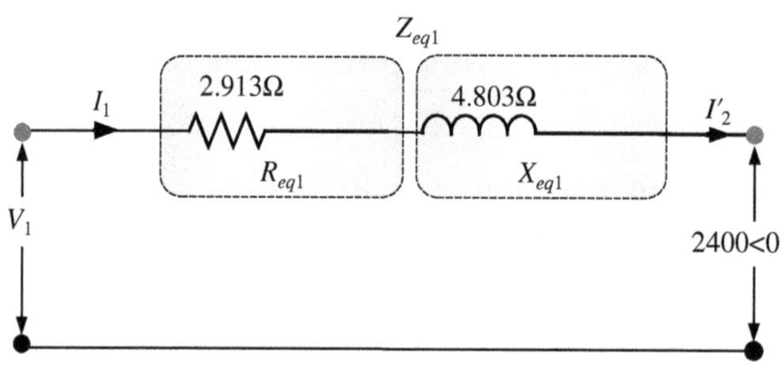

Figure E3.7

(a) For 0.8 pf lagging and full load.
The rated current is

$$I_1 = I_2' = \frac{50000}{2400} = 20.83A$$

$$pf = 0.8$$

$$\theta = \cos^{-1}(0.8) = 36.86^0$$

$$I_1 = 20.83\angle - 36.86$$

negative angle because of lagging power factor

$$\overline{V}_1 = \overline{V}_2' + \overline{I}_1\overline{Z}_{eq1} = 2400\angle 0 + 20.83\angle - 36.86\,(2.913 + j4.803)$$

$$\overline{V}_1 = 2400 + j0 + 20.83\angle - 36.86\,(5.617\angle 58.76)$$

$$\overline{V}_1 = 2400 + j0 + 117\angle 21.9 = 2400 + j0 + 108.55 + j43.64$$

$$\overline{V}_1 = 2508.55 + j43.64 = 2508.93\angle 0.99$$

$$\text{Voltage regulation} = \frac{|V_1| - \left|V_2'\right|_{load}}{\left|V_2'\right|_{load}} = \frac{2508.93 - 2400}{2400} \times 100\% = 4.54\%$$

For unity power factor and full load

$$pf = 1.0$$
$$\theta = \cos^{-1}(1.0) = 0^0$$
$$I_1 = 20.83\angle 0$$

$$\overline{V}_1 = \overline{V}_2' + \overline{I}_1\overline{Z}_{eq1} = 2400\angle 0 + 20.83\angle 0\,(2.913 + j4.803)$$
$$\overline{V}_1 = 2400 + j0 + 20.83\angle 0\,(5.617\angle 58.76)$$
$$\overline{V}_1 = 2400 + j0 + 117\angle 58.76 = 2400 + j0 + 60.68 + j100$$
$$\overline{V}_1 = 2460.68 + j100 = 2462.71\angle 2.32$$

$$\text{Voltage regulation} = \frac{|V_1| - |V_2'|_{load}}{|V_2'|_{load}} = \frac{2462.71 - 2400}{2400} \times 100\% = 2.61\%$$

For 0.6 power factor leading and full load

$$pf = 0.6$$
$$\theta = \cos^{-1}(0.6) = 53.13^0$$
$$I_1 = 20.83\angle 53.13$$

positive angle because of leading power factor

$$\overline{V}_1 = \overline{V}_2' + \overline{I}_1\overline{Z}_{eq1} = 2400\angle 0 + 20.83\angle 53.13\,(2.913 + j4.803)$$
$$\overline{V}_1 = 2400 + j0 + 20.83\angle 53.13\,(5.617\angle 58.76)$$
$$\overline{V}_1 = 2400 + j0 + 117\angle 111.89 = 2400 + j0 - 43.62 + j108.56$$
$$\overline{V}_1 = 2356.38 + j108.56 = 2358.88\angle 2.64$$

$$\text{Voltage regulation} = \frac{|V_1| - |V_2'|_{load}}{|V_2'|_{load}} = \frac{2358.88 - 2400}{2400} \times 100\% = -1.71\%$$

(b) For 0.6 lagging Half full load

$$I_1 = I_2' = \frac{50000}{2400} = 20.83A \times 0.5 = 10.415A$$
$$pf = 0.6$$
$$\theta = \cos^{-1}(0.6) = 53.13^0$$
$$I_1 = 10.415\angle -53.13$$

negative angle because of lagging power factor

$$\overline{V}_1 = \overline{V}_2' + \overline{I}_1\overline{Z}_{eq1} = 2400\angle 0 + 10.415\angle -53.13\,(2.913 + j4.803)$$
$$\overline{V}_1 = 2400 + j0 + 10.415\angle -53.13\,(5.617\angle 58.76)$$
$$\overline{V}_1 = 2400 + j0 + 58.5\angle 5.63 = 2400 + j0 + 58.218 + j5.74$$

$$\overline{V}_1 = 2458.218 + j5.74 = 2458.22\angle 0.133$$

$$\text{Voltage regulation} = \frac{|V_1| - |V_2'|_{load}}{|V_2'|_{load}} = \frac{2458.22 - 2400}{2400} \times 100\% = 2.42\%$$

For unity power factor half load

$$pf = 1.0$$
$$\theta = \cos^{-1}(1.0) = 0^0$$
$$I_1 = 10.415\angle 0$$

$$\overline{V}_1 = \overline{V}_2' + \overline{I}_1\overline{Z}_{eq1} = 2400\angle 0 + 10.415\angle 0\,(2.913 + j4.803)$$
$$\overline{V}_1 = 2400 + j0 + 10.415\angle 0\,(5.617\angle 58.76)$$
$$\overline{V}_1 = 2400 + j0 + 58.5\angle 58.76 = 2400 + j0 + 30.34 + j50$$
$$\overline{V}_1 = 2430.34 + j50 = 2430.85\angle 1.18$$

$$\text{Voltage regulation} = \frac{|V_1| - |V_2'|_{load}}{|V_2'|_{load}} = \frac{2430.85 - 2400}{2400} \times 100\% = 1.28\%$$

3.10.2 Efficiency of Transformer

Efficiency is defined as the ratio of output power in Watts to the ratio of input power in Watts.

$$\eta = \frac{\text{Output Power}}{\text{Input Power}} \times 100\% = \frac{\text{Output Power}}{\text{Output Power} + \text{Losses}} \times 100\%$$
$$\eta = \frac{\text{Output Power}}{\text{Output Power} + \text{Iron/Core Loss} + \text{Copper Loss}} \times 100\%$$
$$\eta = \frac{V_2 I_2 \cos(\theta)}{V_2 I_2 \cos(\theta) + P_{core} + P_{cu}} \times 100\% \tag{3.59}$$

The core loss in a transformer is independent of load. It remains constant from no-load to full load.

The copper loss depends as square of the load

$$P_{cu} = I_2^2 R_{eq2} = I_1^2 R_{eq1} \tag{3.60}$$

Since the copper loss depends upon the load and output also varies with the load, the efficiency also changes with change in the load.

Considering the fraction of load supplied by a transformer also called load factor is given as;

$$x = \frac{I_{\sup plied}}{I_{rated}} \tag{3.61}$$

When the transformer supplies rated load, $x = 1$, it is called 100% loading.

The output power from transformer at any load factor x, can then be written as

$$P_{out} = x\left(V_2 I_2 \cos(\theta)\right) = \text{Load factor} \times \text{Rated output Watt} \tag{3.62}$$

Since the core loss or iron loss does not depend upon the load, it remains constant

$$P_{core,x} = P_{core} \tag{3.63}$$

The copper loss depends upon square of the current and hence square of the load factor and is given as

$$P_{cu,x} = x^2 P_{cu,fl} \tag{3.64}$$

Where $P_{cu,x}$ is the copper loss at any load x while $P_{cu,fl}$ is the full load copper loss. The efficiency of transformer when supplying a load of x is given as

$$\eta = \frac{x\left(V_2 I_2 \cos(\theta)\right)}{x\left(V_2 I_2 \cos(\theta)\right) + P_{core} + x^2 P_{cu,fl}} \times 100\% \tag{3.65}$$

3.10.3 Maximum Efficiency Condition

If the terminal voltage and the load power factor are constant, the maximum efficiency depends upon the load current and occurs when

$$\frac{d\eta}{dI_2} = 0 \tag{3.66}$$

If condition is applied to Eq. (3.65), the condition for maximum efficiency is obtained as

$$P_{core} = x^2 P_{cu,fl} = P_{cu}$$

$$\text{Iron Loss} = \text{Copper Loss}$$

$$x = \sqrt{\frac{P_{core}}{P_{cu,fl}}} \tag{3.67}$$

The maximum efficiency is obtained as

$$\eta_{\max} = \frac{x\left(V_2 I_2 \cos(\theta)\right)}{x\left(V_2 I_2 \cos(\theta)\right) + 2P_{core}} \times 100\% \tag{3.68}$$

If terminal voltage and load current are kept constant, the maximum efficiency occur when

$$\frac{d\eta}{d\theta} = 0 \tag{3.69}$$

Using this condition in Eq. (3.65), the maximum efficiency condition is

$$\theta = 0$$

$$\cos(\theta) = 1 \tag{3.70}$$

The maximum efficiency in a transformer occurs when the copper loss is equal to the core/iron loss and the load power factor is unity (resistive load). The change in efficiency with the change in load current and load power factor is plotted in Figure 3.35.

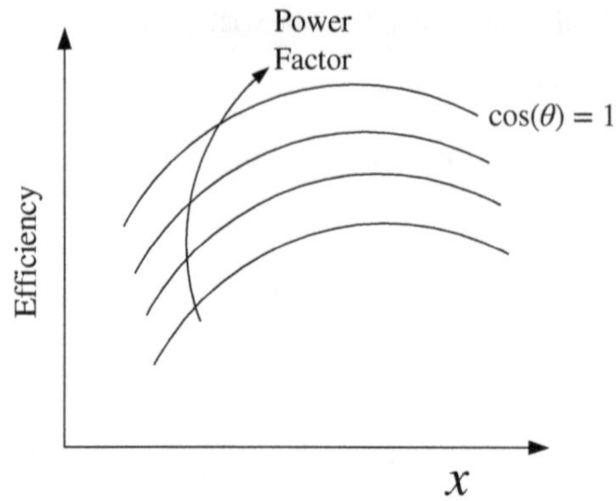

Figure 3.35 Efficiency vs load of a transformer.

Example 3.8 Consider the data of Example 3.6, calculate the

(a) Efficiency for full load 0.8 pf lagging and unity pf
(b) Efficiency for half load 0.6 pf lagging and unity power factor
(c) At what load maximum efficiency occurs
(d) Output power in kVA for maximum efficiency
(e) Maximum efficiency.

Solution

(a) At full load or rated load condition and 0.8 pf lagging
Given the rated kVA = 50 kVA
Core loss (Pc) = 340 W, Full load Copper loss (P_{flcu}) = 950 W
Efficiency is given as

$$\eta = \frac{V_2 I_2 \cos(\theta)}{V_2 I_2 \cos(\theta) + P_{core} + P_{cu}} \times 100\%$$

$$\eta = \frac{(KVA)_{rated} \cos(\theta)}{(KVA)_{rated} \cos(\theta) + P_c + P_{cu,fl}} \times 100\%$$

$$\eta = \frac{\left(50 \times 10^3\right) 0.8}{\left(50 \times 10^3\right) 0.8 + 340 + 950} \times 100\% = \frac{40000}{40000 + 1300} \times 100\%$$

$$\eta = 96.85\%$$

At full load and unity power factor, the efficiency is calculated as

$$\eta = \frac{(KVA)_{rated} \cos(\theta)}{(KVA)_{rated} \cos(\theta) + P_c + P_{cu,fl}} \times 100\%$$

$$\eta = \frac{\left(50 \times 10^3\right) 1}{\left(50 \times 10^3\right) 1 + 340 + 950} \times 100\% = \frac{50000}{50000 + 1300} \times 100\%$$

$$\eta = 97.46\%$$

(b) At half full load ($x = 0.5$) and 0.6 pf lagging, the core loss remains the same while the copper loss is changed as square of the load factor, the efficiency is calculated as;

$$\eta = \frac{x\left(V_2 I_2 \cos(\theta)\right)}{x\left(V_2 I_2 \cos(\theta)\right) + P_{core} + x^2 P_{cu,fl}} \times 100\%$$

$$\eta = \frac{x(KVA)_{rated} \cos(\theta)}{x(KVA)_{rated} \cos(\theta) + P_c + x^2 P_{cu,fl}} \times 100\%$$

$$\eta = \frac{0.5\left(50 \times 10^3\right) 0.6}{0.5\left(50 \times 10^3\right) 0.6 + 340 + (0.5)^2 950} \times 100\% = \frac{15000}{15000 + 577.5} \times 100\%$$

$$\eta = 96.29\%$$

At half full load ($x = 0.5$) and unity power factor the efficiency is

$$\eta = \frac{x(KVA)_{rated} \cos(\theta)}{x(KVA)_{rated} \cos(\theta) + P_c + x^2 P_{cu,fl}} \times 100\%$$

$$\eta = \frac{0.5\left(50 \times 10^3\right) 1}{0.5\left(50 \times 10^3\right) 1 + 340 + (0.5)^2 950} \times 100\% = \frac{25000}{25000 + 577.5} \times 100\%$$

$$\eta = 97.74\%$$

(c) The load at which maximum efficiency is achieved is given as

$$x = \sqrt{\frac{P_{core}}{P_{cu,fl}}} = \sqrt{\frac{340}{950}} = 0.598$$

(d) Power supplied at the maximum efficiency condition is calculated as

$$(KVA)_{\eta_{max}} = x(KVA)_{rated} = 0.598\,(50) = 29.912KVA$$

(e) The maximum efficiency is calculated as

$$\eta_{max} = \frac{x\left(V_2 I_2 \cos(\theta)\right)}{x\left(V_2 I_2 \cos(\theta)\right) + 2P_{core}} \times 100\%$$

$$\eta_{max} = \frac{29.912}{29.912 + 2 * 0.34} \times 100\% = 97.77\%$$

Example 3.9 Given the maximum efficiency of 98.8% at 12 kVA and 0.9 pf. Compute the copper loss at 4.5 kW and 0.85 pf.

Solution

The maximum efficiency is given as

$$\eta_{max} = \frac{x(KVA)_{rated} \cos(\theta)}{x(KVA)_{rated} \cos(\theta) + 2P_{core}} = 0.985$$

$$0.985 = \frac{12 \times 0.9}{12 \times 0.9 + 2P_{core}}$$

$$0.985\,(12 \times 0.9) + 0.985\left(2P_{core}\right) = 12 \times 0.9$$

$$1.97P_{core} = 10.8 - 10.638 = 0.162$$

$$P_{core} = 0.08223kW$$

Load factor for maximum efficiency

$$(KVA)_{\eta_{\max}} = x(KVA)_{rated}$$
$$x = \frac{(KVA)_{\eta_{\max}}}{(KVA)_{rated}}$$

For maximum efficiency

$$P_{core} = x^2 P_{cu,fl} = \left(\frac{(KVA)_{\eta_{\max}}}{(KVA)_{rated}}\right)^2 P_{cu,fl}$$
$$0.08223 = \left(\frac{12}{(KVA)_{rated}}\right)^2 P_{cu,fl}$$
$$P_{cu,fl} = \frac{0.08223 \times (KVA)^2_{rated}}{(12)^2} = 0.00057104 \times (KVA)^2_{rated}$$

Given load is 4.5 kW and pf = 0.85

$$KVA = 4.5/0.8 = 5.2941 \text{ kVA}$$

$$x = \left(5.2941/(\text{KVA})_{\text{rated}}\right)$$

$$P_{cu} = x^2 P_{cu,fl} = \left(\frac{5.2941}{(KVA)_{rated}}\right)^2 P_{cu,fl}$$
$$P_{cu} = \left(\frac{5.2941}{(KVA)_{rated}}\right)^2 \left(0.00057104 \times (KVA)^2_{rated}\right) = 0.016kW$$

Example 3.10 A 100 kVA, 11 kV/440 V 1-phase transformer has maximum efficiency of 99% when it delivers 92 kVA at unity pf and rated voltage. Compute the efficiency at full load and 0.8 pf lagging.

Solution

Efficiency is given as

$$\eta = \frac{x(KVA)_{rated} \cos(\theta)}{x(KVA)_{rated} \cos(\theta) + P_c + x^2 P_{cu,fl}} \times 100\%$$

$$(KVA)_{rated} = 100KVA$$
$$(KVA)_{\eta_{\max}} = x(KVA)_{rated} = 92KVA$$
$$pf = 1$$
$$\eta_{\max} = \frac{92}{92 + 2P_{core}} = 0.985$$

$$0.985\,(92 \times 1) + 0.985\,\left(2P_{core}\right) = 92 \times 1$$
$$1.97P_{core} = 92 - 90.62 = 1.38$$
$$P_{core} = 700.5W$$

$$(KVA)_{\eta_{\max}} = x(KVA)_{rated} = 92KVA$$

$$x = \frac{92}{(KVA)_{rated}} = \frac{92}{100} = 0.92$$

$$P_{cu,fl} = \frac{P_{core}}{(x)^2} = \frac{700.5}{(0.92)^2} = 827.62W$$

$$\eta = \frac{x(KVA)_{rated}\cos(\theta)}{x(KVA)_{rated}\cos(\theta) + P_c + x^2 P_{cu,fl}} \times 100\%$$

$$= \frac{0.92\,(100)\,(0.8)}{0.92\,(100)\,(0.8) + 0.7005 + 0.82762} \times 100\%$$

$$\eta = 97.96\%$$

3.11 All-Day Efficiency or Energy Efficiency

The distribution transformer supplying power to the houses and used in the distribution network are subjected to the varying loading conditions. All-day or energy efficiency is defined for such transformers as

$$\eta_{all,day} = \frac{\text{Output Energy in 24 hours}}{\text{Input Energy in 24 hours}} \times 100\%$$

$$= \frac{\text{Output Energy in 24 hours}}{\text{Output Energy in 24 hours} + \text{Energy Losses in 24 Hours}} \times 100\%$$

$$\eta_{all,day} = \frac{\text{Output Energy in 24 hours}}{\text{Output Energy in 24 hours} + \text{Iron/Core Energy Loss in 24 hours} + \text{Copper Energy Loss in 24 hours}} \times 100\%$$

$$\text{Energy} = \text{Power} \times \text{Time}$$

Knowing the load cycle, the energy efficiency can be computed for a distribution transformer.

Example 3.11 A 50 kVA, 2400/240 V, single-phase 50 Hz transformer has core loss of 340 W and full-load copper loss of 950 W. The load cycle of the transformer is given as

%load	0%	50%	75%	100%	110%
Pf		0.95	1	0.9	0.9
Hours	5	7	5	4	3

Calculate the all-day efficiency

Solution

Energy output = Output Power (kVA) × pf × Time = Load Factor × Rated KVA × pf × Time

Energy Output first 5 hours = 0 × 50 × 5 = 0 kWh

Energy Output for 7 hours = 0.5 × 50 × 0.95 × 7 = 166.25 kWh

Energy Output for 5 hours = 0.75 × 50 × 1 × 5 = 187.5 kWh

Energy Output for 4 hours = $1 \times 50 \times 0.9 \times 4 = 180$ kWh
Energy Output for 3 hours = $1.1 \times 50 \times 0.9 \times 3 = 148.5$ kWh
Total Energy Output = $166.25 + 187.5 + 180 + 148.5 = 682.25$ kWh
Energy Losses = Core Loss × Time + Copper Loss × Time
Energy Core Loss = $0.34 \times 24 = 8.16$ kWh
Energy Copper Loss for 5 hours = $(x)^2 P_{cu,fl} \times \text{Time} = (0)^2(0.95)(5) = 0$ kWh
Energy Copper Loss for 7 hours = $(x)^2 P_{cu,fl} \times \text{Time} = (0.5)^2(0.95)(7) = 1.6625$ kWh
Energy Copper Loss for 5 hours = $(x)^2 P_{cu,fl} \times \text{Time} = (0.75)^2(0.95)(5) = 2.6718$ kWh
Energy Copper Loss for 4 hours = $(x)^2 P_{cu,fl} \times \text{Time} = (1)^2(0.95)(4) = 3.8$ kWh
Energy Copper Loss for 3 hours = $(x)^2 P_{cu,fl} \times \text{Time} = (1.1)^2(0.95)(3) = 3.4485$ kWh
Total Energy Loss = 19.7428 kWh

The all-day efficiency is;

$$\eta_{allday} = \frac{\text{Output Energy}}{\text{Output Energy} + \text{Energy Losses}} \times 100\%$$

$$\eta_{allday} = \frac{682.25}{682.25 + 19.7428} \times 100\% = 97.18\%$$

3.12 Autotransformer

Autotransformers are single-winding transformers that output variable voltage. Both primary and secondary windings are created from one winding. The output voltage is continuously varied by changing the position of the stud. Secondary winding is common between the input and output sides. There is no isolation between the two input and output windings. The two windings are electrically connected. Hence, the power from input to output is transferred by induction and conduction.

Advantages of Autotransformer:

- Low Leakage reactances
- Lower Losses
- Lower Exciting current
- Higher kVA rating
- Variable output voltages.

Disadvantages of Autotransformer:

- The main disadvantage of an autotransformer is that it does not have the primary to secondary winding isolation. Then, an autotransformer cannot safely be used for stepping down higher voltages to much lower voltages suitable for smaller loads.
- If the secondary side winding becomes open-circuited, load current stops flowing through the primary winding stopping the transformer action resulting in the full primary voltage being applied to the secondary terminals.
- If the secondary circuit suffers a short-circuit condition, the resulting primary current would be much greater than an equivalent two wound transformer damaging the autotransformer.

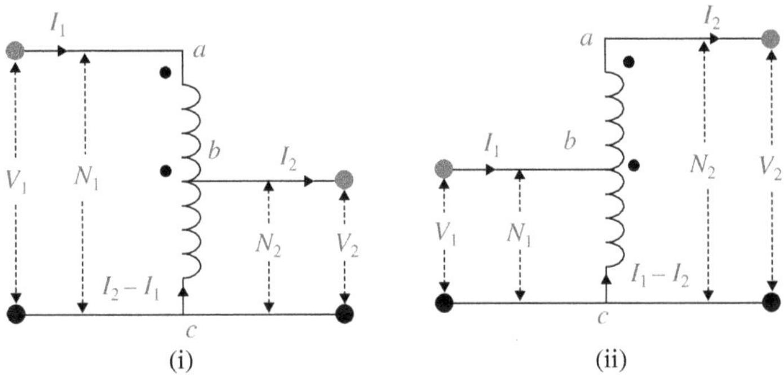

Figure 3.36 Autotransformer, (i) Step-down, (ii) Step-Up.

- Since the neutral connection is common to both primary and secondary windings, earthing of the secondary winding automatically Earth's the primary as there is no isolation between the two windings. Double-wound transformers are sometimes used to isolate equipment from earth.

Considering Figure 3.36 (i), the mmf for the upper and lower portion of windings are calculated as

$$\left.\begin{aligned} F_U &= (N_1 - N_2)\, I_1 = \left(1 - \tfrac{1}{a}\right) N_1 I_1 \\ F_L &= N_2\,(I_1 - I_2) = \tfrac{N_1}{a}\,(I_2 - I_1) \end{aligned}\right\} \Rightarrow F_U = F_L \Rightarrow \frac{I_1}{I_2} = \frac{1}{a} \tag{3.71}$$

$$\frac{V_1}{V_2} = \frac{N_1}{N_2} = a$$

The turn ratios are the same as that of a two-winding transformer.

A two-winding transformer with N_1 and N_2 number of turns can be connected to form an autotransformer as shown in Figure 3.37.

KVA of the two-winding transformer and autotransformer is;

$$\begin{aligned} |S_{tw}| &= E_1 I_1 = E_2 I_2 \\ |S_{auto}| &= V_1 I_s = V_2 I_{load} \end{aligned} \tag{3.72}$$

$$\begin{aligned} |S_{auto}| &= V_1 I_s = (E_1 + E_2)\, I_1 = E_1 I_1 + E_2 I_1 = |S_{tw}| + E_2 I_1 \\ |S_{auto}| &> |S_{tw}| \end{aligned} \tag{3.73}$$

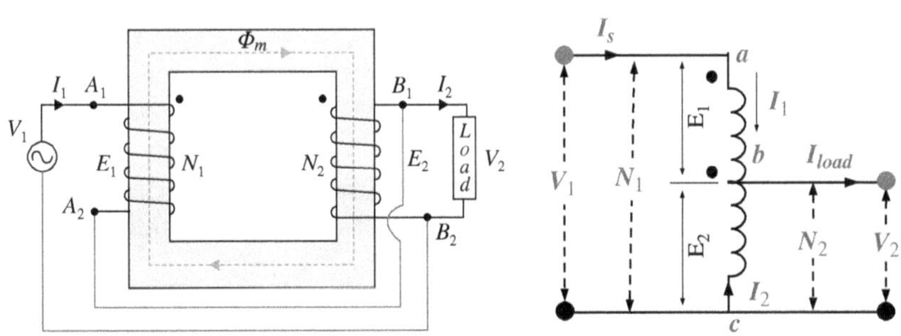

Figure 3.37 Reconnecting a two-winding transformer as an autotransformer.

Example 3.12 A single-phase, 50 kVA, 2400/240 V two-winding transformer is connected as an autotransformer as shown in Figure 3.43 such that more 2400 V at the secondary side. The portion ab is 240 V winding and the portion bc is 2400 V winding. Calculate the kVA rating as an autotransformer (Figure E3.12a).

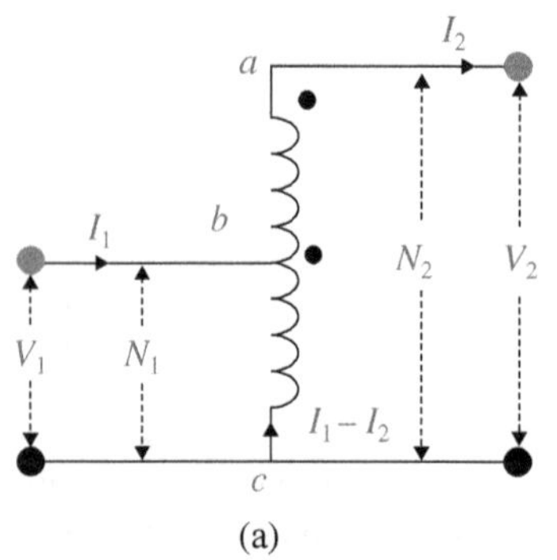

Figure E3.12a

The current in the two windings

$$I_{ab} = \frac{S_{TW}}{V_{ab}} = \frac{50 \times 10^3}{240} = 208.33A$$

$$I_{bc} = \frac{S_{TW}}{V_{bc}} = \frac{50 \times 10^3}{2400} = 20.833A$$

$$S_{Auto} = V_1 I_1 = 2400 \times 229.166 = 550kVA$$

or

$$S_{Auto} = V_2 I_2 = 2640 \times 208.33 = 550kVA$$

Figure E3.12b

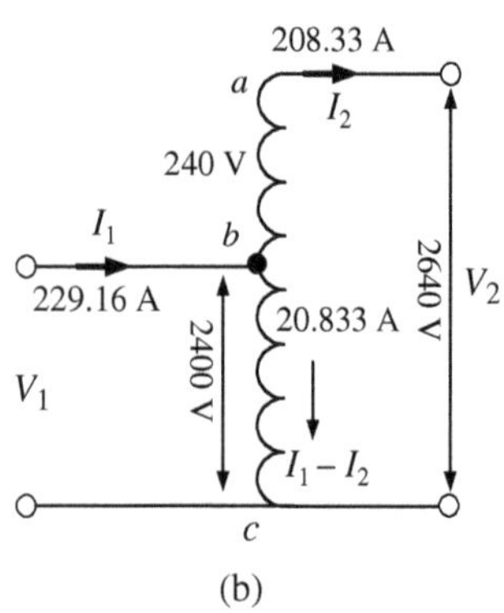

Figure E3.12b

Example 3.13 A single-phase 20 kVA 560/220 V transformer has an efficiency of 97% when it delivers 18 kW at 0.8 pf. The transformer is reconnected as an autotransformer to supply load to a 560 V circuit from a 780 V source.

(a) Show the connection diagram
(b) Determine the kVA of the autotransformer with output voltage equal to 560 V
(c) Calculate the efficiency of the autotransformer for full load at 0.9 pf.

Figure E3.13

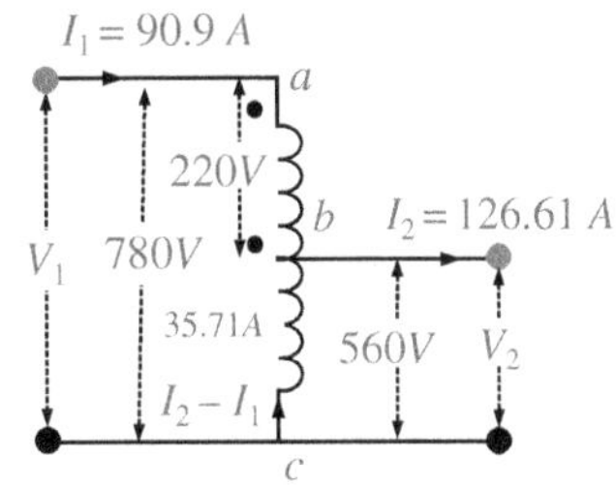

Solution

(a) The currents in windings of the autotransformer are

$$I_{ab} = \frac{20000}{220} = 90.9A$$
$$I_{bc} = \frac{20000}{560} = 35.71A$$

(b) KVA Rating

$$S_{auto} = 780 \times 90.9 = 70.902kVA$$
$$or$$
$$S_{auto} = 560 \times 126.61 = 70.902kVA$$

(c) Power Loss is calculated as

$$\eta = \frac{\text{Output Power in Watts } (P_{out})}{\text{Output Power in Watts } (P_{out}) + \text{Losses } (P_{losses})}$$
$$\eta\, P_{out} + \eta P_{losses} = P_{out}$$
$$P_{losses} = P_{out}\,(1-\eta)\,\frac{1}{\eta} = 18\,(1-0.97)\,\frac{1}{0.97} = 0.5567kW$$

The efficiency is then calculated as

$$\eta = \frac{70.902 \times 0.9}{70.902 \times 0.9 + 0.5567} \times 100\% = 99.13\%$$

Example 3.14 A single-phase 20 kVA 560/220 V two-winding transformer is to be connected as an autotransformer.

(a) Show the possible connections
(b) Calculate the maximum possible KVA.

Solution
The currents in the two windings are;

$$I_{220V} = \frac{20000}{220} = 90.9A$$
$$I_{560V} = \frac{20000}{560} = 35.71A$$

There are four possible connections as shown in Figure E3.14.

Looking at the possible connection type, it is seen that the maximum possible KVA can be achieved in connection a or connection d, and the maximum KVA is

$$S_{auto} = 780 \times 90.9 = 70.902kVA$$

or

$$S_{auto} = 560 \times 126.61 = 70.902kVA$$

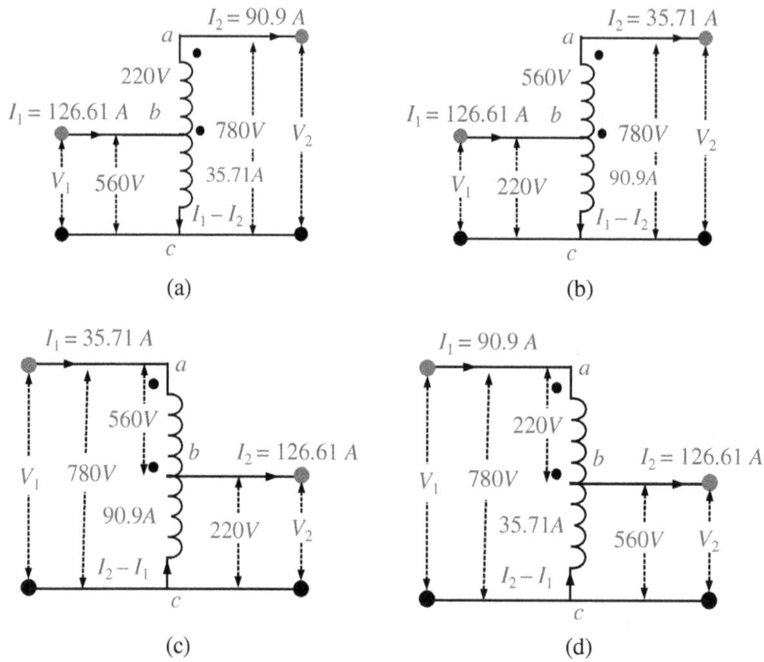

Figure E3.14

3.13 Three-Phase Transformer

Electrical Power is generated, transmitted and distributed in three-phase. During the course of power generation to utilization, three-phase transformers are used at various places to step-up and step-down. There are two possibilities to construct three-phase transformers; (i) Single core with six-windings, and (ii) Three separate single-phase transformers externally connected to form three-phase. The two configurations are shown in Figures 3.38 and 3.39, respectively.

A three-phase transformer constructed from a single core, occupies less space, and costs less. Three-phase transformers made up of three single-phase transformers has higher volume. However, this provides a reliable power supply system because, in the event of failure of one unit, the remaining two healthy transformers can be connected in open-delta and can still supply about 58% three-phase power [3].

Comparison between the three single units of transformers and a single unit of three-phase transformer is given in Table 3.2

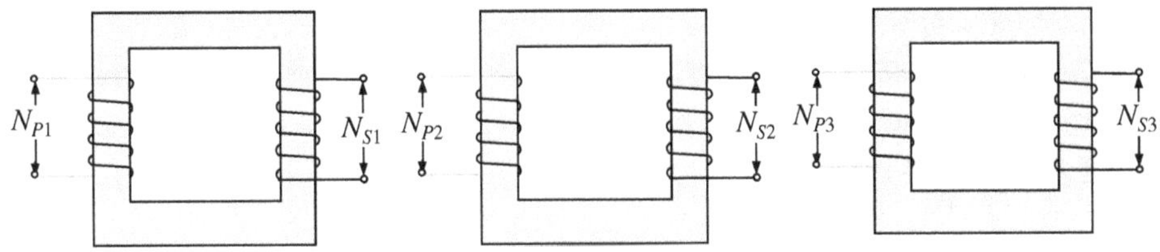

Figure 3.38 Three single-phase transformer to obtain three-phase supply.

Figure 3.39 Single unit of a three-phase transformer.

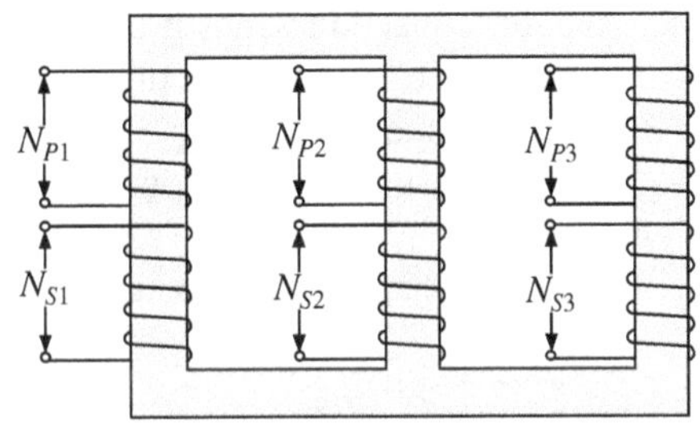

Table 3.2 Comparison between single-phase three units and three-phase single unit.

Three units of Single-phase Transformer	Single units of Three-phase transformer
Requires more space, more cost due to higher auxiliary, separate tanks and more oil is needed	Lower cost, shorter magnetic path, less volume, small tank size, low volume of oil
For star or delta connection on H.V. side six cables are needed to brought outside the tank. Hence six H.V. bushings are required, thus high cost.	The winding connection can be done inside the tank. Thus only 3 bushings are required, thus low cost
It requires more floor area. Thus, more cost especially for indoor sub-station.	It needs less floor space and for indoor sub-station, the building cost is low.
Offer greater flexibility in installation	Less flexibility in installation
If one unit of single-phase transformer is faulted or damaged, it can be replaced easily.	If any phase is faulted or damaged the whole transformer need to be out of service.
Only one single-phase transformer unit is kept as standby, therefor, cost of standby unit is low	One full unit of three-phase unit is needed as a standby, therefore, cost of standby unit is high
Because of more iron, the core loss is high and overall efficiency is low	Because of less iron, the overall core loss is low and thus higher efficiency
Transportation from production to installation area is easy because the units are small in size	Transportation from production to installation area is difficult because of the higher size of the single unit.

The single unit of three-phase transformer is more efficient, less costly, and therefore, more commonly used compared to the three units of single-phase transformers. Nevertheless, in hilly areas, mines and isolated areas where the transportation is difficult, single-phase transformer units are preferred.

The possible connections are

- Input star (Y), output delta (Δ)
- Input delta (Δ), output star (Y)
- Input delta (Δ), output delta (Δ)
- Input star (Y), output star (Y).

3.13.1 Input (Y), Output (Δ)

The connection diagram with primary as star and secondary as delta is shown in Figure 3.40. This type of connection is mostly used in step-down applications. The high-voltage side (star connected) can be grounded, which is desirable, as discussed later in this chapter.

Given the primary line voltage, V and line current I and turn ratio $a = N_1/N_2$. The phase voltage of the primary side is given as $V/\sqrt{3}$ and the phase current is same as the line current (I).

The phase voltage is transformed to the secondary phase voltage as

$$\frac{V_{1,phase}}{V_{2,phase}} = a$$

$$\frac{V/\sqrt{3}}{V_{2,phase}} = a$$

$$V_{2,phase} = \frac{V}{a\sqrt{3}} \tag{3.74}$$

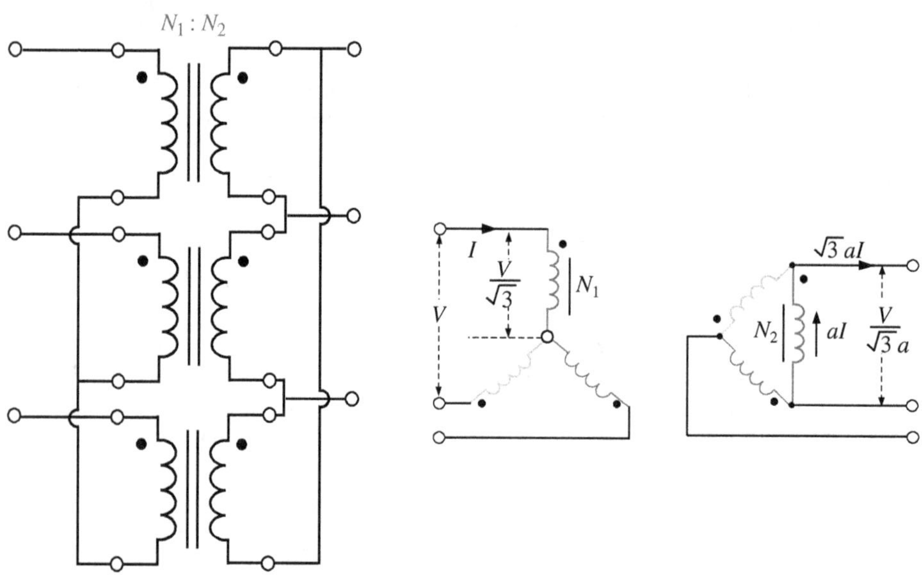

Figure 3.40 Star-delta three-phase transformer connection.

The primary side phase current is transformed to the secondary side phase current as

$$\frac{I_{2,phase}}{I_{1,phase}} = a$$

$$\frac{I_{2,phase}}{I} = a$$

$$I_{2,phase} = aI \tag{3.75}$$

The secondary phase quantities are now transformed to the line quantities as
Line Voltage = Phase Voltage (because of delta connection)

$$V_{2,line} = V_{2,phase} = \frac{V}{a\sqrt{3}} \tag{3.76}$$

$$I_{2,line} = \sqrt{3}I_{2,phase} = \sqrt{3}aI \tag{3.77}$$

The phase shift between the input and output voltages is obtained from the phasor diagram. Different voltages of primary and secondary sides are shown in Figure 3.41 and corresponding phasors in Figure 3.42.

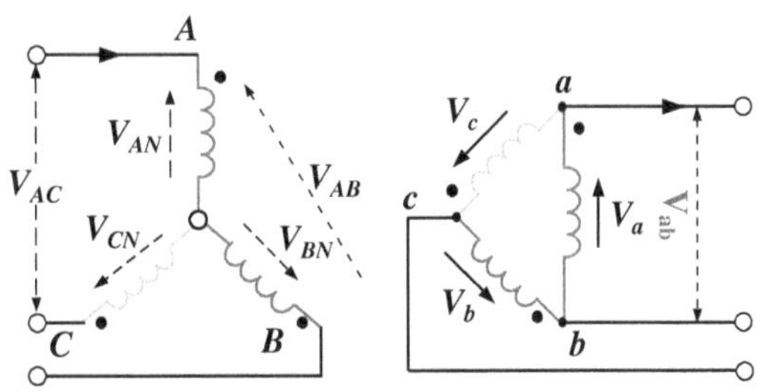

Figure 3.41 Voltages in star-delta connections.

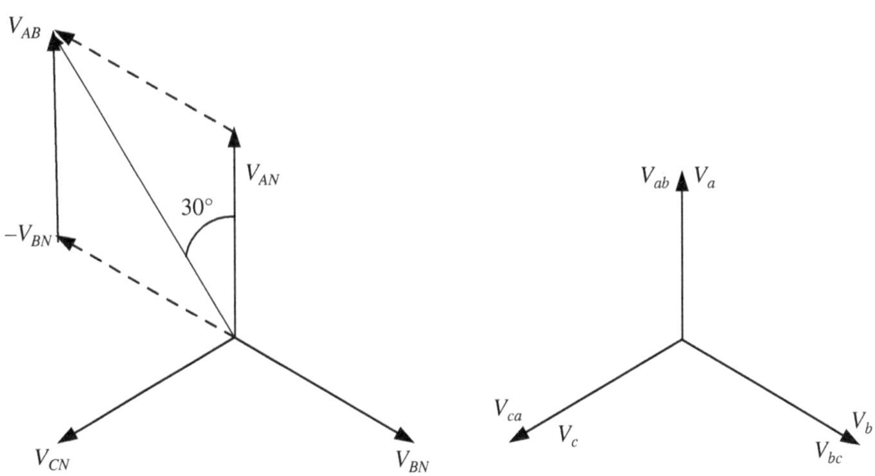

Figure 3.42 Phasor diagram for star-delta connection.

The line voltage of the star side (primary side) is given as;

$$V_{AB} = V_{AN} - V_{BN}$$
$$V_{BC} = V_{BN} - V_{CN}$$
$$V_{CA} = V_{CN} - V_{AN} \tag{3.78}$$

The line voltage of the delta side (secondary) is equal to the phase voltage.

Comparing the line voltage of the primary and secondary sides, it is seen that the primary voltage (star) leads the secondary side (delta) by 30°.

3.13.2 Input Delta (Δ), Output Star (Y)

The connection diagram with input delta and output star is shown in Figure 3.43. This type of transformer connection is used for step-down applications.

Given the primary line voltage, V and line current I and turn ratio $a = N_1/N_2$. The phase voltage of the primary side is the same as line voltage (delta connection) and the phase current is $^I/_{\sqrt{3}}$.

The phase voltage is transformed to the secondary phase voltage as

$$\frac{V_{1,phase}}{V_{2,phase}} = a$$
$$\frac{V}{V_{2,phase}} = a$$
$$V_{2,phase} = \frac{V}{a} \tag{3.79}$$

The primary side phase current is transformed to the secondary side phase current as

$$\frac{I_{2,phase}}{I_{1,phase}} = a$$
$$\frac{I_{2,phase}}{^I/_{\sqrt{3}}} = a$$
$$I_{2,phase} = a^I/_{\sqrt{3}} \tag{3.80}$$

The secondary phase quantities are now transformed to the line quantities as

Line Current = Phase Current (because of delta connection)

$$V_{2,line} = \sqrt{3}V_{2,phase} = \frac{\sqrt{3}V}{a} \tag{3.81}$$

$$I_{2,line} = I_{2,phase} = I \tag{3.82}$$

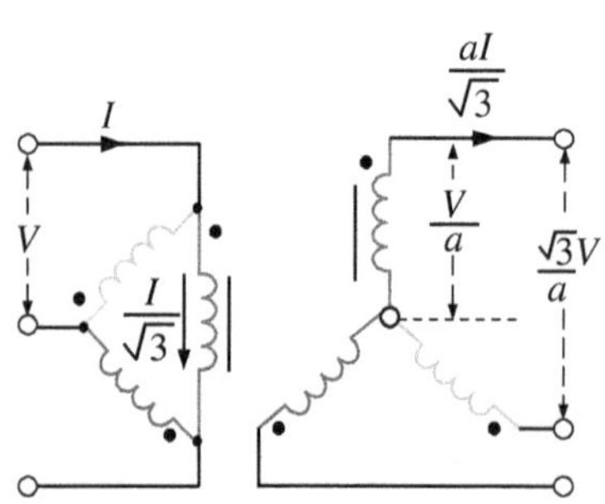

Figure 3.43 Delta-star connection of three-phase transformer.

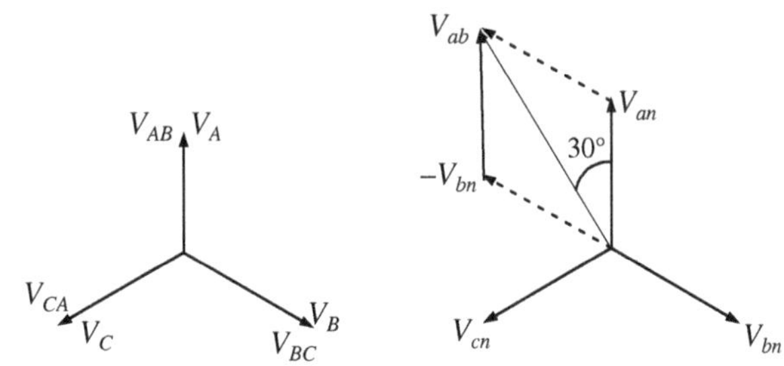

Figure 3.44 Phasor diagram for delta-star connection.

Comparing the line voltage of the primary and secondary sides and it is seen that the primary voltage (star) lags behind the secondary side (delta) by 30° as seen from the phasor diagram of Figure 3.44.

3.13.3 Input Delta (Δ), Output Delta (Δ)

The connection diagram with input delta and output delta is shown in Figure 3.45. This type of transformer connection is used for a reliable power supply system. If one of the transformers is faulty, it can be removed and the whole system run in open-delta mode or V connection. The output power can still be supplied in this mode. The 58% of the rated can be supplied in open-delta mode. The voltage current relationship can be obtained in the same way.

Given the primary line voltage, V and line current I and turn ratio $a = N_1/N_2$. The phase voltage of the primary side is the same as line voltage (delta connection) and the phase current is $^I/_{\sqrt{3}}$.

The phase voltage is transformed to the secondary phase voltage as

$$\frac{V_{1,phase}}{V_{2,phase}} = a$$

$$\frac{V}{V_{2,phase}} = a$$

$$V_{2,phase} = \frac{V}{a} \tag{3.83}$$

The primary side phase current is transformed to the secondary side phase current as

$$\frac{I_{2,phase}}{I_{1,phase}} = a$$

$$\frac{I_{2,phase}}{^I/_{\sqrt{3}}} = a$$

$$I_{2,phase} = a^I/_{\sqrt{3}} \tag{3.84}$$

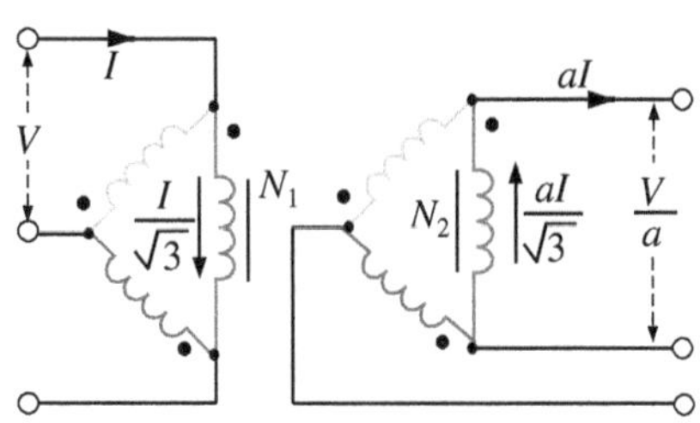

Figure 3.45 Delta-delta connection of three-phase transformer.

The secondary phase quantities are now transformed to the line quantities as

$$V_{2,line} = V_{2,phase} = \frac{V}{a} \tag{3.85}$$

$$I_{2,line} = \sqrt{3}I_{2,phase} = \sqrt{3}{}^{aI}/_{\sqrt{3}} = aI \tag{3.86}$$

There is no phase shift between primary and secondary side voltages.

3.13.4 Input Star (Y), Output Star (Y)

The connection diagram of star primary and star secondary is shown in Figure 3.46. This type of connection is not very commonly in use because of the problem with the exciting current (distorted current with harmonic content).

Given the primary line voltage, V and line current I and turn ratio $a = N_1/N_2$. The phase voltage of the primary side is given as ${}^{V}/_{\sqrt{3}}$ and the phase current is same as the line current (I).

The phase voltage is transformed to the secondary phase voltage as

$$\frac{V_{1,phase}}{V_{2,phase}} = a$$

$$\frac{{}^{V}/_{\sqrt{3}}}{V_{2,phase}} = a$$

$$V_{2,phase} = \frac{V}{a\sqrt{3}} \tag{3.87}$$

The primary side phase current is transformed to the secondary side phase current as

$$\frac{I_{2,phase}}{I_{1,phase}} = a$$

$$\frac{I_{2,phase}}{I} = a$$

$$I_{2,phase} = aI \tag{3.88}$$

The secondary phase quantities are now transformed to the line quantities as;

$$V_{2,line} = \sqrt{3}V_{2,phase} = \sqrt{3}\frac{V}{a\sqrt{3}} = \frac{V}{a} \tag{3.89}$$

$$I_{2,line} = I_{2,phase} = aI \tag{3.90}$$

There is no phase shift between primary and secondary quantities.

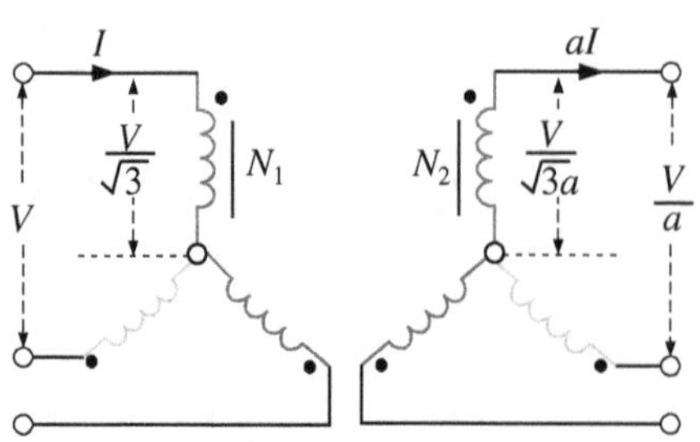

Figure 3.46 Star-star connection of three-phase transformer.

3.14 Single-Phase Equivalent Circuit of Three-Phase Transformer

Under balanced operating conditions, it is sufficient to analyse a three-phase transformer on per-phase basis. Under balanced conditions, each phase has equal magnitude with a phase difference of 120°.

For analytical purposes, the source and load are transformed into star-star equivalent circuit. One phase is then used for the purpose of analysis.

Consider a star-delta transformer supplying a delta connected load as shown in Figure 3.47. The delta secondary and delta load should be converted to equivalent star, keeping the line voltage and currents of secondary unchanged.

Given the line voltage of the primary side as V and the primary side line current as I, the line voltage and current of the secondary side is $V/a\sqrt{3}$ and, respectively. Keeping the same line voltage and line current, the equivalent star on the secondary side is shown in Figure 3.48. The corresponding phase voltage of the star secondary is

$$V_{phase} = \frac{V_{line}}{\sqrt{3}} = \frac{V}{a\sqrt{3}} \bullet \frac{1}{\sqrt{3}} = \frac{V}{3a} \tag{3.91}$$

Phase current = line current.

The equivalent star-star circuit is shown in Figure 3.48.

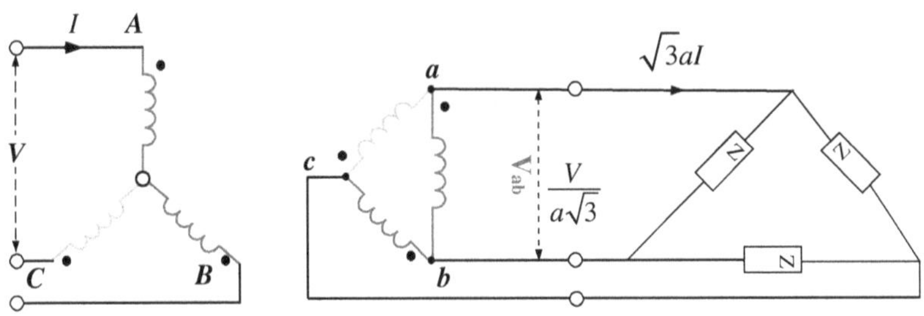

Figure 3.47 Star-delta transformer supplying a delta load.

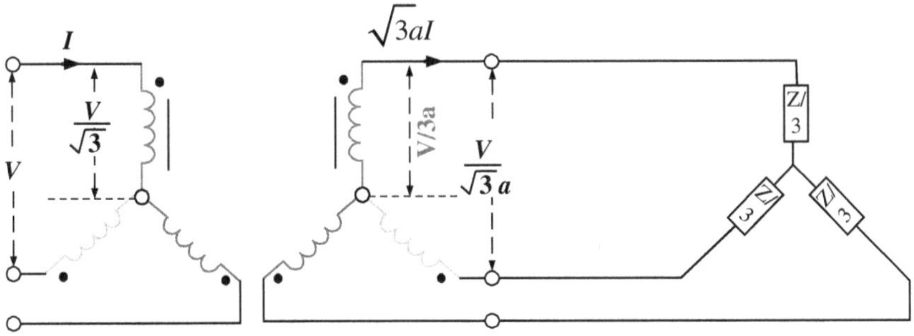

Figure 3.48 Equivalent star-star connection.

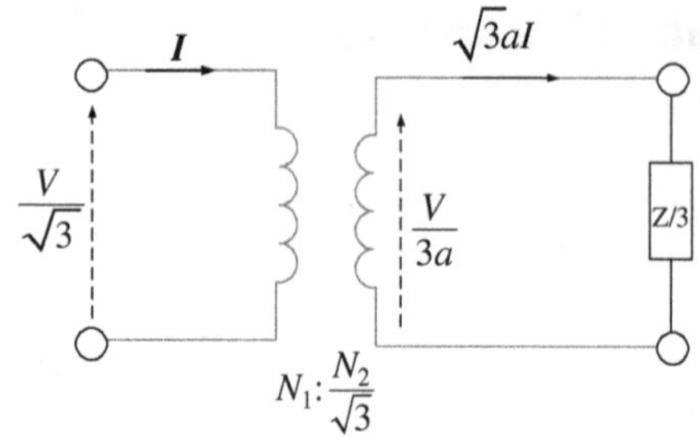

Figure 3.49 Single-phase equivalent.

Per-phase equivalent can be derived as shown in Figure 3.49. The turn ratio should be modified as

$$a' = \frac{V/\sqrt{3}}{V/3a} = \sqrt{3}a = \sqrt{3}\frac{N_1}{N_2} = \frac{N_1}{N_2/\sqrt{3}}$$

$$a' = N_1 : \frac{N_2}{\sqrt{3}} \tag{3.92}$$

Example 3.15 Three units of 40 kVA, 2400/240 V, single-phase transformers are connected to form a three-phase 4200/240 V transformer bank. The equivalent impedance of each transformer referred to low voltage side is $0.011 + j\,0.015\,\Omega$. A three-phase load of 100 kVA, 240 V, 0.8 pf lagging load is supplied by the transformer bank.

(a) Draw the circuit diagram for the transformer connection
(b) Calculate the transformer winding currents
(c) Calculate the primary voltage (line-to-line) required
(d) Calculate the voltage regulation.

Solution

(a) The three-phase transformer bank need to feed a 240 V load and the primary voltage should be 4200 V.
This indicates that the primary should be connected in star so that the line voltage becomes $V_{p(line)} = \sqrt{3}V_{p(phase)} = \sqrt{3}(2400) = 4200V$. The secondary should be connected in delta since $V_{s(line)} = V_{s(phase)} = 240V$. The connection is shown in Figure 3.58.

(b) Currents are calculated as
Secondary line current

$$S_{Load} = \sqrt{3}V_sI_s$$

$$I_s = \frac{S_{load}}{\sqrt{3}V_s} = \frac{100}{\sqrt{3}(240)} = 0.2405kA$$

Secondary phase current

$$I_2 = \frac{I_s}{\sqrt{3}} = \frac{0.2405}{\sqrt{3}} = 0.1668kA$$

Turn Ratio is 2400/240 = 10

The phase and line currents on the primary side

$$I_p = \frac{0.1668}{10} = 16.68A$$

(Figure E3.15a)

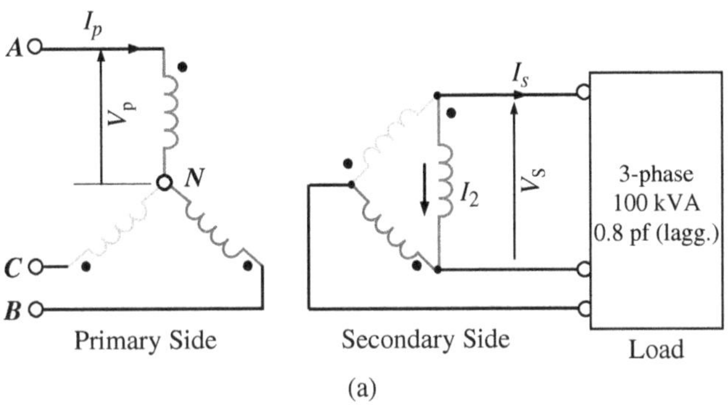

Figure E3.15a

(c) The primary side voltage can be computed by using per-phase equivalent circuit

$$Z_{eq1} = (a)^2\,(0.011 + j0.015) = 1.866\angle 53.75^0$$

Load pf = 0.8

$$\theta = \cos^{-1}(0.8) = 36.9^0$$

(Figure E3.15b)

Figure E3.15b

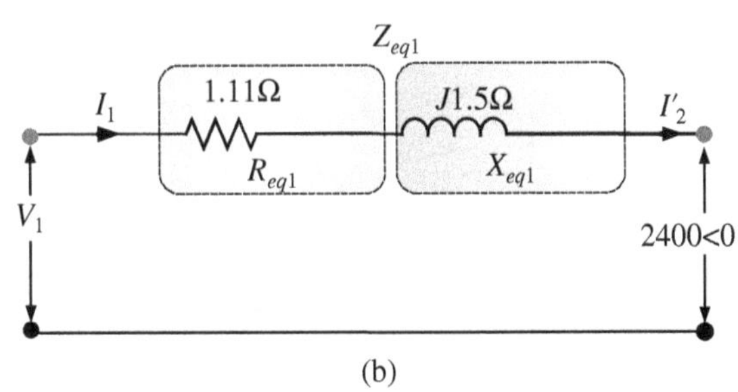

$$I_2' = 16.68\angle -36.9^0$$

The primary side applied voltage is

$$\overline{V}_1 = \overline{V}_2' + \overline{I}_2'\overline{Z}_{eq1} = 2400\angle 0 + 16.68\angle -36.9\,(1.866\angle 53.75)$$

$$\overline{V}_1 = 2400\angle 0 + 31.124\angle 16.85 = 2400 + j0 + 29.787 + j9.022$$

$$\overline{V}_1 = 2429.787 + j9.022 = 2429.80\angle 0.213$$

(d) Voltage Regulation

$$VR = \frac{|V_1| - |V_2'|}{|V_2'|} \times 100\% = \frac{2429.80 - 2400}{2400} \times 100\% = 1.24\%$$

3.15 Open-Delta Connection or V Connection

The major advantage of using delta-delta connected three-phase transformer (constructed from three units single-phase transformer), is the continuity of power supply in the event of failure of one single-phase transformer. Even with the loss of one single-phase transformer, balanced three-phase output is still available from two units of single-phase transformer, however, with reduced output power. A delta-delta three-phase transformer supplying a three-phase load is shown in Figure 3.50.

The KVA supplied by three-phase delta-connected transformer unit is given as

$$(KVA)_\Delta = \sqrt{3}V_L I_L^* \tag{3.93}$$

Where I_L^* is the conjugate of line current. The KVA supplied can be determined in terms of phase quantities.

$$(KVA)_\Delta = \sqrt{3}V_P\left(\sqrt{3}I_P^*\right) = 3V_P I_P^* \tag{3.94}$$

Considering one single-phase transformer is out of service for regular maintenance or due to fault. The remaining two units of single-phase transformer can now still produce balanced three-phase output as shown in Figure 3.51. This is called open-delta or V-connection (due to the shape of the remaining windings). Now there are two currents forming balanced three-phase currents. Look at Figure 3.51, the line current and phase currents are now same

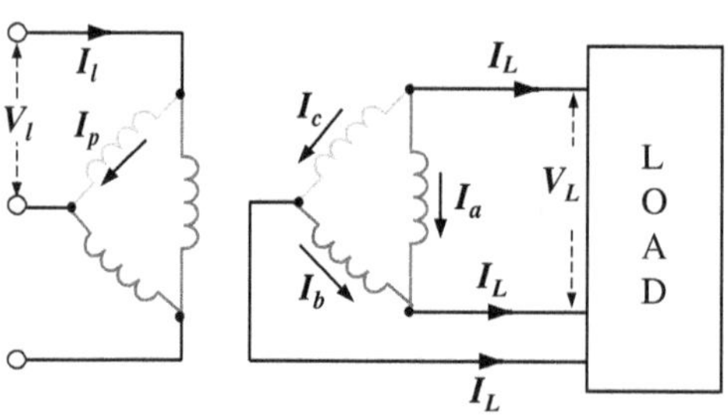

Figure 3.50 Three single-phase transformer connected in delta-delta.

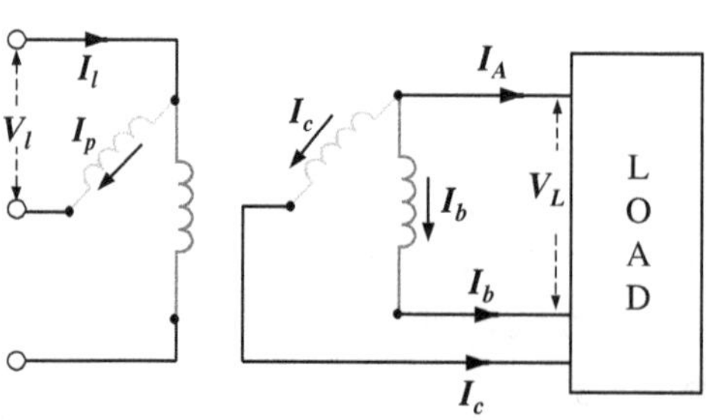

Figure 3.51 Open-delta connected transformer bank.

(I_b and I_c) due to absence of one winding. Thus, the output KVA delivered to the load under open-delta of V-connection is given as

$$(KVA)_V = \sqrt{3}V_L I_P^* = \sqrt{3}V_P I_P^* \tag{3.95}$$

The ratio of power supplied under healthy (Δ) and faulty (V) condition is

$$\frac{(KVA)_V}{(KVA)_\Delta} = \frac{\sqrt{3}V_P I_P^*}{3V_P I_P^*} = \frac{1}{\sqrt{3}} = 0.577 \tag{3.96}$$

The power supplied under open-delta condition is 57.7% of the power supplied during the healthy delta connected transformer bank.

Under open-delta condition, the two transformers operate at the lower KVA $\left(\sqrt{3}V_{phase}I_{Phase}\right)$ compared with the sum of the individual transformer KVA ratings $(2V_{phase}I_{Phase})$. The ratio is called utilization factor or rating factor. Utilization factor for open-delta and closed-delta are different.

Utlization Factor for open-delta connection is;

$$\text{Utlization Factor} = \frac{\text{Actual available KVA ratings}}{\text{Sum of KVA ratings of the installed transformers}}$$

$$= \frac{\sqrt{3}V_{phase}I_{phase}}{2V_{phase}I_{phase}} = \frac{\sqrt{3}}{2} = 0.866$$

Utlization Factor for closed-delta connection is;

$$\text{Utlization Factor} = \frac{\text{Actual available KVA ratings}}{\text{Sum of KVA ratings of the installed transformers}}$$

$$= \frac{\sqrt{3}V_{line}I_{line}}{\sqrt{3}V_{line}I_{line}} = 1$$

Let us consider that transformer 'a' is out of service and transformer 'b' and transformer 'c' are supplying power as open delta. The power supplied by two remaining healthy transformers (b and c) depends upon the load power factor. Let us analyse the power share by the two transformers. Considering a three-phase balanced load supplied by open-delta source as shown in Figure 3.52. Assuming a lagging load power factor angle as ϕ.

The corresponding phasor diagram is shown in Figure 3.53. The load phase voltages are V_{an}, V_{bn}, V_{cn} which are 120° apart. The phase currents are I_a, I_b, I_c lagging by the phase voltages by an angle ϕ.

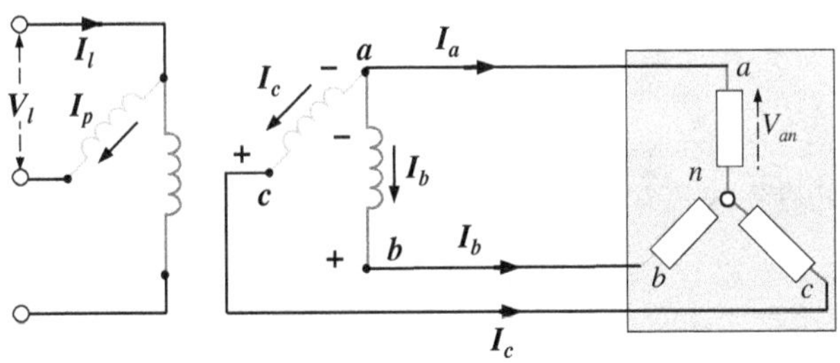

Figure 3.52 Open-delta source supplying a balanced three-phase load.

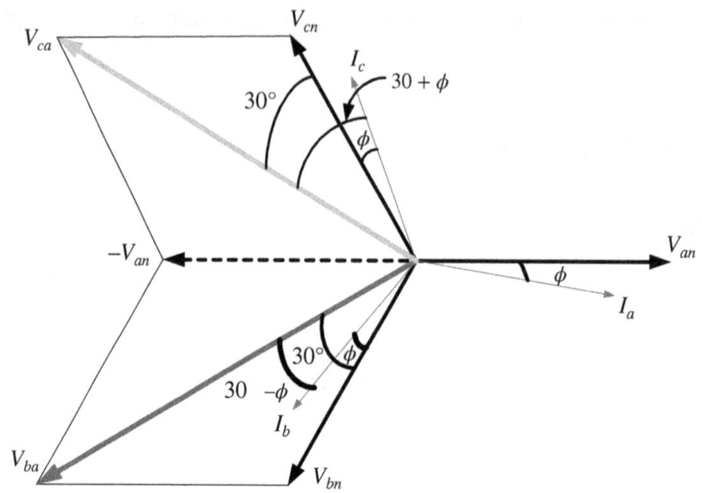

Figure 3.53 Phase diagram of open-delta connection.

The KVA supplied by transformer 'b', is the product of the line voltage (V_{ba}) and conjugate of corresponding current, I_{b}. Look at Figure 3.53, terminal 'b' is positive since the current is flowing out of this terminal. Hence, the voltage across the transformer 'b' winding is V_{ba}.

The line voltage V_{ba} is given as

$$V_{ba} = V_{bn} - V_{an}$$

The current through this transformer (b) secondary winding is given as

$$\bar{I}_b = |I_b| \angle (30 - \phi) - \text{Leading}$$

From the phasor diagram, it is seen that the phase angle between the line voltage V_{ba} and current I_{b} is $30 - \phi$. Hence, the KVA supplied by transformer 'b' is

$$(KVA)_b = V_{ba}\left(I_b \angle 30 - \phi\right)^* = V_{ba} I_b \angle \phi - 30 \tag{3.97}$$

It is further noticed that V_{ba} and I_{b} are phase voltage and phase current of the open-delta transformer, respectively. Hence, the KVA supplied by this transformer is given as

$$(KVA)_b = V_{ba}\left(I_b \angle 30 - \phi\right)^* = V_{ba} I_b \angle \phi - 30 = V_{phase} I_{phase} \angle (\phi - 30)$$

$$(KVA)_b = P + jQ = V_{phase} I_{phase} \cos (\phi - 30) + j V_{phase} I_{phase} \sin (\phi - 30) \tag{3.98}$$

The angle $30 - \phi$ is taken positive because the current I_{b} is leading the line voltage V_{ba} (look at the phasor diagram of Figure 3.53). The angle is then changed to $-30 + \phi$ or $\phi - 30$ because of conjugate.

Similarly, the KVA supplied by transformer 'c', is the product of the line voltage (V_{ca}) and conjugate of corresponding current, I_{c}. Look at Figure 3.53, terminal 'c' is positive since the current is flowing out of this terminal. Hence, the voltage across the transformer 'c' winding is V_{ca}.

The line voltage V_{ca} is given as

$$V_{ca} = V_{cn} - V_{an}$$

The current through this transformer (c) secondary winding is given as

$$\bar{I}_c = |I_c| \angle (30 + \phi) - \text{Lagging}$$

From the phasor diagram, it is seen that the phase angle between the line voltage V_{ca} and current I_c is $30 + \phi$. Hence, the KVA supplied by transformer 'c' is;

$$(KVA)_c = V_{ca}\left(I_c\angle -(30+\phi)\right)^* = V_{ca}I_c\angle\phi + 30 \tag{3.99}$$

The angle $30 + \phi$ is taken as negative because the current I_c is lagging the line voltage V_{ca} (look at the phasor diagram of Figure 3.53). The angle is then changed to $\phi + 30$ because of conjugate.

It is also seen in Figure 3.52 that V_{ca} and I_c are phase voltage and phase current of the open-delta transformer, respectively. Hence, the KVA supplied by this transformer is given as

$$(KVA)_c = V_{ca}\left(I_c\angle -(30+\phi)\right)^* = V_{ca}I_c\angle\phi + 30 = V_{phase}I_{phase}\angle(\phi + 30)$$

$$(KVA)_c = P + jQ = V_{phase}I_{phase}\cos(\phi+30) + jV_{phase}I_{phase}\sin(\phi+30) \tag{3.100}$$

Total KVA supplied under open-delta condition is

$$(KVA)_V = (KVA)_b + (KVA)_c \tag{3.101}$$

Considering a purely resistive load ($\phi = 0$)supplied by the two open-delta connected transformers. In this case, the power shared by the two transformers are

$$(KVA)_{trf-1} = V_{phase}I_{phase}\angle -30 = V_{phase}I_{phase}\cos(-30) + jV_{phase}I_{phase}\sin(-30)$$

$$(KVA)_{trf-1} = \frac{\sqrt{3}}{2}V_{phase}I_{phase} - j\frac{1}{2}V_{phase}I_{phase} \tag{3.102}$$

$$(KVA)_{trf-2} = V_{phase}I_{phase}\angle 30 = V_{phase}I_{phase}\cos(30) + jV_{phase}I_{phase}\sin(30)$$

$$(KVA)_{trf-1} = \frac{\sqrt{3}}{2}V_{phase}I_{phase} + j\frac{1}{2}V_{phase}I_{phase} \tag{3.103}$$

Both transformers are supplying active and reactive powers. However, the load is resistive and does not require reactive power. The reactive powers are thus exchanged between the two transformers.

Similarly, the power shared between the two transformers can be computed for different power factor angles.

Example 3.16 A 100 kVA load is supplied by three single-phase transformer banks connected in delta-delta. The primaries are connected to 11 kV line-to-line voltage and secondaries are connected to 440 V. If one single-phase transformer is removed for service and repair, calculate current windings in lv and hv when operated in open-delta model and how much load the remaining two transformers can supply without being overloaded.

Solution

In healthy conditions, when three transformers are supplying load the line voltage is 440 V and the line current is obtained as

$$(KVA)_\Delta = \sqrt{3}V_L I_L$$

$$I_L = \frac{(KVA)_\Delta}{\sqrt{3}V_L} = \frac{100 \times 1000}{\sqrt{3}(440)} = 131.216A$$

$$a = \frac{11000}{440} = 25$$

The phase current on the secondary side

$$I_P = \frac{I_L}{\sqrt{3}} = \frac{131.216}{\sqrt{3}} = 75.75A$$

The phase currents on the primary side

$$I_p = \frac{I_P}{a} = \frac{75.75}{25} = 3.03A$$

The line current on primary side is

$$I_l = \sqrt{3}I_p = \sqrt{3}(3.03) = 5.24A$$

The currents are shown in the Figure E3.16a

Under open-delta condition, the line current becomes phase currents and also the currents should not exceed the 75.75 A (not to overload the transformer), Figure E3.16b.

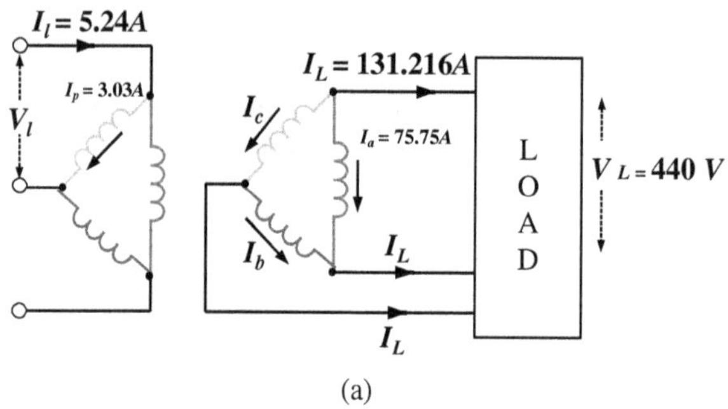

Figure E3.16a

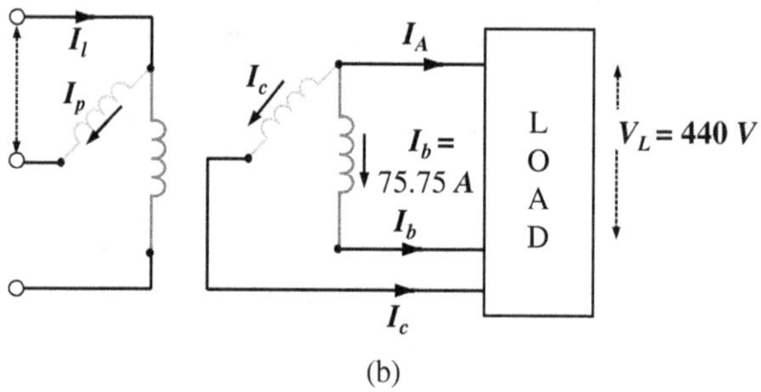

Figure E3.16b

Under open-delta condition, the power supplied to the load is

$$(KVA)_V = \sqrt{3}V_L I_P = \sqrt{3}(440)(75.75) = 57.729kVA$$

Check:

$$\frac{(KVA)_\Delta}{(KVA)_V} = \frac{100}{57.729} = \sqrt{3}$$

3.16 Harmonics in a Single-Phase Transformer

Harmonics are the higher frequency and non-fundamental components of any periodic waveform. Considering periodic current waveform with harmonics. The periodic current waveform is written as sum of fundamental voltage and harmonic currents as

$$i(t) = I_{m1}\sin(\omega t) + I_{m3}\sin(3\omega t) + I_{m5}\sin(5\omega t) + \dots\dots + I_{mn}\sin(n\omega t) \tag{3.104}$$

The periodic current waveform containing fundamental and 3rd harmonic components is shown in Figure 3.54. The harmonic component is assumed in phase with the fundamental. It is seen that the peak of fundamental and peak of 3rd harmonic component are opposite to each other. This causes a dip in the resultant peak. Such resultant current waveform is known as a flat-topped waveform.

If the 3rd harmonic component is in phase opposition with the fundamental component, the resultant waveform is shown in Figure 3.55.The peak of fundamental and peak of 3rd harmonic coincides. This gives a resultant peaky waveform.

In a single-phase transformer, when sinusoidal voltage is applied to the primary winding, current flows that causes sinusoidal flux in the core. This current is called exciting current. The nature of flux is governed by the B-H curve [4].

Case-1: Considering at first linear B-H curve. The X-axis of B-H curve is the exciting current (proportional to H) and the Y-axis is the flux (proportional to B). If the flux is sinusoidal

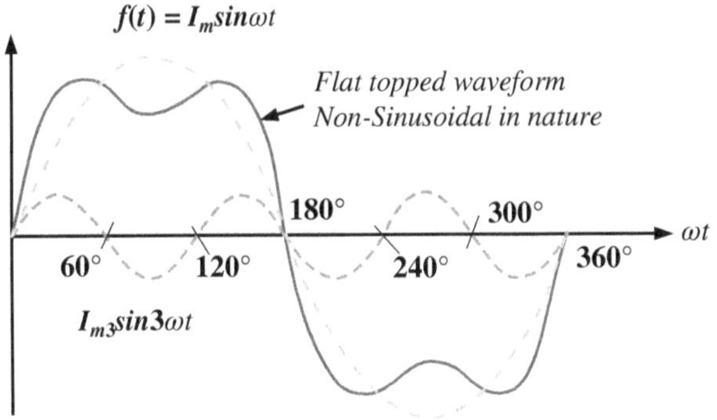

Figure 3.54 Fundamental + 3rd harmonic waveform.

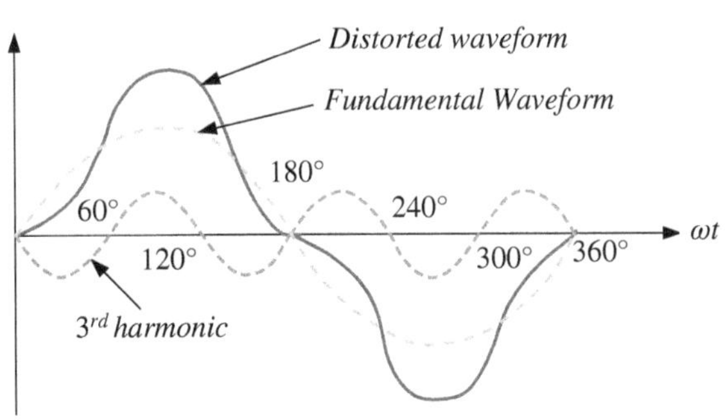

Figure 3.55 Fundamental – 3rd harmonic waveform.

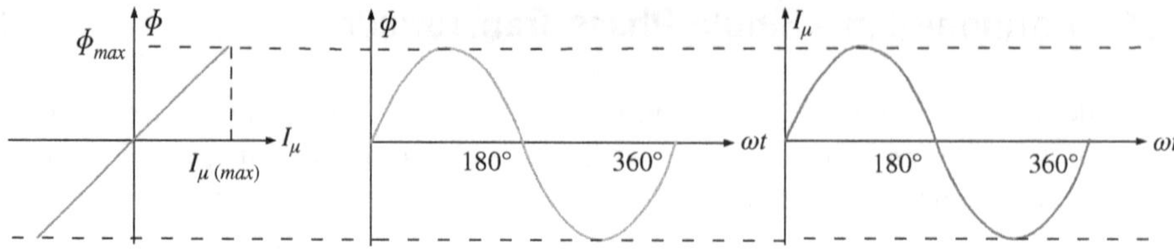

Figure 3.56 Flux and current shapes for linear B-H curve.

the current is also sinusoidal as shown in Figure 3.56. The flux and current both are sinusoidal since the relation between them is linear. However, this is an ideal case which does not exist in reality.

Case-2: Consider a real situation, where the relation between B and H is non-linear or the B-H curve is non-linear. If flux is assumed sinusoidal, the exciting current would be peaky in nature as shown in Figure 3.57. It is seen from the B-H (flux-current) relation, that initially the flux changes at a fast rate while the exciting current changes slowly. In the later part of the curve, the change in flux is slow while the exciting current changes quickly. This variation can be drawn as a peaky waveform. The peaky waveform contains 3rd, 5th and 7th, etc. harmonics. However, the dominant harmonic would be the 3rd. If the flux is sinusoidal

$$\Phi(t) = \Phi_m \sin(\omega t) \tag{3.105}$$

The exciting current can be written as

$$i(t) = I_{m1} \sin(\omega t) - I_{m3} \sin(3\omega t) \tag{3.106}$$

The 3rd harmonic is opposing in nature because the resultant waveform is peaky in nature.

Since the emf induced is given as $e = -N\frac{d\Phi}{dt}$, the nature of emf will be sinusoidal. Thus, to summarize

- The flux is sinusoidal
- The induced emf is sinusoidal
- The excitation current is non-sinusoidal and peaky in nature.

Case-3: Now considering the hysteresis effect of non-linear B-H curve on the shape of flux and excitation current. Assuming sinusoidal flux, the resultant exciting current can be drawn as shown in Figure 3.58. The excitation current remains peaky in nature however, it will have some phase shift. The peaky excitation current and the corresponding sinusoidal flux is shown in the right most figure. Look at the hysteresis loop, in the beginning,

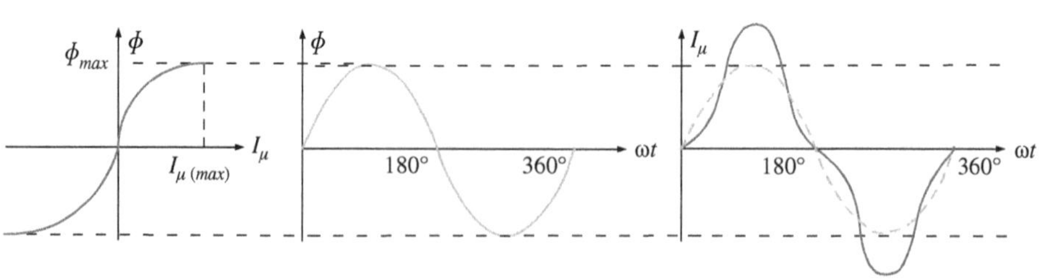

Figure 3.57 Flux and current shapes for non-linear B-H curve.

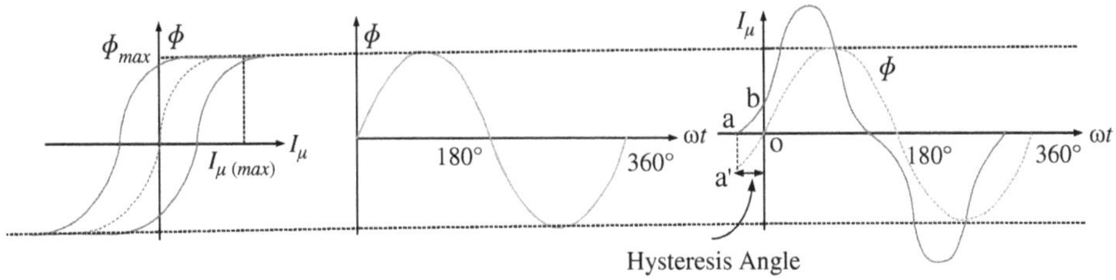

Figure 3.58 Flux and current shapes for non-linear B-H curve with hysteresis.

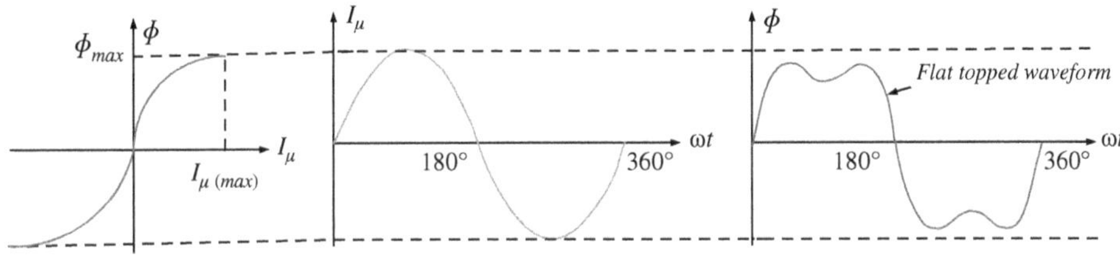

Figure 3.59 Flux and current shapes for non-linear B-H curve with sinusoidal current.

when current is zero (point a), there is a negative flux (point a′). Now the current increase in positive direction reaches point b, the flux also increases and reaches zero (point o). Thus, the flux lags the excitation current by some angle called 'hysteresis angle'.

Case-4: One last case is considered with non-linear B-H curve and sinusoidal current. Mapping the flux corresponding to the sinusoidal current. The flux waveform is found as flat topped as shown in Figure 3.59.

This means that the flux contains harmonics in the same phase as that of fundamental. Although the flux waveform contains several odd harmonics, however, the 3rd harmonic is the most dominant. The flux waveform is written as

$$\Phi(t) = \Phi_{m1}\sin(\omega t) + \Phi_{m3}\sin(3\omega t) + \Phi_{m5}\sin(5\omega t) + \ldots\ldots + \Phi_{mn}\sin(n\omega t) \quad (3.107)$$

Neglecting the 5th, 7th, etc. harmonic the flux waveform is approximated by

$$\Phi(t) = \Phi_{m1}\sin(\omega t) + \Phi_{m3}\sin(3\omega t) \quad (3.108)$$

The corresponding induced voltage is given as

$$e = -N\frac{d\Phi}{dt} = -N\frac{d}{dt}\left(\Phi_{m1}\sin(\omega t) + \Phi_{m3}\sin(3\omega t)\right)$$

$$e = -N\left(\Phi_{m1}\omega\cos(\omega t) + 3\Phi_{m3}\omega\cos(3\omega t)\right)$$

$$e = -e_{m1}\cos(\omega t) - e_{m3}\cos(3\omega t) \quad (3.109)$$

Adding the fundamental and 3rd harmonic in opposing phase leads to resultant peaky waveform. Hence, the voltage waveform for the distorted flux is also peaky in nature as shown in Figure 3.60.

Thus, to summarize:

- The current is sinusoidal
- The corresponding flux is non-sinusoidal with flat topped waveform
- The induced emf is peaky in nature.

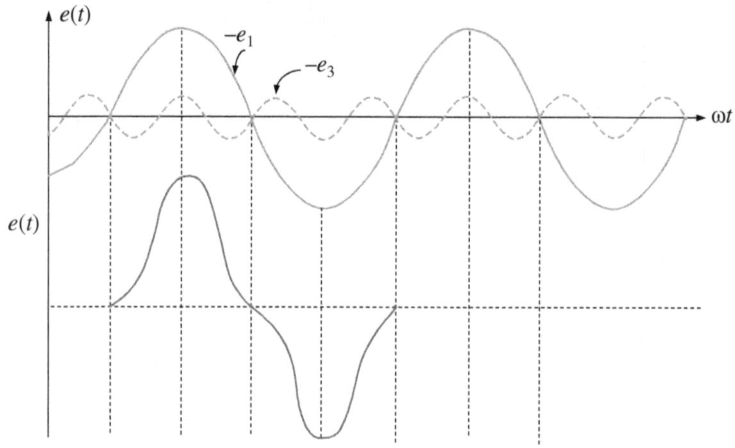

Figure 3.60 Flux and current shapes for non-linear B-H curve with sinusoidal current.

Note:

- If the flux is sinusoidal, the current becomes non-sinusoidal and induced voltage remains sinusoidal.
- If the flux is non-sinusoidal, the current is sinusoidal and the induced voltage is non-sinusoidal.
- Core loss in a transformer is approximately proportional to the square of maximum value of flux density. Thus, the core loss in a flat-topped flux wave is less than for a sine wave flux (since the peak of flat-topped wave is lower than the sine wave).
- In practice, the applied voltage is sine wave, exciting current is peaky, flux is therefore, also peaky. So, the effect of harmonics is to increase the core loss.

3.16.1 Excitation Phenomena in a Single-Phase Transformer

Considering a sinusoidal voltage applied to the primary winding at t = 0 while secondary is open circuited (no-load). Assuming no resistive and inductive voltage drop in the primary winding (ideal case), the induced emf is equal to the applied voltage.

$$v(t) = V_m \sin(\omega t)$$

$$v(t) = e(t) = V_m \sin(\omega t)$$

$$N\frac{d\Phi}{dt} = V_m \sin(\omega t)$$

$$\int d\Phi = \int \frac{V_m}{N} Sin(\omega t)\, dt$$

$$\Phi(\omega t) = -\frac{V_m}{\omega N}\cos(\omega t) + C \tag{3.110}$$

The induced voltage is taken as positive, because the sign is already taken care of in the polarity of Figure 3.61 (induced emf opposes the applied voltage). If the switching takes place at ωt = 0, the flux is also 0 at this instant, i.e. Φ(0) = 0

$$At\ \omega t = 0$$

$$\Phi(\omega t) = -\frac{V_m}{\omega N}\cos(\omega t) + C$$

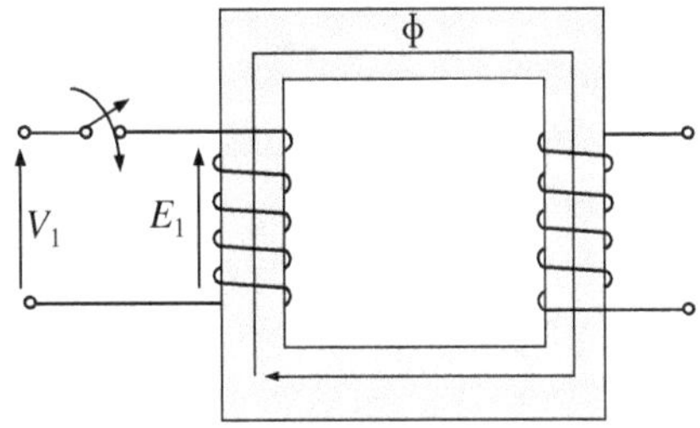

Figure 3.61 Transformer switching.

$$0 = -\frac{V_m}{\omega N}\cos(0) + C$$

$$C = \frac{V_m}{\omega N} \tag{3.111}$$

Substituting the value of integration constant C in Eq. (3.110), the following is obtained.

$$\Phi(t) = -\frac{V_m}{\omega N}\cos(\omega t) + \frac{V_m}{\omega N}$$

$$\Phi(t) = \frac{V_m}{\omega N}[1 - \cos(\omega t)]$$

$$\Phi(t) = \Phi_{max}[1 - \cos(\omega t)] \tag{3.112}$$

Let us examine the flux at different instances of time (at the switching instant, the flux is zero in the core)

$$At\ \omega t = 0$$

$$\Phi(0) = 0$$

$$At\ \omega t = \frac{\pi}{2}$$

$$\Phi\left(\frac{\pi}{2}\right) = \Phi_{max}$$

$$At\ \omega t = \pi$$

$$\Phi(\pi) = 2\Phi_{max} \tag{3.113}$$

Flux become double that of its maximum value, called the 'Doubling Effect' as shown in Figure 3.62. Due to this doubling effect, a large amount of current is drawn by the transformer at the time of switching. This is called an initial inrush of current which is approximately 6–8 times the normal rated current.

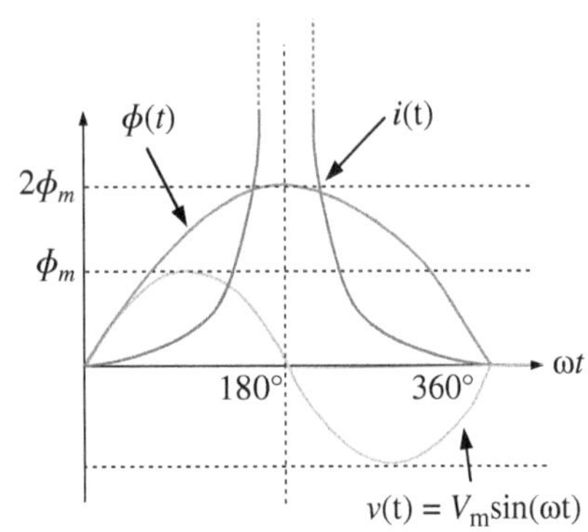

Figure 3.62 Switching transient in a transformer.

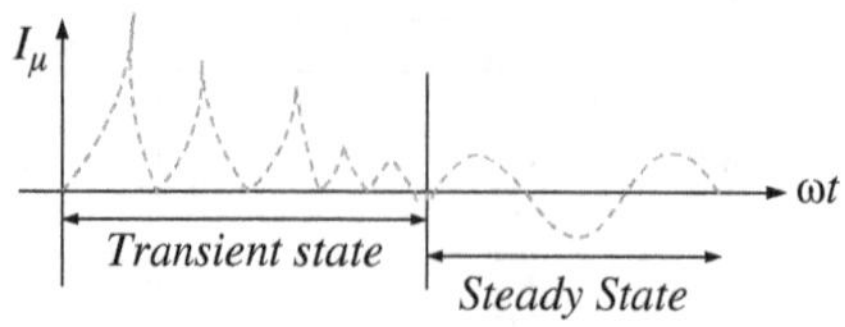

Figure 3.63 Excitation current transformer switched under no-load condition.

It is to be noted that if the transformer is switched under load condition (secondary supplying load), the inrush current will be small and the transient period will be short. This is due to the fact that the secondary flux (flux due to load current) opposes the main flux. This will cause an overall reduction in the flux and the doubling effect will not reach.

The switching transient is shown only for one cycle in Figure 3.62. The transient remains for few cycles followed by steady-state condition. The transient and steady-state currents are shown in Figure 3.63

If switching is done at $\omega t = \pi/2$, the flux is zero at the instant of switching

$$\begin{aligned} &At\ \omega t = {}^{\pi}/_{2} \\ \Phi\left(\frac{\pi}{2}\right) &= -\frac{V_m}{\omega N}\cos\left(\frac{\pi}{2}\right) + C \\ 0 &= -\frac{V_m}{\omega N}\cos\left(\frac{\pi}{2}\right) + C \\ C &= 0 \end{aligned} \tag{3.114}$$

Hence, the flux is given as

$$\Phi(\omega t) = -\frac{V_m}{\omega N}\cos(\omega t) = -\Phi_{\max}\cos(\omega t) \tag{3.115}$$

In this case, maximum flux goes to $\Phi_{\max}$ and no doubling effect is seen. Hence, no inrush current is in the transformer if the switching is done at $\omega t = \pi/2$.

3.16.2 Harmonics in a Three-Phase Transformer

In three-phase transformer, two types of harmonics are present.

1. Order of harmonics is $3n \pm 1$; $n = 2, 4, 6\ldots\ldots$
2. Order of harmonics $3n$; $n = 1, 3, 5, \ldots.$

Let us examine each harmonic component.

For $(3n+1)^{\text{th}}$ harmonics, the most dominant is the 7th.

The fundamental component for *abc* phase sequence (called positive phase sequence) is given as

$$\begin{aligned} I_a &= I_m \sin(\omega t) \\ I_b &= I_m \sin(\omega t - 120) \\ I_c &= I_m \sin(\omega t + 120) \\ I_a + I_b + I_c &= 0 \end{aligned} \tag{3.116}$$

The fundamental component for *acb* phase sequence (called negative phase sequence) is given as

$$I_a = I_m \sin(\omega t)$$
$$I_b = I_m \sin(\omega t + 120)$$
$$I_c = I_m \sin(\omega t - 120)$$
$$I_a + I_b + I_c = 0 \tag{3.117}$$

The 7th harmonics are (positive phase sequence)

$$I_{a7} = I_{m7} \sin(7\omega t)$$
$$I_{b7} = I_{m7} \sin(7\omega t - 7(120)) = I_{m7} \sin(7\omega t - 120)$$
$$I_{c7} = I_{m7} \sin(7\omega t + 7(120)) = I_{m7} \sin(7\omega t + 120)$$
$$I_{a7} + I_{b7} + I_{c7} = 0 \tag{3.118}$$

For $(3n\text{-}1)^{th}$ harmonics, the most dominant is the 5th (negative phase sequence). The 5th harmonics are given as;

$$I_{a5} = I_{m5} \sin(5\omega t)$$
$$I_{b5} = I_{m5} \sin(5\omega t - 5(120)) = I_{m5} \sin(5\omega t + 120)$$
$$I_{c5} = I_{m5} \sin(5\omega t + 5(120)) = I_{m5} \sin(5\omega t - 120)$$
$$I_{a5} + I_{b5} + I_{c5} = 0 \tag{3.119}$$

The $3n$ harmonic is multiple of the 3rd harmonic. The most dominant is 3rd harmonics. The 3rd harmonics components are

$$I_{a3} = I_{m3} \sin(3\omega t)$$
$$I_{b3} = I_{m3} \sin(3\omega t - 3(120)) = I_{m3} \sin(3\omega t)$$
$$I_{c3} = I_{m3} \sin(3\omega t + 3(120)) = I_{m3} \sin(3\omega t)$$
$$I_{a5} + I_{b5} + I_{c5} = 3I_{m3} \sin(3\omega t) \tag{3.120}$$

In a star connection, the flow of 3rd harmonic current depends upon the neutral connection. Applying KCL at the neutral point in isolated neutral (not grounded) connection in Figure 3.64a, gives

$$I_N = 0 = I_a + I_b + I_c$$
$$0 = I_m \sin(\omega t) + I_m \sin(\omega t - 120) + I_m \sin(\omega t + 120) \tag{3.121}$$

The neutral current is zero because the neutral current has no path to flow. If the current contains the 3rd harmonic, the sum of current still remains zero. No 3rd harmonic current can flow in star-connected windings with isolated neutral.

$$I_N = I_a + I_b + I_c + I_{a3} + I_{b3} + I_{c3}$$
$$0 = I_m \sin(\omega t) + I_m \sin(\omega t - 120) + I_m \sin(\omega t + 120) + 3I_{m3} \sin(3\omega t) \tag{3.122}$$

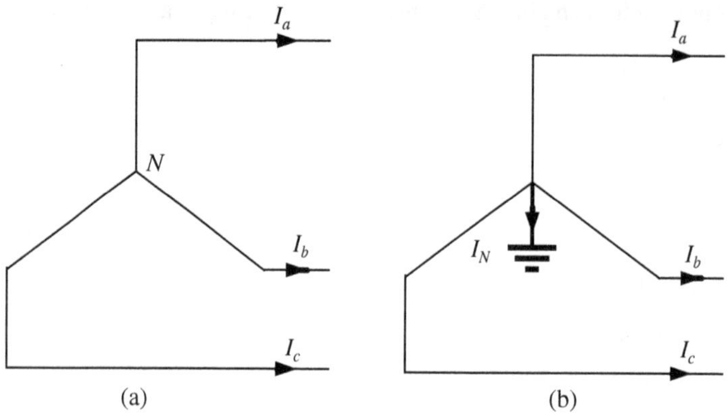

Figure 3.64 Star connection, (a) Isolated neutral, (b) Grounded Neutral.

If the neutral is grounded (Figure 3.64b), the sum of the three-phase current is equal to the neutral current. Applying KCL at the neutral point

$$I_N = I_a + I_b + I_c$$
$$I_N = I_m \sin(\omega t) + I_m \sin(\omega t - 120) + I_m \sin(\omega t + 120) = 0 \tag{3.123}$$

If the current is balanced three-phase sinusoidal, the sum of current is zero. However, if the current contains 3rd harmonic, the neutral current is three times the 3rd harmonic

$$I_N = I_a + I_b + I_c + I_{a3} + I_{b3} + I_{c3}$$
$$I_N = I_m \sin(\omega t) + I_m \sin(\omega t - 120) + I_m \sin(\omega t + 120) + 3I_{m3} \sin(3\omega t)$$
$$I_N = 3I_{m3} \sin(3\omega t) \tag{3.124}$$

Thus, the 3rd harmonic current flows in star-connected windings if the neutral is grounded.

Note: The following observations are made (discussion is given in previous Section 3.16)

1. If the current contains 3rd harmonic, the flux will be sinusoidal and hence the voltage will be sinusoidal. This is the case of star with neutral grounded.
2. If the current does not carry 3rd harmonic and is purely sinusoidal, the flux will be distorted containing the 3rd harmonics. The voltage will also be distorted. This is the case of star connection with isolated neutral.

In delta-connected windings, the phase currents form a loop as shown in Figure 3.65. If the phase current contains 3rd harmonic component, the sum of three currents will flow within the loop. The phase current thus carries 3rd harmonic current.

The 3rd harmonic current in the loop is give as

$$I_{ab3} = I_{bc3} = I_{ca3} = I_{m3} \sin(3\omega t)$$
$$I_{loop} = 3I_{m3} \sin(3\omega t) \tag{3.125}$$

However, the line current will not carry the 3rd harmonic. Since the 3rd harmonic currents are all in phase, they cancel out at the point of connection. Thus, the line current only carries

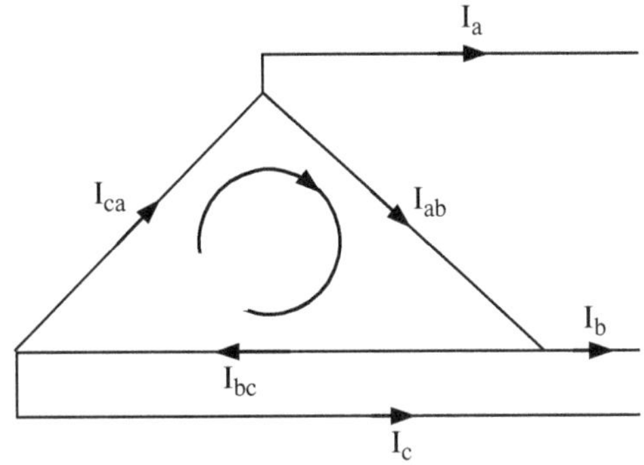

Figure 3.65 Delta connection.

the fundamental while the 3rd harmonic is restricted to the phase loop.

$$I_{a3} = I_{ca3} - I_{ab3} = I_{m3} \sin(3\omega t) - I_{m3} \sin(3\omega t) = 0 \tag{3.126}$$

3.16.2.1 Star-Delta Connection with Grounded Neutral

To examine the harmonics in different connection type, consider a star-delta three-phase transformer with primary star and secondary delta shown in Figure 3.66. Let us assume that the switch SW1 is closed (grounded neutral) and SW2 open (no-load secondary). Since the transformer is operating under no-load, excitation current flows in the primary winding that contains 3rd harmonic component as seen in Section 3.16 (due to non-linear BH curve). The line/phase currents of primary side is

$$I_N = I_A + I_B + I_C + I_{A3} + I_{B3} + I_{C3}$$
$$I_N = I_m \sin(\omega t) + I_m \sin(\omega t - 120) + I_m \sin(\omega t + 120) + 3I_{m3} \sin(3\omega t)$$
$$I_N = 3I_{m3} \sin(3\omega t) \tag{3.127}$$

The neutral line current carries only the 3rd harmonic. Because the exciting current is non-sinusoidal, the flux in the core and hence induced voltages in the windings will be sinusoidal.

Since the secondary is open, the voltage across the secondary is same as induced emf.

$$v_{\Delta 0} = v_A + v_B + v_C = 0 \tag{3.128}$$

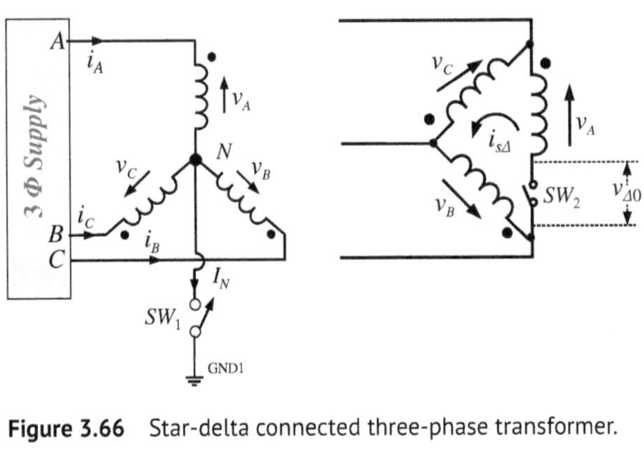

Figure 3.66 Star-delta connected three-phase transformer.

3.16.2.2 Star-Delta Connection without Grounded Neutral

Consider SW1 and SW2 open that means, primary is with isolated neutral connection and secondary is on no-load. In this case, excitation current flowing in the primary winding will be sinusoidal since the 3rd harmonic current has no path to flow. Due to non-linear B-H curve, the flux is now distorted containing 3rd harmonic components. This causes the induced voltage distortion as well. Thus, the primary phase voltages is

$$v_A = v_{A1} + v_{A3}$$
$$v_B = v_{B1} + v_{B3}$$
$$v_C = v_{C1} + v_{C3} \tag{3.129}$$

The line-to-line voltage of the primary side will not contain 3rd harmonic component as they cancel out because 3rd harmonic voltages are in the same phase

$$v_{AB} = v_A - v_B = v_{A1} + v_{A3} - v_{B1} - v_{B3}$$
$$v_{AB} = v_{A1} - v_{B1} \tag{3.130}$$

The secondary side phase voltage also contains 3rd harmonic component (due to transformation from phase of primary). The net voltage in the closed loop of the delta side

$$v_{\Delta 0} = v_A + v_B + v_C = v_{A1} + v_{B1} + v_{C1} + v_{A3} + v_{B3} + v_{C3}$$
$$v_{\Delta 0} = 0 + v_{A3} + v_{B3} + v_{C3} = 3V_m \sin(3\omega t) \tag{3.131}$$

Hence, the voltage across the open delta is the sum of three harmonic voltages.

3.16.3 Summary

In order to flow the 3rd harmonic current, either

- Star neutral is grounded
- A closed path is available (as in delta).

If 3rd harmonic current is allowed to flow in a transformer, then the

- Current will be distorted containing 3rd harmonic
- Flux will be sinusoidal without any harmonics
- Voltage will be sinusoidal without any harmonics.

3.16.4 Star-Star with Isolated Neutral

If it is necessary to use Star-Star connection without grounded neutral, then no path will be available for 3rd harmonics and hence the flux and voltage will be distorted. In such cases, a third winding called tertiary winding in delta connection is used to allow the flow of 3rd harmonics as shown in Figure 3.67. The delta connected winding is also wound on the same core. The tertiary winding provides a closed path for 3rd harmonic to flow and thus making the flux and voltage sinusoidal. The tertiary winding may be used to supply some auxiliary load.

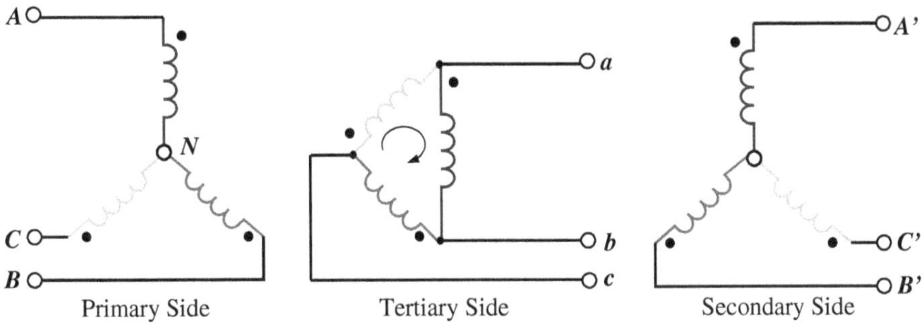

Figure 3.67 Star-star with delta connected tertiary winding.

3.17 Disadvantages of Harmonics in Transformer

Harmonic voltages and harmonic currents have different effects on the transformer behaviour. Harmonic voltages are considered more problematic than harmonic current. Means are adopted to suppress harmonic voltages.

3.17.1 Effect of Harmonic Currents

Losses: The presence of harmonic current incurs additional copper losses and hence increased temperature rise.

Flux is flat-topped when the 3rd harmonic currents are not flowing in the transformer winding and the emf is peaky when the flux is flat topped.

Core loss is composed of hysteresis and eddy current losses. The hysteresis loss is a major portion of the overall core loss. The hysteresis loss is directly proportional to square of the flux and the eddy current loss is directly proportional to the square of the emf. Flat-topped waveform means reduced peak and peaky waveform means higher peak value. Thus, for flat-topped flux waveform hysteresis loss reduces and eddy current loss increases and overall core loss is reduced. If the emf is flat topped and flux is peaky, the core loss is increased.

3.17.2 Electromagnetic Interference

The harmonic currents generate harmonic fluxes around the transmission line conductors. These harmonic fluxes induce emfs in the communication circuit in the vicinity. These cause interference in the communication lines.

3.17.3 Effect of Harmonic Voltages

Insulation Failure: Due to the presence of harmonics, the voltage increases and may exceed the rated voltage of the winding. In such cases, the dielectric losses increase, the life of the insulation is reduced and the transformer efficiency is reduced and insulation failure is possible. For example, in isolated neutral star/star three-phase transformer, the phase voltages are more than the rated voltages contain higher harmonics emfs.

Electrostatic Interference: A communication circuit running in parallel to a transmission line is subjected to interference due to harmonic voltages. The transmission line and communication lines are coupled using stray capacitance. The resonance may occur when the capacitor reactance becomes equal to the inductive reactance at some harmonic frequency.

3.17.4 Summary

The harmonics of order $3n \pm 1$ are present in both star and delta connections. However, their magnitudes are small and are thus neglected. The harmonics of order 3n is summarized in Table 3.3.

3.17.5 Oscillating Neutral Phenomena

Three-phase transformer with independent magnetic circuit (three single-phase transformer forming a bank of three-phase transformer or shell-type construction) exhibit an oscillating neutral point when connected in star-star configuration. The neutral voltage oscillates at three times the fundamental frequency. The effects of oscillating neutral voltages are

- Fluctuation in the phase voltages.
- The maximum voltage across each phase varies with time.
- The relationship between line voltage and phase voltage is no more $V_{line} \neq \sqrt{3}V_{phase}$.

The oscillation of neutral is explained using the Figure 3.68.

Fundamental emf and 3rd harmonic emf are shown for three different instants of time in Figure 3.68. They combine to form the resultant. The 3rd harmonic voltages are co-phasal (i.e. lies in same phase) and rotates at three times the speed of fundamental in the same direction. The resultant of fundamental and 3rd harmonic voltages is shown in the dotted line on the right most side of the Figure 3.68. Since the 3rd harmonic rotates at a speed

Table 3.3 Summary of three-phase transformer connection.

Connection type	3rd Harmonic current	3rd harmonic flux/voltage
	Absent	Present
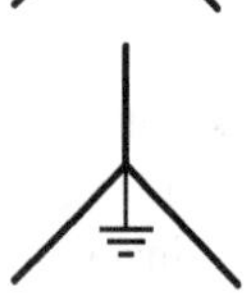	Present and peaky in nature	Absent
	Present in Phase current Absent in line current	Absent

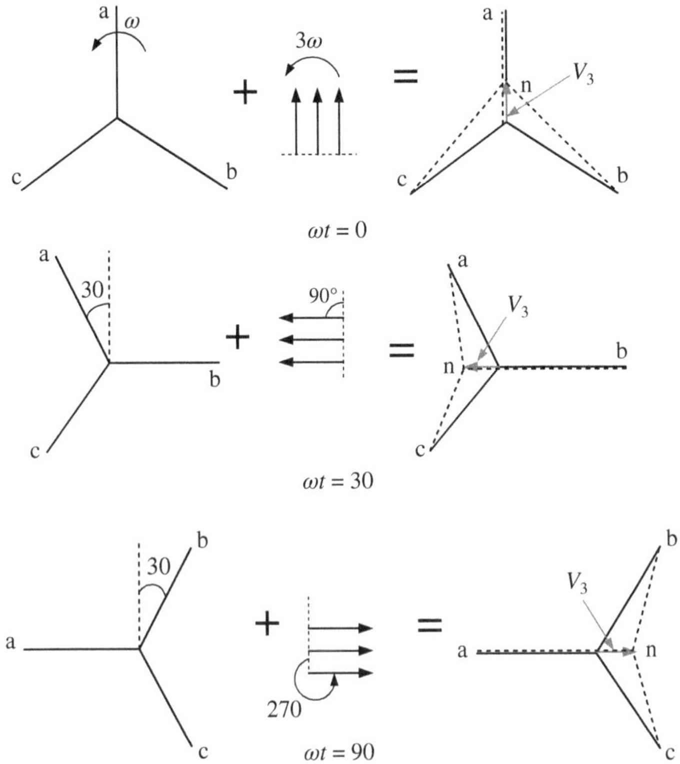

Figure 3.68 Oscillation of neutral point in Y-Y connection.

three times faster than the fundamental, when fundamental is at 30° the 3rd harmonic is at $3 \times 30 = 90°$. It is seen from the figure that the neutral itself changes its position with time forming a circle. This is called oscillating neutral.

The net phase voltage is the sum of the fundamental and 3rd harmonic voltage i.e. $V = V_{m1} + V_{m3}$

The line voltages remain unaffected since the 3rd harmonic cancels out and does not appear in the line. The resultant phase voltage is $\sqrt{(V_1)^2 + (V_3)^2}$. Thus, the relationship between the line and phase voltage of $V_{line} \neq \sqrt{3}V_{phase}$ is no more valid.

Hence, the line voltage is not equal to sqrt(3) time the phase voltage for Star/Star transformer connection with separate magnetic circuit and without grounded neutral (or isolated neutral).

3.18 Open Circuit and Short-Circuit Conditions in a Three-Phase Transformer

This section describes the behaviour of a three-phase transformer when one of the feeders supplying the primary winding gets open-circuited. For simplicity, turn ratio is taken as unity.

For the discussion, the instantaneous polarity is shown for different phases. Consider at first a Delta/Star connection. When one feeder is disconnected, the input to the primary delta becomes single-phase (V). The other two phases must then be V/2 and V/2 to satisfy the KVL in the delta loop as shown in Figure 3.69. The corresponding phases in the secondary is shown in Figure 3.69. Two phases will have V/2 voltage across them and one phase have V. Now applying KVL (considering the polarity of the voltages), the three-line voltages are single-phase and 1.5 V, 1.5 V, and 0. Hence unbalanced voltages are observed at the output secondary lines.

When one feeder becomes open-circuited, sometimes the connection of the primary delta is made in such a way that voltage across one of the primary winding is reduced to zero as shown in Figure 3.69b. In such a connection, the two-feeder yield voltage V across two phases is as shown in Figure 3.69b and one phase voltage is 0 due to connection.

The corresponding phase voltages (V, V, and 0) are marked in the secondary side. The line voltage is obtained by using KVL. The three-line voltages are single-phase and V, V, and 2 V. This produces unbalanced condition at the secondary side output.

Next, consider a Star/Delta three-phase transformer as shown in Figure 3.70. The single phasing is assumed in one phase of the primary. The applied voltage is V, that causes V/2 voltages across the two phases. The corresponding phase voltages are V/2, V/2, and 0 as shown in the Figure 3.70. The secondary line voltages become single-phase with magnitudes, V/2, V/2, and 0.

In case the open-circuit phase is connected with live terminal on primary side as shown in Figure 3.70b, phase voltages of primary become V/2, V/2, and V/2 (use KVL). The corresponding line voltages on secondary side is V/2, V/2, and V/2.

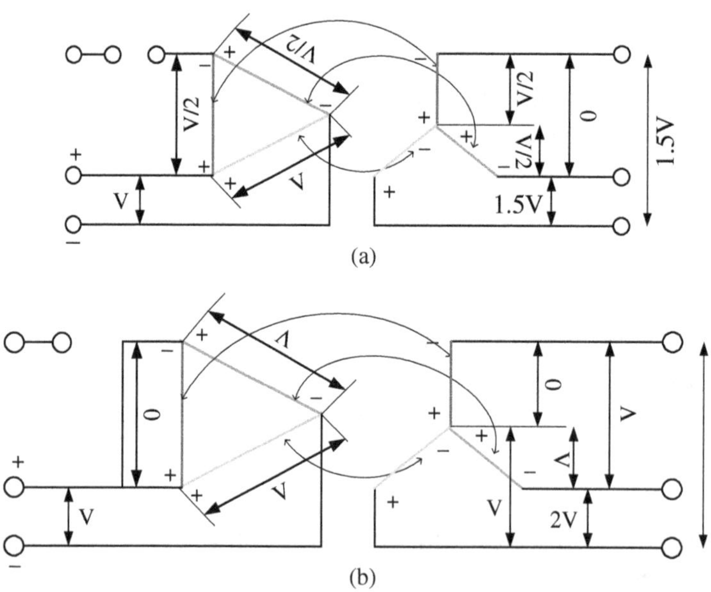

Figure 3.69 Single phasing in delta/star connection.

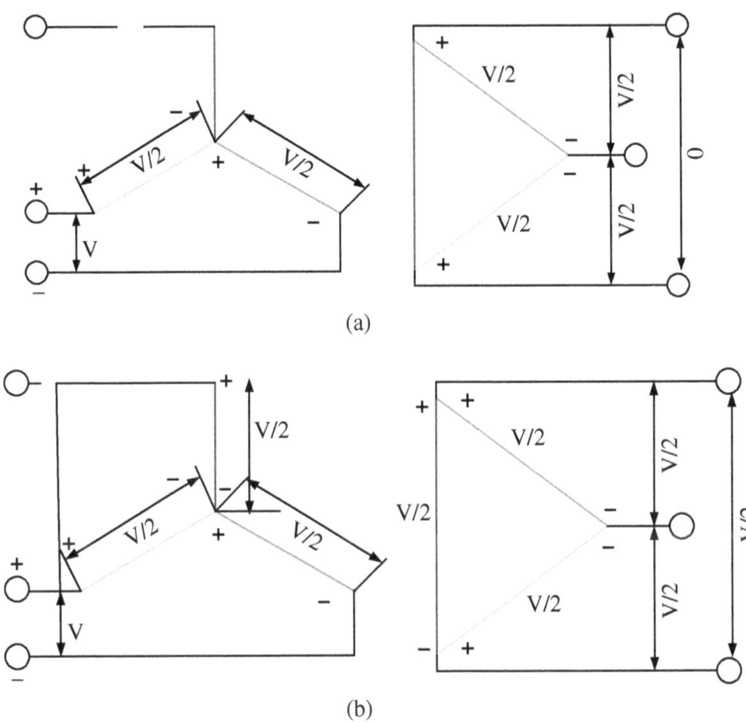

Figure 3.70 Single phasing in star/delta connection.

3.19 Matlab/Simulink Model of a Single-Phase Transformer

Simulink Model of a single-phase transformer is shown in Figure 3.71 using Powersystem blocksets of Matlab/Simulink. In the powersystem toolbox model of transformer with multiple windings are built in. The parameters can be set into the parameter setup box and the desired model can be simulated. The parameters used for simulation purpose is given in Table 3.4. The following model is simply implemented to show the working of a single-phase transformer.

The following parameters have been used for the simulation examples presented in this section.

Table 3.4 Simulation parameter.

Transformer type		Turn ratio	Voltage parameter
Single phase	Step up	1 : 2	Input voltage = 120 V, Output voltage = 240 V
	Step down	1 : 0.5	Input voltage = 120 V, Output voltage = 60 V
Autotransformer	Additive	1 : 2	Input voltage = 120 V, Output voltage = 180 V
	Subtractive	1 : 0.5	Input voltage = 120 V, Output voltage = 60 V
Three-phase transformer		1 : 2	Input voltage = 120 V

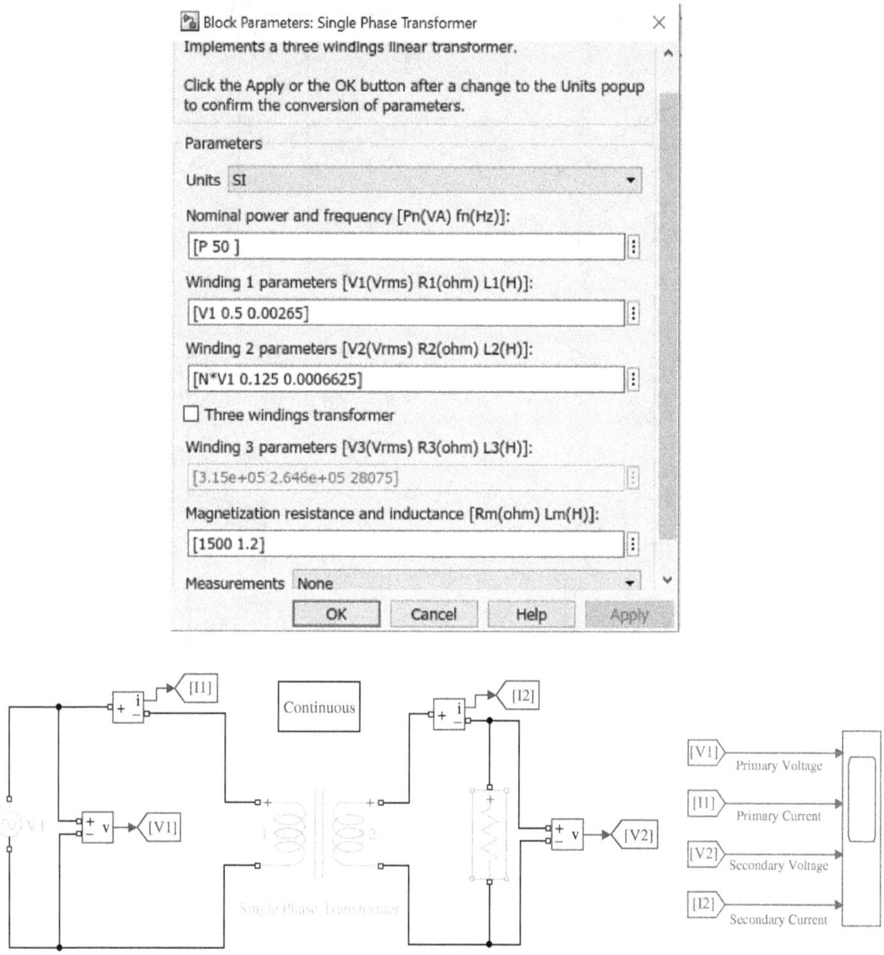

Figure 3.71 Simulink model for single-phase transformer (File Name: M1.slx).

Single Phase Transformer: The Simulink model of a single-phase transformer is depicted in Figure 3.71, Where,

V_1 = Primary side voltage
I_1 = Primary side current
V_2 = Secondary side voltage
I_2 = Secondary side current.

The parameter setup box is also shown where values of input voltage V1, power rating P and turn ratio can be inserted to obtain the output and input voltages and currents, waveforms and values.

The results shown in Figure 3.72, where primary voltage, primary currents and secondary voltage and secondary currents are shown for step-up operation. Step-down operation is shown in Figure 3.73.

(a) Step up
(b) Step down

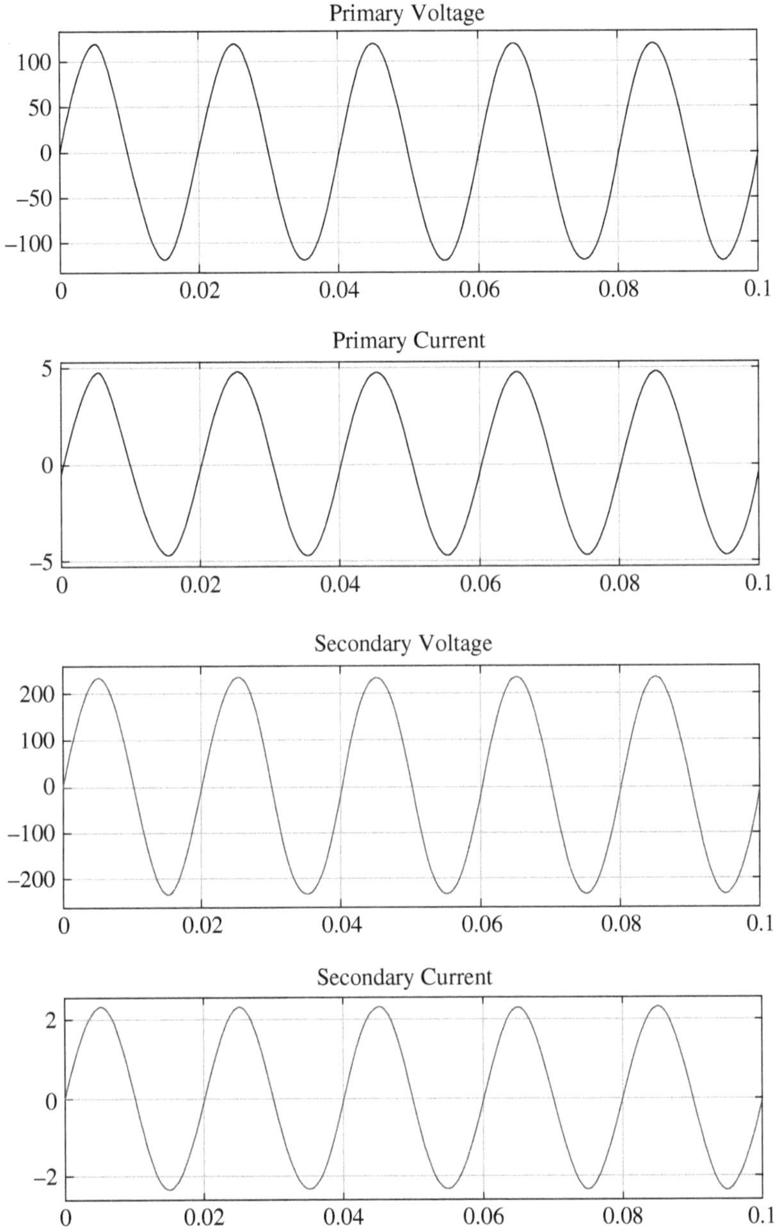

Figure 3.72 Voltage and current waveforms of single-phase step up transformer.

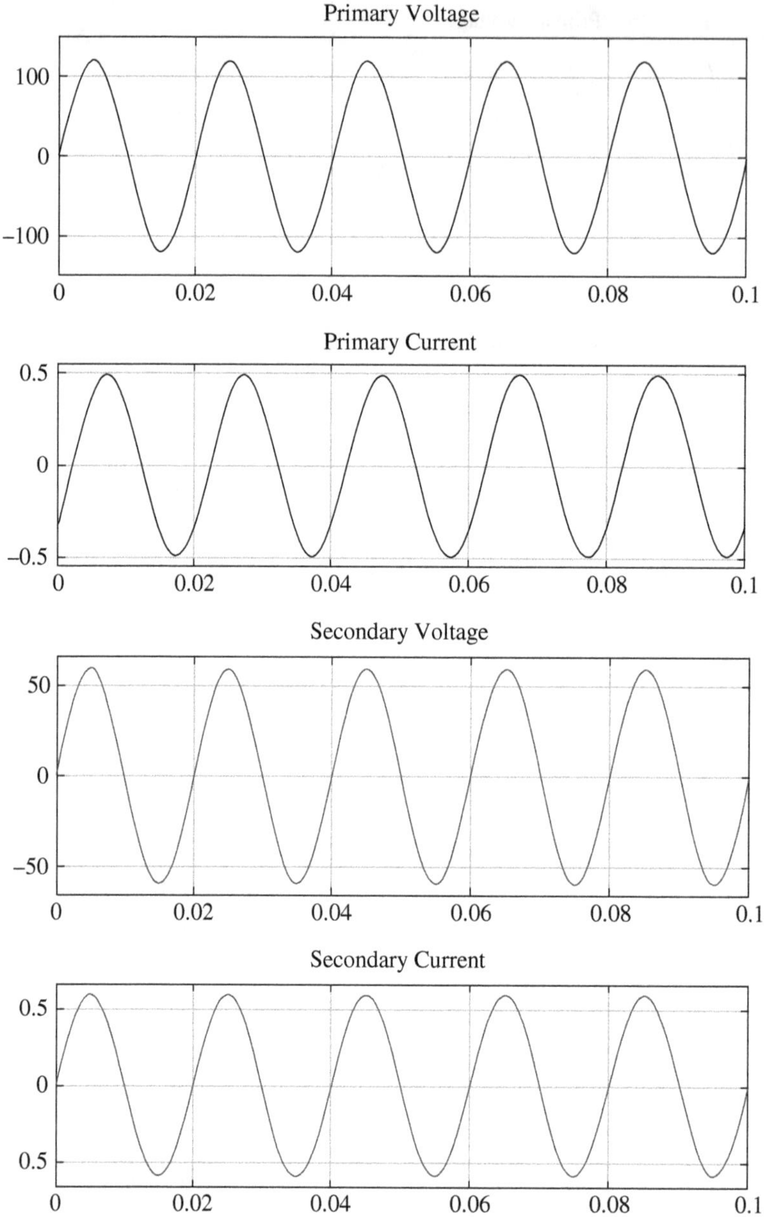

Figure 3.73 Voltage and current waveforms of single-phase step down transformer.

3.20 Matlab/Simulink Model of Testing of Transformer

Approximate equivalent circuit model referred to the primary side is computed using the Matlab/Simulink Model shown in Figure 3.74. Inputs to the code are the test values and the outputs are the equivalent series parameters and shunt parameters. The parameter calculation part is written in embedded Matlab code.

```
function [tf] = test1 (u)
Poc = u(1);
Voc = u(2);
Ioc = u(3);
Psc = u(4);
Vsc = u(5);
Isc = u(6);
N = u(7);
Ic = Poc/Voc;
Im = sqrt(((Ioc)^2)-((Ic)^2));
Rc = Voc/Ic;
Xm = Voc/Im;
Zeq = Vsc/Isc;
Req = Psc/((Isc)^2);
%Req1 is the equivalent resistance reffered to primary side%
Req1 = (N^2)*Req;
Xed = sqrt(((Zeq)^2)-((Req)^2));
%Xeq1 is the equivalent resistance reffered to primary side%
Xeq1 = (N^2)*Xeq;
tf = [Rc, Xm, Req1, Xeq1];
%All the output parameters are reffered to primary side%
```

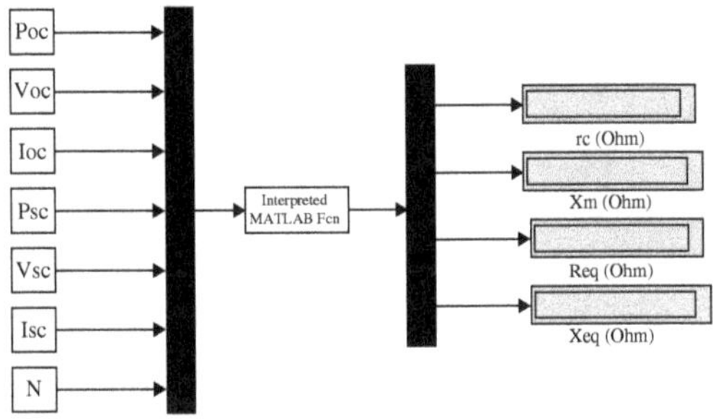

Figure 3.74 OCT and SCT Matlab/Simulink Model (File Name M2.slx).

3.21 Matlab/Simulink Model of Autotransformer

(a) Additive polarity: To build a Simulink model of autotransformer, a single-phase transformer is used and both windings are connected in series such that the voltages of both windings are added. The input voltage is connected to the primary side of the transformer and the output voltage is obtained across both windings. The step-up factor will depend upon the turn ratio of the single-phase transformer. The Simulink model for additive polarity autotransformer model is shown in Figure 3.75. The model yield results as shown in Figure 3.76 (primary and secondary side voltages and currents are depicted).

(b) Subtractive polarity

The Simulink model of an autotransformer for subtractive poliarity is shown in Figure 3.77 and the corresponding result is shown in Figure 3.78.

3.22 Matlab/Simulink Model of Three-Phase Transformer

Figure 3.79 shows the Simulink model for a three-phase transformer. The ratings and types of connection can be changed in the dialogue box (Figure 3.80). The voltage and

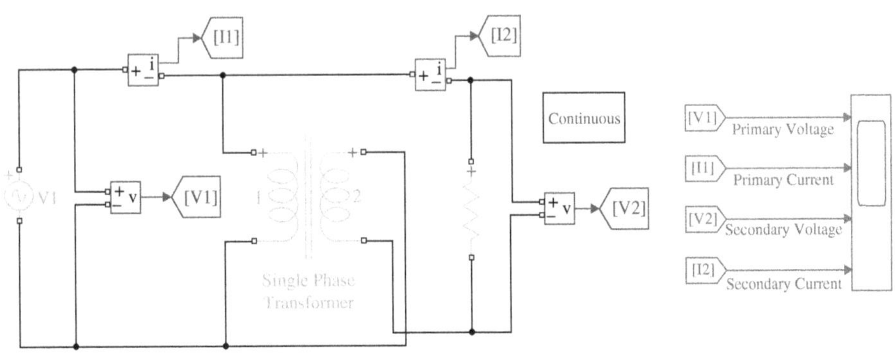

Figure 3.75 Simulink model for auto transformer in additive polarity (File Name M3.slx).

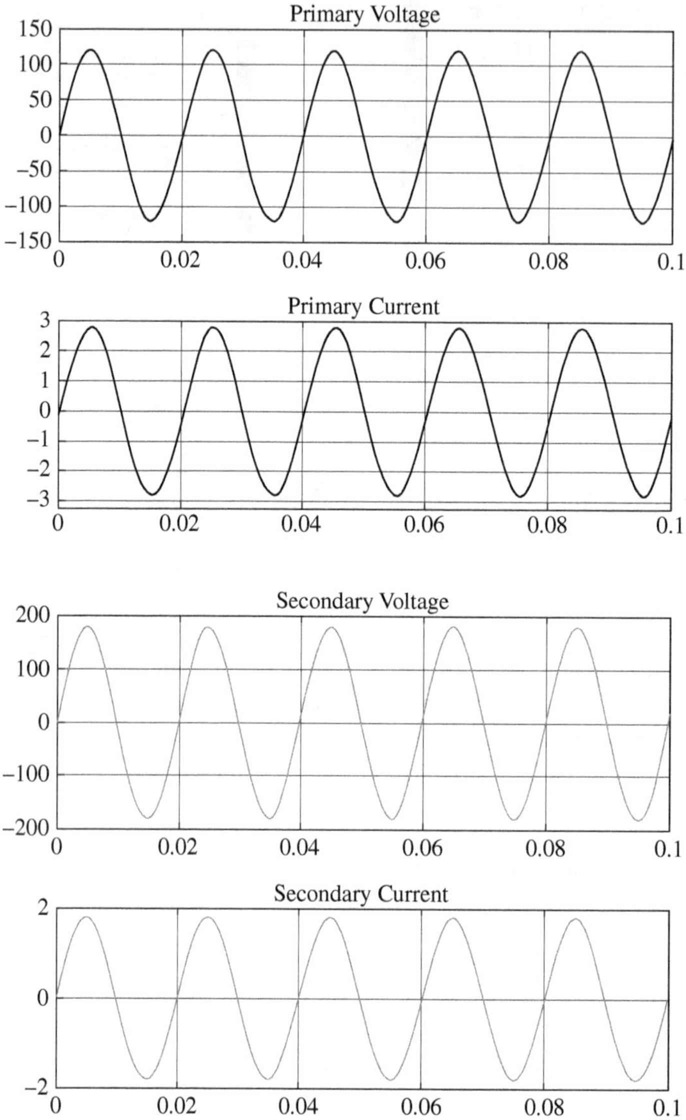

Figure 3.76 Voltage and current waveforms with additive polarity configured auto transformer.

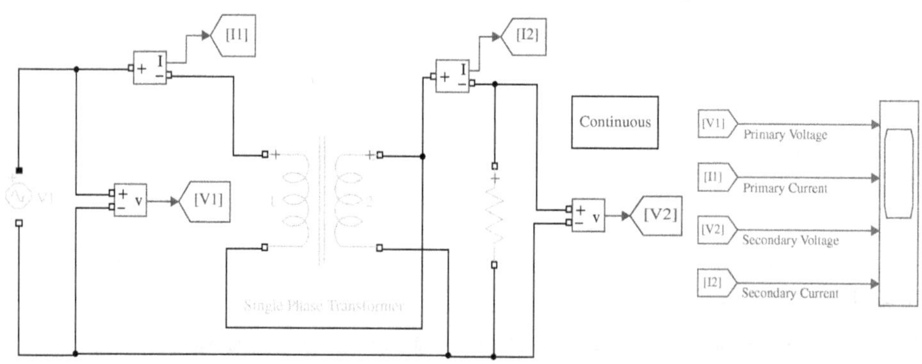

Figure 3.77 Simulink model of auto transformer with subtractive polarity.

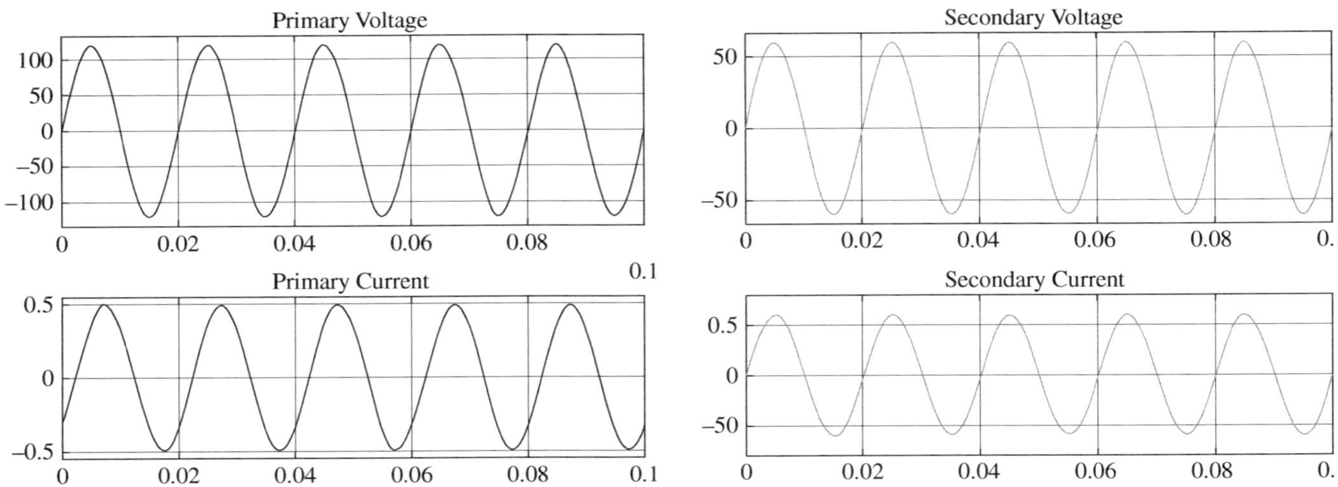

Figure 3.78 Voltage and current waveforms of single-phase auto transformer subtractive polarity.

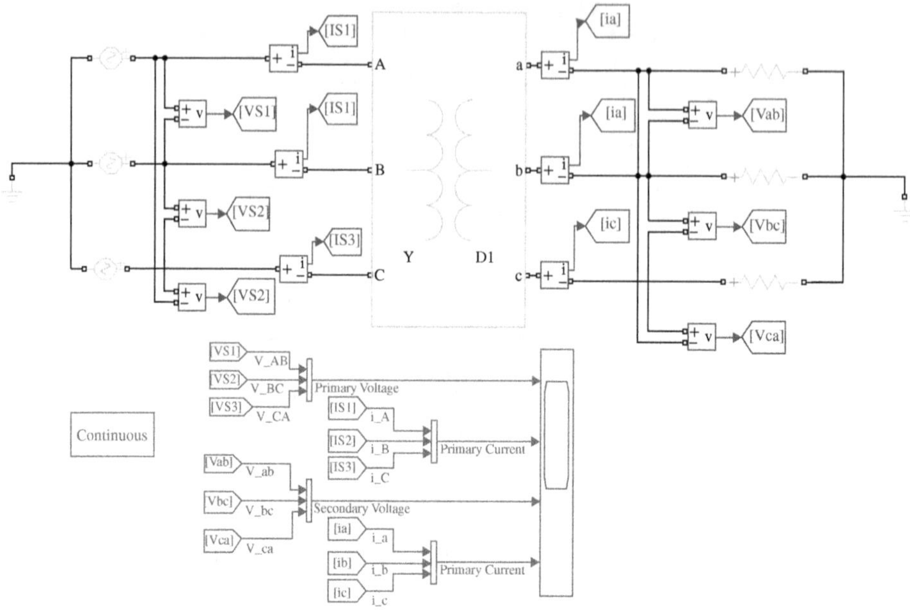

Figure 3.79 Simulink model for three-phase transformer (File Name: M4.slx).

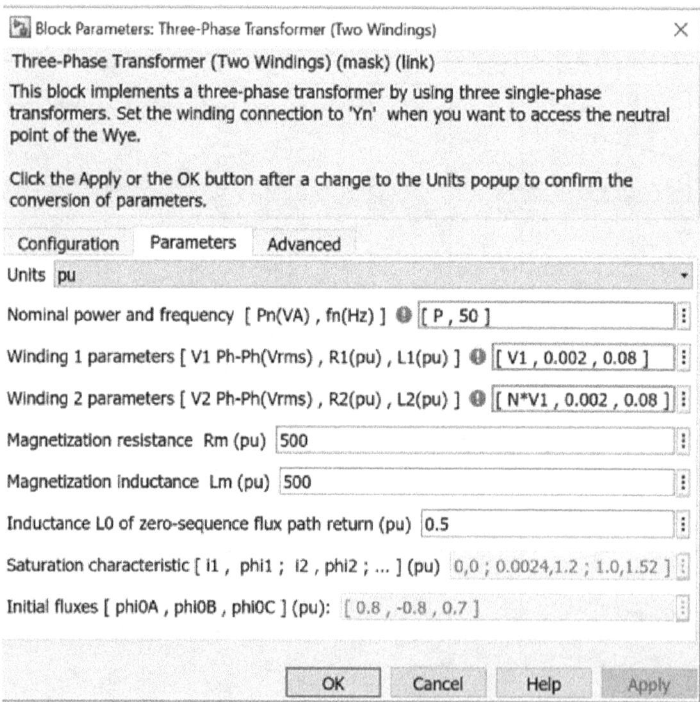

Figure 3.80 Dialogue box to set the parameters for a three-phase transformer.

current waveforms for different connections of three-phase transformers are shown in Figure 3.81–3.86.

(a) Star Star (YY) Connection
(b) Star Delta (ΔY)
(c) Delta Delta (ΔΔ)
(d) Delta Star (ΔY)
(e) Star Delta 11 (YΔ11)
(f) Delta Delta 11(ΔΔ11)

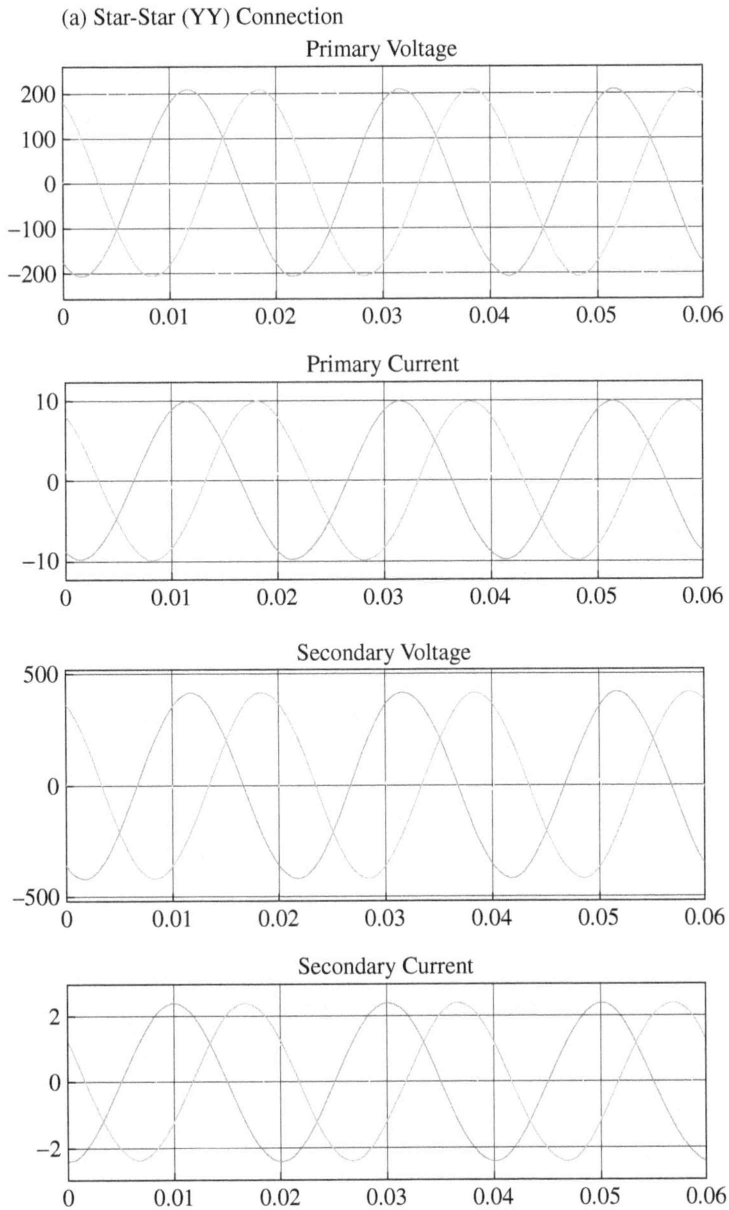

Figure 3.81 Voltage and current waveforms of three-phase transformer with YY connection.

(b) Star-Delta (ΔY):

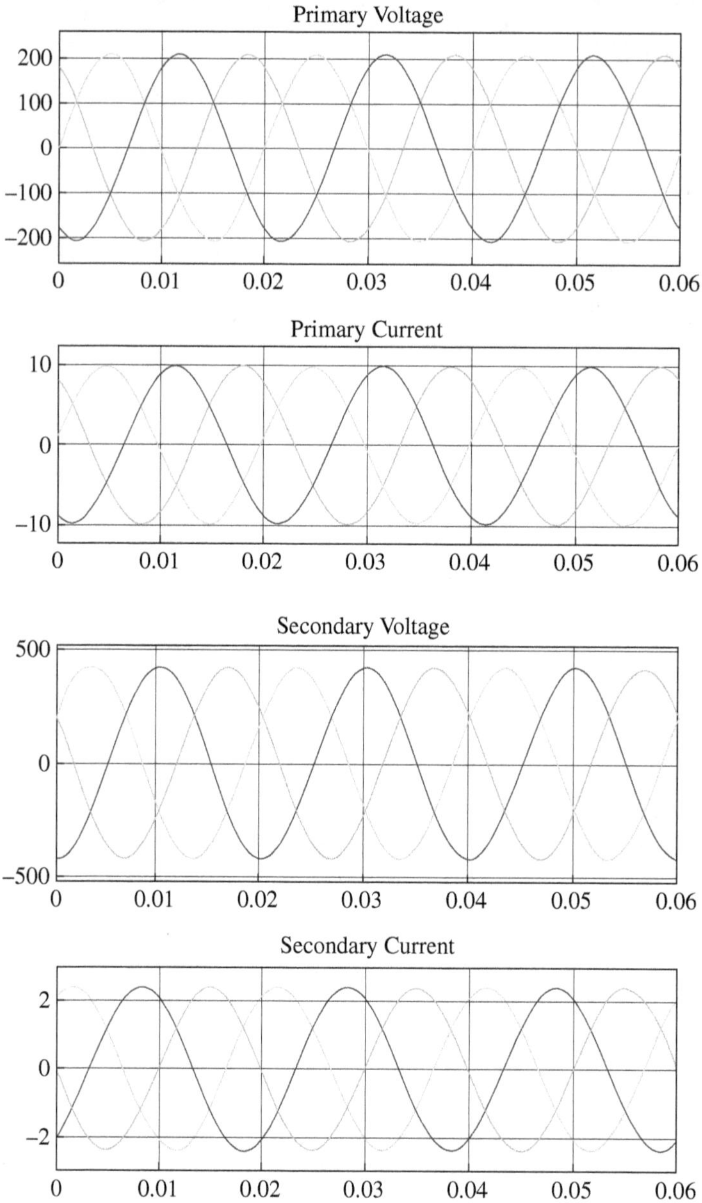

Figure 3.82 Voltage and current waveforms of three-phase transformer with YΔ connection.

(c) Delta delta (ΔΔ)

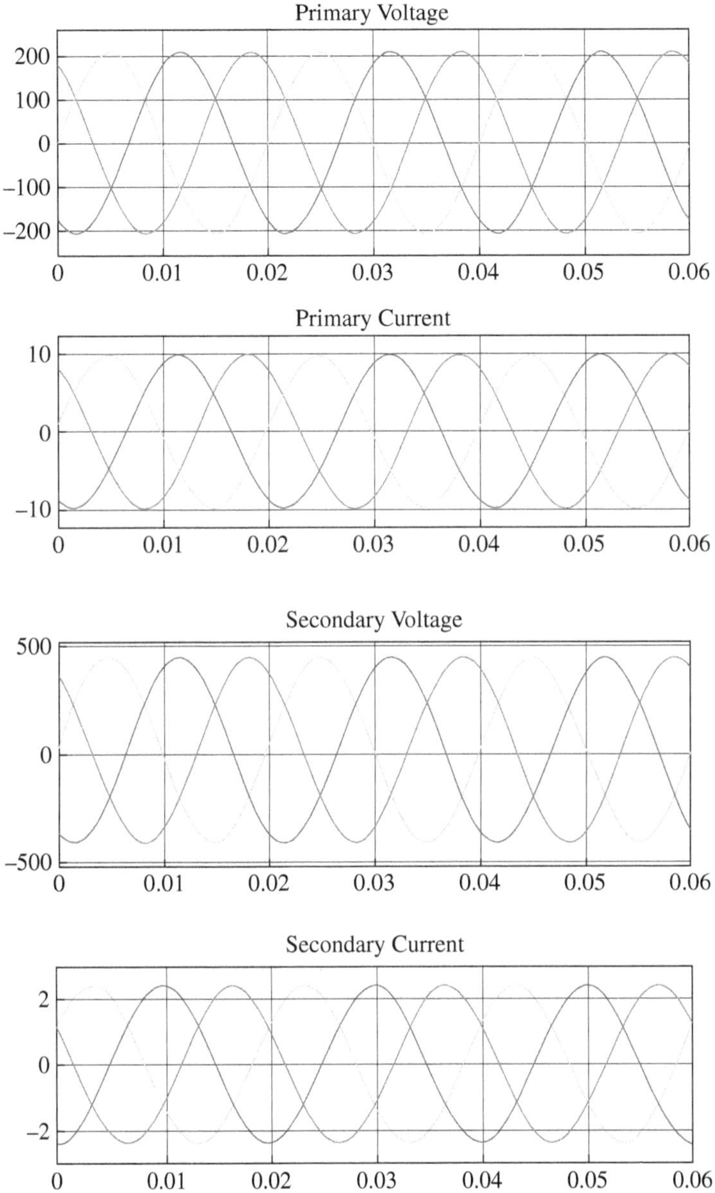

Figure 3.83 Voltage and current waveforms of three-phase transformer with ΔΔ connection.

(d) Delta Star (ΔY)

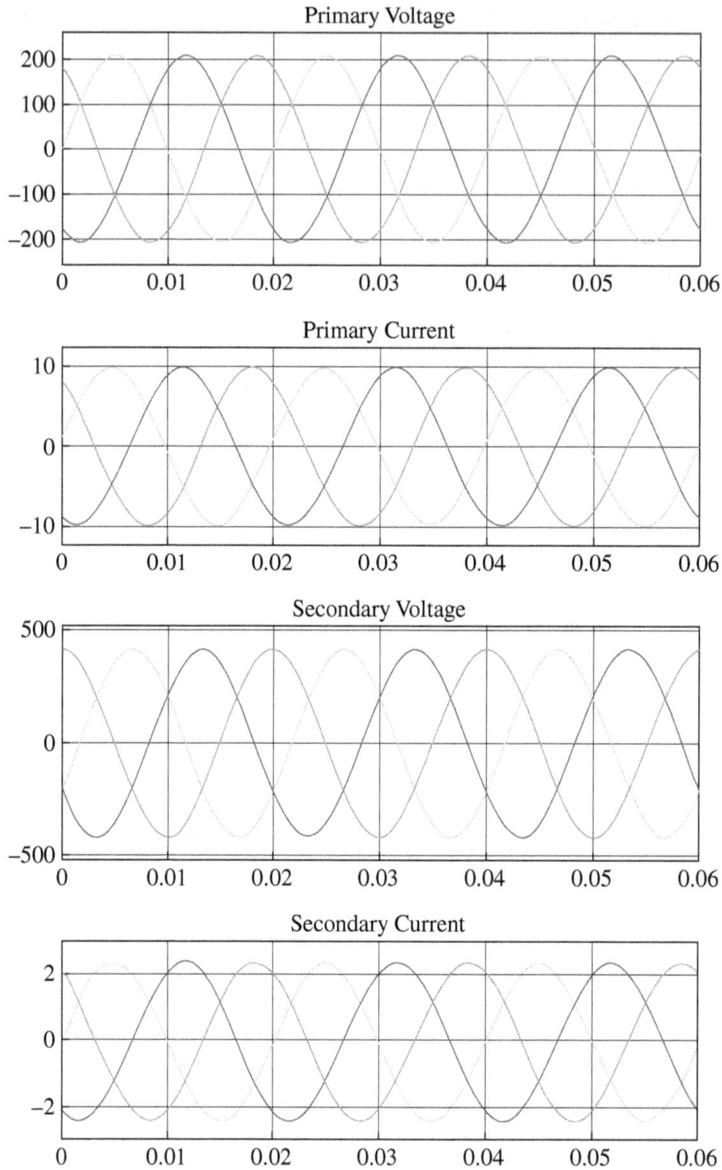

Figure 3.84 Voltage and current waveforms of three-phase transformer with ΔY connection.

(e) Star Delta 11 (YΔ11)

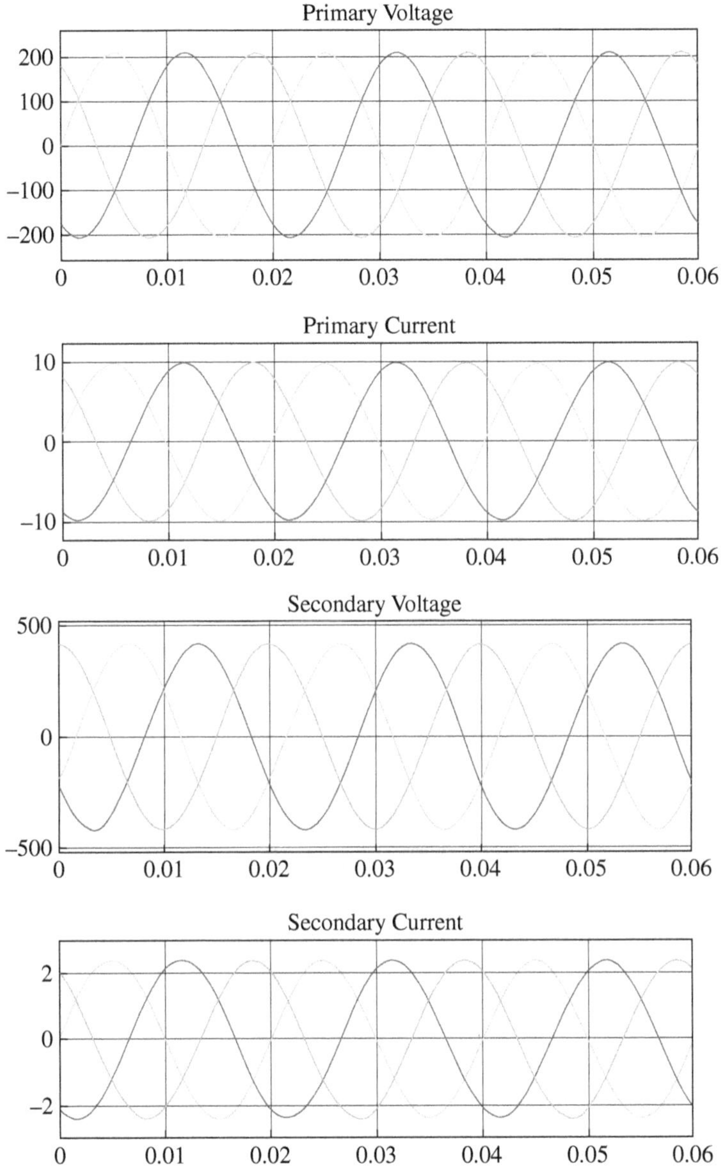

Figure 3.85 Voltage and current waveforms of three-phase transformer with YΔ11 connection.

(f) Delta Delta 11 (ΔΔ11)

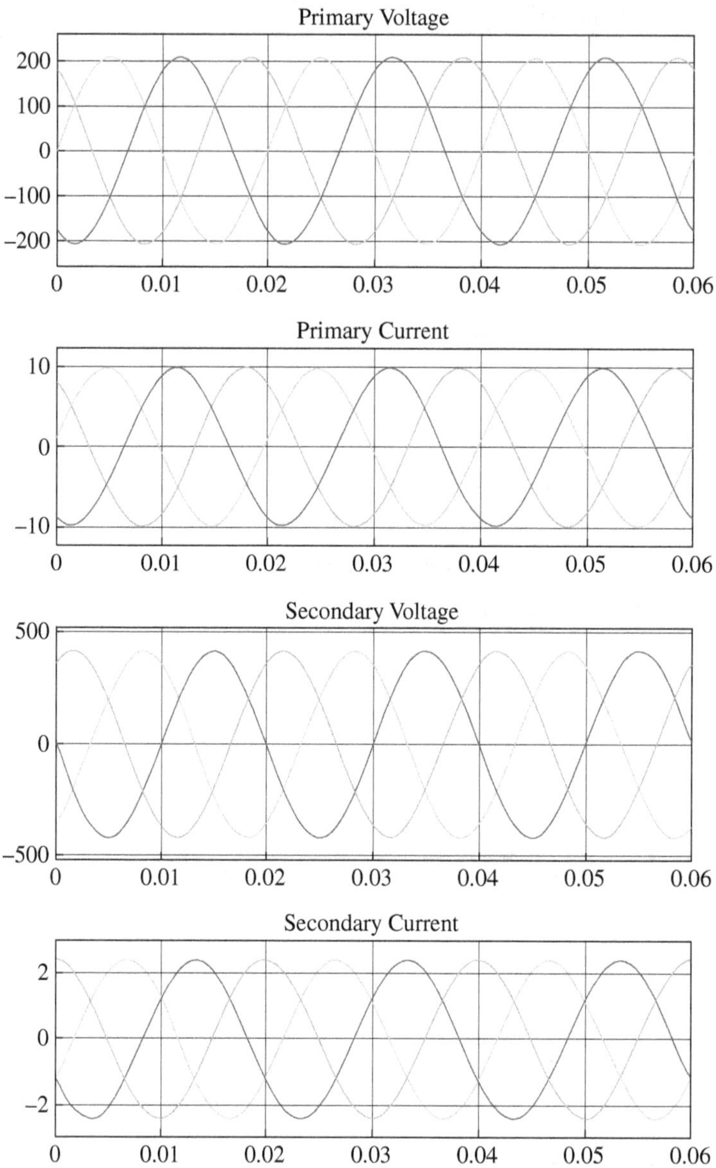

Figure 3.86 Voltage and current waveforms of three-phase transformer with ΔΔ11 connection.

3.23 Supplementary Solved Problems

Problem 3.1 A single-phase transformer is designed to work at 440/220 V and 50 Hz. What will be the effect on the transformer performance if the supply frequency is reduced by 10% to 45 Hz and voltage is increased by 10% to 484 V?

Solution

The emf equation in a single-phase transformer is given by

$$V_1 \simeq E_1 = 4.44 f \phi_m N_1$$

Flux Dessity and flux are related as, $\phi_m = B_m A_c$

Where B_m is flux density and A_c is the area of core

$$V_1 = 4.44 f B_m A_c N_1$$

$$\frac{(V_1)_{45Hz}}{(V_1)_{50Hz}} = \frac{4.44 (B_m)_{45Hz} (45) A_c N_1}{4.44 (B_m)_{50Hz} (50) A_c N_1}$$

$$\frac{484}{440} = \frac{(B_m)_{45Hz} (45)}{(B_m)_{50Hz} (50)}$$

$$\frac{(B_m)_{45Hz}}{(B_m)_{50Hz}} = 1.222$$

$$(B_m)_{45Hz} = 1.222 (B_m)_{50Hz}$$

This shows that the flux density increases if frequency is reduced and the applied voltage is increased simultaneously. The increased flux density will cause higher no-load current, higher core losses and noise. However, if the frequency and voltage are reduced or increased in the same proportion, the flux density will remain constant.

Problem 3.2 A single-phase 50 Hz transformer has equal hysteresis and eddy current loss when rated voltage is applied. If the applied voltage is 50% at 25 Hz, compute the percentage change in the total core loss. (Total core loss is the sum of hyestersis and eddy current loss)

Solution

Hyestersis loss is given as

$$P_h = K_h B_m^{1.6} f$$

Eddy Current loss is given as

$$P_e = K_e B_m^2 f^2$$

Total Core loss is

$$P_{core} = P_h + P_e$$

Induced voltage at the primary side is

$$V_1 \approx E_1 = 4.44 f N_1 \phi_m$$

Since

$$\phi_m = B_m A_c$$

$$B_m = \frac{1}{4.44 N A_c} \left(\frac{V_1}{f}\right) = C \left(\frac{V_1}{f}\right)$$

Substituting the maximum flux density B_m in the loss expressions

$$P_{core} = K_h C\left(\frac{V_1}{f}\right)^{1.6} f + K_e C\left(\frac{V_1}{f}\right)^2 f^2 = K_1 \frac{(V_1)^{1.6}}{(f)^{0.6}} + K_2 (V_1)^2$$

Given the applied voltage and frequency as half of the rated value

$$V_1'' = 0.5V_1; f'' = 0.5f$$

$$P_{core(new)} = K_1 \frac{(V_1)^{1.6}}{(f)^{0.6}} + K_2 (V_1)^2 = K_1 \frac{(0.5V_1)^{1.6}}{(0.5f)^{0.6}} + K_2 (0.5V_1)^2$$

$$P_{core(new)} = 0.5K_1 \frac{(V_1)^{1.6}}{(f)^{0.6}} + 0.25K_2 (V_1)^2 = 0.5P_h + 0.25P_e$$

Since at the rated condition the Hyestersis and eddy current losses are the same i.e.

$$P_h = P_e = \frac{P_{core}}{2}$$

$$P_{core(new)} = 0.5P_h + 0.25P_e = 0.5\frac{P_{core}}{2} + 0.25\frac{P_{core}}{2} = 0.375P_{core}$$

%age Reduction in core loss;

$$= \frac{P_{core(new)} - P_{core}}{P_{core}} \times 100\% = \frac{0.375P_{core} - P_{core}}{P_{core}} \times 100\% = 62.5\%$$

Problem 3.3 In a single-phase transformer, core loss is obtained as 50 W at 25 Hz and 120 W at 50 Hz. Calculate the hysteresis and eddy current losses in the transformer. Assuming the same flux density at both frequencies.

Solution

Hyestersis loss is given as

$$P_h = K_h B_m^{1.6} f$$

Eddy Current loss is given as

$$P_e = K_e B_m^2 f^2$$

Total Core loss is

$$P_{core} = P_h + P_e$$

For constant flux density the core loss is given as

$$P_{core} = P_h + P_e = k_h f + k_e f^2$$

$$50 = k_h (25) + k_e (25)^2 \quad \text{At 25 Hz}$$

$$120 = k_h (50) + k_e (50)^2 \quad \text{At 50 Hz}$$

Solving the two equations two get two unknowns

$$2 = k_h + k_e\,(25)$$
$$k_h = 2 - 25k_e$$
$$120 = k_h\,(50) + k_e(50)^2 \quad \text{At 50 Hz}$$

Problem 3.4 Open-circuit and short-circuit tests are performed on a 30 MVA 132/33 kV three-phase Star/Delta transformer. The test results are given below

Open Circuit Test: 33 kV, 5.2 A, 20.5 kW
Short Circuit Test: 15.1 kV, 131.21 A, 90 kW

Compute the equivalent circuit parameter referred to LV and HV sides

Solution

The given test values are considered as line quantities. The equivalent circuit parameters are calculated on per-phase basis. Thus, all the quantities are converted to per-phase values.

The OCT is conducted on LV side (delta side). The per-phase values are

$$V_{\text{phase}} = 33\,\text{kV}$$

$$I_{phase} = \frac{I_{line}}{\sqrt{3}} = \frac{5.2}{\sqrt{3}} = 3A$$

Per-phase core loss $P_{core} = \frac{P_{core-total}}{3} = \frac{20500}{3} = 6833.33W$

Core-loss component $I_c = I_o \cos\left(\theta_0\right) = \frac{P_{core}}{V_1} = \frac{6833.33}{33000} = 0.207A$

Magnetizing component $I_m = \sqrt{\left(I_o\right)^2 - \left(I_c\right)^2} = \sqrt{(3)^2 - (0.207)^2} = 2.993A$

The shunt branch parameters referred to the LV side

$$R_{c2} = \frac{V_2}{I_{c2}} = \frac{33000}{0.207} = 159.42k\Omega$$
$$X_{m2} = \frac{V_2}{I_{m2}} = \frac{33000}{0.2993} = 110.257k\Omega$$

Rated current on the HV side is

$$S = \sqrt{3}V_1 I_1$$
$$I_1 = \frac{S}{\sqrt{3}V_1} = \frac{30000}{\sqrt{3}\,(132)} = 131.21A$$

The short-circuit test is conducted on HV side. The equivalent resistance and leakage reactance referred to the HV side is computed as

Per-phase applied voltage to the star-connected HV winding $= V_{1(phase)} = \frac{15.1\times1000}{\sqrt{3}}$ $= 8718V$

For star connection the line and phase currents are equal 131.21 A.

$$Z_{eq1} = \frac{V_{sc}}{I_{sc}} = \frac{8718}{131.21} = 66.44\Omega$$

$$per - phase \text{ Power} = \frac{P}{3} = \frac{90}{3} = 30kW$$

$$R_{eq1} = \frac{P_{phase}}{I_{sc}^2} = \frac{30 \times 1000}{(131.21)^2} = 1.742\Omega$$

$$X_{eq1} = \sqrt{Z_{eq1} - R_{eq1}} = \sqrt{(66.44)^2 - (1.742)^2} = 66.46\Omega$$

Referred to the HV side

Referring the LV side to the HV sides;

$$a = \frac{N_1}{N_2} = \frac{V_{1(phase)}}{V_{2(phase)}} = \frac{\left(\frac{132}{\sqrt{3}}\right)}{33} = 2.309$$

$$R_{c1} = (a)^2 R_{c2} = (2.309)^2 159.42 = 850.24k\Omega$$

$$X_{m1} = (a)^2 X_{m2} = (2.309)^2 110.257 = 1357.3k\Omega$$

$$R_{eq1} = 1.742\Omega$$

$$X_{eq1} = 66.46\Omega$$

Referring the HV side to the LV sides

$$R_{eq2} = \left(\frac{1}{a}\right)^2 R_{eq1} = \left(\frac{1}{2.309}\right)^2 1.742 = 0.3267\Omega$$

$$X_{eq2} = \left(\frac{1}{a}\right)^2 R_{eq2} = \left(\frac{1}{2.309}\right)^2 66.46 = 12.46\Omega$$

$R_{c2} = 159.42k\Omega, X_{m2} = 110.257k\Omega$

Problem 3.5 A 200 kVA single-phase transformer has an efficiency of 96% at full load 0.8 pf lagging and a 95% efficiency at half-load unity power factor.

Calculate

(a) Core loss and copper loss at full-load condition
(b) Maximum efficiency at unity power factor

Solution

(a) The efficiency of the transformer at full load is;

$$\eta = \frac{x(KVA)_{rated} \cos(\theta)}{x(KVA)_{rated} \cos(\theta) + P_c + x^2 P_{cu,fl}}$$

$$\eta = 0.96 = \frac{1\,(200)\,(0.8)}{1\,(200)\,(0.8) + P_{Losses}}$$

$$0.96 P_{Losses} + 0.96\,(200)\,(0.8) = (200)\,(0.8)$$

$$P_{Losses} = \frac{160 - 153.6}{0.96} = 6.667kW$$

$$P_{core} + P_{cu,fl} = 6.667 ------(1) \tag{1}$$

The efficiency of the transformer at half full load is

$$\eta = \frac{x(KVA)_{rated}\cos(\theta)}{x(KVA)_{rated}\cos(\theta) + P_c + x^2 P_{cu,fl}}$$

$$\eta = 0.97 = \frac{0.5\,(200)\,(1)}{0.5\,(200)\,(1) + P_{Losses}}$$

$$0.97P_{Losses} + 0.97\,(200)\,(0.5) = (200)\,(0.5)$$

$$P_{Losses} = \frac{100 - 97}{0.97} = 3.09278kW$$

$$P_{core} + P_{cu} = 3.09278$$

$$P_{core} + (0.5)^2 P_{cu,fl} = 3.09278 ------(2)$$

Since

$$P_{cu} = (x)^2 P_{cu,fl} \qquad (2)$$

Solving Eq. (1) and (2) to get the core loss and copper loss

$$P_{core} = 6.667 - P_{cu,fl}$$

$$6.667 - P_{cu,fl} + 0.25P_{cu,fl} = 3.09278$$

$$0.75P_{cu,fl} = 3.5742$$

$$P_{cu,fl} = \frac{3.5742}{0.75} = 4.7656kW$$

$$P_{core} = 6.667 - 4.7656 = 1.9014kW$$

(b) The maximum efficiency occurs at the load factor given as

$$x = \sqrt{\frac{P_{core}}{P_{cu,fl}}} = \sqrt{\frac{1.9014}{4.7656}} = 0.632$$

The maximum efficiency is calculated as

$$\eta_{\max} = \frac{x(KVA)_{rated}\cos(\theta)}{x(KVA)_{rated}\cos(\theta) + 2P_c} \times 100\% = \frac{0.632\,(200)\,1}{0.632\,(200)\,1 + 2\,(1.9014)} \times 100\% = 97.08\%$$

Problem 3.6 Given a 20 KVA 400/100 V two-winding transformer. This is used as an autotransformer to supply a load of 500 V from a 400 V source. When tested as a two-winding transformer at rated load, 0.8 pf lagging its efficiency is 96%. Calculate the efficiency as an autotransformer when operated at the same power factor assuming the same losses.

Solution

The rated current on primary and secondary windings are

$$I_1 = \frac{20000}{400} = 50A$$

$$I_2 = \frac{20000}{100} = 200A$$

The KVA rating as an autotransformer

$$(KVA)_{auto} = 500 \times 200 = 100KVA$$

The circuit diagram of the autotransformer is shown below

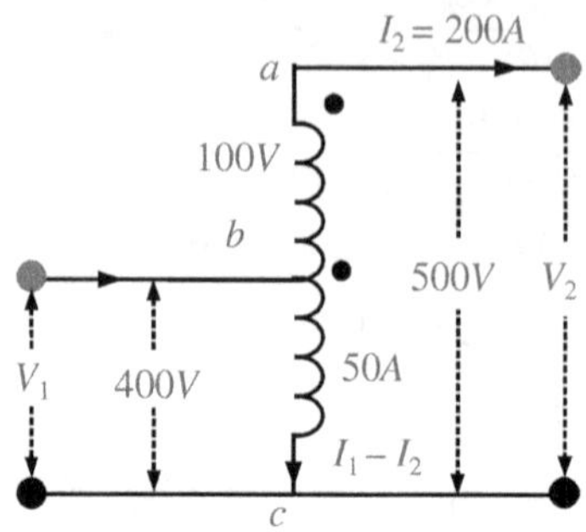

Efficiency as a two-winding transformer

$$\eta_{tw} = \frac{x(KVA)_{rated}\cos(\theta)}{x(KVA)_{rated}\cos(\theta) + P_c + x^2 P_{cu,fl}}$$

$$\eta_{tw} = 0.96 = \frac{1\,(20)\,(0.8)}{1\,(20)\,(0.8) + P_{Losses}}$$

$$0.96P_{Losses} + 0.96\,(20)\,(0.8) = (20)\,(0.8)$$

$$P_{Losses} = \frac{16 - 15.36}{0.96} = 0.6667kW$$

Efficiency as an autotransformer

$$\eta_{auto} = \frac{x(KVA)_{rated}\cos(\theta)}{x(KVA)_{rated}\cos(\theta) + P_c + x^2 P_{cu,fl}} \times 100\%$$

$$\eta_{auto} = \frac{1\,(100)\,(0.8)}{1\,(100)\,(0.8) + P_{Losses}} \times 100\% = \frac{1\,(100)\,(0.8)}{1\,(100)\,(0.8) + 0.6667} \times 100\% = 99.17\%$$

Problem 3.7 Determine the relationship between the KVA ratings of an autotransformer and a two-winding transformer in terms of turn ratio a.

Solution

Consider a two-winding transformer with turn ratio

$$a = \frac{N_1}{N_2}$$

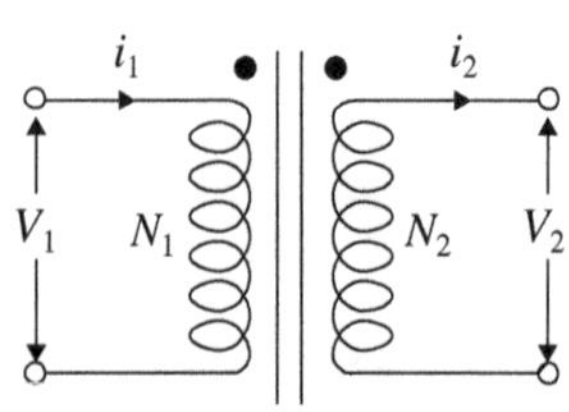

(a)

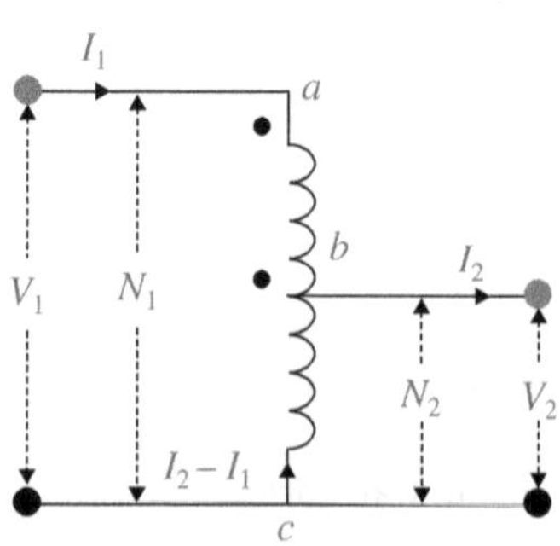

(b)

Considering the figure of the autotransformer the following equation is written

$$S_{auto} = V_1 I_1 = V_2 I_2$$
$$S_{tw} = (V_1 - V_2)\, I_1 = V_2\, (I_2 - I_1)$$

Multiplying and dividing the equation of S_{tw} by V_1, the following is obtained

$$S_{tw} = (V_1 - V_2)\, I_1 = \frac{V_1}{V_1}\,(V_1 - V_2)\, I_1 = \left(1 - \frac{V_2}{V_1}\right)(V_1 I_1)$$

$$S_{tw} = \left(1 - \frac{1}{\frac{V_1}{V_2}}\right)(V_1 I_1) = \left(1 - \frac{1}{a}\right) S_{auto}$$

$$S_{auto} = \left(\frac{a}{a-1}\right) S_{tw}$$

If

$a \gg 1; S_{auto} \approx S_{tw}$

if

$a \approx 1; S_{auto} \gg S_{tw}$

Problem 3.8 A three-phase step-down transformer is supplied from 11 kV source. It draws a line current of 30 A from the supply mains. The magnetizing current and losses are neglected. Per-phase turn ration is 25. Calculate the output voltage, output current and output kVA for following connections

(a) Star/Star
(b) Delta/Delta
(c) Star/Delta
(d) Delta/Star

Solution

(a) Star/Star connection

$$\text{Per-phase primary voltage} = \frac{11000}{\sqrt{3}} = 6350.85V$$

$$\text{Per-phase secondary voltage} = \frac{6350.85}{25} = 254V$$

$$\text{Secondary line voltage} = 254\left(\sqrt{3}\right) = 440V$$

$$\text{Primary side phase/line current} = 30\text{A}$$

$$\text{Secondary side phase/line current} = 30\,(25) = 750A$$

$$\text{Output kVA} = \sqrt{3}V_{line}I_{line} = \sqrt{3}\,(440)\,(750) = 571.576kVA$$

(b) Delta/Delta connection

$$\text{Primary phase/line voltage} = 11000V$$

$$\text{Secondary phase/line voltage} = \frac{11000}{25} = 440V$$

$$\text{Primary phase current} = \frac{30}{\sqrt{3}} = 17.32\text{A}$$

$$\text{Secondary phase current} = 17.32\,(25) = 433.01\text{A}$$

$$\text{Secondary line current} = \sqrt{3}\,(433.01) = 750A$$

$$\text{Output kVA} = \sqrt{3}V_{line}I_{line} = \sqrt{3}\,(440)\,(750) = 571.576kVA$$

(c) Star/Delta

$$\text{Per-phase primary voltage} = \frac{11000}{\sqrt{3}} = 6350.85V$$

$$\text{Per-phase secondary voltage} = \frac{6350.85}{25} = 254V$$

$$\text{Secondary line voltage} = 254V$$

$$\text{Primary side phase/line current} = 30\text{A}$$

$$\text{Secondary side phase current} = 30\,(25) = 750A$$

$$\text{Secondary side line current} = \sqrt{3}\,(750) = 1299.04A$$

$$\text{Output kVA} = \sqrt{3}V_{line}I_{line} = \sqrt{3}\,(254)\,(1299.04) = 571.576kVA$$

(d) Delta/Star

$$\text{Primary phase/line voltage} = 11000V$$

$$\text{Secondary phase voltage} = \frac{11000}{25} = 440V$$

$$\text{Secondary line voltage} = \left(\sqrt{3}\right)440 = 762.1V$$

$$\text{Primary phase current} = \frac{30}{\sqrt{3}} = 17.32\text{A}$$

$$\text{Secondary phase current} = 17.32\,(25) = 433.01\text{A}$$

$$\text{Secondary line current} = 433.01A$$

$$\text{Output kVA} = \sqrt{3}V_{line}I_{line} = \sqrt{3}\,(440)\,(762.1) = 571.576kVA$$

Problem 3.9 A 440 V, three-phase industrial load of 250 kW at 0.85 lagging pf is supplied from 11 kV mains through three identical single-phase step-down transformers. Calculate the voltage, current and kVA ratings of each of the three transformers of they are connected in

(a) Star/Star
(b) Delta/Delta
(c) Star/Delta
(d) Delta/Star

Solution

Load in kVA = 250/0.85 = 294.11 kVA

(a) Star/Star Connection

$$\text{Per-phase secondary voltage} = \frac{440}{\sqrt{3}} = 254\,V$$

$$\text{Per-phase primary voltage} = \frac{11000}{\sqrt{3}} = 6350.85V$$

$$\text{Voltage rating of each transformer} = \frac{6350.85}{254}\text{V}$$

$$\text{Secondary phase current} = \frac{S_{1-phase}}{V_2} = \frac{\left(\frac{294.11}{3}\right)}{254} = 385.97A$$

$$\text{Primary phase current} = \frac{S_{1-phase}}{V_1} = \frac{\left(\frac{294.11}{3}\right)}{6350.85} = 15.44A$$

$$\text{kVA rating of each transformer} = V_{phase}I_{phase} = (254)\,(386.97) = 98.04kVA$$

(b) Delta/Delta connection

$$\text{Per-phase secondary voltage} = 440V$$

$$\text{Per-phase primary voltage} = 11000V$$

$$\text{Voltage rating of each transformer} = \frac{11000}{440}\text{V}$$

$$\text{Secondary phase current} = \frac{S_{1-phase}}{V_2} = \frac{\left(\frac{294.11}{3}\right)}{440} = 222.816A$$

$$\text{Primary phase current} = \frac{S_{1-phase}}{V_1} = \frac{\left(\frac{294.11}{3}\right)}{11000} = 8.9124A$$

$$\text{kVA rating of each transformer} = V_{phase}I_{phase} = (440)\,(222.816) = 98.04kVA$$

(c) Star/Delta Connection

$$\text{Per-phase secondary voltage} = 440\,V$$

$$\text{Per-phase primary voltage} = \frac{11000}{\sqrt{3}} = 6350.85V$$

$$\text{Voltage rating of each transformer} = \frac{6350.85}{440}\text{V}$$

$$\text{Secondary phase current} = \frac{S_{1-phase}}{V_2} = \frac{\left(\frac{294.11}{3}\right)}{440} = 222.81A$$

$$\text{Primary phase current} = \frac{S_{1-phase}}{V_1} = \frac{\left(\frac{294.11}{3}\right)}{6350.85} = 15.44A$$

$$\text{kVA rating of each transformer} = V_{phase}I_{phase} = (440)\,(222.81) = 98.04kVA$$

(d) Delta/Star Connection

$$\text{Per-phase secondary voltage} = \frac{440}{\sqrt{3}} = 254V$$

Per-phase primary voltage $= 11000V$

Voltage rating of each transformer $= \frac{11000}{254}$V

Secondary phase current $= \frac{S_{1-phase}}{V_2} = \frac{\left(\frac{294.11}{3}\right)}{254} = 385.97A$

Primary phase current $= \frac{S_{1-phase}}{V_1} = \frac{\left(\frac{294.11}{3}\right)}{11000} = 8.9124A$

kVA rating of each transformer $= V_{phase}I_{phase} = (254)(385.97) = 98.04kVA$

Problem 3.10 Three-single-phase 50 kVA, 6350/440 V, transformer are connected to form a three-phase 11 000/440 V transformer bank. Given the equivalent impedance of each transformer referred to its low voltage side is 0.02 + j 0.09. Three-phase supply is given to the transformer bank using a feeder with impedance of 0.3 + j1.2. The transformer bank supplies full load at 440 V and 0.8 pf lagging.

(a) Draw the circuit diagram of the transformer connection
(b) Find the single-phase equivalent circuit
(c) Calculate the transformer winding currents
(d) Calculate the sending end voltage of the 3-phase source.

Solution

(a) Since the primary voltage of the three-phase bank is required to be 11 kV. The primaries of the single-phase transformers must be connected in star.

$$V_P = \sqrt{3}(6350) = 11000V$$

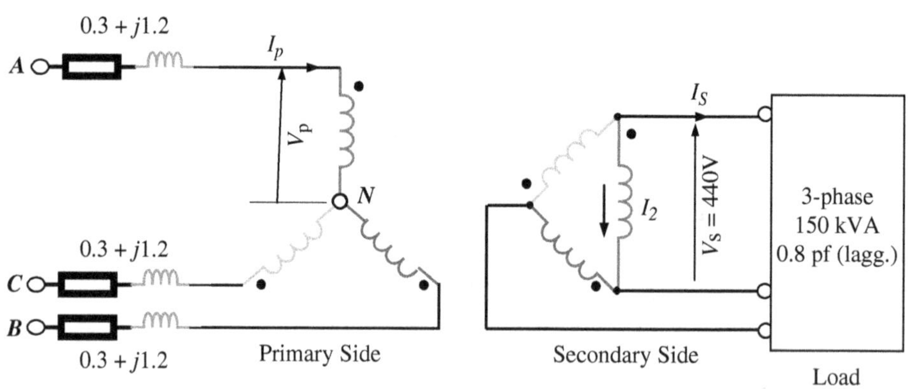

(a)

(b) Equivalent primary side voltage $V_p = \frac{6350}{\sqrt{3}} = 3666.2V$

Equivalent secondary side voltage $V_p = \frac{440}{\sqrt{3}} = 254.0V$

Equivalent turn ratio $a = \frac{11000}{440} = 25$

The equivalent impedance of Δ

$$Z_{eq1(\Delta)} = (0.02 + j0.09) \times 25^2 = 12.5 + j56.25\ \Omega$$

The equivalent impedance of Y

$$Z_{eq1(Y)} = \frac{12.5 + j56.25}{3} = 4.167 + j18.75\Omega$$

The single-phase equivalent circuit is shown below.

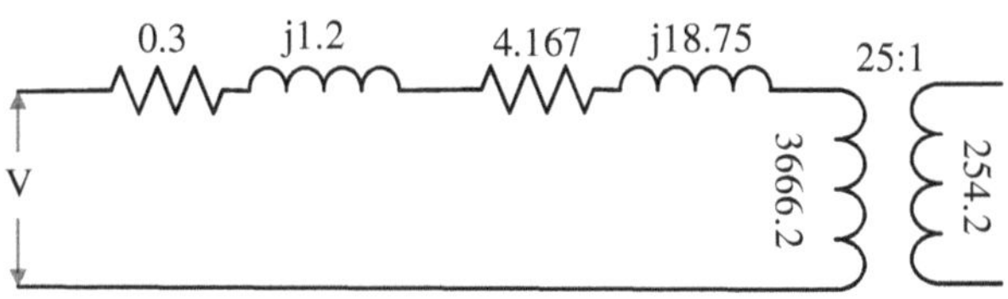

(b)

(c) Transformer winding current

$$I_s = \frac{150 \times 1000}{\sqrt{3} \times 440} = 196.82A$$

$$\emptyset = \cos^{-1}0.8 = 36.9°$$

$$I_p = \frac{196.82}{25} = 7.87\angle - 36.9A$$

$$I_1 = \frac{7.87}{\sqrt{3}} = 4.54A$$

$$I_2 = \frac{196.82}{\sqrt{3}} = 113.63A$$

(d) Sending end voltage

$$V = 3666.2\angle 0 + (4.467 + j18.95) \times 7.87\angle - 36.9 = 4284.8\angle 6.59V$$

Sending end voltage $= \sqrt{3} \times 4284.8V = 7372.5V$

Problem 3.11 Three-single-phase 15 kVA 440/110 V transformer are connected to form a 45 kVA 440/190 V three-phase transformer bank. The equivalent impedance of each transformer referred to high voltage is 0.5 + j 1 Ω. The transformer delivers 25 kW at 0.8 pf lagging.

(a) Draw circuit diagram showing the transformer connections
(b) Calculate the transformer winding and line currents
(c) Calculate the primary side voltage
(d) Calculate the voltage regulation

Solution

(a) The primary side voltage is 440 V in three-phase connection while single-phase voltage is also 440 V. This indicates delta connected primary. While in secondary side the per-phase voltage is 110 V and three-phase voltage should be 190 V, hence star connection is used in the secondary.

The connection diagram is shown in the following figure

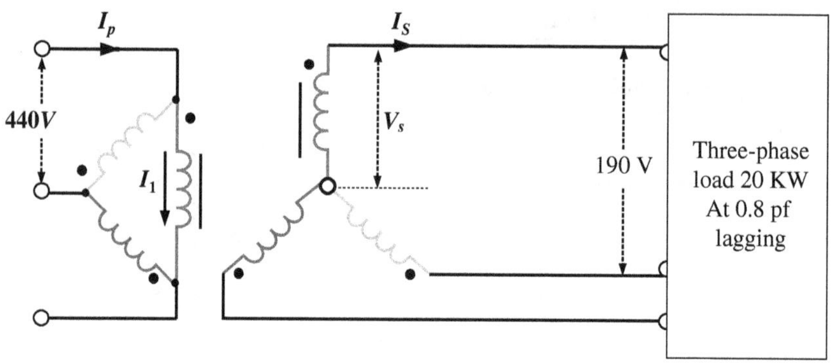

(b) Load KVA = 25/0.8 = 31.25 KVA
Secondary Current $= I_s = \frac{S_L}{\sqrt{3}V} = \frac{31250}{\sqrt{3}(190)} = 94.962A$
Turn ratio $a = \frac{440}{110} = 4$
The primary phase current or winding current is $I_1 = \frac{94.962}{4} = 23.74A$
The primary side-line current $= I_p = \sqrt{3}\,(23.74) = 41.12A$

(c) Primary side voltage (V_p)

$$Z_{eq1} = 0.5 + j1.0 \text{ and } \emptyset = \cos^{-1}0.8 = 36.9°$$

Therefore, $V_p = (440\angle 0) + 41.12\angle -36.9 \times (0.5 + j1.0) = 481.13\angle 2.44V$

(d) Voltage Regulation

$$VR\,(\%) = \frac{V_p - V_1}{V_1} \times 100 = \frac{481.13 - 440}{440} \times 100 = 9.35\%$$

Problem 3.12 A voltage that contain 3rd harmonic component is applied to the primary of a transformer under no-load condition is given as

$$v = 440\sin(\omega t) + 110\sin(3\omega t)$$

The primary winding has 1000 turns and frequency of the fundamental component of the applied voltage is 50 Hz.

(a) Calculate the maximum value of the flux and its shape.
(b) Given the no-load current as $i_0 = 0.5\sin(\omega t - 30) + 0.04\sin(3\omega t - 5)$, calculate the total core loss. Neglect copper losses (no winding resistance).
(c) If the primary voltage does not contain 3rd harmonic component, calculate the % age change in eddy current loss.

Solution

(a) The applied voltage is same as induced emf, neglecting the stator voltage drop

$$v \approx e = -N\frac{d\phi}{dt}$$

$$d\phi = -\frac{1}{N}\int vdt = -\frac{1}{1000}\int (440\sin(\omega t) + 110\sin(3\omega t))\,dt$$

$$\phi = -\frac{110}{1000}\int (4\sin(\omega t) + \sin(3\omega t))\,dt = \frac{0.11}{\omega}\left(4\cos(\omega t) + \frac{1}{3}\cos(3\omega t)\right)$$

$$\phi = \frac{0.11}{2\pi f}\left(4\cos(2\pi ft) + \frac{1}{3}\cos(6\pi ft)\right) = \frac{0.11}{100\pi}\left(4\cos(100\pi t) + \frac{1}{3}\cos(300\pi t)\right)$$

Maximum value occurs at $\omega t = 0$

Maximum value of flux $\phi_m = \frac{0.11}{100\pi}\left(4 + \frac{1}{3}\right) = 1.517mWb$

The waveform is drawn using Matlab/Simulink. Its shows peaky waveform

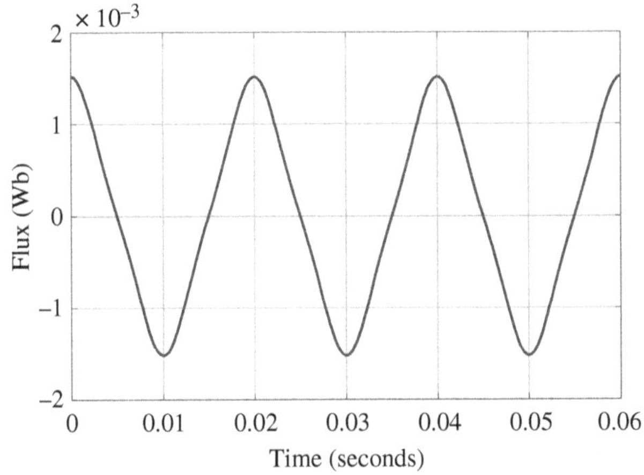

(b) For negligible primary winding resistance, the no-load power supplied to the transformer is equal to the total core less (since copper loss is neglected and no-load condition).
Total Core Loss =

$$V.I.\cos(\theta_1) + V.I.\cos(\theta_3) = \frac{440}{\sqrt{2}}.\left(\frac{0.5}{\sqrt{2}}\right).\cos(30) + \frac{110}{\sqrt{2}}.\left(\frac{0.04}{\sqrt{2}}\right).\cos(5)$$

$$= 95.263 + 2.191 = 97.454W$$

(c) When 3rd harmonic is absent from the applied voltage, then it is given as, $v = 440\sin(\omega t)$
Eddy current loss is given as

$$P_{eddy} = K_e B_m^2 f^2$$

$$P_{eddy} \infty (V)^2$$

Voltage when third harmonic is present

$$V = \sqrt{\left(\frac{440}{\sqrt{2}}\right)^2 + \left(\frac{110}{\sqrt{2}}\right)^2} = \sqrt{(96800) + (6050)} = 320.702V$$

Voltage when third harmonic is absent

$$V' = \frac{440}{\sqrt{2}} = 311.126V$$

%age reduction in eddy current loss

$$= \frac{320.703 - 311.126}{320.701} \times 100\% = 2.986\%$$

(Figure P3.12)

Problem 3.13 The primary of the three identical single-phase transformers are connected in star. The phase voltage, due to the fundamental component of flux ϕ1 is 220 V. The 3rd harmonic flux ϕ3 is 25% of the fundamental. The turn ratio a = N1/N2 = 5.

(a) If the secondary is connected in star, compute (i) phase voltage, and (ii) line voltage of the secondary side
(b) If the secondary is connected in delta, compute (i) the voltage across the open-circuited delta, and (ii) line to line voltage when delta is closed.
(c) If the primary neutral is connected to the source neutral and the secondary is in star, calculate the secondary line and phase voltages.
(d) In part (c) if the secondary winding is arranged in delta, find the voltage across the secondary open-circuited delta.

Solution

In the presence of harmonics, the induced voltage is given as

$$\phi = \phi_{m1} \sin(\omega t) + \phi_{m3} \sin(3\omega t) + \phi_{m5} \sin(5\omega t) + \dots\dots\dots$$

$$e = -N\frac{d\phi}{dt} = -N\omega\left[\phi_{m1} \cos(\omega t) + 3\phi_{m3} \cos(3\omega t) + 5\phi_{m5} \cos(5\omega t) + \dots\dots\right]$$

Given the 3rd harmonic component of flux as 25%, the induced voltage due to 3rd harmonic flux will be 25 × 3 = 75% of that induced due to fundamental. Thus

E3 = (75/100) x 220 = 165 V

(a) Emfs when referred to the secondary side

$$E_1' = 220 \times 5 = 1100V$$

$$E_3' = 165 \times 5 = 825V$$

(*i*) The phase voltage on the secondary side

$$E_{phase} = \sqrt{(E_1')^2 + (E_3')^2} = \sqrt{(1100)^2 + (825)^2} = 1375V$$

(*ii*) The line voltage will not contain 3rd harmonic voltage

$$E_{line} = \sqrt{3}E_1' = \sqrt{3} \times 1100 = 1905.25V$$

(b) (i) The voltage across the open-delta is due to the 3rd harmonic only. This is due to the absence of 3rd harmonic current. Since delta is open ,hence, no closed path is available for 3rd harmonic current to flow. In this case, the voltage will contain 3rd harmonic.

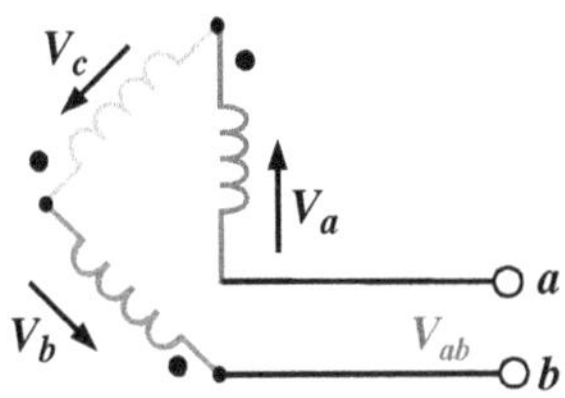

$$V_{ab} = 3E'_3 = 3\,(825) = 2475V$$

(ii) When the delta is closed, the line voltage, will be free from the 3rd harmonic and only fundamental will be seen.

$$E'_1 = 1100V$$

(c) When the primary and source neutrals are connected, the 3rd harmonic no-load current can flow, causing the core flux sinusoidal. Therefore, the secondary phase voltage is free of 3rd harmonic voltage.
The secondary phase voltages = $E'_1 = 1100V$
The secondary line voltage = $\sqrt{3}\,(1100) = 1905.25V$

(d) Since the flux is sinusoidal (as explained in (c)), the 3rd harmonic emf is absent. Hence the voltage across the open-circuit delta is zero.

Problem 3.14 A Delta/Star three-phase power transformer is connected to a Star/Star three-phase distribution. The secondary of the power transformer is grounded and the secondary line voltages are sinusoidal. The two neutrals of the distribution transformer are not grounded. Voltage is measured using voltmeter between the neutral and ground is 2000 V. Compute the effective values of line and phase voltages across the secondary of the distribution transformer, for which the primary to secondary turns is 20.

Solution

The transformer connection is shown in the following Figure

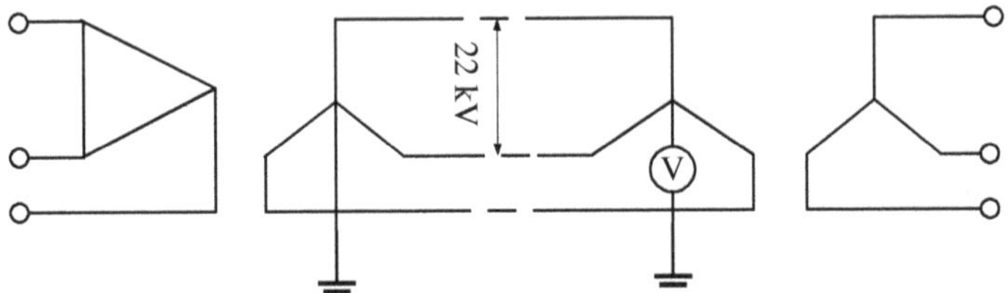

The voltmeter reading of 2000 V, present between the neutral and the ground is due to the 3rd harmonic voltages in the primary phase of the distribution transformer.

The component of this voltage in the secondary phase is given as

$$2000/(N1/N2) = 2000/20 = 100\text{ V}$$

For Distribution transformer

Fundamental component of primary phase voltage $= \frac{22000}{\sqrt{3}} = 12701.70V$
Fundamental component of secondary phase voltage $= \frac{12701.70}{20} = 635.08V$
Effective secondary phase voltage $= \sqrt{(635.08)^2 + (100)^2} = 642.9V$

The line voltage is not affected by the presence of 3rd harmonic content in phase voltage.

$$\text{Line voltage} = \sqrt{3}\,(635.08) = 1100V$$

Problem 3.15 There are five three-phase induction motors each of 20 kW, 440 V rating that are supplied by transformers connected in open-delta from an 11 kV mains. When operated at full load, each motor has an efficiency of 94% and operate at a pf. of 0.85 lagging.

(a) Calculate the KVA rating of each of the two transformers operating in open-delta mode and their turn ratio.
(b) Calculate the line currents on secondary and primary sides.
(c) At what power factor are each of transformers operating?
(d) Calculate the real power supplied by each transformer.
(e) Calculate the available capacity, if a third transformer of the same rating is used to form the closed delta.

Solution

Total load on the transformers = 5 × 20 kW = 100 kW
Total load in terms of kVA = 100/(0.94 × 0.85) = 125.156 kVA

The open-delta must be able to supply 125.156 kVA to the load. In other words, the actual available capacity of the open-delta transformer bank must be 125.156 kVA.

(a) In open-delta condition

$$\frac{\text{Actual Available KVA Rating}}{\text{Sum of the KVA Ratings}} = 0.866$$

$$\text{Sum of the KVA Ratings} = \frac{\text{Actual Available KVA Rating}}{0.866} = \frac{125.156}{0.866}$$

$$= 144.522kVA$$

$$\text{Individual Transformer rating} = \frac{144.522}{2} = 72.261kVA$$

Turn ratio $a = \frac{11000}{440} = 25$

(b) Secondary side-line current = 72.261/440 = 164.23 A
Primary side-line current = 164.23/25 = 6.57 A

(c) Given load power factor = 0.85

$$\theta = \cos^{-1}(0.85) = 31.78$$

$$pf \text{ of Transformer 1} = \cos(30 - \theta) = \cos(30 - 31.78) = 0.99$$

$$pf \text{ of Transformer 2} = \cos(30 + \theta) = \cos(30 + 31.78) = 0.47$$

(d) Real power supplied by Transformer 1 = KVA × pf = 72.261 × 0.99 = 71.538 kW
Real power supplied by Transformer 2 = KVA × pf = 72.261 × 0.47 = 33.962 kW
Hence, the total power supplied by the two transformers = 71.538 + 33.962 = 105.5 kW

(e) When a 3rd transformer is used to form closed delta, the available KVA becomes = 3 × 72.261 = 216.783 kVA

3.24 Summary

This chapter is dedicated to different typed of transformers. Single-phase and three-phase transformers are discussed at length. Additionally, autotransformer is also elaborated with some key numerical problems. Working principle, analysis, testing and performance of single- and three-phase transformer are explained. All limiting conditions of transformer are described with the help of fundamental theorem of electrical engineering. Numerous numerical examples are given throughout the chapter for full understanding of the chapter. Some supplementary numerical examples are also given that cover the entire chapter. Some complex problems are also given. Additionally, some unsolved examples are given for practice by students. Simulink models of single- and three-phase transformers are developed and given in the chapter. Students can develop their own models based on the examples given in this chapter.

3.25 Problems

3.1 A 2000/200 V, 50 KVA transformer has 92 turns in secondary winding calculate:
(a) Primary turns
(b) Secondary full load current, neglecting the loss.

3.2 An ideal transformer has a turns ratio of 10 : 1 and the primary current is 5A when it is supplied at 110 V. Calculate the secondary voltage and current.

3.3 A primary supply of 220 V is given to an ideal single-phase transformer. The secondary is supplying a lighting resistive load of 100 W at 12 V. Compute the turn ratio of the transformer and the current on both primary and secondary sides.

3.4 Calculate full-load primary and secondary currents for a 100 kVA 11 kV/440 V single-phase transformer.

3.5 The number of primary winding turns are 2000 and secondary winding turns are 1000. The load connected to the secondary of the transformer is 10 kVA when supplied by 440 V. Considering the transformer as idea
(a) Compute the load voltage
(b) Calculate the Secondary side current
(c) Calculate the load impedance
(d) Calculate the load impedance referred to the primary side

3.6 A single-phase transformer is wound with 500 primary turns and 100 secondary turns. The primary and secondary resistances are 0.20 Ω and 0.015 Ω, respectively, and the corresponding leakage reactances are 1.5 Ω and 0.05 Ω, respectively.
Determine (a) the equivalent resistance referred to the primary side, (b) the equivalent reactance referred to the primary side, (c) the equivalent impedance referred to the primary side, (d) the equivalent resistance referred to the secondary side, (e) the

equivalent reactance referred to the secondary side, (f) the equivalent impedance referred to the secondary side.

3.7 An ideal 240/60 V step-down transformer which has a primary winding of 50 turns, if the primary current is 1 A, what is the number of secondary windings, what is the turn ratio and what is the secondary current?

3.8 A 20 kVA, 3000/300 V single-phase transformer has the following parameters; $R_1 = 1.5\,\Omega$, $X_1 = 4.5\,\Omega$, $R_2 = 0.003\,\Omega$, $X_2 = 0.006\,\Omega$; calculate

(a) Equivalent resistance referred to HV side
(b) Equivalent reactance referred to HV side
(c) Equivalent impedance referred to HV side
(d) Equivalent resistance referred to LV side
(e) Equivalent reactance referred to LV side
(f) Equivalent impedance referred to LV side.

3.9 A 30 KVA, 2300/220 V, 50 Hz single-phase transformer draw no load current of 10A and consumes 250 W under no-load condition. Calculate two components of no-load current.

3.10 A 60 KVA, 3000/250 V single-phase transformer has the following parameters; $R_1 = 0.5\,\Omega$, $X_1 = 1.5\,\Omega$, $R_2 = 0.02\,\Omega$, $X_2 = 0.3\,\Omega$. Compute the impedance referred to HV side and total copper losses.

3.11 An ideal single-phase 1 kVA, 480/220 V step-down transformer have the open- and short-test circuit tests performed, and the result is the following:

Open-circuit test (OCT) [Secondary side]	Short-circuit test (SCT) [Secondary side]
$V_{oc} = 220\,V$	$V_{sc} = 18\,V$
$I_{oc} = 0.5\,A$	$I_{sc} = 4.5\,A$
$P_{oc} = 35\,W$	$P_{sc} = 65\,W$

Find the equivalent circuit with all the parameters referred to the primary. (Note: $R_1 = 0.87\,\Omega$)

3.12 A 100 kVA, 440/2200 V Single-phase transformer has the following parameters

$$R_1 = 0.012\Omega, R_2 = 0.27\Omega$$
$$X_1 = 0.05\Omega, X_2 = 1\Omega$$
$$R_{c1} = 476\Omega, X_{m1} = 154\Omega$$

The shunt-branch parameters are given with reference to the primary side.
The transformer is supplying a load of 95 KVA at 2200 V and 0.8 pf lagging. Compute the primary side voltage and current for the equivalent circuits of the following figure. Also, compute the voltage regulation in each case.

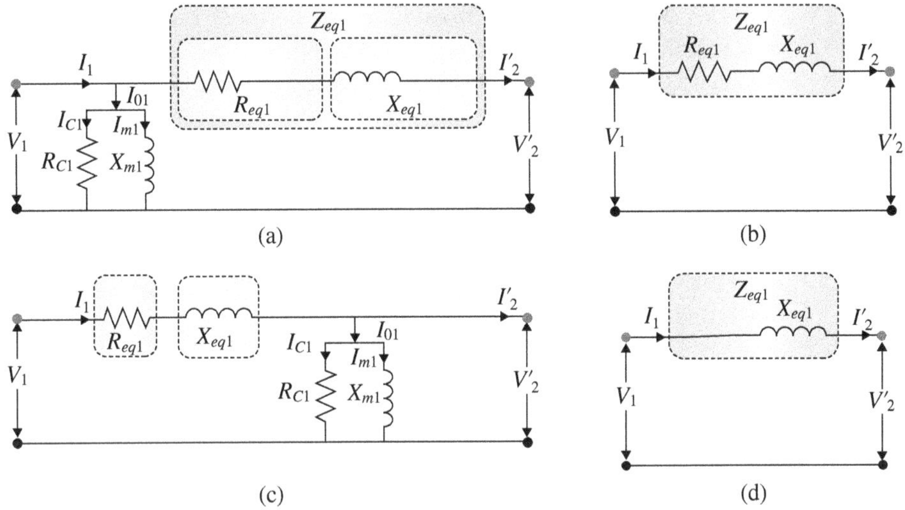

3.13 A 30 KVA, 3000/300 V single-phase 50 Hz, Transformer has the variable and constant losses as 200 W and 100 W, respectively. Calculate the efficiency of the transformer operating at unity power factor and half of the rated load.

3.14 A 50KVA, single-phase transformer has iron loss of 1 kW. The full load copper loss is 1.2 kW. Calculate

(a) Efficiency at full load at 0.8 lagging p.f
(b) Load factor at which maximum efficiency occurs
(c) KVA supplied at maximum efficiency
(d) Maximum efficiency at unity power factor

3.15 A 250KVA, Transformer has iron loss of 2000 W and full load copper loss is 6000 W. If the power factor of the load is 0.8 lagging; Calculate

(a) Full load efficiency
(b) Percentage of full load at which maximum efficiency occurs
(c) Maximum efficiency at 0.9 pf lagging

3.16 The following results were obtained from Open-Circuit Test and Short-Circuit Test on a 5 kVA, 300/200 V, 50 Hz transformer
OC Test: 150 V, 3A, 40 W (on LV side)
SC Test: 10 V, 80 W, 15A (on HV side)
Calculate the full load efficiency at 0.8 power factor.

3.17 The maximum efficiency of a 100 kVA single-phase 50 Hz transformer is 97.08% and operates at a load factor of 95% of full load with unity power factor. Calculate the efficiency of the transformer at full-load condition with unity power factor.

3.18 A 50 KVA, single-phase 2200/220 V transformer gave the following test parameters
OC Test: 200 V, 5A, 150 W
SC Test: 10 V, 250 W, 15A
Calculate the efficiency at full load and 0.9 lagging power factor.

3.19 A 50KVA, 3000/300 V, single-phase 50 Hz transformer has following parameters $R_1 = 8\,\Omega, X_1 = 15\,\Omega, R_2 = 0.5\,\Omega, X_2 = 0.8\,\Omega$; Compute the voltage regulation and the secondary terminal voltage at full load for a power factor of
(a) 0.9 lagging
(b) 0.9 leading and the primary voltage of the transformer is 3000 V (constant)

3.20 Short-circuit test is performed on the HV side of a 20 kVA, 2200/220 V single-phase transformer. It gives the following parameters; 60 V, 2A, 120 W; If the LV side is delivering full load (or rated) current at 0.9 p.f lag and at 220 V. Find the voltage applied to HV side.

3.21 The iron loss is found to be 48 W at 45 Hz and 60 W at 55 Hz in a single-phase transformer and measured at the same peak flux density. Calculate the hysteresis and eddy current losses at 50 Hz.

3.22 A 20KVA, 2200/220 V, single phase transformer gave the following test parameters
OC Test; 220 V, 0.6A, 60 W
SC Test; 80 V, 4A, 40 W
Calculate the efficiency at $(3/4)^{\text{th}}$ of full load at 0.85 p.f.

3.23 A 10 kVA, single-phase transformer has iron loss of 50 W and full load copper loss of 120 W. The daily variation of load on the transformer is given below
8 a.m. to 12 Noon: 2 kW at pf 0.7
12 Noon to 5 p.m.: 3 kW at pf 0.8
5 p.m. to 2 a.m.: 5 kW at pf 0.9
2 a.m. to 8 a.m.: No load
Calculate all-day efficiency of the transformer.

3.24 A 2200/220 V, 400 KVA, 50 Hz, single-phase distribution transformer has the following parameters:
Core loss: 1.5 kW Copper loss: 4 kW at full load.
During the day it is loaded:

% Load	Power factor	Hours
0	—	2
25	0.65 lagging	4
50	0.8 lagging	4
80	0.85 lagging	6
100	1	6
120	0.8 lagging	2

Calculate all-day efficiency of the transformer

3.25 A 1φ, 20 kVA, 2300/230 V, 50 Hz distribution transformer has the following characteristics
Core loss at full voltage = 200 W
Copper loss at half load = 120 W
(a) Determine the efficiency of the transformer when it delivers full load at 0.6 power factor lagging.
(b) Determine the per-unit load fraction at which the transformer efficiency is a maximum. Determine this efficiency if the load power factor is 0.9.
(c) The transformer has the following load cycle
No load for 6 hours
50% full load for 8 hours at 0.8 PF
80% full load for 10 hours at 0.9 PF
Determine the all-day efficiency of the transformer.

3.26 A 150 kVA, 50 Hz, 440/11000 V, single-phase transformer has an efficiency of 97% when supplying full-load current at 0.8 power factor lagging, and an efficiency of 98.5% percent when supplying half-full load current at unity power factor. Find the core losses and copper losses corresponding to full-load current. At what value of load current will the maximum efficiency be attained?

3.27 A 4 kVA, 2200/200 V, single-phase two-winding transformer is used as an autotransformer to raise the supply voltage of 2200 V to an output voltage of 2350 V. The LV winding of the two-winding transformer consists of two equal parts of 150 V each. If both parts of the LV winding are used, Calculate,
(a) kVA output of autotransformer, if rated current of HV is 4 A.
(b) kVA transformed and conducted.

3.28 A 1φ, 20 kVA, 2300/230 V, 50 Hz two-winding transformer is to be used as an autotransformer.
(a) Show the connection that will result in maximum kVA rating.
(b) Determine the voltage ratings of the high- and low-voltage sides.
(c) Determine the kVA rating of the autotransformer. Calculate for both high- and low voltage sides.

3.29 A 1φ, 20 kVA, 440/220 V, 50 Hz transformer has an efficiency of 96% when it delivers 18 kW at 0.9 power factor. This transformer is connected as an autotransformer to supply load to a 440 V circuit from a 660 V source.
(a) Show the autotransformer connection.
(b) Determine the maximum kVA the autotransformer can supply to the 460 V circuit.
(c) Determine the efficiency of the autotransformer for full load at 0.9 power factor.

3.30 A single-phase transformer is designed to operate at 220/110 V, 50 Hz. Calculate the secondary no-load voltage and its frequency if the HV side of the transformer is connected to 220 V, 25 Hz.

3.31 Three single-phase transformers of 25 KVA, 2200/220 V, 50 Hz are connected in Star/Star to form a three-phase transformer bank. The HV side is connected to a 4000 V supply and LV side in open circuited. The Neutral of the primary side is not connected to the supply neutral. The voltage between the primary neutral and the supply neutral is measured as 1100 V.

(a) What is the voltage waveshape between the primary neutral and supply neutral? Neglecting the higher order harmonics more than three.

(b) Calculate the ratio of (i) phase voltages of the two sides, and (ii) Line voltages of the two side.

(c) Determine the ratio of the RMS line-to-line voltages and the RMS line to neutral voltage on each side.

3.32 Two identical single-phase 200 kVA, 220/440 V are supplying a three-phase load at 440 V in open-delta mode. The power factor of the load is 0.7 lagging. Calculate the following:

(a) The maximum secondary side-line current without overloading the transformer.

(b) The amount of real power delivered by each transformer.

(c) The primary side-line current.

(d) A third identical transformer is now added to complete the closed-delta. Calculate the percentage increase in real power supplied. Assume that the load voltage and pf remain same as 440 V and 0.7 pf lagging.

3.33 Two single-phase transformer each rated for 200 kVA, 11KV/400 V are supplying power in open-delta connection mode

a) Determine the load KVA that can be delivered by the connection scheme without overloading either transformer

b) The connection scheme supplies a three-phase balanced load of 200 kVA at 400 V, 0.866 pf lagging. Determine (i) the transformer currents on LV and HV side.

3.34 If 500KVA load at 0.8 pf lagging is supplied by a bank of single-phase transformers connected in Delta/Delta. Each single-phase transformer is rated at 250KVA, 2200/220 V. If fault occurred in one of the transformers and is removed from the service calculate for open-delta connection

a) Primary and Secondary Line currents

b) KVA load carried by each transformer

c) Real Power Supplied by each transformer

d) Percentage increase in load on each transformer when one transformer is removed.

3.35 Three identical single-phase transformers connected in Delta/Delta supply a balanced three-phase load. Each transformer carries a rated current of 200 A at rated voltage

a) One transformer develop fault and is removed for service. For the same temperature rise, determine in each transformer and % age reduction in in KVA capacity of the open-delta.

3.36 A load of 20 kW at 0.8 pf lagging is to be supplied to 230 V by transformer connected in open-delta from 11 kV mains

a) Determine the KVA rating of each of the two transformers and their turn rations
b) Determine the line currents in both HV and LV sides
c) At what pf is each transformer operating?

3.37 A three-phase 11 kV alternator delivers 100 MVA to a three-phase 132 kV transmission line using three identical single-phase transformer banks. Calculate the voltage, current and KVA ratings of each of the transformers if they are connected in (a) Star/Star, (b) Delta/Delta, (c) Star/Delta (d) Delta/Star.

3.38 The voltage applied to the primary of a single-phase transformer at no-load is given as

$$v = 220\cos(\omega t) + 100\cos(3\omega t) + 40\cos(5\omega t)$$

If the primary has 100 turns and the supply frequency is 50 Hz.

(a) Calculate the maximum value of the flux and its shape.
(b) Given the no-load current as $i_0 = 0.45\cos(\omega t - 30) + 0.04\cos(3\omega t - 10) + 0.01\cos(5\omega t - 5)$, calculate the total core loss. Neglect copper losses (no winding resistance).
(c) If the primary voltage does not contain 3rd harmonic component, calculate the % age change in eddy current loss.

3.39 The magnetizing current of a single-phase transformer at rated voltage and rated frequency is given by

$$i = 5\cos(\omega t) + \cos\left(3\omega t + \theta_3\right) + 0.3\cos\left(5\omega t + \theta_5\right)$$

If such transformers are connected with their primary in delta and secondary in star, then calculate the rms value of (a) primary side-phase current (b) primary side-line current.

3.40 The primary side-phase voltage of a 100 kVA, 11 kV/440 V, 50 Hz, three-phase transformer contain 3rd harmonic and is given as

$$v = 220\cos(\omega t) + 100\cos(3\omega t)$$

Calculate the effective value of secondary line-to-line voltage and secondary phase voltage, and waveforms for voltages and currents (no-load and full-load condition (a) Star/Star, and (b) Star/Delta)

References

1 Kulkari, S.V. and Khaparde, S.A. (2005). *Transformer Engineering-Design and Practice.* Marcel Decker.

2 Iqbal, A., Moinuddin, S., Khan, M.R. et al. A novel three-phase to five-phase transformation using special transformer connection. *IEEE Trans. On Power Delivery* 25 (3): 1637–1644.

3 Sen, P.C. (2013). *Principles of Electric Machines and Power Electronics*, 3e. Wiley.

4 Fitzgerald, A.E., Kingsley, C. Jr., and Umans, S.D. (2003). *Electric Machinery*, 6e. Singapore: Mc Graw Hill.

4

Fundamentals of Rotating Electrical Machines and Machine Windings

4.1 Preliminary Remarks

This chapter gives the fundamental concepts on how to understand the basic operating principle of electric machines. An electric machine either operates as a generator or a motor. When operating as a generator, it converts mechanical power to electrical and vice-versa when operating as a motor. Magnetic field and working currents are used to get the machines working. The working principle is explained using a single loop conductor placed in a magnetic field produced by a pair of magnetic poles. Operation of both AC and DC machines are explained throughout the chapter. Detail on machine winding is given in this chapter. Focus is on DC winding with numerous solved examples for clear understanding. Production of mmf and their analysis is presented in conjunction with operation of AC machines. The effect of harmonics and method of elimination or reduction of harmonics by using measures in machine winding is explained in this chapter.

4.2 Generator Principle

An electric generator is an electric machine that converts mechanical energy into electrical energy. An electric generator works on the principle that whenever flux is cut by a conductor, an emf is induced which will cause a current to flow if the conductor circuit is closed. The direction of induced emf (and hence current) is given by Fleming's right-hand rule. Therefore, the essential components of a generator are

(a) a magnetic field
(b) conductor or a group of conductors
(c) motion of conductor w. r. t. magnetic field.

4.2.1 Simple Loop Generator

Consider a single turn loop ABCD rotating clockwise in a uniform magnetic field with a constant speed as shown in Figure 4.1. As the loop rotates, the flux, linking the coil sides AB and CD changes continuously. Hence, the emf induced in these coil sides also changes but the emf induced in one coil side adds to that induced in the other.

Electrical Machine Fundamentals with Numerical Simulation using MATLAB/SIMULINK, First Edition.
Atif Iqbal, Shaikh Moinoddin, and Bhimireddy Prathap Reddy.

Companion website: www.wiley.com/go/iqbal/electricalmachine

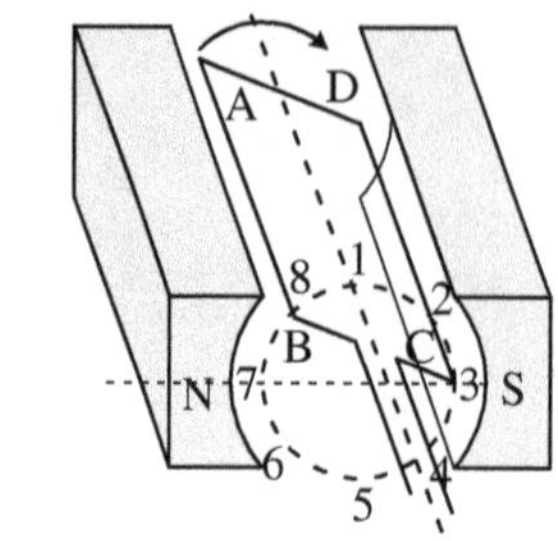

Figure 4.1 Single turn loop rotating in uniform magnetic field [1].

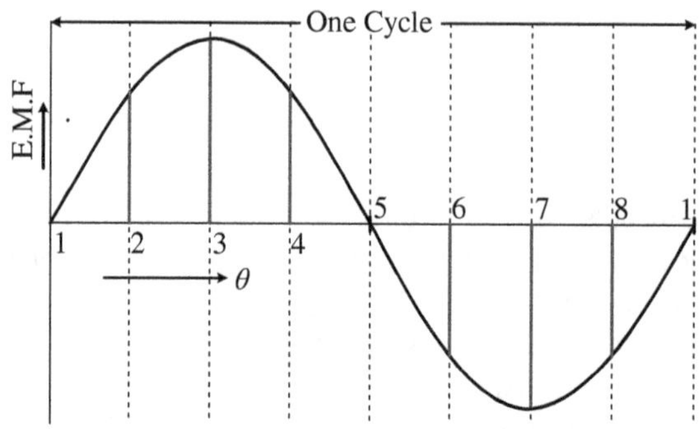

Figure 4.2 Generated emf [1].

(i) When the loop is in position no. 1 (see Figure 4.1), the generated emf is zero because the coil sides (AB and CD) are cutting no flux but are moving parallel to it.
(ii) When the loop is in position no. 2, the coil sides are moving at an angle to the flux and, therefore, a small emf is generated as indicated by point 2 in Figure 4.2.
(iii) When the loop is in position no. 3, the coil sides (AB and CD) are at right angle to the flux and are, therefore, cutting the flux at a maximum rate. Hence at this instant, the generated emf is maximum as indicated by point 3 in Figure 4.2.
(iv) At position 4, the generated emf is less because the coil sides are cutting the flux at an angle.
(v) At position 5, no magnetic lines are cut and hence induced emf is zero as indicated by point 5 in Figure 4.2.
(vi) At position 6, the coil sides move under a pole of opposite polarity and hence the direction of generated emf is reversed. The maximum emf in this direction (i.e. reverse direction, see Figure 4.2) will be when the loop is at position 7 and zero when at position 1. This cycle repeats with each revolution of the coil.

Note that emf generated in the loop is alternating in nature. It is because any coil side, say AB, has emf in one direction when under the influence of N-pole and in the other direction when under the influence of S-pole. If a load is connected across the ends of the loop, then alternating current will flow through the load. The alternating voltage, generated in the loop, can be converted into direct voltage by a device called a commutator. By using a commutator, the same AC generator can be converted to a DC generator. A commutator is made up from copper in the form of a semi-circular ring. Two commutator segments are joined with an insulation between them (Figure 4.3). A commutator is a mechanical rectifier that converts AC signal to DC.

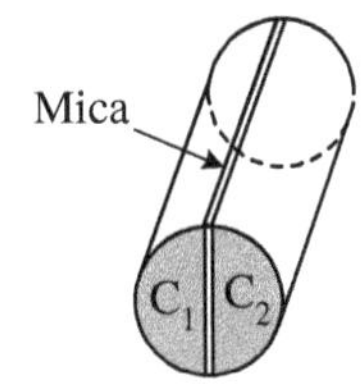

Figure 4.3 Two segment commutator [1].

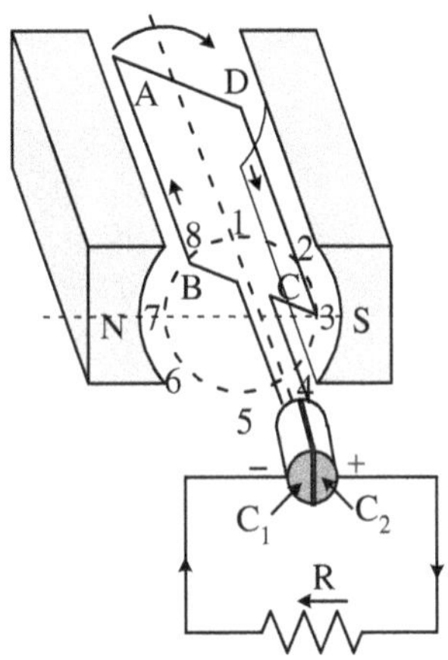

Figure 4.4 Direction of current from 0° to 180° [1].

4.2.2 Action of Commutator

If, somehow, connection of the coil side to the external load is reversed at the same instant the current in the coil side reverses, the current through the load will be direct current. This is what a commutator does. Figure 4.3 shows a commutator with two segments C_1 and C_2. It consists of a cylindrical metal ring cut into two halves or segments C_1 and C_2, respectively separated by a thin sheet of mica. The commutator is mounted on, but insulated from, the rotor shaft. The ends of coil sides AB and CD are connected to the segments C_1 and C_2, respectively as shown in Figure 4.4. Two stationary carbon brushes rest on the commutator and lead current to the external load. With this arrangement, the commutator at all times connects the coil side under S-pole to the +ve brush and that under N-pole to the −ve brush.

(i) In Figure 4.4, the coil sides AB and CD are under N-pole and S-pole, respectively. Note that segment C_1 connects the coil side AB to point P of the load resistance R and the segment C_2 connects the coil side CD to point Q of the load. Also, note the direction of current through load. It is from Q to P.
(ii) After half a revolution of the loop (i.e. 180° rotation), the coil side AB is under S-pole and the coil side CD goes under N-pole as shown in Figure 4.5. The currents in the coil sides now flow in the reverse direction but the segments C_1 and C_2 have also moved through 180° i.e. segment C_1 is now in contact with +ve brush and segment C_2 in contact with −ve brush terminal. Note that a commutator has reversed the coil connections to the load i.e. coil side AB is now connected to point Q of the load and coil side CD to the point P of the load. Also note the direction of current through the load. It is again from Q to P.

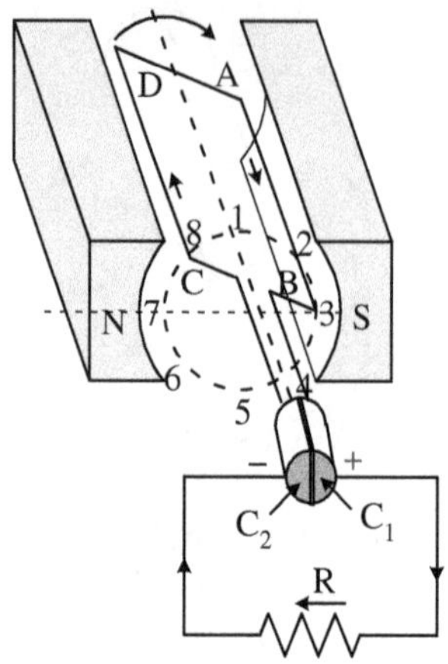

Figure 4.5 Direction of current from 180° to 360° [1].

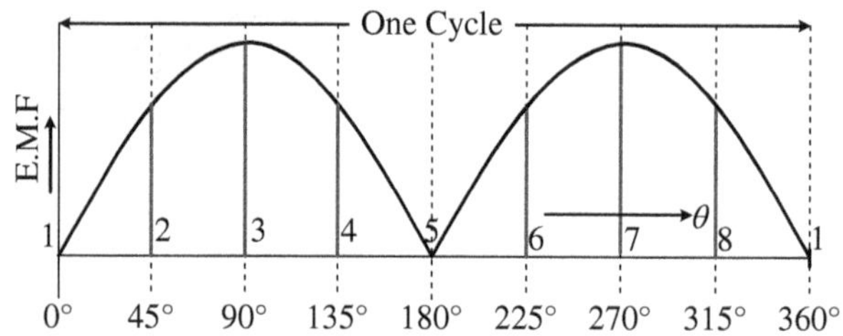

Figure 4.6 Output voltage during one cycle of rotation of the loop [1].

Thus, the alternating voltage generated in the loop will appear as direct voltage across the brushes. The reader may note that emf generated in the armature winding of a DC generator is alternating in nature. It is that by the use of a commutator that generated AC, emf is converted into direct voltage. The purpose of brushes is simply to lead current from the rotating loop or winding to the external stationary load.

The variation of voltage across the brushes with the angular displacement of the loop will be as shown in Figure 4.6. This is not a steady direct voltage but has a pulsating character. It is because the voltage appearing across the brushes varies from zero to maximum value and back to zero twice for each revolution of the loop. A pulsating direct voltage such as is produced by a single loop is not suitable for many commercial uses. What we require is the steady direct voltage. This can be achieved by using a large number of coils connected in series. The resulting arrangement is known as armature winding.

4.2.3 Force on a Conductor

4.2.3.1 DC Motor Principle

A machine that converts DC electrical power into mechanical power is known as a DC motor. Its operation is based on the principle that when a current carrying conductor is placed in a magnetic field, the conductor experiences a mechanical force. The direction of this force is given by Fleming's left-hand rule and magnitude is given by

$$\mathrm{F} = \mathrm{B}Il \text{ newtons} \tag{4.1}$$

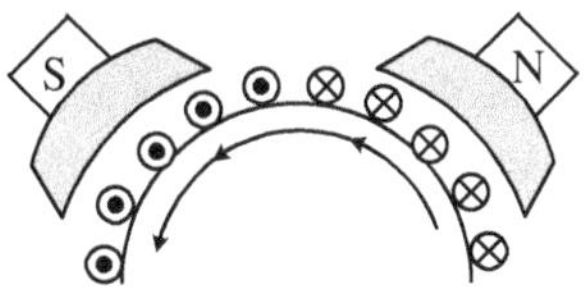

Figure 4.7 Half portion of multi-polar DC motor [1].

Basically, there is no constructional difference between a DC motor and a DC generator. The same DC machine can be run as a generator or motor.

4.2.3.2 Motor Action

4.2.3.2.1 Working of DC Motor

Consider a part of a multi-polar DC motor as shown in Figure 4.7. When the terminals of the motor are connected to an external source of DC supply:

(i) the field magnets are excited developing alternate N and S poles
(ii) the armature conductors carry currents. All conductors under N-pole carry currents in one direction while all the conductors under S-pole carry currents in the opposite direction.

Suppose the conductors under N-pole carry currents into the plane of the paper and those under S-pole carry currents out of the plane of the paper as shown in Figure 4.7. Since each armature conductor is carrying current and is placed in the magnetic field, mechanical force acts on it. Referring to Figure 4.7 and applying Fleming's left-hand rule, it is clear that force on each conductor is tending to rotate the armature in an anticlockwise direction. All these forces add together to produce a driving torque which sets the armature rotating. When the conductor moves from one side of a brush to the other, the current in that conductor is reversed and, at the same time, it comes under the influence of the next pole which is of opposite polarity. Consequently, the direction of force on the conductor remains the same.

4.3 Machine Windings

Machine winding is made up of several coils. Coils are made from a number of conductors turns. The section below elaborates the terms related to machine windings.

4.3.1 Coil Construction

4.3.1.1 Coil Construction: Distributed Winding

This type of winding is made up of several full-pitched or short pitched coils connected in series/parallel fashion [2].

A single turn coil is formed by two joined conductors. The distance between two coil sides is called coil span or coil pitch. The distance between two pole sides is called pole pitch.

If the coil pitch is equal to the pole pitch, it is a full pitched coil.

If the coil pitch is less than the pole pitch, it is called a short-pitched coil. A short-pitched coil is purposely made to eliminate certain harmonics. This is discussed later in the chapter.

The coils are placed in the slots along the periphery. The resultant emf, produced by the winding, is lower than the equivalent concentrated winding. Taking into account the reduction in the overall emf due to the distribution nature of the winding, a factor called 'distribution factor' is used in emf calculation. The emf and air gap mmf is sinusoidal in nature due to distribution of the winding along the periphery of the air gap.

4.3.1.2 Coil Construction: Concentrated Winding

Multiturn coils is placed in two slots is called concentrated winding [2]. The mmf waveform produced by such winding is a square signal. The amount of copper used is lower compared to distributed winding. The overhang is not required in this type of winding. Thus, reduced copper means reduced cost, volume, and copper losses and thus greater efficiency.

Induced emf in concentrated windings is higher than distributed windings. However, due to square waveform of mmf and emf, the harmonic content is higher.

4.3.1.3 Coil Construction: Conductor Bar

In fabricating higher power ratings electric machines, it is generally required to bundle several individual conductors to form the conductor bar [2]. The conductor used is insulated using varnish. The insulation used between the conductor bar and stator core is called main insulation. To fabricate stator windings, several conductor bars are connected to each other at the machine ends.

Conductor bars are prefabricated with either rectangular- or circular-shaped copper wires, coated with insulation material. In fabricating conductor bars, the individual conductor can either be twisted around each other or maintained parallel to each other.

The annealed copper conductors, generally circular shape, are used for winding small- and medium-rating electrical machines.

Field Coil Formation: Field coils are wound with insulated copper wire whose diameter and number of turns depend on the exciting voltage and machine capacity. The wire can be wound on a wooden former that consists of the inner dimensions of the coil.

Stator/rotor/armature Coil Formation: Diamond-shaped wooden former is constructed to the required dimensions, length and width of the coil. The coils are wound over the former with the help of the coil-winding machine. All coils are wound identically. The number of turns depend on the voltage rating of the machine whereas the conductor size depends on the current rating.

4.3.2 Revolving (Rotor) Winding

The winding that rotates with either the rotor of an induction machine or the armature of a DC machine, is called a revolving winding.

4.3.3 Stationary (Stator) Winding

The winding wound either on the stator of an induction machine or on the field system of a DC machine is called a stationary winding.

4.3.4 DC Armature Windings

There are two types of DC armature windings, namely the lap and wave windings. For development of DC armature windings, few pitches related to the types of DC armature windings are Back pitch, Front pitch and Winding pitch.

4.3.4.1 Pole Pitch (Y_p)

It may be defined as:

(i) The periphery of the armature divided by the number of poles of the generator i.e. the distance between two adjacent poles.
(ii) It is equal to the number of armature conductors (or armature slots) per pole. If there are 48 conductors and 4 poles, the pole pitch is 48/4 = 12.

4.3.4.2 Coil Pitch or Coil Span (Y_{cs})

It is the distance, measured in terms of armature slots (or armature conductors) between two sides of a coil. It is, in fact, the periphery of the armature spanned by the two sides of the coil.

If the pole span or coil pitch is equal to the pole pitch, then winding is called ***full-pitched.*** It means that coil span is 180 electrical degrees. In this case, the coil sides lie under opposite poles, hence the induced emfs of two sides are additive. Therefore, maximum emf is induced in the coil as a whole; it being the sum of the emfs induced in the two coil sides. For example, if there are 36 slots and 4 poles, then coil span is 36/4 = 9 slots. If number of slots is 35, then $Y_S = 35/4 = 8$ because it is customary to drop fractions.

If the coil span is less than the pole pitch, then the winding is fractional-pitched. In this case, there is a phase difference between the emfs in the two sides of the coil. Hence, the total emf round the coil which is the vector sum of emfs in the two coil sides, is less in this case compared to that in the first case.

4.3.4.3 Back Pitch (Y_b)

Back Pitch, Y_b: It is the distance between the two active sides of the same coil under adjacent opposite poles. For double layer winding

$$Y_b = \frac{2C}{P} \pm b \tag{4.2}$$

where b = any integer or fraction, added or subtracted with $\frac{2C}{P}$, that will give the value of Y_b an odd integer.

The distance, measured in terms of the armature conductors, which a coil advances on the back of the armature is called back pitch and is denoted by Y_b.

For simplex, duplex, and triplex

Figure 4.8 shows lap winding and Figure 4.9 shows wave winding.

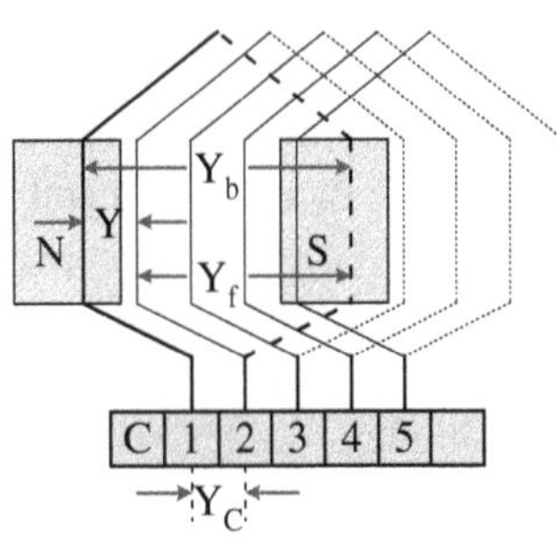

Figure 4.8 Lap winding coil showing back pitch, front pitch and resultant pitch.

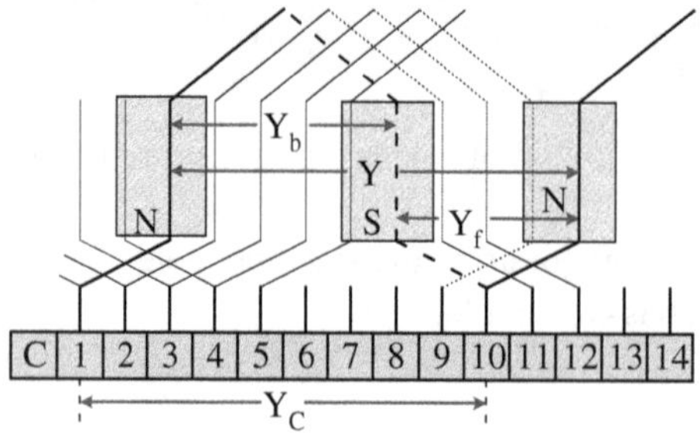

Figure 4.9 Wave winding coil showing back pitch, front pitch and resultant pitch.

4.3.4.4 Front Pitch (Y_f)

Front Pitch, Y_f: It is the distance between two coil sides connected to the same commutator segment. It should be an odd integer.

The number of armature conductors spanned by a coil on the front (or commutator end of an armature) is called the front pitch and is designated by Y_F. Alternatively, the front pitch may be defined as the distance (in terms of armature conductors) between the second conductor of one coil and the first conductor of the next coil which are connected together at the front i.e. commutator end of the armature. Both front and back pitches for lap and wave-winding are shown in Figures 4.8 and 4.9.

4.3.4.5 Resultant Pitch (Y)

It is the distance between the beginning of one coil and the beginning of the next coil to which it is connected (Figures 4.8 and 4.9).

Winding Pitch or coil pitch, Y: It is the distance between starting ends of two consecutive coils expressed in terms of coil sides.

For a double-layer winding, winding pitch should be an even integer.

$$Y = \pm 2\,m \text{ For lap winding} \tag{4.3}$$

$$Y = \frac{2C \pm 2m}{P/2} \text{ For wave winding} \tag{4.4}$$

where $m = 1, 2, 3$ for simplex, duplex and triplex, respectively. P = number of poles and C = number of coils. + sign indicates progressive and – sign indicates retrogressive winding.

As a matter of precaution, it should be kept in mind that all these pitches, although normally stated in terms of armature conductors, are also sometimes given in terms of armature slots or commutator bars because a commutator is, after all, an image of the winding.

4.3.4.6 Commutator Pitch (a)

It is the distance (measured in commutator bars or segments) between the segments to which the two ends of a coil are connected. From Figures 4.8 and 4.9 it is clear that for lap winding, Y_C is the ***difference*** of Y_b and Y_f whereas for wave winding it is the *sum* of Y_b and Y_f. Obviously, commutator pitch is equal to the number of bars between coil leads. In general, Y_C equals the 'plex' of the lap-wound armature. Hence, it is equal to 1, 2, 3, 4 etc. for simplex-, duplex, triplex–and quadruplex etc. lap-windings.

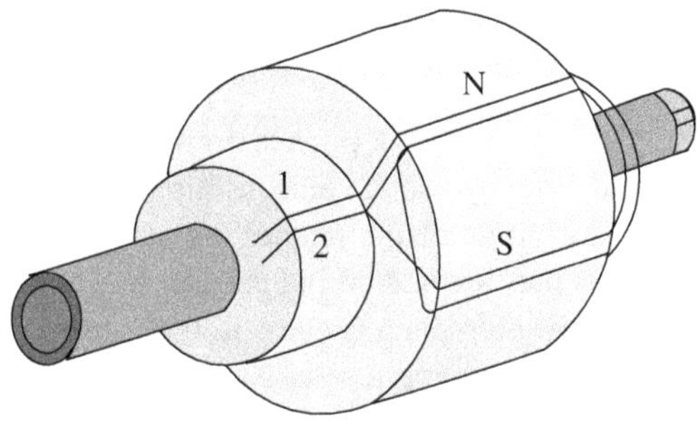

Figure 4.10 Lap winding showing the relative position of one winding element or coil of two turns on the armature of a four-pole machine [3].

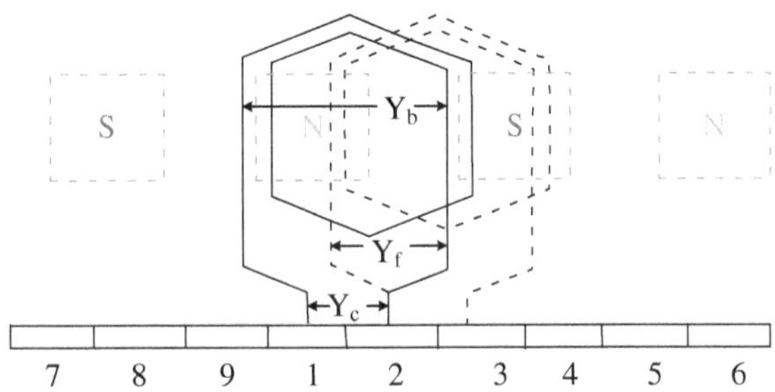

Figure 4.11 Winding diagram for the lap winding of Figure 4.10 showing how the finish end of one winding element or coil of two turns is joined to the commutator and to the start end of the next coil under the same pair of poles [3].

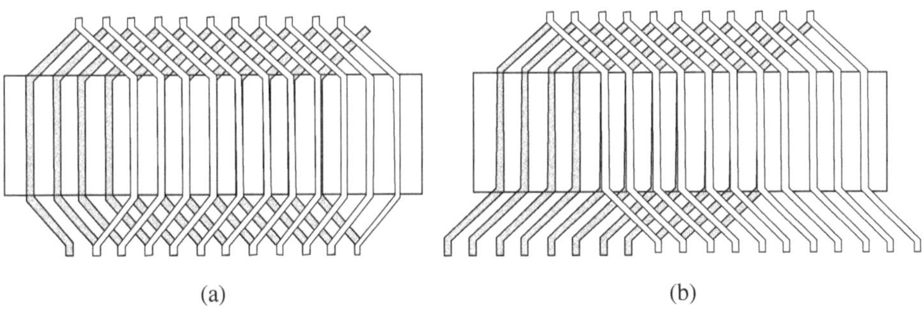

Figure 4.12 (a) Lap winding showing both end connections of each coil side bent towards center of the coil. (b) Wave winding showing end connections of each coil side bent in opposite directions [3].

4.3.5 Lap Winding

4.3.5.1 Lap Multiple or Parallel Windings

A lap winding is so called from the lapping back of the coils or winding elements as they are wound on the armature core [3]. This is shown in Figures 4.10 and 4.11. A better name is a multiple or parallel winding because of the fact that such a winding consists of as many circuits as there are field poles and these circuits are connected in parallel between the brushes. The number of brushes required, therefore, equals the number of armature sections in parallel and also equals the number of poles, i.e. $a = p$. A lap winding can be distinguished from a so-called wave winding from the appearance of the end connections of coils. In the wave winding the front and rear ends of the coils lead in opposite directions while in the lap winding, they continue in the same direction around the armature, as shown in Figure 4.12.

4.3.5.2 Formulas for Lap Winding

The following rules and restrictions govern the assembling and use of a lap winding:

1. The front and back pitches (in winding spaces or coil sides) must both be odd numbers and differ by two or some multiple thereof.
2. The front and back pitches are of opposite sign (one positive and the other negative) since they are laid off in opposite directions on the armature.
3. The winding pitch is equal to the algebraic sum of the front and back pitches. That is, where the front pitch is +9 and the back pitch 7, the algebraic sum is (9–7) or 2.
4. The total number of armature conductors or inductors must be a multiple of the number of slots on the armature.
5. The number of armature slots may be odd or even.
6. The number of current collecting points or brushes on the commutator must equal the number of poles.
7. The maximum emf between two consecutive coil sides (top and bottom) in the same slot, is equal to or a little less than the terminal voltage.
8. The end of one coil is joined to the commutator and to the start of another coil (usually the next) under the same pair of poles.

The following formulas apply in Lap windings:

$$\text{Front pitch} = y_f = \frac{N \pm b}{2p} \pm 2 \tag{4.5}$$

$$\text{Back pitch} = y_b = \frac{N \pm b}{2p} \tag{4.6}$$

$$\text{Winding pitch} = \mathrm{y} = \text{algebraic sum of } \mathrm{y}_1 \text{ and } \mathrm{y}_2 = \pm 2 \tag{4.7}$$

$$\text{Commutator pitch} = \mathrm{y_c} = \pm 1 \tag{4.8}$$

In the formulas for front and back pitches, N is the number of coil sides in the winding; p the number of pairs of poles and b any even number which will make both y_1 and y_2 odd whole numbers and about equal to the pole pitch. The value of b should be taken to be as large as allowable but not too large because then, there is a risk of the two sides of a coil coming under the influence of poles of the same polarity so that the induced emfs would oppose each other. For $b = 0$, the back pitch y_2 becomes equal to the pole pitch. When b is positive, y_2 becomes greater than the pole pitch, while with b negative, y_2 becomes less than the pole pitch. In most cases b is taken as negative so that y_2 is equal to or less than the pole pitch.

The current (i) which flows in each conductor of a lap winding is, $i = \frac{I}{2p}$ Where I is the total armature current or that flowing in the external circuit. The value of (i) should be considered when selecting the proper size of wire in making up or ordering coils. The only practical lap winding from the standpoint of symmetry is the one in which the number of pairs of circuits (current paths) in parallel equals the number of pairs of poles or $a = p$.

A lap winding (Figure 4.13) is symmetrical when $S \div a$ and $S \div p$ is a whole number. Here, S equals the number of slots, a the pairs of circuits in parallel and p, the pairs of poles. With four-pole motors it is not always found that $S \div a$ and $S \div p$ are whole numbers,

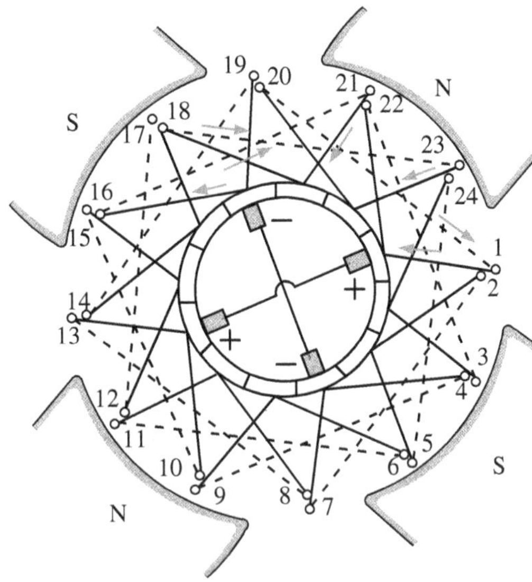

Figure 4.13 Lap winding for a four-pole machine showing the direction of current flow in one of the four current paths for the brush positions as illustrated [3].

since manufacturers sometimes make the number of commutator bars and slots odd so as to make the armature suitable for a two-circuit wave winding as well as a lap winding.

In lap windings where the commutator pitch (y_c) equals one, when $a = p$, the common factor of the number of commutator bars, and y_c will always be unity and there will be a single winding.

In a double-layer winding with two coil sides per slot, the number of slots equals the number. of coils in the winding. When there are more than two coil sides in a slot, the number of coils, $C = SN_s \div 2$ with no idle coils; where S equals the number of slots and N_s equals the number of coil sides per slot. More than two coil sides per slot greatly reduce the number of slots required. Such a winding is symmetrical only when $S \div a$ is a whole number.

For a lap winding the potential pitch, $y_p = K \div p$, where K equals the number of commutator bars and p is the number of pairs of poles.

When equalized rings are needed with lap winding, it is good practice to use one ring for every 6 or 12 commutator bars.

4.3.5.3 Multiplex, Single, Double, and Triple Windings

In cases where it is necessary that the armature carries a very heavy current, more than one winding may be used on the armature with an equal number of commutator bars for each. Therefore, both lap and wave windings may be made up with one, two or three entirely separate windings. In the case of a double lap winding, the sides of a winding element or coil of one winding are sandwiched between the coil sides of the other and likewise, the commutator bars of one winding are sandwiched between those of the other. Each brush must be thick enough to always touch two commutator bars so that both windings will always be connected to the brushes and both deliver or receive current evenly. Three windings so sandwiched make up a triple winding.

A single lap winding always has the same number of current paths between brush sets as there are field poles while a double lap winding has twice the number of current paths as there are field poles. The triple lap winding has three times the number of current paths as there are field poles. A single wave winding always has two current paths between brush sets while the double wave winding has four current paths between brush sets and the triple wave winding has six such paths between brush sets.

An armature has one winding (single) when the number of commutator bars K and commutator pitch y_c have no common factor; it is double when their common factor is two, and it is triple when their common factor is three.

4.3.5.4 Meaning of the Term Re-entrant

A winding is often said to be single or double re-entrant. In the case of a single winding, this means that the winding closes on itself or returns to the beginning point after being traced through all the coils upon passing once around the armature core. A winding is doubly re-entrant if it only re-enters itself after making two passages around the coils of the armature. A single winding may be either single re-entrant or double re-entrant.

In the case of double and triple windings, the term re-entrant is sometimes used in an improper sense. For this reason, it is advisable to specify types of armature windings by the number of separate windings used. Thus, a single winding as a single-closed; two windings as double-closed and three windings as triple-closed etc. A winding made up of two single windings, each of which re-enters itself, will therefore be a double-closed winding not a 'double-re-entrant' one.

4.3.5.5 Multiplex Lap Windings

As explained under the heading of 'Multiplex Windings', a lap winding may be made up of one, two or three separate windings in order to handle heavy armature currents. In the case of a double lap winding, two similar windings insulated from each other are placed in the armature slots with the even numbered commutator bars connected to one winding and the odd numbered bars connected to the other winding. In the same way for a triple lap winding one-third of the commutator bars provided would be connected to each winding.

The formulas which apply in multiplex lap windings are as follows

$$\text{Front pitch} = y_f = \frac{N \pm b}{2p} \pm 2m$$

$$\text{Back pitch} = y_b = \frac{N \pm b}{2p}$$

$$\text{Winding pitch} = \text{y} = \text{y}_\text{b} - \text{y}_\text{f} = \pm 2\,\text{m}$$

$$\text{Commutator pitch} = y_c = \frac{y_b - y_f}{2} = \pm m$$

In these formulas m is 2 for a double winding and 3 for a triple winding, etc. When the number of commutator bars is exactly divisible by m, the windings will be entirely separate from each other.

A double lap winding will have $2 \times (2p)$ current paths between brushes, where $2p$ is the *number of poles*. That is, each winding for a four-pole machine will have four current paths so that a double lap winding on a four-pole machine will have eight current paths between brushes.

In order that the two sides of a winding element or a coil may move simultaneously under field poles of opposite polarity, the total number of coil sides that make up the complete winding divided by the number of the field poles, that is $N \div 2p$, is the approximate value for both the front and back pitches as in the case of the single lap winding. Under such conditions the electromotive forces induced in the two sides of a winding element add up. The smallest front or back pitch, to satisfy this condition, is the distance across a single pole face and the largest front or back pitch is the distance from one pole tip to the nearest pole tip of the same polarity. If the front and back pitches are much less than $N \div 2p$, then a chorded winding results.

Since y_f and y_b must be odd numbers and approximately equal to $N \div 2p$, where $2p$ is the number of field poles, this value will help in determining the value of b to be used in the formula for front pitch (y_f) and back pitch (y_b).

The multiplex lap winding is largely confined to windings for small- and medium-sized machines carrying large currents and using coils made up of wire rather than copper strips or bars. This is because that in such a winding for a large multipolar machine the bars would be too thin and a mechanical construction would result which would make another type of winding advisable, probably a series-parallel design.

Example 4.1 ***Two-layer Simplex Lap Winding***

Write down the winding table for a two-layer simplex lap-winding for a four-pole DC generator having **(*a*)** 20 slots and **(*b*)** 13 slots. What are the back and front pitches as measured in terms of armature conductors?

Solution

(a) The number of commutator segments = 20
Number of conductors (Z) or coil sides $20 \times 2 = 40$; pole pitch = 40/4 = 10
Now remembering that (*i*) Y_b and Y_f have to be odd and (*ii*) have to differ by 2, we get for a progressive winding $Y_b = 11$; $Y_f = -9$ (retrogressive winding will result if $Y_b = 9$ and $Y_f = -11$).
Obviously, commutator pitch $Y_C = -1$.
[Otherwise, for progressive winding $Y_f = \frac{Z}{P} - 1 = \frac{40}{4} - 1 = 9$ and $Y_b = \frac{Z}{P} + 1 = \frac{40}{4} + 1 = 11$]
The simple winding table is given as under

Back connections		Front connections
1 to (1 + 11) = 12	→	12 to (12 − 9) = 3
3 to (3 + 11) = 14	→	14 to (14 − 9) = 5
5 to (5 + 11) = 16	→	16 to (16 − 9) = 7
7 to (7 + 11) = 18	→	18 to (18 − 9) = 9
9 to (9 + 11) = 20	→	20 to (20 − 9) = 11
11 to (11 + 11) = 22	→	22 to (22 − 9) = 13
13 to (13 + 11) = 24	→	24 to (24 − 9) = 15
15 to (15 + 11) = 26	→	26 to (26 − 9) = 17
17 to (17 + 11) = 28	→	28 to (28 − 9) = 19
19 to (19 + 11) = 30	→	30 to (30 − 9) = 21
21 to (21 + 11) = 32	→	32 to (32 − 9) = 23
23 to (23 + 11) = 34	→	34 to (34 − 9) = 25
25 to (25 + 11) = 36	→	36 to (36 − 9) = 27
27 to (27 + 11) = 38	→	38 to (38 − 9) = 29
29 to (29 + 11) = 40	→	40 to (40 − 9) = 31
31 to (31 + 11) = 42 = (42 − 40) = 2	→	2 to (40 − 9) = 33
33 to (33 + 11) = 44 = (44 − 40) = 4		4 to (44 − 9) = 35
35 to (35 + 11) = 46 = (46 − 40) = 6		6 to (46 − 9) = 37
37 to (37 + 11) = 48 = (48 − 40) = 8		8 to (48 − 9) = 39
39 to (39 + 11) = 50 = (50 − 40) = 10		10 to (10 − 9) = 1

The winding ends here because we come back to the conductor from where we started.

We will now discuss the developed diagram which is one that is obtained by imagining the armature surface to be removed and then laid out flat so that the slots and conductors can be viewed without the necessity of turning round the armature in order to trace out the armature windings. Such a developed diagram is shown in Figure 4.14.

Figures 4.15–4.17 are sequence, derived and radial diagrams for the developed winding of Figure 4.14.

(b) No. of conductors = 26; $Y_b = 7; Y_f = -5$
The number of commutator segments = 13
Number of conductors or coil sides 13 × 2 = 26; pole pitch = 26/4 = 6
Now remembering that (*i*) Y_b and Y_f have to be odd and (*ii*) have to differ by 2, we get for a progressive winding $Y_b = 7$; $Y_f = -5$ (retrogressive winding will result if $Y_b = 5$ and $Y_f = -7$).

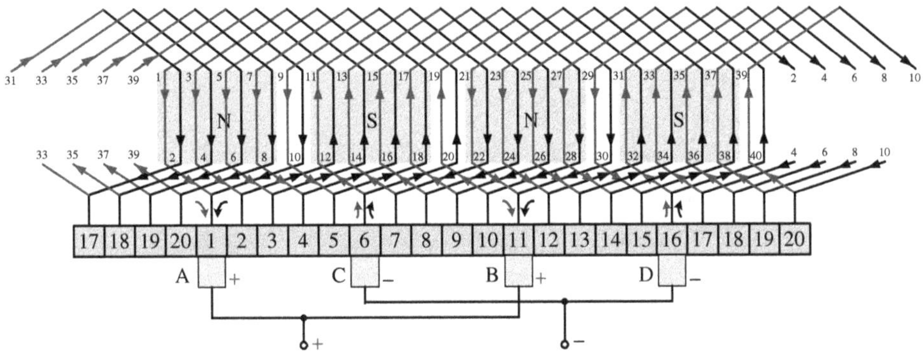

Figure 4.14 Simplex lap-winding for 20 slots, 2-layer.

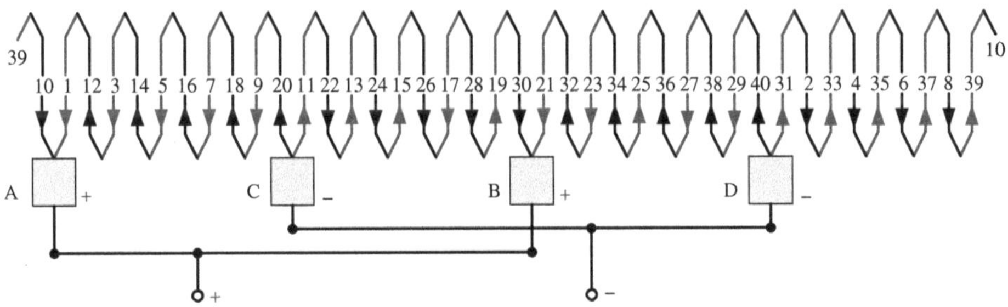

Figure 4.15 Sequence diagram for winding of Figure 4.14.

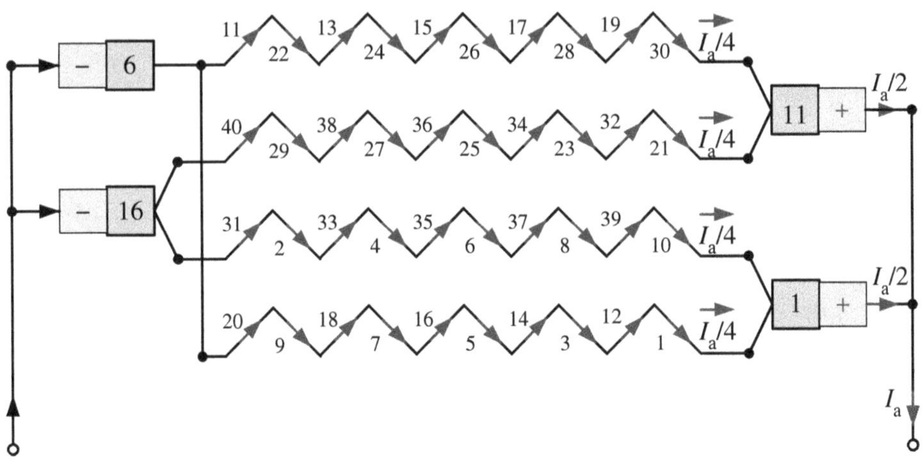

Figure 4.16 Derived diagram for Example 4.1(a).

Obviously, commutator pitch $Y_C = -1$.

[Otherwise, for progressive winding $Y_f = \frac{Z}{P} - 1 = \frac{26}{4} - 1 = 5$ and $Y_b = \frac{Z}{P} + 1 = \frac{32}{4} + 1 = 7$]

The simple winding table is given as under

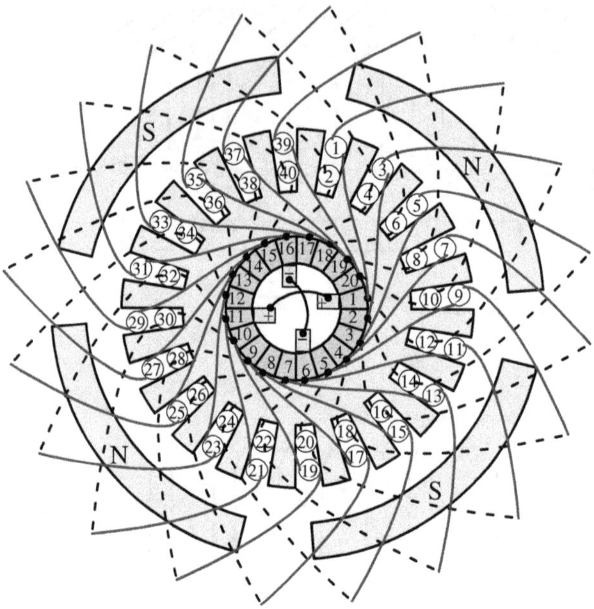

Figure 4.17 Radial view of Figure 4.14.

Back connections		*Front connections*
1 to (1 + 7) = 8	→	8 to (8 − 5) = 3
3 to (3 + 7) = 10	→	10 to (10 − 5) = 5
5 to (5 + 7) = 12	→	12 to (12 − 5) = 7
7 to (7 + 7) = 14	→	14 to (14 − 5) = 9
9 to (9 + 7) = 16	→	16 to (16 − 5) = 11
11 to (11 + 7) = 18	→	18 to (18 − 5) = 13
13 to (13 + 7) = 20	→	20 to (20 − 5) = 15
15 to (15 + 7) = 22	→	22 to (22 − 5) = 17
17 to (17 + 7) = 24	→	24 to (24 − 5) = 19
19 to (19 + 7) = 26	→	26 to (26 − 5) = 21
21 to (21 + 7) = 28 = (28 – 26) = 2	→	2 to (28 − 5) = 23
23 to (23 + 7) = 30 = (30 – 26) = 4	→	4 to (30 − 5) = 25
25 to (25 + 7) = 32 = (32 − 26) = 6	→	6 to (6 − 5) = 1

The winding ends here because we come back to the conductor from where we started.

We will now discuss the developed diagram which is one that is obtained by imagining the armature surface to be removed and then laid out flat so that the slots and conductors

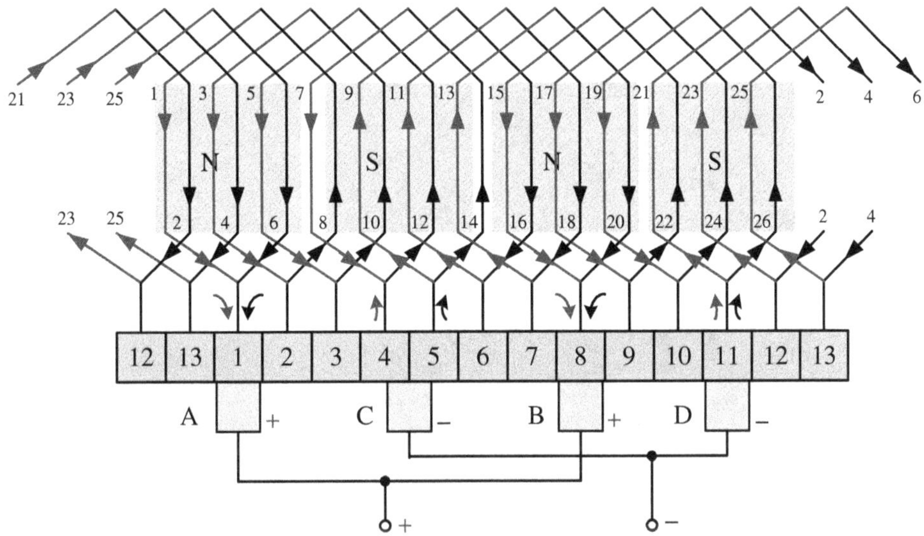

Figure 4.18 Simplex lap-winding for 13 slots, 2-layer.

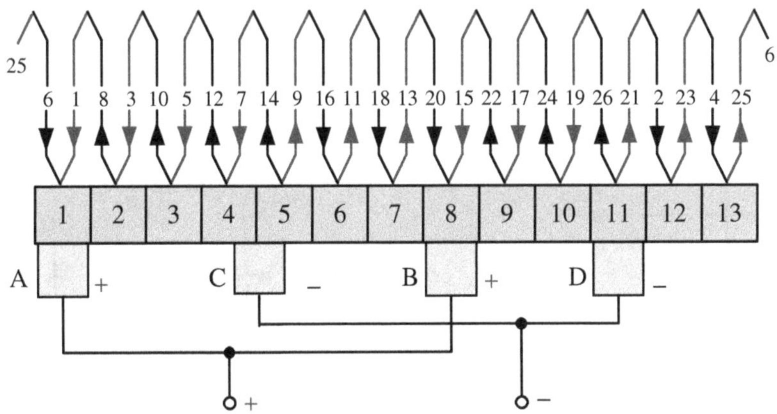

Figure 4.19 Sequence diagram for winding of Figure 4.18.

can be viewed without the necessity of turning round the armature in order to trace out the armature windings. Such a developed diagram is shown in Figure 4.18.

Figures 4.19–4.21 are sequence, radial and derived diagrams for the developed winding of Figure 4.18.

Example 4.2 ***Four-layer Duplex Lap Winding***

Give the winding details for a 4 pole, 36 slots, 72 segments DC generator armature with 8 parallel paths. Draw winding, sequence and derived diagrams for each circuit.

Solution

As the number of parallel paths is 8, which is twice the number of poles, the winding is duplex lap type.

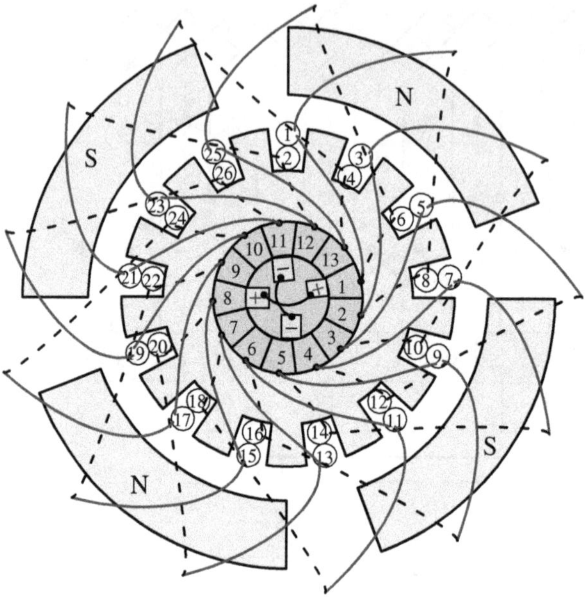

Figure 4.20 Radial view of Figure 4.18.

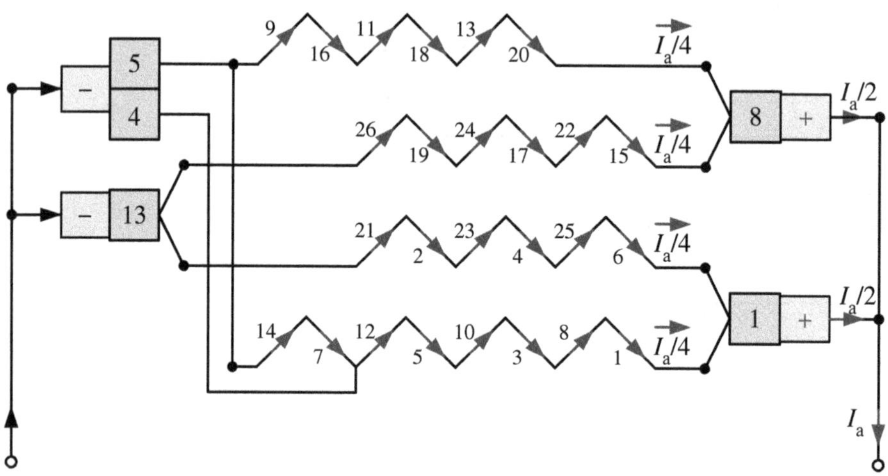

Figure 4.21 Derived diagram for Example 4.1(b).

Number of coils = C = 72

Coil sides per slot $N_s = \frac{2\times72}{36} = 4$

Back pitch = $Y_b = \frac{2C}{P} = \frac{2\times72\pm4}{4} = 37$ or 35

We select $Y_b = 37$ as this satisfies the relation $\frac{Y_b-1}{\text{coil sides per slot}} = \frac{37-1}{4} = 9$ is an integer.

Now for a progressive duplex lap winding

$$Y_c = +2 \ \textit{and} \ Y = 2Y_c = +4$$

Front pitch $= Y_f = Y_b - 4 = 37 - 4 = 33$

As the number of segments in Y_c is an even number, the winding is doubly re-entrant.

Pole pitch = $\frac{\text{slots}}{\text{poles}} = \frac{36}{4} = 9$. Therefore, a total of 36 sides are under each pole.

Winding table

Circuit 1		Circuit 2	
Back connections ($Y_b = 37$)	Front connections ($Y_f = 33$)	Back connections ($Y_b = 37$)	Front connections ($Y_f = 33$)
1 → 38	38 → 5	3 → 40	40 → 7
5 → 42	42 → 9	7 → 44	44 → 11
9 → 46	46 → 13	11 → 48	48 → 15
13 → 50	50 → 17	15 → 52	52 → 19
17 → 54	54 → 21	19 → 56	56 → 23
21 → 58	58 → 25	23 → 60	60 → 27
25 → 62	62 → 29	27 → 64	64 → 31
29 → 66	66 → 33	31 → 68	68 → 35
33 → 70	70 → 37	35 → 72	72 → 39
37 → 74	74 → 41	39 → 76	76 → 43
41 → 78	78 → 45	43 → 80	80 → 47
45 → 82	82 → 49	47 → 84	84 → 51
49 → 86	86 → 53	51 → 88	88 → 55
53 → 90	90 → 57	55 → 92	92 → 59
57 → 94	94 → 61	59 → 96	96 → 63
61 → 98	98 → 65	63 → 100	100 → 67
65 → 102	102 → 69	67 → 104	104 → 71
69 → 106	106 → 73	71 → 108	108 → 75
73 → 110	110 → 77	75 → 112	112 → 79
77 → 114	114 → 81	79 → 116	116 → 83
81 → 118	118 → 85	83 → 120	120 → 87
85 → 122	122 → 89	87 → 124	124 → 91
89 → 126	126 → 93	91 → 128	128 → 95
93 → 130	130 → 97	95 → 132	132 → 99
97 → 134	134 → 101	99 → 136	136 → 103
101 → 138	138 → 105	103 → 140	140 → 107
105 → 142	142 → 109	107 → 144	144 → 111
109 → 2	2 → 113	111 → 4	4 → 115
113 → 6	6 → 117	115 → 8	8 → 119
117 → 10	10 → 121	119 → 12	12 → 123
121 → 14	14 → 125	123 → 16	16 → 127
125 → 18	18 → 129	127 → 20	20 → 131
129 → 22	22 → 133	131 → 24	24 → 135
133 → 26	26 → 137	135 → 28	28 → 139
137 → 30	30 → 141	139 → 32	32 → 143
141 → 34	34 → 1	143 → 36	36 → 3

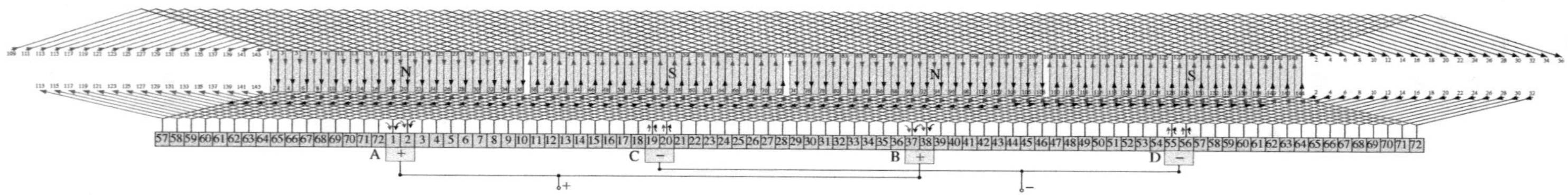

Figure 4.22 Duplex lap winding, 4 pole, 36 slot DC Generator.

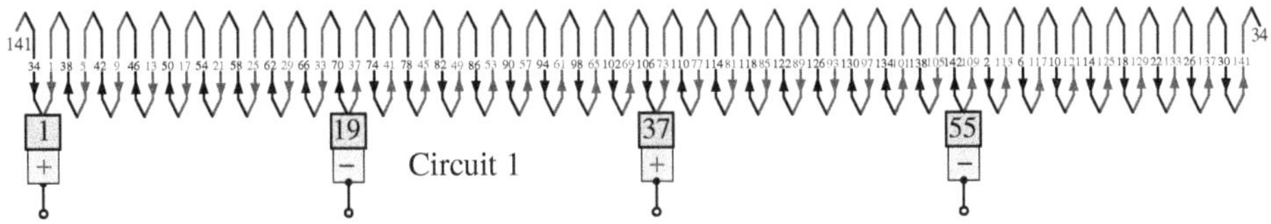

Figure 4.23 Sequence diagram for circuit 1.

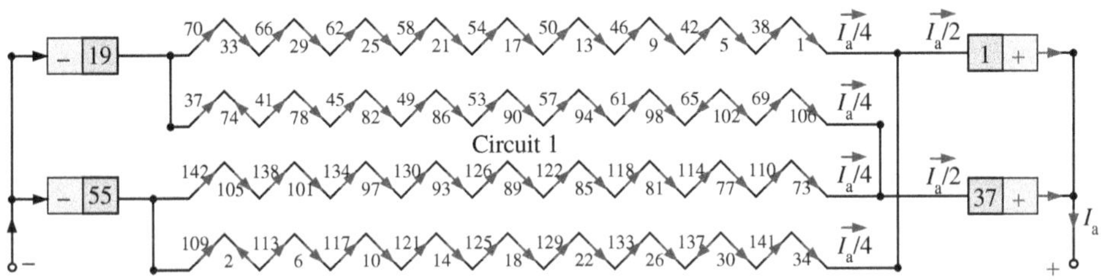

Figure 4.24 Derived diagram for circuit 1.

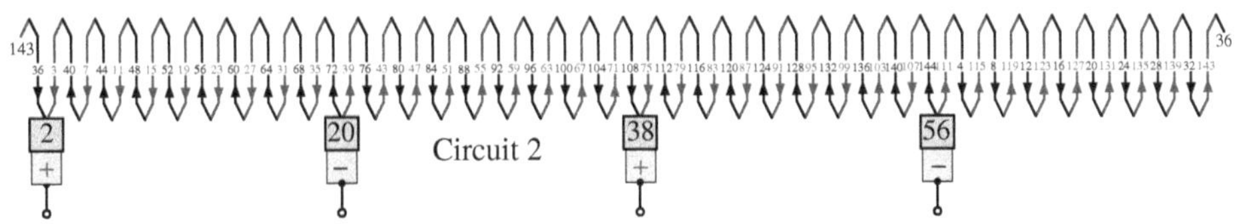

Figure 4.25 Sequence diagram for circuit 2.

Winding layout for 144 conductors with direction of induced voltage/current and 72 commutator segments is shown in Figure 4.22.

There are nine slots under each pole. Hence assume, slots 1–9 and 19–27 are under North Pole and slots 10–18 and 25–36 are under South Pole. Therefore, coil sides 1–36 and 73–108 are under North Pole while coil sides 37–72 and 109–144 are under South Pole.

Figure 4.22 shows the developed winding diagram for progressive lap winding. Figure 4.23 shows the sequence diagram for circuit-1 of the lap winding from which we conclude that it has four parallel paths because of four poles and four brushes. Observing the sequence diagram, the positive brushes position are at segment number 1 and 37 and the negative brushes can be placed at segment number 19 and 55. Figure 4.24 is the derived diagram for cicuit-1, which shows four parallel paths. Similarly, Figure 4.25 shows the sequence diagram for circuit-2 of the lap winding from which we conclude that it has four parallel paths because of four poles and four brushes. Observing the sequence diagram (Figure 4.25) the positive brushes position are at segment number 2 and 35 and the negative brushes can be placed at segment number 20 and 56. Figure 4.26 is the derived diagram for cicuit-2, which shows four parallel paths.

Example 4.3 ***Single-layer Duplex Lap Winding***

Develop the duplex single layer lap winding diagram for a DC machine having pole pitch = 6 and poles = 4.

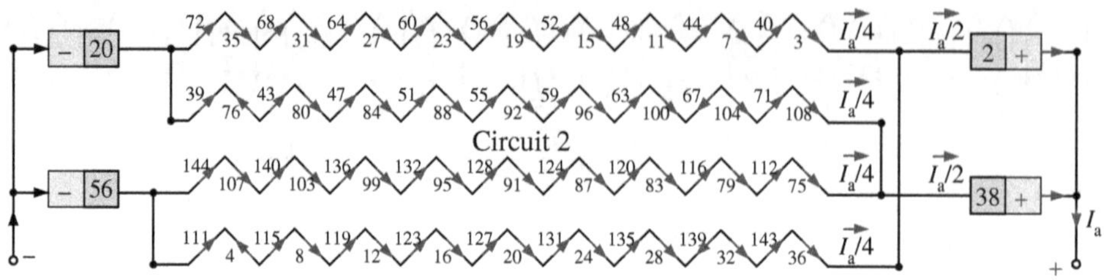

Figure 4.26 Derived diagram for circuit 2.

Solution

Number of slots = 6 × 4 = 24.

It is duplex lap winding, so, number of parallel paths = $a = 2 \times 4 = 8$

Number of coils = C = 24/2 = 12

Number of coil sides = N = 2C = 24 because it is single layer.

Number of commutators = K = 12

Back pitch = $Y_b = \frac{2C \pm b}{P} = \frac{2\times12\pm4}{4} = 7$ or 5

Front pitch = $Y_f = \frac{2C\pm b}{P} \pm 2m = \frac{2\times12\pm4}{4} \pm 4 = 11$ or 3, 1 or 9

Winding pitch = $Y = Y_b - Y_f = \pm 2\,m$

Commutator pitch = $Y_c = \frac{Y_b - Y_f}{2} = \pm m = \pm 2$

Conditions of symmetry are satisfied because (i) $\frac{2S}{A}$ = integer = $\frac{2\times24}{8} = 6$ where A = Number of parallel paths and (ii) $\frac{2S}{P}$ = integer = $\frac{2\times24}{4} = 12$

Slots per pole = 24/4 = 6

Since there is a common factor of 2 between commutator pitch and commutators, the winding is doubly re-entrant.

Winding table:

Circuit 1		Circuit 2	
Back connections ($Y_b = 7$)	Front connections ($Y_f = -3$)	Back connections ($Y_b = 7$)	Front connections ($Y_f = -3$)
1 → 8 →	8 → 5	3 → 10 →	10 → 7
5 → 12 →	12 → 9	7 → 14 →	14 → 11
9 → 16 →	16 → 13	11 → 18 →	18 → 15
13 → 20 →	20 → 17	15 → 22 →	22 → 19
17 → 24 →	24 → 21	19 → 2 →	2 → 23
21 → 4 →	4 → 1	23 → 6 →	6 → 3

There are six slots under each pole. Hence assume, slots 1–6 and 13–18 are under North Pole and slots 7–12 and 19–24 are under South Pole. Therefore, coil sides 1–6 and 13–18 are under North Pole while coil sides 7–12 and 19–24 are under South Pole.

Figure 4.27 shows the developed winding diagram for progressive duplex lap winding. The width of brushes is such that they should make contact with two commutator segments

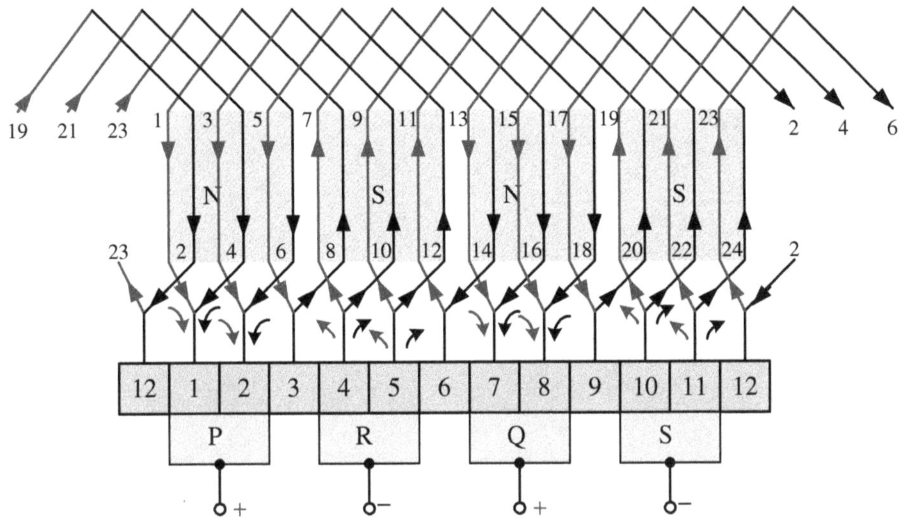

Figure 4.27 4 pole, 24 slot, single layer duplex lap winding diagram.

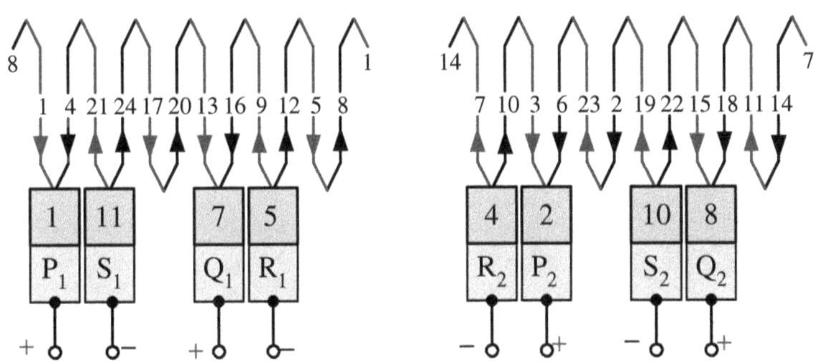

Figure 4.28 Sequence diagram for Figure 4.27.

at a time. Figure 4.28 shows the sequence diagram Figure 4.27 of the duplex lap winding from which we conclude that it has eight parallel paths because of duplex lap type. Observing the sequence diagram, the positive brushes position are at segment number 1, 2 and 7, 8 and the negative brushes can be placed at segment number 4, 5 and 11, 12. Figure 4.29 is the derived diagram for Figure 4.28, which shows eight parallel paths. One can observe that there is unequal voltage across the parallel branches since there are two coil sides in four branches and four coil sides in the remaining four branches. So, there shall be circulating current within the branches. Therefore, duplex lap type winding is not advisable in single layer 24 slot, four-pole machine.

4.3.6 Wave Winding

The wave winding is so called from the zig zag or wave path that the winding takes through the slots of the armature, as shown in Figures 4.30 and 4.31. This type of winding is more definitely described as a series or two-circuit winding because of the fact that half of the

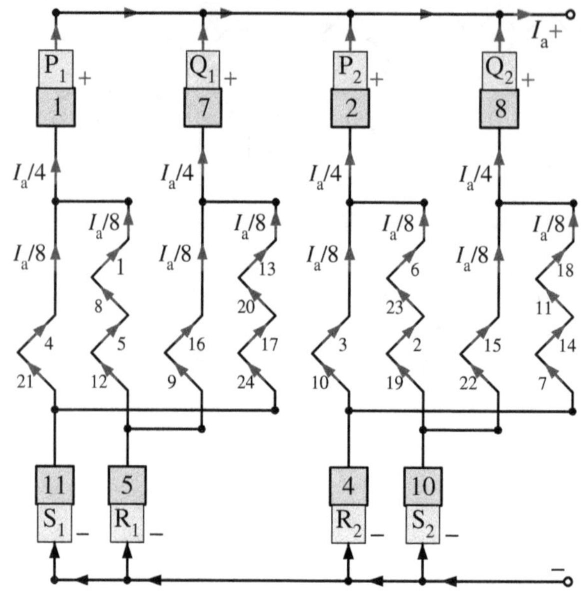

Figure 4.29 Derived diagram for Figure 4.28.

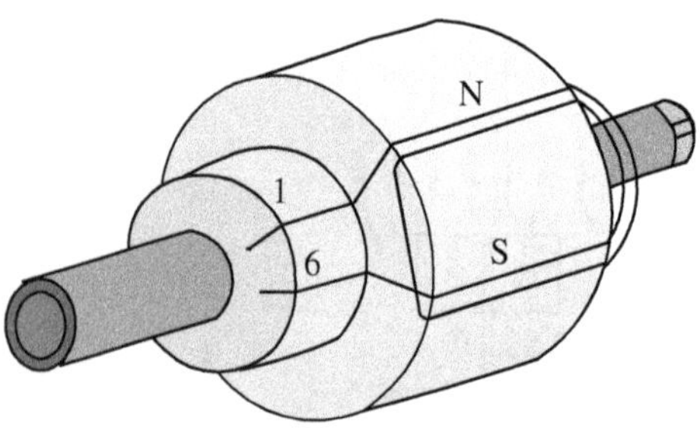

Figure 4.30 Wave winding showing relative position of one winding element or coil of two turns on the armature of a four-pole machine [3].

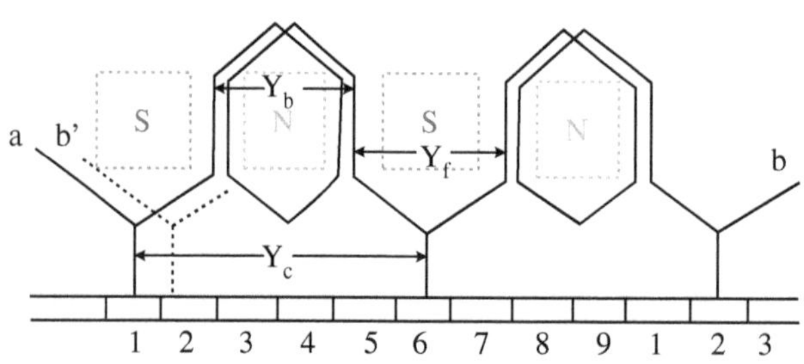

Figure 4.31 Winding diagram, for wave winding of Figure 4.30 showing how the finish end of one coil is joined to the commutator and connected to the start of the next coil under the next pair of poles. b' is a continuation of b [3].

armature coils or sections are connected in series and the two halves are connected in parallel. Therefore, this winding has only two current paths in parallel between brushes regardless of the number of poles. Only two sets of brushes are required for a machine of any number of poles, but improved commutation is brought about when the number of brushes equals the number of field poles. The wave winding is used in small- and medium-sized machines where it is desired to keep the number of coils as small as possible.

4.3.6.1 Formulas for Wave Winding

The following rules and restrictions govern the assembling and use of a wave winding:

1. The front and back pitches (in winding spaces or coil sides) must both be odd.
2. The front and back pitches may be equal or may differ by two or some multiple thereof. The former condition is usually the case.
3. The front and back pitches are of the same sign since they are laid off in the same direction.
4. The winding pitch is equal to the sum of the front and back pitches.
5. The commutator pitch and number of commutator bars must not have a common factor.
6. The number of current collecting points or brushes on the commutator is always two for any number of poles but the number of sets of brushes may equal the number of poles.
7. The maximum emf between two consecutive coil sides (top and bottom) in the same slot, is equal or nearly equal to the terminal voltage of the machine.
8. The finish of one coil is joined to a commutator bar and to the start of another coil which lies under the next pair of poles.

The following formulas apply in wave winding:

$$\text{Winding pitch} = y = y_{\mathrm{f}} + y_{\mathrm{b}} = \frac{N \pm 2}{p} \tag{4.9}$$

Front pitch $= y_{\mathrm{f}}$
Back pitch $= y_{\mathrm{b}}$
Number pairs of poles $= p$

$$\text{Commutator pitch} = \mathrm{y_c} = \frac{y_f + y_b}{2} = \frac{K \pm 1}{p} \tag{4.10}$$

$$\text{Number commutator bars} = \mathrm{K} = \frac{N}{2} = p \times y_c \pm 1 \tag{4.11}$$

The number of coil sides (N) must be such that $\frac{N\pm2}{p}$ will be an even number with y_1 and y_2 both odd numbers. y_{f} and y_{b} are usually taken equal. Under these conditions, the back pitch (y_{b}) is nearly equal to the pole pitch and the sum of the front and back pitches is nearly equal to double the pole pitch. If y_{b} is reduced, then y_{f} must be increased so that $y_{\mathrm{f}} + y_{\mathrm{b}}$ will be constant. This is what happens in a chorded or fractional pitch winding.

Figure 4.32a shows the brush position of four-pole wave winding.

The current (i) which flows in each conductor of a wave winding is always one-half the total armature current, $i = I \div 2$. This value of (i) should be considered when selecting the proper size of conductors for coils of wave windings. The closed winding formula for a wave winding is $y_{\mathrm{c}} = (\mathrm{K} \pm a) \div p$. Where y_{c} is the commutator pitch, K the number of commutator bars, a the pairs of parallel circuits in the winding and p the number of pairs of poles. When $a = 1$, it is only possible to make a single winding. The highest common factor of y_{c} and K gives the type of winding, whether single, double, etc. The number of coil sides in a wave winding equals twice the number of commutator bars.

The number of slots without idle coils must satisfy the formula, $S = (2\,\mathrm{K}) \div \mathrm{N_s}$. Where K is the number of commutator bars and $\mathrm{N_s}$ the number of coil sides per slot. A wave winding is symmetrical when $\mathrm{K} \div a$, $S \div a$ and $p \div a$, are whole numbers. Where K is the number of commutator bars, a, the number of pairs of parallel circuits in the winding, S the number of slots, and p the number of pairs of poles.

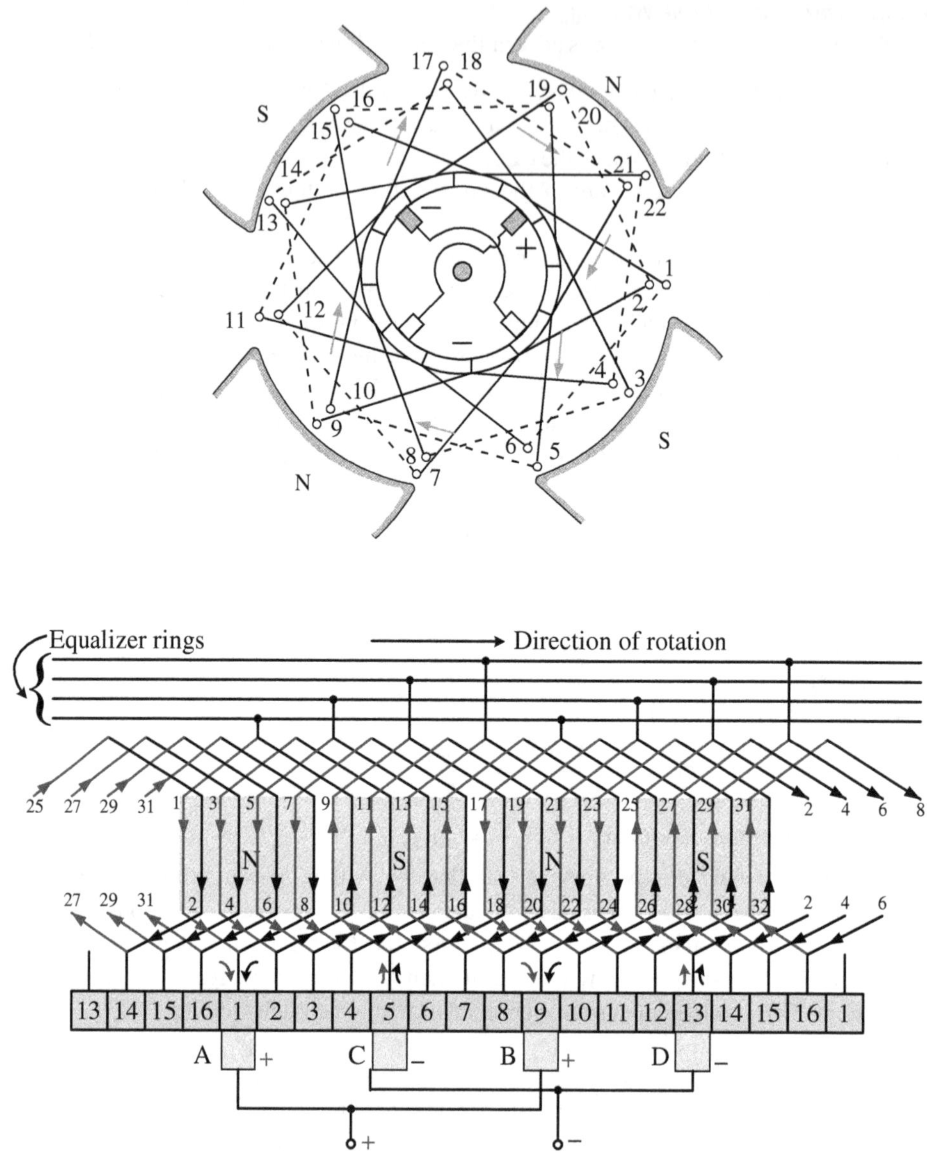

Figure 4.32 (a) Wave winding for a four-pole machine showing the direction of current flow in one of the two current paths for the brush position as illustrated [3]. (b) Developed winding diagram [4].

When equalizer rings are needed with a wave winding, the use of one ring to every 15–20 commutator bars is good practice.

4.3.6.2 Multiplex Wave or Series-Parallel Winding

In a single wave or series winding, the number of armature circuits in parallel is always equal to 2, while in a lap or multiple winding the number of circuits in parallel is equal to the number of field poles. The series-parallel winding is a design of wave winding in which the number of armature circuits in parallel may be larger than 2 and yet smaller than the number of field poles. It is especially suitable for large multipolar armatures which use winding elements made up of copper bars.

4.3.6.3 Formulas for Series-Parallel Winding

The following rules govern the assembling and use of this winding

1. The front and back pitches (in winding spaces or coil sides) must be odd numbers. They may be equal to or different from each other although they are usually equal.
2. The winding should be symmetrical as shown by the number of field poles divided by the number of armature circuits in parallel being a whole number. If this is not the case, the final winding should be made up of a sufficient number of independent windings to make the number of field poles divided by the number of armature circuits in parallel a whole number for each of the independent windings.
3. It is necessary that the number of commutator bars divided by one-half the number of armature circuits in parallel shall be a whole number. If the conditions of (2) are fulfilled, the condition named here is likewise satisfied at the same time.
4. If the number of commutator bars and the commutator pitch have no common factor, the winding is a single one; if the common factor is 2, the winding is double or if it is equal to m, the winding is multiplex of m separate windings. The common factor must, however, be smaller than half the number of circuits. An illustration will bring out these conditions. In the case of an eight-pole machine with 27 slots, 54 commutator bars, 4 conductors per slot, and a commutator pitch of 13, the two series windings are singly closed forming one winding, since 54 and 13 have no common factor. For a machine of four-poles, with 20 slots, 40 commutator bars, 4 conductors per slot, commutator pitch of 18, the winding may be made up of two series windings, the values 40, 18 and 4 have a common factor of 2. This makes the winding doubly closed and consists of two series windings with each in turn consisting of two other series windings, each of the latter forming one continuous winding.
5. For large machines using series-parallel windings, it has been found advisable in most cases to use equipotential connections and connect about every fourth or eighth commutator section. (See heading 'Equipotential Connectors'.)
6. The number of current collecting points or brushes on the commutator is equal to the number of armature sections in parallel.

The following formulas apply in the case of a series–parallel winding:

Front pitch $= y_f$

Back pitch $= y_b$

$$\text{Winding pitch} = y = y_f + y_b = \frac{N \pm 2a}{p} \tag{4.12}$$

$$\text{Commutator pitch} = y_c = \frac{K \pm a}{p} = \frac{C \pm a}{\frac{P}{2}} = \frac{2(C \pm a)}{P} \tag{4.13}$$

$$\text{Number commutator bars} = \mathrm{K} = \frac{N}{2} \tag{4.14}$$

Where p is number of *pairs of poles*, P is the number of poles and N is the number of coil sides in the winding. N must be such a value that $\frac{N \pm 2a}{p}$ is a whole number.

4.3.7 Symmetrical Windings

All armature windings should be made symmetrical if possible. A winding is symmetrical when the total number of slots divided by one-half the number of armature sections in parallel is a whole number ($2S/a$). If this is not the case, the emfs induced in different circuits will produce circulating currents. A winding is also symmetrical if the number of field poles divided by the number of armature sections in parallel is a whole number (P/a) or the number of commutator bars is divisible by one-half the number of armature sections in parallel ($2K/a$). To be sure that any winding is symmetrical, all three of these conditions should be fulfilled.

4.3.7.1 Possible Symmetrical Windings for DC Machines of a Different Number of Poles

The following are the most suitable windings for machines of different number of poles

1. For a two-pole machine either a two-circuit lap or wave winding can be used. The former is usually preferred.
2. For a four-pole machine either a two-circuit wave or a four-circuit lap winding can be used.
3. For a six-pole machine, we are limited to the two-circuit wave and six-circuit lap windings, since the four-circuit wave winding ($a = 2, p = 3$) is unsymmetrical. This is probably the most objectionable restriction of all, because so many cases arise where four circuits are desirable in a six-pole machine.
4. For an eight-pole machine, wave windings with two or four circuits and an eight-circuit lap winding can be used. This is the first instance of a symmetrical wave winding with $a > 1 < p$ and can be used with great advantage in many cases.
5. For a 10-pole machine there are again only two symmetrical windings the two-circuit wave and 10-circuit lap windings.
6. For a 12-pole machine, the possibilities are much greater, for wave windings with two, four or six circuits and a lap winding with 12 circuits can be used.
7. The list can be continued further, if desired, but it is plainly seen that certain very important advantages are open to the designer by making use of wave windings with more than two circuits in 8, 12, and 16-pole machines. It is necessary, however, to observe the other conditions of symmetry, namely s/a must be a whole number and idle coils must be avoided.

4.3.8 Equipotential Connectors (Equalizing Rings)

In a symmetrical winding, that is, a winding with identical K/a phase systems, there are always coils at the same potential, and these can be joined together, if desired. If no dissymmetry whatever were present, however, there would be little object in making such connections, unless they are needed as phase tappings to obtain an alternating emf. There are many cases of dissymmetry in a machine, apart from those due to the arrangement of the winding already mentioned. Thus, the magnetic material may not be uniform, the pole-shoes may not be properly spaced, and the gap may not be uniform, and so on. In general, perfect symmetry must be regarded as an unattainable ideal in practice. One can pay attention to the following conditions.

Owing to these dissymmetries, the emf induced in the several armature circuits varies, and causes equalizing currents to flow through the brushes. These equalizing currents, if of sufficient strength, produce sparking and in any case, load the brushes in an undesirable manner. In other words, the flux through each parallel coil is not the same due to non-uninform air gap, magnetic material non uniformity etc. Due to different flux passing through different parallel coil, the emf induced is different. This causes circulating current among the parallel connected coils. The current through one brush is higher than the other causing sparking in brush. It is interesting to note that, even when no equalizing rings are present, the positive and negative collecting rings act as equalizers, and the tendency of these equalizing currents is to neutralize the inequalities in the magnetic field to which they are due. When equalizing rings are used, however, large equalizing currents will flow along these and strongly damp out any inequalities in the field, and so reduce the difference of potential between corresponding points in the winding. Consequently, the brushes are relieved, and are so much better able to perform their proper function of collecting the current.

Equalizing rings must not be regarded as in any way essential, and many machines work quite well without them. Nevertheless, they add a certain factor of safety which the manufacturer is often glad to purchase at so small a cost, for they are not only surer of their machine passing the test satisfactorily but also knows that after-effects, like wear of the bearings, cannot give rise to such serious trouble as when equalizers are absent. Consequently, equalizers are seen on many large machines with lap-wound armatures, or with wave windings with more than two circuits. Equalizing rings should not have an extremely low resistance. Such practice requires not only an excessive amount of copper but leads to considerable loss and heating in the winding. All that is necessary is to provide an alternative path of negligible resistance compared to that of the brushes and, for this purpose, it is usually sufficient to make the section of the rings about half that of the conductors.

Regarding the number of equalizing rings, much depends on the opinion of the designer. With lap windings, one ring for every 6–12 commutator segments is common practice, but this is scarcely feasible when $p > a$, for here the potential pitch, $y_p = K/a$, may be fairly large and the number of rings becomes prohibitive (The potential pitch is the number of commutator bars between successive equipotential points in the winding which can be joined together when equalizing rings are needed.). In such cases (wave windings with more than two circuits), one ring for every 15–20 commutator segments may suffice. There is no need to make the pitch between all the rings the same, but designers generally prefer to split up y_p into a whole number of parts. In this case, the tappings form a symmetrical poly-phase system of pressures.

For making equalizer connections, the number of coils should be multiples of the number of poles and, at the same time, the number of coils per pole should be divisible by a small number 2, 3, or 4.

For instance, consider a four-pole, 16-slot double layer, lap wound armature winding having 32 coil sides. Its developed winding diagram is shown in Figure 4.32b. Here, the number of coil sides per pole is eight. The equalizer connections are shown in Figure 4.32b. The following points are worth noting [4]:

1. That every third coil is connected to an equalizer. Each coil-side (or coil) connected to the equalizer occupies the same position relative to the poles such that the same emf is induced in all the coils at every instant.
2. The number of connections to each equalizing ring is equal to the number of pair of poles.
3. The number of equalizing rings is equal to the number of coils under one pair of poles (i.e. 8/2 = 4).
4. Here, alternate coils have been connected to the equalizing ring and, as such, it is said that the winding is equalized by 50%. If all the coils had been connected to the equalizers, it would have been said that the winding is 100% equalized.

Equalizer rings not needed in wave winding. The armature coils forming each of the two parallel paths are under the influence of all pole-pairs so that the effect of the magnetic circuit asymmetry is equally present in both the parallel paths resulting in equal parallel-path voltages. Thus, equalizer rings are not needed in a wave winding.

4.3.9 Applications of Lap and Wave Windings

In multipolar machines, for a given number of poles (P) and armature conductors (Z), a wave winding has a higher terminal voltage than a lap winding because it has more conductors in series [1]. On the other hand, the lap winding carries more current than a wave winding because it has more parallel paths.

In small machines, the current-carrying capacity of the armature conductors is not critical and in order to achieve suitable voltages, wave windings are used. On the other hand, in large machines suitable voltages are easily obtained because of the availability of a high number of armature conductors and the current carrying capacity is more critical. Hence, in large machines, lap windings are used.

Note: In general, a high-current armature is lap-wound to provide a large number of parallel paths and a low-current armature is wave-wound to provide a small number of parallel paths (Table 4.1) where $m = 1, 2, 3$ for simplex, duplex and triplex, respectively. P = number of poles and C = number of coils. + sign indicates progressive and – sign indicates retrogressive winding.

Example 4.4 ***Two-layer Simplex Wave Winding***

With a simplex two-layer wave winding having 26 conductors and four-poles, write down the winding table. What will the front and back pitches of the winding be?

Solution

Here, $Y_A = \frac{26\pm2}{4} = 7$ or 6. If we take 7, then the pitches would be: $Y_b = 8$ and $Y_f = 6$ or $Y_b = 6$ and $Y_f = 8$, which are even numbers. Taking $Y_A = 6$, we have $Y_b = 5$, $Y_f = 7$. Incidentally, if $Y_A = Y_C$ is taken as 6.

Table 4.1 Difference between lap and wave winding.

S. No.	Terms	Lap winding	Wave winding
1	Coil-span (Y_{cs})	$Y_{cs} = \frac{S}{P}$ (lower integer)	$Y_{cs} = \frac{S}{P}$ (lower integer)
2	Back-pitch (Y_b)	$Y_b = N_s Y_{cs} + 1$	$Y_b = N_s Y_{cs} + 1$
3	Commutator pitch (Y_c)	$Y_c = \pm 1$ (+ for progressive, – for retrogressive)	$Y_c = \frac{2(C \pm 1)}{P}$ (must be integral) (+ for progressive, – for retrogressive)
4	Front-pitch (Y_f)	$Y_f = Y_b \pm 2$ (+ for progressive, – for retrogressive)	$Y_f = 2Y_c - Y_b$
5	Resultant pitch (Y)	$Y = \pm 2\,m$	$Y = \frac{2C \pm 2m}{P/2}$
6	Parallel paths (A)	$A = P$	$A = 2$
7	Conductor current (I_c)	$I_c = I_a/A$	$I_c = I_a/2$
8	Number of brushes	$A = P$	Number of brushes = 2 (Large machine use *P* brushes)
9	Dummy coils	not needed	may be needed
10	Equalizer rings	needed	may not be needed

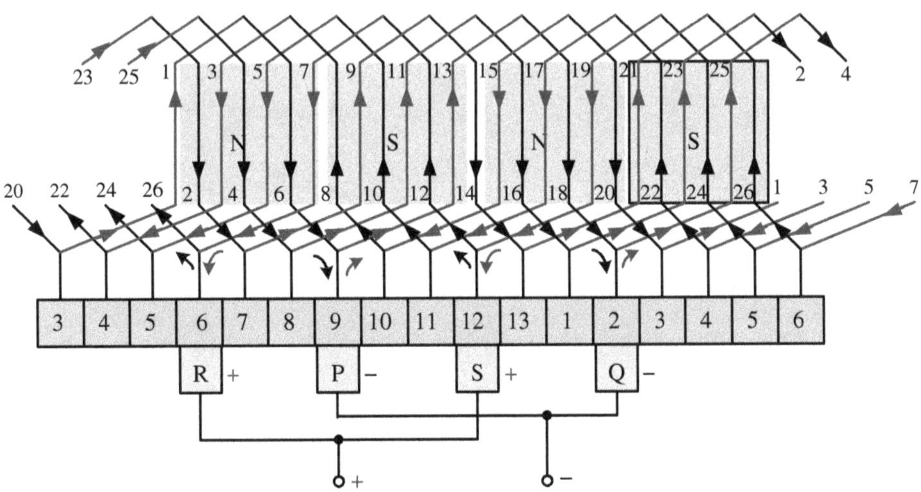

Figure 4.33 4 pole, 13 slot, 2-layer simplex wave winding diagram.

As shown in Figures 4.33 and 4.35, conductor No. 5 is taken to conductor No. 5 + 7 = 12 at the back and is joined to commutator segment 5 at the front. Next, the conductor No. 12 is joined to commutator segment 5 + 7 = 12 ($\because Y_C = 7$) to which is joined conductor No. 12 + 7 = 19. Continuing this way, we return to conductor No. 5 from where we started. Hence, the winding closes in on itself.

The simple winding table is as under:

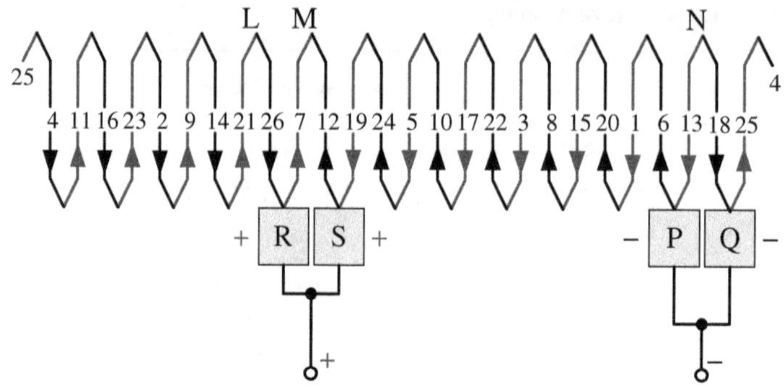

Figure 4.34 Sequence diagram for winding of Figure 4.33.

Back connections		*Front connections*
1 to (1 + 5) = 6	→	6 to (6 + 7) = 13
13 to (13 + 5) = 18	→	18 to (18 + 7) = 25
25 to (25 + 5) = 30 = (30 − 26) = 4	→	4 to (4 + 7) = 11
11 to (11 + 5) = 16	→	16 to (16 + 7) = 23
23 to (23 + 5) = 28 = (28 − 26) = 2	→	2 to (2 + 7) = 9
9 to (9 + 5) = 14	→	14 to (14 + 7) = 21
21 to (21 + 5) = 26	→	26 to (26 + 7) = 33 = (33 − 26) = 7
7 to (7 + 5) = 12	→	12 to (12 + 7) = 19
19 to (19 + 5) = 24	→	24 to (24 + 7) = 31 = (31 − 26) = 5
5 to (5 + 5) = 10	→	10 to (10 + 7) = 17
17 to (17 + 5) = 22	→	22 to (22 + 7) = 29 = (29 − 26) = 3
3 to (3 + 5) = 8	→	8 to (8 + 7) = 15
15 to (15 + 5) = 20	→	20 to (20 + 7) = 27 = (27 − 26) = 1

As we come back to the conductor No. 1 from where we started, the winding becomes closed at this stage.

Figure 4.35 shows the radial view of Figure 4.33 whereas Figures 4.36 and 4.37 show the derived diagrams from Figure 4.34.

Brush Position

Location of brush position in a wave winding is slightly difficult. In Figure 4.33 conductors are supposed to be moving from left to right over the poles. By applying Fleming's Right-hand rule, the directions of the induced emfs in various armature conductors can be found. The directions shown in the figure have been found in this manner. In Figure 4.34 the equivalent ring or spiral or sequence diagram is shown which is very helpful in understanding the formation of various parallel paths in the armature. It is seen that the winding is electrically divided into two portions. One portion consists of conductors lying between points N and L and the other of conductors lying between N and M. In the first portion, the general trend of the induced emfs is from left to right whereas in the second portion it is

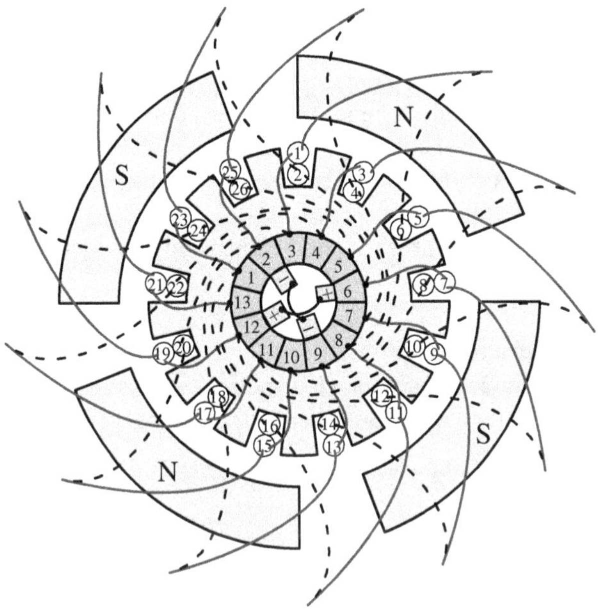

Figure 4.35 Radial view of Figure 4.33.

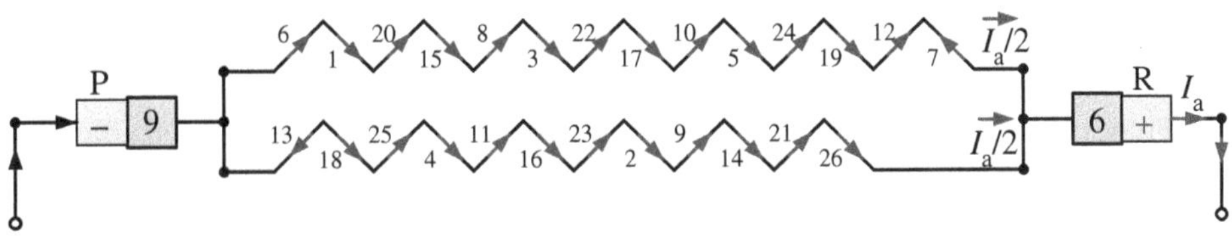

Figure 4.36 Derived diagram from Figure 4.34 when brushes P and R are considered.

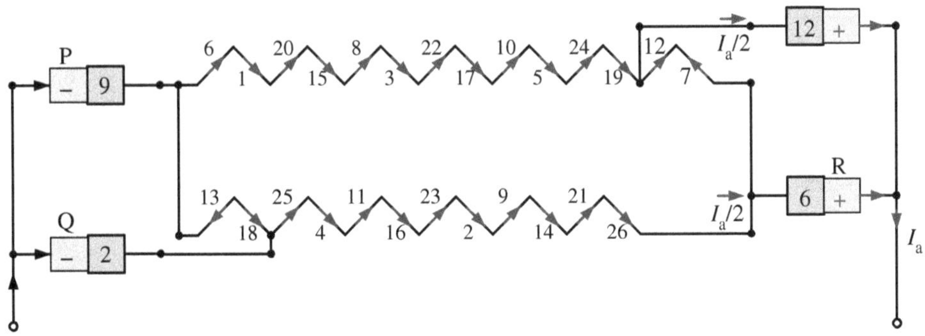

Figure 4.37 Derived diagram from Figure 4.34 when all four brushes considered.

from right to left. Hence, in general, there are only two parallel paths through the winding, so that two brushes are required; one positive and one negative.

From the equivalent ring diagram, it is seen that point *N* is the separating point of the emf induced in the two portions of the winding. Hence, this fixes the position of the negative brush. But, as it is at the back and not at the commutator end of the armature, the negative brush has two alternative positions i.e. either at point *P* or *Q*. These points on the equivalent diagram correspond to commutator segments No. 9 and 2.

Now, we will find the position of the positive brush. It is found that there are two meeting points of the induced emf, i.e. points L and M but both these points are at the back or non-commutator end of the armature. These two points are separated by one loop only, namely, the loop composed of conductors 6 and 12, hence the middle point R of this loop fixes the position of the positive brush, which should be placed in touch with commutator segment No. 6. We find that for one position of the +ve brush, there are two alternative positions for the −ve brush.

Taking the +ve brush at point R and negative brush at point P, the winding is seen to be divided into the following two paths.

Example 4.5 ***Two-layer Simplex Wave Winding***

Draw the winding diagram in spread and radial form for a four-pole, 13 slot, simplex wave connected DC generator with a commutator having 13 segments. The number of coil sides per slot is 2. Indicate the position of brushes.

Solution

This example is similar to Example 4.4, but solved to eliminate the condition of short circuiting of coils by brushes.

Number of coils $= \mathrm{C} = \frac{1}{2} \times 2 \times 13 = 13$

Number of coils sides $= \mathrm{N} = 2\mathrm{C} = 2 \times 13 = 26$

Back pitch $= Y_b = \frac{2C \pm K}{P} = \frac{2\times13\pm2}{4} = 7$ or 6. Take $Y_b = 7$.

For progressive windings with commutator pitch $= Y_c = \frac{C\pm1}{P/2} = \frac{13+1}{4/2} = 7$

Total winding pitch $\mathrm{Y} = 2\mathrm{Y_c} = 2 \times 7 = 14$

∴ Front pitch $= Y_f = Y_b - 2 = 5$; but when we take $Y_f = 5$, brush position is difficult to find, so take $Y_f = Y_b = 7$.

Slots/pole $= \frac{13}{4} = 3\frac{1}{4}$. Therefore, there are three slots each under three poles and four slots under the fourth pole.

The simplex winding table is as shown below:

Back connections		*Front connections*
1 to (1 + 7) = 8	→	8 to (8 + 7) = 15
15 to (15 + 7) = 22	→	22 to (22 + 7) = 29 = (29 − 26) = 3
3 to (3 + 7) = 10	→	10 to (10 + 7) = 17
17 to (17 + 7) = 24	→	24 to (24 + 7) = 31 = (31 − 26) = 5
5 to (5 + 7) = 12	→	12 to (12 + 7) = 19
19 to (19 + 7) = 26	→	26 to (26 + 7) = 33 = (33 − 26) = 7
7 to (7 + 7) = 14	→	14 to (14 + 7) = 21
21 to (21 + 7) = 28 = (28 − 26) = 2	→	2 to (2 + 7) = 9
9 to (9 + 7) = 16	→	16 to (16 + 7) = 23
23 to (23 + 7) = 30 = (30 − 26) = 4	→	4 to (4 + 7) = 11
11 to (11 + 7) = 18	→	18 to (18 + 7) = 25
25 to (25 + 7) = 32 = (32 − 26) = 6	→	6 to (6 + 7) = 13
13 to (13 + 7) = 20	→	20 to (20 + 7) = 27 = (27 − 26) = 1

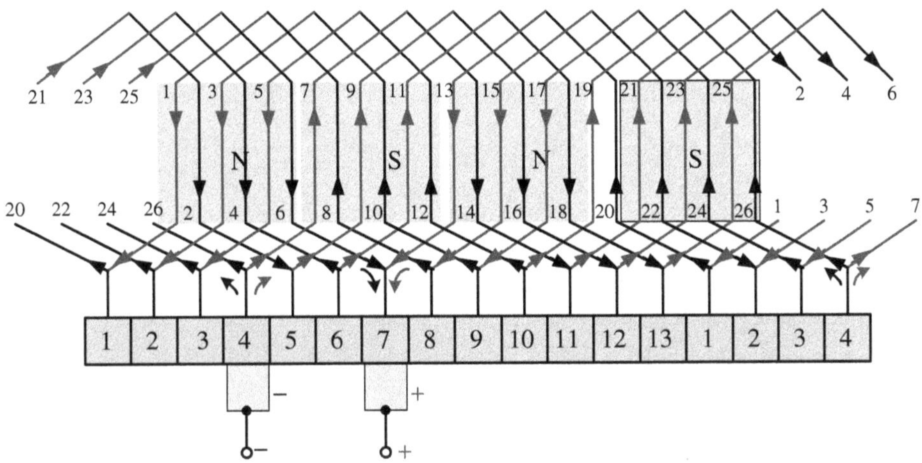

Figure 4.38 Simplex wave winding.

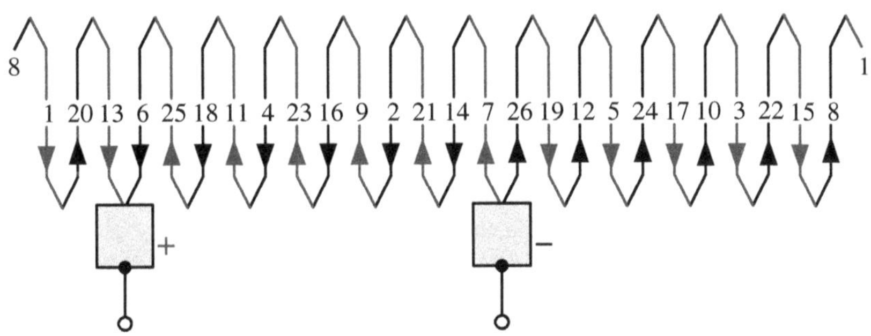

Figure 4.39 Sequence diagram for winding of Figure 4.38.

Winding layout for 26 conductors with direction of induced voltage/current and 13 commutator segments is shown in Figure 4.38. The sequence diagram (Figure 4.39) of the wave winding, from which, we conclude that it has two parallel paths irrespective of the number of poles and has a minimum of two brush sets although it is possible to put as many brush sets as the number of poles.

The various connections are made as per the table and also the connections to the various commutator segments are drawn. The 13 slots with two coil sides each and the commutator with 13 segments are drawn as shown in Figure 4.39. On inspection, it is found that the two coil sides connected to each of the segments 4 and 7 carry current in the same direction and, therefore, the brushes are placed on these two segments. Starting at commutator segment 4 and progressing round the winding in a clockwise direction. The diagram (Figure 4.39) includes all coil sides. Going round the winding, it is found that there are only two reversals of emf, and, therefore, two parallel paths. This is always true whatever the number of poles may be. The radial and derived diagrams are shown in Figures 4.40 and 4.41.

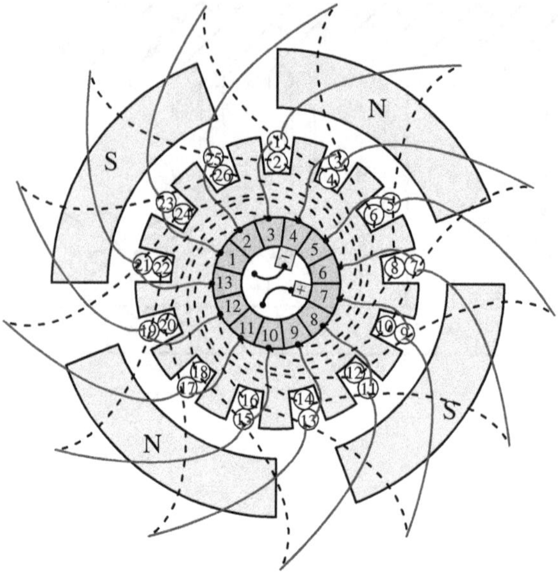

Figure 4.40 Simplex wave winding of a 4 pole, 13 slots, 13 commutator segments DC. generator.

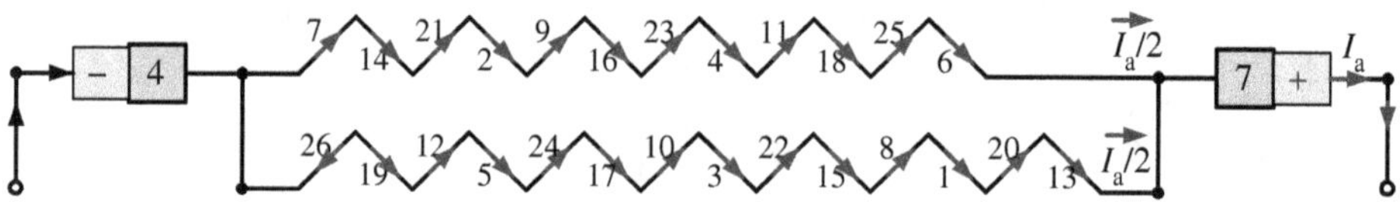

Figure 4.41 Derived diagram for winding of Example 4.5.

Example 4.6 ***Two-layer Simplex Wave Winding***
A four-pole, simplex wave wound armature has 25 slots and 25 coils. The commutator has 25 segments. Draw the winding diagram in spread and radial form.

Solution

$$\text{Slots per pole} = \frac{25}{4} = 6\frac{1}{4}$$

$$\text{Number of coil sides} = 2 \times 25 = 50$$

$$\text{Coil sides per slot} = \frac{50}{25} = 2$$

$$\text{Winding pitch} = Y = \frac{2C \pm 2}{P/2} = \frac{2 \times 25 \pm 2}{4/2} = 26 \text{ or } 24$$

We take Y = 26. We have $Y = Y_b + Y_f$. This give $Y_b = Y_f = 13$ (because Y_b and Y_f have to be odd integer).

The choice of $Y_b = 13$, gives the coil span equal to six slots which is nearly equal to slots per pole.

For progressive winding commutator pitch = $Y_c = \frac{C+1}{P/2} = \frac{25+1}{4/2} = 13$ segments.

The table of the winding to facilitate the winding connections is given below:

Back connections		*Front connections*
1 to (1 + 13) = 14	→	14 to (14 + 13) = 27
27 to (27 + 13) = 40	→	40 to (40 + 13) = 53 = (53 – 50) = 3
3 to (3 + 13) = 16	→	16 to (16 + 13) = 29
29 to (29 + 13) = 42	→	42 to (42 + 13) = 55 = (55 – 50) = 5
5 to (5 + 13) = 18	→	18 to (18 + 13) = 31
31 to (31 + 13) = 44	→	44 to (44 + 13) = 57 = (57 – 50) = 7
7 to (7 + 13) = 20	→	20 to (20 + 13) = 33
33 to (33 + 13) = 46	→	45 to (46 + 13) = 59 = (59 – 50) = 9
9 to (9 + 13) = 22	→	22 to (22 + 13) = 35
35 to (35 + 13) = 48	→	48 to (48 + 13) = 61 = (61 – 50) = 11
11 to (11 + 13) = 24	→	24 to (24 + 13) = 37
37 to (37 + 13) = 50	→	50 to (50 + 13) = 63 = (63 – 50) = 13
13 to (13 + 13) = 26	→	26 to (26 + 13) = 39
39 to (39 + 13 = 52 = (52 – 50) = 2	→	2 to (2 + 13) = 15
15 to (15 + 13) = 28	→	28 to (28 + 13) = 41
41 to (41 + 13) = 54 = (54 – 50) = 4	→	4 to (4 + 13) = 17
17 to (17 + 13) = 30	→	30 to (30 + 13) = 43
43 to (43 + 13) = 56 = (56 – 50) = 6	→	6 to (6 + 13) = 19
19 to (19 + 13) = 32	→	32 to (32 + 13) = 45
45 to (45 + 13) = 58 = (58 – 50) = 8	→	8 to (8 + 13) = 21
21 to (21 + 13) = 34	→	34 to (34 + 13) = 47
47 to (47 + 13) = 60 = (60 – 50) = 10	→	10 to (10 + 13) = 23
23 to (23 + 13) = 36	→	36 to (36 + 13) = 49
49 to (49 + 13) = 62 = (62 – 50) = 12	→	12 to (12 + 13) = 25
25 to (25 + 13) = 38	→	38 to (38 + 13) = 51 = (51 – 50) = 1

There are six slots each under three poles while there are seven slots under the fourth pole. Hence assume, slots 1–6 and 13–18 are under North Pole and slots 7–12 and 19–25 are under South Pole. Therefore, coil sides 1–12 and 25–36 are under North Pole while coil sides 13–24 and 37–50 are under South Pole.

Figure 4.42 shows the developed winding diagram for progressive wave winding. Figure 4.43 shows the sequence diagram of the wave winding from which we conclude that it has two parallel paths irrespective of the number of poles and has a minimum of two brush sets although it is possible to put as many brush sets as the number of poles.

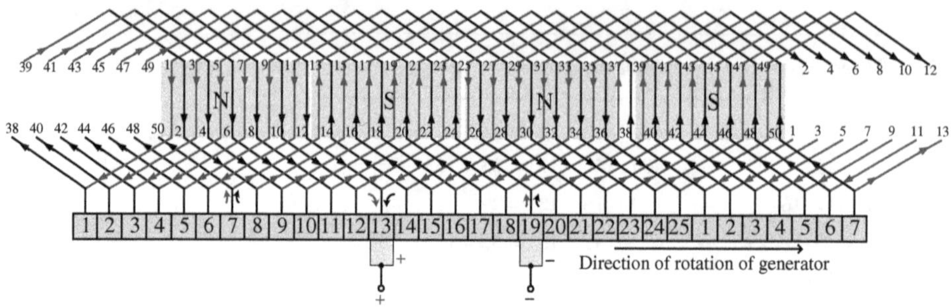

Figure 4.42 Simplex wave winding.

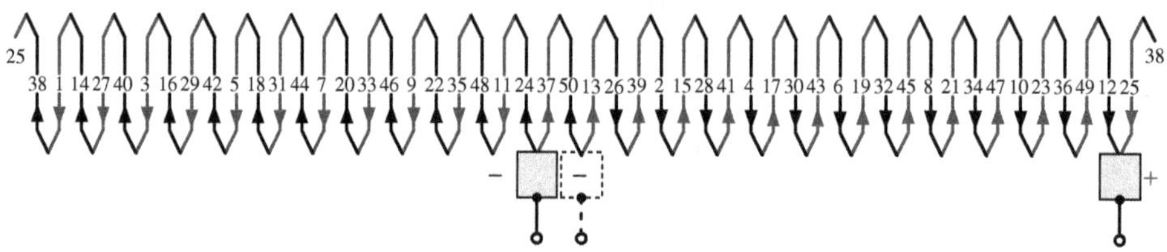

Figure 4.43 Sequence diagram for winding of Figure 4.42.

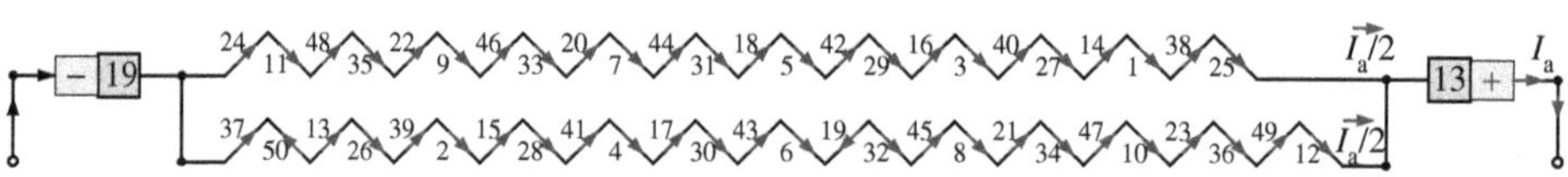

Figure 4.44 Derived diagram for winding of Figure 4.42.

Observing the sequence diagram, it can be seen that the positive brush position is at segment number 13 and the negative brush can be placed at segment number 7 or 19. Here, it is placed at segment number 19. It is observed in Figure 4.44 that there are 24 conductors in series in the upper branch whereas in the lower branch 26 conductors are in series but conductor number 50 has opposite polarity giving 24 effective conductors to make the voltage same across the parallel branches.

Example 4.7 *Two-layer Simplex Wave Winding*

Draw a developed diagram of a simplex two-layer wave-winding for a four-pole DC generator with 30 armature conductors.

Solution

$$\text{Number of coils} = \text{C} = \frac{30}{2} = 15$$

$$\text{Number of slots} = 15$$

$$\text{Number of commutator segments} = 15$$

$$\text{Number of coils sides} = 2\text{C} = 2 \times 15 = 30$$

$$\text{Back pitch} = Y_b = \frac{2C \pm b}{P} = \frac{2 \times 15 \pm 2}{4} = 8 \text{ or } 7.\text{Take odd value } Y_b = 7.$$

For progressive windings with commutator pitch = $Y_c = \frac{C\pm1}{P/2} = \frac{15+1}{4/2} = 8$
Total winding pitch $Y = 2Y_c = 2\times8 = 16$
$\therefore$ Front pitch $= Y_f = Y - Y_b = 16 - 7 = 9$;
Slots/pole $= \frac{15}{4} = 3\frac{3}{4}$. Therefore, there are four slots each under three poles and three slots under the fourth pole.
The simplex winding table is as shown below:

Back connections		*Front connections*
1 to (1 + 7) = 8	→	8 to (8 + 9) = 17
17 to (17 + 7) = 24	→	24 to (24 + 9) = 33 = (33 − 30) = 3
3 to (3 + 7) = 10	→	10 to (10 + 9) = 19
19 to (19 + 7) = 26	→	26 to (26 + 9) = 35 = (35 − 30) = 5
5 to (5 + 7) = 12	→	12 to (12 + 9) = 21
21 to (21 + 7) = 28	→	28 to (28 + 9) = 37 = (37 − 30) = 7
7 to (7 + 7) = 14	→	14 to (14 + 9) = 23
23 to (23 + 7) = 30	→	30 to (30 + 9) = 39 = (39 − 30) = 9
9 to (9 + 7) = 16	→	16 to (16 + 9) = 25
25 to (25 + 7) = 32 = (32 − 30) = 2	→	2 to (2 + 9) = 11
11 to (11 + 7) = 18	→	18 to (18 + 9) = 27
27 to (27 + 7) = 34 = (34 − 30) = 4	→	4 to (4 + 9) = 13
13 to (13 + 7) = 20	→	20 to (20 + 9) = 29
29 to (29 + 7) = 36 = (36 − 30) = 6	→	6 to (6 + 9) = 15
15 to (15 + 7) = 22	→	22 to (22 + 9) = 31 = (31 − 30) = 1

Winding layout for 30 conductors with direction of induced voltage/current and 15 commutator segments is shown in Figure 4.45.

There are four slots each under three poles while there are three slots under the fourth pole. Hence assume, slots 14 and 9–12 are under North Pole and slots 5–8 and 13–15 are under South Pole. Therefore, coil sides 1–8 and 17–24 are under North Pole while coil sides 9–16 and 25–30 are under South Pole.

Figure 4.45 shows the developed winding diagram for progressive wave winding. Figure 4.46 shows the sequence diagram of the wave winding from which we conclude that it has two parallel paths irrespective of the number of poles and has a minimum of two brush sets although it is possible to put as many brush sets as the number of poles. Observing the sequence diagram, it can be seen that the positive brushes position is at segment number 1 and/or 9 and the negative brushes can be placed at segment number 5 and/or 13. When two brushes are employed i.e. brushes at segments 1 and 5, actually there are 14 coil sides in series but 12 coils are effectively in series because coil side

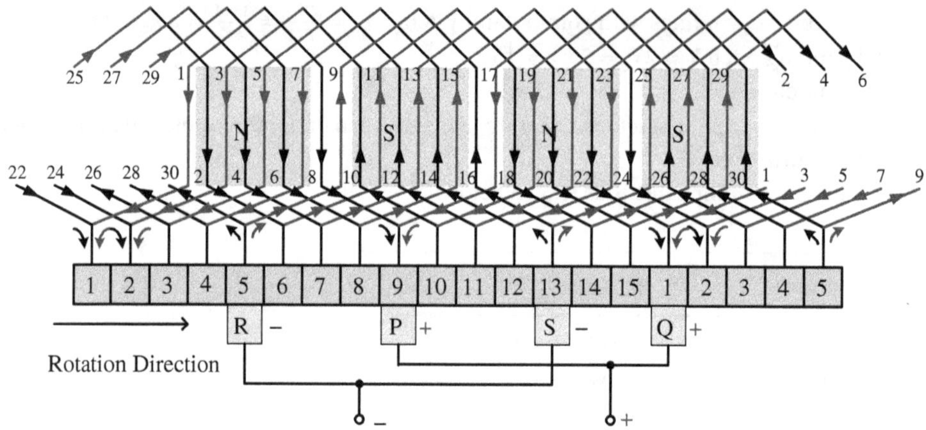

Figure 4.45 Simplex wave winding.

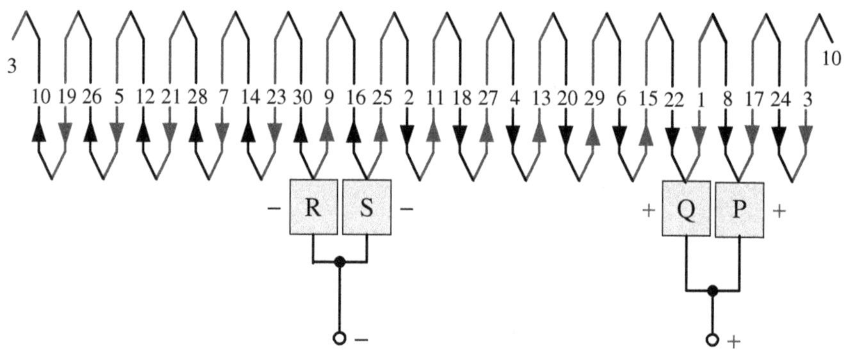

Figure 4.46 Sequence diagram.

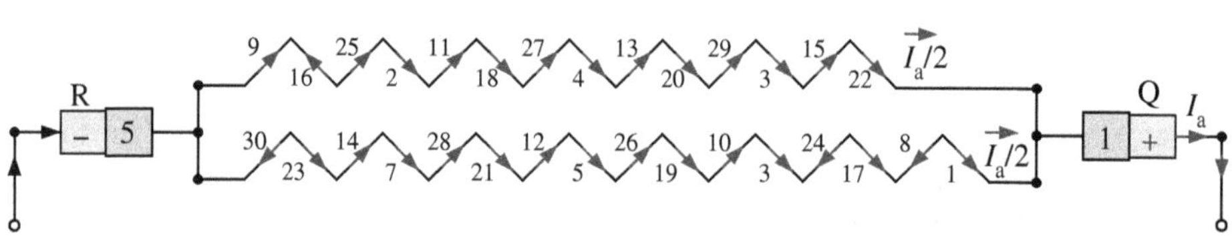

Figure 4.47 Derived diagram when 2 brushes are used.

number 16 in the upper branch and coil side numbers 30 and 8 in the lower branch have opposite polarity of induced emf (Figure 4.47). When four brushes are used, the negative brushes short circuit the coil side number 9 and 16 while the positive brushes short circuit the coil side number 1 and 8 (see Figure 4.48). Here, only 10 sides in each branch are effectively in series because side number 8 in the lower branch has opposite polarity of induced emf.

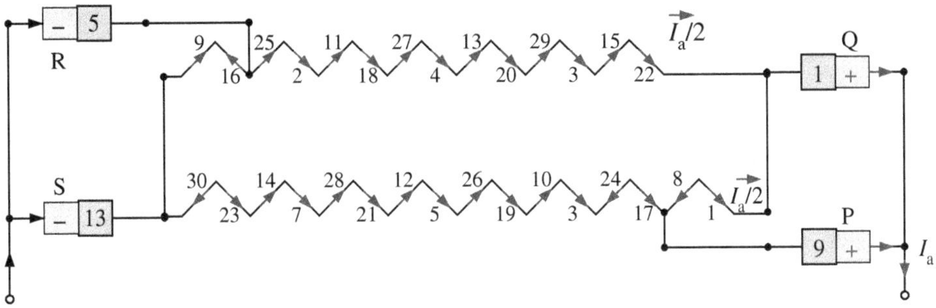

Figure 4.48 Derived diagram when 4 brushes are used.

Example 4.8 *Single-layer Triplex Lap Winding*

Develop the triplex single layer parallel winding diagram for a DC machine having pole pitch = 6 and poles = 4.

Solution

Number of slots = 6 × 4 = 24.

It is triplex lap winding, so, number of parallel paths = $a = 3 \times 4 = 12$

Number of coils = C = 24/2 = 12

Number of coil sides = 2C = 24 because it is single layer.

Number of commutators = C = 12

Back pitch = $Y_b = \frac{2C \pm b}{P} = \frac{2\times12\pm4}{4} = 7$ or 5

Front pitch = $Y_f = \frac{2C\pm b}{P} \pm 2m = \frac{2\times12\pm4}{4} \pm 6 = 13$ or $1, 11$ or -1

Winding pitch = $Y = Y_b - Y_f = \pm 2\,m$

Commutator pitch = $Y_c = \frac{Y_b - Y_f}{2} = \pm m = \pm 3$

Conditions of symmetry are satisfied because (i) $\frac{2S}{a}$ = integer = $\frac{2\times24}{12} = 4$ and (ii) $\frac{2S}{p}$ = integer = $\frac{2\times24}{4} = 12$

Slots per pole = 24/4 = 6

Since there is a common factor of 3 between commutator pitch and commutators, the winding is triply re-entrant.

Winding table:

Circuit 1								Circuit 2							
Back connections ($Y_b = 7$)				Front connections ($Y_f = -3$)				Back connections ($Y_b = 7$)				Front connections ($Y_f = -3$)			
1	→	8	→	8	→	5		3	→	10	→	10	→	7	
5	→	12	→	12	→	9		7	→	14	→	14	→	11	
9	→	16	→	16	→	13		11	→	18	→	18	→	15	
13	→	20	→	20	→	17		15	→	22	→	22	→	19	
17	→	24	→	24	→	21		19	→	2	→	2	→	23	
21	→	4	→	4	→	1		23	→	6	→	6	→	3	

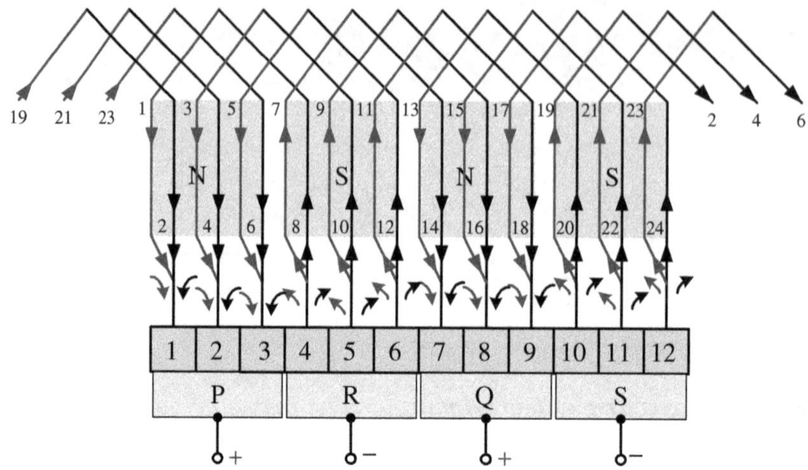

Figure 4.49 4 pole, 24 slot, single layer triplex lap winding diagram.

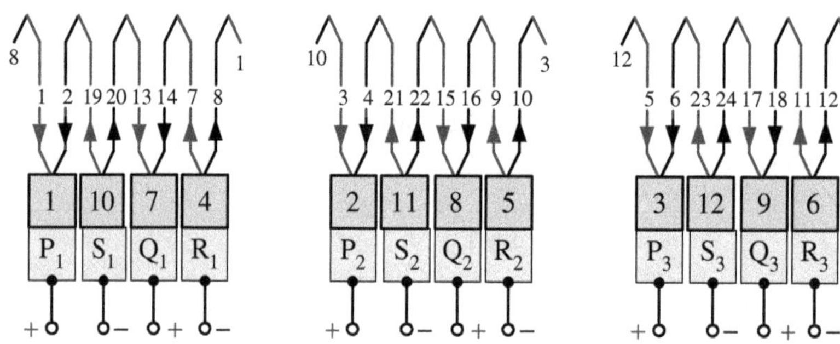

Figure 4.50 Sequence diagram for Figure 4.49.

There are six slots under each pole. Hence assume, slots 1–6 and 13–18 are under North Pole and slots 7–12 and 19–24 are under South Pole. Therefore, coil sides 1–6 and 13–18 are under North Pole while coil sides 7–12 and 19–24 are under South Pole.

Figure 4.49 shows the developed winding diagram for progressive triplex lap winding. The width of brushes is such that they should make contact with three commutator segments at a time. Figure 4.50 shows the sequence diagram Figure 4.49 of the triplex lap winding from which we conclude that it has 12 parallel paths because of triplex lap type. Observing the sequence diagram, the positive brushes position are at segment number 1, 2, 3 and 7, 8, 9 and the negative brushes can be placed at segment number 4, 5, 6 and 10, 11, 12. Figure 4.51 is the derived diagram for Figure 4.50, which shows 12 parallel paths.

Example 4.9 ***Single-layer Duplex Wave Winding***

Develop the duplex single layer series winding diagram for a DC machine having pole pitch = 6 and poles = 4.

Solution

Number of slots = $6 \times 4 = 24$.

It is duplex wave winding, so, number of parallel paths = $a = 4$

Number of coils = $C = 24/2 = 12$

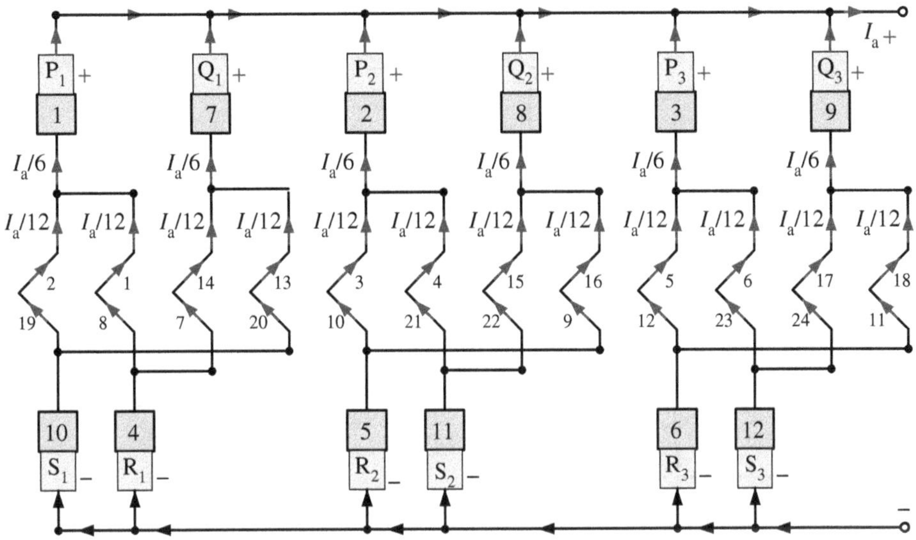

Figure 4.51 Derived diagram for Figure 4.50.

Number of coil sides = N = 2C = 24 because it is single layer.

Number of commutators = C = 12

Back pitch = $Y_b = \frac{2C}{P} \pm b = 7$

Commutator pitch = $Y_c = \frac{C\pm2}{P/2} = \frac{12\pm2}{4/2} = 7$ or 5

Average Winding pitch = $Y = 2Y_c = 2\times7 = 14$

Front Pitch $= Y_f = Y - Y_b = 14 - 7 = 7$

Conditions of symmetry are satisfied because (i) $\frac{2S}{A}$ = integer = $\frac{2\times24}{4} = 12$ where A is number of parallel paths and (ii) $\frac{P}{a}$ = integer = $\frac{4}{4} = 1$

Since Y_c is an odd integer, the winding is singly re-entrant duplex wave type.

Winding table:

Back connections ($Y_b = 7$)				**Front connections ($Y_f = 7$)**		
1	→	8	→	8	→	15
15	→	22	→	22	→	5
5	→	12	→	12	→	19
19	→	2	→	2	→	9
9	→	16	→	16	→	23
23	→	6	→	6	→	13
13	→	20	→	20	→	3
3	→	10	→	10	→	17
17	→	24	→	24	→	7
7	→	14	→	14	→	21
21	→	4	→	4	→	11
11	→	18	→	18	→	1

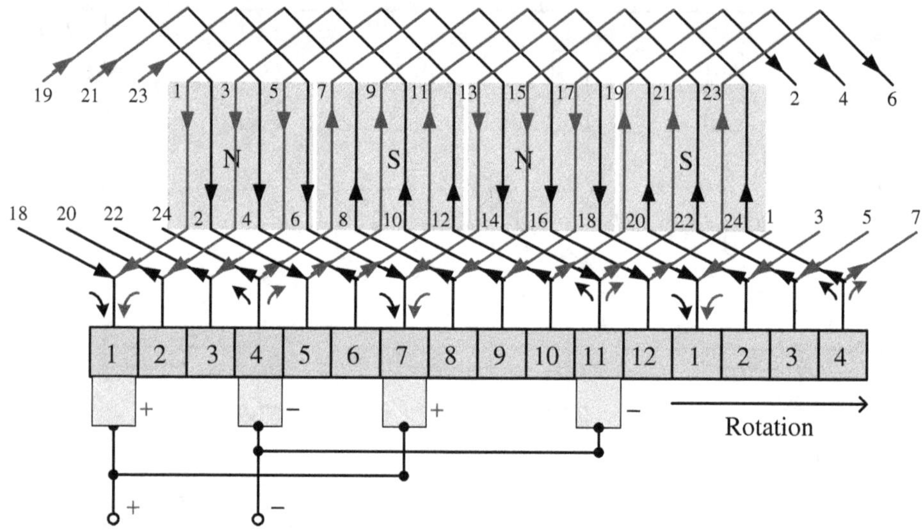

Figure 4.52 Duplex wave winding diagram.

Figure 4.53 Radial duplex wave winding diagram.

There are six slots under each pole. Hence assume, slots 1–6 and 13–18 are under North Pole and slots 7–12 and 19–24 are under South Pole. Therefore, coil sides 1–6 and 13–18 are under North Pole while coil sides 7–12 and 19–24 are under South Pole.

Figure 4.52 shows the developed winding diagram for progressive duplex wave winding. Figure 4.53 shows the radial view of Figure 4.52. Figure 4.54 shows the sequence diagram

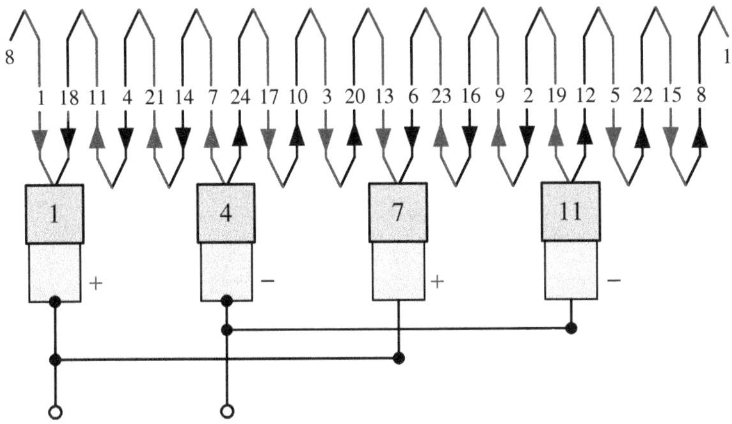

Figure 4.54 Sequence diagram for Figure 4.52.

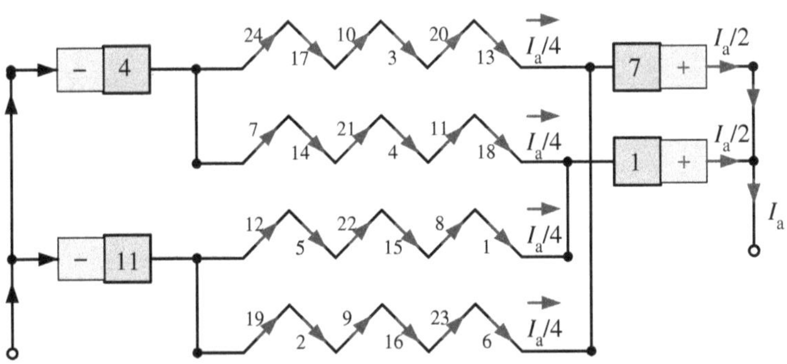

Figure 4.55 Derived diagram for Figure 4.54.

Figure 4.52 of the duplex wave winding from which we conclude that it has four parallel paths because of duplex wave type. Observing the sequence diagram, the positive brushes position is at segment number 1 and 7 and the negative brushes can be placed at segment number 4 and 11. Figure 4.55 is the derived diagram for Figure 4.54, which shows four parallel paths.

Example 4.10 ***Double-layer Duplex Wave Winding***

Draw winding diagram for 6 pole, 28 slots, 28 commutator segments, 4 paths double layer wave winding of an armature of DC generator.

Solution

As the number of paths is 4, while number of poles is 6, it is a duplex wave winding.

Number of coils = C = number of slots = 28

Number of coil sides = 2 × 28 = 56

Back pitch = $Y_b = \frac{2\times C}{P} \pm b = \frac{2\times 28}{6} \pm b = 9$

For duplex wave winding commutator pitch = $Y_c = \frac{C\pm 2}{P/2} = \frac{28\pm 2}{6/2} = 10$

Even numbered commutator pitch gives that the winding doubly re-entrant.

Winding pitch = Y = $2Y_c = 2\times 10 = 20$

Front pitch = $Y_f = Y - Y_b = 20 – 9 = 11$
The winding is asymmetrical because $P/a = 6/4$ = not an integer.
Slots per pole = $\frac{28}{6} = 4\frac{2}{3}$
5, 5 and 4 slots are under three north poles as well as under three south poles.
Winding table for duplex double layer wave winding

Circuit 1									Circuit 2								
Back connections ($Y_b = 9$)				Front connections ($Y_f = 11$)					Back connections ($Y_b = 9$)				Front connections ($Y_f = 11$)				
1	→	10	→	10	→	21			3	→	12	→	12	→	23		
21	→	30	→	30	→	41			23	→	32	→	32	→	43		
41	→	50	→	50	→	5			43	→	52	→	52	→	7		
5	→	14	→	14	→	25			7	→	16	→	16	→	27		
25	→	34	→	34	→	45			27	→	36	→	36	→	47		
45	→	54	→	54	→	9			47	→	56	→	56	→	11		
9	→	18	→	18	→	29			11	→	20	→	20	→	31		
29	→	38	→	38	→	49			31	→	40	→	40	→	51		
49	→	2	→	2	→	13			51	→	4	→	4	→	15		
13	→	22	→	22	→	33			15	→	24	→	24	→	35		
33	→	42	→	42	→	53			35	→	44	→	44	→	55		
53	→	6	→	6	→	17			55	→	8	→	8	→	19		
17	→	26	→	26	→	37			19	→	28	→	28	→	39		
37	→	46	→	46	→	1			39	→	48	→	48	→	3		

There are five slots under each two north poles and two south poles. There are four slots under each one north pole and one South Pole. Therefore, coil sides 1–10, 21–28 and 39–48 are under North Poles while coil sides 11–20, 29–38 and 49–56 are under South Poles.

Figure 4.56 shows the developed winding diagram for progressive duplex wave winding. Observing Figure 4.56, there are six brushes out of which two brushes (P and Q) have width equal to two commutator segments while other four brushes (R, S, T, and U) have width equal to one commutator segments. Figure 4.57 shows the sequence diagram for circuit 1 from Figure 4.56 of the duplex wave winding from which we conclude that it has two parallel paths because of wave type winding. Figure 4.58 shows the sequence diagram for circuit 2 from Figure 4.56 of the duplex wave winding from which we conclude that it has two parallel paths because of wave type winding. Observing the sequence diagram, the positive brushes position is at segment numbers 1, 2 (P), 11 (R), and 20 (T) and the negative brushes are placed at segment numbers 6 (S). 15, 16 (Q), and 25 (U). Eight coil sides are commuted. The coil side number 1 and 10 by brushes P and R, 29 and 38 by brushes Q and U. 39 and 48 by brushes T and P, and coil side number 11 and 20 by brushes S and Q. Figure 4.59 is the derived diagram for Figures 4.57 and 4.58, which show four parallel paths.

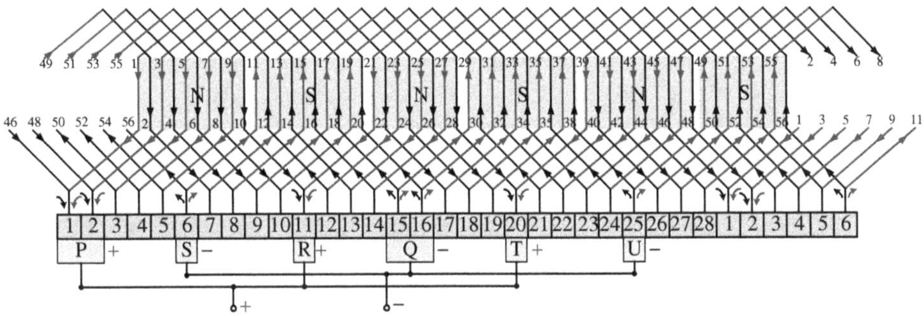

Figure 4.56 Duplex wave winding diagram.

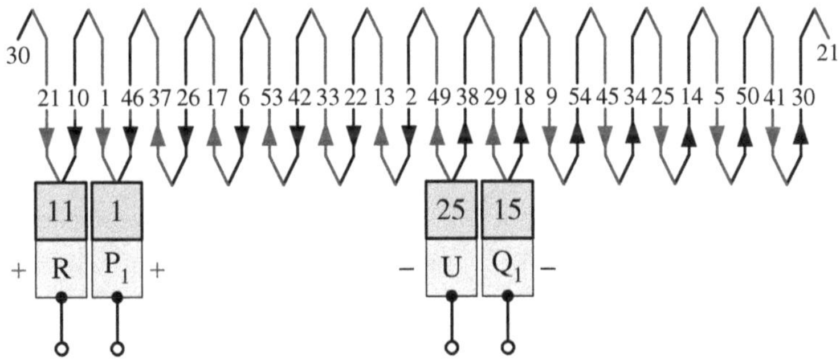

Figure 4.57 Sequence diagram for circuit 1 of Figure 4.56.

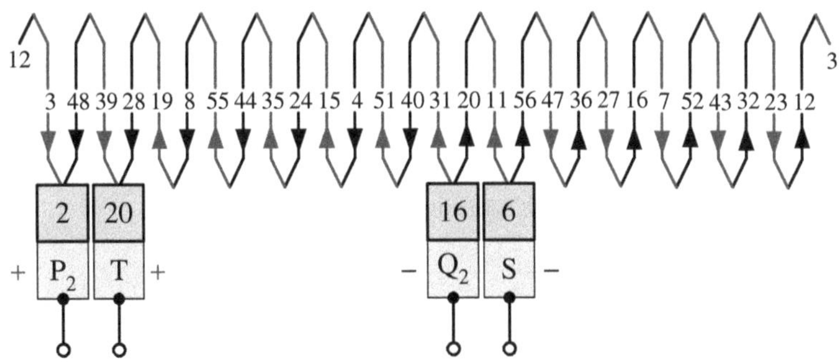

Figure 4.58 Sequence diagram for circuit 2 of Figure 4.56.

Since this is an asymmetrical winding and every time four coils are getting commutated, there will be spark at brushes which enhances wear and tear of brushes along with heating of the commutated coils. So, it is not advisable to use asymmetrical wave winding.

Example 4.11 ***Double-layer Duplex Wave Winding***

Draw winding diagram for four pole, 28 slots, 28 commutator segments, four paths double layer wave winding of an armature of DC generator.

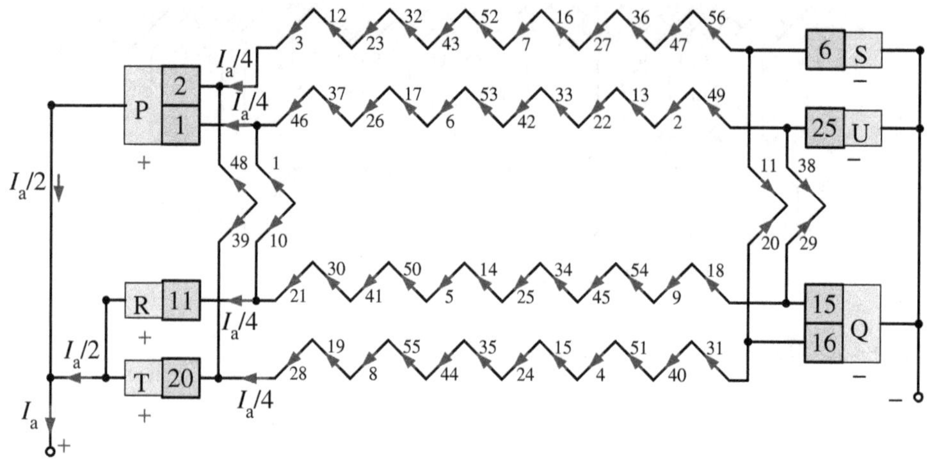

Figure 4.59 Derived diagram for Figures 4.57 and 4.58.

Solution

As the number of paths is four, while number of poles is four, it is a duplex wave winding.

Number of coils = C = number of slots = 28

Number of coil sides = 2 × 28 = 56

Back pitch = $Y_b = \frac{2\times C}{P} \pm b = \frac{2\times 28}{4} \pm b = 15$

For duplex wave winding commutator pitch = $Y_c = \frac{C\pm 2}{P/2} = \frac{28\pm 2}{4/2} = 15$ or 13. Take $Y_c = 15$

Odd numbered commutator pitch gives that the winding singly re-entrant.

Winding pitch = $Y = 2Y_c = 2\times 15 = 30$

Front pitch = $Y_f = Y - Y_b = 30–15 = 15$

The winding is symmetrical because $P/A = 4/4$ = an integer and $2S/A = 2\times 28/4$ = an integer.

Slots per pole = $\frac{28}{4} = 7$, where A is the number of parallel paths.

Seven are under two north poles as well as under two south poles.

Winding table for duplex double layer wave winding

Back connections ($Y_b = 15$)				Front connections ($Y_f = 15$)		
1	→	16	→	16	→	31
31	→	46	→	46	→	5
5	→	20	→	20	→	35
35	→	50	→	50	→	9
9	→	24	→	24	→	39
39	→	54	→	54	→	13
13	→	28	→	28	→	43
43	→	2	→	2	→	17
17	→	32	→	32	→	47

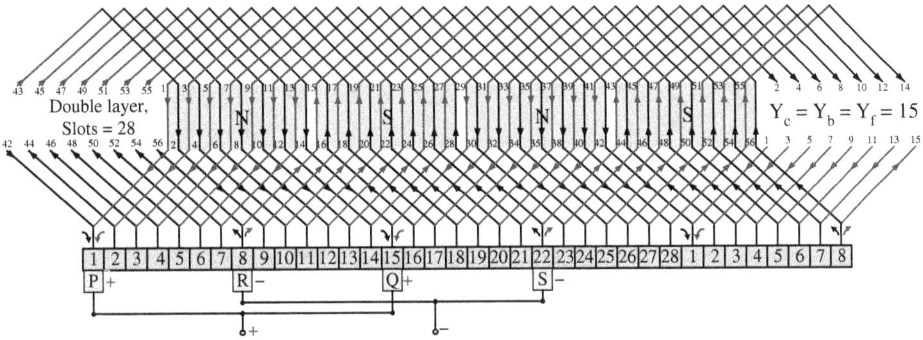

Figure 4.60 Duplex wave winding diagram.

Back connections ($Y_b = 15$)				Front connections ($Y_f = 15$)		
47	→	6	→	6	→	21
21	→	36	→	36	→	51
51	→	10	→	10	→	25
25	→	40	→	40	→	55
55	→	14	→	14	→	29
29	→	44	→	44	→	3
3	→	18	→	18	→	33
33	→	48	→	48	→	7
7	→	22	→	22	→	37
37	→	52	→	52	→	11
11	→	26	→	26	→	41
41	→	56	→	56	→	15
15	→	30	→	30	→	45
45	→	4	→	4	→	19
19	→	34	→	34	→	49
49	→	8	→	8	→	23
23	→	38	→	38	→	53
53	→	12	→	12	→	27
27	→	42	→	42	→	1

There are seven slots under each two north poles and two south poles. Therefore, coil sides 1–14, and 29–42 are under North Poles while coil sides 15–28, and 43–56 are under South Poles.

Figure 4.60 shows the developed winding diagram for progressive duplex wave winding. Observing Figure 4.60, there are four brushes out of which two brushes (P and Q) are positive while other two brushes (R and S) are negative.

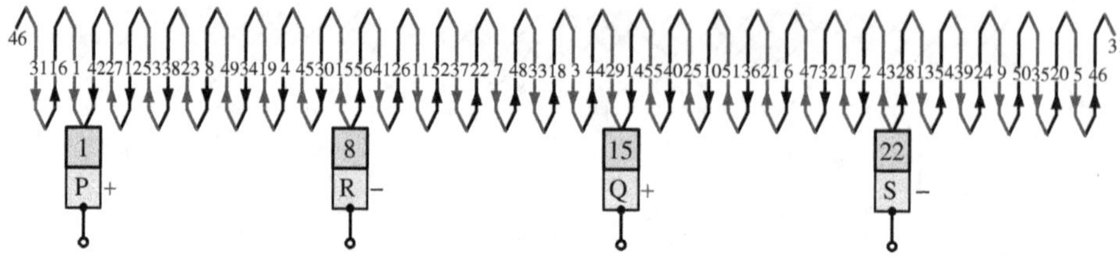

Figure 4.61 Sequence diagram for Figure 4.60.

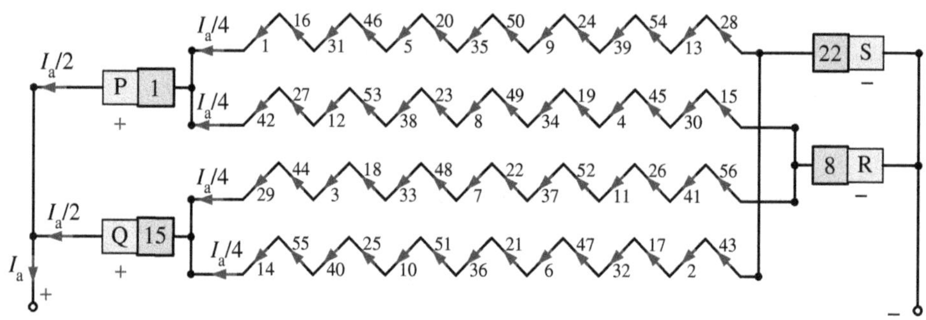

Figure 4.62 Derived diagram for Figure 4.61.

Figure 4.61 shows the sequence diagram for Figure 4.60 of the duplex wave winding from which we conclude that it has four parallel paths because of duplex wave type winding. Observing the sequence diagram, the positive brushes position is at segment numbers 1 (P), and 15 (Q) and the negative brushes are placed at segment numbers 8 (R) and 22 (S). Figure 4.62 is the derived diagram for Figure 4.61, which show four parallel paths.

The same problem solved when we take $Y_c = 13$

For duplex wave winding commutator pitch $= Y_c = \frac{C\pm 2}{P/2} = \frac{28\pm 2}{4/2} = 15$ or 13. Take $Y_c = 13$

Odd numbered commutator pitch gives that the winding singly re-entrant.

Winding pitch $= Y = 2Y_c = 2\times 13 = 26$

Front pitch $= Y_f = Y - Y_b = 26{-}15 = 11$

Winding table for duplex double layer wave winding

Back connections ($Y_b = 15$)				Front connections ($Y_f = 11$)		
1	→	16	→	16	→	27
27	→	42	→	42	→	53
53	→	12	→	12	→	23
23	→	38	→	38	→	49
49	→	8	→	8	→	19
19	→	34	→	34	→	45
45	→	4	→	4	→	15
15	→	30	→	30	→	41

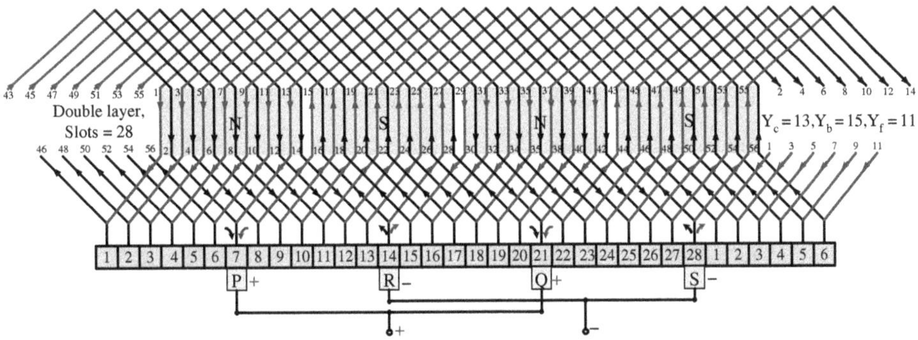

Figure 4.63 Duplex wave winding diagram.

Back connections ($Y_b = 15$)				Front connections ($Y_f = 11$)		
41	→	56	→	56	→	11
11	→	26	→	26	→	37
37	→	52	→	52	→	7
7	→	22	→	22	→	33
33	→	48	→	48	→	3
3	→	18	→	18	→	29
29	→	44	→	44	→	55
55	→	14	→	14	→	25
25	→	40	→	40	→	51
51	→	10	→	10	→	21
21	→	36	→	36	→	47
47	→	6	→	6	→	17
17	→	32	→	32	→	43
43	→	2	→	2	→	13
13	→	28	→	28	→	39
39	→	54	→	54	→	9
9	→	24	→	24	→	35
35	→	50	→	50	→	5
5	→	20	→	20	→	31
31	→	46	→	46	→	1

There are seven slots under each two north poles and two south poles. Therefore, coil sides 1–14, and 29–42 are under North Poles while coil sides 15–28, and 43–56 are under South Poles.

Figure 4.63 shows the developed winding diagram for progressive duplex wave winding.

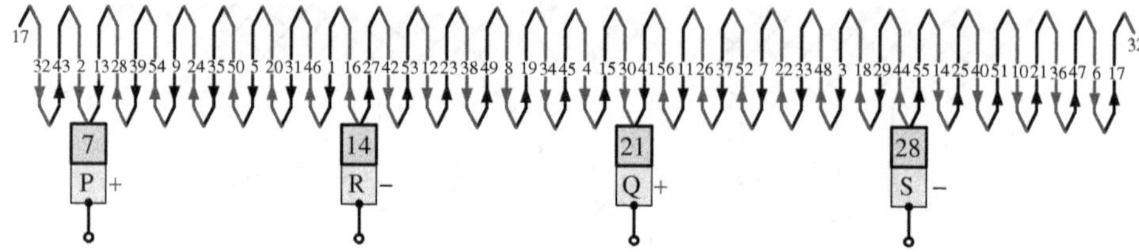

Figure 4.64 Sequence diagram for Figure 4.63.

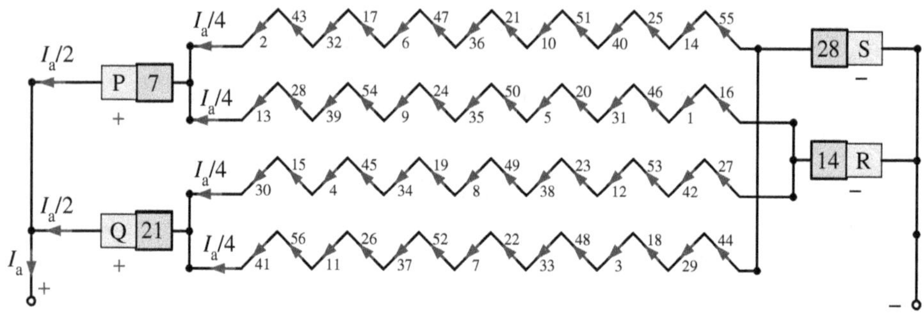

Figure 4.65 Derived diagram for Figure 4.64.

Observing Figure 4.63, there are four brushes out of which two brushes (P and Q) are positive while other two brushes (R and S) are negative. Figure 4.64 shows the sequence diagram for Figure 4.63 of the duplex wave winding from which we conclude that it has four parallel paths because of duplex wave type winding. Observing the sequence diagram, the positive brushes position is at segment numbers 7 (P), and 21 (Q) and the negative brushes are placed at segment numbers 14 (R) and 28 (S). Figure 4.65 is the derived diagram for Figure 4.64, which show four parallel paths.

Example 4.12 ***Double-layer Triplex Wave Winding***

Draw winding diagram for six pole, 30 slots, 30 commutator segments, six paths double layer wave winding of an armature of DC generator.

Solution

As the number of paths is six, while number of poles is six, it is a triplex wave winding.

Number of coils = C = number of slots = 30

Number of coil sides = 2 × 30 = 60

Back pitch = $Y_b = \frac{2 \times C}{P} \pm b = \frac{2 \times 30}{6} \pm b = 11$

For triplex wave winding commutator pitch = $Y_c = \frac{C \pm 3}{P/2} = \frac{30 \pm 3}{6/2} = 11$ or 9. Take $Y_c = 11$

Odd numbered commutator pitch gives that the winding singly re-entrant.

Winding pitch = $Y = 2Y_c = 2 \times 11 = 22$

Front pitch = $Y_f = Y - Y_b = 22 - 11 = 11$

The winding is symmetrical because $P/A = 6/6$ = an integer and $2S/A = 2 \times 30/6$ = an integer.

Slots per pole = $\frac{30}{6} = 5$, where A is the number of parallel paths.

Five slots are under three north poles as well as under three south poles.

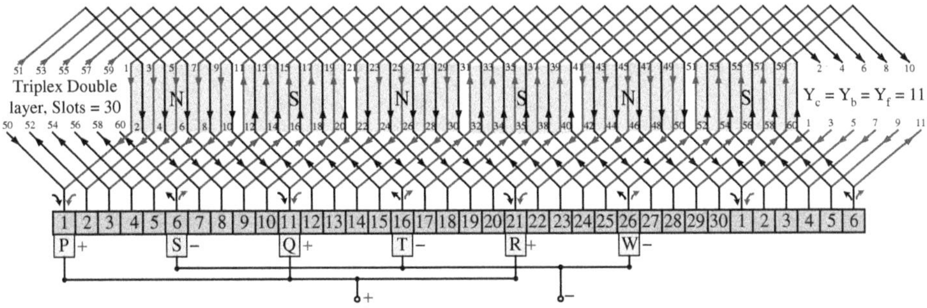

Figure 4.66 Triplex wave winding diagram.

Winding table for triplex double layer wave winding

Back connections ($Y_b = 11$)				Front connections ($Y_f = 11$)			Back connections ($Y_b = 11$)				Front connections ($Y_f = 11$)		
1	→	12	→	12	→	23	23	→	34	→	34	→	45
45	→	56	→	56	→	7	7	→	18	→	18	→	29
29	→	40	→	40	→	51	51	→	2	→	2	→	13
13	→	24	→	24	→	35	35	→	46	→	46	→	57
57	→	8	→	8	→	19	19	→	30	→	30	→	41
41	→	52	→	52	→	3	3	→	14	→	14	→	25
25	→	36	→	36	→	47	47	→	58	→	58	→	9
9	→	20	→	20	→	31	31	→	42	→	42	→	53
53	→	4	→	4	→	15	15	→	26	→	26	→	37
37	→	48	→	48	→	59	59	→	10	→	10	→	21
21	→	32	→	32	→	43	43	→	54	→	54	→	5
5	→	16	→	16	→	27	27	→	38	→	38	→	49
49	→	60	→	60	→	11	11	→	22	→	22	→	33
33	→	44	→	44	→	55	55	→	6	→	6	→	17
17	→	28	→	28	→	39	39	→	50	→	50	→	1

If we assume $Y_c = 9$, then Y_f will be 7. With $Y_b = 11$ and $Y_f = 7$, the winding table does not cover all coil sides. There are five slots under each three north poles and three south poles. Therefore, coil sides 1–10, 21–30 and 41–50 are under North Poles while coil sides 11–20, 31–40, and 51–60 are under South Poles. Figure 4.66 shows the developed winding diagram for progressive duplex wave winding.

Observing Figure 4.66, there are six brushes out of which three brushes (P, Q, and R) are positive while other three brushes (R, S, and T) are negative. Figure 4.67 shows the sequence diagram for Figure 4.66 of the triplex wave winding from which we conclude that it has six parallel paths because of triplex wave type winding. Observing the sequence diagram, the positive brushes position is at segment numbers 1 (P), 11 (Q), and 21 (R) and the negative

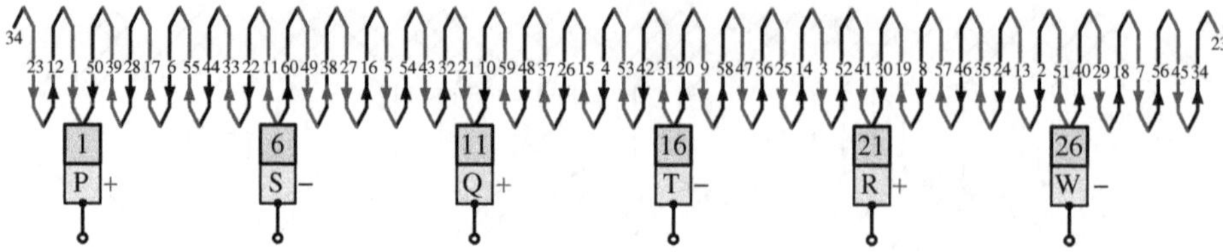

Figure 4.67 Sequence diagram for Figure 4.66.

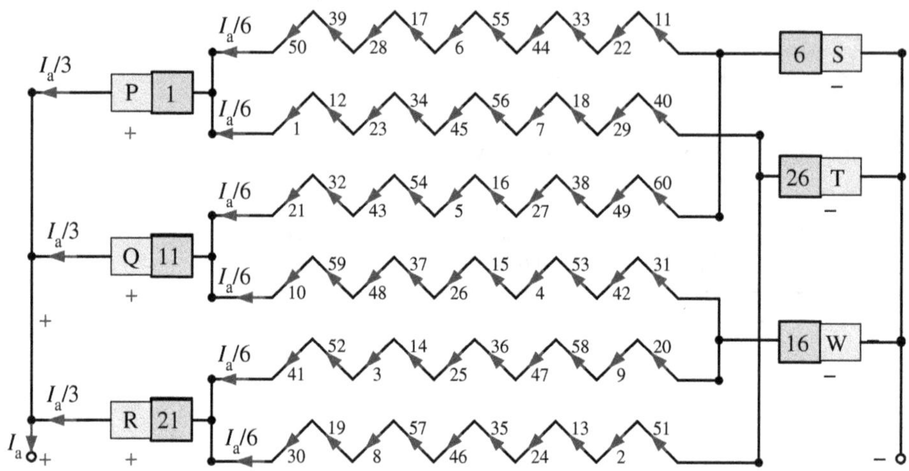

Figure 4.68 Derived diagram for Figure 4.67.

brushes are placed at segment numbers 6 (S), 16 (T), and 26 (W). Figure 4.68 is the derived diagram for Figure 4.67, which show six parallel paths.

4.3.10 Dummy or Idle Coils

4.3.10.1 Dummy Coils

In a lap winding $y_c = \pm 1$ irrespective of the number of armature coils so that coils can always be chosen to completely fill all the slots

$$\left(C = \frac{1}{2} N_s S\right) \tag{4.15}$$

Where N_s = Number of coil sides per slot, S = Number of slots.

In a simplex wave winding from Eq. (4.13), the number of coils must fulfil the condition

$$C = \frac{P}{2} y_c \pm 1 \tag{4.16}$$

where y_c is *commutator pitch,* while at the same time C must also be governed by

$$\left(C = \frac{1}{2} N_s S\right) \tag{4.17}$$

For a certain-design value of P and the choice of S restricted by manufacturing considerations (availability of a certain notching gear for armature stamping), the values of C as obtained from Eqs. (4.16) and (4.17) may not be the same. In such a situation, the number of coils C' is dictated by

$$C' = \frac{1}{2}N_s S$$

in such a manner that

$$C' > C$$

and y_c is so selected that $(C' - C)$ is the least possible. Electrically, only C coils (Eq. (4.16)) are needed, but C' coils are accommodated in the armature slots to ensure dynamic (mechanical) balancing of the armature. The difference $(C' - C)$ is called *dummy coils* and are placed in appropriate slots with their ends electrically insulated.

As an example, if $P = 4n$ (multiple of four), C can only be odd (Eq. (4.16)), while C' may be even if an even number of slots are used. In this case, then at least one dummy coil would be needed.

In a simplex wave winding, the average pitch $\mathrm{Y_A}$ (or commutator pitch $\mathrm{Y_C}$) should be a whole number. Sometimes, the standard armature punching available in the market have slots that do not satisfy the above requirement so that more coils (usually only one more) are provided than can be utilized. These extra coils are called dummy or dead coils. The dummy coil is inserted into the slots in the same way as the others to make the armature dynamically balanced, but it is not a part of the armature winding [1].

Let us illustrate the use of dummy coils with a numerical example. Suppose the number of slots is 22 and each slot contains 2 conductors. The number of poles is four. For simplex wave wound armature

$$Y_A = \frac{Z \pm 2}{P} = \frac{2 \times 22 \pm 2}{4} = \frac{44 \pm 2}{4} = 11\frac{1}{2} \quad \text{or} \quad 10\frac{1}{2}$$

Since the results are not whole numbers, the number of coils (and hence segments) must be reduced. If we make one coil dummy, we have 42 conductors and

$$Y_A = \frac{Z \pm 2}{P} = \frac{2 \times 21 \pm 2}{4} = \frac{42 \pm 2}{4} = 11 \quad \text{or} \quad 10$$

This means that armature can be wound only if we use 21 coils and 21 segments. The extra coil or dummy coil is put in the slot. One end of this coil is taped, and the other end connected to the unused commutator segment (segment 22) for the sake of appearance. Since only 21 segments are required, the two (21 and 22 segments) are connected together and considered as one.

4.3.11 Whole-Coil Winding and Half-Coil Winding

When the coils of an alternating current winding are connected so that there are as many coils per phase as there are poles, the winding is called 'whole-coiled' (Figure 4.69a). When the coils are connected so that there is only one coil per phase per pair of poles, the winding

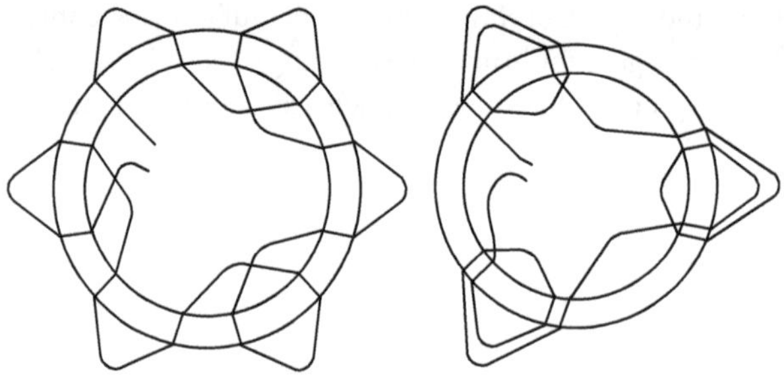

(a) Whole-coiled winding (b) Half-coiled winding

Figure 4.69 A six-pole stator with whole-coiled and half-coiled windings. The whole-coiled winding, (a), has as many coils per phase as there are poles. The half-coiled winding (b) has only one coil per phase per pair of poles [3].

is called 'half-coiled' (Figure 4.69b) [3]. The main difference between these two connections is in the method of making the end connections for the coils. In the 'whole-coiled' winding each slot contains two coil sides. It is not, however, strictly a double-layer winding, as the coil sides are placed side-by-side and not one above the other. In the 'half-coiled' winding, however, each coil may have twice the number of turns of a 'whole-coiled' winding or the two coils under a north or south pole of the latter type may be connected in series and taped together to form one coil in case of a change in connections.

The 'half-coiled' winding has the advantage that, when used with large generators, the armature frame may be split into two sections for shipment or repair, without disturbing many of the end connections.

4.3.12 Concentrated Winding

A winding with only one slot per pole per phase is called a concentrated winding. The uni-slot (all turn sides in one slot) or concentrated winding gives the largest possible emf from a given number of inductors in the winding. That is for a definite fixed speed and field strength in an alternator, the concentrated winding requires a lesser number of inductors than a distributed winding, but increases the number of turns per coil.

4.3.13 Distributed Winding

An armature winding which has its inductors of any one phase under a single pole placed in several slots, is said to be distributed. When these inductors are bunched together in one slot per pole, per phase, the winding is called concentrated. It is usual, in a distributed winding, to distribute the series inductors in any phase of the winding among two or more slots under each pole. This tends to diminish armature reactance and gives a better emf wave, besides offering a better distribution of the heating due to armature copper loss than in concentrated windings.

4.4 Electromotive Force (emf) Equation

4.4.1 emf Equation of an Alternator [1]

Let
Z = No. of conductors or coil sides in series per phase
ϕ = Flux per pole in Webers
P = Number of poles
N = Rotor speed in rpm.

In one revolution (i.e. 60/N second), each stator conductor is cut by pf Webers i.e.
$d\phi = P\phi$; $dt = 60/N$
∴ Average emf induced in one stator conductor

$$= \frac{d\phi}{dt} = \frac{P\phi}{60/N} = \frac{P\phi N}{60}\text{volts}$$

and

$$N = \frac{120f}{P}$$

where f is the frequency of the emf
Since there are Z conductors in series per phase,

$$\therefore \text{Average emf/phase} = \frac{P\phi N}{60} \times Z = \frac{P\phi Z}{60} \times \frac{120f}{P} = 2f\phi Z\,\text{volts}$$

rms value of emf/phase = Average value/phase × form factor

$$= 2f\phi Z \times 1.11 = 2.22 f\phi Z\,\text{volts}$$

$$\therefore E_{\text{r.m.s.}}/\text{phase} = 2.22 f\phi Z\,\text{volts} \tag{4.18}$$

If K_p and K_d are the pitch factor and distribution factor of the armature winding, then,

$$E_{\text{r.m.s.}}/\text{phase} = 2.22 K_p K_d\, f\phi Z\,\text{volts} \tag{4.19}$$

Sometimes, the turns (T) per phase rather than conductors per phase are specified, in that case, Eq. (4.19) becomes:

$$E_{\text{r.m.s.}}/\text{phase} = 4.44 K_p K_d\, f\phi T\,\text{volts} \tag{4.20}$$

The line voltage will depend upon whether the winding is star or delta connected.

4.4.1.1 Winding Factor (Coil Pitch and Distributed Windings)

4.4.2 Winding Factors

The armature winding of an alternator is distributed over the entire armature. The distributed winding produces nearly a sine waveform and the heating is more uniform. Likewise, the coils of armature winding are not full-pitched i.e. the two sides of a coil are not at corresponding points under adjacent poles. The fractional pitched armature winding requires less copper per coil and, at the same time, waveform of output voltage is near to sinusoid. The distribution and pitching of the coils affect the voltages induced in the coils. We shall discuss two winding factors

(i) Pitch factor (K_p), also known as chord factor
(ii) Distribution factor (K_d), also called breadth factor

4.4.2.1 Pitch Factor or Coil Pitch (Pitch Factor (K_p) or Coil Span Factor [K_c])

4.4.2.1.1 Pitch Factor (K_p)

A coil whose sides are separated by one pole pitch (i.e. coil span is 180° electrical) is called a full-pitch coil. With a full-pitch coil, the emfs induced in the two coil sides are in phase with each other and the resultant emf is the arithmetic sum of individual emfs. However, the waveform of the resultant emf can be improved by making the coil pitch less than a pole pitch. Such a coil is called short-pitch coil. This practice is only possible with double-layer type of winding. The emf induced in a short-pitch coil is less than that of a full-pitch coil. The factor by which emf per coil is reduced is called pitch factor K_p. It is defined as

$$K_p = \frac{\text{e.m.f.induced in full} - \text{pitch coil}}{\text{e.m.f.induced in short} - \text{pitch coil}} \tag{4.21}$$

Expression for K_p. Consider a coil AB which is short-pitch by an angle β electrical degrees as shown in Figure 4.70. The emfs generated in the coil sides A and B differ in phase by an angle β and can be represented by phasors E_A and E_B respectively as shown in Figure 4.71. The diagonal of the parallelogram represents the resultant emf E_R of the coil.

Since $E_A = E_B$, $E_R = 2E_A \cos \beta/2$

Pitch factor

$$K_p = \frac{\text{e.m.f.in full} - \text{pitch coil}}{\text{e.m.f.in short} - \text{pitch coil}} = \frac{2E_A \cos\left(\frac{\beta}{2}\right)}{2E_A} = \cos\left(\frac{\beta}{2}\right) \tag{4.22}$$

For a full-pitch winding, $K_p = 1$. However, for a short-pitch winding, $K_p < 1$. Note that β is always an integer multiple of the slot angle α.

$K_{p1} = \cos\left(\frac{\beta}{2}\right)$ for the fundamental and $K_{pn} = \cos\left(\frac{n\beta}{2}\right)$ for n^{th} harmonic. In general, the coils should be chorded by an angle = π/n for elimination of nth harmonic.

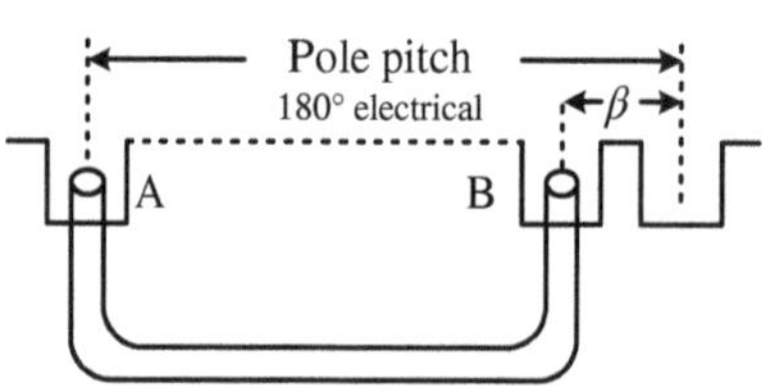

Figure 4.70 Coil AB which is short-pitch by an angle β electrical degrees

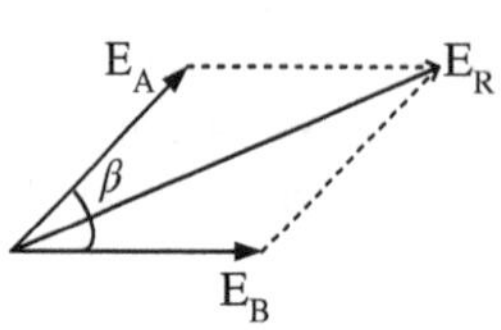

Figure 4.71 Coil sides A and B differ in phase by an angle β and can be represented by phasors E_A and E_B respectively

4.4.3 Distribution Factor (Breadth Factor (K_b) or Distribution Factor (K_d))

4.4.3.1 Distribution Factor (K_d)

A winding with only one slot per pole per phase is called a concentrated winding. In this type of winding, the emf generated/phase is equal to the arithmetic sum of the individual coil emfs in that phase. However, if the coils/phase are distributed over several slots in space (distributed winding), the emfs in the coils are not in phase (i.e. phase difference is not zero) but are displaced from each by the slot angle α (The angular displacement in electrical agrees between the adjacent slots is called slot angle). The emf/phase will be the phasor sum of coil emfs. The distribution factor K_d is defined as

$$K_d = \frac{\text{emfwith distributed winding}}{\text{emfwith concentrated winding}} = \frac{\text{phasor sum of coil emfs/phase}}{\text{arithmetic sum of coil emfs/phase}} \tag{4.23}$$

Note that numerator is less than denominator so that $K_d < 1$.

Expression for K_d: Let

$$\alpha = \text{slot angle} = \frac{180^\circ \text{electrical}}{\text{No.of slots/pole}} \tag{4.24}$$

n = slots per pole per phase

The distribution factor can be determined by constructing a phasor diagram for the coil emfs. Let n = 3. The three coils emfs are shown as phasors AB, BC, and CD (see Figure 4.72a) each of which is a chord of circle with centre at O and subtends an angle α at O. The phasor sum of the coil emfs subtends an angle nα (Here n = 3) at O. Draw perpendicular bisectors of each chord such as Ox, Oy, etc (see Figure 4.72b).

$$K_d = \frac{AD}{n \times AB} = \frac{2 \times Ax}{n \times (2Ay)} = \frac{Ax}{n \times Ay} = \frac{OA \times \sin\left(\frac{n\alpha}{2}\right)}{n \times OA \times \sin\left(\frac{\alpha}{2}\right)} = \frac{\sin\left(\frac{n\alpha}{2}\right)}{n\,\sin\left(\frac{\alpha}{2}\right)} \tag{4.25a}$$

Note that nα = σ is the phase spread. When the n coils form a pole phase group, n = q = slots per pole per phase. Therefore, distribution factor for fundamental from (4.25a)

$$K_{d1} = \frac{\sin\left(\frac{\sigma}{2}\right)}{q\,\sin\left(\frac{\sigma}{2q}\right)} \tag{4.25b}$$

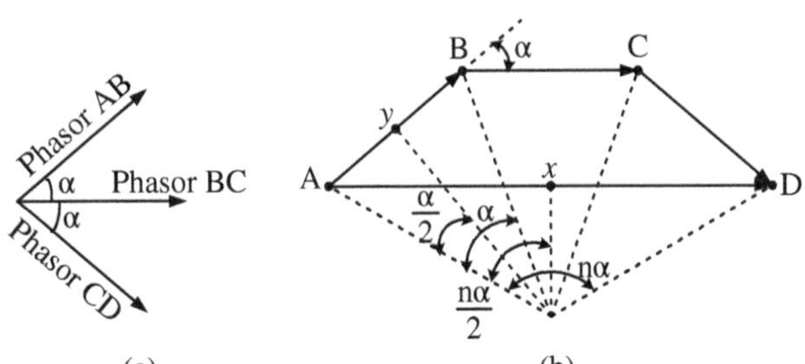

Figure 4.72 Phasor diagram for the coil emfs [1].

distribution factor for nth harmonic

$$K_{dn} = \frac{\sin\left(n\frac{\sigma}{2}\right)}{q\,\sin\left(\frac{n\sigma}{2q}\right)} \tag{4.25c}$$

Winding factor is defined as the product of distribution factor (K_d) and pitch factor (K_p). It is denoted by the symbol K_w.

$$\therefore K_w = K_d \times K_p$$

4.5 Magnetomotive Force (mmf) of AC Windings

Magnetomotive force and flux in the rotating machine [5].

4.5.1 mmf and Flux in Rotating Machine

Current flowing in winding produces magnetic flux. The flux produced produces emf in the winding and the interaction of current and flux generates useful torque.

4.5.2 Main Air-Gap Flux (Field Flux)

The magnetic flux produced by current links throughout the machine, however, its effect depends primarily on the distribution of flux density round the air-gap. Thus, it is important to analyse the flux available in the air-gap. The flux density waveform can acquire any shape deepening upon the winding placement and current waveform. The *main flux* of an AC machine is determined by the fundamental component of air-gap flux density waveform, and the radial line where the fundamental density is a maximum is called the *axis of the flux*. The main flux is then completely defined by a magnitude and a direction [5].

4.5.3 mmf of a Coil [5]

4.5.3.1 mmf

mmf produced due to current is assumed to be sinusoidal as shown in Figure 4.73. Developed diagram is shown in Figure 4.73 where circular shape is cut out and shown in a plane. A two-pole machine is assumed; hence, half of the coils carry current in one direction and other half in the opposite direction shown by dot and cross signs in Figure 4.63. Current distribution in conductors is known and hence the mmf can be found. One closed path is shown as ACDFGH that include iron portion and air-gaps. Since the reluctance of air-gap is much higher than that of iron portion, mmf can be assumed to concentrated across the air-gap. mmf wave is crossing the air gap at two points A and F. At A, the mmf is zero and hence all the mmf is then assumed to be concentrated at point F. Thus, mmf distribution round the air gap can be drawn for any value of current flowing in the coil. Therefore, although mmf is line integral round a closed path, a value can be associated with each point along the air-gap circumference, giving the space distributed mmf waveform of the

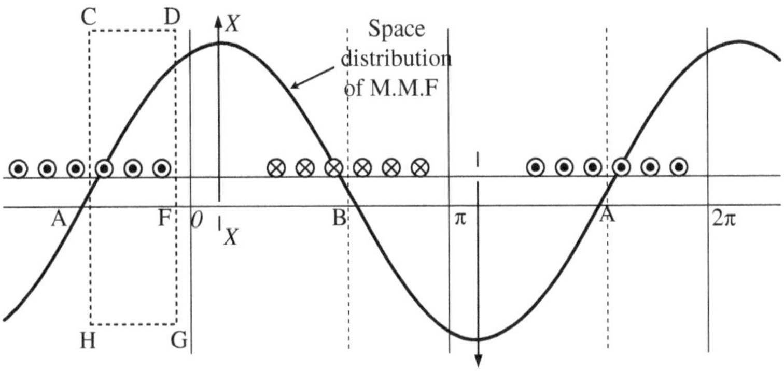

Figure 4.73 Distribution of current and mmf in a coil [5].

machine. If the conductors are assumed to be located at points around the air-gap circumference, the mmf waveform is a stepped curve, but the fundamental component, which, by symmetry, is zero at the points A and B, can be drawn as in Figure 4.73. The radial line at the point of maximum mmf *(XX* in Figure 4.72) is called the *axis of the mmf,* and since this depends only on the conductor distribution, it is also the *axis of the coil.* The curve shows the instantaneous magnitude of the mmf, which depends on the instantaneous value of the current [5].

4.5.3.2 mmf of Distributed Windings

mmf of distributed AC windings [6].

It has been seen earlier that the armature of a practical machine has distributed winding wound for the same number of poles as the field winding. As the armature carries current, the resultant field of its current-carrying coils has the same number of poles as the field winding. Our approach will be to find the mmf space distribution of the current-carrying armature by superimposing the mmf space waves of individual coils.

4.5.3.3 mmf Space Wave of a Single Coil

A cylindrical rotor machine with small air gap is shown in Figure 4.74a [6]. The stator is imagined to be wound for two-poles with a single N-turn full-pitch coil carrying current i in the direction indicated. The figure shows some flux lines of the magnetic field set up. A north and corresponding south pole are induced on the stator periphery. The magnetic axis of the coil is from the stator north to the stator south. Each flux line radially crosses the air gap twice, normal to the stator and rotor iron surfaces and is associated with constant mmf Ni. On the assumption that the reluctance of the iron path is negligible, half the mmf ($Ni/2$) is consumed to create flux from the rotor to stator in the air gap and the other half is used up to establish flux from the stator to rotor in the air-gap. mmf and flux radially outwards from the rotor to the stator (south pole on stator) will be assumed to be positive and that from the stator to rotor as negative.

The physical picture is more easily visualized by the developed diagram of Figure 4.74b where the stator with the winding is laid down flat with rotor on the top of it. It is seen that the mmf is a rectangular space wherein mmf of $+ Ni/2$ is consumed in setting flux from the rotor to stator and mmf of $– Ni/2$ is consumed in setting up flux from the stator to the rotor.

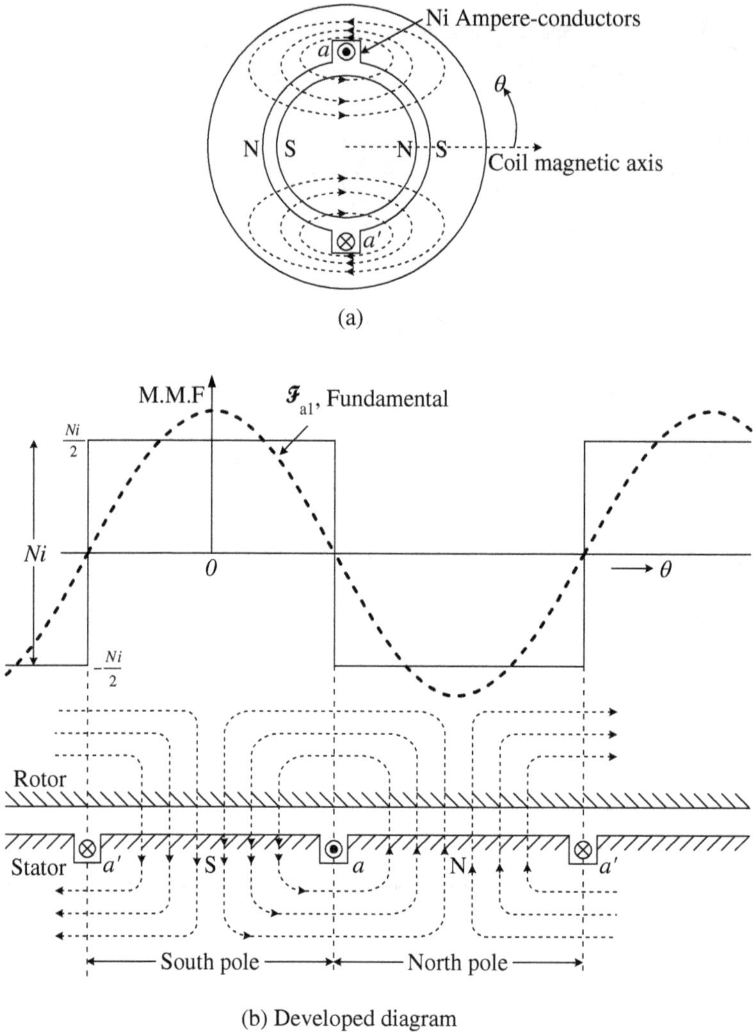

Figure 4.74 mmf space wave of a single coil [6] (a) A cylindrical rotor machine with small air-gap, (b) Developed diagram.

It has been imagined here that the coil-sides occupy a narrow space on the stator and the mmf changes abruptly from $-Ni/2$ to $+Ni/2$ at one slot and in reverse direction at the other slot. The mmf change at any slot is

$$Ni = \text{ampere} - \text{conductors/slot}$$

and its sign depends upon the current direction.

With the mmf space wave of a single coil being rectangular, it can be split up into its fundamental and harmonics. It easily follows from the Fourier series analysis that the fundamental of the mmf wave as shown in Figure 4.74b is

$$\mathcal{F}_{a1} = \frac{4}{\pi}\left(\frac{Ni}{2}\right)\cos(\theta) = F_{1p}\cos(\theta) \tag{4.26}$$

where θ is the electrical angle measured from the magnetic axis of the coil which coincides with the positive peak of the fundamental wave. From Eq. (4.26)

$$\mathscr{F}_{a1}\,(\text{peak}) = F_{1p} = \frac{4}{\pi}\left(\frac{Ni}{2}\right) \tag{4.27}$$

For a rectangular wave, the normalized peaks of the harmonic waves are:

$$F_{3p} = \frac{4}{3\pi}, F_{5p} = \frac{4}{5\pi}, F_{7p} = \frac{4}{7\pi}, \ldots..$$

Needless to say, that there cannot even be any harmonics.

From now onwards, only the fundamental mmf wave of a current-carrying coil will be considered, neglecting its associated harmonic space waves whose amplitudes are small. Furthermore, as will soon be seen in a distributed winding, the rectangular mmf waves of individual phase group coils add up to produce an mmf wave closer to a sine wave, i.e. the harmonics tend to cancel out.

It may be noted here that with the assumption of the narrow air-gap, the mmf distribution will be the same if the coils were located in the rotor slots instead.

4.5.3.4 mmf Space Wave of One Phase of a Distributed Winding [6]

Consider now a basic two-pole structure with a round rotor, with five slots/pole/phase (SPP) and a 2-layer winding as shown in Figure 4.75. The corresponding developed diagram is shown in Figure 4.76a along with the mmf diagram which now is a stepped wave–obviously closer to a sine wave than the rectangular mmf wave of a single coil (Figure 4.74b). Here, since SPP is odd (5), half the ampere-conductors of the middle slot of the phase group a and a' contribute towards establishment of south pole and half towards north pole on the stator. At each slot, the mmf wave has a step jump of $2N_c i_c$ ampere-conductors where N_c = coil turns (equal to conductors/layer) and i_c = conductor current.

Now F_{1p}, the peak of the fundamental of the mmf wave, has to be determined. Rather than directly finding the fundamental of the stepped wave, one can proceed by adding the fundamentals of the mmfs of individual slot-pairs (with a span of one pole-pitch). These fundamentals are progressively out of phase (space phase as different from time phase)

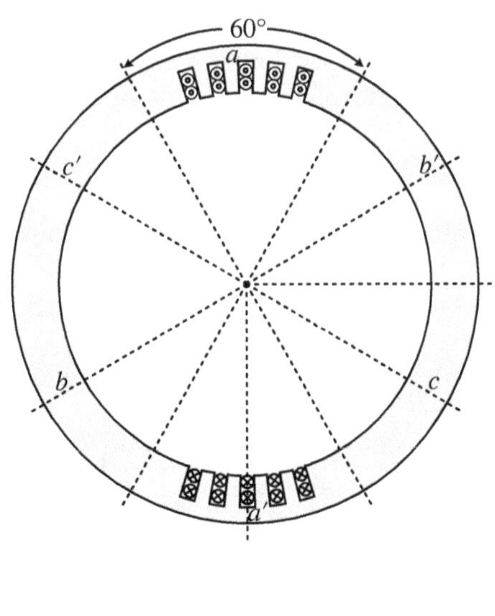

Figure 4.75 A 3-phase, 2-pole structure with two-layer winding [6].

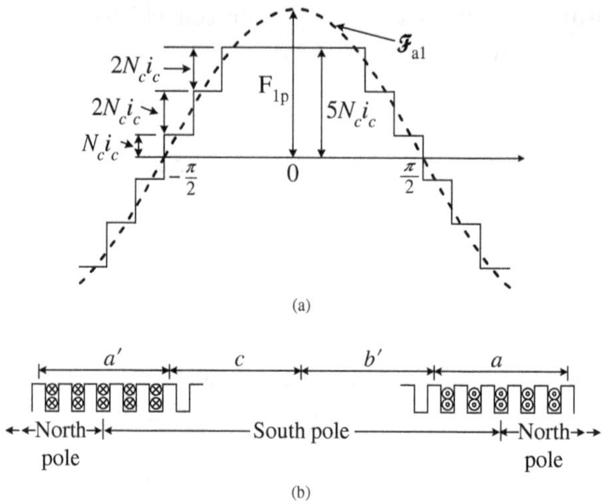

Figure 4.76 Developed diagram and mmf wave of the machine of Figure 4.75 [6].

with each other by the angle γ. This addition is easily accomplished by defining the breadth factor K_b, which will be the same as in the case of the generated emf of a coil group.

Let

N_{ph}(series) = series turns per parallel path of a phase
A = number of parallel paths of a phase

Then

AT/parallel path = $N_{ph}(\text{series})i_c$

$$\text{AT/phase} = A\left(N_{ph}(\text{series})\, i_c\right) = N_{ph}(\text{series})\, i_a \tag{4.28}$$

where i_a = phase current = Ai_c

It now follows that

$$\text{AT/pole/phase} = \left(\frac{N_{ph}(\text{series})}{P}\right) i_a \tag{4.29}$$

provided the winding is a concentrated one. The fundamental peak of the concentrated winding is then

$$F_{1P}\left(\text{concentrated winding}\right) = \frac{4}{\pi}\left(\frac{N_{ph}(\text{series})}{P}\right) i_a \tag{4.30}$$

Since the actual winding is a distributed one, the fundamental peak will be reduced by the breadth factor K_b. Thus

$$F_{1P}\left(\text{distributed winding}\right) = \frac{4}{\pi}K_b\left(\frac{N_{ph}(\text{series})}{P}\right) i_a \tag{4.31}$$

Hence

$$\mathcal{F}_{a1} = \frac{4}{\pi}K_b\left(\frac{N_{ph}(\text{series})}{P}\right) i_a \cos(\theta) \tag{4.32}$$

where the pole axis is taken as the angle reference (Figure 4.76b).

The effect on the mmf wave of short-pitched coils can be visualized by Figure 4.77 in which the stator has two short-pitched coils ($a_1a'_1, a_2a'_2$) for phase a of a two-pole structure (The arrangement of coils for 3-phase two-pole, two-layer winding, with $S = 12$, $m = 2$ and coil-pitch = 5 (chorded by one slot) is shown in Figure 4.78. It is seen from Figure 4.78 that the coil-sides in a given slot do not belong to the same phase. This is in contrast to Figure 4.75 depicting a full-pitch winding.). The mmf of each coil establishes one pole. From the developed diagram of Figure 4.77b it is seen that the mmf wave is rectangular but of shorter space length than a pole-pitch. The amplitude of the fundamental peak gets reduced by a factor K_p, called the *pitch factor*, compared to the full-pitch rectangular mmf wave. It can be shown by Fourier analysis that

$$F_{1P} = \frac{2}{\pi}\int_{-\frac{\pi}{2}+\frac{\theta_{sp}}{2}}^{\frac{\pi}{2}-\frac{\theta_{sp}}{2}} F_{\text{pole}}\cos\theta \, d\theta = \frac{4}{\pi}F_{\text{pole}}\cos\left(\frac{\theta_{sp}}{2}\right)$$

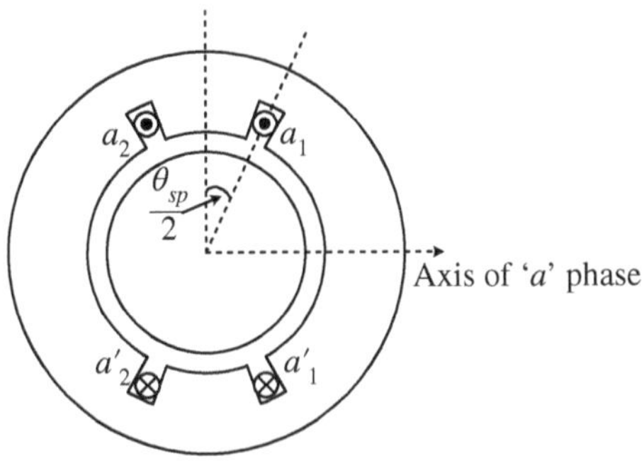

(a) Stator with short-pitched cpol at phase 'a' (single-layer winding)

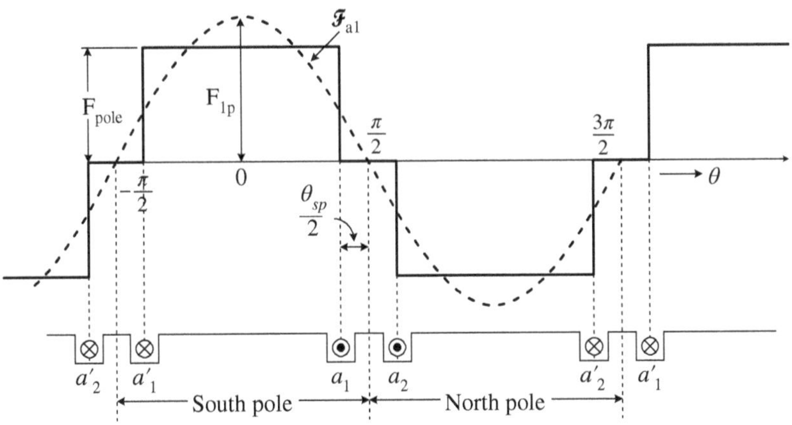

(b) Effect of short-pitched coil on mmf wave

Figure 4.77 [6] (a) Stator with short-pitched coil at phase 'a' (single-layer winding) (b) Effect of short-pitched coil on mmf wave.

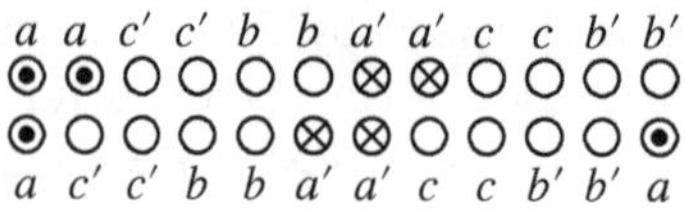

Figure 4.78 Two-layer short-pitch winding [6].

Hence

$$K_p = \cos\left(\frac{\theta_{sp}}{2}\right) \tag{4.33}$$

It is not surprising that the same result is obtained for K_p for the space mmf wave as for the generated emf in a short-pitched coil.

In general, when the winding is both distributed and short-pitched, the fundamental space mmf of phase a as in Eq. (4.32) generalizes to

$$\mathcal{F}_{a1} = \frac{4}{\pi}K_w\left(\frac{N_{ph}\,(\text{series})}{P}\right)i_a\cos(\theta) \tag{4.34}$$

where $K_w = K_bK_p = \textit{winding factor}$

As in the case of the induced emf, distribution of stator winding and short-pitching, both help reducing harmonics in the space mmf wave. In the analysis from now onwards, it will be assumed that the space mmf wave created by each phase of the stator winding when carrying current is purely sinusoidal. Also, K_b and K_p are defined in the same manner as for the induced emf.

When the phase a carries sinusoidal current

$$i_a = I_m\cos\omega t;\ \left(I_m = \text{maximum value of the phase current}\right) \tag{4.35}$$

the mmf wave is given by

$$\mathcal{F}_{a1} = \frac{4}{\pi}K_w\left(\frac{N_{ph}\,(\text{series})}{P}\right)I_m\cos(\omega t)\cos(\theta) = F_m\cos(\omega t)\cos(\theta) \tag{4.36}$$

where

$$\mathcal{F}_m = \frac{4}{\pi}K_w\left(\frac{N_{ph}\,(\text{series})}{P}\right)I_m = \frac{4\sqrt{2}}{\pi}K_w\left(\frac{N_{ph}\,(\text{series})}{P}\right)I \tag{4.37}$$

where $I = \frac{I_m}{\sqrt{2}}$ = rms value of phase current.

As per Eq. (4.36), the mmf of one phase is a *standing wave* (*pulsating wave*) in space whose peak always coincides with the phase axis while the peak amplitude varies sinusoidally with time. This is illustrated in Figure 4.79, where a half-cycle of pulsation is indicated.

4.6 Harmonic Effect [7]

4.6.1 The Form Factor and the emf per Conductor

If ϕ_a is the flux per pole and p the number of poles, then one armature conductor cuts $\phi_a p$ lines of force per revolution or $\phi_a p \times ([\text{rpm}]/60)$ lines per second and the average emf in one conductor $\phi_a p \times ([\text{rpm}]/60) \times 10^{-8}$ V.

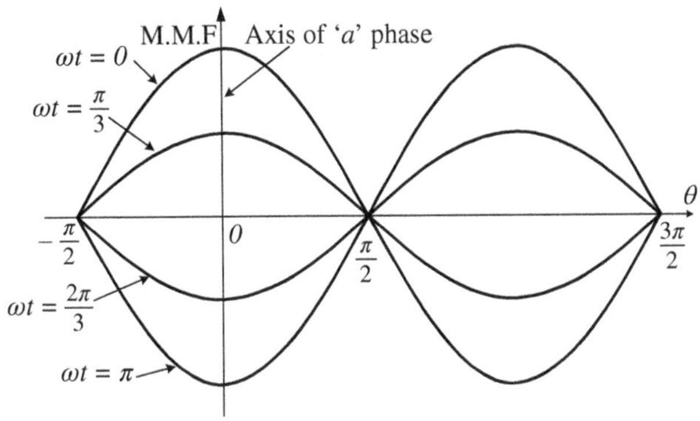

Figure 4.79 Standing (pulsating) mmf wave caused by sinusoidal current in phase *a* coils [6].

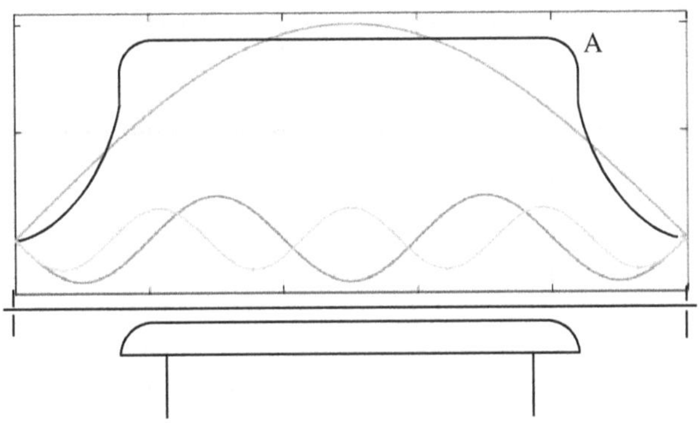

Figure 4.80 E. M. F. wave of an alternator [7].

The form factor of an emf wave is defined as the ratio

$$\frac{\text{effective voltage}}{\text{average voltage}} \tag{4.38}$$

and for a sine wave of emf this value $= \frac{\frac{1}{\sqrt{2}}E_{max}}{\frac{2}{\pi}E_{max}} = 1.11$

The effective emf per conductor $= \phi_a p \frac{rpm}{60} \times 10^{-8} \times \text{form factor} = 1.11\phi_a p \frac{rpm}{60} \times 10^{-8}$ for sine wave emf

$$= 2.22\phi_a f \times 10^{-8} \tag{4.39}$$

since f, the frequency $= \frac{p \times rpm}{120}$

4.6.2 The Wave Form

Figure 4.80 shows the shape of pole face that is in general use and curve A shows the distribution of flux in the air gap under such a pole face. The emf in a conductor is proportional to the rate of cutting lines of force and has, therefore, a waveform of the same shape as the curve of flux distribution.

Curve A is not a sine wave but can be considered as the resultant of a number of sine waves consisting of a fundamental and harmonics. The frequency and magnitude of these harmonics depend principally on the ratio of pole arc to pole-pitch. In Figure 4.80 this ratio is 0.65 and the fundamental and harmonics which go to make up the flux distribution curve are shown to scale, higher harmonics than the seventh being neglected in this case.

4.6.3 Problem Due to Harmonics

The fundamental and harmonics that go to make up an emf wave act as if each had a separate existence. If the circuit to which this emf is applied consists of an inductance L in series with a capacity C so that the impedance of the circuit

$$= 2\pi fL - \frac{1}{2\pi fC} \tag{4.40}$$

then the current in this circuit consists of a fundamental $= \frac{E_1}{2\pi f_1 L - \frac{1}{2\pi f_1 C}}$

and harmonics of the form $\frac{E_n}{2\pi f_n L - \frac{1}{2\pi f_n C}}$

where E_n is the effective value of the nth harmonic. If now, f_n has such a value that

$$2\pi f_n L = \frac{1}{2\pi f_n C} \tag{4.41}$$

so that the circuit is in resonance at this frequency, then the nth harmonic of current will be infinite, and the nth harmonic of emf across L and C individually will also be infinite.

The above is an ideal case; in ordinary circuits, the current cannot reach infinity on account of the resistance that is always present, nevertheless, if the circuit is in resonance at the frequency of the fundamental or of any of the harmonics, dangerously high voltages will be produced, between different points in the circuit. The constants of a circuit seldom give trouble at the fundamental frequency, trouble is generally due to high-frequency harmonics. It is desirable then to eliminate harmonics from the emf wave of a generator as far as possible and several of the methods adopted are described below.

4.6.4 Elimination or Suppression of Harmonics

4.6.4.1 Shape of Pole Face

The pole face is sometimes shaped as shown in Figure 4.81, that is, the air gap is varied from a minimum under the centre of the pole to a maximum at the pole tip, so as to make the flux distribution curve approximately a sine curve; then the emf wave from each conductor will be approximately a sine wave.

4.6.4.2 Use of Several Slots per Phase per Pole

Figure 4.82 shows part of a three-phase machine which has six slots per pole or two slots per phase per pole. The emf generated in a conductor in slot A is represented at any instant

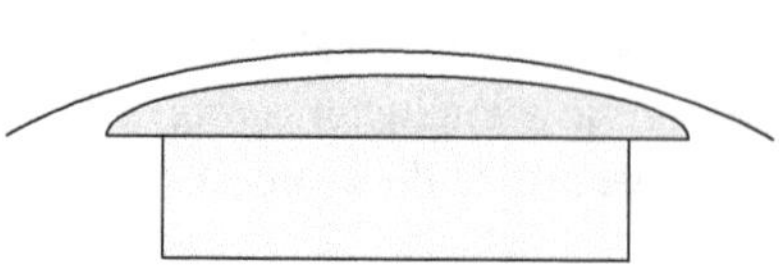

Figure 4.81 Pole face shaped to give a sine wave e. m. f. [7].

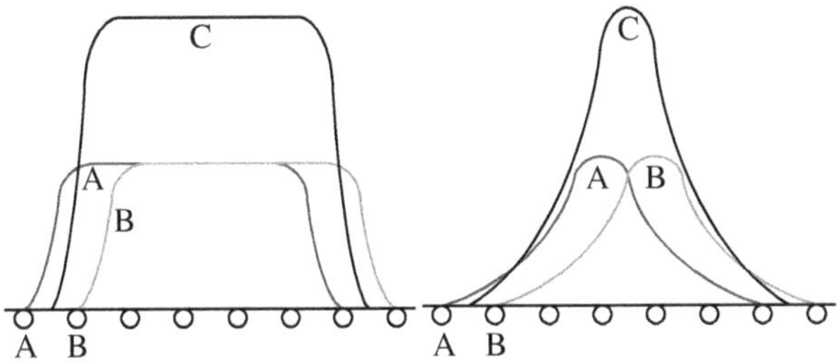

Figure 4.82 E. M. F. wave of a three-phase alternator with two slots per phase per pole [7].

by curve A, and that in a conductor in slot B by curve B, which is out of phase with curve A by the angle corresponding to one slot-pitch, or 30°.

When the conductors in slots A and B are connected in series so that their emfs add up, the resultant emf at any instant is given by curve C, which is obtained by adding together the ordinates of curves A and B. C is more nearly a sine curve than either A or B.

If the fundamental and harmonics that go to make up curves A and B are known, then those which go to make up curve C can be readily found as follows: The fundamental of the resultant wave C is the vector sum of the fundamentals of A and B, which are θ electrical degrees apart, and the nth harmonic of C is the vector sum of the nth harmonics of A and B which are $(n \times \theta)$ electrical degrees apart. For example, curves A and B, Figure 4.83 are 30° out of phase with one another and each consists of a fundamental and a third harmonic as shown. C_1 is the resultant of the fundamentals A_1 and B_1 and C_3, the third harmonic of curve C, is the resultant of the two third harmonics A_3 and B_3.

4.6.4.3 Use of Short-Pitch Windings

Figure 4.84 shows a short-pitch coil. The emf waves in the conductors A and B are out of phase with one another by θ degrees, but each has the same shape as the curve of flux distribution. The problem of finding the resultant emf wave is, therefore, the same as that discussed in the last article.

If $n\theta$, the phase angle between the nth harmonics of curves A and B, becomes equal to 180°, then the corresponding harmonic is eliminated from the resultant curve C since the two harmonics which go to make it up are equal and opposite. If, for example, there are nine slots per pole and the coil is one slot short, then the angle θ is 20 electrical degrees, and the 9th harmonic is eliminated from the voltage wave of the coil; in general, if the pitch of the coil be shortened by $\frac{1}{n}$ of the pole-pitch then, as shown in Figure 4.85, the nth harmonic will be eliminated.

It may be pointed out here that an even harmonic is seldom found in the emf wave of an alternator, because the resultant of a fundamental and an even harmonic gives an unsymmetrical curve, as shown in Figure 4.86, where the resultant curve is made up of a fundamental and a second harmonic. If then the emf wave is symmetrical, it may be assumed that no even harmonics are present.

The various advantages of short pitch coils are:

1. The length required for the end connections of coils is less, that is, in active length of winding is less. So, less copper is required.
2. As copper required is less; it is economical.

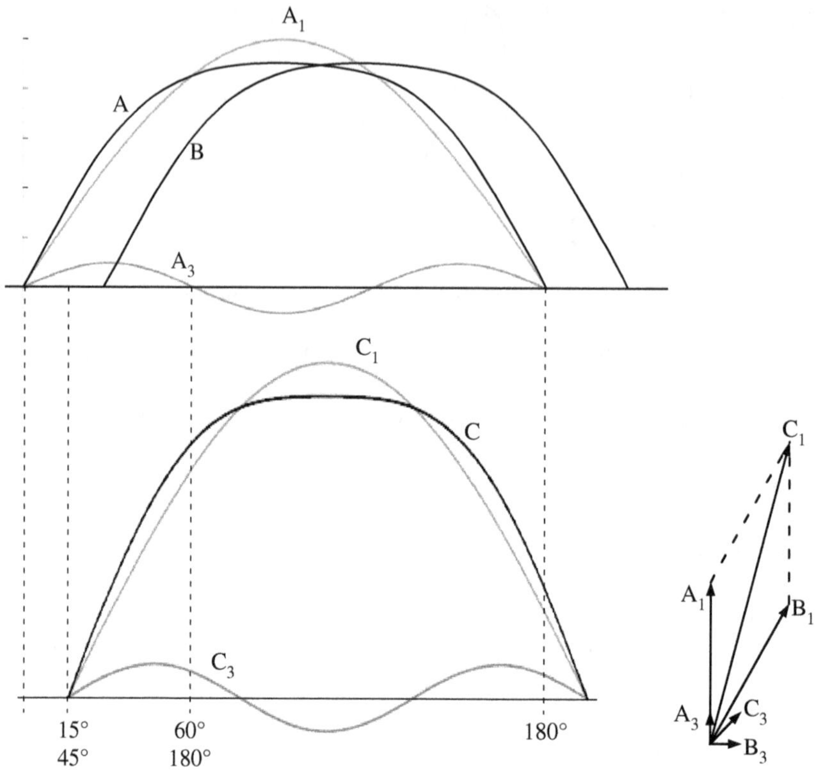

Figure 4.83 The vector diagram for the fundamental and the harmonics. [7].

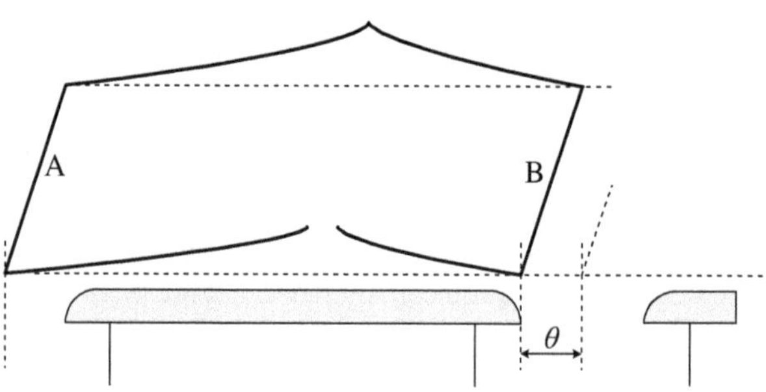

Figure 4.84 Short-pitch coil [7].

3. Short pitching eliminates high-frequency harmonics which distort the sinusoidal nature of emf Hence, waveform of induced emf is more sinusoidal due to short pitching.
4. As high-frequency harmonics get eliminated, eddy current and hysteresis losses which depend on frequency also get eliminated.

The efficiency increases due to reduced losses.

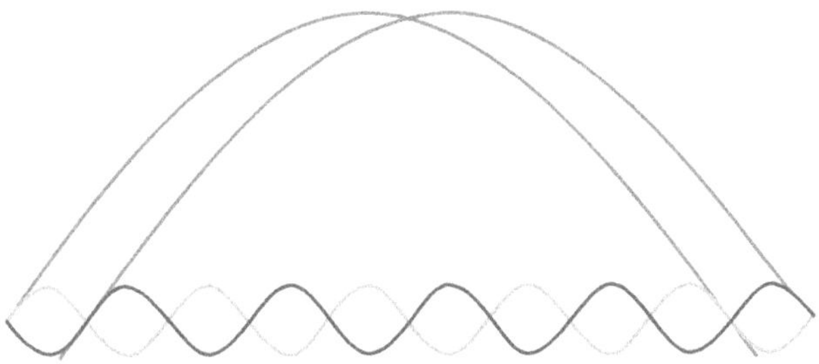

Figure 4.85 Elimination of harmonics from, the e. m. f. wave [7].

Figure 4.86 Unsymmetrical waves due to even harmonics [7].

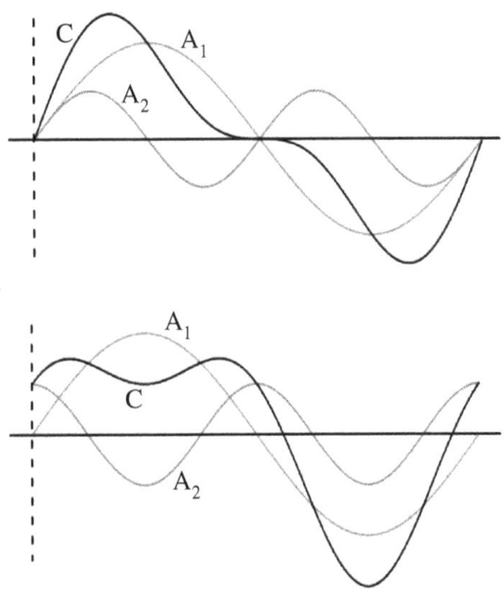

4.6.4.4 Effect of the Y- and Δ -Connection on Harmonics

The fundamentals of the three emfs are 120° out of phase with one another and are represented by vectors in diagram A, Figure 4.87. The nth harmonics are $(n \times 120)$ degrees out of phase with one another.

When the phases are Y-connected the terminal F_2 is brought to the potential of terminal F_1 and the resultant fundamental between S_1 and S_2 is represented by the vector S_1S_2 and = 1.73 times the fundamental in one phase. In the case of the third harmonic the emfs are $(3 \times 120) = 360°$ out of phase with one another and are represented by vectors in Figure 4.87b; the resultant third harmonic between S_1 and S_2 is zero so that, in a Y-connected alternator, no third harmonic, nor any harmonic which is a multiple of three, is found in the terminal voltage wave. This can be observed in Figure 4.87e that third harmonics and its multiple are eliminated because they all are in phase. Mathematically, third harmonics can present as:

$$V_{S13} = V_{max} \sin(3\omega t)$$

$$V_{S23} = V_{max} \sin\left(3\omega t - 3\frac{2\pi}{3}\right) = V_{max} \sin(3\omega t - 2\pi)$$

$$V_{S33} = V_{max} \sin\left(3\omega t - 3\frac{4\pi}{3}\right) = V_{max} \sin(3\omega t - 4\pi)$$

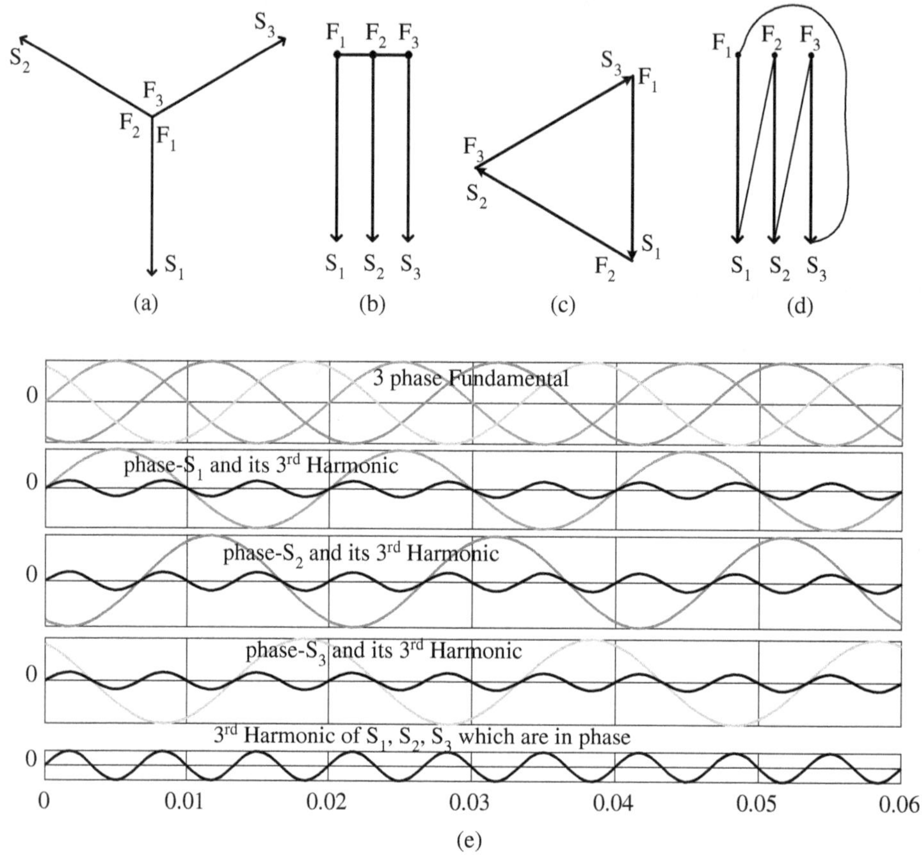

Figure 4.87 Effect of the Y- and Δ-connection on the third harmonic [7].

When the phases are Δ-connected, any harmonic in the voltage wave of one phase will also be found in that of the terminal voltage; a greater objection to the use of this connection for alternators is that the harmonics cause circulating currents to flow in the closed circuit produced by the Δ-connection. Figure 4.87c shows the voltage vector diagram for the fundamentals in the emf wave of each phase; the three vectors are 120° out of phase with one another and the resultant voltage in the closed circuit due to the fundamentals is zero.

Figure 4.87d shows the voltage vector diagram for the third harmonic in the emf wave of each phase, the three vectors are 360° out of phase with one another and the resultant voltage in the closed circuit due to the third harmonics is three times the value of the third harmonic in one phase. A circulating current will flow in the closed circuit, of triple frequency and of a value $= \frac{3E_3}{3Z_3}$ where E_3 is the effective value of the third harmonic in each phase and z_3 is the impedance per phase to the third harmonic.

4.6.4.5 Harmonics Produced by Armature Slots

Figure 4.88 shows two positions of the pole of an alternator relative to the armature. In position A, the air-gap reluctance is a minimum and in position B is a maximum. The flux per pole pulsates, due to this change in reluctance, once in the distance of a slot-pitch, or $2a$

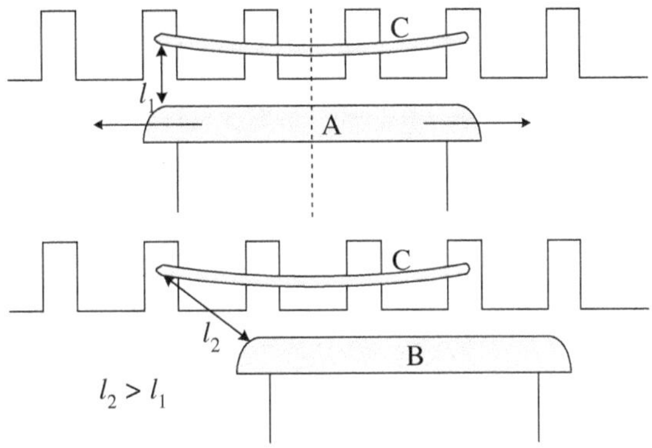

Figure 4.88 Variation of the air gap reluctance [7].

times in the distance of two pole-pitches, where a = slots per pole, and the emf generated in each coil by the main field goes through one cycle while the pole moves, relative to the armature, through the distance of two pole-pitches, therefore, the frequency of the flux pulsation = $2af$.

The flux per pole then consists of a constant value ϕ_a, and a superimposed alternating flux which has a frequency of $2a$ times the fundamental frequency of the machine, or at any instant the flux per pole

$$= \phi_p = \phi_a + \phi_1 \cos 2a\theta$$

where $\phi_a + \phi_1$ is the maximum value of the pulsating flux and θ is the angle moved through from position A, Figure 4.88, in electrical degrees; therefore, the flux threading coil C, which is a full-pitch coil

= $\phi_a + \phi_1$ when the coil is in position A

= $[\phi_a + \phi_1 \cos 2a\theta]\cos\theta$ when the coil has moved through θ electric degrees relative to the pole, and the emf in coil C at any instant

$$= -T\frac{d\left(\phi_a + \phi_1 \cos(2a\theta)\right)\cos(\theta)}{dt} = -T\frac{d\left(\phi_a + \phi_1 \cos(2a\theta)\right)\cos(\theta)}{d\theta} \times \frac{d\theta}{dt}$$

$$= -2\pi f\left[-\sin(\theta)\left\{\phi_a + \phi_1\cos(2a\theta)\right\} - \cos(\theta)\left\{2a\phi_1 \sin(2a\theta)\right\}\right]T$$

$$= 2\pi f\left[\phi_a \sin(\theta) + \phi_1 \sin(\theta)\cos(2a\theta) + 2a\phi_1\cos(\theta)\sin(2a\theta)\right]T$$

$$= 2\pi f\left[\phi_a \sin(\theta) + \frac{\phi_1}{2}\{\sin(2a+1)\theta - \sin(2a-1)\theta\}\right.$$

$$\left. + \frac{2a\phi_1}{2}\{\sin(2a+1)\theta + \sin(2a-1)\theta\}\right]T$$

$$= 2\pi f T\left[\phi_a \sin(\theta) + \frac{\phi_1 + 2a\phi_1}{2}\sin(2a+1)\theta + \frac{\phi_1 - 2a\phi_1}{2}\sin(2a-1)\theta\right] \quad (4.42)$$

so that, due to the variation of the air gap reluctance as the poles move past the armature slots, two harmonics are produced which have frequencies of $(2a + 1)$ and of $(2a - 1)$ times that of the fundamental, respectively.

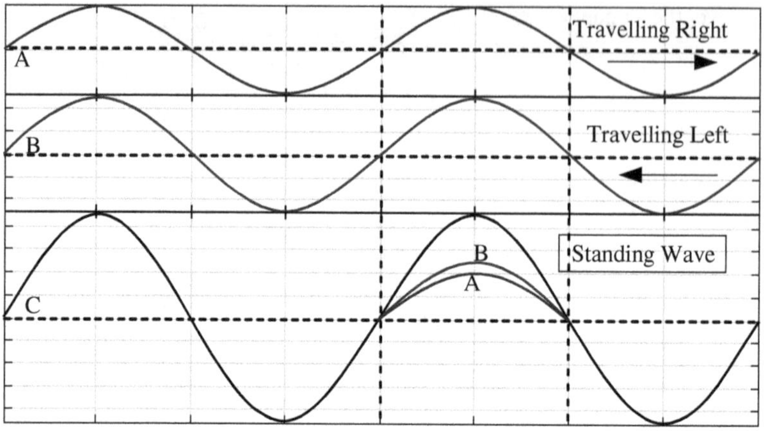

Figure 4.89 Resolution of a stationary wave into two progressive waves [7].

A convenient physical interpretation of the above result is as follows: A and B, Figure 4.89, are two equal trains of waves which are constant in magnitude, and move in opposite directions at the same speed. The resultant of two such wave trains superimposed on one another is a series of stationary waves as shown at C.

A stationary wave, such as that of the alternating magnetic field in the air gap of an alternator due to a variation in the air gap reluctance can, therefore, be exactly represented by two progressive waves of constant value which move in opposite directions through the distance of two pole-pitches while the alternating wave goes through one cycle.

Consider both of these waves or fields to exist separately from the main field then, when the pole moves with the constant flux ϕ_a, through a distance y relative to the armature, one of these constant progressive fields moves through a distance $2ay$ relative to the pole, or through a distance $(2a + 1)y$ relative to the armature, while the other moves through a distance $-2ay$ relative to the poles, or through a distance $(2a - 1)y$ relative to the armature. If then the fundamental frequency of the generated emf is f, the two other fields will generate emfs of frequencies $= (2a + 1)f$ and $(2a - 1)f$, respectively.

To keep down the value of these harmonics, the reluctance of the air gap under the poles should be made as nearly constant as possible for all positions of the pole relative to the armature.

4.7 Basic Principles of Electric Machines

Although a far greater percentage of the electrical machines in service are AC machines, the DC machines are of considerable industrial importance. The principal advantage of the DC machine, particularly the DC motor, is that it provides a fine, simple and flexible speed control. Flux and torque are independently controllable contrary to AC machine where advanced control approaches such as vector control and direct torque control schemes are adopted to achieve the same functionality. Nevertheless, DC generators are not as common as they used to be, because direct current, when required, is mainly obtained from an AC supply by the use of power electronic converters. Yet, an understanding of DC generator is important because it represents a logical introduction to the behaviour of DC motors. Indeed, many DC motors in industry actually operate as DC generators for a brief period [8].

DC motors are rarely used in general purpose applications because DC supply is not readily available. However, for special applications such as in steel mills, mines and electric trains, it is advantageous to convert alternating current into direct current in order to use DC motors. The reason is that speed/torque characteristics of DC motors are much more superior to that of AC motors. Therefore, it is not surprising to note that for industrial drives, DC motors are as popular as 3-phase induction motors. Like DC generators, DC motors are also of three types viz., series-wound, shunt-wound and compound wound. The use of a particular motor depends upon the mechanical load it has to drive [8].

4.7.1 AC Rotating Machines

AC machines are motors that convert AC electric energy to mechanical energy and generators that convert mechanical energy to AC electric energy [8]. The two major classes of AC machines are synchronous and induction machines. The field current of synchronous machines (motors and generators) is supplied by a separate DC power source while the field current of induction machines is supplied by magnetic induction (transformer action) into the field windings.

AC machines differ from DC machines by having their *armature windings* almost always located on the stator while their *field windings* are located on the rotor. A set of three-phase AC voltages is induced into the stator armature windings of an AC machine by the rotating magnetic field from the rotor field windings (generator action). Conversely, a set of three-phase currents flowing in the stator armature windings produces a rotating magnetic field within the stator. This magnetic field interacts with the rotor magnetic field to produce the torque in the machine (motor action).

4.7.1.1 The Rotating Magnetic Field

The main principle of ac machine operation is this: A three-phase set of currents, flowing in an armature winding, each of equal magnitude and differing in phase by 120°, produces a rotating magnetic field of constant magnitude. The stator shown in Figure 4.90 has three coils, each 120° apart. The currents flowing in the stator are given by

$$i_{aa'}(t) = I_M \sin(\omega t) \text{ A} \tag{4.43a}$$

$$i_{bb'}(t) = I_M \sin\left(\omega t - 120^o\right) \text{ A} \tag{4.43b}$$

$$i_{cc'}(t) = I_M \sin\left(\omega t - 240^o\right) \text{ A} \tag{4.43c}$$

The resulting magnetic flux densities are

$$B_{aa'}(t) = B_M \sin(\omega t) \angle 0^o \text{Wb/m}^2 \tag{4.44a}$$

$$B_{bb'}(t) = B_M \sin\left(\omega t - 120^o\right) \angle 120^o \text{Wb/m}^2 \tag{4.44b}$$

$$B_{cc'}(t) = B_M \sin\left(\omega t - 240^o\right) \angle 240^o \text{Wb/m}^2 \tag{4.44c}$$

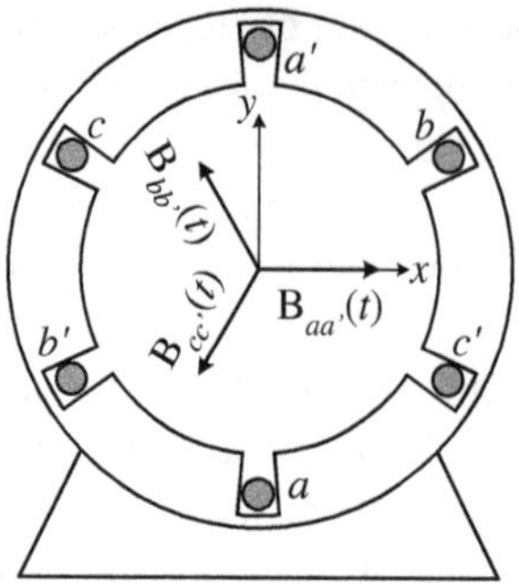

Figure 4.90 A simple three-phase stator. Currents in this stator are assumed positive if they flow into the unprimed and out of the primed ends of the coils [8].

The directions of these fluxes are given by the right-hand rule i.e. when the fingers of the right-hand curl in the direction of the current in a coil, the thumb points in the direction of the resulting magnetic flux density [8].

An examination of the currents and their corresponding magnetic flux densities at specific times is used to determine the resulting net magnetic flux density. For example, at time $\omega t = 0°$, the magnetic field from coil aa' will be

$$\mathbf{B}_{aa'} = 0 \tag{4.45a}$$

The magnetic field from coil bb' will be

$$\mathbf{B}_{bb'} = B_M \sin\left(-120°\right) \angle - 120° \tag{4.45b}$$

and the magnetic field from coil cc' will be

$$\mathbf{B}_{cc'} = B_M \sin\left(-240°\right) \angle - 240° \tag{4.45c}$$

The total magnetic field from all three coils added together will be

$$\mathbf{B}_{net} = \mathbf{B}'_{aa} + \mathbf{B}'_{bb} + \mathbf{B}'_{cc} = 0 + \left(-\frac{\sqrt{3}}{2}B_M\right)\angle 120^o - \frac{\sqrt{3}}{2}B_M\angle 240^o = 1.5B_M\angle - 90^o \tag{4.46}$$

As another example, look at the magnetic field at time $\omega t = 90°$. At that time, the currents are

$$i_{aa'}(t) = I_M \sin\left(90^o\right) \text{A} \tag{4.47a}$$

$$i_{bb'}(t) = I_M \sin\left(-30^o\right) \text{A} \tag{4.47b}$$

$$i_{cc'}(t) = I_M \sin\left(-150^o\right) \text{A} \tag{4.47c}$$

and the magnetic fields are

$$B_{aa'} = B_M\angle 0^o \text{Wb/m}^2 \tag{4.48a}$$

$$B_{bb'} = -0.5B_M\angle 120^o \text{Wb/m}^2 \tag{4.48b}$$

$$B_{cc'} = -0.5B_M\angle 240^o \text{Wb/m}^2 \tag{4.48c}$$

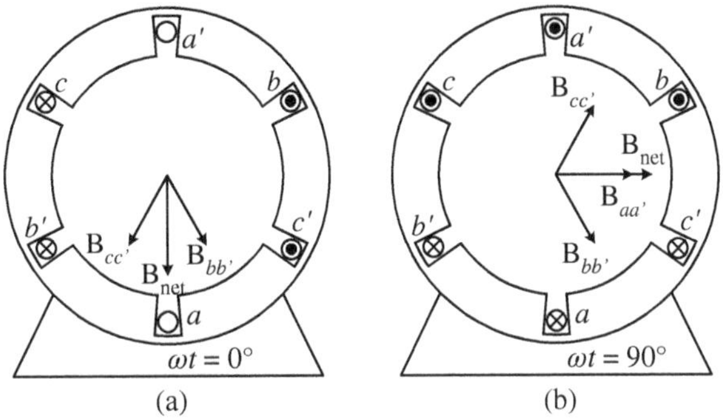

Figure 4.91 (a) The vector magnetic field in a stator at time $\omega t = 0°$. (b) The vector magnetic field in a stator at time $\omega t = 90°$ [8].

The resulting net magnetic field is

$$\begin{aligned} \boldsymbol{B}_{net} &= \boldsymbol{B}'_{aa} + \boldsymbol{B}'_{bb} + \boldsymbol{B}'_{cc} = B_M\angle 0^o + (-0.5)\,B_M\angle 120^o + (-0.5)\,B_M\angle 240^o \\ &= 1.5B_M\angle 0^o \end{aligned} \tag{4.49}$$

The resulting magnetic flux is shown in Figure 4.91. Notice that the direction of the magnetic flux has changed, but its magnitude remained constant. The magnetic flux is rotating counter-clockwise while its magnitude remained constant [8].

At any time, t, the magnetic flux has the same magnitude 1.5 BM. It continues to rotate at angular velocity ω. A proof of this concept is presented in Chapter 2.

4.7.1.2 The Relationship between Electrical Frequency and the Speed of Magnetic Field Rotation

Figure 4.92 illustrates that the rotating magnetic field in the stator can be represented as a north and a south pole. The flux leaves the stator at the North Pole and enters the stator at the South Pole. The magnetic poles complete one complete revolution around the stator surface for each electrical cycle of the applied current. Therefore, the mechanical angular speed of rotation in revolutions per second is equal to the electrical frequency in hertz:

$$f_e = f_m \quad \text{two poles} \tag{4.50a}$$

$$\omega_e = \omega_m \quad \text{two poles} \tag{4.50b}$$

where f_m and ω_m are the mechanical speed of rotation in revolutions per second and radians per second, respectively. Both f_e and ω_e are the electrical frequency (speed) in hertz and radians per second, respectively.

The windings on the two-pole stator shown in Figure 4.90 occur in the order (taken counter-clockwise)

$$a - c' - b - a' - c - b$$

If this pattern is repeated twice within the stator, the pattern of windings becomes

$$a - c' - b - a' - c - b' - a - c' - b - a' - c - b'$$

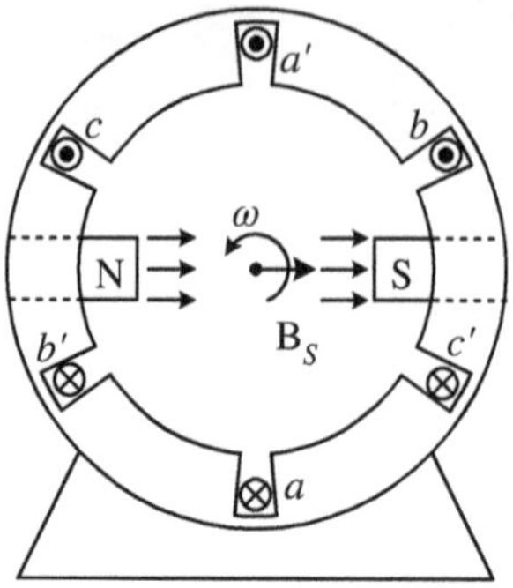

Figure 4.92 The rotating magnetic field in a stator represented as moving north and south stator poles [8].

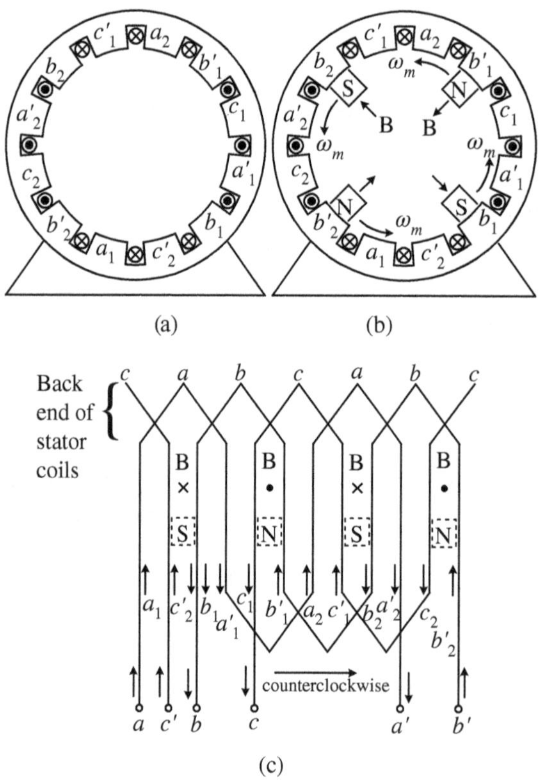

Figure 4.93 (a) A simple four-pole stator winding. (b) The resulting stator magnetic poles. Notice that there are moving poles of alternating polarity every 90° around the stator surface. (c) A winding diagram of the stator as seen from its inner surface, showing how the stator currents produce north and south magnetic poles [8].

Figure 4.93 illustrates the *two* north poles and *two* south poles that are produced in the stator when a three-phase set of currents is applied to the stator.

In this stator, the pole moves around half the stator surface in one electrical cycle. Since the mechanical motion is 180° for a complete electrical cycle (360°), the electrical angle θ_e is related to the mechanical angle θ_m by

$$\theta_e = 2\theta_m \text{ four poles} \tag{4.51a}$$

Therefore, for a four-pole stator, the electrical frequency is double the mechanical frequency of rotation

$$f_e = 2f_m \text{ four poles} \tag{4.51b}$$

$$\omega_e = 2\omega_m \text{ four poles} \tag{4.51c}$$

In general, if P is the number of magnetic poles on the stator, then there are $P/2$ repetitions of the windings. The electrical and mechanical quantities of the machine are related by

$$\theta_e = \frac{P}{2}\theta_m \tag{4.52a}$$

$$f_e = \frac{P}{2}f_m \tag{4.52b}$$

$$\omega_e = \frac{P}{2}\omega_m \tag{4.52c}$$

Since the mechanical frequency $f_m = \frac{n_m}{60}$, the electrical frequency in hertz is related to the mechanical speed of the magnetic fields in revolutions per minute by

$$f_e = \frac{P}{2}\left(\frac{n_m}{60}\right) = \frac{Pn_m}{120} \tag{4.53}$$

4.7.1.3 Reversing the Direction of the Magnetic Field Rotation

The direction of the magnetic field's rotation is reversed when the current in any two of three coils is swapped. Therefore, it is possible to reverse the direction of rotation of an AC motor by just switching any two of the three phases [9].

4.7.1.4 The Induced Voltage in AC Machines

Just as a rotating magnetic field can be produced by three-phase set of currents in a stator, a three-phase set of voltages in the coils of a stator can be produced by a rotating magnetic field.

4.7.1.5 The Induced Voltage in a Coil on a Two-Pole Stator

Figure 4.94 illustrates a *stationary* coil with a *rotating* magnetic field moving in its centre. The induced voltage in a wire is given by

$$e_{\text{ind}} = \left(\vec{v} \times \vec{B}\right) \cdot \vec{l} \tag{4.54}$$

where $\mathbf{v}$ = velocity of wire *relative to magnetic field*

$\mathbf{B}$ = magnetic flux density of field
$\boldsymbol{l}$ = length of wire

This equation was derived for a *wire moving within a stationary magnetic field*. In AC machines, the magnetic field is moving, and the wire is stationary.

Figure 4.95 illustrates the velocities and vector magnetic field from the point of view of a moving wire and a stationary magnetic field. The voltages induced in the sides of the coil are

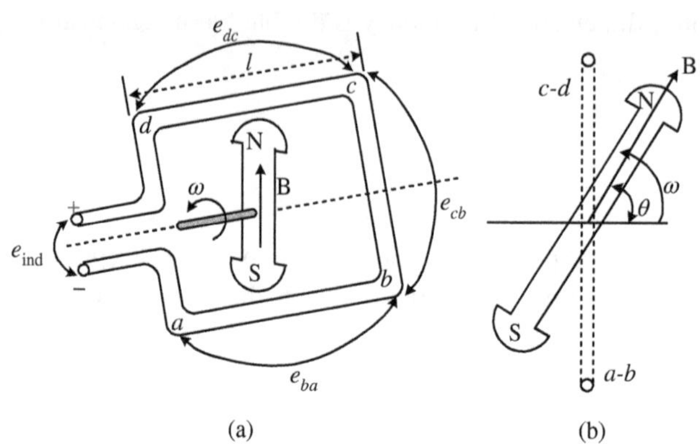

Figure 4.94 A rotating magnetic field inside a fixed coil: (a) Perspective view; (b) end view [8].

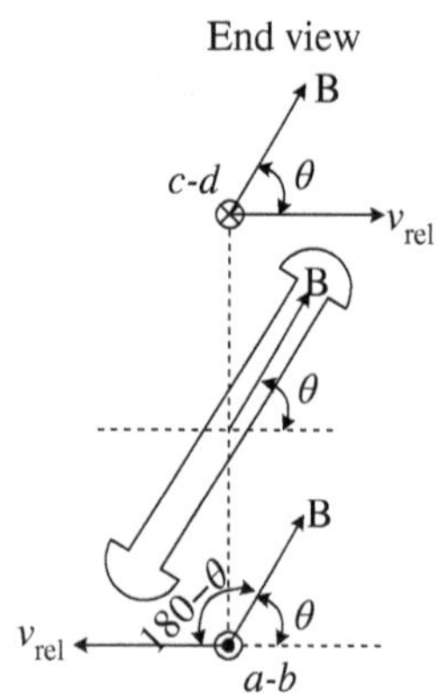

Figure 4.95 The magnetic fields and velocities of the coil sides as seen from a frame of reference in which the magnetic field is stationary [8].

1. *Segment ab*. The angle between **v** and **B** in segment *bc* is 180° – θ, while the quantity **v** × **B** is in the direction of ***l***, so

$$e_{\text{ind}} = \left(\vec{v} \times \vec{B}\right) \cdot \vec{l} = vBl \sin\left(180^o - \theta\right) \text{ directed } into \text{ page} \tag{4.55}$$

   The direction of e_{ba} is given by the right-hand rule. By trigonometric identity, Sin (180° – θ) = sin θ. So

$$e_{ba} = vBl \sin\theta \tag{4.56}$$

2. *Segment bc*. The voltage on segment *bc*is zero, since the vector quantity **v** × B is perpendicular to ***l***.

$$e_{cb} = \left(\vec{v} \times \vec{B}\right) \cdot \vec{l} = 0 \tag{4.57}$$

3. *Segment cd*. The angle between υ and *B* in segment *cd* is θ, while the quantity **v** × **B** is in the direction of ***l***. So

$$e_{dc} = vBl \sin\theta \text{ directed } out\ of \text{ page} \tag{4.58}$$

4. *Segment da*. The voltage on segment *da* is zero, for the same reason as in segment *bc*:

$$e_{ad} = \left(\vec{v} \times \vec{B}\right) \cdot \vec{l} = 0 \tag{4.59}$$

The total voltage induced within a single turn coil is given by

$$e_{\text{ind}} = 2vBl\sin\theta \tag{4.60}$$

Since angle $\theta = \omega_e t$, the induced voltage can be rewritten as

$$e_{\text{ind}} = 2vBl\sin\omega_e t \tag{4.61}$$

Since the cross-sectional area A of the turn is $2rl$ and the velocity of the end conductors is given by $\upsilon = r\omega_m$, the equation can be rewritten as

$$e_{\text{ind}} = 2\left(r\omega_m\right)Bl\sin\omega_e t = (2rl)\,B\omega_m\sin\omega_e t = AB\omega_m\sin\omega_e t \tag{4.62}$$

The maximum flux passing through the coil is $\phi = AB$. For a two-pole stator $\omega_m = \omega_e = \omega$, the induced voltage is

$$e_{\text{ind}} = \phi\omega\sin\omega t \tag{4.63}$$

This equation describes the voltage induced in a single-turn coil; if the coil (phase) has N_c turns of wire in it, the total induced voltage will be

$$e_{\text{ind}} = N_c\phi\omega\sin\omega t \tag{4.64}$$

4.7.1.6 The Induced Voltage in a Three-Phase Set of Coils

Figure 4.96 illustrates three coils each of N_c turns placed around the rotor magnetic field. The voltage induced in each has the same magnitude but differs in phase by 120°. The resulting voltages in the three phases are

$$e_{aa'}(t) = N_c\phi\omega\sin\omega t\ \text{V} \tag{4.65a}$$

$$e_{bb'}(t) = N_c\phi\omega\sin\left(\omega t - 120^o\right)\text{V} \tag{4.65b}$$

$$e_{cc'}(t) = N_c\phi\omega\sin\left(\omega t - 240^o\right)\text{V} \tag{4.65c}$$

Therefore, a set of three-phase currents generates a rotating uniform magnetic field within the stator of the machine, and a uniform magnetic field induces a set of three-phase voltages in such a stator.

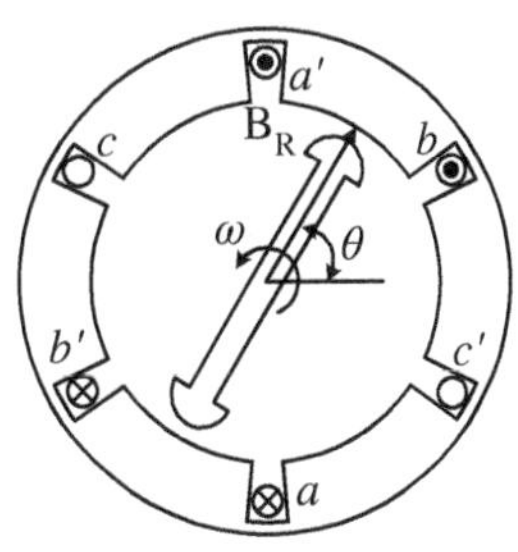

Figure 4.96 The production of three-phase voltages from three coils spaced 120° apart. [8].

4.7.1.7 The rms Voltage in a Three-Phase Stator

The peak voltage in any phase is

$$E_{max} = N_c \phi \omega \tag{4.66}$$

Since $\omega = 2\pi f$, the rms voltage in any phase is

$$E_{rms} = E_A = \frac{2\pi}{\sqrt{2}} N_c \phi f = E = \sqrt{2}\pi N_c \phi f = 4.44288 N_c \phi f \tag{4.67}$$

The rms voltage at the terminals of the machine depends on whether the stator is Y- or Δ- connected. If the machine is Y-connected, the terminal voltage is 3 times E_A. In Δ-connected machines, the terminal voltage is the same as E_A.

4.7.2 The Induced Torque in an AC Machine

During normal operation of AC machines (motors and generators), there are two magnetic fields: a magnetic field from the rotor and another from the stator. A torque is induced in the machine due to the interaction of the two magnetic fields.

A synchronous machine is illustrated in Figure 4.97. A magnetic flux density $\mathbf{B}_R$ is produced by the rotor, and a magnetic flux density $\mathbf{B}_S$ is produced by the stator. The induced torque in a machine (motors and generators) is given by

$$\vec{\tau}_{\text{ind}} = k\vec{B}_R \times \vec{B}_S \tag{4.68a}$$

$$\tau_{\text{ind}} = kB_R B_S \sin(\gamma) \tag{4.68b}$$

where τ_{ind} = induced torque in machine

$\mathbf{B}_R$ = rotor flux density
$\mathbf{B}_S$ = stator flux density
γ = angle between $\mathbf{B}_R$ and $\mathbf{B}_S$

The net magnetic field in the machine is the vector sum of the fields from the stator and rotor

$$\vec{B}_{\text{net}} = \vec{B}_R + \vec{B}_S \tag{4.69}$$

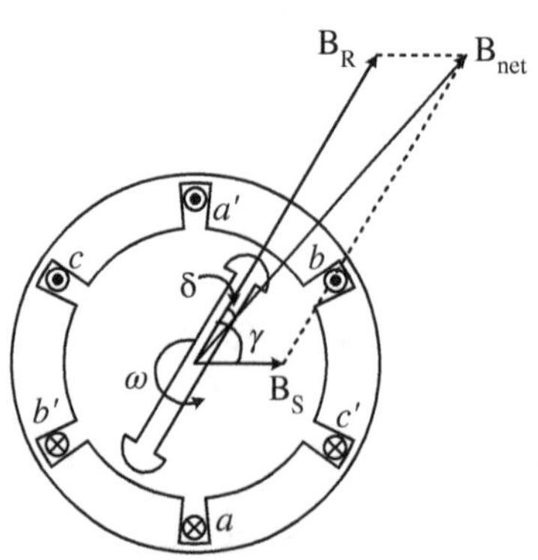

Figure 4.97 A simplified synchronous machine showing its rotor and stator magnetic fields [8].

The induced torque can be expressed as

$$\vec{\tau}_{\text{ind}} = k\vec{B}_R \times \vec{B}_{\text{net}} \tag{4.70}$$

The magnitude of the torque is

$$\tau_{\text{ind}} = kB_R B_{\text{net}} \sin(\delta) \tag{4.71}$$

The magnetic fields of the synchronous machine shown in Figure 4.97, are rotating in a counter-clockwise direction. What is the direction of the induced torque on the rotor of the machine?

By applying the right-hand rule to the equation of the induced torque, we see that the induced torque is clockwise. It is opposing the direction of rotation of the rotor. Therefore, this machine is working as a generator [8].

4.8 Summary

This chapter is dedicated to the development of the fundamental concept of rotating electric machines. The basic operating principle of electric generator and motor is well explained. The operating principle is explained with the help of fundamental laws of electrical engineering. Fundamental equations are derived that will help readers to follow the machine working in subsequent chapters. mmf waveform, produced by concentrated and distributed winding, is elaborated that help to understand the behaviour of machines. Detailed DC machine winding is given in this chapter with numerous solved examples. AC winding is explained in Chapter 9. As such, both AC and DC windings are covered in this book.

Problems

4.1 Develop the duplex double layer parallel winding diagram for a DC machine with 16 slots and four brushes.

4.2 Develop the duplex double layer lap winding diagram for a DC machine with 13 slots, four brushes and $Y_b < Y_f$.

4.3 Develop the duplex double layer lap winding table for a DC machine. The number of conductors in the machine is 64 and pole is four.

4.4 Develop the triplex four-layer lap winding table for a DC machine. The number of conductors in the machine is 128 and pole is four.

4.5 A six-pole DC armature with lap connected simplex double layer winding has 144 coil sides. Determine the winding pitches and connections to the six equalizing rings.

4.6 A six-pole DC machine has 18 armature slots. Develop a double layer lap winding diagram for a DC machine and mark the poles. Draw a sequence diagram and indicate the position of the brushes and also show the direction of induced emf.

References

1 Mehta, V.K. and Mehta, R. (2006). *Principle of Electrical Machines*. S Chand; Reprint Edn. 2006 edition (1 December 2014).

2 Tong, W. (2014). *Mechanical Design of Electric Motors*. CRC Press.

3 Braymer, D.H. (1920). *Armature Winding and Motor Repair*, 1e. Mcgraw-Hill Book Company, Inc.

4 Sahdev, S.K. (2018). *Electrical Machines*. Cambridge University Press.

5 Adkins, B. and Harley, R.G. (1975). *The General Theory of Alternating Current Machines: Application to Practical Problems*. Springer-Science+Business Media, B.V.

6 Nagrath and Kothari (2004). *Electrical Machines*, 4e. New Delhi: Tata McGraw Hill Education Private Limited.

7 Gray, A. (1913). *Electrical Machine Design*, 1e, eighth imprint. Mcgraw-Hill Book Company, Inc.

8 Kiameh, P. (2003). *Electrical Equipment Handbook - Troubleshooting and Maintenance*, 1e. McGraw-Hill Professional.

9 Chapman, S.J. (1991). *Electric Machinery Fundamentals*, 2e. New York: McGraw-Hill.

5

DC Machines

5.1 Preliminary Remarks

DC Machines are an electro-mechanical energy conversion device that uses DC power as input and/or DC output. A DC machine is a general term that refers to both the DC motor and DC generator. A DC generator takes DC power as input and generates DC power as output. The DC motor takes in DC power and outputs mechanical power. DC machine construction has two parts, a stator and a rotor separated by an air gap. Any rotating machine has two kinds of magnetic field in order to produce rotation. One is called the main field and the other is called the armature field. Main field is on the stator and armature is on the rotor in a DC machine. In case of a DC motor, DC supply is given to both main field and armature. While in a DC generator, DC power is supplied to the main field and the usable electrical power is collected from the armature winding and supplied to DC electrical loads. Hence, the DC electrical load is connected to the armature of the DC generator. The stator has projected poles where concentrated windings are wound and DC power supply is given to produce main field. The armature winding is DC that is distributed and placed on rotor. The armature and main field are either supplied from two separate DC power supplies called separately excited DC machine or the windings are connected together, called self-excited DC machine and supplied by a single DC power supply. When the field and armature windings are connected in series, it is called DC series Machine. When the field and armature windings are connected in parallel, it is called DC shunt Machine. When the DC machine has both series and parallel (shunt) field winding, they are called DC compound machines. If the stator contains a permanent magnet, it is called a DC permanent Magnet Machine. DC machines has an integral part called a 'commutator' (split slip ring structure made of copper strips) that is used as a mechanical rectifier to convert the internally generated AC to output DC in a DC generator. In a DC motor, the commutator makes the torque unidirectional by changing the direction of current. The DC machine is considered a simple machine with flexible and easy control characteristics. DC machines are used in a variety of applications. DC motors are mostly used in applications such as Traction, Lathe Machine, Centrifugal Pumps, Cranes, Compressors, Vacuum Cleaner, Weaving Machine, Spinning Machine, Lifts, Conveyors, Pumps, Fans, etc. DC generators are rarely used in modern times, because Power Electronic Converters are readily available that efficiently and effectively convert the AC power to DC Power. However, there exists a small

Electrical Machine Fundamentals with Numerical Simulation using MATLAB/SIMULINK, First Edition.
Atif Iqbal, Shaikh Moinoddin, and Bhimireddy Prathap Reddy.

Companion website: www.wiley.com/go/iqbal/electricalmachine

ripple in the DC voltage obtained using Power Electronic Converters. This chapter gives a fundamental understanding of DC machines. Construction, operation and control are discussed in the chapter. Some numerical examples are provided to make understanding more clear.

5.2 Construction and Types of DC Generator

Figure 5.1 shows the structure of a basic four-pole DC machine. Two main components of a DC machine are stator and rotor. Stator is the stationary part and rotor is the rotating part.

5.2.1 Construction of DC Machine

Below are the basic construction parts of the DC machine.

i. **Yoke**: A DC-machine's external frame is called a yoke. The material of the yoke is with iron or steel cast iron. It not only provides the entire assembly with mechanical support, but also carries the magnetic flux generated by the field winding.
ii. **Pole shoes and poles**: Poles are connected by bolts or welding to the yokes. They are winding on the field and have pole shoes tied to them. Pole shoes serve two purposes, i.e. (a) support field coils and (b) evenly distribute the air gap current.
iii. **Winding in the field**: The winding of the field is typically made of copper. The field coils are formerly wound and mounted on each pole in a row. They are wound in such a way that they form alternating North and South poles when energized.
iv. **Armature core**: An Armature core is a DC rotor. It is cylindrical with slots to bear the winding of the armature. The armature is constructed of thin laminated circular steel disks to reduce the loss. The air ducts can be supplied for axial air flow for cooling purposes. Armature is connected to the shaft of the rotor.
v. **Armature winding**: Usually, the armature winding is a wound copper spiral that rests in armature slots. The armature conductors are insulated both from each other and from the reinforcement core. Winding of the armature can be done by either of the two

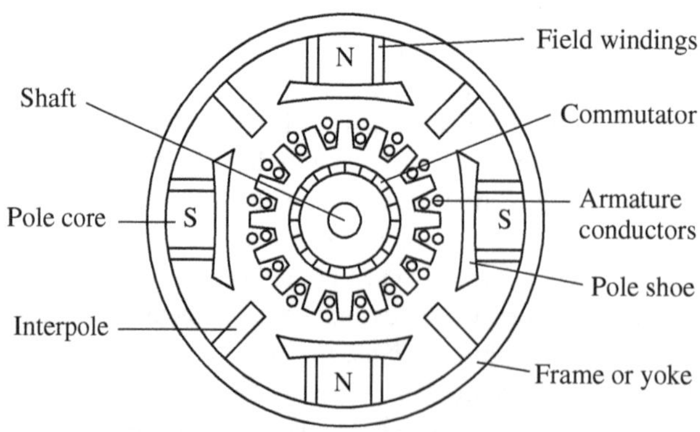

Figure 5.1 Structure of the DC machine.

methods, i.e. lap winding or wave winding. Usually, double layer or wave windings are used. Winding a double layer means that every frame slot has two different spindles.

vi. **Commutator and brushes**: The physical relation to the winding of the armature is rendered by a change arrangement. In a DC generator, the function of the commutator is to collect the current generated in the armature conductors. For the case of a DC motor, the commutator helps to provide energy to the armature conductors. A commutator consists of a set of isolated copper segments separated from each other. Each segment is connected to an armature coil. Brushes are usually made from graphite or carbon. They rest on the commutator segments and slide on the segments when the commutator turns and keeps contact with the current. So, the commutator and brush unit of the DC motor is concerned with transmitting the power from the static electrical circuit to the mechanically rotating region or the rotor.

5.2.2 Types of DC Generator

Based on the placement of the filed winding, the DC generators can be categorized as

(i) Separately excited DC generator
(ii) Self-excited DC generator
  (a) Shunt DC generator
  (b) Series DC generator
  (c) Compound DC generators
    - Long-shunt compound DC generator
    - Short-shunt compound DC generator

In a separately excited generator field, winding is energized from a separate DC voltage source to generate main magnetic field. The flux produced is proportional to field current as long as the machine operates in unsaturated condition. The schematic diagram of the separately excited DC generator is shown in Figure 5.2a. In the category of self-excited DC generators, the external DC source for the field winding is omitted. One category of the self-excited DC generator is the series winding connected DC generator. The field winding is intended to be connected in series and naturally designed for the rated armature current. The winding of the series field consists of a few turns and a large cross-sectional area (since the series field carries a large armature current, the resistance should be low). Obviously, due to a residual field, there will be virtually no voltage or very small voltage under no load conditions ($I_a = 0$). However, the field is strengthened because the load develops rated voltage across the armature, and the voltage is increasing. Series DC generators are not used to delivering continuous power as the field current in this type of generator depends on the armature current. The series generator found application in the DC transmission system to boost voltage. The configuration of the series generator is shown in Figure 5.2b. Another category of the self-excited DC generator is shunt DC generator in which the field winding is in parallel to the armature circuit. The shunt DC generator is shown in Figure 5.2c. As the field winding is in parallel to the armature circuit, the generated voltage appears across it. The shunt field winding consists of a coil with a large number of turns and a smaller cross-sectional area.

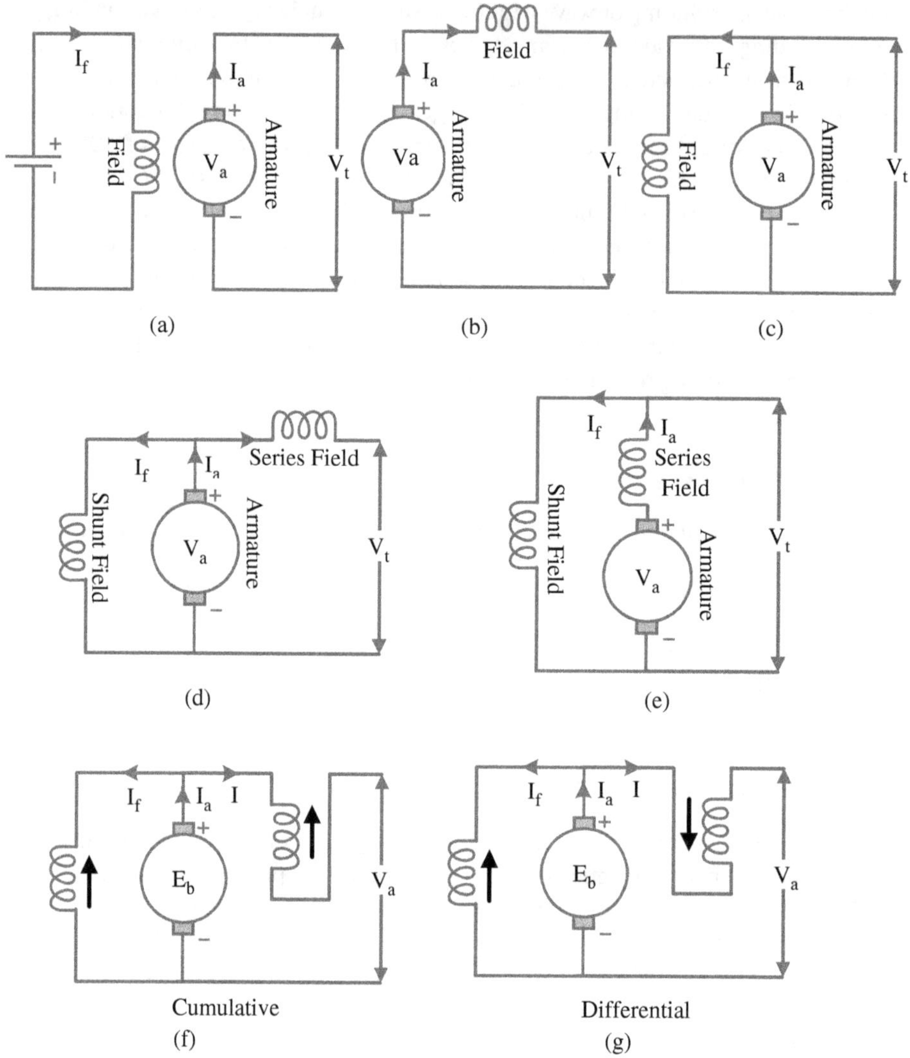

Figure 5.2 Classification of DC generators.

In the compound type of DC generator, two independent field coils are wound on the field poles in a compound generator and act as series and shunt-field winding. When the magnetic flux produced by the series winding supports the flux produced by the shunt winding, the compound generator is termed as cumulatively compound as shown in Figure 5.2f. The cumulative compounded generators may be over compounded, flat compounded and under compounded. We can obtain the desired terminal voltage by compensating the drop due to the armature reaction and ohmic drop in the line. Cumulative compound generators are generally used for lighting, power supply purpose and for heavy power services because of their constant voltage property. When the series field flux opposes the shunt-field flux, then the generator works as differentially compounded system as shown in Figure 5.2g. This differential process of flux has a net effect of producing a much lower terminal electromotive force (emf) than any generator of comparable size or number of poles and windings.

Both types of compound generators, i.e. cumulative compound and differential compound can be constructed in two ways. One is a generator of short-shunt compound in which the series winding is connected to the combination of shunt generator, and another a generator of long-shunt compound, in which the series generator is made parallel through a shunt winding. The long- and short-shunt generators are shown in Figures 5.2d,e, respectively.

5.3 Principle of Operation of DC Generator

Under Faraday's electromagnetic induction rules, when a conductor is put in a particular magnetic field (OR a conductor is moved in a magnetic field), the conductor produces an emf. The magnitude of emf induced can be determined from the DC generator emf equation given in Eq. (5.3). The induced current flows through the conductor if there is a closed path. In a DC generator, field coils produce an emf and the armature conductors are rotated into the field. The armature conductors placed in rotor are rotated by means of an external prime mover. Thus, an electromagnetically induced emf is generated in the armature conductors. Fleming's right-hand rule provides the direction of the induced current. According to Fleming's right-hand rule as shown in Figure 5.4, the direction of induced current changes whenever the direction of motion of the conductor changes. When the armature-coil ABCD is rotated clockwise, an emf is induced in the coil and a current flows in the direction ABCD (Fleming's right-hand rule), as shown in Figure 5.4. In the external circuit, the current flows from XX to X. For half the revolution, U is in contact with XX and W with X as shown in Figure 5.3a. But as soon as the coil passes the vertical, W comes in contact with XX and U with X and remains so during the next half revolution (Figure 5.3b). Although the induced emf in the coil is reversed and the current in the coil flows in the direction DCBA, but, in the external circuit, the current still flows from XX to X, and therefore, a unidirectional current is obtained at the output terminals.

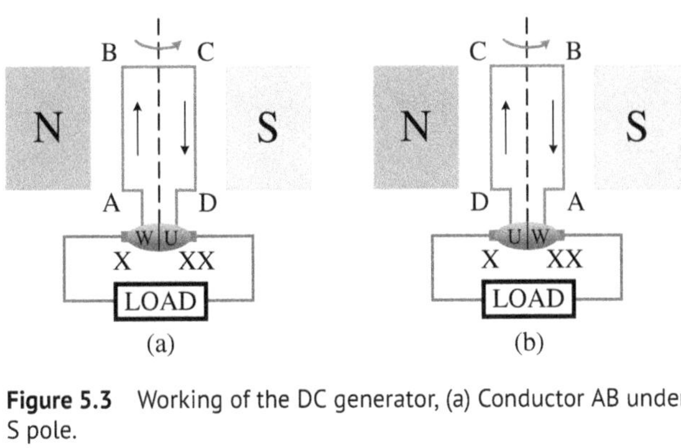

Figure 5.3 Working of the DC generator, (a) Conductor AB under N pole, (b) Conductor AB under S pole.

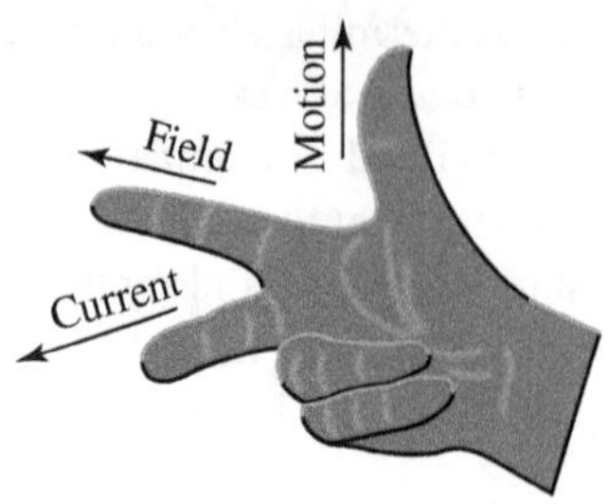

Figure 5.4 Fleming's right-hand rule.

5.3.1 Voltage Build-Up in a DC Generator

When a DC shunt generator is powered at a constant speed without closing its shunt circuit, a very low residual voltage (1 to 5 V) occurs at its terminals due to the residual magnetism inherent in the iron. When the field circuit is properly connected, the residual voltage pushes a small exciting current through the field circuit and thus increases the field strength. The voltage produced increases due to the increased field strength. This process will be repeating until the actual terminal voltage is reached. Once the terminal voltage is reached, then the winding will be saturated and hence there will not be any further increase in flux also, the voltage gets constant. Many modern generators are constructed in such a way that the voltage can rise to about 125% of the rated value without any resistance in series with the field. The condition is defined in Figure 5.5 by point d. The straight-line Od is called a field resistance line, since its slope V / I_f is equal to R_f, the field circuit resistance including the field rheostat. When the voltage has built up, it can be changed to any desired value between points 'd' and 'e' by increasing the field circuit's resistance through its rheostat. Generators are typically designed to produce the average voltage at point b, or at a slightly higher field current. The resistance line will have a different slope for any value of the field-circuit resistance, such as 'Ob' or 'Oc'.

The voltage build-up of the generator will fail if the resistance is so high that the slope of the resistance line is equal to or greater than the right portion of the magnetization curve and will drop to the residual value '*oO*' as shown in Figure 5.5.

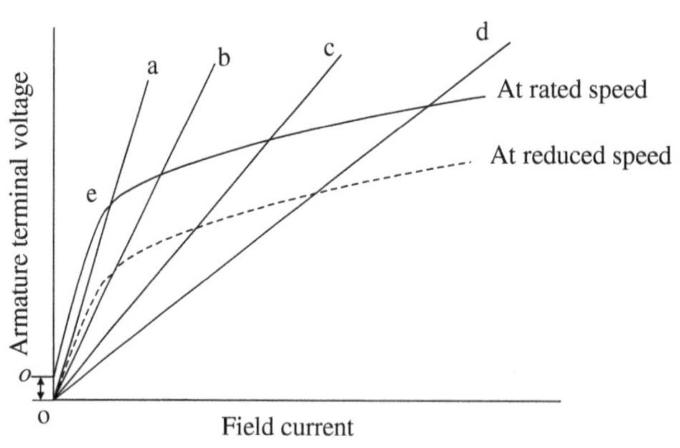

Figure 5.5 Generator self-excitation curve.

For any of the following reasons a generator may fail to build up:

1. Too high a resistance to field-circuiting; including open circuit, which will reduce the field current to a very small value.
2. Low speed.
3. Loss of the remaining magnetism.
4. Incorrect rotation direction.
5. Connected generator terminals with too low resistance to external circuit.
6. Reversed Shunt-field terminals.

If the generator speed is decreased, the magnetization curve will decrease proportionately in height as shown by the dashed curve in Figure 5.5. Whereas when the field resistance line is 'Ob', the generator will build up at the rated speed; this line is too steep to allow build-up at the reduced speed. If the direction of rotation is wrong, the residual voltage would push the current in a direction through the field coils to demagnetize the field, so there can be no build-up. If the generator is connected to a load circuit with too low a resistance, by its ++magnetizing action in the armature, the load-circuit current caused by residual voltage can prevent build-up. The residual magnetism can be restored by connecting the field circuit to any suitable direct current source.

5.3.2 Function of Commutator

The commutator is simply a mechanical rectifier and is shown in Figure 5.6. The voltage generated in an armature of a DC generator, placed in a rotating magnetic, is alternating in nature. The commutation on the DC machine, or more specifically, the commutation in the DC generator, is the process where the generated alternating current in the winding is converted into a direct current of the DC machine. This change of current from the rotating armature of a DC machine to stationary brushes requires continuous contact between the commutator segments and the brushes. When the frame is turning, the bobbins underneath one pole (let it be N pole) are rotated between a positive brush and the consecutive negative brush and the current runs through the commutator. The coil is then shortened to a very short span of time with the help of a brush. It is called the commutating time. As depicted and explained through Figure 5.3a,b, this reverse current phenomenon is called the process of commutation. From the brush terminal, a direct current is obtained. Figure 5.7a–c illustrates the change of brush segment. For that flow of current in coil *A*, three different instants are shown. For a satisfactory commutation, the direction of flow of current in coil *A* must completely reversed as shown in Figure 5.7a,c, by the time the brush moves from segment 2 to segment 3. As in Figure 5.7a, coil *A* carries a current of I/2 and it flows in the direction of motion. As the brush position is changed from 2 to 3, as shown in Figure 5.7b and the current through coil *A* is zero and the current is supplied by other coils adjacent to both sides of coil *A*. As shown in Figure 5.7c, now the brush position is at segment 3 and the direction of the current to coil *A* is reversed. Figure 5.7 shows the current-time graph for coil A undergoing the commutation. The horizontal line AB represents a constant current of I/2 up to the beginning of commutation. From the finish of commutation, it is represented by another horizontal line CD on the opposite side of the zero line and the same distance from it as AB i.e. the current has reversed exactly (−I/2). The way in which current changes from

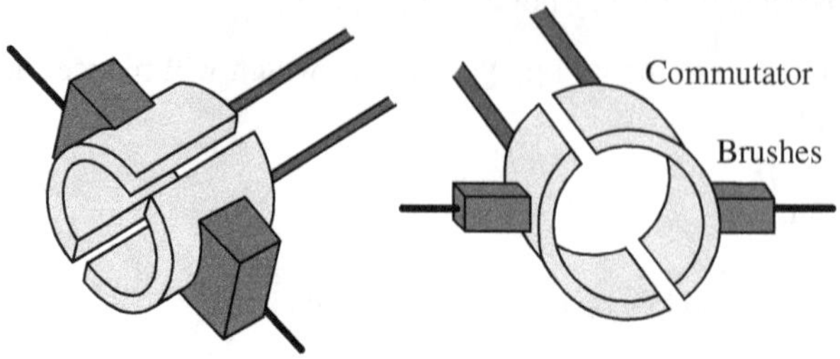

(a)

(b)

Figure 5.6 (a) and (b) Commutator.

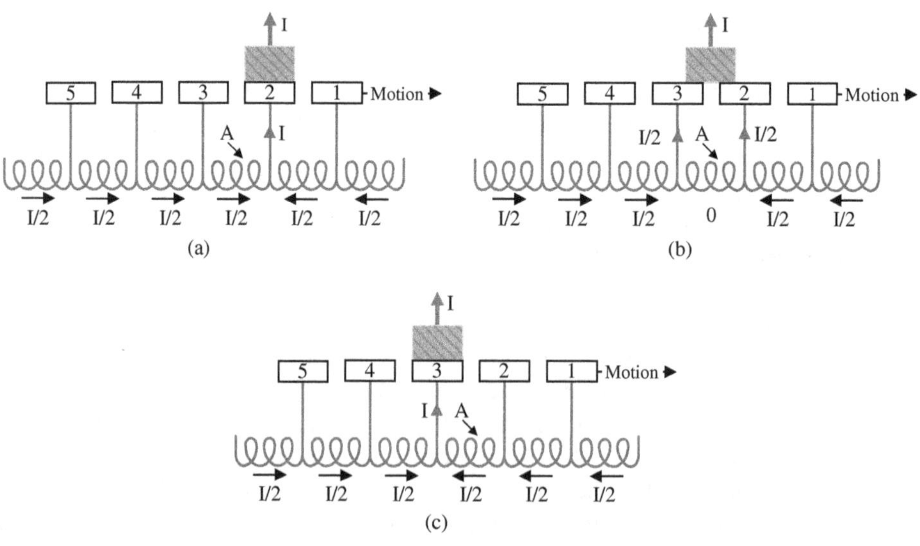

Figure 5.7 Commutation process.

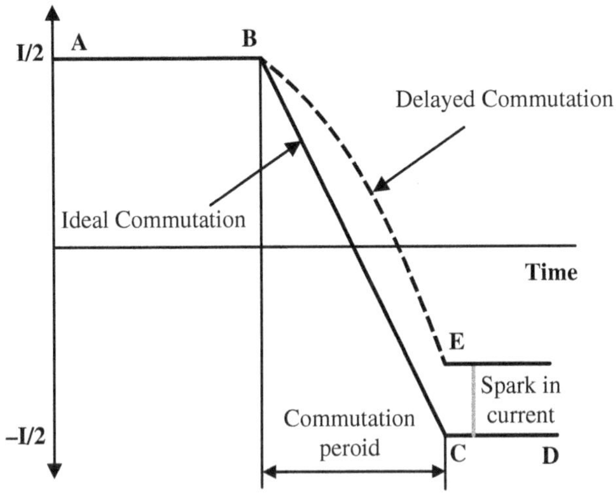

Figure 5.8 Type of commutation.

B to C depends upon the conditions under which the coil undergoes commutation. If the current changes at a uniform rate (i.e. BC is a straight line), then it is called ideal commutation as shown in Figure 5.7. Under such conditions, no sparking will take place between the brush and the commutator contact surface.

The ideal commutation (i.e. straight-line change of current) cannot be attained in practice. This is mainly because the armature coils have appreciable inductance. When the current in the coil undergoing commutation changes, self-induced emf is produced in the coil. This is generally called reactance voltage. This reactance voltage opposes the change of current in the coil undergoing commutation. The result is the change of current in the coil undergoing commutation. This is illustrated in Figure 5.8. The straight-line BC represents the ideal commutation whereas the curve BE represents the change in current when self-inductance of the coil is taken into account. This results in sparking similar to when any other current-carrying circuit is broken. The sparking results in overheating of commutator-brush contact and causes damage to both.

5.4 Commutation Problem and Solution

To make the commutation satisfactory, one must ensure that the current that flows through the coil completely reversing during the commutation period reaches its full value. When commutation takes place, the coil undergoing commutation is short-circuited by the brush. The brief period when the coil remains short-circuited is known as the commutation period. Three principal ways to improve commutation are available and discussed below.

5.4.1 Brush Shifting

When the armature of a DC machine carries current, the magnetic and mechanical neutrals do not coincide. Sparking will occur at the brushes unless they occupy positions that short circuit coil sides in the neutral zone. Therefore, the brushes, must be shifted to the magnetic neutral. The effect of the armature reaction depends upon the value of armature current. The brushes, therefore, must be shifted back or forth continuously as the value of the load current changes or the brushes must be located in a compromise position that represents the best average load.

5.4.2 Commutating Poles

The limitation of brush shift has led to the use of inter poles in almost all the medium- and large-sized DC machines. Inter poles are long but narrow poles placed in the inter polar axis. They have the polarity of succeeding pole (coming next in sequence of rotation) in generator action and proceeding (which has passed behind in rotation sequence) pole in motor action. The inter pole is designed to neutralize the armature reaction mmf in the inter polar axis. This is because the direction of armature reaction mmf is in the inter polar axis. It also provides commutation voltage for the coil undergoing commutation such that the commutation voltage completely neutralizes the reactance voltage ($L\,{}^{di}/_{dt}$). Thus, no sparking takes place. Inter polar windings are always kept in series with armature, so inter polar winding carries the armature current; therefore, it works satisfactorily irrespective of load, the direction of rotation or the mode of operation. Inter poles are made narrower to ensure that they influence only the coil undergoing commutation and its effect does not spread to the other coils. The base of the inter poles is made wider to avoid saturation and to improve response.

5.4.3 Compensating Windings

At heavy loads, the cross-magnetizing armature reaction may cause very high flux density in the trailing pole tip in generator action and the leading pole tip in the motor action. Consequently, the coil under this tip may develop induced voltage high enough to cause a flashover between the associated adjacent commutator segments particularly, because this coil is physically close to the commutation zone (at the brushes) where the air temperature might already be high due to the commutation process. This flashover may spread to the neighbouring commutator segments, ultimately leading to a complete fire over the commutator surface from brush to brush. Also, when the machine is subjected to rapidly fluctuating loads, then the voltage $L\,{}^{di}/_{dt}$, that appears across the adjacent commutator segments may reach a value high enough to cause flashover between the adjacent commutator segments. This would start from the centre of the pole as the coil below it possesses the maximum inductance. This may again cause a similar fire as described above. This problem is more acute while the load is decreasing in generating action and increasing in motor action as then, the induced emf and voltage $L\,{}^{di}/_{dt}$ will support each other. The above problems are solved by use of compensating winding. Compensating winding as shown in Figure 5.9, consists of conductors embedded in the pole face that run parallel to the shaft and carry an armature current in a direction opposite to the direction of current in the armature conductors under that pole arc. With complete compensation the main field is restored. This also reduces armature circuit's inductor and improves system response. Compensating winding functions satisfactorily irrespective of the load, direction of rotation and mode of operation. Obviously, it is of help in commutation as the inter polar winding is relieved from its duty to compensate for the armature mmf under the pole arc.

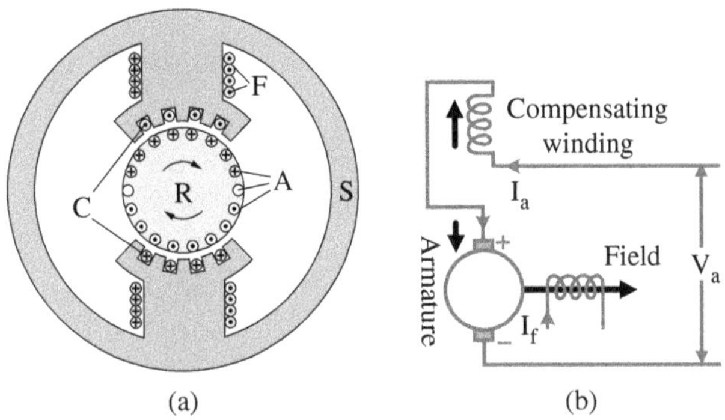

Figure 5.9 Cross-section of DC machine with compensation windings. A = armature windings, C = compensation windings, F = field windings, R = rotor, S = stator.

5.5 Types of Windings

Armature windings are mainly of two types – lap winding and wave winding. Lap winding is the winding in which successive coils overlap each other. It is named 'Lap' winding because it doubles or laps back with its succeeding coils. In this winding, the finishing end of one coil is connected to one commutator segment and the starting end of the next coil situated under the same pole and connected with the same commutator segment.

Here, it is seen from Figure 5.10a,b. The finishing end of coil 1 and starting end of coil 2 are both connected to the commutator segment 2 and both coils are under the same magnetic pole that, here, is N pole. In lap type winding, if the connection is started from slot 1, then connections overlap each other as winding proceeds till the starting point is reached again. Due to such a connection, the number of the parallel paths is equal to the number of poles of the machine. A large number of parallel paths indicate high current carrying capacity of the machine; hence, lap winding is preferred for high current rating generators.

Wave winding is another type of armature winding. In this winding, the end of one coil is connected to the starting of another coil of the same polarity as shown in Figure 5.11. In this type of winding, the coil side progresses forwards around the armature to another coil side and goes on successively passing through N and S pole till it returns to a conductor lying under the starting pole. This winding forms a wave with its coil, hence the name wave winding, and it also travels like a progressive wave. Since we connect the coils in series here, we also call it series winding. This type of winding always travels ahead avoiding overlapping. With this type of winding, the total number of conductors are divided into two parallel paths, irrespective of the number of poles. As the number of parallel paths is less, it is preferable for low current, high voltage generators.

5.6 emf Equations in a DC Generator

Take a DC generator whose armature coil is excited to produce an air gap flux density distribution and the armature has a prime mover at a constant speed. Let us assume that:

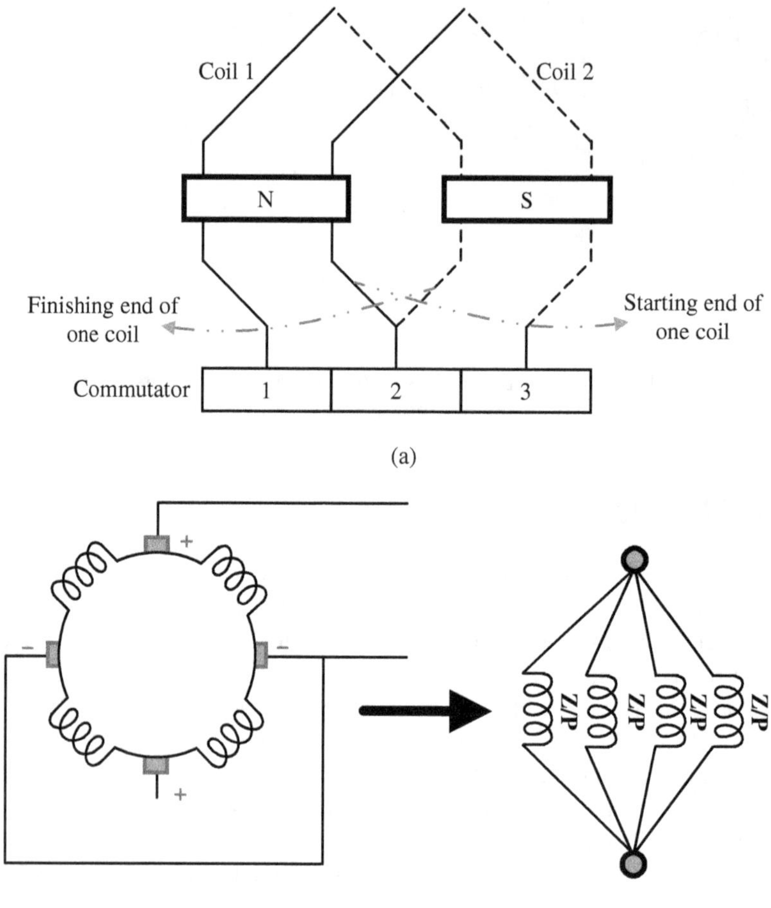

(b)

Figure 5.10 (a) and (b) Lap winding.

P = Number of poles of DC generator
N = Speed of primary mover in rpm
Z = total number of conductors in the armature = Number of slots × number of conductor per slot
A = Number of parallel paths
Φ = Flux produced by each pole in weber (W_b)
Total flux produced by all the poles = $\phi \times P$
Time taken for one revolution = $\frac{60}{N}$

Therefore, according to Faraday's law of electromagnetic induction, the induced emf is given as

$$e = \frac{d\Phi}{dt} = \frac{total\,flux}{time} \tag{5.1}$$

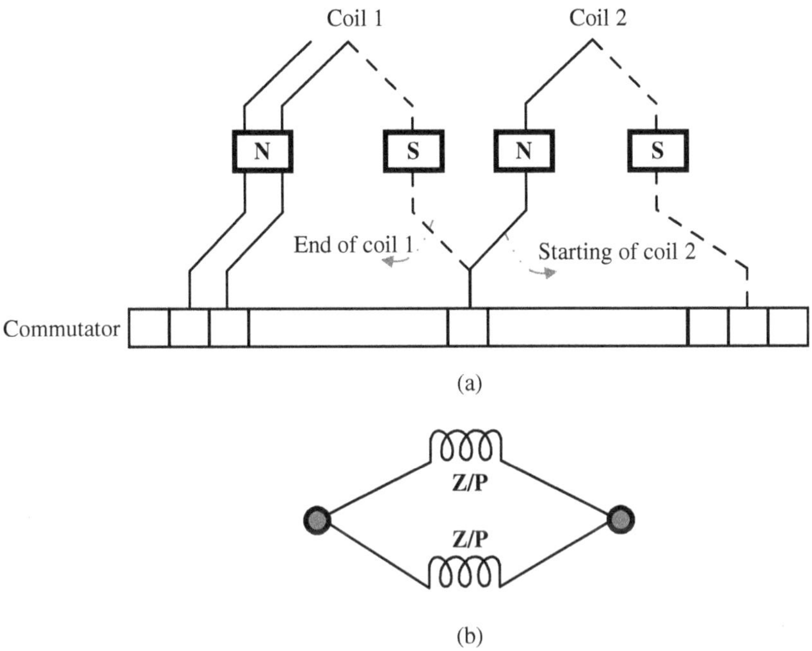

Figure 5.11 Wave winding.

The induced emf due to one conductor is given as

$$e = \frac{\Phi \times P}{{}^{60}\!/_{N}} = \frac{NP\Phi}{60} \tag{5.2}$$

As the number of conductors per path is Z/A, the induced emf of the generator is given as

$$e = \frac{NP\Phi}{60} \times \frac{Z}{A} = \frac{NP\Phi Z}{60A} \tag{5.3}$$

$$\Rightarrow N = \frac{60\omega_m}{2\pi}$$

$$E = \frac{NP\Phi Z}{60A} = \frac{P\Phi Z}{60A} X \frac{60\omega_m}{2\pi} = k\Phi\omega_m$$

$$k = \frac{PZ}{2\pi A} \tag{5.4}$$

A = 2 for Wave winding and
A = P (number of poles) for Lap Winding

5.7 Brush Placement in a DC Machine

Each time commutator segment is passed through the brush, the coil is short-circuited, so the brushes should be placed along the point where the emf induced in the armature is minimum, and that point is along the Magnetic Neutral Axis or MNA or q-axis. MNA may be defined as the axis along which no emf is generated in the armature conductors as they move parallel to the flux lines. Geometrical Neutral Axis (GNA) may be defined as the axis which is perpendicular to the stator field axis. In older DC machines, the brushes were mounted on an adjustable ring to change the position of the brushing. It was usually a 'hit and miss' test method for the location of the brush. The brush position in the generator affects terminal voltage and therefore torque in the motor. It affects switching in both the generator and motor. Brushes are always placed along the MNA because the reversal of current in the armature conductors takes place along this axis. Although in newer DC machines, the brushes are attached to the geometric neutral axis of the machine. Inter poles are used to achieve good switching, terminal voltage, and torque. Inter poles compensate for the armature reactions and improve the commutation.

5.8 Equivalent Circuit of DC Generator

The DC generator armature can be represented by an electric circuit equivalent. Three series-connected elements E_b, R_a, and V_b can be represented. Element E_b is the generated voltage, R_a is the armature resistance, and V_b is the brush contact voltage drop (brush has resistance and hence a small voltage drop of 1–2 V takes place across the brush). The equivalent circuit of the armature of a DC generator is shown in Figure 5.12.

Based on the equivalent equation of the DC generator, the terminal voltage is calculated by applying KVL.

$$V_t = E_b - I_a R_a - V_b \tag{5.5}$$

As the brush drop is of very low magnitude, the terminal voltage is modified as:

$$V_t = E_b - I_a R_a \tag{5.6}$$

5.9 Losses of DC Generator

A DC generator transforms mechanical power into electrical power, and a DC motor transforms mechanical power. Therefore, the input power for a DC generator is mechanical and the output power is electric. On the other hand, input power for a DC motor is in the form of electrical power, and output power is mechanical. In a practical computer, the entire input power cannot be transformed to output power, because some power in the conversion cycle is lost. This leads to reduced machine efficiency. The output is the ratio of output power

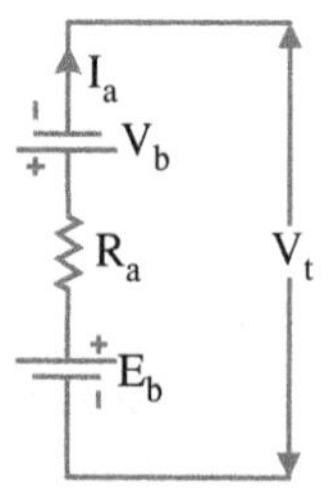

Figure 5.12 Equivalent circuit of the DC generator.

and input power. Therefore, it is important to research the losses in rotating DC machines with greater efficiency.

Copper losses are winding losses that occur during the current flow through the winding. Those losses are due to the winding resistance. Copper losses occur in the armature and field resistance. Thus, copper losses categorize into three parts; loss of armature, loss of winding field and loss of brush contact resistance. The losses of copper are proportional to the current square which flows through the winding.

$$P_{arm\_losses} = I_a^2 R_a$$
$$P_{field\_losses} = I_f^2 R_f \tag{5.7}$$

where P_{arm_losses} and P_{field_losses} are the losses of the armature and field resistance respectively.

As the core of the armature is made of iron and rotates in a magnetic field, a small current is also induced in the core itself. Due to this current, eddy current loss and hysteresis loss occur in the iron core of the armature. Iron losses are also referred to as core losses or magnetic losses.

Mechanical losses are another type of loss due to friction in bearings and switches. The loss of air friction of the rotating armature also contributes to this. These losses represent about 10–20% of total-load losses.

In addition to the losses described above, there may be small losses present which is referred to as stray losses or miscellaneous losses. These losses are difficult to account for. They are usually due to inaccuracies in the design and modelling of the machine. Most of the time, stray losses are assumed to be 1% of the total load.

Example 5.1 A shunt generator delivers 25 A at 230 V and the resistance of the shunt field and armature are 100 and 0.1 Ω respectively, shown in Figure E5.1. Calculate the generated emf.

Solution

Field winding current is given as

$$I_{sh} = \frac{230}{100} = 2.3\,A$$

Load current I = 25 A

Total Armature current $I_a = 25 + 2.3 = 27.3$ A

Armature voltage drop $= I_a R_a = 25.3 \times 0.1 = 2.53$ V

Figure E5.1 DC Shunt generator model for Example 5.1.

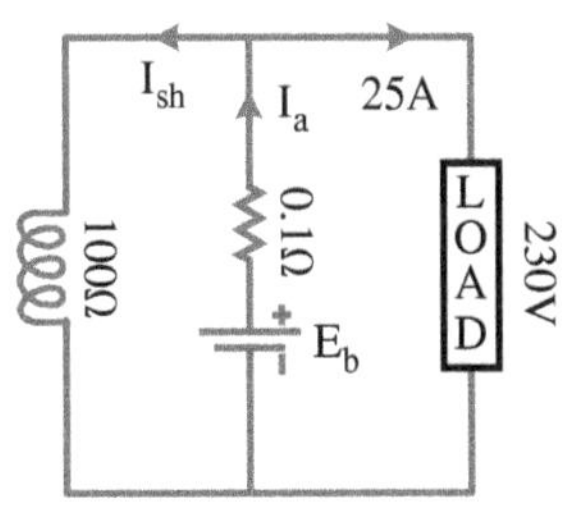

Therefore, the generated emf (E) is given as

$$E = V + I_a R_a = 230 + 2.53 = 232.53\ V$$

Example 5.2 A four-pole generator with wave-wound armature winding has 21 slots, each slot containing 50 conductors. What will the voltage generated in the machine be when driven at 1000 rpm assuming the flux per pole to be 5.0 mWb?

Solution
The generated emf is given as

$$E = \frac{NP\Phi Z}{60A}$$

Here

$$\Phi = 5 \times 10^{-3}\ Wb$$

$$Z = 21 \times 50 = 1050$$

$$A = P = 4$$

$$N = 1000\ rpm$$

Therefore

$$E = \frac{1000 \times 4 \times 5 \times 10^{-3} \times 1050}{60 \times 4} = 87.5\ V$$

Example 5.3 A four-pole DC shunt generator with 428 wave-connected armature conductors and running at 1500 rpm supplies a load of 10 Ω resistance at terminal voltage of 230 V. The armature resistance is 0.2 Ω and the field resistance is 150 Ω. Find the armature current, the induced emf and the flux per pole.

Solution
Field winding current is given as

$$I_{sh} = \frac{230}{150} = 1.53A$$

Load current

$$I_L = \frac{230}{10} = 23A$$

Therefore, armature current is given as

$$I_a = I_L + I_{sh} = 23 + 1.53 = 24.53A$$

Induced emf is given as

$$E = 230 + (0.2 \times 24.53) = 234.9V$$

Now

$$E = \frac{NP\Phi Z}{60A}$$

$$\Rightarrow \Phi = \frac{60EA}{NPZ} = \frac{60 \times 234.9 \times 2}{1500 \times 4 \times 428} = 0.01097 = 10.97mWb$$

Example 5.4 A 5 kW, 100 V DC series generator has a series field resistance of and armature 0.05 Ω resistance of 0.1 Ω. Assuming the generator is supplying at rated speed and current. Find (a) armature current, (b) generated voltage.

Solution

(a) Output current = armature current = $I_a = I_L = \frac{5\times10^3}{100} = 50A$
(b) Generated voltage = $\mathrm{E = V + I_a\,(R_a + R_s) = 100 + 50 \times (0.1 + 0.05) = 107.5\,V}$

Example 5.5 When a generator is being driven at 1000 rpm, the generated emf is 110 V. Determine the generated emf (a) if the field flux is decreased by 15% with the speed remaining unchanged, and (b) if the speed is reduced to 900 rpm, the field flux remaining unchanged.

Solution

(a) As the generated emf is directly proportional to the field flux, the generated emf with the 10% decrease in field flux will be: $E_g = 110 \times (1.0 - 0.1) = 99V$

(b)
$$\frac{E_{g1}}{E_{g2}} = \frac{N_1}{N_2}$$
$$\Rightarrow E_{g2} = \frac{N_2}{N_1} \times E_{g1} = \frac{900}{1000} \times 110 = 99V$$

Example 5.6 Assuming constant field excitation, calculate the no-load voltage of a separately excited generator whose armature voltage is 120 V at a speed of 1500 rpm, when

(a) The speed is increased to 1800 rpm.
(b) The speed is reduced to 1200 rpm.

Solution

$$\frac{E_{g1}}{E_{g2}} = \frac{N_1}{N_2}$$
$$\Rightarrow E_{g2} = \frac{N_2}{N_1} \times E_{g1}$$

With $E_{g1} = 120V, N_1 = 1500rpm$

(a) $E_{g2} = \frac{1800}{1500} \times 120 = 144V$
(b) $E_{g2} = \frac{1200}{1500} \times 120 = 96V$

Example 5.7 A 5-kW 200-V self-excited generator, when delivering rated load, has an armature-circuit voltage drop that is 3% of the terminal voltage and a shunt-field current equal to 5% of rated load current. Calculate the resistance of the armature circuit and that of the field circuit.

V = 250 V, rated value

$$\text{rated load current, } I = \frac{5000}{200} = 25A$$

$$\text{field current, } I_f = 0.05 \times 25 = 1.25A$$

From the equation, $I_a = I + I_f$,

$$I_a = 25 + 1.25 = 26.25\text{ A}$$

$$I_a R_a = 0.03 \times 200 = 6V$$

$$R_a = \frac{6}{26.25} = 0.2286\Omega$$

$$I_f R_f = 200V$$

$$\Rightarrow R_f = \frac{200}{1.25} = 133.3\Omega$$

Example 5.8 A 200-kw 220-V compound generator is delivering 600 amp at 220 V. The shunt-field current is 10 amp. The armature resistance is 0.005 Ω, and the series-field resistance is 0.003 Ω. The stray power at this load is 3000 watts. The generator is connected long-shunt. Determine generator efficiency at this load.

Solution

Output = 220 × 600 = 132 000 W.

Shunt-field loss = 220 × 10 = 2200 W.
Armature loss = $610^2 \times 0.005 = 1860.5$ W.
Series-field loss = $610^2 \times 0.003 = 1116.3$ W.
Stray power = 3000 W.
Stray-load loss, 0.01 × 132 000 = 1320 W.
Total loss = 9496.8 W.

$$\eta = \frac{P_o}{P_o + losses} \times 100\% = \frac{132000}{132000 + 9496.8} \times 100\% = 93.29\%$$

Example 5.9 A certain 100-V shunt generator has an armature and brush resistance of 0.05 Ω at full load of 60 A. The resistance of the shunt field is 55 Ω, and the stray power losses are found to be 630 W. Calculate the full-load efficiency of the generator.

Solution

The total armature current is the sum of the load current plus the field current.

$$\text{Field current } I_f = \frac{100}{55} = 1.818A$$

$$\therefore I_a = 60 + 1.818 = 61.818A$$

$$I_a^2 R_a = 61.818^2 \times 0.05 = 191 \text{ watts}$$

$$I_f^2 R_f = 1.818^2 \times 55 = 181.8 \text{ watts}$$

Total copper loss = 372.8 watts

Stray power loss = 630 watts

Total losses = 1002.8 W

$$\eta = \frac{P_o}{P_o + losses} \times 100\% = \frac{100 \times 60}{100 \times 60 + 1002.8} \times 100\% = 85.7\%$$

Example 5.10 A 10 kW, 230 V, 1750 rpm shunt generator was run light as a motor to determine its rotational losses at its rated load. The applied voltage across the armature, V_a, computed for the test, was 245 V, and the armature current drawn was 2 A. The field resistance of the generator was 230 Ω and the armature circuit resistance measured 0.2 Ω. Calculate:

(a) The rotational (stray power) losses at full load
(b) The full-load armature circuit loss and the field loss
(c) The generator efficiency at (1/4), (1/2), and (3/4) of the rated load; at the rated load, and at 1(1/4) times the rated load.

Solution

(a) Rotational loss $= V_aI_a - I_a{}^2R_a = (245 \times 2) - (2^2 \times 0.2) = 490 - 0.8 = 489.2$ W
Note that 490 W may be used with negligible error because of negligible armature loss.
(b) At the rated load, $I_L = (W/V_t) = \{(10{,}000\text{ W}) / (230\text{ V})\} = 43.5$ A
$I_a = I_f + I_L = \{(230\text{ V}) / (230\ \Omega)\} + 43.5 = 44.5$ A
The full-load armature loss

$$I_a{}^2R_a = (44.5)^2 \times 0.2 = 376\text{ W}$$

The field loss

$$V_fI_f = 230\text{ V} \times 1\text{ A} = 230\text{ W}$$

The efficiency at any load, for a generator is given by the following equation:

$$\eta = \frac{\text{Output at that load}}{(\text{Output at the load} + \text{Rotational loss} + \text{Electric loss at that load})}$$

Efficiency at (1/4) load

$$\eta_{1/4} = \frac{10000/4}{(10000/4 + 489.2 + 376/16 + 230)} = 77\%$$

Efficiency at (1/2) load

$$\eta_{1/2} = \frac{10000/2}{(10000/2 + 489.2 + 376/4 + 230)} = 86.2\%$$

$= 86.2\%.$

Efficiency at (3/4) load

$$\eta_{3/4} = \frac{10000 \times 3/4}{(10000 \times 3/4 + 489.2 + 376 \times 9/16 + 230)} = 89\%$$

Efficiency at full load

$$\eta_{fl} = \frac{10000}{(10000 + 489.2 + 376 + 230)} = 90.1\%$$
$$= \left[(10{,}000) / \{10{,}000 + 489.2 + (376 + 230)\}\right] \times 100$$
$$= 90.1\%.$$

5.10 Armature Reaction

Armature reaction is the interaction of armature flux (that is produced due to armature current when machine is loaded) with the main field flux, as shown in Figure 5.13a–c.

emf is induced by cutting the magnetic field lines in the armature conductors. There is an axis (or, you might say, a plane) along which the armature conductors move parallel to the flux lines and thus do not cut the flux lines while they are on that plane. MNA may be defined as the axis along which the armature conductors do not generate emf as they move in parallel with the flux lines. Brushes are always placed along the MNA because the reversal of current occurs along this axis in the armature conductors. GNA may be defined as an axis perpendicular to the axis of the stator field.

The flux per pole alone is decided by the field current in an unloaded DC machine. The uniform distribution of the force lines is upset when the armature carries current due to

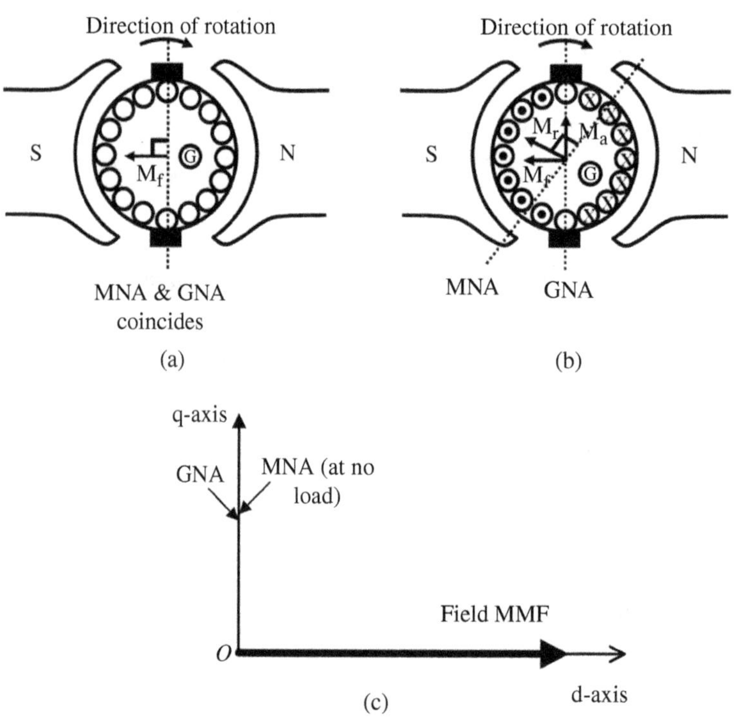

Figure 5.13 MNA and GNA of the DC machine.

loading. Flux lines are concentrated in one half of the pole and are much less in the other half. One can argue that flux per pole during loading is the same as in no-loading operation because the increase in the flux in one half is compensated for by the decrease in the other half. Since the emf generated and the torque produced by the machine are determined by the flux per pole, there is apparently no effect as regards machine performance due to an armature reaction. In fact, this is almost true if the machine is loaded slightly or moderately. However, at rated armature current the increase of flux in one half of the pole is rather less than the decrease in the other half due to the presence of *saturation*. In other words, the flux per pole will decrease during sufficient loading of the machine. This will have a direct impact on emf as well as the developed torque affecting the machine's performance. Apart from this, there is a certain amount of flux along the q axis (brush axis) of the machine due to distortions in the flux distribution as shown in Figure 5.13c. This makes it difficult to commutate. In the following sections, the armature reaction is given in some details.

5.10.1 No-Load Operation

When the DC machine operates under no-load condition whatsoever, the armature current is zero. Under such a scenario, the torque produced is zero, and the machine runs at a constant no-load speed. In the absence of any armature current, the flux per pole is only determined by the field current inside the machine and lines of force are distributed uniformly under a pole as shown in Figure 5.14a.

5.10.2 Loaded Operation

Once a resistance is connected across the armature and power is supplied to the resistance, the generator is loaded. The direction of the current in the conductor (whether crossing or pointing) is determined by the fact that the direction of the developed torque is opposite to the direction of rotation. Therefore, it is obvious to see that flux per pole, formed in the generator, should be determined not only by the field winding mmf alone, but also by the armature mmf and the armature now carrying present. By superimposing the no-load field lines and the armature field lines one can get the resultant field lines pattern as shown in Figures 5.14b and 5.15. The tip of the pole which is seen by a moving conductor first during the course of rotation is called the *leading pole tip* and the tip of the pole which is seen later

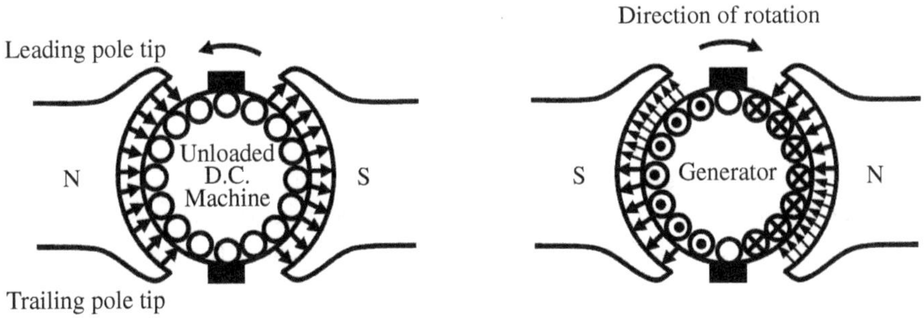

Figure 5.14 (a) Flux lines during no load conditions and (b) flux lines for loaded generator.

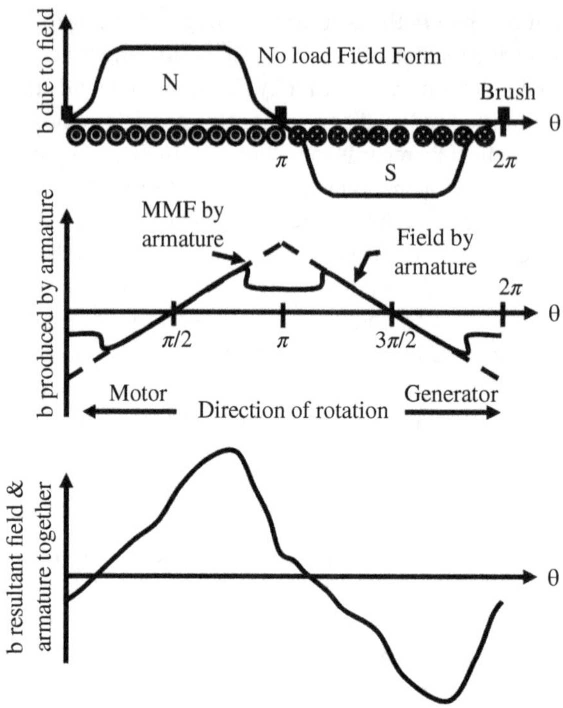

Figure 5.15 Effect of armature reaction.

is called the *trailing pole tip*. In case of generator mode, the lines of forces are concentrated near the trailing edge thereby producing torque in the opposite direction of rotation. How the trapezoidal no-load field becomes distorted along the air gap of the generator is shown in Figure 5.15. In this figure, notice that the distribution of armature mmf is triangular in nature, and the distribution of flux density due to armature current is obtained by dividing armature mmf with air gap reluctance. The reluctance is constant and small at any point under the pole. This means that, at any point within the pole, the reluctance is constant and low. This means that the density of the armature flux must essentially follow the pattern of the armature mmf. However, the reluctance in the q-axis field is very high, giving rise to a small resulting flux of polarity the same as the main pole in the q-axis behind.

5.11 Principle of Operation of a DC Motor

The DC motor operating principle is based on the Lorentz force equation. An input DC supply is given to the armature winding placed on the rotor. The coil is about to carry current in the windings when this supply is turned ON. Because of this current in the coil, the Lorentz force is exerted on the coil. As it will exert some force on a coil that carries the current under the magnetic field. Using Fleming's left-hand rule, the direction of force is found. As shown in, Figure 5.16, the current flows inside the coil side are AB and inside whereas CD is outside. According to the current direction and by Fleming's rule force at the side, AB is upwards and at side CD, it's downwards. This force allows the coil within the magnetic field to develop to a torque. But this force exerted to the magnetic field is not

Figure 5.16 Operation of a DC motor.

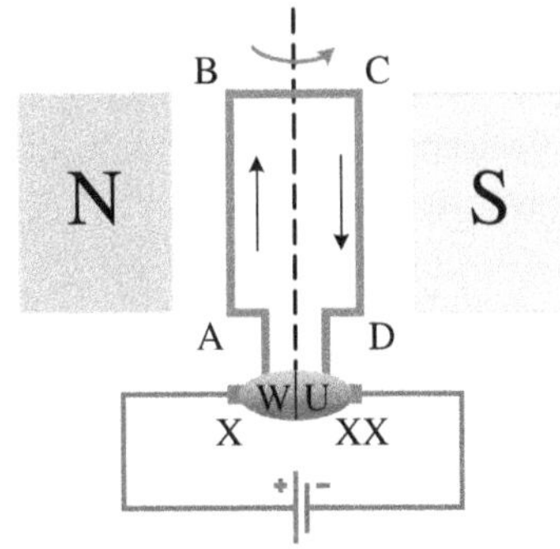

constant when the coil is at 90°. When the armature coil is perpendicular to the main field, the torque induces zero in the coil. The zero torque signifies that the motor stops rotating. To avoid this, the multiple number of coils is placed together in such a way that a uniform force is developed which, in turn, results in unidirectional torque. So, if one of their coils is perpendicular to the field, then the torque is induced by the other coils. And the rotor keeps on moving. Also, for obtaining the continuous torque, the arrangement is maintained in such a way that the current direction in the coils is reversed whenever the coils cut the MNA of the magnet. This is done with the help of a commutator.

5.11.1 Equivalent Circuit of a DC Motor

An equivalent circuit may represent the armature of a DC Motor. It can be represented by three elements E_b, R_a, and V_b that are connected in series. The element E_b is the back emf, R_a is the resistance of the armature and V_b is the drop in the voltage of the brush contact. The equivalent circuit of a DC motor armature is displayed below in Figure 5.17. In the DC motor, the armature current flow depends upon the armature voltage V_a. By applying the KVL law to the equivalent circuit, the following is obtained;

$$V_a = E_b + I_aR_a + V_b \tag{5.7a}$$

The above Eq. (5.7a) is called the basic Motor Equation. The motor's Back emf is seen to always be lower than its applied terminal voltage V_a. The brush contact drop is small about 1–2 V. If the brush contact drop is neglected, then Eq. (5.7a) is rewritten as

$$V_a = E_b + I_aR_a \tag{5.7b}$$

Figure 5.17 Equivalent circuit of the DC motor.

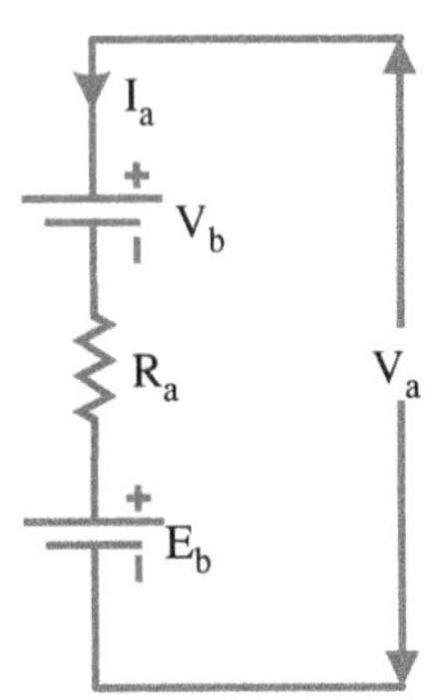

5.12 emf and Torque Equations of DC Motor

Neglecting the voltage drop due to the brushes, the voltage equation of a DC motor is given as

$$V_a = E_b + I_a R_a \tag{5.8}$$

To obtain the torque of the motor, multiply the armature current I_a on both sides of the above equations

$$I_a V_a = I_a E_b + I_a^2 R_a \tag{5.9}$$

as $I_a^2 R_a$ represents the total power loss in the armature resistance, the effective mechanical power P_m required to produce the desired torque is given as

$$P_m = E_b I_a \tag{5.10}$$

The relation between the electromagnetic torque T_e and mechanical power P_m with the rotational speed ω (in rad/sec) is given as

$$P_m = T_e \omega_m \tag{5.11}$$

Also, E_b and ω are given as

$$E_b = \frac{NP\Phi Z}{60A}$$

$$\omega_m = \frac{2\pi N}{60} \tag{5.12}$$

From Eq. (5.11), the following is obtained

$$T_e = \frac{P_m}{\omega_m} = \frac{E_b I_a}{\omega_m} = \frac{k\Phi\omega_m}{\omega_m} I_a$$

$$T_e = k\Phi I_a \tag{5.13}$$

where $k = \frac{PZ}{2\pi A}$ is called as armature constant.

5.13 Types of DC Motor

Based on the placement of the field winding, the DC motors can be categorized as

5.13.1 Separately Excited DC Motor

As the name suggests, the DC power supply is given separately for field and armature windings in the case of a separately excited DC motor. The main distinguishing fact in these types of DC motors is that, as the field winding is energized from a separate external source of DC current, the armature current does not flow through the field windings as shown in Figure 5.18. From the DC motor torque equation, we know $T_e = k\Phi I_a$. So, in this case the

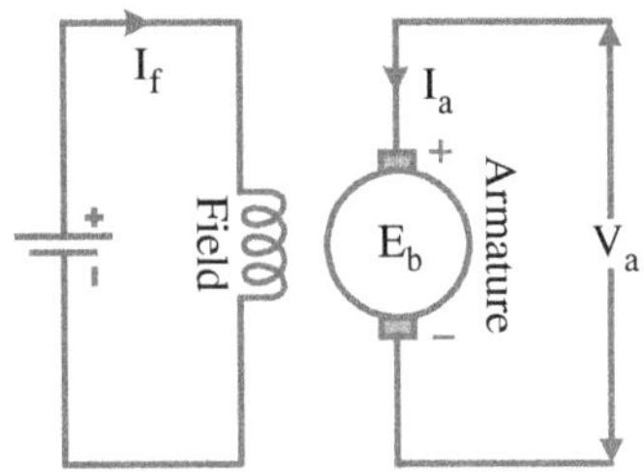

Figure 5.18 Separately excited DC motor.

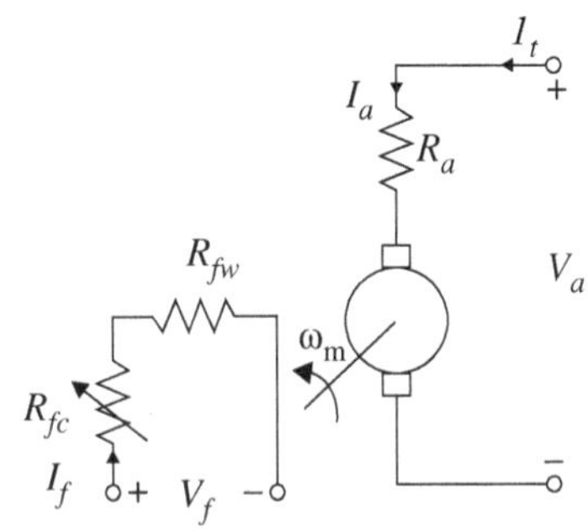

Figure 5.19 Separately excited DC motor with field resistance for speed control.

torque can be varied by varying field flux ϕ, independent of the armature current I_a. Considering the brush voltage drop, the terminal equation for the separately excited DC motor is given as shown in Figure 5.19 below

$$V_a = E_b + I_a R_a + V_b \tag{5.14a}$$

If the brush voltage drop is neglected, the terminal equation modifies as

$$V_a = E_b + I_a R_a \tag{5.14b}$$

$I_t = I_a$; terminal current is the same as the armature current

$$V_f = I_f R_f$$

- R_{fw}: resistance of field winding
- R_{fc}: resistance of control rheostat used in field circuit
- $R_f = R_{fw} + R_{fc}$: total field resistance
- R_a: resistance of armature circuit

$$E_b = k\Phi\omega_m \tag{5.15}$$

$$T_e = k\Phi I_a \tag{5.16}$$

5.13.2 Self-Excited DC Motor

5.13.2.1 Shunt DC Motor

In the shunt motor, the field is connected in parallel (shunt) to the armature windings. The shunt-connected motor provides good regulation of speed. The field winding may be excited or connected separately to the same source as the armature. An advantage of a separately excited shunt field is the ability of a variable speed drive to provide independent armature

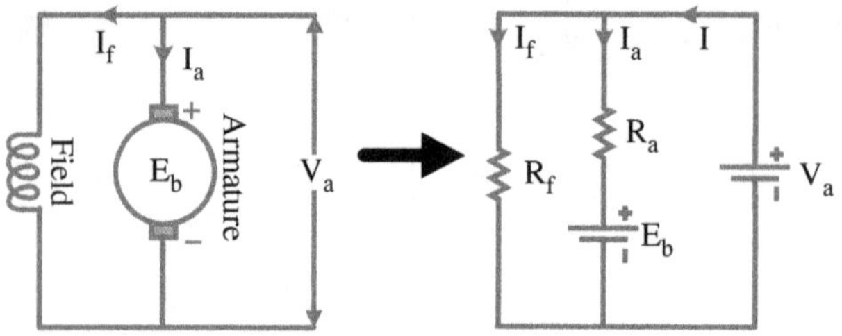

Figure 5.20 Shunt DC motor with its equivalent circuit.

and field control. The shunt-connected motor provides for simplified reversal control. This is particularly advantageous for regenerative drives. Figure 5.20 shows the schematic and its equivalent circuit of the DC shunt motor. The terminal equations of the DC shunt motor are the same as that of the separately excited DC motor and are given in (5.14a and 5.14b)

The different equations for the shunt DC motor are summarized in (5.17).

$$\begin{aligned} V_f &= R_f I_f = V_a \\ E_b &= V_a - I_a R_a \\ E_b &= K_a \Phi \omega_m \\ T_e &= K_a \Phi I_a \\ I &= I_t - I_f \end{aligned} \tag{5.17}$$

5.13.2.2 Series DC Motor

In a series of DC motors, the field is connected in series to the armature. The field is wound with a few turns of large wire because it must carry the full armature current. A characteristic of series motor is that the motor develops a large amount of starting torque. However, the speed varies greatly between no load and full load. Series motors cannot be used where constant speed is required under varying loads. A series motor should never be started at no load. With no mechanical load on the series motor, the current is low, the counter-emf produced by the field winding is weak, and so the armature must turn faster to produce sufficient counter-emf to balance the supply voltage. The speed of the DC motor is given in (5.18).

$$N \propto \frac{E_b}{\Phi \propto I_a} \tag{5.18}$$

At the beginning, the value of ϕ is almost zero, which leads to the value of speed being very high. The motor can be damaged by over-speed. This is called a runaway condition. Some load must always be connected to a series-connected motor. Series-connected motors are generally not suitable for use in most variable speed drive applications.

The terminal equation for the DC series motor from Figure 5.21 is given as

$$V_a = E_b + I_a\left(R_a + R_s\right) + V_b \tag{5.19a}$$

If the brush voltage drop is neglected, the terminal equation modifies as

$$V_a = E_b + I_a\left(R_a + R_s\right) \tag{5.19b}$$

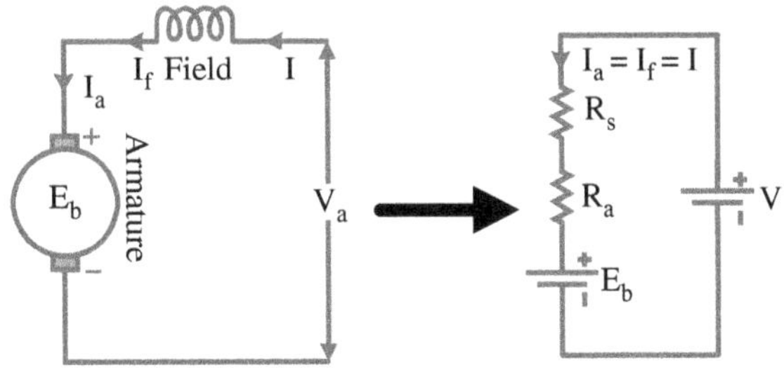

Figure 5.21 Series DC motor with its equivalent circuit.

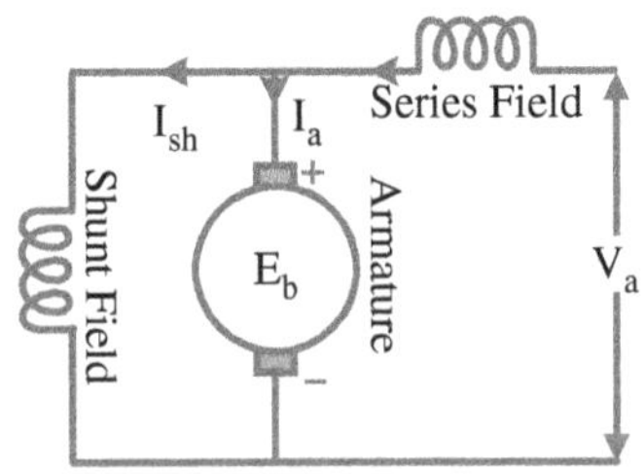

Figure 5.22 Compound DC motor.

The different equations for the shunt DC motor are summarized in (5.20).

$$
\begin{aligned}
V_a &= E_b + I_a\left(R_a + R_s\right)\\
I &= I_a = I_f\\
E_b &= K_a \Phi_s \omega_m\\
T_e &= K_a \Phi_s I
\end{aligned}
\tag{5.20}
$$

a. Compound DC motor

Compound DC motors, like shunt DC motors, have a shunt field which is excited separately as shown in Figure 5.22. Compound DC motors have a good starting torque but may experience control problems in variable speed drive applications.

5.14 Characteristics of DC Motors

Every motor has its characteristics to determine its performance. By this, one can conclude where it can be applied, and one can also take the necessary precautions to avoid any damage. For example, a series motor should not be started directly by applying a very light or no load, as heavy currents flow to the windings that damage the winding mechanism. So, it is possible to avoid this type of damage by knowing its performance characteristics so that any necessary precautionary methods can be taken to avoid such instances.

The three characteristic curves generally considered important for the DC motors are

(i) torque vs. armature current
(ii) speed vs. armature current, and
(iii) speed vs. torque.

Torque and speed play a crucial role in motor operation. Torque is generally produced by main flux produced by field winding, which interacts with the flux of the armature produced by armature current. The interaction of these two fluxes develops a unidirectional torque which allows rotation of the motor, t_m. The speed with which the rotor should rotate is another main concern. The rotor speed is generally dependent on the back emf and flux produced inside the motor. The torque-to-speed relationship explains how we can draw a characteristic curve between them. These different types have been classified, based on the arrangement of the winding, and consequently the characteristics also differ.

5.14.1 Separately Excited and DC Shunt Motor

Basically, the field winding in a shunt motor is connected to the winding of the armature. When a current is supplied via a DC source, flux is generated which interacts with the flux of the armature to produce a unidirectional torque. The torque-to-speed relationship of a shunt motor is determined by

$$T \propto \Phi \times I_a \tag{5.21}$$

The flux Φ is constant in the case of separately excited DC motor and DC shunt motor. The torque is therefore directly proportional to the current of the armature. The speed vs. armature current (torque) characteristics of the separately excited DC motor and shunt motor can be drawn by determining the relation between the speed and armature current.

$$V_a = E_b + I_a R_a \tag{5.22}$$

$$E_b = k\Phi\omega_m$$

$$\omega_m = \frac{E_b}{k\Phi} = \frac{V_a - I_a R_a}{k\Phi} \tag{5.23}$$

$$\omega_m = \frac{V_a}{k\Phi} - \frac{R_a}{k\Phi} I_a \tag{5.24}$$

For no-load condition, $I_a = 0$, the no-load speed is obtained as

$$\omega_{mo} = \frac{V_a}{k\Phi} \tag{5.25}$$

As the armature current increases, the speed drops from its no-load value with a slope of $-\frac{R_a}{k\Phi}$ as shown in Figure 5.23.

A similar relationship is obtained between speed and torque as

$$T_e = k\Phi I_a$$

$$I_a = \frac{T_e}{k\Phi} \tag{5.26}$$

Current is substituted in Eq. (5.24) to obtain the relationship between speed and torque

$$\omega_m = \frac{V_a}{k\Phi} - \frac{R_a}{k\Phi}\frac{T_e}{k\Phi}$$

$$\omega_m = \frac{V_a}{k\Phi} - \frac{R_a}{(k\Phi)^2} T_e \tag{5.27}$$

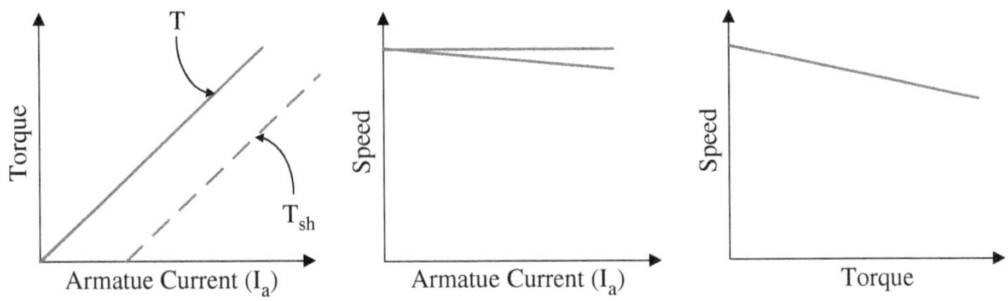

Figure 5.23 Characteristics of the DC shunt motor.

Under no-load condition, Te = 0, the no-load speed is seen as

$$\omega_{mo} = \frac{V_a}{k\Phi} \tag{5.28}$$

As the load torque increases, the speed drops from its no-load value with a slope of $-\frac{R_a}{(k\Phi)^2}$ as shown in Figure 5.23.where E_b is constant as given in (5.12). A slight drop in the characteristic can be found due to the voltage drop in the armature resistance. A similar curve to the speed vs. armature current can be found for speed vs. torque as the torque is directly proportional to the armature current. Figure 5.23 shows the characteristics curve for the separately excited and shunt motor.

5.14.2 DC Series Motor

The torque is directly proportional to the product of the armature current and the field flux. In DC series motors, field winding is connected to the armature in series. As the field current and eventually the flux is directly proportional to the armature current, i.e.

$$T \propto \Phi \times I_a \propto I_a^2 \tag{5.29}$$

Equations (5.19a and 5.19b) are valid up to the saturation point. Thus, before the magnetic saturation of the field, the torque-armature current curve is a parabola for smaller values of I_a. After the magnetic saturation of the field, the flux is independent of the armature current I_a. Thus, the torque varies proportionally to I_a only. Therefore, after magnetic saturation, the torque-armature current curve becomes a straight line.

The speed vs. armature current (torque) characteristics of a series DC motor can be drawn by determining the relationship between the speed and armature current.

$$V_a = E_b + I_a\left(R_a + R_s\right)$$

$$E_b = k\Phi\omega_m$$

$$\Phi = KI_a$$

$$E_b = kK\omega_m I_a = K_1\omega_m I_a$$

$$\omega_m = \frac{E_b}{K_1 I_a} = \frac{V_a - I_a\left(R_a + R_s\right)}{K_1 I_a} = \frac{V_a}{K_1 I_a} - \frac{\left(R_a + R_s\right)}{K_1} \tag{5.30}$$

For no-load condition, Ia = 0, the speed becomes infinite. Hence, a DC series motor is never allowed to start no-load condition. For a loaded condition, the curve is shown in Figure 5.24.

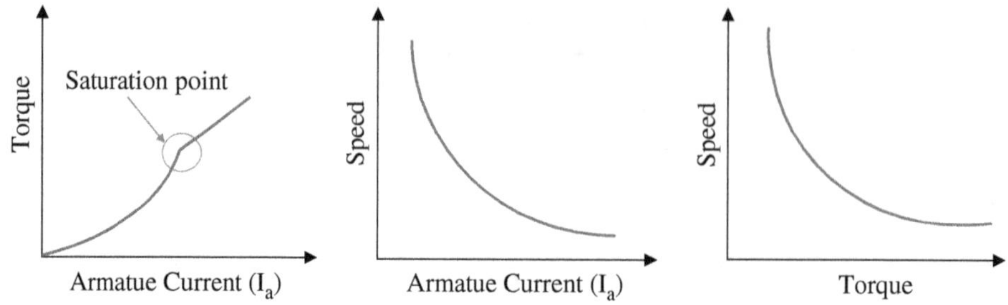

Figure 5.24 Characteristics of the DC shunt motor.

Similarly, the speed-torque curve is reached by obtaining the relationship between the speed and torque. Considering the speed–current relationship of Eq. (5.30), replacing the current with torque.

$$T_e = k\Phi I_a = kK(I_a)^2 = K_1(I_a)^2$$

$$I_a = \sqrt{\frac{T_e}{K_1}} \tag{5.31}$$

Substitute the current into Eq. (5.30)

$$\omega_m = \frac{V_a}{K_1 I_a} - \frac{(R_a + R_s)}{K_1} = \frac{V_a}{\sqrt{K_1}}\frac{1}{\sqrt{T_e}} - \frac{(R_a + R_s)}{K_1} \tag{5.32}$$

The variation of the speed with respect to the armature current can be given as

$$N \propto \frac{E_b}{\Phi \propto I_a} \tag{5.33}$$

As the speed of the DC series motor is inversely proportional to the armature current, the increase in the load current decreases the speed of the motor, however, for low value of load current, the speed becomes very high which must be avoided. Therefore, the series motor must be started with a mechanical load connected to it. A similar curve is obtained for the speed vs. torque relationship.

5.14.3 Compound Motor

Compound motor characteristics basically depend on whether the motor is cumulatively compound or differential compound. Cumulative compound motor can produce large quantities of torque at low speeds much like series motor. However, even at light or no load, it does not have series motor disadvantages. The shunt field winding produces the definite flux, and the series flux helps to increase the overall flux level by the shunt field flux. As a result, cumulative compound motor can run at acceptable speed and does not operate at extremely high speeds such as series motor, on light or without load conditions. Since two fluxes oppose each other in differential compound motor, the resulting flux decreases as the load or armature current rises, thus the system runs at a higher speed with load rise. This property is dangerous as the motor can attempt to run at extremely high speed on full load. Nevertheless, differential compound motor is rarely used. The

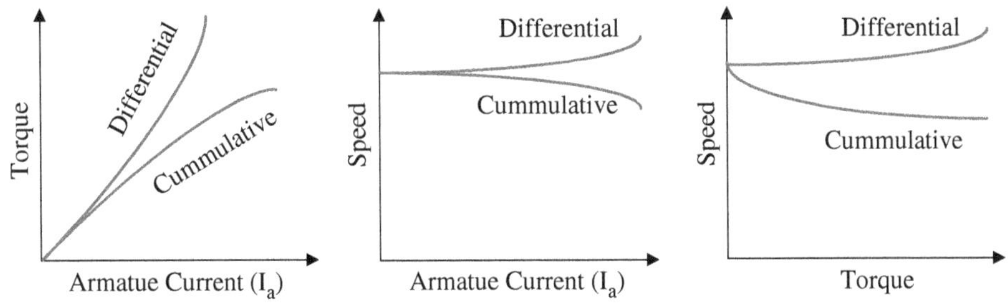

Figure 5.25 Characteristics of the DC compound motor.

specific characteristics of both the combined compound motor forms and the differential are shown in the Figure 5.25.

5.15 Starting of a DC Motor

The basic operating equation of a DC motor are given as

$$V_a = E_b + I_a R_a$$

$$I_a = \frac{V_a - E_b}{R_a} \tag{5.34}$$

When the motor is at rest, the back emf is zero, i.e.

$$E_b = \frac{N\Phi PZ}{60A}$$

$$E_b \propto N \tag{5.35}$$

Thus, at the time of starting, armature current can be given as

$$I_{a,st} = V_a/R_a. \tag{5.36}$$

The resistance of the armature in practical DC machines is very low, usually about 0.5 Ω. Therefore, during initialization/start, a large current flows through the armature. The current is large enough to damage the circuit of the armature. Because of this excessive starting current, the fuses will blow out and the winding armature and/or switching brush system may get harmed. A very high starting torque will be produced (as the torque is directly proportional to the current of the armature), and this high starting torque will cause enormous centrifugal force which will throw off the winding armature. Many loads connected to the same source/feeder can experience a voltage drop in the terminal. Because of their high rotor inertia, the DC motor can pick up speed very slowly. Therefore, building up the back emf gradually causing the high starting current level sustained for quite a while. That can result in severe damage. A suitable DC motor starter needs to be used to avoid this. Nevertheless, very small DC motors can be started directly by connecting them to the supply using a switch or a contactor. It does no damage as they gain speed easily due to low inertia of the rotor. In this scenario, because of the rapid rise in the back emf, the broad starting current will die down quickly.

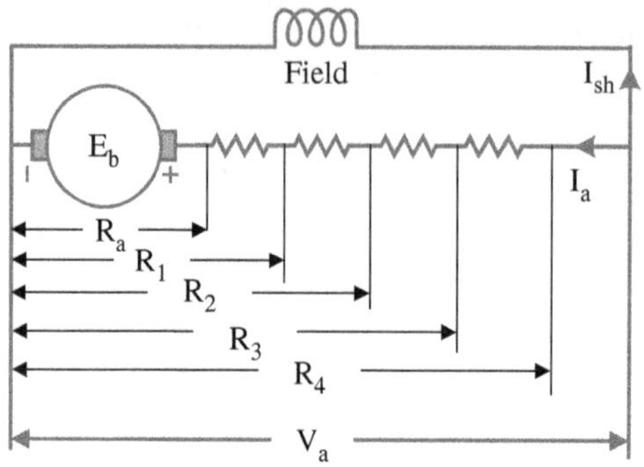

Figure 5.26 Design of a starter.

5.15.1 Design of a Starter for a DC Motor

Figure 5.26 shows a four-step starter system which is designed to reduce the starting armature current to a low value and gradually reduces the resistance to the final resistance of the armature. The motor then reaches its rated current and works normally.

Initially, now starting armature current $I_a = V_a/R_4$; which comes out to be very less. Let the values of each resistance from the armature side be X_1, X_2, X_3, and X_4, ohms. Therefore

$$R_1 = R_a + X_1$$
$$R_2 = R_a + X_1 + X_2$$
$$R_3 = R_a + X_1 + X_2 + X_3$$
$$R_4 = R_a + X_1 + X_2 + X_3 + X_4$$

Normally, motors are designed with 50% overload to start, advancing the starter one step when armature current has fallen to a definite lower value. Either this lower current limit can be fixed, or the number of steps can be fixed. In the former case, the number of steps is chosen to match the upper and lower current limits, whereas in the latter case, the lower current limit depends on the number of steps specified. It can be shown that the resistances in circuits from geometric progression on successive studs, having a common ratio equal to lower current limit by upper current limit i.e. I_2/I_1

$$\frac{Lower\,Current\,Limit\ (I_1)}{Upper\,Current\,Limit\ (I_2)} = \frac{R_4}{R_3} = \frac{R_3}{R_2} = \frac{R_2}{R_1} = \frac{R_1}{R_a} = k \tag{5.37}$$

For example, let the upper current limit be set to 2 times the lower current limit, i.e. $k = 2$ and the value of armature resistance R_a is 0.5 Ω.

Therefore

$$\frac{R_1}{R_a} = k = 2$$
$$\Rightarrow R_1 = 2 \times 0.5 = 1.0\Omega$$

Similarly

$$R_2 = 2.0\Omega$$
$$R_3 = 4.0\Omega$$
$$R_4 = 8.0\Omega$$

From the above values, the actual values of the step resistance will be

$$X_1 = 0.5\Omega$$
$$X_2 = 1.0\Omega$$
$$X_3 = 2.0\Omega$$
$$X_4 = 4.0\Omega$$

Example 5.11 A 110 V, a 1000 rpm DC motor has a full load armature current of 20 A. It is proposed to design a starter which restricts the maximum armature current during the start to 40 A. For design purposes, the minimum current during the starting is to be restricted to 20 A, the series resistance in the armature circuit is being cut out in steps when the current falls to 20 A. Calculate the maximum series resistance used in the starter and the amount of resistance cut out during each of the first two steps. The armature resistance of DC motor is 0.25 Ω.

Solution
As the maximum armature current needs to be restricted at 40 A, the armature resistance R_1 can be calculated as

$$R_1 = \frac{110}{40} = 2.75\Omega$$

The value of external resistance to be used at the start = 2.75 Ω – 0.25 Ω = 2.50 Ω.

$$\text{Now, } k = \frac{\text{Upper current limit}}{\text{Lower current limit}} = \frac{40}{20} = 2$$

$$R_2 = \frac{R_1}{k} = \frac{2.75}{2} = 1.375\Omega$$
$$R_3 = \frac{R_2}{k} = \frac{1.375}{2} = 0.6875\Omega$$

The resistance cut out in first step = $R_1 - R_2$ = 2.75 Ω – 1.375 Ω = 1.375 Ω.
The resistance cut out in second step = $R_2 - R_3$ = 1.375 Ω – 0.6875 Ω = 0.6875 Ω.

5.15.2 Types of Starters

To prevent the dangers mentioned above when starting a DC motor, the starting current must be reduced. Therefore, a starter is used to start a DC motor. There are different types of DC motor starters, such as a three-point starter, four-point starter, no-load release coil starter, thyristor controller starter, etc. The basic principle behind each DC motor starter is to add external resistance to the armature winding during start-up.

5.15.2.1 Three-Point Starter

One way to restrict the inrush current is to reduce the voltage applied to the motor, and then gradually bring it up to the rated voltage. It obviously requires a variable-voltage DC supply. While these supplies are also available in the laboratory, they are usually not available at work. We must therefore restrict the current by other means. One method is to reduce the voltage by connecting an external series resistance. The motor can be started easily without

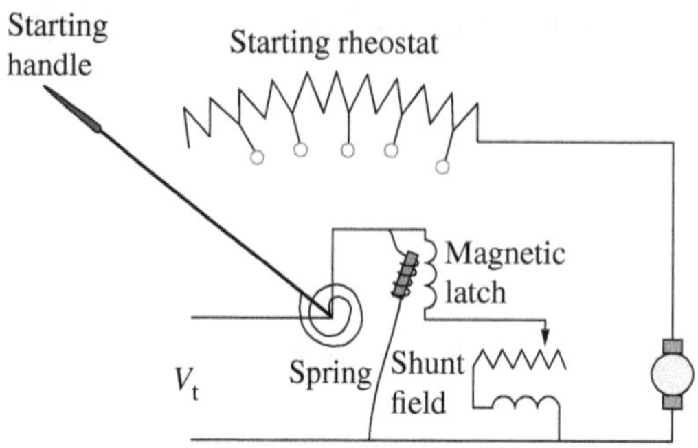

Figure 5.27 Three-point starter.

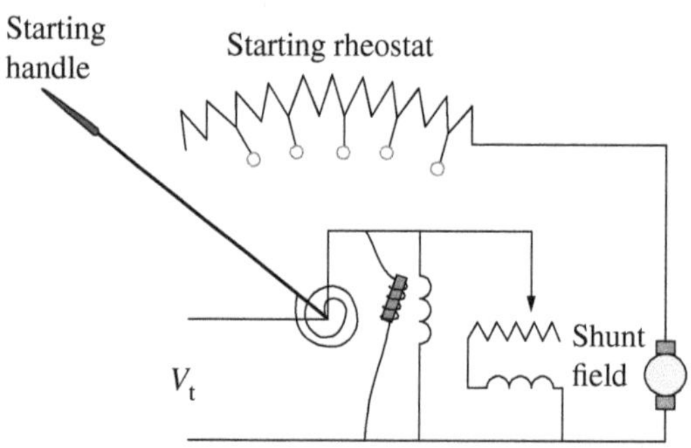

Figure 5.28 Four-point starter.

unnecessary armature current by putting a large resistance in the circuit at start-up and increasing it in steps as the speed rises. This is achieved using both manual and automatic starters. Figure 5.27 displays a circuit diagram for the manual starter. The arm is powered by the spring and is rotated in the direction of the clockwise direction, slowly reducing the resistance of the armature as the motor accelerates. This starter is a three-point starter. The electromagnet holding the starter in the place of run is in the circuit of ground. This form of starter can be used for shunt and compound motors and the starter drops out if the field is lost, preventing the motor from runaway. The downside of this type of starter is that if field resistance control is used to weaken the field for increased motor speed it can drop out the electromagnet. Therefore, this type of starter cannot be used for the DC series motor.

5.15.2.2 Four-Point Starter

Figure 5.28 displays a circuit diagram for four-point starters. Here, the electromagnet is directly connected to the voltage line. It does not drop out on the shunt machine at low field current values, and thus does not protect against runaway. The fuses in the motor circuit, however, would possibly blow before the motor reached a dangerous speed. Also, it can be used to start series motors.

5.16 Speed Control of a DC Motor

We frequently want to control the speed of a DC motor. Speed control is frequently needed in numerous applications. This requires change in drive speed known as a DC motor speed control. Speed control of a DC motor is done either by the operator manually, or by an automatic control unit. The speed of a DC motor (N) is equal to

$$N = \frac{V_a - I_a R_a}{k_a \Phi} \tag{5.38}$$

The speed of the three types of DC motors-, series- and compound-regulated by changing the quantities on the right side of the above equation.

5.16.1 Separately Excited and DC Shunt Motor

For controlling the speed of the shunt motor, there are essentially two methods, namely:

(i) Varying armature resistance: This is the most widely employed method. Here, the controlling resistance is directly connected to the motor supply in series, as shown in the Figure 5.29. A variable resistance is added to the armature circuit in armature resistance control. Field is connected directly across the supply so the flux is not changed due to series resistance variations. The power loss in the control resistance of the DC series motor can be neglected as this control method is used to reduce the speed under light-load condition for a long portion of time. This speed control method is most cost effective for constant torque. This speed control method is used for motor-driving cranes, hoists, trains etc. of the DC series.

(ii) Flux control method: The speed of the DC motor is inversely proportional to the flux. Thus, speed can be increased by decreasing the flux, and vice versa. To control the flux, a resistance is added as shown in the circuit diagram of Figure 5.30, in series with field winding. Adding more resistance with field winding in series will increase the speed as flux decreases. As field current is relatively small in shunt motors, the loss of the external resistance is small. So this method is quite effective. Although speed can be increased above the rated value by reducing flux with this method, it places a limit on maximum speed as weakening field flux beyond a limit will adversely affect the switching process. Reducing the field flux also reduces the electromagnetic torque produced by motor.

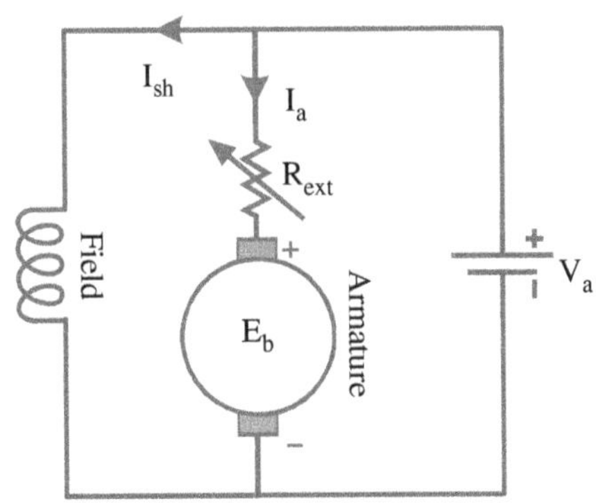

Figure 5.29 Varying armature resistance method of speed control.

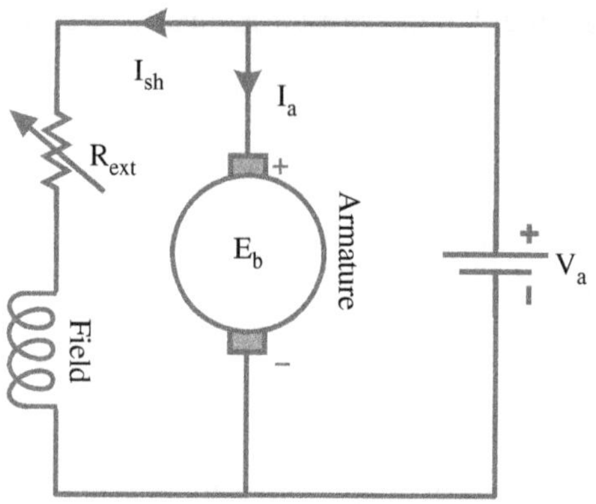

Figure 5.30 Flux control method of speed control.

5.16.2 DC Series Motor

Speed adjustment by field control of a DC series motor may be done by

(i). Field diverter method: In this method, an external resistance is connected in parallel to the series field which works as a diverter for the armature current as shown in Figure 5.31. Here, the field flux is produced by a portion of the motor current around the series field by diverting some portion of the armature current through the diverter. Reducing the resistance to the diverter means less field current, therefore less flux is more speed. This method gives above-normal speed, and the method is used in electrical drives where speed should rise sharply as soon as load is lowered.

(ii). Tapped field control: This is another way to increase the speed by reducing the flux, and is done by reducing the number of field winding turns through which current flows as shown in Figure 5.32. In this method, a number of field winding tappings are brought out. This method is used in the electric traction.

Example 5.12 A 25-KW, 125 V separately excited DC machine is operated at a constant speed of 3000 rpm with a constant field current such that the open-circuit armature voltage

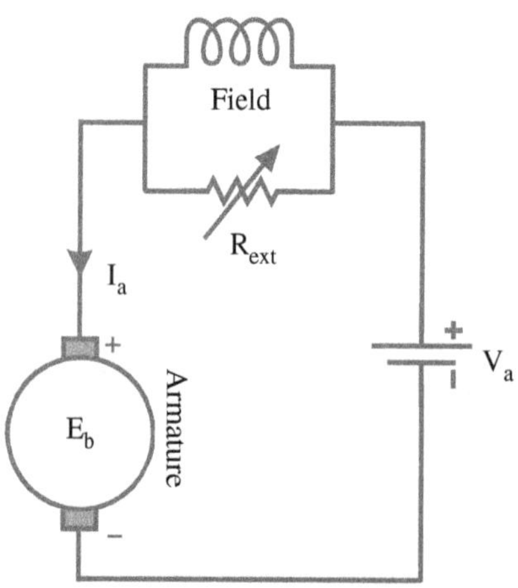

Figure 5.31 Field diverter method of speed control.

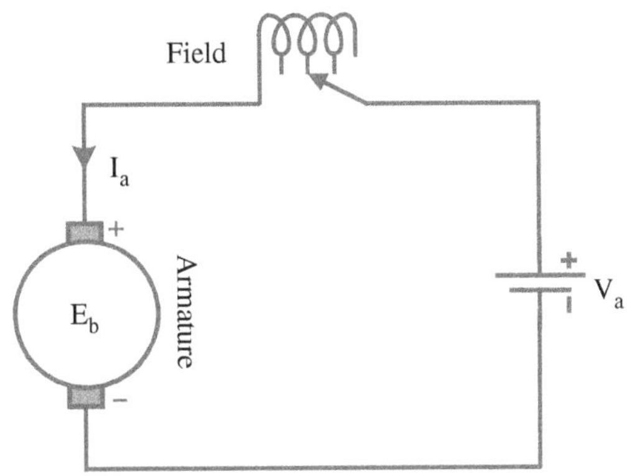

Figure 5.32 Field diverter method of speed control.

is 125 V. Design a DC motor starter to restrict the starting current to twice rated current of DC motor.

Solution

$$I_{a|rated} = P/V_{rated} = 25\,k/125 = 200\,A$$

$$I_{a|starting} = Vt/Ra = 125/Ra \text{ Assume } Ra = 0.1\,\Omega \rightarrow 1250.1 = 1250\,A$$

$$\text{(This value is very high and undesired)}$$

Given 200% of rated value

$$I = Vt/\,(Rae + Rae1) = 2\,(200) = 1250/\,(0.1 + Rae1)\,;\, Rae1 = 0.2125\,\Omega$$

At t = t2⁻

$$Ia = 200\,A$$

$$Ea2 = Vt - Ia\,(Ra + Rae1) = 125 - 200\,(0.1 + 0.2125) = 62.5\,V$$

At t = t2+

$$Ia = 400\,A$$

$$400 = (Vt - Ea)\,/\,(Ra + Raext) = (125 - 62.5)\,/\,(0.1 + Rae2)\,;\, Rae2 = 0.05625\,\Omega$$

At t = t3⁻

$$Ia = 200\,A$$

$$Ea3 = Vt - Ia\,(Ra + Rae1) = 125 - 200\,(0.1 + 0.05625) = 93.75\,V$$

At t = t3⁺

$$400 = (Vt - Ea)\,/\,(Ra + Raext) = (125 - 93.75)\,/\,(0.1 + Rae3)\,;\, Rae3 = -0.021875\,\Omega$$

Negative value means it is not appropriate

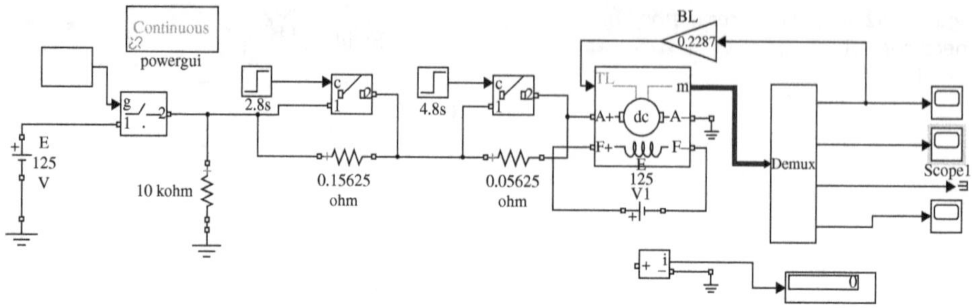

Figure E5.12 Simulink model of starter.

$$R_1 = R_{ae1} - R_{ae2} = 0.2125 - 0.056251 = 0.15625\,\Omega$$

$$R_2 = R_{ae2} - R_{ae3} = 0.05625 - 0 = 0.05625\,\Omega$$

Simulink Model of starter shown in Figure E5.12.

5.17 Solved Examples

Example 5.13 A 1 kW, 200 V DC shunt motor has armature resistance of 0.2 Ω and shunt field resistance of 200 Ω. At no load, the motor draws 3 A from 200 V supply and runs at 1000 rpm. Calculate the total loss of the machine at full load.

Solution

Current in the shunt branch $I_{sh} = \frac{V_{in}}{R_{sh}} = \frac{200}{200} = 1A$

Cu-loss in the shunt branch $I_{sh}^2 R_{sh} = 1^2 \times 200 = 200W$

No load input to motor $P_{nl} = 200 \times 3 = 600W$

No load current in the armature branch $I_{ao} = 3 - 1 = 2A$

No load Cu-loss in the armature resistance $I_{ao}^2 R_a = 2^2 \times 0.2 = 0.8W$

Constant or standing losses of the motor = 600–0.8 = 599.2 W

Full load current $I_{fl} = \frac{1000}{200} = 5A$

Full load cu-loss in the armature = $I_a^2 R_a = 5^2 \times 0.2 = 5W$

Therefore, total losses at full load = 200 + 599.2 + 5 = 804.2 W

Example 5.14 A two-pole series motor with a lap wave winding with 500 conductors. Total resistance is 1 Ω. When supplied from a 200 V, motor supplies a load of 6.4 kW and takes 40 A with a flux of 5 mWb. Compute the developed torque.

Solution

Total resistance which is sum of armature resistance and series field resistance

$$R = Ra + Rse = 1\,\Omega$$

Applied voltage $V_a = 200\,V$

$$Ia = 40\,A;\ P = 10\,kW\ (load)$$

$\Phi = 5$ mWb

E = Va − Ia Ra = 200 − 40 X1 = 160 V

$$E = \frac{N\Phi PZ}{60A}$$

$$N = \frac{60AE}{\Phi PZ} = \frac{60X2X160}{5X10^{-3}X2X500} = 3840rpm$$

Speed in rad/sec

$$\omega_m = \frac{2\pi N}{60} = \frac{2\pi\,(3850)}{60} = 403.17rad/\sec$$

Torque Developed

$$T = \frac{P}{\omega_m} = \frac{6.4X1000}{403.17} = 15.87Nm$$

Example 5.15 A commutatively compound generator is supplying power to a DC machine. Suddenly, if the prime mover of the generator fails and the speed drops how the generator will behave?

Solution

Due to failure of the prime mover, the speed drops. When the speed drops, the back emf will reduce. The generator now behaves as a motor since ($E_b < V$).

When the machine becomes a motor, it will start drawing current from the load. Now the direction of series field will reverse. The overall flux will be subtractive. The machine becomes a differential compound motor.

Example 5.16 A DC motor takes an armature current of 50 A at 400 V. The armature circuit resistance is 0.1 Ω. The machine has four poles and the armature is lap-connected with 264 conductors. The flux per pole is 0.08 Wb. Calculate

(i) the speed and
(ii) the gross torque developed by the armature

Solution

(i)

$$E_b = V_a - I_a R_a = 400 - 50 \times 0.1 = 395V$$

$$\phi = 0.08Wb$$

$$Z = 264$$

$$E_b = \frac{\phi ZNP}{60A}$$

$$395 = \frac{0.08 \times 264 \times N \times 4}{60 \times 4}$$

$$\Rightarrow N = 1122.2rpm$$

(ii)

$$T_a = \frac{0.159 \times \phi Z P I_a}{A}$$
$$= \frac{0.159 \times 0.08 \times 264 \times 4 \times 50}{4} = 167.9N - m$$

Example 5.17 Determine the developed torque and shaft torque of 220-V, four-pole series motor with 800 wave-connected conductors supplying a load of 8.2 kW by taking 45 A from the mains. The flux per pole is 25 mWb and its armature circuit resistance is 0.6 Ω.

Solution

Developed torque or gross torque is the same thing as armature torque

$$T_a = \frac{0.159 \times \phi Z A P}{A}$$
$$= \frac{0.159 \times 25 \times 10^{-3} \times 800 \times 45 \times 4}{2} = 286.2N - m$$
$$E_b = V_a - I_a R_a = 220 - 45 \times 0.6 = 193V$$
$$E_b = \frac{\phi Z N P}{60A}$$
$$193 = \frac{25 \times 10^{-3} \times 800 \times N\pi \times 4}{2}$$
$$N = 4.825 rps$$

Also

$$2\pi N T_{sh} = P_o$$
$$2\pi \times 4.825 \times T_{sh} = 8200$$
$$\Rightarrow T_{sh} = 270.5\, N - m$$

Example 5.18 A 125-V shunt motor has an armature whose resistance is 0.2 Ω. Assuming a voltage across the brush contacts of 5 V, what armature current will flow

(a) when the back emf is 110 V?
(b) if the motor load is increased so that the back emf drops to 96 V?

Solution

$$I_a = \frac{V_a - E_b}{R_a}$$

where

I_a	= armature current
V_a	= applied voltage
E_b	= back emf
R_a	= armature resistance

(a) $I_a = \frac{V_a - E_b}{R_a} = \frac{125 - 5 - 110}{0.5} = 20A$

(b) $I_a = \frac{V_a - E_b}{R_a} = \frac{125 - 5 - 96}{0.5} = 48A$

Example 5.19 A 250 V DC shunt motor has an armature resistance of 0.04 Ω and the field resistance is 100 Ω. Calculate;

(a) The value of resistance to be added to the field circuit to increase speed from 1000 rpm to 1400 rpm when supply current is 150 A.
(b) With this field resistance, calculate the speed when the supply current is reduced to 100 A.

Solution

(a)

$$V_a = 250\text{ V}, I_a = 150\text{ A}, Ra = 0.04\,\Omega, R_{sh1} = 100\,\Omega; N_1 = 1000\text{ rpm}, R_{sh2} = ?,$$
$$N_2 = 1400\text{ rpm}$$

Field current

$$I_{sh1} = \frac{V_a}{R_{sh1}} = \frac{250}{100} = 2.5A$$
$$I_{a1} = I_{t1} - I_{sh1} = 150 - 2.5 = 147.5\text{ A}$$
$$E_{b1} = V_a - I_a R_a = 250 - 147.5 \times 0.04 = 244.1V$$
$$E_b = \frac{N\Phi PZ}{60A} = K_n \Phi N$$
$$\Phi \infty I_{sh} = KI_{sh}\ (\text{Linear Magnetization})$$
$$E_b = K_n K I_{sh} N$$
$$K_n K = \frac{E_b}{I_{sh} N} = \frac{244.1}{2.5\,(1000)} = 0.09764 V/Arpm$$

When the field resistance is added
Assuming the line current and armature current are the same

$$I_t \approx I_a$$
$$E_{b2} = V_a - I_a R_a = 250 - 150 \times 0.04 = 244V$$
$$E_b = K_n K I_{sh} N$$
$$I_{sh2} = \frac{E_{b2}}{K_n K N} = \frac{246}{0.09764\,(1400)} = 1.8A$$
$$I_{sh2} = \frac{V_a}{R_{sh2}}; R_{sh2} = \frac{V_a}{I_{sh2}} = \frac{250}{1.8} = 138.88\Omega$$
$$R_{ext} = R_{sh2} - R_{sh1} = 138.88 - 100 = 38.88\Omega$$

(b) Va = 250 V, Ish = 1.8 A, Rsh = 38.88 Ω, N =?

$$I_a = I_t - I_{sh} = 100 - 1.8 = 98.2A$$

$$E_b = V_a - I_a R_a = 250 - 98.2 \times 0.04 = 246.07V$$

$$E_b = K_n K I_{sh} N$$

$$N = \frac{E_b}{K_n K I_{sh}} = \frac{246.07}{0.09764\,(1.8)} = 1400.1rpm$$

Example 5.20 The armature of a 200 V shunt motor has a resistance of 0.5 Ω. If the armature current is not to exceed 60 amp, calculate: (a) the resistance that must be inserted in series with the armature at the instant of starting; (b) the value to which this resistance can be reduced when the armature accelerates until E_b is 160 V; (c) the armature current at the instant of starting if no resistance is inserted in the armature circuit. (Assume a 2-V drop at the brushes.)

Solution

(a)
$$I_a = \frac{V_a - E_b}{R_a}$$

$$\Rightarrow R'_a = \frac{V_a - E_b}{I_a} = \frac{220 - 2 - 0}{60} = 3.63\Omega$$

External resistance $R = 3.63 - R_a = 3.63 - 0.5 = 3.13\Omega$

(b)
$$\Rightarrow R'_a = \frac{V_a - E_b}{I_a} = \frac{220 - 2 - 160}{60} = 0.966\Omega$$

$$R = 0.966 - R_a = 0.966 - 0.5 = 0.466\Omega$$

(c)
$$I_A = \frac{220 - 2}{0.5} = 436A$$

Example 5.21 Find the condition for maximum power developed in a DC machine. Calculate the power and efficiency under this condition.

Solution

Power Developed by a DC machine is given as

$$P = E_b I_a = \left(V_a - I_a R_a\right) I_a = V_a I_a - I_a^2 R_a$$

For maximum power

$$\frac{dP}{dI_a} = 0 \Rightarrow V_a - 2I_a R_a = 0$$

$$I_a = \frac{V_a}{2R_a}$$

Substituting the current equation to the back emf equation, the condition for maximum power

$$E_b = V_a - I_a R_a = V_a - \left(\frac{V_a}{2R_a}\right) R_a = \frac{V_a}{2}$$

Thus, the back emf is half of the applied armature voltage and is the condition of maximum power.

Maximum Power

$$P_{\max} = E_b I_a = \left(\frac{V_a}{2}\right)\left(\frac{V_a}{2R_a}\right) = \frac{V_a^2}{4R_a}$$

Input power under maximum output power condition

$$P_{in} = V_a I_a = V_a \left(\frac{V_a}{2R_a}\right) = \frac{V_a^2}{2R_a}$$

Efficiency

$$\eta = \frac{P}{P_{in}} \times 100\% = \frac{\left(\frac{V_a^2}{4R_a}\right)}{\left(\frac{V_a^2}{2R_a}\right)} \times 100\% = 50\%$$

Example 5.22 A 220 V DC Series Motor takes 50 A when running at 1500 rpm and delivering rated output power. What resistance should be added in armature to reduce the speed to 1000 rpm at the rated output power? Total resistance is given as 0.5 Ω

Solution

Total resistance

$$R_{total} = R_a + R_{se} = 0.5\Omega$$
$$V_a = 220V, I_a = 50A, N = 1500rpm$$
$$E_b = V_a - I_a R_{total} = 220 - 50 \times 0.5 = 195V$$

$$E_b = \frac{N\Phi PZ}{60A} = K_n \Phi_{se} N$$
$$\Phi_{se} \infty I_{se} = K_{se} I_{se} \text{ (Linear Magnetization)}$$
$$E_b = K_n K_{se} I_{se} N$$
$$K_n K_{se} = \frac{E_b}{I_{se} N} = \frac{195}{50 \times 1500} = 0.0026V/Arpm$$

Torque

$$T = K_a \Phi_{se} I_a$$

T = rated constant

I_a = rated Constant

Va = 220 V, Ia = 50 A, N = 1000 rpm

After adding external resistance, total resistance is

$$R'_{total} = R_a + R_{ext}$$

$$E_b = K_n K_{se} I_{se} N' = 0.0026 \times 50 \times 1000 = 130V$$

$$130 = 220 - 50R'_{total}$$

$$R'_{total} = 1.8\Omega$$

$$R_{ext} = 1.8 - 0.5 = 1.3\Omega$$

Example 5.23 A DC shunt motor runs at 1500 rpm when rated voltage is applied. If the voltage is reduced to half of its rated value, find the new speed of the machine. Neglect the armature resistance.

Solution

$$E_b = V_a - I_a R_a$$

$$K_n \Phi N = V_a - I_a R_a$$

$$N = \frac{V_a - I_a R_a}{K_n \Phi}$$

$$N = \frac{V_a - I_a R_a}{K_n \Phi} = \frac{V_a}{K_n \Phi}$$

$$V'_a = \frac{V_a}{2}; I_f = \frac{V_a}{R_f}$$

$$I'_f = \frac{V'_a}{R_f} = \frac{\left(\frac{V_a}{2}\right)}{R_f} = \frac{V_a}{2R_f} = \frac{I_f}{2}$$

$$\Phi' \infty I_f; \Phi' = \frac{\Phi}{2}$$

$$N' = \frac{V'_a}{K_n \Phi'} = \frac{\left(\frac{V_a}{2}\right)}{K_n \frac{\Phi}{2}} = \frac{V_a}{K_n \Phi} = N = 1500rpm$$

Example 5.24 A 120 V DC shunt motor has an armature resistance of 0.2 Ω and a brush volt drop of 2 V. The rated full-load armature current is 75 A. Calculate

(i) the current at the instant of starting, and the % of full load.
(ii) Calculate the various values (taps) of starting resistance to limit the current in the motor of Part (i) to
  (a) 150% rated load at the instant of starting.
  (b) 150% rated load, when the counter emf is 25% of the armature voltage, V.
  (c) 150% rated load, when the counter emf is 50% of the armature voltage, V_a.
(iii) Find the counter emf at full load, without starting resistance.

Solution

(i) $I_{st} = \frac{V_a - BD}{R_a} = \frac{120-2}{0.2} = 590A$ (when back emf is zero)

$$\text{Percent full load} = \frac{590}{76} \times 100 = 786\%$$

(ii) $I_a = \frac{V_a - (E_b + BD)}{R_a + R_s}$

Solving for R_s,

$$R_s = \frac{V_a - (E_b + BD)}{I_a} - R_a$$

(a) At starting, E_b is zero

$$R_s = \frac{V_a - BD}{I_a} - R_a$$

$$= \frac{120 - 2}{1.5 \times 75} - 0.2 = 0.85\Omega$$

(b) $R_e = \frac{V_a - (E_b + BD)}{I_a} - R_a = \frac{120-30-2}{1.5 \times 75} - 0.2 = 0.582\Omega$ $R_e = \left[\left\{V_a - \left(E_c + BD\right)\right\}/I_a\right] -$

(c) $R_s = \frac{V_a - (E_b + BD)}{I_a} - R_a = \frac{120-60-2}{1.5 \times 75} - 0.2 = 0.316\Omega$

(iii) $E_b = V_a - (I_a R_a + BD) = 120 - (75 \times 0.2) + 2 = 103V$

Example 5.25 Consider a DC series motor used in traction application (constant power load). It is running at rated load and rated voltage. For speed control, armature voltage control is used. If the reduced is dropped to 25% of the rated value, calculate the applied voltage to be applied.

Solution

$$E_b = V_a - I_a R_a$$

$$K_n \Phi N = V_a - I_a R_a$$

$$N = \frac{V_a - I_a R_a}{K_n \Phi}$$

$$N = \frac{V_a - I_a\left(R_a + R_{se}\right)}{K_n \Phi} = \frac{V_a}{K_n \Phi}; (R \to 0)$$

$$\Phi \infty I_a$$

$$N = \frac{V_a}{K_n K_{se} I_a}$$

Mechanical Power Developed

$$P = E_b I_a$$

$$E_b \approx V_a; (R \rightarrow 0)$$

$$P = V_a I_a$$

$$I_a = \frac{P}{V_a}$$

$$N = \frac{V_a}{K_n K_{se} I_a} = \frac{(V_a)^2}{K_n K_{se} P}$$

At reduced voltage

$$N_1 = \frac{V_a}{K_n K_{se} I_a} = \frac{(V_a)_1^{\ 2}}{K_n K_{se} P}$$

$$\frac{N_1}{N} = \frac{(V_a)_1^{\ 2}}{(V_a)^2} = \frac{(0.5)^2}{(1)} = 0.25$$

$$N_1 = 0.25N$$

Example 5.26 A 250 V DC shunt motor has armature circuit resistance of 0.6 Ω and shunt field resistance of 120 Ω. The machine is connected to a supply of 250 V. The motor is operating as a generator and then as a motor separately. The line current of the machine in both cases is 50 A. Calculate the ration of speed as a generator to the speed as a motor.

Solution

Shunt field current

$$I_f = \frac{250}{125} = 2A$$

As a generator

The armature current at no load is $I_a = 50 + 2 = 52$ A

$$E = V_a + I_a R_a = 250 + 52 \times 0.6 = 281.2V$$

The induced emf for generator is given as $E_g = k_a \times \phi \times N_g = 281.2V$

As a motor

The armature current at no load is $I_a = 50–2 = 48$ A

$$E = V_a - I_a R_a = 250 - 48 \times 0.6 = 221.2V$$

The back emf for motor is given as $E_m = k_a \times \phi \times N_m = 221.2V$

$$\therefore \frac{N_g}{N_m} = \frac{E_g}{E_m} = \frac{281.2}{221.2} = 1.27$$

5.18 Matlab/Simulink Model of a DC Machine

5.18.1 Matlab/Simulink Model of a Separately/ Shunt DC Motor

According to Kirchhoff's voltage law, the electrical equation of the shunt DC motor is described as [1].

$$R_a i_a + L_a \frac{di_a}{dt} + E_b = V_a \tag{5.39}$$

Also,

$$E_b = L_{af} \omega I_f \tag{5.40}$$

where L_{af} is the mutual inductance between field and armature coil, ω is the speed and I_f represents the field current.

Taking Laplace transform of the above equation we get

$$I_a(s) = \frac{V_a - E_b}{R_a + sL} \tag{5.41}$$

The equation for the mechanical system is given by

$$J\frac{d\omega}{dt} + B\omega = T \tag{5.42}$$

where J = Moment of inertia of motor and load and B = Frictional coefficient of motor and load

Taking Laplace transform of the above equation, we get:

$$\omega(s) = \frac{T(s)}{B + sJ} \tag{5.43}$$

The equation of the developed torque T_e is given by

$$T_e = L_{af} I_a I_f \tag{5.44}$$

From the above equations, the block diagram for the DC motor is given in Figure 5.33 as:

The simulation results are shown in Figure 5.34 for a DC shunt motor. The machine is allowed to run freely for t = 0 to t = 3 and then a load torque of T_L = 20 Nm is applied at t = 3 seconds. The armature current goes to peak value and then reduces at steady state. To feed the load demand the current increases again when load is applied.

The torque generated by machine also goes to peak value and then reaches a minimum value equal to the no-load losses. The electromagnetic torque increases to compensate for the load torque as shown in Figure 5.34b.

The machine initially accelerates and reaches steady-state value. Once the load is applied, the speed drops and continues to run at a lower speed, shown in Figure 5.34c.

5.18.2 Matlab/Simulink Model of a DC Series Motor

The different equations for the DC series motor are given as

$$(R_a + R_f)\, i_a + (L_a + L_f)\frac{di_a}{dt} + E_b = V_a \tag{5.45}$$

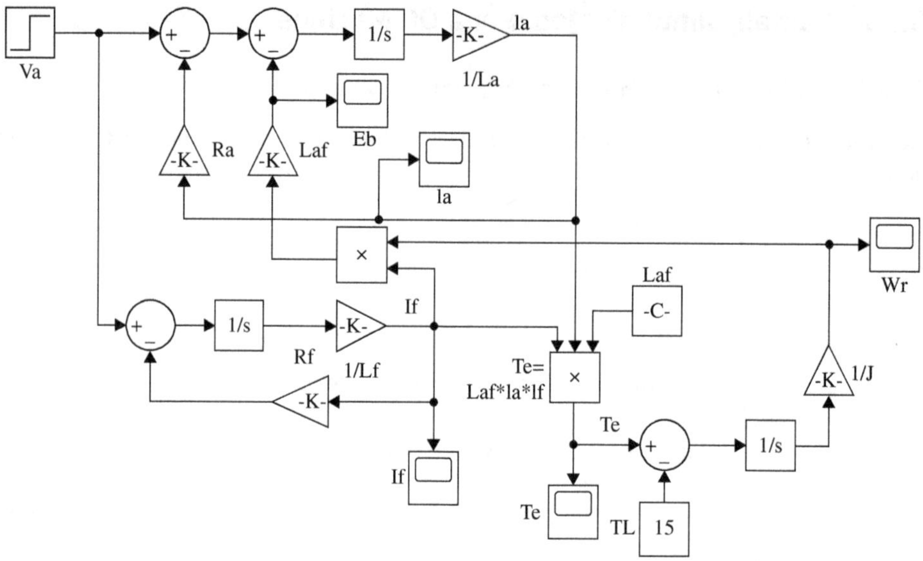

Figure 5.33 MATLAB model for the separately/ shunt DC motor [File Name: DC Motor Shunt.slx].

Taking the Laplace transform of (5.4)

$$I_a(s) = \frac{V_a - E_b}{(R_a + R_f) + s(L_a + L_f)} \tag{5.46}$$

The equation of torque is given by

$$T_e = L_{af} I_a^2 \tag{5.47}$$

From (5.40, 5.42, 5.43, 5.45 and 5.47), the Matlab model for the DC series motor is shown in Figure 5.35.

The simulation results are shown in Figure 5.36 for a DC series motor. The machine is allowed to run freely for t = 0 to t = 100 and then a load torque of T_L = 20 Nm is applied at t = 50 seconds. The armature current goes to peak value and then reduces at steady state. To feed the load demand, the current increases again when load is applied.

5.18.3 Matlab/Simulink Model of a Compound DC Motor

For a compound motor, the electrical equations are given as

$$(R_a + R_{fs}) i_a + (L_a + L_{fs}) \frac{di_a}{dt} + L_{af}\omega (I_{fs} + I_{fsh}) = V_a \tag{5.48}$$

where R_{fs}, L_{fs}, I_{fs}, and I_{fsh} represents the series field resistance, series field inductance, series field current, and shunt field current respectively. Similarly, the torque equation is modified as

$$T_e = L_{af} (I_{fs} + I_{fsh}) I_a \tag{5.49}$$

From the above equations, the block diagram for the DC compound motor is given in Figure 5.37 as

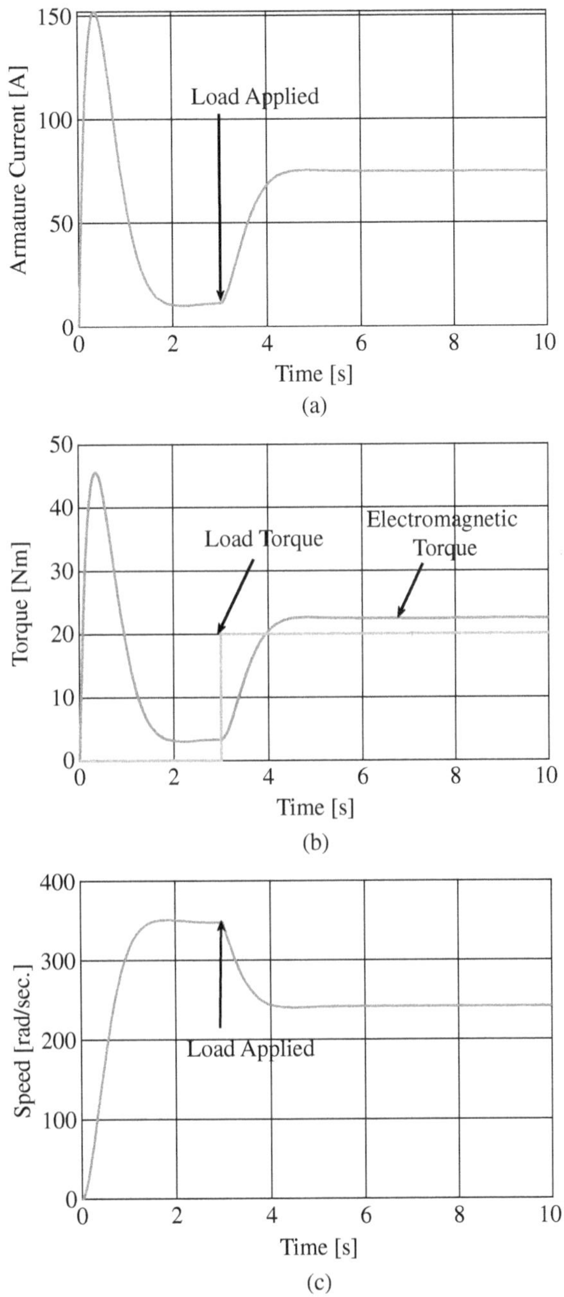

Figure 5.34 Simulation results of a DC shunt motor, (a) armature current, (b) torque, and (c) speed.

The simulation results are shown in Figure 5.38 for a DC compound motor. The machine is allowed to run freely for t = 0 to t = 100 and initially, the load torque of $T_L = 5\,Nm$ is applied. At t = 20 seconds, it is changed to 15 Nm and it gradually increases to 20 Nm till t = 100 seconds. The armature current goes to peak value and then reduces at steady state. To

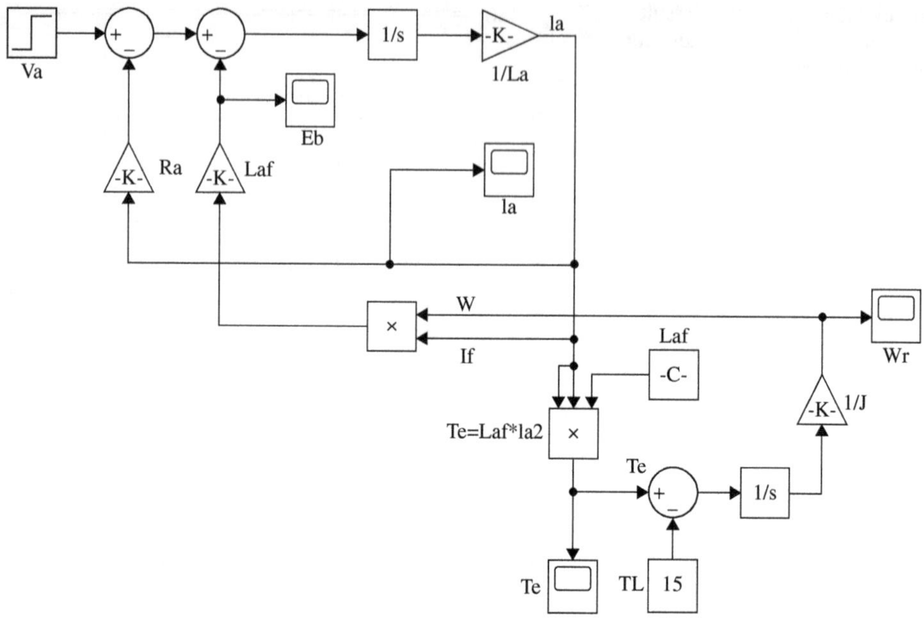

Figure 5.35 MATLAB model for the DC series motor [File Name: DC Motor Series.slx].

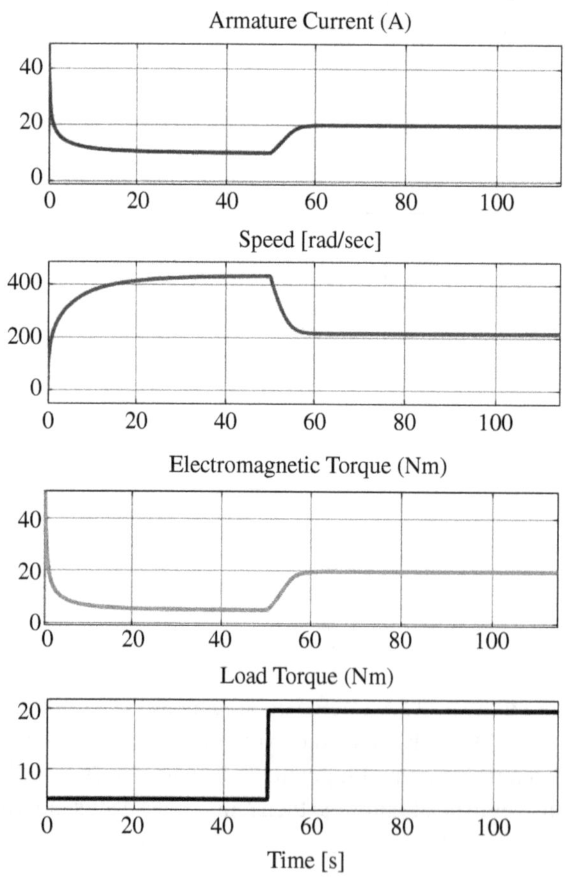

Figure 5.36 Simulation results of a DC series motor.

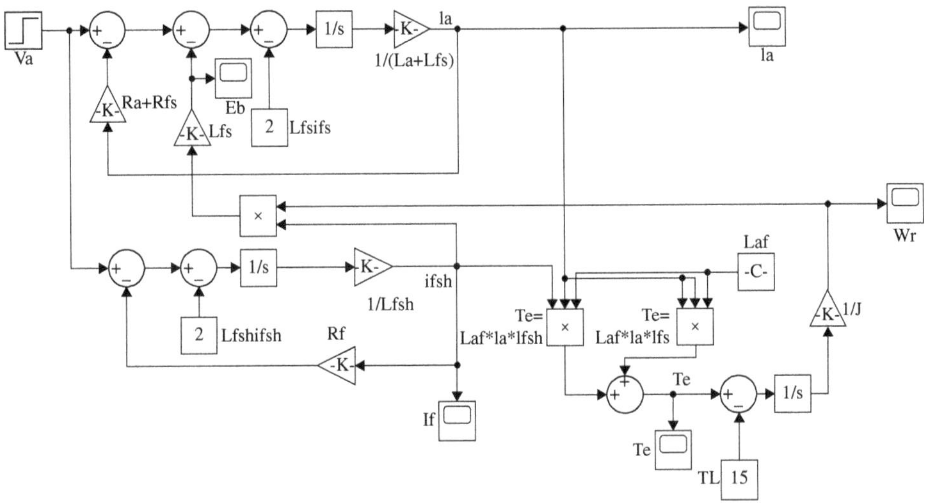

Figure 5.37 MATLAB model for the DC compound motor [File Name: DC Motor Compound.slx].

Figure 5.38 Simulation results of a DC compound motor.

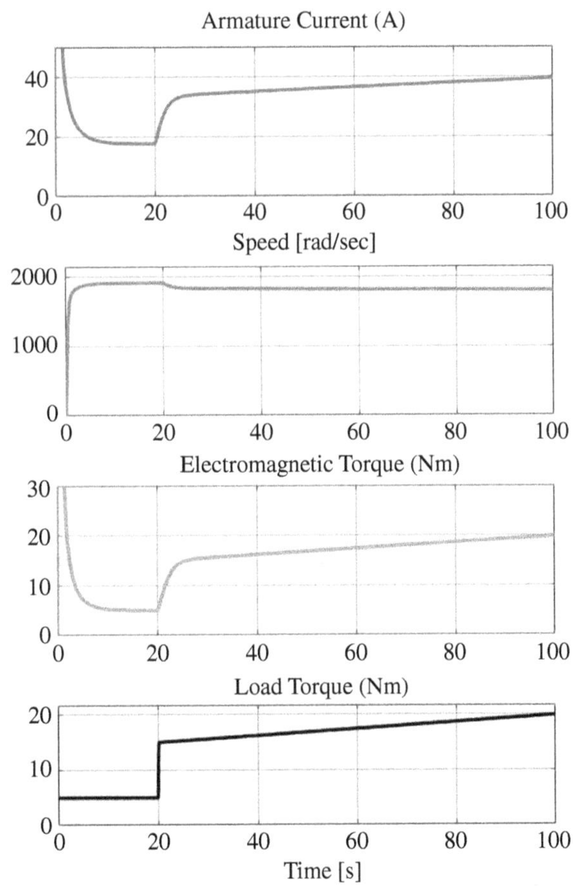

feed the load demand, the current increases again when load is applied. The corresponding plot of speed and electromagnetic torque are also shown in Figure 5.38.

5.19 Summary

This chapter is dedicated to DC machines (both motor and generator). The fundamental operation and control of DC machine is discussed. DC motor and generators and constructionally the same. It is the direction of power/current flow that make this machine operate either as a motor or as a generator. Commutator in a DC machine plays an important role in the operation of a DC machine. Commutator works as a mechanical rectifier that converts AC to DC. However, if the commutator process (the reversal of current) is not appropriate, sparking takes place at the contact point of brush and commutator. This makes this machine vulnerable for use in hazardous environments such as chemical factories. However, the DC series motor characteristics matches the requirement of traction and hence is widely used for such applications. DC shunt and separately excited motors offer constant speed operation and hence are used in such applications where speed drop is not tolerated. Due to widespread applications of Power Electronic Converters, DC generators are rarely used since AC-DC conversion is easily done by means of such power converters. Speed control and starting mechanism are also discussed in the chapter. Matlab/Simulink model is provided for students to work on more advanced control of machines. A large number of solved and unsolved examples are given for clear understanding of the subject.

Problems

5.1 An eight-pole, lap winging, DC machine has 700 active conductors running at a speed of 1500 rpm. The flux per pole is 25 mWb. Calculate the induced armature voltage.

5.2 Calculate the induced armature voltage with the data of problem 1 when the armature is a wave wound.

5.3 If the armature in Problem 5.1 is designed to carry a maximum current of 50 A, what is the maximum electromagnetic power and torque developed?

5.4 A DC machine has a six-pole, wave wound armature with 50 slots and 10 conductors per slot. If the induced armature voltage is 400 V at 1000 rpm, determine the flux per pole.

5.5 A 50 kW, 220 V shunt generator has armature resistance of 0.05 Ω and field resistance of 60 Ω. If the generator operated at the rated voltage, calculate the induced voltage at (a) full load, and (b) half load. Neglect the brush contact drop.

5.6 The generator of Problem 5.5 has a total mechanical and core loss of 0.5 kW. Calculate (a) the generator efficiency at full load and (b) the horsepower output from the prime mover to drive the generator at this load.

5.7 A 25 kW, 200 V short-shunt compound generator has the following data: Ra = 0.05 Ω, Rse = 0.02 Ω, and Rf = 120 Ω. Calculate the induced armature voltage at rated load and terminal voltage. Take 3 V as a total brush contact drop.

5.8 A separately excited DC generator has a constant loss of Pc, and operates at a voltage V with armature current of Ia, the armature resistance is Ra. At what value of Ia, is the generator efficiency at its maximum?

5.9 An eight-pole lap-connected armature of a DC shunt generator is required to supply the loads connected in parallel
2.5 kW motor at 200 V, and 1 kW Lighting load also at 200 V. The Generator has an armature resistance of 0.1 Ω and a field resistance of 200 Ω. The armature has 100 conductors in the slots and runs at 1200 rpm. Allowing 2 V per brush for contact drops and neglecting friction, find
(a) Flux per pole
(b) Armature-current per parallel path.

5.10 A separately excited generator, when running at 900 rpm supplied 100 A at 250 V. What will the load current be when the speed drops to 800 rpm if it is unchanged? Given that the armature resistance = 0.05 Ω and brush drop = 1 V.

5.11 An eight-pole, DC shunt generator with a shunt field resistance of 200 Ω and an armature resistance of 0.5 Ω has 458 wave-connected conductors in its armature. The flux per pole is 0.01 Wb. If a load resistance of 5 Ω is connected across the armature terminals and the generator is driven at 1500 rpm, calculate the power absorbed by the load.

5.12 A 200 V a DC machine has an armature resistance of 0.8 Ω. If the full-load armature current is 35 A, find the induced emf when the machine acts as (i) generator (ii) motor.

5.13 A 400 V, shunt motor has armature resistance of 0.55 Ω and field resistance of 100 Ω. Determine the back emf when giving an output of 5 kW at 90% efficiency.

5.14 A 120 V DC shunt motor runs at 1000 rpm when the armature current is 25 A. Calculate the speed if the torque is doubled. Given Ra = 0.5 Ω.

5.15 A four-pole, 200-V shunt motor has a 440 lap-wound conductor. It takes 22 A from the supply mains and develops output power of 2 kW. The field winding takes 1 A. The armature resistance is 0.1 Ω and the flux per pole is 30 mWb. Calculate (i) the speed and (ii) the torque developed in newton-metre.

5.16 Find the load and full-load speeds for a four-pole, 200 V, and 10 kW, shunt motor having the following data:
Field–current = 2 amp, armature resistance = 0.05 Ω,
Flux per pole = 0.1 Wb, number of armature-conductors = 100, Two-circuit wave connection, full load current = 100 amp, No load current = 4 A. Neglect armature reaction.

5.17 Determine the torque established by the armature of a four-pole DC motor with 442 conductors, two paths in parallel, 50 mWb of pole-flux and the armature current is 30 A.

5.18 A DC series motor takes 20 A at 200 V and runs at 1000 rpm. If the armature and field resistance are 0.3 Ω and 0.2 Ω respectively and the iron and friction losses are 100 W, find the torque developed in the armature. What will the output of the motor be?

5.19 A 200-V shunt motor runs at 1500 rpm at no-load and takes 10 A. The total armature and shunt field resistances are respectively 0.5 Ω and 200 Ω. Calculate the speed when loaded and taking 20 A. Assume the flux to be constant.

5.20 The armature circuit resistance of a 10 kW, 200 V series motor is 0.5 Ω, the brush voltage drop is 2 V, and the series field resistance is 0.1. When the motor takes 10 A, speed is 1000 rpm. Calculate the speed when the current is 15 A.

5.21 A 240 V, DC shunt motor draws 15 A while supplying the rated load at a speed of 80 rad/s. The armature resistance is 0.5 W and the field winding resistance is 80 W. Calculate
(a) The net voltage across the armature resistance at the time of plugging.
(b) The external resistance to be added in the armature circuit to limit the armature current to 125% of its rated value.

5.22 A 220 V shunt motor delivers 10 hp at the shaft at 1020 rpm. If the motor has an efficiency of 92% at this load, determine
(a) The total input power.
(b) The line current.

5.23 A 220 V, 1000 rpm DC motor has a full load armature current of 55 A. It is proposed that a starter be designed which restricts the maximum armature current during the start to 85 A. For design purposes, the minimum current during the starting is to be restricted to 30 A, the series resistance in the armature circuit is being cut out in steps when the current falls to 30 A. Calculate the maximum series resistance used in the starter and the amount of resistance cut out during each of the first three steps. The armature resistance of the DC motor is 0.1 Ω.

5.24 A 5 kW 220 V, DC shunt motor draws a line current of 4.2 A while running at a no-load speed of 1100 rpm from a 220 V power supply. It has an armature resistance of 0.15 Ω and a field resistance of 110 Ω. Estimate the efficiency of the motor at rated load.

5.25 A 120 V shunt motor has an armature resistance of 0.2 Ω and takes a current of 35 A at full load. By how much must the main flux be reduced to raise the speed by 40% if the developed torque is constant?

5.26 A separately excited DC motor runs at 1500 rpm under no-load with 200 V applied to the armature. The field voltage is maintained at its rated value. The speed of the motor, when it delivers a torque of 5 Nm, is 1400 rpm as shown in the Figure P5.26. The rotational losses and armature reaction are neglected. Calculate

Figure P5.26 Speed and torque characteristics of separately excited DC motor.

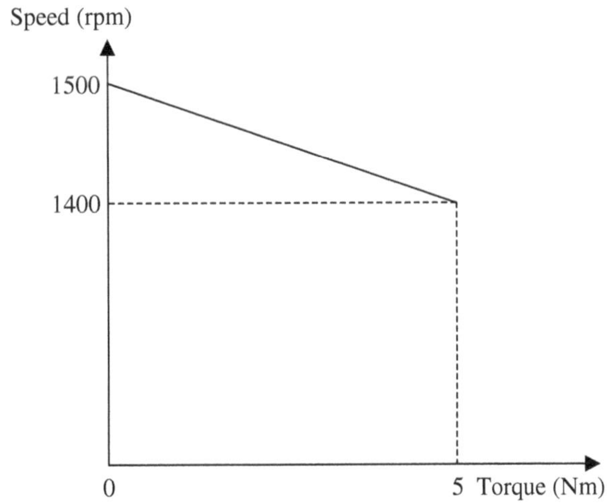

The armature voltage for the motor to deliver a torque of 2.5 Nm at 1400 rpm
The armature resistance of the motor.

5.27 The armature resistance of a permanent magnet DC motor is 0.8 Ω. At no load, the motor draws 1.5 A from a supply voltage of 25 V and runs at 1500 rpm. Find the efficiency of the motor while it is operating on load at 1500 rpm drawing a current of 3.5 A from the same source.

5.28 A 220 V DC machine supplies 20 A at 200 V as a generator. The armature resistance is 0.2 Ω. If the machine is now operated as a motor at same terminal voltage and current but with a flux increased by 10%, calculate the ratio of speed of motor to the speed of generator.

5.29 An eight-pole, DC generator has a simple wave wound armature containing 32 coils of six turns each. Its flux per pole is 0.06 Wb. The machine is running at 250 rpm. Calculate the included armature voltage.

5.30 A DC generator produces 240 V while operating at 1200 rpm with a magnetic flux of 0.02 Wb. The same generator is operated at 1000 rpm with a magnetic flux of 0.05 Wb. Disregarding armature resistance, what new voltage does the generator produces?

5.31 A DC generator operates at 1000 rpm with a magnetic flux of 0.3 Wb at the poles. The armature constant of the generator is 1 V-min/Wb. Calculate the generated voltage.

5.32 The armature of a DC motor draws 50 A while generating a magnetic flux of 0.03 Wb. The armature constant of the motor is 1 V-min/Wb. Calculate the developed torque.

5.33 A three-phase, 440 V, 50 Hz, four-pole slip ring induction motor is fed from the rotor side through an autotransformer and the stator is connected to a variable resistance as shown in the Figure P5.33.

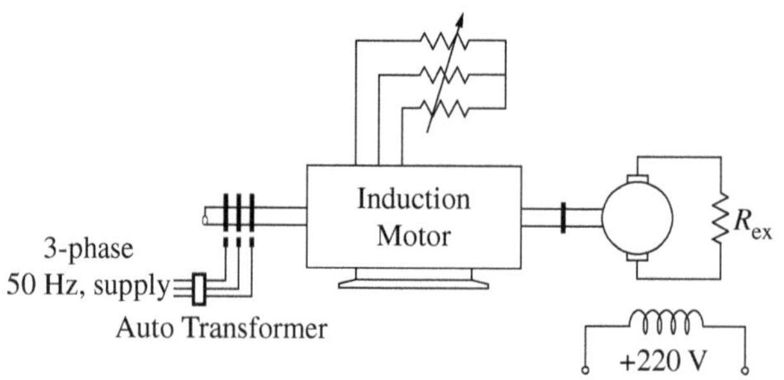

Figure P5.33 Autotransformer and the stator are connected to a variable resistance.

The motor is coupled to a 220 V, separately excited DC generator feeding power to fixed resistance of 10 W. A two-watt meter method is used to measure the input power to the induction motor. The variable resistance is adjusted such that the motor runs at 1410 rpm and the following readings were recorded W1 = 1800 W, W2 = −200 W.

Calculate

(a) The speed of rotation of stator magnetic field with respect to rotor structure.

(b) Neglecting all losses of both the machines, the DC generator power output and the current through resistance (Rex)

5.34 A 240 V, DC shunt motor draws 15 A while supplying the rated load at a speed of 80 rad/s. The armature resistance is 0.5 W and the field winding resistance is 80 W.

Calculate

The net voltage across the armature resistance at the time of plugging.

The external resistance to be added in the armature circuit to limit the armature current to 125% of its rated value.

5.35 A 5 kW, 200 V DC shunt motor has armature resistance of 0.1 Ω and shunt field resistance of 100 Ω. At no load, the motor draws 6 A from 200 V supply and runs at 1000 rpm. Calculate the total loss of the machine at full load.

5.36 A separately excited DC generator has an armature circuit resistance of 0.1 Ω and the total brush-drop is 2 V. When running at 1000 rpm, it delivers a current of 100 A at 250 V to a load of constant resistance. If the generator speed drop to 700 rpm, with field-current unaltered, find the current delivered to load.

5.37 A DC motor takes an armature current of 110 A at 480 V. The armature circuit resistance is 0.2 Ω. The machine has six poles and the armature is lap-connected with 864 conductors. The flux per pole is 0.05 Wb. Calculate

(i) the speed

(ii) the gross torque developed by the armature

5.38 Determine the developed torque and shaft torque of 220-V, four-pole series motor with 800 wave-connected conductors supplying a load of 8.2 kW by taking 45 A from the mains. The flux per pole is 25 mWb and its armature circuit resistance is 0.6 Ω.

5.39 The armature winding of a 200-V, four--pole, series motor is lap-connected. There are 280 slots and each slot has four conductors. The current is 45 A and the flux per pole is 18 mWb. The field resistance is 0.3 Ω; the armature resistance 0.5 Ω and the iron and friction losses total 800 W. The pulley diameter is 0.41 m. Find the pull in newton at the rim of the pulley.

5.40 A 50-h.p. (37.3 kW), 460-V DC shunt motor running light takes a current of 4 A and runs at a speed of 660 rpm. The resistance of the armature circuit (including brushes) is 0.3 Ω and that of the shunt field circuit 270 Ω. Determine when the motor is running at full load

(i) the current input

(ii) the speed. Determine the armature current at which efficiency is maximum. Ignore the effect of armature reaction.

5.41 When operated from a 230-V DC supply, a DC series motor operates at 975 rpm with a line current of 90 A. Its armature-circuit resistance is 0.11 Ω and its series-field resistance is 0.08 Ω. Due to saturation effects, the flux produced by an armature current of 30 A is 48% of that at an armature current of 90 A. Find the motor speed when the armature voltage is 230 V and the armature current is 30 A.

5.42 A series-connected DC motor has an armature resistance of 0.5 Ω and field-winding resistance of 1.5 Ω. In driving a certain load at 1200 rpm, the current drawn by the motor is 20 A from a voltage source of 220 V. The rotational loss is 150 W. Find the output power and efficiency.

5.43 A 230 V, 10 hp. DC shunt motor delivers power to a load at 1200 r/min. The armature current drawn by the motor is 200 A. The armature circuit resistance of the motor is 0.2 Ω and the field resistance is 115 Ω. If the rotational losses are 500 W, what is the value of the load torque?

5.44 A 500-V, 37.3 kW, 1000 rpm DC shunt motor has, on full-load, an efficiency of 90%. The armature circuit resistance is 0.24 Ω and there is total voltage drop of 2 V at the brushes. The field current is 1.8 A. Determine (i) full-load line current (ii) full-load shaft torque in N-m and (iii) total resistance in motor starter to limit the starting current to 1.5 times the full-load current.

5.45 Find the load and full-load speeds for a four-pole, 220-V, and 20-kW, shunt motor having the following data
Field–current = 5 amp
Armature resistance = 0.04 Ω
Flux per pole = 0.04 W
Number of armature-conductors = 160,
Two-circuit wave-connection
Full load current = 95 amp
No load current =9 A. Neglect armature reaction

5.46 A cutting tool exerts a tangential force of 400 N on a steel bar of diameter 10 cm which is being turned in a simple lathe. The lathe is driven by a chain at 840 rpm from a 220 V DC Motor which runs at 1800 rpm. Calculate the current taken by the motor if its efficiency is 80%. What size is the motor pulley if the lathe pulley has a diameter of 24 cm?

5.47 The no-load speed of a 230 V separately excited DC motor is 1400 rpm. The armature resistance drop and the brush drop are neglected. The field current is kept constant at rated value. Calculate the torque of the motor for an armature current of 8 A.

5.48 A 15 kW, 230 V DC shunt motor has armature circuit resistance of 0.4 Ω and field circuit resistance of 230 Ω. At no load and rated voltage, the motor runs at 1400 rpm and the line current drawn by the motor is 5 A. At full load, the motor draws a line current of 70 A. Calculate the full load speed of the motor in rpm.

5.49 A shunt generator us supplying a load of 4 kW. If armature resistance is 0.2 Ω and the induced emf is 250 V, Calculate the load resistance.

5.50 A 220 V, 10 kW DC shunt motor is operating at a speed of 1000 rpm when taking 11 A from supply. Its armature resistance is 2 Ω and the field resistance is 220 Ω. Calculate the developed torque.

5.51 A DC series motor is driving a load which is proportional to the cube of speed. The total resistance of the motor is 2 Ω and motor takes 5 A and runs at 1000 rpm when operating from 200 V supply. Calculate the resistance to be inserted in series with armature to reduce the operating speed to 600 rpm.

Reference

1 Krause, P., Wasynczuk, O., Sudhoff, S., and Pekarek, S. (2013). *Analysis of Electric Machinery and Drive Systems*. IEEE Press.

6

Three-Phase Induction Machine

6.1 Preliminary Remarks

Machine is a general term that is used for both motor and generator. Electrical generators convert mechanical power into electrical power while motors convert electrical power to mechanical power. Both conversions occur via magnetic field. Hence magnetic field is the basic mechanism that transforms one form of energy into another. Amongst several types of electrical motors, induction motors are most widely used in industries and household applications. Induction motors are called the 'workhorse' of industries. Nearly 60% of the total electrical energy generated is consumed by induction machinery [1]. The popularity of the induction machine in industrial application is attributed to the many advantages that this machine offers. They are cheap due to 'off the shelf' availability, simple in design, rugged and maintenance free due to the absence of carbon brushes, run at almost constant speed from no-load to full-load condition. Three-phase induction is a single-fed machine as it requires only one power source for the stator and no power source is needed for the rotor. Rotor voltage is induced due to magnetic field produced by the stator. The principle of operation is simple based on Faraday's law of electromagnetic induction. The motor is self-starting, meaning it starts rotating as soon as supply is switched on due to the inherent production of the starting torque. The characteristics of this machine can be augmented when power electronic converters are used as a supply source.

This chapter is dedicated to the fundamental concepts associated with the operation and characteristics of a three-phase induction motor. The constructional detail is presented that helps to visualize the design of the machine. The steady-state analysis is presented along with the starting and speed control properties. The machine equivalent circuit is presented which is then used for further analysis of the machine characteristics. The testing procedure is also provided that helps in determining the machine parameters. At the end of the chapter, basic the concept of high-performance control of three-phase induction machine is given. Matlab/Simulink model given in this chapter helps in understanding the performance and characteristics of a three-phase induction motor.

Electrical Machine Fundamentals with Numerical Simulation using MATLAB/SIMULINK, First Edition.
Atif Iqbal, Shaikh Moinoddin, and Bhimireddy Prathap Reddy.

Companion website: www.wiley.com/go/iqbal/electricalmachine

6.2 Construction of a Three-Phase Induction Machine

A three-phase induction machine has two parts; stationary called stator and the rotating, called rotor. The stator and rotor are separated by a uniform air-gap. The length of the air gap should be kept as minimum as mechanically possible. This is in done in order to keep the magnetizing current low (air gap offers high reluctance to the flow of flux). Lower magnetizing current means small no-load current and hence better power factor. A block diagram of a three-phase induction machine is shown in Figure 6.1. The stator has slots to accommodate three-phase windings. The rotor has two different structures; (i) squirrel cage type and (ii) wound rotor type or slip ring type.

6.2.1 Stator

Stator is made up from a number of thin laminations of high-grade steel. The lamination has a number of slots in it to accommodate three-phase windings. The laminations have different types of slots; open slot, and semi-closed slots. The slots are chosen depending upon the types of design and requirement. In an open-slot machine, the winding is accessible and easier for maintenance. The advantage of using open slot is lower leakage reactance. But the disadvantage is higher tooth pulsation and high noise. In semi-closed slots, the magnetizing current is low and the machine has quitter operation. Three-phase windings in the stator are connected either in star or in delta. A typical stator lamination is shown in Figure 6.2.

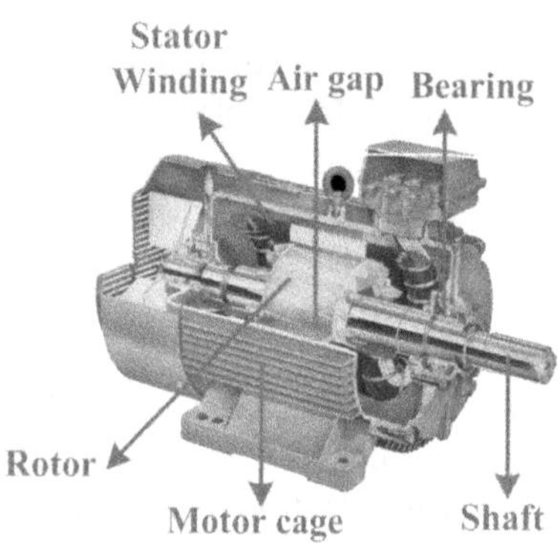

Figure 6.1 Three-phase induction motor.

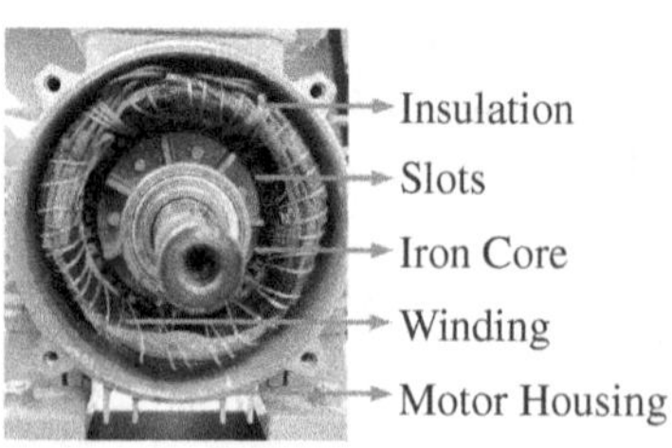

Figure 6.2 A typical stator of an induction motor.

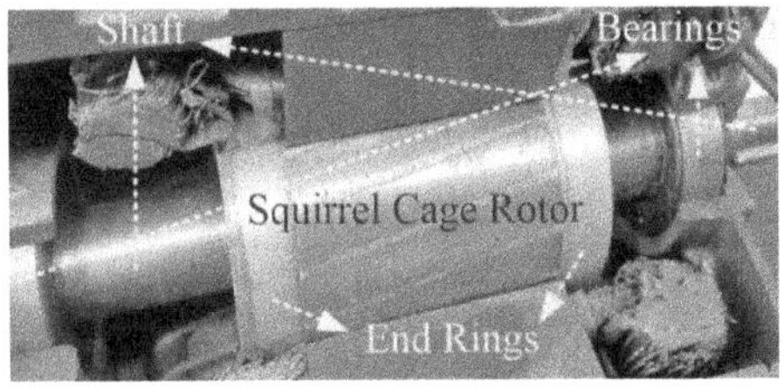

Figure 6.3 Squirrel cage rotor in a three-phase induction motor.

6.2.2 Stator Frame

Stator and rotor are inserted inside the stator frame. The stator frame is a strong body that hold the stator, rotor and the shaft. Stator frame has axial fins that increases the surface area to make cooling better. A block of stator frame is shown in Figure 6.3.

6.2.3 Rotor

Rotor are of two types

a) Squirrel cage type
b) Slip ring type or wound rotor type

The ***Squirrel cage*** type of rotor is the most common in an induction machine. The rotor has copper or aluminium bars that are cast in a solid piece and shorted through end rings as shown in Figure 6.3. The rotor is in the shape of a cage and the bars (that behave as conductor) are slightly skewed. The bars are of the same shape as that of the fur on the back of a squirrel and hence the name is given as 'squirrel cage'. The rotor structure assumes the same number of poles as that of the stator. The construction is highly rugged and strong and require no maintenance.

A ***Wound rotor*** has the same shape as that of the stator and carries a three-phase winding similar to a stator. The three terminals of winding have to be out and should be accessible from the external stationary world (rotor is rotating). From a total of six terminals of three-phase rotor winding, three terminals are joined together to form star point inside and three terminals are taken out of the rotor. The rotor winding should be short circuited through external resistance (which is stationary) to improve its performance, as discussed later. This is only possible if the winding terminals are available. Hence, three copper rings called 'slip rings' are placed on the rotor shaft to which the three winding terminals are connected. External resistances are connected to these slip rings through carbon brushes. A three-phase wound rotor is shown in Figures 6.4 and 6.5. The wound rotor motor is used for high-power applications (MW range). This is also used as a double-fed induction generator for wind energy generation.

It is be to noted, that both the squirrel cage and wound rotor should be short-circuited either directly or through external resistance. External resistance can only be connected in a wound rotor type machine. The machine will not produce any torque if the rotor is left open. This is explained in the next section.

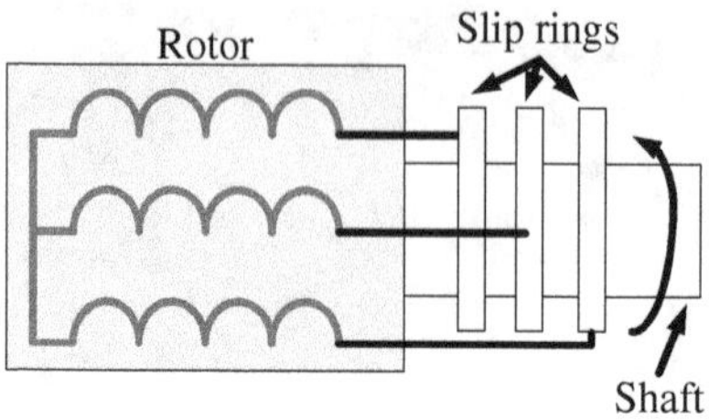

Figure 6.4 Rotor circuit of a wound rotor type three-phase induction motor.

Figure 6.5 Wound rotor of a three-phase induction motor.

6.3 Principle Operation of a Three-Phase Induction Motor

The principle of operation of a three-phase induction motor depends upon Faraday's law of electromagnetic induction. When balanced, three-phase stator winding (120° spatial phase difference) is supplied with a balanced three-phase supply (120° time-phase difference), rotating magnetic field is produced in the air gap (as discussed in Chapter-2). This rotating magnetic field rotates at synchronous speed given as

$$N_s = \frac{120f}{P} \text{ rpm} \tag{6.1}$$

The frequency of the generated voltage depends upon the number of field poles and on the speed at which the field poles are rotated. One complete cycle of voltage is generated in an armature coil when a pair of field poles (one north and one south pole) passes over the coil.

Let

P = Total number of field poles
p = pair of field poles
N = speed of the field poles in rpm
n = speed of the field poles in rps

f = frequency on the generated voltage in Hz

obviously, $\frac{N}{60}=\boldsymbol{n}$

and $\frac{P}{2}=p$

In one revolution of the rotor, an armature coil is cut by $\frac{P}{2}$ north poles and $\frac{P}{2}$ south poles. Since one cycle is generated in an armature coil when a pair of field poles pass over the coil, the number of cycles generated in one revolution of the rotor will be equal to the number of pairs of poles. That is

number of cycles per revolutions = p

also, number of revolutions per second = n

now frequency = number of cycles per second

$$= \frac{\text{number of cycles}}{\text{revolutions}} \times \frac{\text{revolutions}}{\text{seconds}}$$

$$f = p \times n$$

Since n = N/60 and p = P/2

$$f = \frac{PN}{120}$$

Hence

$$f = \frac{N_s P}{120}\ \text{Hz}$$

where f is the frequency of the stator supply and P is the number of poles of the stator winding. The speed is given in rpm.

The rotating magnetic field cuts the rotor conductors. Since the rotor is stationary at the beginning and the magnetic field is rotating, emf is induced in the rotor conductor as per Faraday's law. This induced emf causes current to flow in the rotor since the rotor is short-circuited. If the rotor is open-circuited, no current can flow in the rotor. The rotor induced voltage and current have a different frequency (much lower) than the stator supply frequency. This is given in next section. The induced current in the rotor conductor will cause its own flux according to the Ampere's law (a current carrying conductor produces its own magnetic field). The rotor flux also rotates at synchronous speed. Hence, there will now be two fluxes rotating at synchronous speed and have a phase difference of δ. The interaction of the two fluxes will generate electromagnetic torque in a three-phase induction machine and hence the motor starts moving. The general expression of electromagnetic torque in a round rotor machine is given as

$$T_e = K\phi_s\phi_r \sin(\delta) \tag{6.2}$$

where ϕ_s, ϕ_r are the stator and rotor fluxes, respectively. The stator and rotor flux vectors are shown in Figure 6.6. The angle between the two fluxes should remain fixed in order to have a positive average torque. If the angle between the two fluxes changes, then the torque will be sinusoidal and the average value of any sine functions is zero. Hence, in order to have positive average torque, δ should be constant, and this is only possible if the speed of rotation of the fluxes are same.

It is to be noted that the speed of the rotor never reaches synchronous but it is slightly less than synchronous. If, somehow, the rotor is allowed to run at synchronous speed then

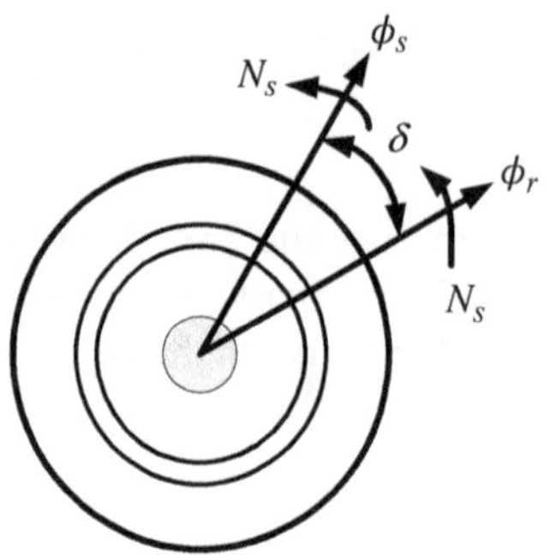

Figure 6.6 Stator and rotor flux vectors.

the rotating magnetic flux (produced from the stator) will appear stationary to the rotor because both rotating stator field and the rotor are now moving at the same speed. If there is no relative speed between the rotor and the stator flux then rate of change of flux will appear to be zero and hence emf cannot be induced in the rotor. Then, no current in the rotor and no torque will be produced. Furthermore, the motor will only operate if the rotor circuit is short-circuited. If the rotor is open-circuited then no current can flow in the rotor and hence no second magnetic field can be produced.

6.3.1 Slip in an Induction Motor

Slip is defined as the per unit or percentage difference between the synchronous speed and actual rotor speed. The difference between synchronous speed and actual speed is called slip speed. Mathematically it is written as

$$s = \frac{N_s - N}{N_s} \text{ p.u.} \tag{6.3}$$

$$s = \frac{N_s - N}{N_s} \times 100\%$$

$$\text{Slip speed} = N_s - N \tag{6.4}$$

where N is the actual speed of three-phase induction machine.

From Eq. (6.3) the following relation can be obtained

$$N = (1 - s)N_s \tag{6.5}$$

Under full-load condition the slip value is 2–5%. When the motor is not rotating or it is stationary, the slip is 1 (since $N = 0$) and the value of slip is 0 if the motor is allowed to run at synchronous speed ($N = Ns$). When an induction machine works as a generator, the speed of rotation of rotor is higher than synchronous value i.e. the speed is super synchronous. Thus, in generating mode, the slip will be negative. During the braking of a motor, the speed of the rotation of rotor is in opposite direction to that of the field. Thus, the slip becomes more than unity. Hence, the following values of slip are obtained

Motoring mode	$0 \le s \le 1$
Generating mode	$s < 0$
Braking/Plugging mode	$s > 1$

6.3.2 Frequency of Rotor Voltage and Current

The frequency of the voltage and current is given as

$$
\begin{aligned}
f_r &= \frac{P}{120} \times (\text{rotor speed with respect to rotating field}) \\
&= \frac{P}{120}(N_s - N) \\
&= \frac{P}{120}\frac{(N_s - N)}{N_s} \times N_s \\
&= \frac{PN_s}{120} \times \frac{(N_s - N)}{N_s} \\
&= sf
\end{aligned} \tag{6.6}
$$

It is to be noted that the speed of the rotor is obtained with respect to the speed of the rotating field. The rotating field's speed is synchronous (Ns) and the speed of rotor is actual (N), hence, the relative speed is the difference between these speeds. Thus, the rotor voltage and current frequency is the product of stator voltage/current frequency times the slip. At start condition, s = 1, and the rotor voltage/current and stator voltage/current frequency are equal. Under full-load condition, the slip is small (say s = 0.04, and assume f = 50 Hz) then the rotor voltage/current frequency will be $f_r = 0.04 \times 50 = 2\,\text{Hz}$. Hence, under full-load condition the rotor voltage/current frequency is very small. Due to the low frequency in the rotor circuit, the iron loss (that depends upon the frequency) is low. This is the reason why the rotor is not laminated (since the hysteresis and eddy current losses will be small due to low frequency).

Example 6.1 A three-phase 50 Hz, 440 V, six-pole Induction Motor runs at 920 rpm under full-load condition. Compute the following;

(i) Speed of the rotating field
(ii) The slip under start condition and under full-load condition
(iii) Frequency of the rotor current under start condition and full-load
(iv) Slip Speed
(v) Speed of the stator field with respect to (a) Stator Structure, (b) Rotor Structure
(vi) Speed of the rotor field with respect to (a) Stator Structure, (b) Rotor Structure
(vii) Rotor induced voltage at full-load condition (assuming the stator induced voltage is the same as the applied voltage) and the turn ratio (N1/N2 = 2)

Solution

(i) Speed of the rotating field is equal to synchronous (Ns):

$$N_s = \frac{120f}{P} = \frac{120 \times 50}{6} = 1000\ rpm$$

Slip is given as: $S = \frac{N_s - N}{N_s}$

(ii) At start N = 0, therefore S = 1

At full load: $S = \frac{N_s - N}{N_s} = \frac{1000-940}{1000} = 0.06$

(iii) Rotor current frequency is given as $f_r = sf$, At standstill s = 1, therefore $f_r = f = 50$ Hz
At full load, s = 0.06 hence $f_r = 0.06 \times 50 = 3$ Hz
(iv) Slip Speed is given as $N_{slip} = N_s - N = 1000 - 940 = 60 rpm$
(v) Stator field speed wrt stator = $N_s - 0 = 1000 - 0 = 1000$ rpm (since the stator is stationary and the stator field rotates at N_s)
Stator field speed wrt rotor = $N_s - N = 1000 - 940 = 60$ rpm
(vi) Rotor field speed wrt stator = $N_s - N = 1000 - 0 = 1000$ rpm (since the rotor field is rotating at Ns and the stator is stationary)
Rotor field speed wrt rotor = $N_s - N = 1000 - 960 = 40$ rpm
(vii) Solution?

6.3.3 Induction Machine and Transformer

Both an induction machine and transformer work on the principle of electromagnetic induction. The induction machine is a rotating machine while a transformer is a stationary machine.

In a transformer, voltage is applied to the primary windings that causes induced voltage in the secondary windings. In an induction motor, voltage is applied to the stator and voltage is induced in the rotor winding/conductors. The frequency of the induced voltage in a transformer has same frequency as that of applied voltage of primary. However, in an induction motor, the frequency of rotor voltage is different from applied voltage, due to the rotation of the rotor.

6.4 Per-phase Equivalent Circuit of a Three-Phase Induction Machine

The analysis of the induction motor is done by obtaining its per-phase equivalent circuit. This is used for analysis as it is assumed that all three phases are identical with a phase shift of ±120°. However, if the supply is unbalanced or the windings are not identical then the analysis should be done for each phase separately. Supply is considered three-phase balanced (three phases supplied have equal magnitude and phase displaced in time by ±120°) and the winding is also considered as three-phase balanced (three phases have an equal number of turns and phase displaced in space by ±120°).

The rotating air-gap field produced due to three-phase current supplied to three-phase balanced stator winding rotates at synchronous speed. This rotating field induces a balanced three-phase emf in stator. This counter emf is the difference between the applied voltage and the stator winding voltage drop due to stator winding resistance and leakage reactance.

$$E_1 = V_1 - I_1 \left(R_1 + jX_1\right) \tag{6.7}$$

where

$$E_1 = 4.44 K_{w1} f N_1 \phi_m = 4.44 N_{1eff} f \phi_m \tag{6.8}$$

where K_w is the stator winding factor with $0 \le K_w \le 1$ (usually the value ranges from 0.9 to 0.98). This winding factor arises due to short pitching and distribution of the stator winding. N_1 is the number of turns per phase and ϕ_m is the flux per pole.

The resultant air-gap field in the machine is due to stator and rotor currents. The stator current has two components; load component (I'_2) and no-load component (I_Φ). The no-load current component has two components; magnetizing and core-loss. The magnetizing component and core-loss components are accounted for using shunt branch with R_c and X_m. The current through the shunt branch depends upon the applied voltage. Since the applied voltage is constant equal to the rated value, these currents are constant. The iron/core loss is constant and does not depend upon the load on the machine.

The rotor equivalent circuit has induced emf sE_2 (E_2 is standstill induced emf), rotor winding resistance per-phase is R_2 and the rotor leakage reactance is sX_2. Induced emf in the rotor is given as

$$E_2 = 4.44K_{w2}fN_2\phi_m = 4.44N_{2eff}f\phi_m \tag{6.9}$$

At standstill, the stator and rotor currents frequencies are same. Under running condition, the rotor induced emf is

$$E_{2s} = 4.44K_w f_2 N_2 \phi_m = sE_2 \tag{6.10}$$

The leakage reactance is also a function of frequency and hence the slip, thus it is given as

$$X_{2s} = 2\pi f_2 L_2 = 2\pi sfL_2 = sX_2 \tag{6.11}$$

The equivalent circuit of a three-phase induction motor on a single-phase basis is shown in Figure 6.7. Here, $a = N_{1eff}/N_{2eff}$.

To obtain an equivalent circuit with all parameters referred to one side, the following procedure is done. Apply Kirchhoff's voltage law (KVL) on the rotor side circuit

$$sE_2 = I_2\left[R_2 + jsX_2\right] \tag{6.12}$$

Dividing both sides by slip s, the following is obtained

$$E_2 = I_2\left[\frac{R_2}{s} + jX_2\right] \tag{6.13}$$

Recall the following relationship between the induced emf on stator side and rotor side

$$\frac{E_1}{E_2} = \frac{N_{1eff}}{N_{2eff}} = a \tag{6.14}$$

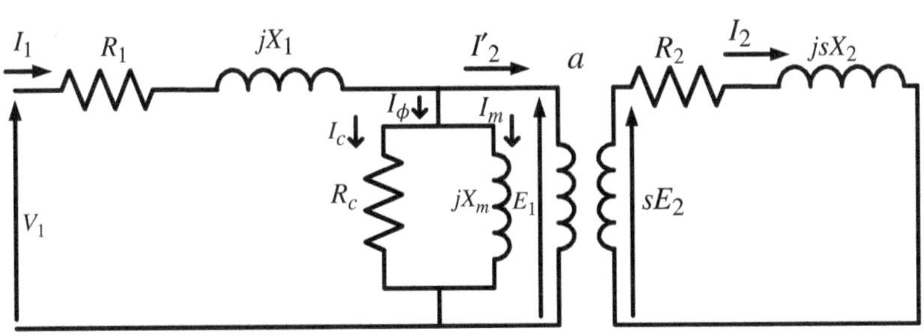

Figure 6.7 Equivalent circuit of a three-phase induction motor.

$$\frac{I_2}{I_2'} = a \tag{6.15}$$

Replacing E_2 From (6.14) and I_2 from (6.15), in Eq. (6.13), the following is obtained

$$\frac{E_1}{a} = aI_2' \left[\frac{R_2}{s} + jX_2\right] \tag{6.16}$$

Multiplying both sides by the turn ratio 'a', the following expression is obtained

$$E_1 = a^2 I_2' \left[\frac{R_2}{s} + jX_2\right] \tag{6.17}$$

$$E_1 = I_2' \left[\frac{a^2 R_2}{s} + ja^2 X_2\right] \tag{6.18}$$

The above equation can be re-written as

$$E_1 = I_2' \left[\frac{R_2'}{s} + jX_2'\right] \tag{6.19}$$

In the above Eq. (6.19), the following relations hold

$R_2' = a^2 R_2$, Rotor resistance referred to the stator side

$X_2' = a^2 X_2$, Rotor leakage reactance referred to the stator side

$I_2' = \frac{I_2}{a}$, Rotor current referred to the stator side

Eq. (6.12, 6.19) are equivalent. Hence, sE_2 can be replaced by E_1, R_2 can be replaced by R_2' and X_2 can be replaced by X_2'. By this substitution of the parameters, the equivalent circuit will change as shown in Figure 6.8. (Look at the rotor circuit in the right-hand side).

Since the voltages on the stator side and rotor side are same, they can be joined together to form a single-sided equivalent circuit as shown in Figure 6.9.

The rotor side resistance can be written as

$$\frac{R_2'}{s} = \frac{R_2'}{s} + R_2' - R_2' = R_2' + R_2'\left[\frac{1}{s} - 1\right] \tag{6.20}$$

Thus, one resistance can be shown as two resistances, one is fixed R_2' that represents the rotor copper loss while the other one is variable resistance since slip 's' is variable and this part represents the mechanical power developed by the machine. The overall per-phase equivalent circuit of three-phase induction motor is shown in Figure 6.10. The applied voltage V_1 is per phase value and the currents are also per phase.

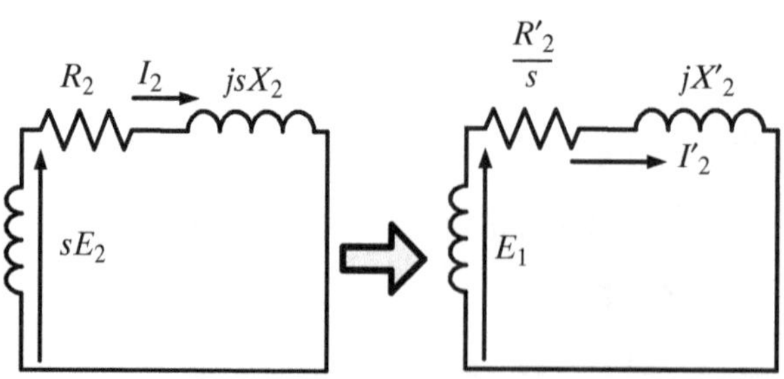

Figure 6.8 Rotor side equivalent circuit.

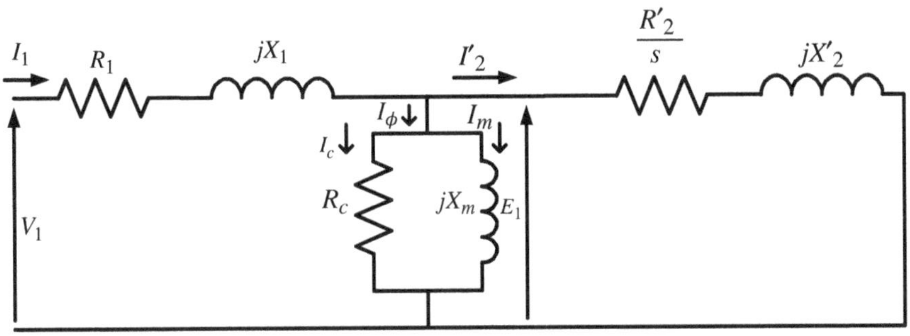

Figure 6.9 Per-phase equivalent circuit of a three-phase induction motor referred to the stator side.

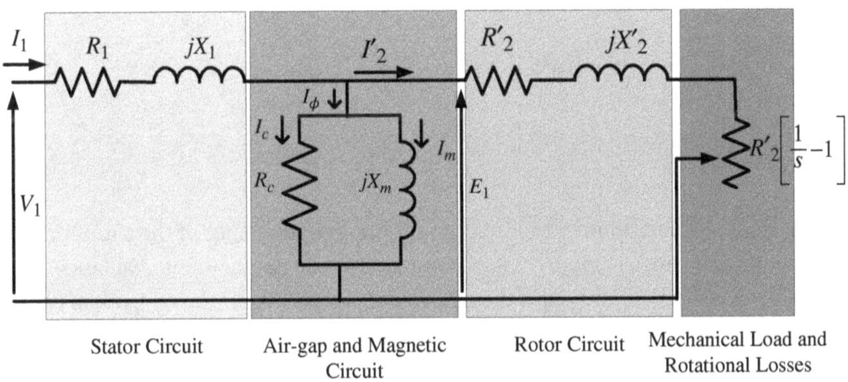

Figure 6.10 Per-phase exact equivalent circuit of a three-phase induction motor referred to the stator side.

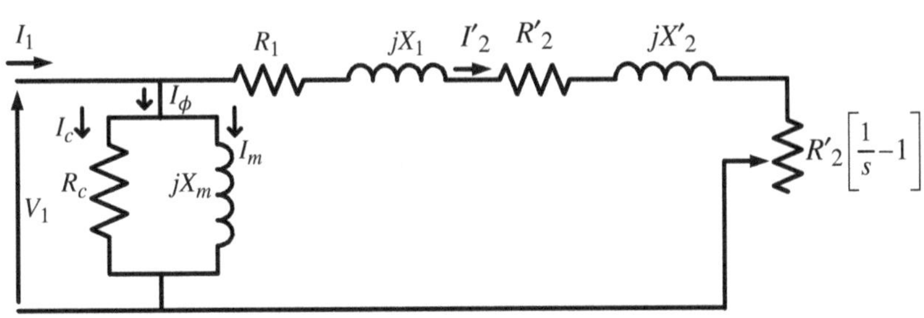

Figure 6.11 Per-phase approximate equivalent circuit of a three-phase induction motor.

The above circuit can be approximated by moving the shunt branch to the source side as shown in Figure 6.11. The shunt branch can be moved to the source side, assuming that the voltage drop at the stator resistance and leakage reactance is small i.e. assuming the applied voltage V_1 and induced voltage E_1 same.

After obtaining the approximate equivalent circuit as shown in Figure 6.11, the following relations can be derived

Impedance of series branch $Z_{se} = \left(R_1 + \frac{R'_2}{s}\right) + j\left(X_1 + X'_2\right)$

Hence, the rotor current referred to the stator side can be obtained as

$$I_2' = \frac{V_1}{Z_{se}} = \frac{V_1}{\left(R_1 + \frac{R_2'}{s}\right) + j\left(X_1 + X_2'\right)} = |I_2'| < -\varphi \tag{6.21}$$

$$|I_2'| = \frac{|V_1|}{\sqrt{\left(R_1 + \frac{R_2'}{s}\right)^2 + \left(X_1 + X_2'\right)^2}} \tag{6.22}$$

$$< -\varphi = -\tan^{-1}\frac{\left(X_1 + X_2'\right)}{\left(R_1 + \frac{R_2'}{s}\right)} \tag{6.23}$$

$$\begin{aligned} I_1 &= I_\varphi + I_2' \\ I_\varphi &= I_m + I_c = \frac{V_1}{jX_m} + \frac{V_1}{R_c}, Also \\ I_m &= I_\varphi \sin(\omega t) \\ I_c &= I_\varphi \cos(\omega t) \end{aligned} \tag{6.24}$$

Another approximation is done by neglecting the core-loss resistance of the shunt branch as the value of this resistance is much higher compared to the magnetizing reactance. This simplification gives a simpler analysis of the equivalent circuit. This is the equivalent circuit that is recommended by IEEE.

The whole equivalent circuit as shown in Figure 6.12 can be represented as a single source and an impedance load.

Impedance of the rotor circuit is given as

$$Z_r = \frac{R_2'}{s} + jX_2' \tag{6.25}$$

This is in parallel to the shunt magnetizing branch

$$Z_1 = Z_r || jX_m$$

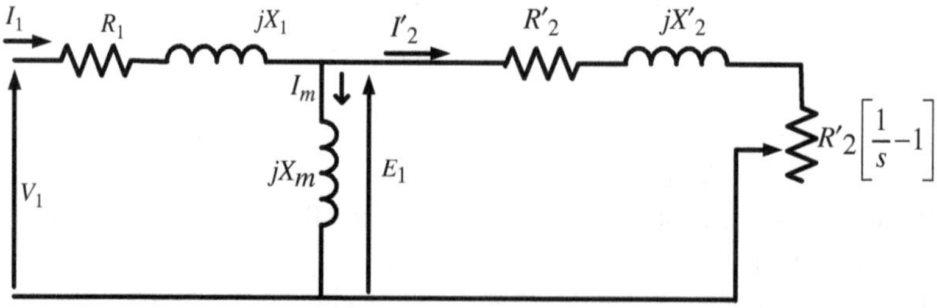

Figure 6.12 Per-phase approximate equivalent circuit of a three-phase induction motor (IEEE Recommended).

Figure 6.13 Single impedance equivalent circuit.

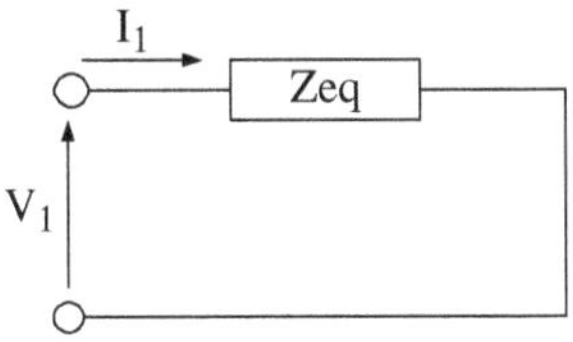

$$\left.\begin{aligned} Z_1 &= \frac{\frac{R_2'X_m^2}{s} + jX_m\left\{\left\{\frac{R_2'}{s}\right\}^2 + X_2'^2 + X_2'X_m\right\}}{\left[\frac{R_2'}{s}\right]^2 + \{X_2'+X_m\}^2} \\ &= \frac{\frac{R_2'X_m^2}{s}}{\left\{\frac{R_2'}{s}\right\}^2 + \{X_2'+X_m\}^2} + j\frac{\left\{\frac{R_2'}{s}\right\}^2 X_m + X_m X_2'^2 + X_2'X_m^2}{\left\{\frac{R_2'}{s}\right\}^2 + \{X_2'+X_m\}^2} \end{aligned}\right\} \tag{6.26}$$

The final equivalent impedance is obtained as

$$Z_{eq} = Z_s + Z_1$$

$$Z_{eq} = (R_1 + jX_1) + \frac{\frac{R_2'X_m^2}{s} + jX_m\left\{\left(\left(\frac{R_2'}{s}\right)^2 + X_2'^2 + X_2'X_m\right)\right\}}{\left(\frac{R_2'}{s}\right)^2 + (X_2'+X_m)^2}$$

$$Z_{eq} = \frac{\left[R_1\left\{\left(\frac{R_2'}{s}\right)^2 + (X_2'+X_m)^2\right\} + \frac{R_2'X_m^2}{s}\right] + j\left[X_m\left\{\left(\left(\frac{R_2'}{s}\right)^2 + X_2'^2 + X_2'X_m\right)\right\} + X_1\left\{\left(\frac{R_2'}{s}\right)^2 + (X_2'+X_m)^2\right\}\right]}{\left(\frac{R_2'}{s}\right)^2 + (X_2'+X_m)^2} \tag{6.27}$$

With this, the final equivalent circuit can be shown as in (Figure 6.13);
Stator and rotor currents can be obtained as

$$I_1 = \frac{V_1}{Z_{eq}} \tag{6.28}$$

Magnetizing branch current is obtained as

$$I_m = \frac{E_1}{X_m}; \quad E_1 = V_1 - I_1\left(R_1 + jX_1\right) \tag{6.29}$$

By using the current divider rule, rotor current can be obtained as

$$I_2' = \frac{jX_m}{\frac{R_2'}{s} + jX_2'} I_1 \tag{6.30}$$

Example 6.2 The equivalent circuit parameter of a three-phase 10 kW, four-pole, 50 Hz, 440 V star-connected induction machine is given as:

$R_1 = 0.2\,\Omega$, $R_2{}' = 1.25\,\Omega$, $X_1 = 1.5\,\Omega$, $X_2{}' = 2.6\,\Omega$, $X_{\mathrm{m}} = 100\,\Omega$, $R_{\mathrm{c}} = 150\,\Omega$

The machine runs at 1440 rpm under full-load condition. Determine under full-load condition:

(i) The rotor current
(ii) The no-load current
(iii) The core-loss component of no-load current
(iv) The magnetizing component of no-load current
(v) The stator current, and
(vi) Input Power Factor under rated-load condition
(vii) Input Power Factor under no-load condition.

Use an approximate equivalent circuit with shunt branch at the source side.

Solution

The equivalent circuit is shown below in (Figure E6.2).

The slip is $S = \frac{N_s - N}{N_s} = \frac{1500-1440}{1500} = 0.04$

(i) Rotor resistance $\frac{R_2'}{s} = 1.25/0.04 = 31.25\,\Omega$

Equivalent impedance $Z_1 = R_1 + \frac{R_2'}{s} + j(X_1 + X_2')\,\Omega$
$= 0.2 + 31.25 + j\,(1.5 + 2.6)\,\Omega$
$= 31.45 + j4.1 = 31.7161\angle 7.427\Omega$

Applied Voltage $= V_1 = \frac{440}{\sqrt{3}} = 254.034V$

Rotor Current: $I_2' = \frac{V_1}{Z_1} = \frac{254.034\angle 0}{31.7161\angle 7.427} = 8\angle -7.427A$

(ii) Shunt Branch equivalent impedance $Z_{sh} = R_c \,||\, jX_m = (100*150\angle 90)/(150+j100) = 83.205\angle 56.31$

Shunt Branch Current (No-load current): $I_\phi = \frac{V_1}{Z_{sh}} = \frac{254.034\angle 0}{83.205\angle 56.31} = 3.053\angle -56.31A$

(iii) Core loss component: $|I_c| = 3.053\cos(-56.31) = 1.6935$ A

(iv) Magnetizing component $|I_m| = 3.053\sin(56.31) = 2.54$ A

(v) Stator current:

$$\bar{I}_1 = \bar{I}_\phi + \bar{I}_2' = 3.053\angle -56.31 + 8\angle -7.427 = 1.6935 - j2.54 + 7.933 - j1.034$$

$$\bar{I}_1 = 10.27\angle -20.368$$

(vi) Input Power Factor under rated-load condition = cos(20.368) = 0.9374

(vii) Input Power Factor under no-load condition = cos(56.31) = 0.554

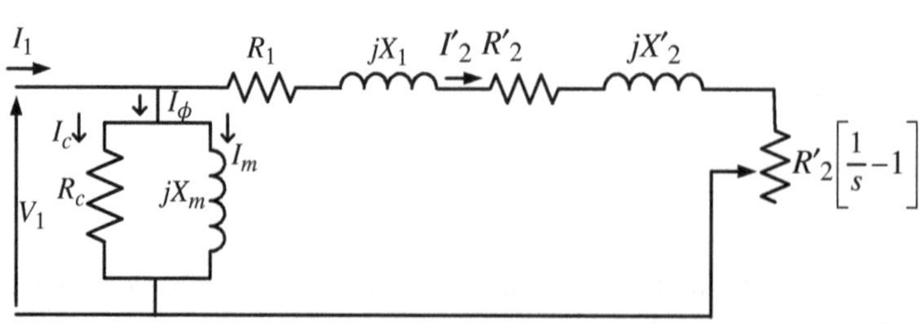

Figure E6.2 Equivalent circuit of three-phase induction motor.

6.5 Power Flow Diagram in a Three-Phase Induction Motor

The input power to an induction motor is three-phase electrical power given as

$$P_{in} = \sqrt{3}V_L I_L \cos(\theta) = 3V_1 I_1 \cos(\theta) \tag{6.31}$$

The electrical power is input to the stator side and out of this input there are losses in the stator core and winding. The power loss in stator winding is called 'copper loss' P_{scl}. The power loss in the core is called 'stator iron loss or stator core loss' P_{si}. This loss is comprised of hysteresis and eddy current losses. The hysteresis loss is attributed to the magnetic property of the stator core. The eddy current loss is due to the eddy current that is caused by the induced emf on the stator core. These losses are given as

$$P_{scl} = 3(I_1)^2 R_1 \tag{6.32}$$

$$P_{si} = 3(I_c)^2 R_c \tag{6.33}$$

The power that remained after subtracting the stator losses from the input power, is transferred to the rotor side and is called, 'Air-gap power' P_{g}. This power is also called 'synchronous watt' since this power is transferred to the rotor through air-gap crossing the magnetic field that is rotating at synchronous speed. Air-gap power is the total power input to the rotor (power dissipated in the total rotor resistance which is $\frac{R_2'}{s}$) and hence is given as

$$P_g = P_{in} - P_{scl} - P_{si} \tag{6.34}$$

$$P_g = 3(I_2')^2 \frac{R_2'}{s} \tag{6.35}$$

The air-gap power is the input power to the rotor. In the rotor, power is lost in the winding (wound rotor type) or conductors (squirrel cage type) resistance, called 'rotor copper loss' P_{rcl}. A small amount of power is lost in the core of the rotor but is negligible. Some unknown losses that occur in an induction motor are called 'stray losses'. The 'stray-load loss' appears under load condition. The modified equivalent circuit considering the 'stray-load loss' is discussed in the next section. The power that remains after the losses in the rotor is 'mechanical power developed', P_{m}, this is also called 'gross mechanical power'.

$$P_m = P_g - P_{rcl} \tag{6.36}$$

$$P_m = 3(I_2')^2 R_2' \left[\frac{1}{s} - 1\right] \tag{6.37}$$

$$P_{rcl} = 3(I_2')^2 R_2' \tag{6.38}$$

The 'mechanical torque developed' or 'electromagnetic torque' is given as

$$T_e = \frac{P_m}{\omega_r} = \frac{P_m}{\left(\frac{2\pi N}{60}\right)} \text{ Nm} \tag{6.39}$$

And,

$$T_e = \frac{P_g}{\omega_s} = \frac{P_g}{\left(\frac{2\pi N_s}{60}\right)} \tag{6.40}$$

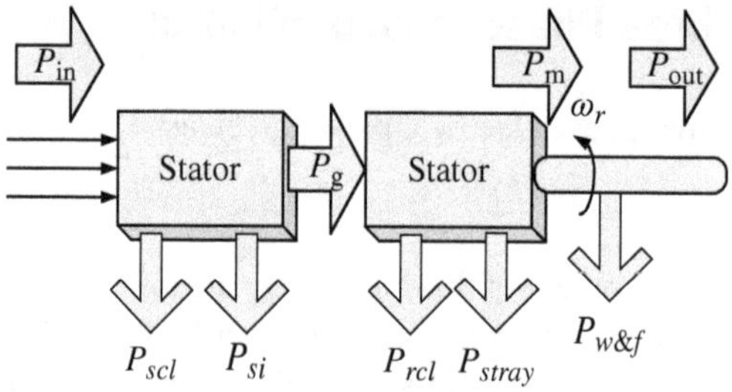

Figure 6.14 Power flow diagram in a three-phase induction motor.

Out of the 'mechanical power developed' some is lost as mechanical losses, that are 'friction and windage losses'. Friction losses occur because the rotor is moving and there are bearings and friction occurs in the rotation. Windage losses are the extra power that the motor exerts to rotate against the air friction. Thus, the net mechanical power available at the shaft is given as

$$P_{out} = P_m - P_{wcf} \tag{6.41}$$

The output torque is given as

$$T_{out} = \frac{P_{out}}{\omega_r} = \frac{P_{out}}{\left(\frac{2\pi N}{60}\right)} \tag{6.42}$$

The overall power flow from electrical input to mechanical output in a three-phase induction motor is shown in Figure 6.14.

6.6 Power Relations in a Three-Phase Induction Motor

A simple relationship can be derived between air-gap power, rotor copper loss and mechanical power developed. Let us rewrite these power equations once more

$$P_g = 3\left(I_2'\right)^2 \frac{R_2'}{s} \tag{6.43}$$

From Eqs. (6.37) and (6.38)

$$P_m = 3\left(I_2'\right)^2 R_2' \left[\frac{1}{s} - 1\right]$$
$$P_{rcl} = 3\left(I_2'\right)^2 R_2' \tag{6.44}$$

The air-gap can be written in terms of rotor copper loss from Eq. (6.35, 6.44) as

$$P_g = \frac{P_{rcl}}{s} \tag{6.45}$$

$$P_{rcl} = sP_g \tag{6.46}$$

It is seen that the rotor copper losses are proportional to the slip. The relationship between mechanical power developed and air-gap power is given from Eq. (6.43) and Eq. (6.44) as

$$\begin{aligned} P_m &= 3\left(I_2'\right)^2 R_2' \left[\frac{1}{s} - 1\right] = 3\left(I_2'\right)^2 \frac{R_2'}{s} - 3\left(I_2'\right)^2 R_2' = P_g - P_{rcl} \\ &= P_g - sP_g \\ &= (1 - s) P_g \end{aligned} \tag{6.47}$$

Thus one can write the following relations

$$P_g : P_{rcl} : P_m = 1 : s : (1 - s) \tag{6.48}$$

6.7 Steps to Find Powers and Efficiency

Step 1: Find the equivalent impedance (depending upon the type of equivalent circuit you use, the impedance equation will be different)
If the exact equivalent circuit is used, the impedance is:

$$\dot{Z}_{eq} = R_1 + jX_1 + \frac{jX_m\left(\frac{R_2'}{s} + jX_2'\right)}{\frac{R_2'}{s} + j\left(X_m + X_2'\right)}$$

Step 2: Determine the stator current

$$\dot{I}_1 = \frac{V_1}{\dot{Z}_{eq}} = |\dot{I}_1| \angle\phi$$

Step 3: Calculate the power factor

$$\cos\phi$$

Step 4: Calculate the input power:

$$P_{in} = 3V_1 I_1 \cos\phi$$

Step 5: Calculate the stator copper loss

$$P_{scl} = 3I_1^2 R_1$$

Step 6: Calculate air-gap power (which is the total input to the rotor)

$$P_g = P_{in} - P_{scl} = 3\frac{I_2'^2 R_2'}{s} = T_{dev}\omega_s$$

Step 7: Calculate the rotor copper loss

$$P_{rcl} = 3I_2^{2\prime} R_2' = sP_g$$

Step 8: Calculate the developed power and torque

$$P_m = 3I_2'^2 R_2' \frac{1-s}{s} = (1-s)P_g \Rightarrow T_e = \frac{P_m}{\omega_m}$$

Step 9: Calculate the output power and torque

$$P_{out} = P_m - P_{wcf} \Rightarrow T_{out} = \frac{P_m}{\omega_r}$$

Step 10: Calculate the efficiency

$$\eta = \frac{P_{out}}{P_{in}} \times 100\%$$

Example 6.3 A three-phase 50 Hz, four-pole, 10 kW, star-connected 440 V induction motor has windage and friction loss of 0.3 kW. The full-load speed is 1450 rpm. Given stator losses as 0.5 kW. Calculate the following

(i) Mechanical Power Developed
(ii) Torque Developed
(iii) Air-gap power
(iv) Rotor Copper Loss
(v) Input power
(vi) Efficiency

Solution

Synchronous speed $N_s = \frac{120f}{P} = \frac{120 \times 50}{4} = 1500rpm$

Slip $S = \frac{N_s - N}{N_s} = \frac{1500-1450}{1500} = 0.033$

(i) Mechanical Power Developed $P_m = P_{out} - P_{W\&f} = 10 - 0.3 = 9.7kW$
(ii) Torque Developed $T_e = \frac{P_m}{\omega_m} = \frac{P_m}{2\pi N/60} = \frac{9.7}{2\pi 1450/60} = 63.88Nm$
(iii) Air-gap Power $P_g = \frac{P_m}{1-s} = \frac{9.7}{1-0.033} = 10.031kW$
(iv) Rotor Copper Loss $P_{rcl} = sP_g = 0.033 \times 10.031 = 0.331kW$
(v) Input Power $P_{in} = P_g + P_{stator} = 10.031 + 1 = 11.531kW$
(vi) Efficiency $\eta = \frac{10}{11.531} \times 100\% = 86.72\%$

Example 6.4 The equivalent circuit parameter of a three-phase four-pole, 50 Hz, 440 V star-connected induction machine is given as:

$R_1 = 0.25\,\Omega$, $R_2' = 1.5\,\Omega$, $X_1 = 1.8\,\Omega$, $X_2' = 2.9\,\Omega$, $X_m = 150\,\Omega$, the windage & Friction loss = 0.3 kW.

The machine runs at 1440 rpm under full-load condition. Using the IEEE Recommended circuit, under full-load condition, determine

(i) Power Developed
(ii) Torque Developed
(iii) Output/Shaft power
(iv) Rotor Copper Loss
(v) Air-gap Power
(vi) Stator Copper Loss
(vii) Input power
(viii) Input Power Factor
(ix) Efficiency

Solution

The IEEE recommended equivalent circuit is shown in (Figure E6.4):

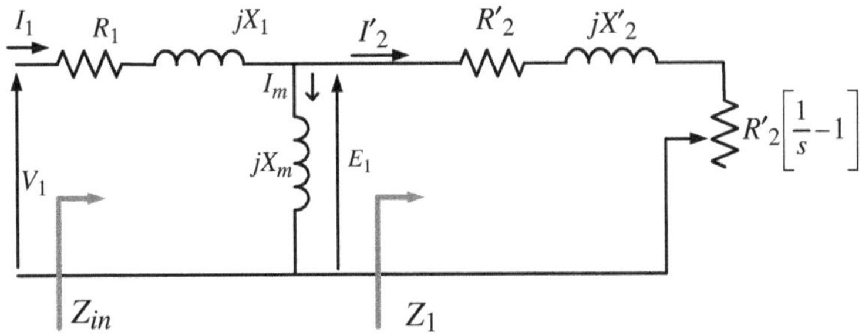

Figure E6.4 Equivalent circuit of three-phase induction motor (IEEE recommended).

Synchronous speed $N_s = \frac{120f}{P} = \frac{120\times50}{4} = 1500rpm$
Slip $S = \frac{N_s-N}{N_s} = \frac{1500-1440}{1500} = 0.04$
Equivalent Impedance

$$Z_1 = \frac{R_2'}{s} + jX_2' = 1.5/0.04 + \text{j}\,2.9 = 37.5 + \text{j}\,2.9 = 37.612 < 4.422$$

$$Z_{\text{in}} = \left(R_1 + jX_1\right) + \left(jX_{\text{m}} \parallel Z_1\right) = \left(0.25 + \text{j}\,1.8\right) + \text{j}150 \parallel 37.612 < 4.422$$
$$= 34.293 + j\,13$$

$$Z_{\text{in}} = 36.674 < 20.76$$

Stator current

$$I_1 = \frac{V_1}{Z_{in}} = \frac{440/\sqrt{3}\angle 0}{36.674\angle 20.76} = 6.927\angle - 20.76A$$

Using the current divider rule to find the rotor current

$$I_2' = \frac{jX_m}{\frac{R_2'}{S} + j\left(X_m + X_2'\right)} I_1 = \frac{j150}{\frac{1.5}{0.04} + j\,(150 + 2.9)} 6.927\angle - 20.76$$
$$= \frac{150\angle 90}{37.5 + j152.9} 6.927\angle - 20.76 = \frac{150\angle 90}{157.431\angle 76.22} 6.927\angle - 20.76$$
$$= 6.601\angle - 6.98A$$

(i) Power Developed $P_m = 3I_2'^2 R_2' \frac{1-s}{s} = 3(6.601)^2 1.5\left(\frac{1-0.04}{0.04}\right) = 4.705kW$
(ii) $T_e = \frac{P_m}{\omega_m} = \frac{P_m}{2\pi N/60} = \frac{4.705}{2\pi 1440/60} = 31.2\ Nm$
(iii) Output Power $P_{\text{out}} = P_{\text{m}} - P_{\text{w\&f}} = 4.705 - 0.3 = 4.405\,\text{kW}$
(iv) Rotor Copper Loss $P_{rcl} = 3\left(I_2'\right)^2 R_2' = 3X(6.601)^2 \times 1.5 = 196.08W$
(v) Air-gap Power $P_{rcl} = sP_g$ $P_g = \frac{P_{rcl}}{s} = \frac{196.08}{0.04} = 4.901kW$
(vi) Stator Copper Loss $P_{scl} = 3(I_1)^2 R_1 = 3(6.927)^2 0.25 = 35.98W$
(vii) Input Power $P_{in} = \sqrt{3}V_1 I_1 \cos(\theta) = \sqrt{3} \times 440 \times 6.927 \cos(20.76) = 4.936kW$
(viii) Input Power Factor $\cos(20.76) = 0.935$
(ix) Efficiency $\eta = \frac{P_{out}}{P_{in}} \times 100\% = \frac{4.405}{4.936} \times 100\% = 89.24\%$

6.8 Per-Phase Equivalent Circuit Considering Stray-Load Losses

In the equivalent circuit of an induction machine shown in the previous section, stray-load losses are neglected. In this section, stray-load losses are considered and the equivalent circuit is modified. Traditionally, to consider the stray-load losses two approaches are used; a resistance R_σ is connected across the rotor leakage reactance, a resistance $\frac{R_\delta}{s}$ is connected across the rotor leakage reactance [2–5]. The following assumptions are made

$$R_\sigma = k\omega \tag{6.49}$$

where k is a constant and ω is the rotor angular speed.

$$R_\delta = k\omega_{sl} \tag{6.50}$$

where ω_{sl} is the slip angular speed. The stray-load loss is shown in the literature [2–5] equal to

$$P_{stray} = k\omega\left(I_2'\right)^2 \tag{6.51}$$

The modified equivalent circuit considering the stray-load losses is shown in Figure 6.15a,b. Further analysis of the equivalent circuit of Figure 6.15a,b is left to the reader.

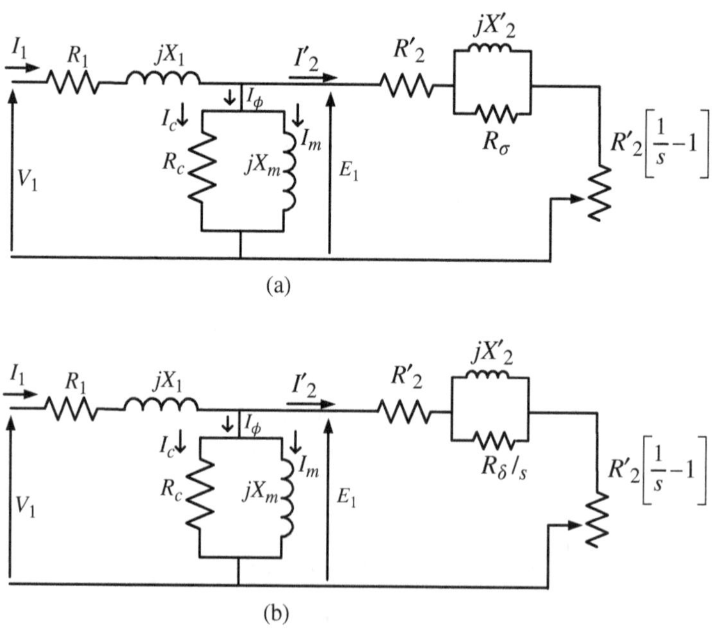

Figure 6.15 (a) Per-phase equivalent circuit-I considering stray-load losses. (b) Per-phase equivalent circuit-II considering stray-load losses.

6.9 Torque and Power using Thevenin's Equivalent Circuit

Analysis of the exact equivalent circuit is quite cumbersome and hence for simple analysis of the equivalent circuit, Thevenin's theorem is applied which states that the whole circuit can be represented by Thevenin's equivalent voltage and Thevenin's equivalent impedance. Considering the rotor impedance as load (shown in Figure 6.16a), equivalent impedance (seen from the load side, making the voltage source short-circuited, as shown in Figure 6.16b) and equivalent voltage (open circuit voltage across the magnetizing branch) are computed. The overall equivalent circuit is shown in Figure 6.16c.

Thevenin's equivalent impedance is obtained as shown in (Figure 6.16b)

$$Z_{TH} = Z_1 \parallel jX_m = \frac{(R_1 + jX_1)\, jX_m}{R_1 + j(X_1 + X_m)} \tag{6.52}$$

Thevenin's equivalent voltage (voltage across the shunt magnetizing branch) is obtained using voltage divider rule as shown in (Figure 6.16c)

$$\begin{aligned} V_{TH} &= \frac{jX_m}{R_1 + j(X_1 + X_m)} V_1 \\ |V_{TH}| &= \frac{X_m}{\sqrt{R_1^2 + (X_1 + X_m)^2}} |V_1| \end{aligned} \tag{6.53}$$

The current drawn by the induction motor can be computed from Thevenin's equivalent circuit as shown in Figure 6.16d as

$$\begin{aligned} I_1 = I_2' &= \frac{V_{TH}}{\left(R_{TH} + \frac{R_2'}{s}\right) + j(X_{TH} + X_2')} \\ |I_1| = |I_2'| &= \frac{|V_{TH}|}{\sqrt{\left(R_{TH} + \frac{R_2'}{s}\right)^2 + (X_{TH} + X_2')^2}} \end{aligned} \tag{6.54}$$

Air-gap power is given from Eq. (6.35)

$$P_g = 3(I_2')^2 \frac{R_2'}{s}$$

Substituting the expression of rotor current in Eq. (6.35)

$$P_g = \frac{3|V_{TH}|^2 \frac{R_2'}{s}}{\left(R_{TH} + \frac{R_2'}{s}\right)^2 + (X_{TH} + X_2')^2} \tag{6.55}$$

Electromagnetic torque developed by induction motor is obtained as

$$T_e = \frac{P_g}{\omega_s} = \frac{3|V_{TH}|^2 \frac{R_2'}{s}}{\omega_s \left[\left(R_{TH} + \frac{R_2'}{s}\right)^2 + (X_{TH} + X_2')^2\right]} \tag{6.56}$$

where $\omega_s = \frac{2\pi N_s}{60}$ rad/sec (synchronous angular speed).

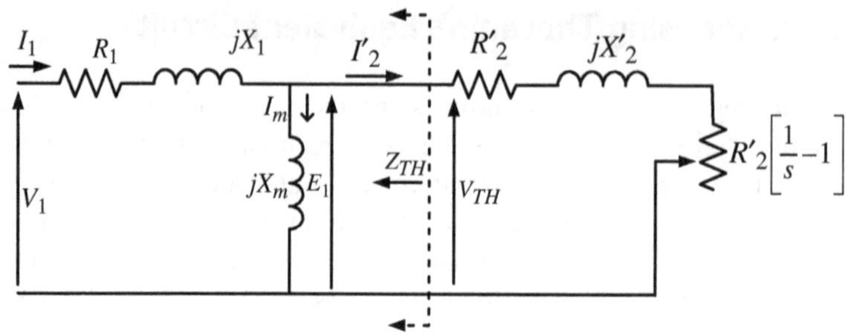

(a)

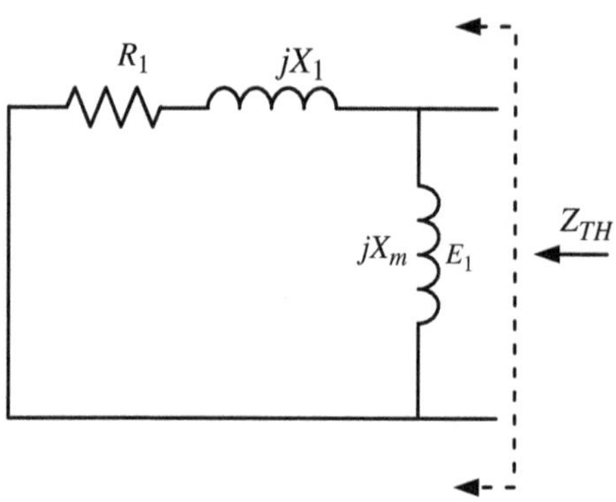

(b)

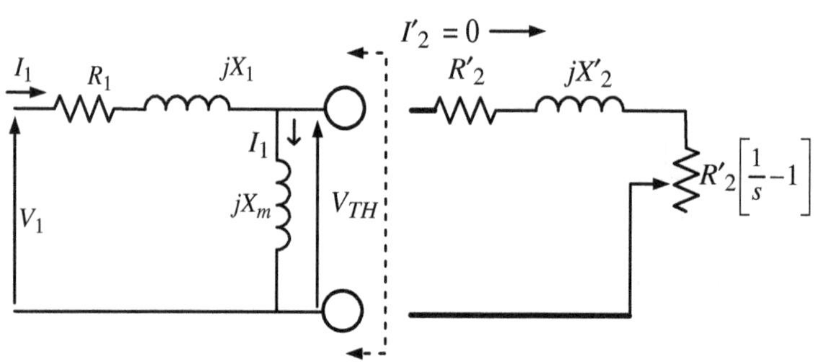

(c)

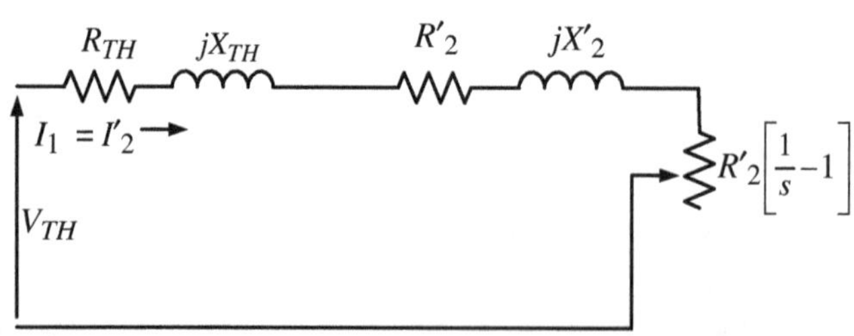

(d)

Figure 6.16 (a) Equivalent circuit for obtaining Thevenin's equivalent. (b) Thevenin's equivalent impedance. (c) Thevenin's equivalent voltage determination. (d) Overall Thevenin's equivalent circuit.

Example 6.5 A three-phase 240 V, 50 Hz, 10 kW, four-pole induction motor has the following equivalent circuit parameters.

$$R_1 = 0.1\,\Omega, X_1 = 0.7\,\Omega, X_2' = 1.5\,\Omega, R_2' = 0.25\,\Omega, X_\mathrm{m} = 20\,\Omega$$

For a slip of 2% calculate the following

(i) Speed of the motor
(ii) Stator Current
(iii) Power Factor
(iv) Power Developed
(v) Output Power
(vi) Output Torque
(vii) Air-gap power
(viii) Input Power
(ix) Efficiency

Use Thevenin's Equivalent Circuit. Given the windage and friction loss as 0.4 kW.

Solution

(i) Synchronous speed $N_s = \frac{120f}{P} = \frac{120X50}{4} = 1500rpm$
Speed of the machine

$$N = N_s(1 - S) = 1500\,(1 - 0.02) = 1470rpm$$

(ii) To calculate the stator current, let us find Thevinin's Equivalent circuit

$$Z_{TH} = Z_1 \parallel jX_m = \frac{(R_1 + jX_1)\,jX_m}{R_1 + j\,(X_1 + X_m)} = \frac{20\angle 90X\,(0.1 + j0.7)}{0.1 + j\,(0.7 + 20)}$$

$$Z_{TH} = \frac{20\angle 90X0.7071\angle 81.87}{20.7\angle 98.72} = 0.6832\angle 82.15 = 0.0933 + j0.6768$$

$$|V_{TH}| = \frac{X_m}{\sqrt{R_1^2 + (X_1 + X_m)^2}}\,|V_1| = \frac{20}{\sqrt{(0.1)^2 + (2.07)^2}}X\frac{240}{\sqrt{3}} = 133.87V$$

$$I_1 = I_2' = \frac{V_{TH}}{\left(R_{TH} + \frac{R_2'}{s}\right) + j\,(X_{TH} + X_2')}$$

$$= \frac{133.87\angle 0}{\left(0.0933 + \frac{0.25}{0.02}\right) + j\,(0.6768 + 1.5)} = \frac{133.87\angle 0}{12.5933 + j2.1768}$$

$$I_1 = I_2' = 10.475\angle - 9.8A$$

(iii) Power Factor $\cos(\theta) = \cos(9.8) = 0.985 lagging$
(iv) Developed Power $P_m = 3I_2'^2R_2'\frac{1-s}{s} = 3(10.475)^2 0.25\left(\frac{1-0.02}{0.02}\right) = 4.0324kW$
(v) Output Power $P_{out} = P_m - P_{w\&f} = 4032.416 - 400 = 3632.416W$
(vi) Output Torque $T_e = \frac{P_{out}}{\omega_m} = \frac{36321.416}{\frac{2\pi 1470}{60}} = 23.6Nm$
(vii) Air-gap Power $P_g = \frac{P_m}{1-S} = \frac{3632.41}{1-0.02} = 3706.54 Watt$

(viii) Input Power $P_{in} = 3I_1^2\left(R_{TH} + \frac{R_2'}{s}\right) = 3(10.475)^2\left(0.0933 + \frac{0.25}{0.02}\right) = 4145.42 Watt$

(ix) Efficiency $\eta = \frac{P_{out}}{P_{in}} X100\% = \frac{3632.416}{4145.42} X100\% = 87.62\%$

6.10 Torque-Speed Characteristics

Torque-slip (speed) characteristics can be drawn using Eq. (6.56), when keeping applied voltage and the machine's parameters constant while varying the slip (speed). Approximate torque equations can be derived assuming the stator resistance and leakage inductance zero.

$$T_e \cong \frac{3|V_{TH}|^2 \frac{R_2'}{s}}{\omega_s\left[\left(\frac{R_2'}{s}\right)^2 + (X_2')^2\right]} \tag{6.57}$$

In the low speed region (high slip), the effect of rotor resistance is negligible when compared with the rotor leakage reactance and this can be neglected from the denominator to simplify the torque equations

$$T_e \cong \frac{3|V_{TH}|^2 \frac{R_2'}{s}}{\omega_s\left[(X_2')^2\right]} = \frac{3|V_{TH}|^2 R_2'}{s\omega_s (X_2')^2} \tag{6.58}$$

The torque is inversely proportional to the slip in low-speed region, considering constant applied voltage, supply frequency and rotor parameters.

For high-speed value (low slip), the effect of resistance component is much higher when compared with the leakage reactance and thus the denominator of the Eq. (6.58) can only be represented by resistance and thus torque equation can be further simplified as

$$T_e \cong \frac{3|V_{TH}|^2 \frac{R_2'}{s}}{\omega_s\left[\left(\frac{R_2'}{s}\right)^2\right]} = \frac{3|V_{TH}|^2 s}{\omega_s R_2'} \tag{6.59}$$

Thus, the torque is directly proportional to the slip for high-speed region considering applied voltage, rotor resistance and supply frequency constant.

Torque-Speed curve of a three-phase induction machine can be drawn as shown in Figure 6.17. The speed increases to the right of the curve while slip increases to the left of the curve. Torque developed at the start of an induction machine (Slip = 1) is called 'Starting torque, T_{st}' as labelled in Figure 6.17. The developed torque reaches maximum at a slip s_{maxT} and then reduces to zero at synchronous speed when slip becomes zero. The torque-speed curve has a positive slope before it reaches maximum torque value and the curve has a negative slope afterwards. During normal steady-state condition, an induction machine operates at the negative slope of the curve, i.e. the positive slope region is an unstable region while a negative slope region is the stable region of operation. This is elaborated further in the next sub section. The same curve can be represented as shown in Figure 6.18 if the abscissa and ordinates are interchanged.

A three-phase induction motor develops starting torque when supply is switched on and thus it is said that the motor is self-starting. The torque becomes zero when the machine

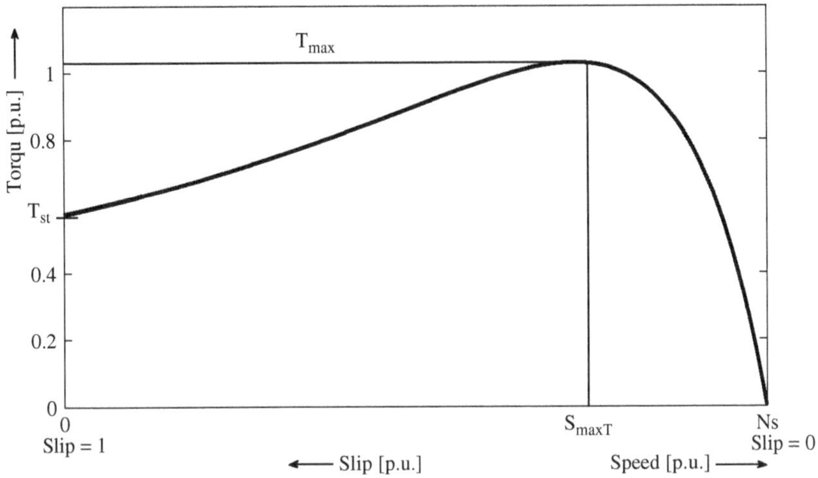

Figure 6.17 Torque-speed curve of a three-phase induction motor.

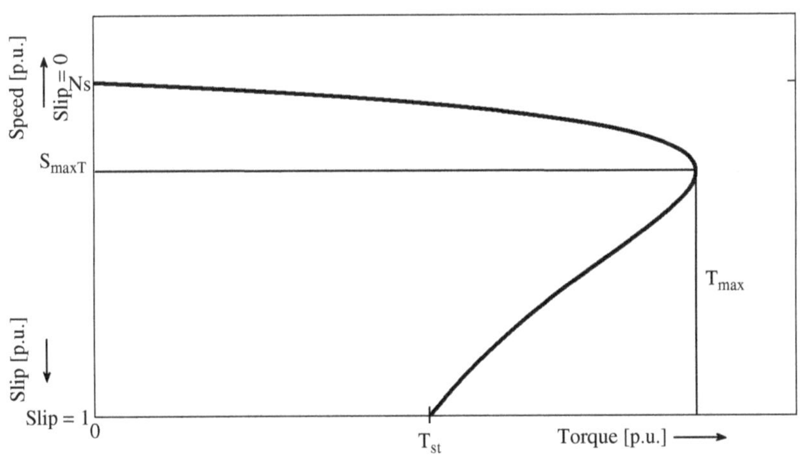

Figure 6.18 Speed-torque curve of a three-phase induction motor.

runs at synchronous speed because there is no relative speed between the rotor and the stator generated rotating field (slip = 0). Thus, an induction machine cannot run at synchronous speed. If, somehow, the motor is allowed to run at a speed higher than the synchronous speed (motor is coupled to a prime-mover), the motor behaves as a generator. The torque becomes negative and hence the machine absorbs mechanical power and generates electrical power.

The starting torque is slightly higher than the full-load torque, so that the motor can start and accelerate when started at full-load. Maximum torque is two to three times the rated full-load torque.

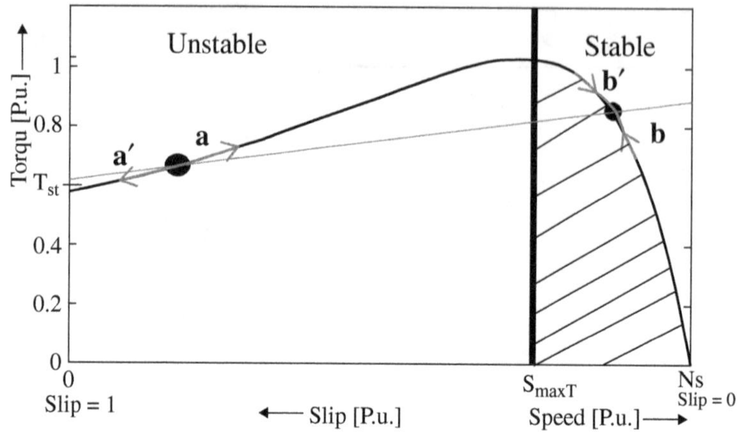

Figure 6.19 Torque-speed curve of motor and load torque showing stable and unstable operation region.

The expression for the starting torque can be obtained by putting the slip =1 in the torque (6.56).

$$T_{st} = \frac{3|V_{TH}|^2 R_2'}{\omega_s\left[(R_{TH} + R_2')^2 + (X_{TH} + X_2')^2\right]} \tag{6.60}$$

It is to be noted that the negative slope region of the torque-speed curve of an induction machine is the stable region of operation. The positive slope region is the unstable region of operation. In other words, the point of intersection between the load curve and motor curve in the negative slope side gives a stable operating point. This can be understood from Figure 6.19 where motor torque and load curve are shown in the same figure. The load curve is intersecting the motor torque curve at two points; one on the positive slope side and the other on the negative slope side. To prove that the point of intersection on the negative side gives stable operation, consider a small perturbation about the operating points. First, let us consider the positive slope point of intersection. If the perturbation moves the operating point to a′, then the motor generated torque becomes less than the load torque and, in this situation, the speed of the motor should fall. Looking at the curve, it can be said that the operating point will hence move as shown by the arrow (towards lower speed). Next, consider perturbation has moved the operating point to a, at this point, the motor torque becomes higher than the load torque and the speed should increase, moving the operating point further away from the original operating point (shown by arrow). Therefore, it is seen that the behaviour is divergent and any small perturbation may cause the operating point to shift from its original position and will never come back to it. This is called an unstable operating condition. Next, consider the operating point on the negative slope region. Any small perturbation may move the operating point to either b or b′. Consider point b′, at this point, motor generated torque is larger than the load torque and hence the motor accelerates (speed increases) causing the operating point to shift back to the original operating point (speed increases towards the right side). Similar behaviour can be analysed at point b. It seen that the behaviour is convergent which indicates a stable operation. The perturbation will bring back the operating point to its original position.

6.10.1 Condition for Maximum Torque

Condition for maximum torque is obtained by using Maximum Power Transfer Theorem. Torque is maximum when the air-gap power is maximum. Air-gap power is the power that is consumed in $\frac{R_2'}{s}$. Thus, maximum torque is developed when the power consumed by $\frac{R_2'}{s}$ is maximum. According to the Maximum Power Transfer theorem, source impedance must be equal to the load impedance for maximum power transfer. Considering Thevenin's equivalent circuit as shown in Figure 6.18c, condition of maximum power transfer can be written as

$$Z_{load} = \frac{R_2'}{s} = Z_{source} = R_{TH} + j\left(X_{TH} + X_2'\right) \tag{6.61}$$

$$\frac{R_2'}{S_{\max T}} = \sqrt{R_{TH} + j\left(X_{TH} + X_2'\right)}$$

$$S_{\max T} = \frac{R_2'}{\sqrt{R_{TH}^2 + \left(X_{TH} + X_2'\right)^2}} \tag{6.62}$$

From Eq. (6.62), it is seen that the slip for maximum torque is directly proportional to the rotor resistance. Increasing rotor resistance, the slip for maximum torque also increases (i.e. the maximum torque occurs at lower speed).

The expression for maximum torque is obtained by substituting the slip Eq. (6.62) into the torque Eq. (6.56). The resulting maximum torque equation is obtained as

$$T_{\max} = \frac{3V_{TH}^2}{2\omega_s\left(R_{TH} + \sqrt{R_{TH}^2 + \left(X_{TH} + X_2'\right)^2}\right)} \tag{6.63}$$

Another method of deriving the equation of maximum torque is to differentiate torque Eq. (6.56) with respect to slip and putting it equal to zero (i.e. $\frac{dT_e}{ds} = 0$). It is seen from above Eq. (6.63), that the maximum torque is proportional to the square of the applied voltage and independent of rotor resistance. Approximate torque equations can be obtained by neglecting Thevenin's equivalent resistance and leakage reactance.

In the slip-ring type of induction machine, the torque-speed curve can be modified by changing the rotor resistance. Rotor resistance can be increased by adding external rotor resistance and this is only possible in a slip-ring type induction machine. No external resistance can be added to squirrel-cage type of induction machine as no external terminal is available.

Furthermore, note that rotor resistance can only be increased and hence a set of characteristics for a slip-ring type induction motor is shown in Figure 6.20 for different values of rotor resistance. It is observed, as shown in Figure 6.20, that the slip for the maximum torque rises with an increase in the rotor resistance while maximum torque remains constant. The starting torque also rises with an increase in the rotor resistance. A situation is reached when starting torque becomes equal to the maximum torque. Modification in the

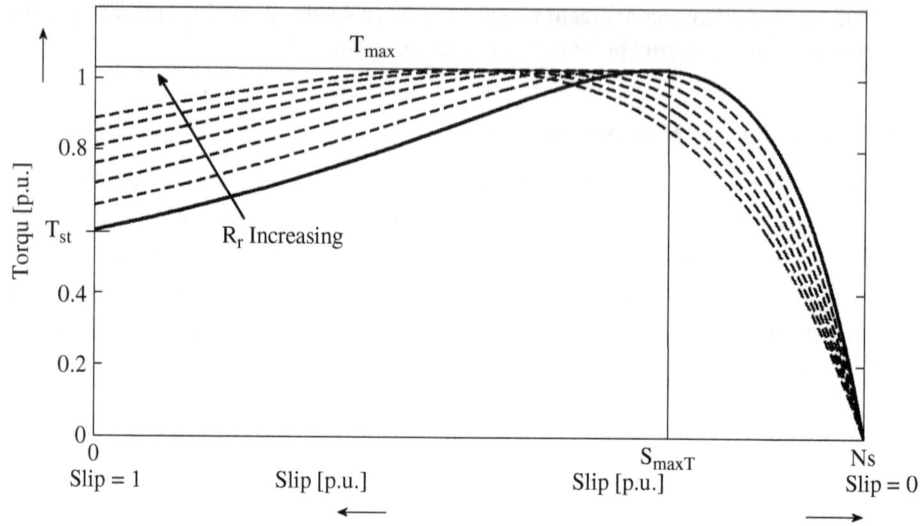

Figure 6.20 Torque-speed curve for various rotor resistance values.

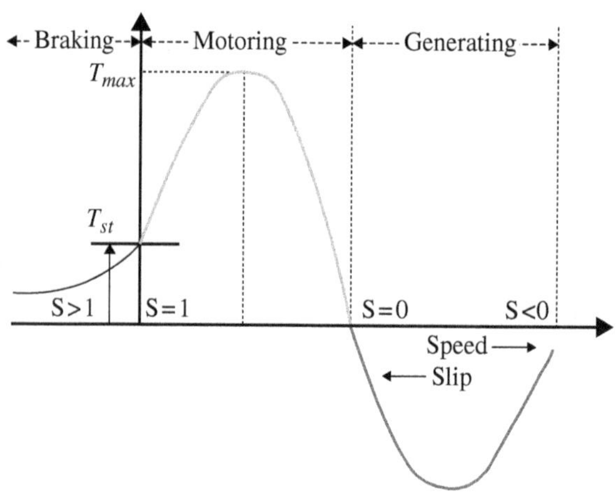

Figure 6.21 Complete torque-speed curve of a three-phase induction machine.

torque-speed curve is important to understand, as one can take advantage of this for starting a heavy load. The starting torque can be increased by inserting external rotor resistance to start heavy load. Once the motor reaches the steady-state condition, the external resistance can be removed and then the speed for which the maximum torque occurs will increase and normal operation can continue.

Overall torque-speed curve of an induction machine is shown in Figure 6.21. The characteristic is shown for all three modes of operation of an induction machine. For speed more than the synchronous value (called super synchronous speed), the induction machine operates as an induction generator and slip is negative. During braking or plugging, the induction machine rotates in the opposite direction and the slip is greater than 1.

6.10.2 Condition for Maximum Torque at Starting

Maximum torque and starting torque can be made equal by adjusting the value of rotor resistance. Considering Eq. (6.62) i.e. the condition for maximum torque

$$s_{\max T} = \frac{R_2'}{\sqrt{R_{TH}^2 + (X_{TH} + X_2')^2}}$$

For starting torque equal to maximum torque $s_{\max T} = 1$

$$R_2' = \sqrt{R_{TH}^2 + (X_{TH} + X_2')^2} \tag{6.64}$$

The approximate condition or value of rotor resistance can be obtained by neglecting Thevenin's equivalent resistance and leakage reactance and is given as

$$R_2' \cong X_2' \tag{6.65}$$

Hence, the rotor resistance and rotor leakage reactance should be equal for maximum torque to occur at start condition. Normally, the rotor resistance is considerably lower than the reactance value and hence the resistance should be increased to achieve the condition of Eq. (6.65).

6.10.3 Approximate Equations

All the above derived equations can be approximated by neglecting the stator impedance ($R_{TH} \approx 0, X_{TH} \approx 0$) from Eqs. (6.60), (6.62), and (6.63).

$$T_{st} \approx \frac{3|V_1|^2 R_2'}{\omega_s\left[(R_2')^2 + (X_2')^2\right]}$$

$$s_{\max T} \approx \frac{R_2'}{X_2'}$$

$$T_{\max} \approx \frac{3|V_1|^2}{2\omega_s X_2'}$$

Example 6.6 For machine of Example 6.5, calculate the following

(i) Slip and speed for maximum torque
(ii) Maximum Torque
(iii) Starting Torque
(iv) External rotor resistance value for Maximum Torque to occur at the starting condition

Solution

(i) Slip and speed for the maximum torque is given as

$$s_{\max T} = \frac{R_2'}{\sqrt{R_{TH}^2 + (X_{TH} + X_2')^2}} = \frac{0.25}{\sqrt{(0.0933)^2 + (0.6768 + 1.5)^2}}$$

$$= \frac{0.25}{2.178} = 0.1147$$

$$N_{\max T} = N_s\left(1 - S_{\max T}\right) = 1500\,(1 - 0.1147) = 1328rpm$$

(ii) Maximum Torque

$$T_{\max} = \frac{3V_{TH}^2}{2\omega_s\left(R_{TH} + \sqrt{R_{TH}^2 + \left(X_{TH} + X_2'\right)^2}\right)}$$

$$V_{TH} = 133.8\,\text{V}, R_{TH} = 0.0933\,\Omega, X_{TH} = 0.678\,\Omega, X_2' = 1.5\,\Omega, R_2' = 0.25\,\Omega$$

$$\omega_s = \frac{2\pi N_s}{60} = \frac{2\pi 1500}{60} = 157.08rad/\sec$$

$$T_{\max} = \frac{3(133.8)^2}{2 \times 157.08\left(0.0933 + \sqrt{(0.0933)^2 + (0.678 + 1.5)^2}\right)} = \frac{53707.32}{714.18} = 75.2Nm$$

(iii) Starting torque is given as

$$T_{st} = \frac{3\left|V_{TH}\right|^2 R_2'}{\omega_s\left[\left(R_{TH} + R_2'\right)^2 + \left(X_{TH} + X_2'\right)^2\right]}$$

Substituting the parameter values

$$T_{st} = \frac{3 \times (133.8)^2 \times 0.25}{157.08 \times \left[(0.0933 + 0.25)^2 + (0.678 + 1.5)^2\right]} = \frac{13426.83}{763.65} = 17.6Nm$$

(iv) For maximum torque to occur at the starting condition, Put $s_{maxT} = 1$ and the external resistance added is R_{ext}.

$$s_{\max T} = \frac{R_2' + R_{ext}}{\sqrt{R_{TH}^2 + \left(X_{TH} + X_2'\right)^2}} = 1$$

$$R_{ext} = \sqrt{R_{TH}^2 + \left(X_{TH} + X_2'\right)^2} - R_2' = \sqrt{(0.0933)^2 + (0.678 + 1.5)^2} - 0.25$$

$$R_{ext} = 2.18 - 0.25 = 1.93\Omega/phase$$

Example 6.7 A four-pole three-phase induction motor develops maximum torque at 1200 rpm when operated at 50 Hz. Rotor resistance is given as 0.8 Ω/phase. Calculate the speed at which it develops maximum torque when supplied with a variable frequency source of 30 Hz.

Solution

$$R_2' = 0.8\Omega, R_1 = X_1 = 0, P = 4$$

$f = 50$ Hz operation

$$N_s = \frac{120f}{P} = \frac{120x50}{4} = 1500rpm$$

Speed for T_{max}, $N_m = 1200\ rpm$

$$s = \frac{N_s - N}{N_s} = \frac{1500 - 1200}{1500} = 0.2$$

$$s = \frac{R_2}{X_2} = \frac{R_2}{2\pi f L_2} = 0.2$$

$$\pi L_2 = 0.04$$

At $f = 30\ \text{Hz}$, $N_S = \frac{120 \times 50}{6} = 1000\ r.p.m.$

$$N_s = \frac{120f}{P} = \frac{120x30}{4} = 900rpm$$

Slip for T_{max}

$$s = \frac{R_2}{X_2} = \frac{R_2}{2\pi f L_2} = \frac{0.8}{2X30X0.04} = 0.333$$

$$\pi L_2 = 0.04$$

$N = (1 - s)N_s = (1 - 0.333)\text{X}900 = 600\ \text{rpm}$

Example 6.8 Develop the Matlab/Simulink Model to determine the performance of a three-phase induction motor

Consider the Inputs

Rated input voltage, number of poles, frequency, rated power, circuit parameters, and mechanical losses.

For any given speed, the simulation model should output: the input power, stator copper losses, iron loss, air-gap power, mechanical power developed, output power and efficiency.

Test the model for the following data

A 400 kW, 3.3 kV, three-phase, 50 Hz, four-pole machine running at 1420 rpm. The mechanical losses are 1 kW. The per-phase equivalent circuit parameters are given as

$R_1 = 0.6\ \Omega$, $R_2 = 1.1\ \Omega$, $X_1 = 2\ \Omega$, $X_2 = 3\ \Omega$, $X_m = 150\ \Omega$.

Solution

```
function [Pin,Pscl,Pg,Pmd,Po,eff]= fcn(Vin,Pl,f,Nr,Pml,R1,R2,
  X1,X2,Xm)
 % Pin: Input Power
 % Pscl: Stator copper loss
 % Pi: Iron loss
 %Pg: Airgap Power
 %Pmd: Mechanical Power developed
 % Po: Output power
 % eff: Efficiency
 % Vin: Input voltage
 % Pl: Number of poles
 % f: frequency
 % Pr: Rated power
 % R1,R2,X1,X2,Xm: Equivalent Circuit parameters
 Ns = (120*f/Pl); % Synchronous speed
```

```
  s = (Ns-Nr)/Ns; % Slip
  Z2 = complex(R2/s, X2);
  Zp = ((Xm*Z2)/(Xm+Z2));
  Z1 = complex(R1, X1);
  Zin = Z1+Zp;
  Vp = Vin/sqrt(3);
  Iin = Vp/Zin;
  theta = angle(Iin);
  Pin = (3*Vp*abs(Iin)*cos(theta))/1000; % Input Power in KW
  Pscl = (3*abs(Iin)^2*R1)/1000; % stator copper loss in KW
  E2 = Iin*Zp;
  I2 = E2/Z2;
  Prcl = (3*abs(I2)^2*R2)/1000; % Rotor copper loss in KW
  Pmd = (3*(abs(I2)^2)*(1-s)*R2/s)/1000; % Mechanical power
developed in KW
  Pg = Pmd+Prcl;
  Po = Pmd - Pml; % Output power in KW
  eff = (Po/Pin)*100; % Efficiency
```

The Simulink diagram of the three-phase IM is shown in Figure E6.8.

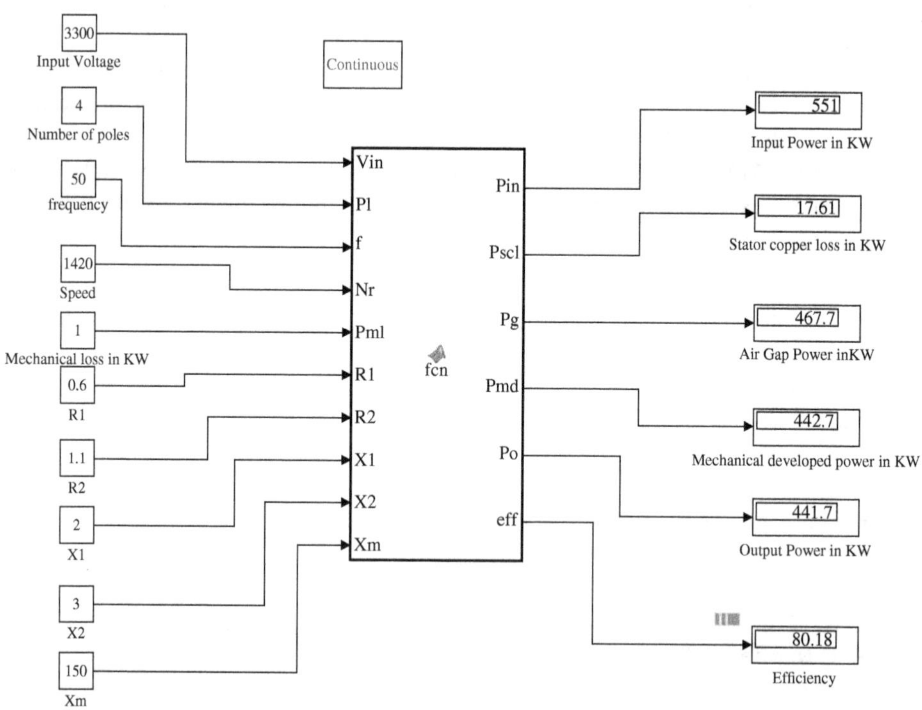

Figure E6.8 Simulink model for obtaining performance of the machine.

6.11 Losses in a Three-Phase Induction Machine

Different types of electrical losses that occurs in an induction machine are

- Copper losses or resistive losses
- Magnetic losses or Iron losses
- Mechanical losses
- Stray losses

These losses are described below.

6.11.1 Copper Losses or Resistive Losses

This type of loss occurs due to the inherent property of the conductors/windings. Both stator and rotor have conductors and hence this loss occurs in both stator and rotor. The copper loss is given as

$$P_{cu} = I^2 R \tag{6.66}$$

The resistance of the conductor is the result of collision of electrons and hence this is expected to rise with an increase in temperature. The resistance changes with the temperature of the conductor. The fractional change in the resistance is proportional to the change in temperature and the proportionality constant is the temperature coefficient

$$\frac{\Delta R}{R_0} = \alpha \Delta T \tag{6.67}$$

where R_0 is the initial resistance at a temperature T_0 degree Celsius and R_1 is the resistance at temperature T_1 degree Celsius. α is the temperature coefficient.

$$\Delta R_0 = R_1 - R_0; \Delta T = T_1 - T_0 \tag{6.68}$$

From Eqs. (6.67) and (6.68), the following relation can be written

$$R_1 = R_0 \left[1 + \alpha \left(T_1 - T_0\right)\right] \tag{6.69}$$

The temperature coefficient values for different metals are

For Aluminium $\alpha = 3.9 \times 10^{-3}/^{o}C$
For Copper $\alpha = 3.9 \times 10^{-3}/^{o}C$

Other factors affect the value of the resistance of induction machine windings: the skin effect and proximity effect. Both are associated with non-uniform distribution of current in the cross-section area of the conductor, the first is caused by irregular distribution of the lines of magnetic fluxes through the conductor which increases the inductance in the centre of the conductor reducing the current flow in this region. The second one is due to distortion in magnetic fields caused by the proximity between conductors; also causing distortions in the current densities. Copper losses also vary with the square of the load current.

Typically out of total losses in an induction machine, 37% occur due to copper losses in the stator windings, while 18% happen in the rotor winding.

6.11.2 Magnetic Losses

The loss that occurs in the stator and rotor cores due to hysteresis properties and eddy currents are termed as Core Loss, Iron Loss or Magnetic Loss. About 20% of the total loss that occurs in an induction machine is due to magnetic losses.

Hysteresis loss takes place due to the inherent property of the magnetic material of remanence and coercivity forming a hysteresis loop. Losses occur due to this hysteresis loop which appears in the form of heat. This loss depends mainly upon the magnitude of the flux density and the frequency of the applied voltage. This loss can be reduced by using thinner lamination and adding small a percentage of Si in the iron.

The eddy current loss occurs due to the flow of eddy currents in the cores of stator and rotor. The rate of change of flux causes induced emf in the stator and rotor cores that causes current to flow in the cores in the form of eddies. Due to the resistance of the cores, loss takes place, called eddy current loss. This type of loss depends upon the square of the frequency of the current, magnitude of the flux density and lamination thickness. This loss can be reduced by lamination of the core.

6.11.3 Mechanical Losses

Mechanical losses are also termed as Windage and Friction losses. Rotor rotates against the air friction and hence has to put more effort to overcome the air resistance and this is called Windage loss. The Friction loss occurs due to the bearings and mechanical rotation. In small induction machines, the mechanical losses are usually about 9% of total losses. The main factors that affect the windage losses are the rotors′ peripheral velocity, the smoothness of the rotor and stator surfaces and the length of the air gap. The smaller the air gap, the bigger the windage losses but, on the other hand, increasing the air gap increases the need for magnetizing current.

6.11.4 Stray-Load Losses

These losses are very difficult to model and quantify. Although the stray-load losses have been the object of several studies and analysis, the phenomena that govern these losses are still under discussion, in particular, from the measurement point of view.

$$P_{Stray} = \left(P_{Electrical} - P_{Mechanical}\right) - P_{Conventional} \tag{6.70}$$

with,

$$P_{Conventional} = \left(P_{JS} + P_{JR} + P_{Iron} + P_{Mec}\right) \tag{6.71}$$

In order to limit the stray-load losses, some possible solutions are represented by the rotor bar insulation (to limit the inter-bar leakage currents), by using a double-layer winding with low space harmonic contributions, by a reduction of the high-frequency flux variations in the motor teeth, etc. It is important to remark that a strong reduction of the stray-load losses can involve a significant motor efficiency increment. Typically, stray-load loss constitutes 16% of the total loss.

6.12 Testing of a Three-Phase Induction Motor

Equivalent circuit parameters of an induction machine are computed from data of voltage, current and power obtained by conducting several tests. The testing procedure for a three-phase induction machine is documented in IEEE standard 112–1991, 112–1996, 112–2004. Summary of some common testing procedures are described in the following sub sections.

Generally, the following tests are conducted

- No-load test
- Blocked rotor test
- DC test to determine stator resistance
- Load test.

6.12.1 No-Load Test

No-load is conducted to determine the shunt-branch parameters of the equivalent circuit and core losses. Machine is allowed to run under no-load condition and rated voltage is applied. Wattmeter, voltmeter and ammeters are connected to record, power, voltage and current, respectively as shown in Figure 6.22. Under no-load condition, the machine runs at high speed very close to the synchronous value and, thus, the slip is very close to zero, the current drawn by motor is small and is called 'no-load current'. The motor under no-load condition develops small torque to overcome the no-load losses. No-load losses are comprised mainly of the core loss and rotational losses (windage and friction loss). Windage losses occur since the motor has to overcome the air friction against the rotation, while friction losses occur due to the rotation of the shaft on bearings and lubricant. A small amount of stator copper loss also occurs but it is neglected.

- As the current drawn by motor is small compared to the rated full-load current, the voltage drop across the stator impedance is small. Hence, the voltage across the shunt impedance branch can be considered the same as the applied voltage.
- Since slip is very close to zero, the rotor-side resistance $\left[R_2'\left(\frac{1}{s}-1\right)\right]$ becomes very high and offers open-circuit and hence can be removed from the equivalent circuit.
- The stator voltage drop is small hence the stator-side impedance is also neglected.

Thus, the equivalent circuit under no-load condition is simple with the shunt impedance branch consisting of stator core loss branch (R_c) and magnetizing reactance (X_m) as shown in Figure 6.23.

Measured Power (Core losses) 'Pc' = $W_1 \pm W_2$ (Neglecting windage and friction losses)

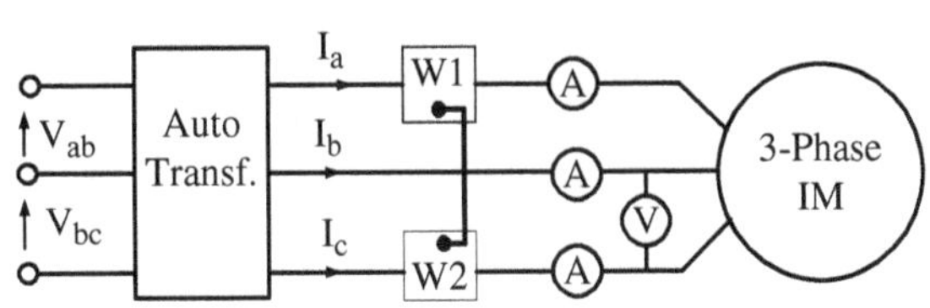

Figure 6.22 No-load test circuit diagram.

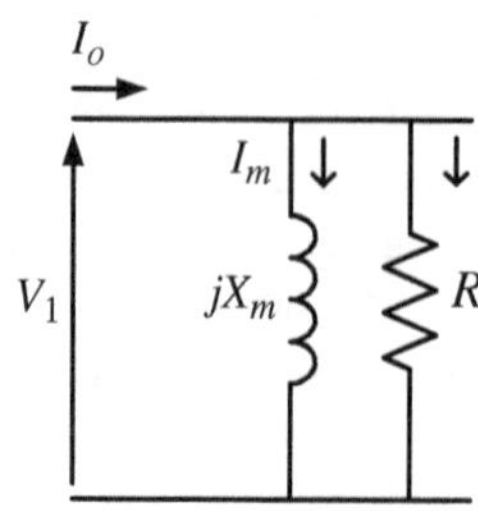

Figure 6.23 Equivalent circuit under no-load condition.

Measured Voltage (rated value) = V_1
Measured Current (No-load current) = I_o

$$\cos(\theta_o) = \frac{P_c}{\sqrt{3}V_1 I_o} \tag{6.72}$$

$$\overline{I_o} = \overline{I_c} + \overline{I_m} = \frac{V_1}{R_c} + \frac{V_1}{jX_m}$$

$$|I_o| \angle\theta_o = |I_o|\cos(\theta_o) + j|I_o|\sin(\theta_o) = \frac{V_1}{R_c} - j\frac{V_1}{X_m} \tag{6.73}$$

$$|I_o|\cos(\theta_o) = \frac{V_1}{R_c}$$

$$R_c = \frac{V_1}{|I_o|\cos(\theta_o)}$$

$$|I_o|\sin(\theta^o) = \frac{V_1}{X_m}$$

$$X_m = \frac{V_1}{|I_o|\sin(\theta_o)} \tag{6.74}$$

The core-loss resistance and the magnetizing reactance are calculated from (6.64).

6.12.2 Blocked Rotor Test

This test is conducted to determine the series parameters of an induction machine. Rotor is blocked from rotation using mechanical means and reduced voltage is applied so that rated current flows in the stator. The circuit diagram for this test is similar to that shown in Figure 6.22 but with a difference of variable frequency source as shown in Figure 6.24.

The reading of voltmeter gives reduced applied voltage; the ammeter gives rated current and the wattmeter gives the total copper loss. This is similar to the short-circuit test of a transformer.

There is a problem with this test because of different values of rotor current frequencies in actual operation and test condition. Under blocked rotor condition, the rotor current frequency is same as the stator current frequency. However, under actual running condition,

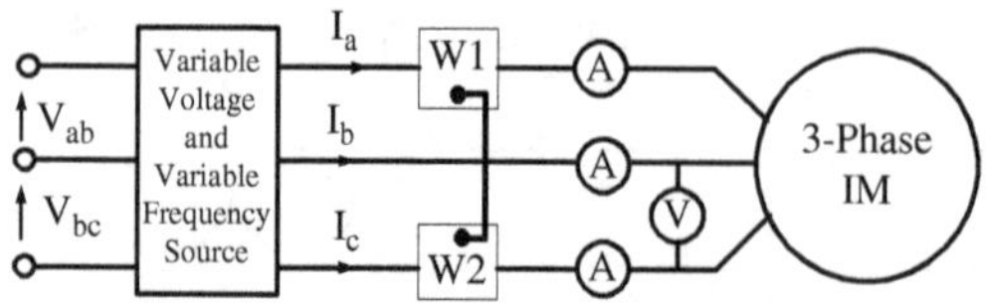

Figure 6.24 Circuit diagram for blocked rotor test.

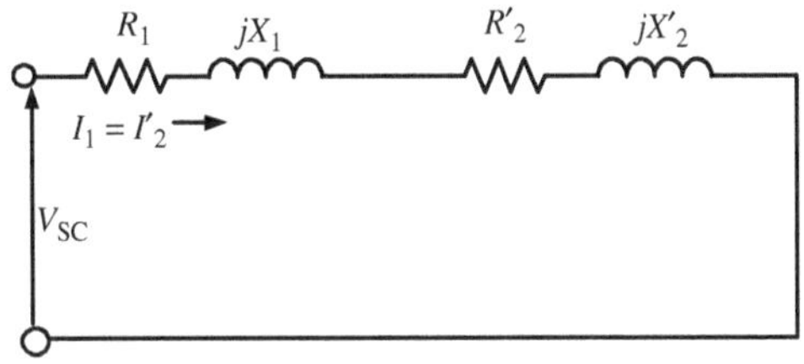

Figure 6.25 Equivalent circuit under blocked-rotor condition.

the rotor current frequency is of the order of 2–4 Hz. Thus, the test frequency should match the actual running frequency. According to IEEE standard 112, the frequency of the applied voltage under blocked rotor condition is kept 25% of the normal line frequency (in case of 50 Hz, it is 12.5 Hz). Assume that the test frequency is f_{test}. Under blocked rotor condition, the slip s = 1, which makes the equivalent circuit component $R'_2\left(\frac{1}{s}-1\right)=0$ and hence the equivalent circuit can be drawn as shown in Figure 6.25.

The following relation is used to compute the series parameters of the equivalent circuit

$$P_{sc} = \sqrt{3}V_{sc}I_1\cos(\theta) \tag{6.75}$$

$$\cos(\theta) = \frac{P_{sc}}{\sqrt{3}V_{sc}I_1} \tag{6.76}$$

$$|Z_{sc}| = \frac{V_{sc}}{I_1}$$

$$Z_{sc} = R_{sc} + jX'_{sc} = |Z_{sc}|\cos(\theta) + j|Z_{sc}|\sin(\theta) \tag{6.77}$$

$$R_{sc} = R_1 + R'_2$$

$$R'_2 = R_{sc} - R_1$$

$$X^*_{sc} = X^*_1 + X^*_2 \tag{6.78}$$

where R_1 is known from the DC test and X^*_1, X^*_2 are leakage inductances at test frequency value. Since the reactance is directly proportional to the frequency, the leakage reactance can be found at the stator-side frequency using the following relation

$$X_{sc} = \frac{f_{rated}}{f_{test}}X^*_{sc} = X_1 + X'_2 \tag{6.79}$$

$$X_1 = X'_2 = 0.5X_{sc} \tag{6.80}$$

Thus, the series parameters of the equivalent circuit can be computed using Eqs. (6.78) and (6.79). The separation of the reactance in Eq. (6.80) is done using equal division. However, for some design, they may be divided in different ratios.

6.12.3 DC Test

To compute the stator-winding resistance DC test is conducted. Since under DC test condition, the stator current is only limited by the stator-winding resistance. No induced emf in the rotor and hence no rotor resistance is seen. Also, all the inductances become short circuited due to DC current. Hence, in the overall induction motor circuit, only stator resistance is available, that can be computed using simple voltage-current relationship. The circuit diagram for DC test is given in Figure 6.26. DC voltage is applied such that

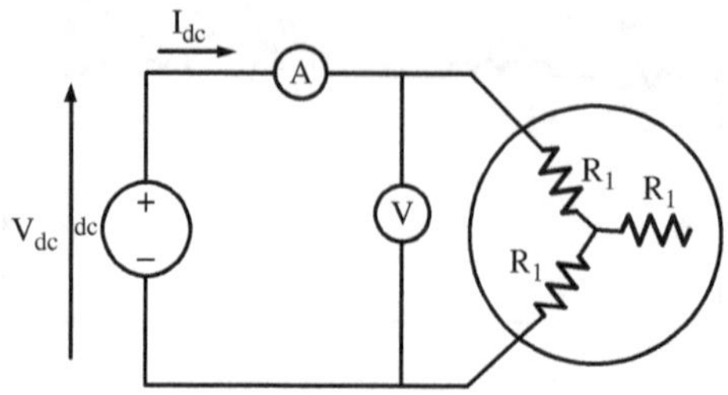

Figure 6.26 Circuit diagram for DC test to determine stator resistance.

rated current starts flowing through the stator. The reading of voltmeter and ammeter is recorded. The stator-winding resistance is then computed using the following relationship for star-connected stator windings

$$2R_1 = \frac{V_{dc}}{I_{dc}}; R_1 = \frac{0.5V_{dc}}{I_{dc}} \tag{6.81}$$

If the winding is delta connected then the resistance is computed as

$$\begin{aligned} R_1 \parallel 2R_1 &= \frac{V_{dc}}{I_{dc}} \\ R_1 &= \frac{1.5V_{dc}}{I_{dc}} \end{aligned} \tag{6.82}$$

The stator resistance computed in this way do not consider the heating and skin effects due to AC current. Hence, the correction factor should be applied to obtain the stator resistance when AC current flows. This can be done using IEEE standard 112. A correction factor of 1.15 to 1.3 is used to compute stator-winding resistance under AC condition. Hence

$$R_{1ac} = 1.15 \text{ to } 1.3 \times R_1 \tag{6.83}$$

6.12.4 Load Test

To determine the efficiency of a three-phase induction motor, load test is conducted. Efficiency of motor is defined as the ratio of output power to input power in Watts

$$\begin{aligned} \text{Efficiency} &= \frac{\text{output power mechanical (Watt)}}{\text{input power electrical (Watt)}} \\ &= \frac{\text{input power electrical (Watt)-losses (Watt)}}{\text{input power electrical (Watt)}} \\ &= \frac{\text{output power mechanical (Watt)}}{\text{input power electrical (Watt)} + \text{Losses (Watt)}} \end{aligned} \tag{6.84}$$

It is seen from Eq. (6.84), that output power and input power have to be recorded alternatively either output power or input power should be known along with the losses. To conduct this test, the machine is loaded and the load is varied from zero to the rated value. Several readings (voltage, current, power and power factor) are taken for variable loads. Typically, the efficiency curve is shown in Figure 6.27. The maximum efficiency depends upon the design class however, a typical range is between 80% and 90%. The efficiency is maximum around 70–90% of the rated load.

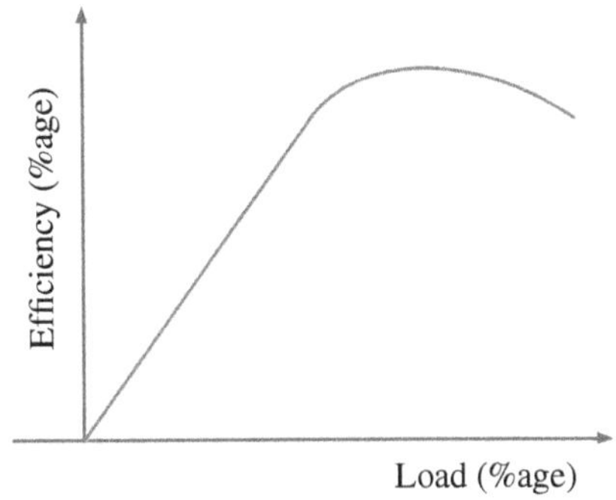

Figure 6.27 A typical efficiency curve of a three-phase induction motor.

Example 6.9 A three-phase, 50 hp, 2.2 kV, four-pole, 50 Hz, star-connected induction motor is tested with the following test results

No-Load Test
Voltmeter Reading: 2200 V
Ammeter Reading: 4.3 A
Wattmeter Reading: 1.5 kW
Blocked-rotor Test
Voltmeter Reading: 250 V
Ammeter Reading: 20 A
Frequency: 20 Hz
Wattmeter Reading: 8500 W
The DC resistance per stator phase is 2.5 Ω. Calculate

(i) The no-load rotational losses
(ii) Parameters of equivalent circuit

Solution

(i) Under no-load condition, there is no current through the rotor circuit. So, the losses are stator copper loss and mechanical rotational loss. From the total input power, if stator copper loss is deduced, the remaining is the mechanical losses.
No-load power Losses (W_{nl}) = 1500 W
No-load rotational Loss (W_{rot}) = W_{nl} – Stator copper loss
$W_{rot} = W_{nl} - 3I_0^2R_1 = 1500 - 3*(4.2)^2 * 2.5 = 1367.7\,W$
(ii) Using (6.74)
Per-phase voltage $= 2200/\sqrt{3} = 1270.17$

$$\cos(\theta_o) = \frac{Pc}{\sqrt{3}V_1I_0} = \frac{1500}{\sqrt{3} * 2200 * 4.3} = 0.09154$$

$$\theta_o = 84.747^0$$

$$R_c = \frac{V_{per-phase}}{|I_o|\cos(\theta_0)} = \frac{1270.17}{4.3 \times \cos(84.747^o)} = \frac{1270.17}{0.39368} = 3226.39\Omega$$

$$X_m = \frac{V_{per-phase}}{|I_o|\sin(\theta_0)} = \frac{1270.17}{4.3 \times \sin(84.747^o)} = \frac{1270.17}{4.2819} = 296.63\Omega$$

Using (6.76)–(6.80) blocked rotor test conditions to determine the series parameters
Per-phase voltage = $250/\sqrt{3}$=144.33 V

$$Z_{sc} = \frac{V_{sc}}{I_{sc}} = \frac{144.33}{20} = 7.2168\Omega$$

$$\cos(\theta_{sc}) = \frac{P_{sc}}{\sqrt{3}V_{sc}I_{sc}} = \frac{8500}{\sqrt{3}\times 250\times 20} = 0.9815$$

$$\theta_{sc} = 11.0394^0$$

$$R_{sc} = R_1 + R_2' = Z_{sc}\times\cos(\theta_{sc}) = 7.2168\times 0.9815 = 7.0833\Omega$$

$$R_2' = 7.0833 - R_1 = 7.0833 - 2.5 = 4.5833\Omega$$

$$X_{sc}^* = X_1 + X_2' = Z_{sc}\times\sin(\theta_{sc}) = 7.2168\times 0.19148 = 1.3819\Omega$$

$$X_{sc} = \frac{f_{rated}}{f_{test}}X_{sc}^* = X_1 + X_2' = \frac{50}{20}\times 1.3819 = 3.4547\Omega$$

$$X_1 = X_2' = 1.7273\Omega$$

Example 6.10 Develop the Simulink Model to compute the equivalent circuit parameters of a three-phase induction motor from tests (No-load and blocked rotor tests).

User Inputs

DC Test Data: V_{dc} and I_{dc}

No-load test data: Voltage, Current and Power

Blocked rotor test data: Frequency, Voltage, Current and Power

Assume $X_1 = X_2$

Output from the code

Equivalent circuit parameters: R_1, R_2, X_1, X_2, and X_m

Test the code for the following case with star-connected stator winding of a 110 KW 2200 V four-pole, 50 Hz squirrel cage motor:

DC test: V_{dc} = 20 V, I_{dc} = 5.5 A

No-load test data: Line voltage 2.2 kV, Line current 7 A, Three-phase power 2.6 kW

Blocked rotor test done at 15 Hz: Line Voltage 300 V, Line Current 50 A, Three-phase Power 15 KW

Solution

MATLAB Code:

```
function [R1,R2,X1,X2,Xm] = Induction_Motor(Vdc,Idc,VLLNL,
  ILLNL,PNL,FNL,VLLBR,ILLBR,PBR,FBR)
%DC Test of Induction Motors
V1=VLLNL./sqrt(3);
ZNL=V1./ILLNL;
RNL=PNL./(3.*(ILLNL.^2));
XNL=sqrt((ZNL.^2)-(RNL.^2));
R1=Vdc./(2.*(Idc));
V1=VLLBR./sqrt(3);
ZBL=V1./ILLBR;
RBL=PBR./(3.*(ILLBR.^2));
```

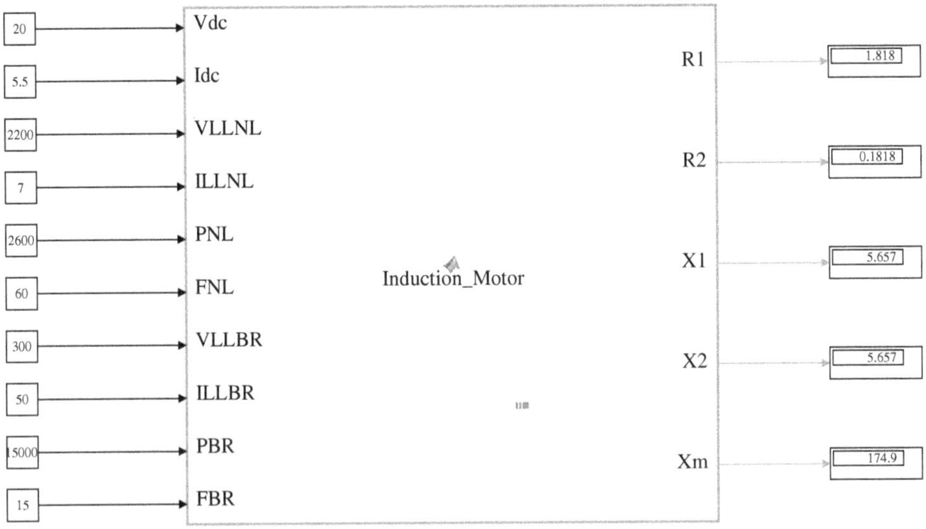

Figure E6.10 Simulink Block for No Load (NL), Blocked Rotor (BR) test for induction motors.

```
R2=RBL-R1;
XBL1=sqrt((ZBL.^2)-(RBL.^2));
XBL2=XBL1.*(FNL./FBR);
X1=XBL2./2;
X2=X1;
Xm=XNL-X2;
```

The Simulink Block for No Load (NL), Blocked Rotor (BR) test for the Induction Motor is shown in Figure E6.10.

Simulink Model

6.12.5 International Standards for Efficiency of Induction Machines

The efforts to reduce the energy consumption begin with its main load, the induction motor. There is world agreement that this objective can be achieved by motor efficiency regulation. The efficiency is of paramount importance nowadays because electrical motors are major consumers of electricity in the modern industrial society; they consume approximately 69% of electricity in the industrial sector. Three-phase, low-voltage squirrel cage induction motors are the most commonly used electric motors in industry.

In the actual electric energy market, energy saving policies are more and more important. In fact, cost and availability of the electric energy can vary in a complex manner. For these reasons, electric energy consumers are interested in using apparatus with high efficiencies to reduce their electric consumption. It is important to observe that the induction motors can be considered as greater users of electrical energy. In the European Union, the electric motors used in the industrial field typically consume 60–70% of the total absorbed electrical energy. In the commercial sector, this percentage is up to 35%. Due to these reasons, for the

Table 6.1 European Cemep efficiency lower limits.

Rated Power	4 KW	7.5 KW	11 KW	15 KW
Eff1 lower limit	88.3	90.1	91.0	91.8
Eff2 lower limit	84.2	87.0	88.4	89.4
Eff3 typical value	80.0	83.0	85.0	87.0

industrial induction motors, the European protocol (the CEMEP committee agreement in Europe) defines the following efficiency classes.

- Eff1 class: 'high efficiency' motors
- Eff2 class: 'energy efficiency' motors
- Eff3 class: 'standard efficiency' motors (motors now in production) (Table 6.1).

6.12.6 International Standards for the Evaluation of Induction Motor Efficiency

The efficiency values provided by the manufacturers have to be determined in accordance with well-known international standards. The most important world references are as follows:

- IEEE 112 – method B (U.S. standard)
- IEC 34-2 (European standard)
- JEC 37 (Japanese standard).

***IEEE 112-method B*:** This standard is the most important in the industrial field because it is applicable to horizontal-axis polyphase squirrel-cage induction motors with power in the range 1–190 kW. Method B requires three tests. In particular, they are as follows.

1. ***Thermal test at the rated load*** – the machine works at the rated load until the main motor temperatures (stator winding, stator lamination core, and external frame), each measured 30 minutes, do not change less than 1^{0} C. At the end of this test, the stator-winding resistance has to be measured.
2. ***No-load test*** – The motor, supplied with the rated voltage and frequency, runs without mechanical load until the bearings are stabilized (between two consecutive measures spaced over 30 minutes, do not change by more than 3%), then a variable voltage test is performed.
3. ***Variable-load test at rated conditions*** – with the motor in steady-state thermal condition at rated load, the motor is loaded with six decreasing load torques (from 150% down to 25% of the rated torque). The winding temperature has to change not more than 10^{0} C with reference to the rated one. Using these tests, it is possible to determine all the motor loss contributions and to calculate the motor efficiency.

***IEC 34–2*:** This standard provides several methods and procedures for the efficiency measurements in accordance with the type and the machine size, with the desired accuracy, etc. These methods can be subdivided into two categories.

Table 6.2 Efficiency classification of induction motor.

Denomination	Pre-EPAct (USA), Padrao (Brasil), IE1 (IEC/Europe), EPAct (NEMA/USA), High-Efficiency (Brasil), IE2 (IEC/Europe), Premium (NEMA/USA),IE3 (IEC/Europe)			Super Premium	*Factors for Improving Efficiency*
Motor	Standard	High Efficiency	Premium	IE4 (IEC/Europe)	• Increase the amount of active material • Use double-layer windings • Utilize high-performance lamination materials • Use termic treatment in the rotor surface • Optimize the air gap dimensions • Improve the efficiency of fan assembly • Increase the rate of heat transfer between active parts and frame • Use high-efficiency bearings • Optimize fabrication process • Rotor with copper busbar
Average Efficiency Level (%)	88.7	90.5	93.3	In analyses	

- ***Direct method*** – The absorbed and provided power at the motor shaft are directly measured.
- ***Indirect method*** – The motor losses are measured by suitable tests and the efficiency is evaluated measuring the motor absorbed power.

***JEC37*:** This standard is less restrictive than the USA and European ones. The efficiency evaluation through the Japanese standard can be considered as an indirect method. JEC 37 neglects the stray-load losses. For this reason, the obtained efficiencies are generally higher. Furthermore, no thermal correction of the joule losses is specified. Because it is very difficult to find the measurement procedures prescribed by the Japanese standard, it is reasonable to evaluate the machine efficiency using the results of the tests required by the other standards (Table 6.2).

6.13 Starting of a Three-Phase Induction Motor

Three-phase induction motors are self-starting machine i.e. the motor starts rotating as soon as electric supply is given. Small- and medium-sized machines are started directly on-line

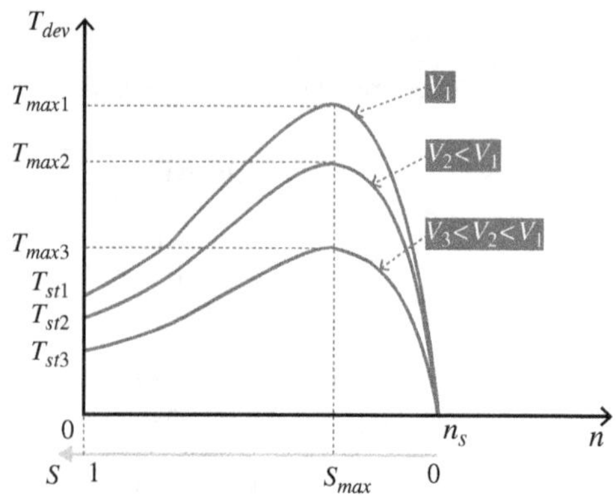

Figure 6.28 Torque-speed characteristics for variable stator voltage.

i.e. electric supply is fed directly from the grid. At start condition, the rotor is short-circuited since slip s = 1, as shown in Figure 6.19 and hence application of full-rated voltage causes a large current to flow into the machine. This large current drawn by motor at start condition is called 'inrush current'. For small- and medium-sized machines such large current is allowed, however, for large machines (order of hundreds of kWs) this is not permitted by the utility companies. Sudden large current drawn by motors may dip the voltage of the feeder supplying them, which will affect the other connected loads. Hence for large machines, starters are used between the utility grid and stator of induction machines. Starters are the devices that are used to feed motors at start conditions. The major aim of the starters is to limit or reduce the starting current. Hence, the basic principle of starters is to apply reduced voltage at start and gradually increase the voltage to the rated voltage.

Starting by reduced voltage application changes the torque-speed characteristics as shown in Figure 6.28.

With reduced voltage starting

- Starting current is reduced since the current is directly proportional to the applied voltage. The starting current is given as (6.54)

$$I_2' = \frac{V_{TH}}{\left(R_{TH} + \frac{R_2'}{s}\right) + j\left(X_{TH} + X_2'\right)}$$

$$|I_2'|_{st} = \frac{|V_{TH}|}{\sqrt{\left(R_{TH} + R_2'\right)^2 + \left(X_{TH} + X_2'\right)^2}}; s = 1 \tag{6.85}$$

- The starting torque is also reduced since it is proportional to square of the applied voltage. This implies that heavy load cannot be started. The starting torque is given as Eq. (6.56), repeated here;

$$T_{st} = \frac{3|V_{TH}|^2 R_2'}{\omega_s\left[\left(R_{TH} + R_2'\right)^2 + \left(X_{TH} + X_2'\right)^2\right]} \tag{6.86}$$

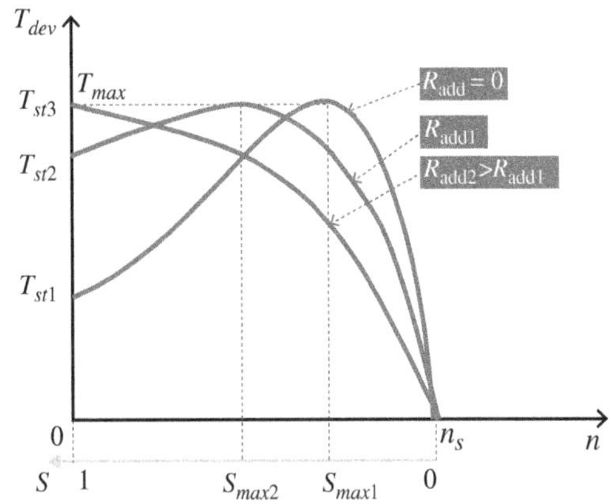

Figure 6.29 Torque-speed characteristics for variable rotor resistance.

□ The maximum torque is also reduced which implies that the acceleration of the motor is reduced. The maximum torque is given as Eq. (6.63), repeated here

$$T_{\max} = \frac{3V_{TH}^2}{2\omega_s\left(R_{TH} + \sqrt{R_{TH}^2 + \left(X_{TH} + X_2'\right)^2}\right)} \tag{6.87}$$

□ Speed or slip at maximum torque remains unchanged since smaxT does not depend upon the applied voltage.

External Resistance Starting

This method of starting is only applicable to slip ring or wound rotor type induction machine. The three terminals of the rotor are available for external connections. Three-phase resistance bank can be connected to the rotor terminals to increase rotor resistance. By increasing the external rotor resistance, the torque-speed characteristics can be modified as shown in Figure 6.29. Three curves are shown, one without the external resistance, and two with external resistances. The following can be deduced from the characteristics

□ Starting current is reduced (see Eq. (6.85))
□ Starting torque is increased (see Eq. (6.86))
□ Maximum torque is unchanged which indicates that Motor acceleration is high (see Eq. (6.87))
□ Speed at maximum torque is reduced (see Eq. (6.66))

Basic types of starters are

- Direct On-line Start
- Line resistance Start
- Auto-transformer Start
- Star-Delta Start

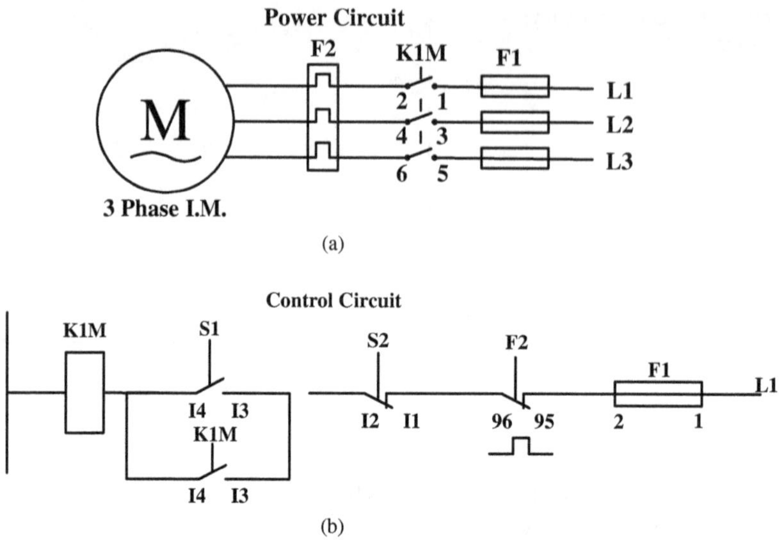

Figure 6.30 Direct on-line starter, (a) power circuit, and (b) control circuit.

6.13.1 Direct-on-Line Start

Small induction motors can be started direct-on-line by connecting to the grid through Direct on-line starter as shown in Figure 6.30. Motor is connected directly to the grid and full voltage is applied to the motor terminals. These types of starters are simple, cheap, and easy to maintain and troubleshoot. The disadvantages are

- Motor draw very high starting current (Typically five to eight times the full-load current of the motor)
- High-starting current causes higher thermal stress on the motor, thereby reducing its life.
- High current causes large voltage drop in the feeder supplying induction motor, affecting other customers connected to the same lines.
- High current yields unnecessarily high starting torque, even when not required by the load, thereby increasing mechanical stress on the mechanical systems such as rotor shaft, bearings, gearbox, coupling, chain-drive equipment, etc. leading to premature failure and plant downtimes.

In the starter, there are two parts: the power circuit and control circuit. In the power circuit the motor is connected to the line through fuses (F1) and (F2). The contactor main coils (1–2, 3–4, and 5–6) are connected to the motor terminals through these fuses. The energizing coil is in the control part that is connected across the start button (S1) which is normally open. There is a stop button which is normally closed. As soon as the start button is pressed, the main contactor energizing coil (C1) is energized closing the contactors in the main power circuit. The motor terminals are then directly connected to the line. To stop the motor, the stop button S2 is pressed that de-energizes the main contactors coil and hence the motor terminals are disconnected from the supply.

Torque and Current relationship in Direct on Line starting

Torque can be found using Eq. (6.35)

$$P_g = 3(I_2')^2 \frac{R_2'}{s}$$

$$T = \frac{P_g}{\omega_s} = \frac{3(I_2')^2 \frac{R_2'}{s}}{\omega_s} = \frac{3(I_2')^2 R_2'}{2\pi f s} \tag{6.88}$$

Under full-load condition:

$$I_2' = I_{2fl}; \quad s = s_{fl}; \quad T = T_{fl}$$

Hence, the torque equation under full-load condition can be written as

$$T_{fl} = \frac{3I_{2fl}^2 R_2'}{2\pi f s_{fl}} \tag{6.89}$$

Under starting condition:

$$I_2' = I_{2st}; \quad s = 1; \quad T = T_{st}$$

$$T_{st} = \frac{3I_{stl}^2 R_2'}{2\pi f} \tag{6.90}$$

Therefore, the ratio of starting to full-load torque can be obtained from Eqs. (6.89) and (6.90) as:

$$\frac{T_{st}}{T_{fl}} = \left(\frac{I_{2st}}{I_{2fl}}\right)^2 s_{fl} \tag{6.91}$$

6.13.2 Line Resistance Start

To apply reduced voltage at start, resistances are inserted between the grid supply and motor stator terminals as shown in Figure 6.31.

Full resistance is inserted at the start condition and is gradually reduced to zero. Most of the voltage is dropped across the line resistances and the voltage applied to the motor is reduced. When the resistance is cut-out of the circuit, the voltage applied to the motor terminal is increased. The advantage of this scheme is that it is cheap, simple, and a reliable solution. The disadvantage of the method is additional losses in the resistance circuit.

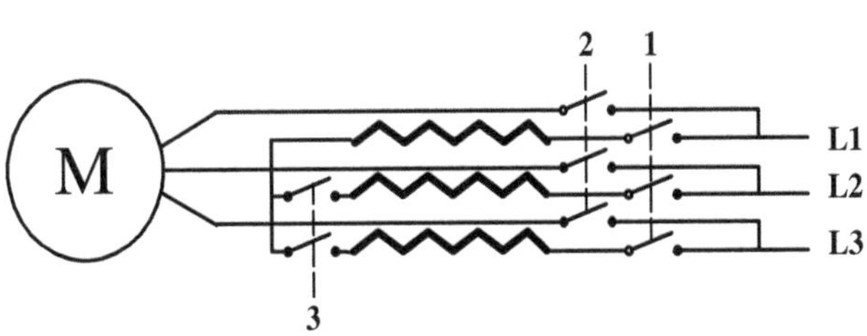

Figure 6.31 Line resistance starter.

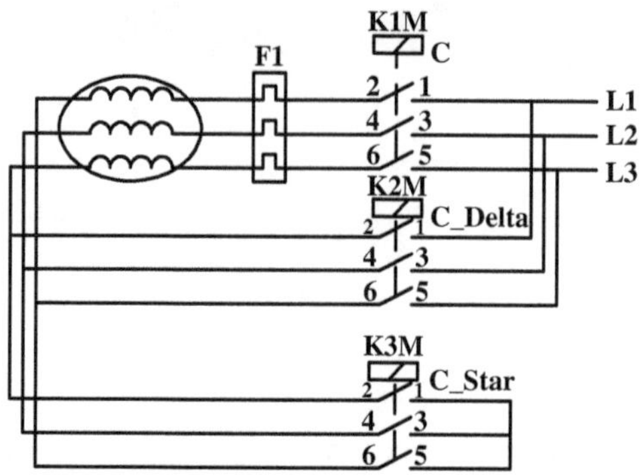

Figure 6.32 Star-Delta starting of three-phase induction motor.

6.13.3 Star-Delta Starter

This is one of the most popular starting methods of a three-phase induction motor. The motor is started as star-connected and once it reaches steady-state condition, is changed to delta connection. When motor is started as star, the applied voltage is $V_L/\sqrt{3}$ and hence the starting current is also reduced. The current is reduced to one-third when compared with the delta connection. When the motor is connected as delta, full-line voltage is applied to the motor terminals. Star-delta starting is shown in Figure 6.32. The torque produced by induction motor varies with square of the applied voltage and, hence, in star-delta starter the starting torque is reduced by one-third compared to direct delta starting. As seen in Figure 6.32, three contactors are shown, C is the main contactor, C_star is for star starting, and C_delta for delta running contactor. The fuse is to protect from over-current.

When C and C_star are energized together, the machine starts as star and once it reaches a certain speed, C_delta is energized.

Let us define the following terms

I_{stY} = Starting current in star connection

$I_{st\Delta}$ = Line Starting current in delta connection

Z_{eqs} = Per-phase Impedenace referred to rotor side at start

V_L = Line voltage

$$\text{Starting line/phase current when started as star} = I_{stY} = \frac{V_L}{\sqrt{3}Z_{eqs}} \tag{6.92}$$

$$\text{Starting phase current when started in delta} = I_{st\Delta} = \frac{V_L}{Z_{eqs}} \tag{6.93}$$

Ratio of starting line currents for star and delta connections can be obtained by dividing Eqs. (6.92) and (6.93); note that Eq. (6.92) gives phase current and to obtain line current this should be multiplied by $\sqrt{3}$

$$\frac{I_{stY}}{I_{st\Delta line}} = \frac{\frac{V_L}{\sqrt{3}Z_{eqs}}}{\sqrt{3}\frac{V_L}{Z_{eqs}}} = \frac{1}{3} \tag{6.94}$$

This shows that when the motor is started in star, it draws one-third of the starting current that it would draw when started in delta.

The ratio of torques for star and delta starting can be obtained in a similar fashion. It is to be noted that the torque is proportional to the square of the applied voltage. Hence, the ratio of starting torques

$$\frac{\text{Starting torque with star connection}}{\text{Starting torque with delta connection}} = \frac{\left(\frac{V_L}{\sqrt{3}}\right)^2}{V_L^2} = \frac{1}{3} \tag{6.95}$$

This shows that the starting torque with star connection is one-third of that obtainable with delta connection.

The ratio of starting torque with star connection and full-load torque with delta winding (running condition is with delta winding) is computed as

$$\frac{T_{st\gamma}}{T_{fl\Delta}} = \frac{\frac{3(I_{st\gamma})^2 R_2'}{2\pi f}}{\frac{3\left(I'_{fl\Delta phase}\right)^2 R_2'}{2\pi f s_{fl}}} = \left(\frac{I_{st\gamma}}{I_{fl\Delta phase}}\right)^2 s_{fl} \tag{6.96}$$

6.13.4 Autotransformer Starter

The motor can be started by applying reduced voltage at the starting and then gradually increasing the voltage when the motor picks up speed. An autotransformer is inserted between the motor and supply. The supply is given to the autotransformer as shown in Figure 6.33. Initially, the output voltage of the autotransformer is zero which is gradually increased (manually or automatically) to full rated value. Since the applied voltage is

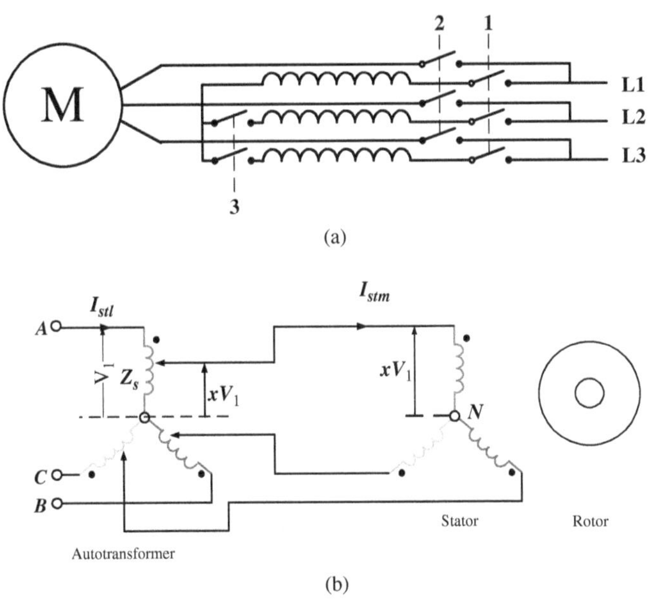

Figure 6.33 Autotransformer starters for three-phase induction motor (a) Schematic, (b) Autotransformer supplying motor.

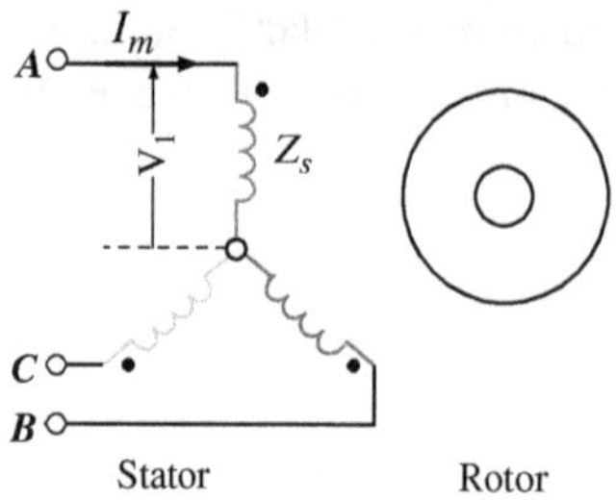

Figure 6.34 Direct starting.

gradually increased, the starting current is limited to safe value. This is a simple method of starting. However, the autotransformer is bulky and expensive.

The starting current drawn by an induction machine when started on direct supply with full rated voltage of V_1 per phase is shown in Figure 6.34.

$$I_m = \frac{V_1}{Z_s} \tag{6.97}$$

If an autotransformer with a tapping of x, is used to supply the induction machine, the applied voltage per phase is xV_1. The starting current drawn is

$$I_{stm} = \frac{xV_1}{Z_s} \tag{6.98}$$

Referring the motor current to the line side (autotransformer side)

Voltage on the line side is V_1, while voltage on the motor side is xV_1, thus the following relationship is obtained

$$\begin{aligned} V_1 I_1 &= V_2 I_2 \\ V_1 I_{stl} &= xV_1 I_{stm} \\ I_{stl} &= xI_{stm} \end{aligned} \tag{6.99}$$

Put the value of I_{stm} from Eq. (6.98), in Eq. (6.99)

$$I_{stl} = xI_{stm} = x\left(\frac{xV_1}{Z_s}\right) = \frac{x^2V_1}{Z_s} \tag{6.100}$$

$$\frac{\text{Starting Current with Auto-transformer}}{\text{Starting Current with direct on-line}} = \frac{I_{stl}}{I_m} = \frac{(x^2V_1/Z_s)}{(V_1/Z_s)} = x^2 \tag{6.101}$$

Torque developed is proportional to the square of the applied voltage. The ratio of the torque with direct on-line starting and with autotransformer starting is obtained as

$$\frac{\text{Starting Torque with Auto-transformer}}{\text{Starting Torque with direct on-line}} = \frac{T_{stl}}{T_m} = \frac{(xV_1)^2}{(V_1)^2} = x^2 \tag{6.102}$$

The ratio of the starting torque to the full-load torque is obtained as follows

$$\frac{T_{st}}{T_{fl}} = \left(\frac{I_{stl}}{I_{fl}}\right)s_{fl} = x^2\left(\frac{I_{stm}}{I_{fl}}\right)s_{fl} \tag{6.103}$$

6.14 Speed Control of Induction Machine

Induction machines are used for numerous variable speed applications in industries. To change the speed of an induction machine, the synchronous speed can be changed from Eq. (6.1).

$$N_s = \frac{120f}{P}$$

The synchronous speed can be changed by changing

- The frequency of the applied voltage
- The number of stator winding poles

The motor slip can also be varied by changing

- Applied stator voltage
- Varying the rotor resistance (by connecting external resistance)
- Applying voltage in the rotor side (rotor voltage injection scheme and slip energy recovery scheme)

6.14.1 By Varying the Frequency of the Supply

The synchronous speed or the speed of the rotating field in an induction machine is directly proportional to the frequency of the applied voltage. Thus, by varying the frequency of the applied voltages, the synchronous speed can be changed which causes the motor to rotate at close to the new synchronous speed. The speed of the motor can be changed continuously by applying variable frequency supply to the stator terminals of induction motor.

The frequency of the applied voltage can be varied by using a power electronic converter to supply the machine as shown in Figure 6.35. The frequency of the supply voltage can easily be varied by modulating the power converter. For detail of the methods used to vary the frequency from the power converter, readers may look into the power electronics courses.

The supply frequency affects many parameters of the machine such as the developed torque, the starting torque, and the maximum torque. The frequency term affects the following

$$\omega_s = 2\pi\frac{N_s}{60} \tag{6.104}$$

$$X_2' = 2\pi f L_{l2}' \tag{6.105}$$

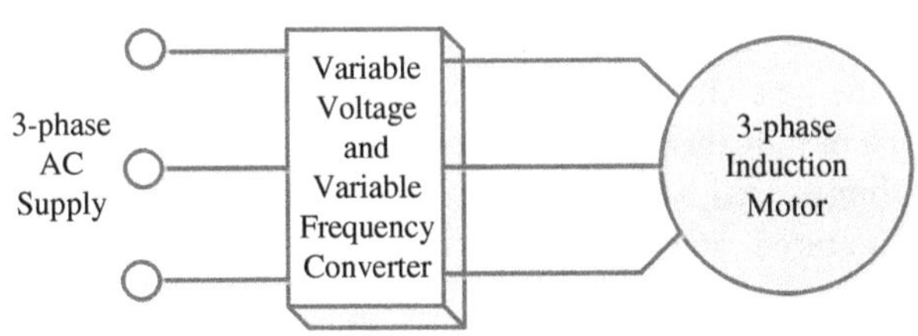

Figure 6.35 Power converter with variable frequency output supplying a three-phase induction motor.

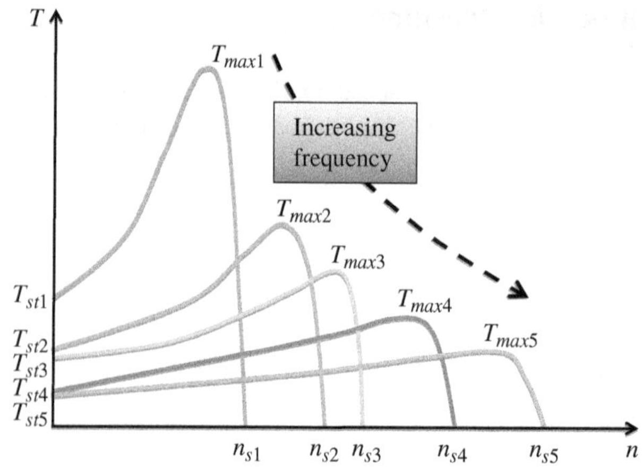

Figure 6.36 Torque vs. Speed curve for variable frequency supply.

Now, considering the torque equations, it is noticed that the torques are affected by the change in frequency due to the $\omega_s$and X_2' (from Eqs. (6.59), (6.60), (6.63), and (6.62)).

$$T_e = \frac{P_g}{\omega_s} = \frac{3|V_{TH}|^2 \frac{R_2'}{s}}{\omega_s \left[\left(R_{TH} + \frac{R_2'}{s}\right)^2 + (X_{TH} + X_2')^2\right]}$$

$$T_{st} = \frac{3|V_{TH}|^2 R_2'}{\omega_s \left[(R_{TH} + R_2')^2 + (X_{TH} + X_2')^2\right]}$$

$$T_{\max} = \frac{3V_{TH}^2}{2\omega_s \left(R_{TH} + \sqrt{R_{TH}^2 + (X_{TH} + X_2')^2}\right)}$$

$$S_{\max T} = \frac{R_2'}{\sqrt{R_{TH}^2 + (X_{TH} + X_2')^2}}$$

The set of torque-speed curves for different frequency of stator voltages is shown in Figure 6.36. It is seen that the maximum torque and starting torque reduces with increase in frequency (increasing speed). The speed at which the maximum torque occurs also increases with increase in the frequency (speed).

6.14.2 Pole Changing Method

This is one of the oldest methods of speed control of a three-phase induction motor. Speed is controlled in steps when using the pole changing method. The pole changing method is applicable to a squirrel-cage type of induction motor. In a modern drive motor system, the pole changing method is no longer used because of the step speed change. The pole changing methods are classified into three categories

1. Multiple numbers of windings
2. Consequent poles method
3. Pole Amplitude Modulation

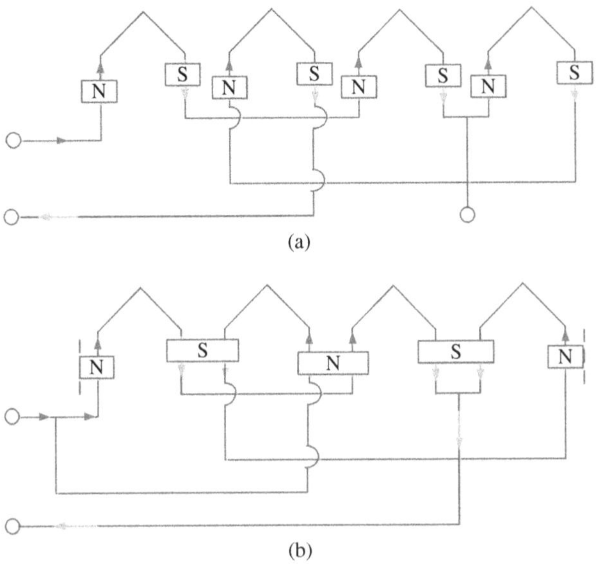

Figure 6.37 (a) Eight-pole connection. (b) four-pole connection.

6.14.2.1 Multiple Numbers of Windings

Two windings are placed in the stator slots corresponding to two different pole numbers. Two different speeds are then possible by using two different pole number windings. One set of three-phase winding is energized at one time.

Only two discrete speed values are possible in this method. No continuous speed change is possible. This method is used for a squirrel-cage type of machine only since the rotor induces the same number of poles as that of the stator. In wound rotor type, winding of a particular pole number (the same as the stator winding number of poles) is placed in the rotor slots; hence, the rotor should also be wound for two different pole numbers. Due to two windings being used in the stator of the machine, it becomes heavy and costly.

6.14.2.2 Consequent Pole Method

In this method of speed control, stator windings are divided into several coil groups. The supply is given in such as a way that two possible pole pairs are created. As can be seen, this is explained with the help of Figure 6.37a,b. Considering North Pole is the for current going up and South Pole for the current going down. It is seen that eight poles are formed, as seen in Figure 6.37a and four poles are formed as shown in Figure 6.37b. In general, the poles are changed in the ratio of 1 : 2 in this method. Hence, two discrete speed are possible.

Combining the above two methods, fours speed are possible. If two windings of two different poles are used in the machine, each winding can be configured for two possible poles. Hence in total, four speeds are possible; e.g. considering two windings with four poles and six poles, then the possible pole numbers that can be generated using consequent pole method are four, eight, six, and 12.

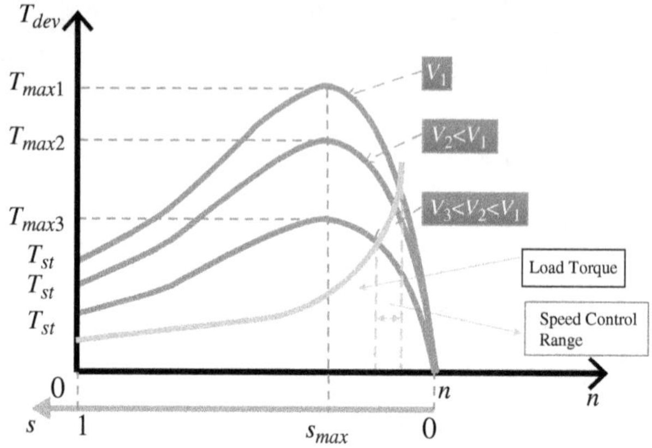

Figure 6.38 Torque-speed curve for variable voltage application.

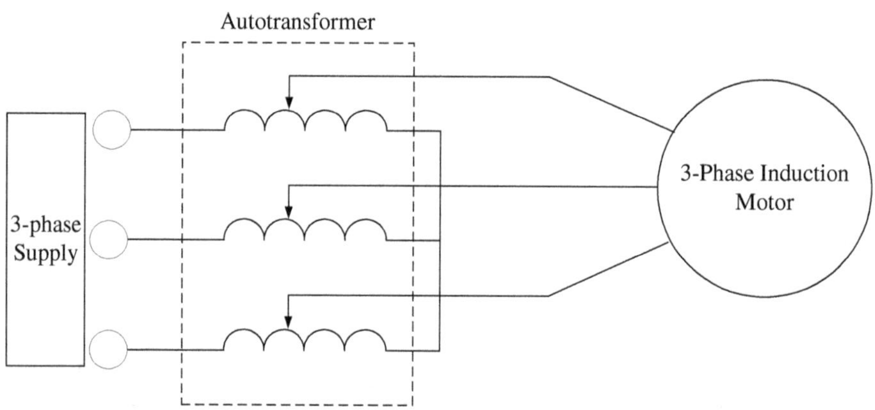

Figure 6.39 Autotransformer for variable voltage supply.

6.14.3 Stator Voltage Control

Stator voltage is varied to change the speed of an induction machine. The torque is proportional to the square of the applied voltage and hence by changing the voltage, the torque is altered. However, voltage can only be reduced below the rated value. The set of torque-speed curve for different applied voltages is shown in Figure 6.38. The following is observed

- ➢ The maximum torque is reduced by a factor of square of the applied voltage
- ➢ The starting torque is also reduced
- ➢ The current drawn by the machine increases
- ➢ Limited/small speed range is achieved.

Stator voltage is varied by

- ➢ Using autotransformer in the supply as shown in Figure 6.39.
- ➢ Using an AC voltage regulator connected in each phase as shown in Figure 6.40.

While using autotransformer, the input voltage to the stator of the induction machine is sinusoidal. When an AC voltage regulator is used, the current is distorted due to the switching action of the semiconductor switches. To obtain an AC voltage regulator, two

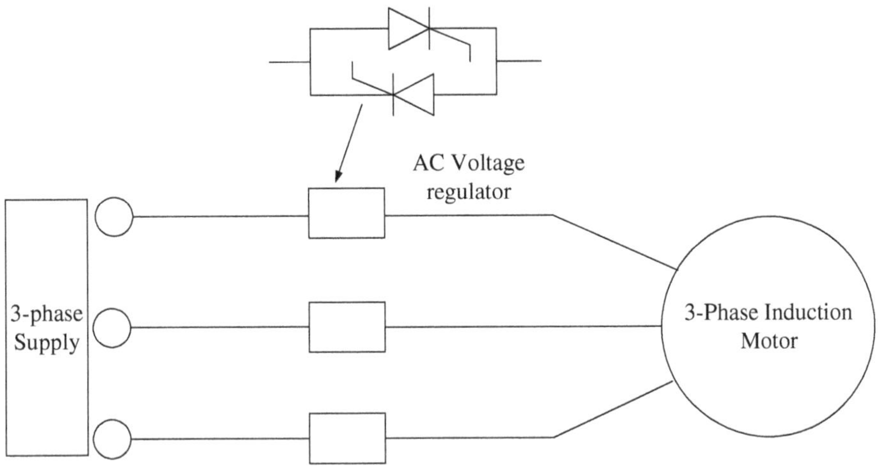

Figure 6.40 AC voltage regulator for variable voltage supply.

switches (such as thyristor) are connected back-to-back. One thyristor conducts in the positive half cycle of the AC wave, while the second operates in the negative half cycle and hence bi-directional power flow is possible.

6.14.3.1 Voltage/Frequency = Constant Control

Stator voltage control is not used alone and, similarly, the frequency control alone is not used for speed control of an induction machine. To understand this, let us consider the emf equation of an induction machine

$$E = 4.44fN_{ph}K_w\phi \tag{6.106}$$

where N_{ph} is the number of stator winding turns per phase, K_w is the winding factor due to short pitching and distribution of the winding and the value is less than 1, and ϕ is flux per pole. If the stator voltage drop is assumed negligible, then the applied voltage is approximately equal to the induced emf

$$V \approx E = 4.44fN_{ph}K_w\phi$$
$$V \approx kf\phi \tag{6.107}$$

If applied voltage is kept and only the frequency is varied, then the flux will also vary. If the frequency is increased, the flux will decrease. If flux decreases then the torque capability of the machine decreases and the machine is underutilized. If the frequency is decreased, then the flux is increased to keep the product constant (V = constant). Since the machine core is already designed to operate at the knee point of the B–H curve, any further increase in the flux leads to the saturation of the core. The saturation will cause distortion in current, increased harmonic losses and a rise in core temperature.

A similar observation is made for voltage variation alone keeping the frequency constant. A decrease in the applied voltage will cause a reduction in the flux.

Hence, the voltage is also varied when frequency is changed in such a way that the ratio of voltage and frequency is constant so that flux remains constant.

$$\frac{V}{f} \approx \text{constant} \tag{6.108}$$

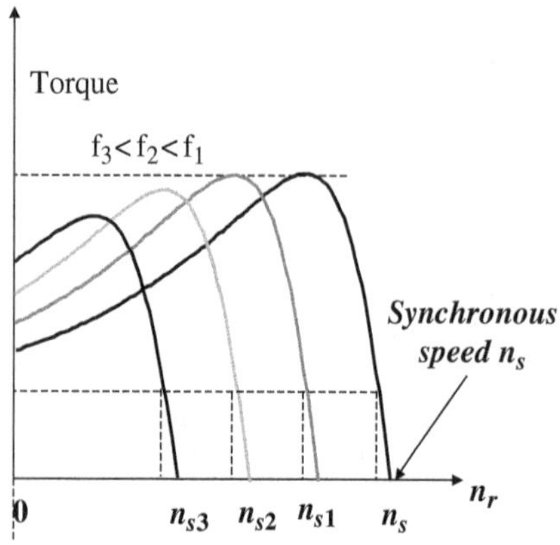

Figure 6.41 Torque-speed curve for v/f = constant condition.

This method is applicable to speed control from zero to base speed region. For speed control above the base speed, voltage cannot be increased further beyond the rated value and hence only frequency is increased. This will cause reduced flux and is called field-weakening mode. A set of torque-speed curve for V/f = constant control is shown in Figure 6.41. It is seen from the characteristics that a wide range of speed control is achieved with relatively constant maximum torque.

6.14.3.2 Rotor Resistance Variation

Rotor resistance is changed (increased) for speed control of the three-phase induction machine. This method is applicable to wound rotor type machines because of the availability of three terminals in the rotor in which the external resistances are connected. The advantages of this method: simple control, the starting current is reduced and starting torque is increased. The disadvantages are: low efficiency due to continuous rotor resistance loss, the control is slow and the unbalancing problem if the external resistances are unequal.

The rotor resistance is also varied by connecting a chopper circuit in the rotor circuit and the duty cycle of the chopper is changed which effectively changes the rotor resistance of the rotor circuit.

6.14.3.3 Rotor Voltage Injection Method

Speed of an induction machine is controlled by injecting a voltage of slip frequency in the rotor by an additional variable frequency source connected to the rotor circuit. The injected voltage should have a frequency equal to the slip frequency and the phase angle has no restriction. By changing the phase of the injected voltage, the effective resistance is varied and thus speed is varied. When the injected voltage is in phase with the rotor induced emf, the effective rotor resistance decreases. When the injected voltage is in opposite phase with the rotor induced emf, the effective rotor resistance increases. This speed control method offers a wide range of speed control in the base speed region and in the field-weakening region. The voltage injection method principle is used in Kramer drive and Scherbius drive. Details of these methods are found in electric drive system books.

6.14.3.4 Cascade Connection of Induction Machines

This method of speed control is done by connecting the two induction machines called the main induction machine and auxiliary induction machine on the same shaft as shown in

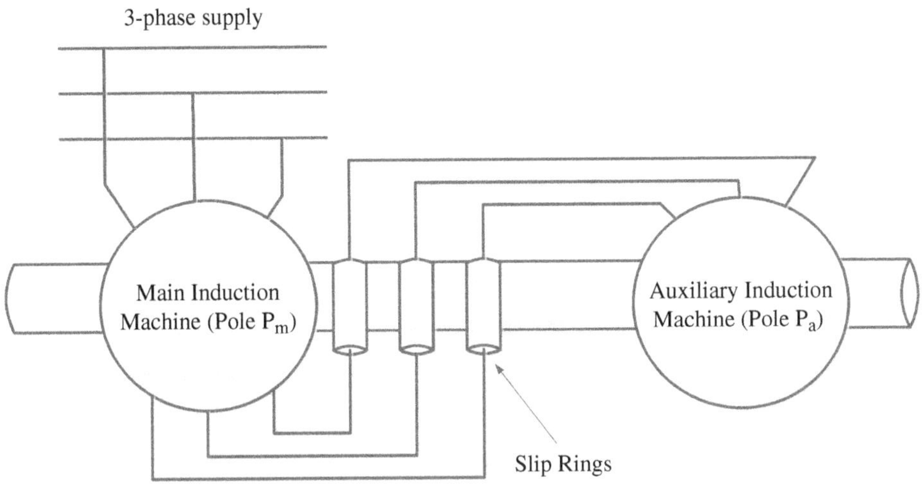

Figure 6.42 Cascade connection of two induction motors.

Figure 6.42. This is called a cascade or tandem connection. The speed of both machines will remain the same as they are mounted on the same shaft. The main induction machine is supplied by three-phase mains and the auxiliary machine is supplied from the rotor side of the main machine. The main motor should be of wound-rotor type while the auxiliary motor is squirrel-cage type. The speed of both motors will either correspond to the sum of the poles or the difference between the poles.

Let the following

f_m = frequency of supply to the stator of the main motor
f_a = frequency of supply to the stator of the auxiliary motor
P_m = Number of poles of the main motor
P_a = Number of poles of the auxiliary motor
N_{sm} = Synchronous speed of the main motor
N_{sa} = Synchronous speed of the auxiliary motor
s_m = Slip of the main motor
s_a = Slip of the auxiliary motor
N = actual speed of the cascaded motors.

The frequency of the rotor voltage and current of the main motor is

$$f_{mr} = s_m \times f_m = f_a \tag{6.109}$$

The supply frequency of the auxiliary motor is the same as the rotor voltage frequency of the main motor.

To find the speed of the cascaded motors, let us do the following

$$N_{sa} = \frac{120 f_a}{P_a} \tag{6.110}$$

Substitute Eq. (6.109) in Eq. (6.110), and the following is obtained

$$N_{sa} = \frac{120 f_a}{P_a} = \frac{120 S_m f_m}{P_a}$$

$$N_{sa} = \frac{120f_m}{P_a}\left[\frac{N_{sm} - N}{N_{sm}}\right] \tag{6.111}$$

Under no-load condition:

$$N_{sm} \approx N \tag{6.112}$$

Substituting Eq. (6.111) in Eq. (6.112)

$$N = \frac{120f_m}{P_a}\left[\frac{N_{sm} - N}{N_{sm}}\right]$$

$$N = \frac{120f_m}{P_a}\left[1 - \frac{N}{N_{sm}}\right]$$

$$N = \frac{120f_m}{P_a}\left[1 - \frac{N}{\frac{120f_m}{P_m}}\right]$$

$$N = \frac{120f_m}{P_a} - \frac{NP_m}{P_a}$$

$$N\left[1 + \frac{P_m}{P_{am}}\right] = \frac{120f_m}{P_a}$$

$$N = \frac{120f_m}{P_m + P_a} \tag{6.113}$$

If the supply terminals of the auxiliary motor are interchanged, the field speed will reverse and the overall speed of the cascaded motors will be

$$N = \frac{120f_m}{P_m - P_a} \tag{6.114}$$

Hence, speed can be varied because of the change in the number of poles.

6.14.3.5 Pole-Phase Modulation for Speed Control

Wider speed and torque range are the major requirements of a drive for applications such as ship propulsion, more electric aircraft and high-power traction etc. Regular motors need to be oversized to have greater speed and torque range which, in turn, increases the volume and cost of the drive. Pole-changing techniques are adopted to achieve an extended range of speed-torque characteristics of the IM without increasing the motor size, i.e. Pole amplitude modulation (PAM), Pole-phase modulation (PPM) and multiple stator windings.

The conventional pole-changing techniques have many limitations such as

- A higher number of the stator winding terminals are required
- De-energization is required before connecting the winding connections
- Mechanical switches are required to change the terminal connections
- Poor utilization of copper
- Auxiliary or separate windings are required for attaining three or four speed ratios.

Because of the above disadvantages, adopting the pole-changing techniques with rearrangeable stator windings for high-power applications is impracticable. In the late 1960s,

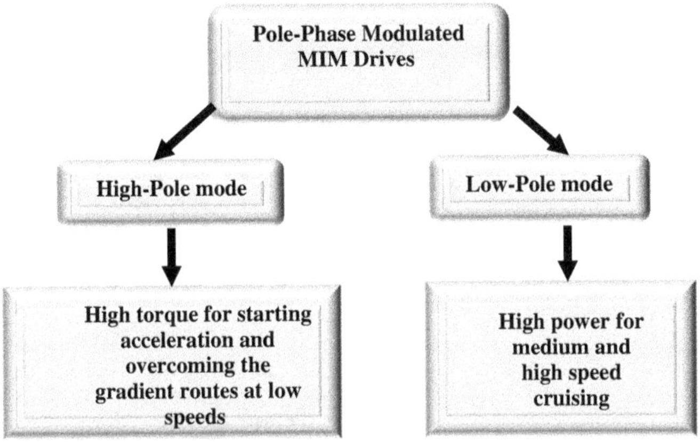

Figure 6.43 The block diagram of the PPMIM drives.

mechanical switches were used for changing the number of poles of the IM. Due to enhancement in power semiconductor technology and microcontrollers, the pole-changing techniques are becoming a cost-effective and efficient solution for achieving a wider range of speed and torque. The power electronic switches are a substitute for mechanical switches, which are controlled by microcontrollers. The continuous changing of the number of poles is possible by varying the supply voltage excitation to respective windings with the help of the power electronic converters called the PPM. The PPM drives deliver a high torque for hill climbing and faster acceleration in high-pole and less-phase operation. Whereas in high-phase and low-pole mode, the drive delivers high speed with constant power for steady-run operation. The block diagram of the PPM drives is shown in Figure 6.43.

PPM is an effective way to obtain the extended speed torque characteristics, where the phase belt is adjusted by changing the excitation of power electronic inverter. In PPM, the number of phases and poles vary perpetually in a constant ratio, i.e. slots/pole/phase. Let us consider Q as the number of stator slots and m is the number of phases and $2p$ is the number of poles, the generalization of PPM is

$$Q = 2p_1 q_1 m_1 = 2\,p_2 q_2 m_2 \tag{6.115}$$

$$k = \frac{p_2}{p_1} = \frac{q_1 m_1}{q_2 m_2} \tag{6.116}$$

where, p_1, m_1 and p_2, m_2 are the number of pole pairs and number of phases of the PPM1 (for a synchronous speed) and PPM2 (for another synchronous speed) respectively. The q_1 and q_2 are phase belt of the PPM1 and PPM2 windings respectively. Eq. (6.116) gives the pole ratio (k), which is a positive integer and it should be always ≥ 1 if the windings are wound for low pole mode.

Consider two PPM models which satisfy the generalization of PPM equations

Model 1: A 36-slot nine-phase pole phase modulation induction motor (PPMIM) drive is considered, that is capable of operating at nine-phase two-pole mode and three-phase six-pole mode with a phase belt of two stator slots. The winding configuration for forming two poles with nine-phase winding is shown Figure 6.44a. For forming the six-pole, the excitation of the phase windings has changed accordingly, as shown in Figure 6.44b.

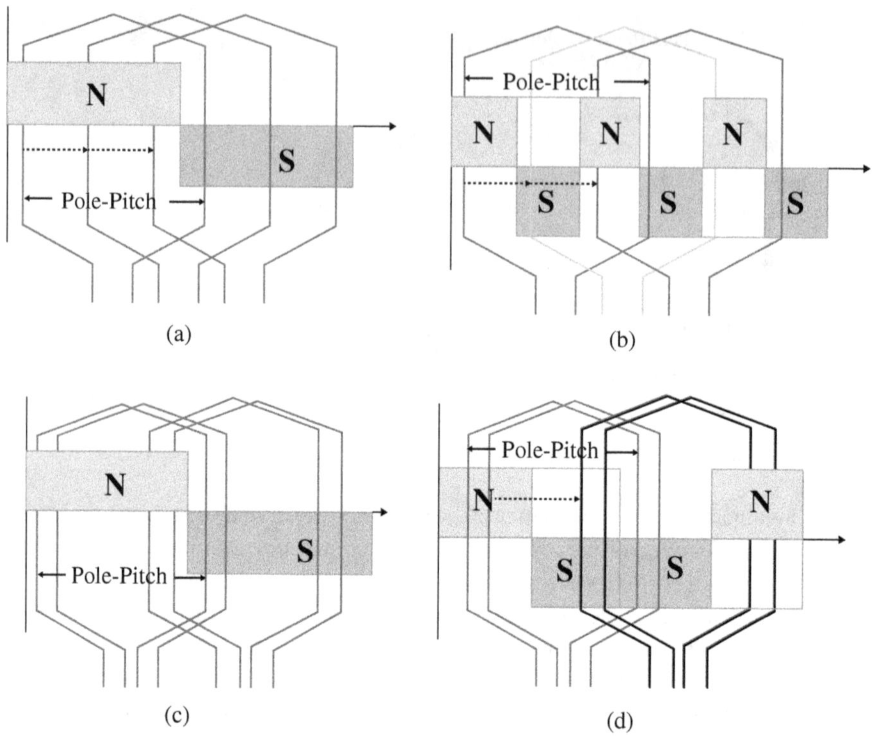

Figure 6.44 9-phase PPMIM Winding arrangement for (a) two-pole winding, (b) six-pole winding; 12-phase PPMIM Winding arrangement; (c) two-pole winding, (d) four-pole winding.

Model 2: A 48-slot 12-phase PPMIM drive is considered, that is capable of operating at 12-phase two-pole mode and six-phase four-pole mode with a phase belt of two stator slots. The winding configuration for forming two poles with 12-phase winding is shown Figure 6.44c. For forming the four-pole, the excitation of the phase windings has changed accordingly, as shown in Figure 6.44d.

From Figure 6.44a,b, it is clear that the nine-phase 36-slot PPMIM satisfies the generalization of PPM equations and it can be possible to implement practically with uniform pole distributions. Whereas, from Figure 6.44c,d, it is observed that the 12-phase 48-slot PPMIM is also satisfies the generalization of PPM equations but the formation of a number of poles is non-uniform (unequal pole widths), which leads to unequal flux distributions and high-torque pulsations.

If the machine wound for $p1$ number of pole pairs with $p1 < p2$ then the pole ratio (k) must be an odd positive integer and it should be always $\geq 2n+1$ (where n = 0, 1, 2, 3...). With this revised generalization, the unequal width of pole formations is avoided. Based on the revised generalization, a 1 : 3 speed ratio of nine-phase PPMIM with 36 and 72 slots, a 1 : 3 : 5 speed ratio of 15-phase PPMIM with 60 slots and a 1 : 3 : 9 : 15 speed ratio of 45-phase PPMIM with 90 slots have presented.

A. **Two Speed Ratios of nine-Phase PPMIM Drive**
   A 36-slot nine-phase IM has been designed based on the generalization of PPM for achieving the two-speed ratios (1 : 3). This nine-phase PPMIM drive is capable of operating at nine-phase four-pole mode with a synchronous speed of 1500 rpm and three-phase 12-pole mode with a synchronous speed of 500 rpm. The phase belt (slots/pole/phase) is constant for both pole phase combinations, i.e. $q = 1$. The winding diagrams for these two pole-phase combinations have given in Table 6.3.
B. **Three Speed Ratios of 15-Phase PPMIM Drive**
   Based on a revised generalization PPM, a 60-slot 15-phase IM is presented in [115] for achieving the three-speed ratios (1 : 3 : 5). This 15-phase PPMIM drive is capable of operating at 15-phase four-pole mode with a synchronous speed of 1500 rpm, 5-phase 12-pole mode with a synchronous speed of 500 rpm and three-phase 20-pole mode with a synchronous speed of 300 rpm. The phase belt (slots/pole/phase) is constant for all three pole-phase combinations, i.e. $q = 1$. The winding diagrams for these two pole-phase combinations are given in Table 6.4.

Advantages and disadvantages of the PPMIM drives:

- Continuous changing of the poles and phase with the help of power electronics converters will deliver better results compared to conventional pole changing techniques.
- In high-pole operations, the PPMIM drive is able to supply the high initial torque for fast acceleration and hill climbing.
- The rated torque of the PPMIM drive in the high pole is k times greater than the low-pole mode. For example, in three-phase 12-pole mode, the nine-phase PPMIM drive is able to supply three times that of the rated torque as compared to nine-phase four-pole mode.
- In low-pole mode, the PPMIM drive is able to supply high speed which is k times greater than the high-pole mode.
- The PPMIM drives are capable of generating three or four speed and torque variations. So, the mechanical gearbox system in conventional IC Engine based vehicles can be eliminated with the PPMIM drives, which will reduce the size and weight of the vehicle.

Even although the PPMIM drives have several advantages (as mentioned above), on the other hand, they also have a few disadvantages, for example

- To achieve the high starting torque in high-power applications, the PPMIM drives must operate in high-pole and low-phase mode, which increases the magnitude of space harmonics in the air gap.
- This increased magnitude of space harmonics results in higher torque pulsations.
- The magnetizing inductance of the machine is inversely proportional to the square of the number of poles, which lowers the power factor.

6.15 Matlab/Simulink Modelling of the Three-Phase Induction Motor

Matlab/Simulink is used extensively for simulation of electric machines. Torque and current computation for steady-state condition is done using the equations directly modelled into Simulink blocks.

Table 6.3 PPM details of the 1 : 3 speed 9-phase IM with 36 slots.

9-Phase 4-pole Operation

Slot Number	1	2	3	4	5	6	7	8	9	10	11	12	13	14	15	16	17	18	19	20	21	22	23	24	25	26	27	28	29	30	31	32	33	34	35	36
Phase Current	a	*f	b	*g	c	*h	d	*i	e	*a	f	*b	g	*c	h	*d	i	*e	a	*f	b	*g	c	*h	d	*i	e	*a	f	*b	g	*c	h	*d	i	*e
Slot Angle	0	20	40	60	80	100	120	140	160	180	200	220	240	260	280	300	320	340	0	20	40	60	80	100	120	140	160	180	200	220	240	260	280	300	320	340
Poles	N									S									N									S								

3-Phase 12-pole Operation

Slot Number	1	2	3	4	5	6	7	8	9	10	11	12	13	14	15	16	17	18	19	20	21	22	23	24	25	26	27	28	29	30	31	32	33	34	35	36
Phase Current	R	*B	Y	*R	B	*Y	R	*B	Y	*R	B	*Y	R	*B	Y	*R	B	*Y	R	*B	Y	*R	B	*Y	R	*B	Y	*R	B	*Y	R	*B	Y	*R	B	*Y
Slot Angle	0	60	120	180	240	300	0	60	120	180	240	300	0	60	120	180	240	300	0	60	120	180	240	300	0	60	120	180	240	300	0	60	120	180	240	300
Poles	N			S			N			S			N			S			N			S			N			S			N			S		

Slot Angle in electrical degrees

Slot angle in electrical degrees.

Table 6.4 PPM details of the 1 : 3 : 5 speed 15-phase IM with 60 slots.

9-Phase 4-pole Operation

Slot Number	1	2	3	4	5	6	7	8	9	10	11	12	13	14	15	16	17	18	19	20	21	22	23	24	25	26	27	28	29	30
Phase Current	a	*i	b	*j	c	*k	d	*l	e	*m	f	*n	g	*o	h	*a	i	*b	j	*c	*k	*d	l	*e	m	*f	n	*g	o	*h
Slot Angle	0	12	24	36	48	60	72	82	96	108	120	132	144	156	168	180	192	204	216	228	240	252	264	276	288	300	312	324	336	348
Poles									N														S							

5-Phase 12-pole Operation

Slot Number	1	2	3	4	5	6	7	8	9	10	11	12	13	14	15	16	17	18	19	20	21	22	23	24	25	26	27	28	29	30
Phase Current	p1	*p4	p2	*p5	p3	*p1	p4	*p2	p5	*p3	p1	*p4	p2	*p5	p3	*p1	p4	*p2	p5	*p3	p1	*p4	p2	*p5	p3	*p1	p4	*p2	p5	*p3
Slot Angle	0	36	72	108	144	180	216	252	288	324	0	36	72	108	144	180	216	252	288	324	0	36	72	108	144	180	216	252	288	324
Poles			N					S					N					S					N					S		

3-Phase 20-pole Operation

Slot Number	1	2	3	4	5	6	7	8	9	10	11	12	13	14	15	16	17	18	19	20	21	22	23	24	25	26	27	28	29	30
Phase Current	R	*B	Y	*R	B	*Y	R	*B	Y	*R	B	*Y	R	*B	Y	*R	B	*Y	R	*B	Y	*R	B	*Y	R	*B	Y	*R	B	*Y
Slot Angle	0	60	120	180	240	300	0	60	120	180	240	300	0	60	120	180	240	300	0	60	120	180	240	300	0	60	120	180	240	300
Poles		N			S			N			S			N			S			N			S			N			S	

Slot Angle in electrical degress

Slot angle in electrical degrees.

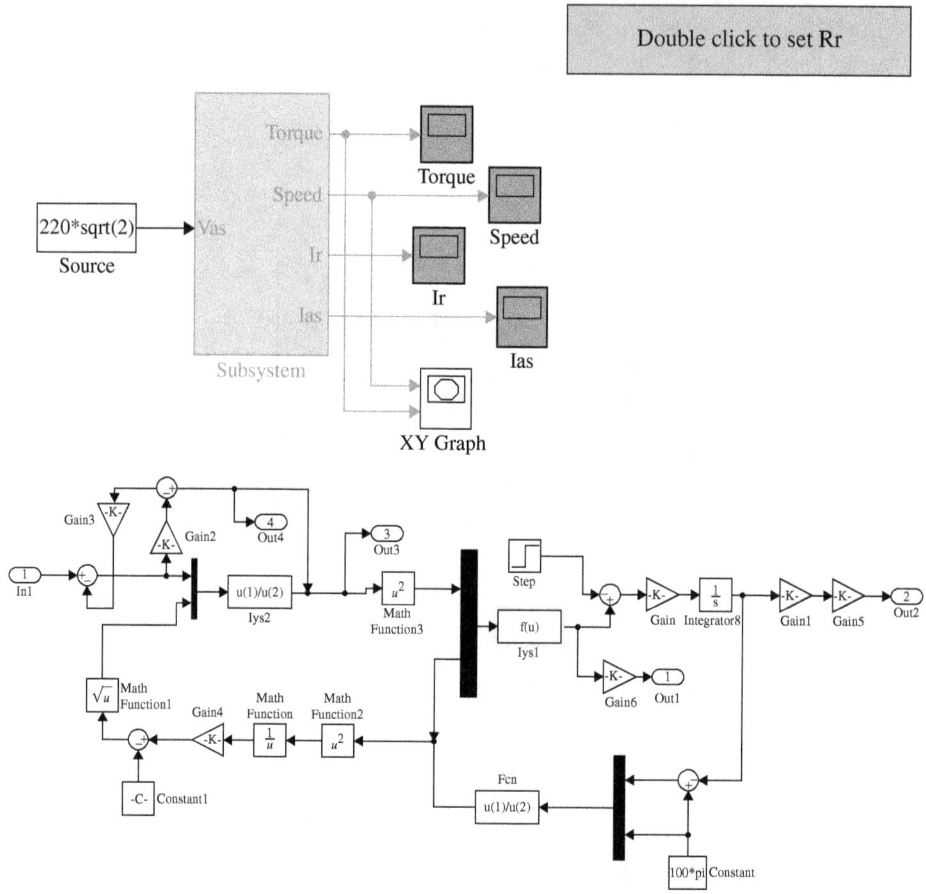

Figure 6.45 Computing torque-speed curve under steady-state condition.

6.15.1 Plotting Torque-Speed Curve under Steady-State Condition

To obtain the torque vs. speed curve, the model is shown in Figure 6.45. Torque vs. speed can easily be plotted from this model for different conditions of rotor resistances and voltages (by changing the applied voltage). Torque, speed, magnitude of stator and rotor currents are plotted.

Torque vs. Speed curve can be plotted using the following Matlab/Simulink model shown in Figure 6.46.

By running this model, the Torque-Speed curve can be obtained as shown in Figure 6.47.

6.15.2 Dynamic Simulation of Induction Machine

In steady-state analysis, all electrical transients are neglected during load changes and stator frequency variations. Such variations arise in applications involving variable-speed drives. The variable-speed drives are converter-fed from finite sources, unlike the utility sources,

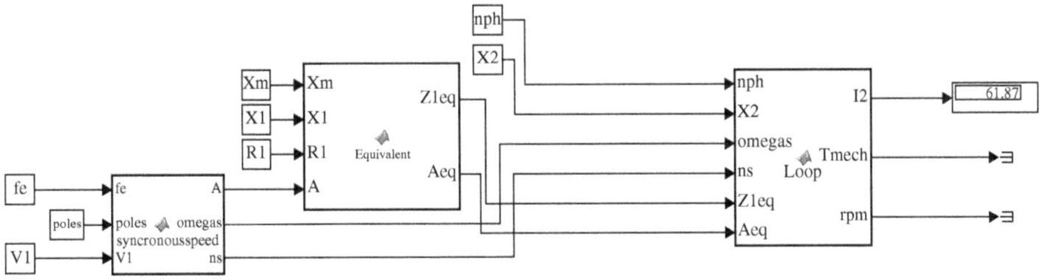

Figure 6.46 Computing torque-speed curve under steady-state condition-Sub Blocks.

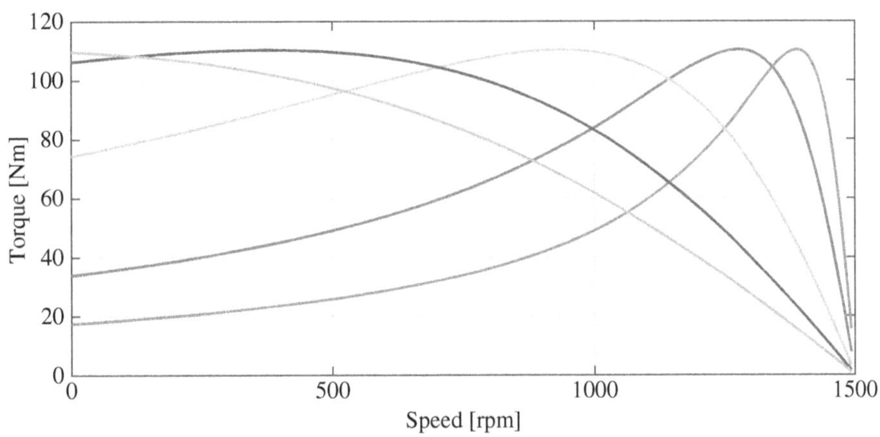

Figure 6.47 Torque-speed curve under steady-state condition.

due to the limitations of the switch ratings and filter sizes. This results in them being incapable of supplying large transient power.

Hence, one needs to evaluate the dynamic simulation of a Three-phase Induction Machine. Dynamic simulation is one of the key steps in the validation of the design process of the motor-drive systems, eliminating inadvertent design mistakes and the resulting errors in the prototype construction and testing and hence the need for dynamic models of the induction machine. The dynamic model of the induction machine is direct-, quadrature-, and zero-sequence axes (known as d-q-0 axes) and is derived from phase variable model in the natural reference frame (a, b, and c reference frame).

The reference frame refers to the point of reference in respect of which the equations are derived. The reference frame theory is used in the modelling of an induction machine to simplify the equations. In the natural a, b, and c reference frame the equations contain rotor-position dependent inductances. To get rid of the position-dependent inductance terms, three-phase to two-phase transformation is done and the equations are found using reference frame theory. Detail is given in several books on machine modelling [6, 7]. Here, detail is not given but the brief theory is given that will help in developing a dynamic Simulink model of an induction machine.

The reference frames are broadly classified into the following categories as shown in Figure 6.48.

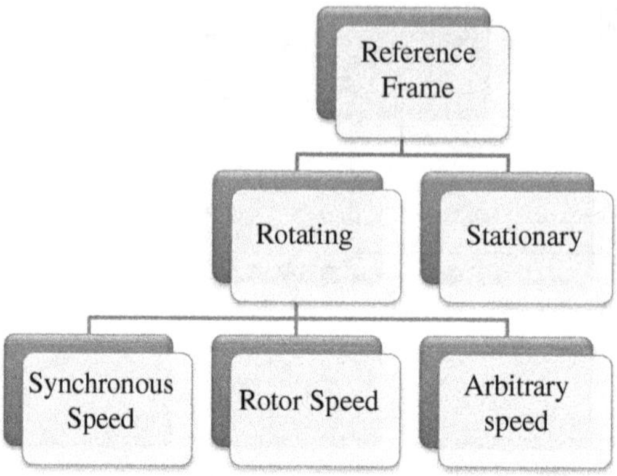

Figure 6.48 Classification of reference frame.

- Stationary reference frame means the reference point is fixed on the stator and the rotational reference frame means the reference point is rotating.
- The rotational speed of the reference frame can be placed on the rotating air-gap field, called the synchronous reference frame.
- When the rotational speed of the reference frame is fixed on the rotor it is known as the rotor reference frame.
- When the rotational speed of the reference frame is rotating at any arbitrary speed, it is called the arbitrary rotating reference frame.

The arbitrary rotating reference frame is the most general form of representation. The following is obtained

$$\omega_a = 0 \rightarrow \text{The stationary reference frame}$$

$$\omega_a = \omega_s \rightarrow \text{The synchronous reference frame}$$

$$\omega_a = \omega_r \rightarrow \text{The rotor reference frame}$$

where ω_a is the arbitrary speed of the reference frame.

In this book, the stationary reference frame model is developed. However, following this concept, other models can be derived.

The transformation of three-phase voltages into two-phase d-q and zero sequence in the stationary reference frame is given as [6]

$$\begin{bmatrix} v_{ds} \\ v_{qs} \\ v_{0s} \end{bmatrix} = \sqrt{\frac{2}{3}} \begin{bmatrix} 1 & 1 & 1 \\ 0 & -\sqrt{3}/2 & \sqrt{3}/2 \\ 1/\sqrt{2} & 1/\sqrt{2} & 1/\sqrt{2} \end{bmatrix} \begin{bmatrix} v_{as} \\ v_{bs} \\ v_{cs} \end{bmatrix} \tag{6.117}$$

The inverse transformation is done as follows

$$\begin{bmatrix} v_{as} \\ v_{bs} \\ v_{cs} \end{bmatrix} = \sqrt{\frac{2}{3}} \begin{bmatrix} 1 & 0 & 1/\sqrt{2} \\ 1 & -\sqrt{3}/2 & 1/\sqrt{2} \\ 1 & \sqrt{3}/2 & 1/\sqrt{2} \end{bmatrix} \begin{bmatrix} v_{ds} \\ v_{qs} \\ v_{0s} \end{bmatrix} \tag{6.118}$$

Nomenclature

r_s	Stator resistance per phase
r_r	Rotor resistance per phase
L_s	Stator inductance per phase
L_r	Rotor inductance per phase
L_m	Magnetizing inductance per phase
ω_r	Rotor Speed
ψ_s	Stator Flux Linkage
ψ_r	Rotor Flux Linkage
p	Derivative operator d/dt

The stator voltage equations in dqo is given as:

$$\begin{aligned} v_{ds} &= r_s i_{ds} + p\psi_{ds} \\ v_{qs} &= r_s i_{qs} + p\psi_{qs} \end{aligned} \tag{6.119}$$

The rotor voltage equation in dqo is given as (the rotor is short-circuited hence the rotor voltages are zero):

$$\begin{aligned} v_{dr} &= 0 = r_r i_{dr} + p\psi_{dr} + \omega_r \psi_{qs} \\ v_{qr} &= 0 = r_r i_{qr} + p\psi_{qr} - \omega_r \psi_{ds} \end{aligned} \tag{6.120}$$

The stator flux linkage equations are

$$\begin{aligned} \psi_{ds} &= L_s i_{ds} + L_m i_{dr} \\ \psi_{qs} &= L_s i_{qs} + L_m i_{qr} \end{aligned} \tag{6.121}$$

The rotor flux linkage equations are

$$\begin{aligned} \psi_{dr} &= L_r i_{dr} + L_m i_{ds} \\ \psi_{qr} &= L_s i_{qr} + L_m i_{qs} \end{aligned} \tag{6.122}$$

The torque equation is given as

$$T_e = \frac{3}{2}\frac{P}{2} L_m \left(i_{dr} i_{qs} - i_{qr} i_{ds} \right) \tag{6.123}$$

Mechanical equation relating the speed and torque is given as

$$T_e = J\frac{d\omega_r}{dt} + B_m \omega_r + T_L \tag{6.124}$$

where J is the inertia, B_m is the viscous friction and T_L is the load torque.

Substituting Eqs. (6.121) and (6.122) in (6.119) and (6.116), to obtain the equations in the matrix form as

$$\begin{bmatrix} v_{ds} \\ v_{qs} \\ v_{dr} \\ v_{qr} \end{bmatrix} = \begin{bmatrix} r_s + pL_s & 0 & L_m p & 0 \\ 0 & r_s + pL_s & 0 & L_m p \\ L_m p & \omega_r L_m & r_r + pL_r & \omega_r L_r \\ -\omega_r L_m & L_m p & -\omega_r L_r & r_r + pL_r \end{bmatrix} \begin{bmatrix} i_{ds} \\ i_{qs} \\ i_{dr} \\ i_{qr} \end{bmatrix} \tag{6.125}$$

The above equations can be used to develop a Simulink model of the three-phase induction motor.

Another approach is described here, by using Eqs. (6.119) and (6.120), the following is obtained

$$\psi_{ds} = \int \left(v_{ds} - r_s i_{ds}\right) dt$$

$$\psi_{qs} = \int \left(v_{qs} - r_s i_{qs}\right) dt \tag{6.126}$$

$$\psi_{dr} = \int \left(v_{dr} - r_r i_{dr} - \omega_r \psi_{qr}\right) dt$$

$$\psi_{qr} = \int \left(v_{qr} - r_r i_{qr} + \omega_r \psi_{dr}\right) dt \tag{6.127}$$

Determining the currents in terms of the flux linkages using Eqs. (6.120) and (6.122)

$$i_{ds} = \frac{1}{L_s}\psi_{ds} - \frac{L_m}{L_s} i_{dr}$$

$$i_{dr} = \frac{1}{L_r}\psi_{dr} - \frac{L_m}{L_r} i_{ds} \tag{6.128}$$

Putting i_{dr} in the equation of i_{ds}, the following is obtained

$$i_{ds} = \left(\frac{1}{L_s L_r - L_m^2}\right)\left[L_r \psi_{ds} - \psi_{dr} L_m\right]$$

$$i_{qs} = \left(\frac{1}{L_s L_r - L_m^2}\right)\left[L_r \psi_{qs} - \psi_{qr} L_m\right] \tag{6.129}$$

Similarly, for the q-axis current, the following expression is deduced

$$i_{qs} = \frac{1}{L_s}\psi_{qs} - \frac{L_m}{L_s} i_{qr}$$

$$i_{qr} = \frac{1}{L_r}\psi_{qr} - \frac{L_m}{L_r} i_{qs} \tag{6.130}$$

Putting i_{qr} in the equation of i_{qs}, the following is obtained

$$i_{dr} = \left(\frac{1}{L_s L_r - L_m^2}\right)\left[L_r \psi_{dr} - \psi_{ds} L_m\right]$$

$$i_{qr} = \left(\frac{1}{L_s L_r - L_m^2}\right)\left[L_r \psi_{qr} - \psi_{qr} L_m\right] \tag{6.131}$$

Currents from Eqs. (6.129) and (6.131) are used to compute the torque in Eq. (6.123). After computing the torque, speed is calculated using Eq. (6.125).

The Simulink using Eqs. (6.119) and (6.131) as shown in Figure 6.49. The sub-blocks are given in Figures 6.50 and 6.51.

Three-phase applied voltages are transformed to two-phase voltages using the transformation equations given in (6.108).

Results obtained using the Simulink model are shown in Figure 6.52. The response shows ripple in the torque during start-up and then settlse to the steady-state condition. The speed rises and reaches its steady-state value of 1500 rpm (four pole and 50 Hz). The speed of the

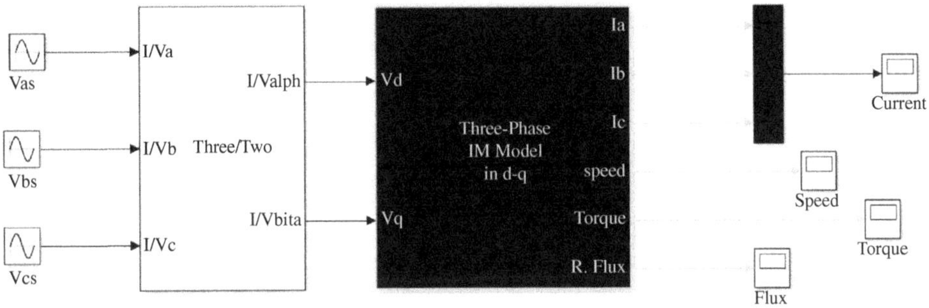

Figure 6.49 Dynamic model of a three-phase induction motor.

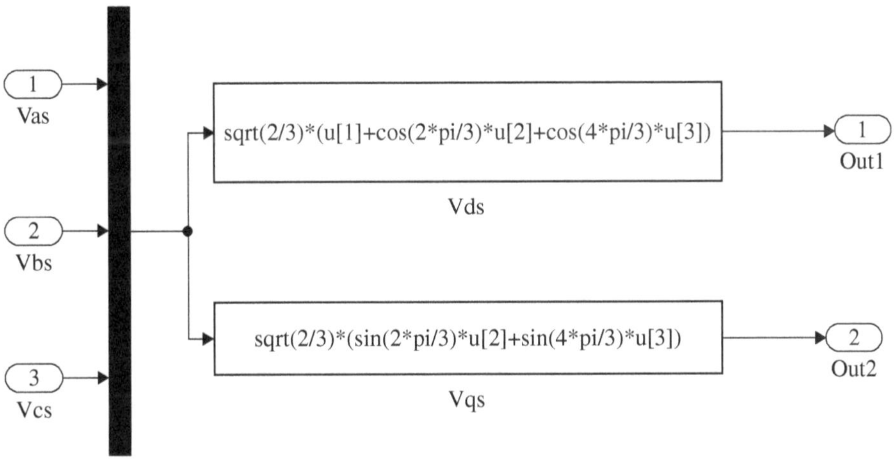

Figure 6.50 Three-phase to two-phase transformation.

machine is the same as the synchronous value because the model does not consider losses in the machine. In starting, the motor draws a large amount of current called initial inrush. The stator current then settles to the steady-state value (which is same as no-load current). The stator current under steady state is equal to the no-load condition which is required to keep the machine energized. Losses are not considered hence the torque under steady state is also zero. The rotor flux also increases and reaches the steady-state value.

6.16 Practice Examples

Example 6.11 A 12 pole 3 ϕ alternator driver at speed of 500 rpm supplies power to an eight- pole 3 ϕ induction motor. If the slip of motor is 0.03p.u, calculate the speed.

Solution

Frequency of supply from alternator, $f = PN/120$

$= 12 \times 500/120$

$= 50\ \text{Hz}$

where $P =$ No. of poles on alternator

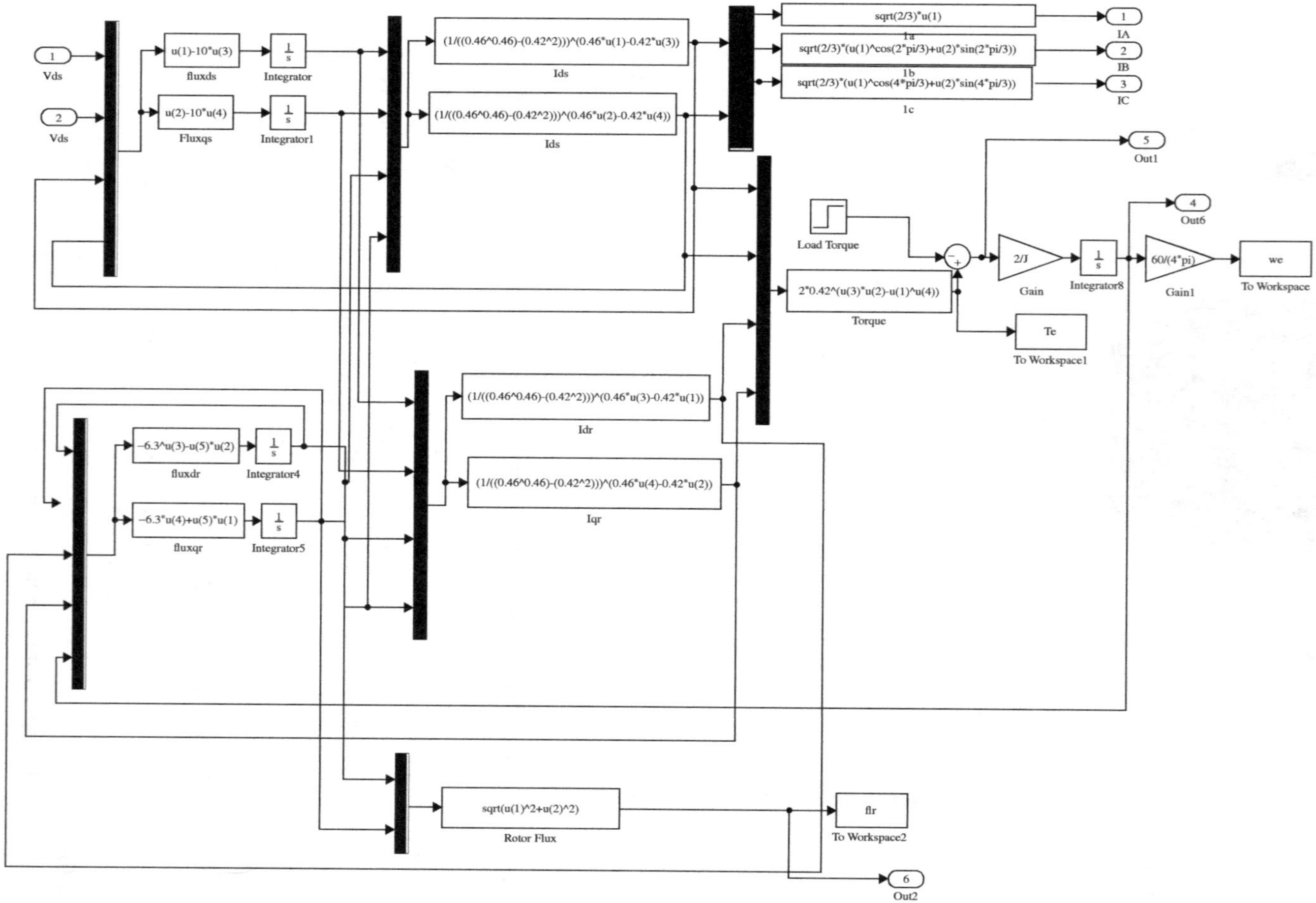

Figure 6.51 Model of three-phase induction motor.

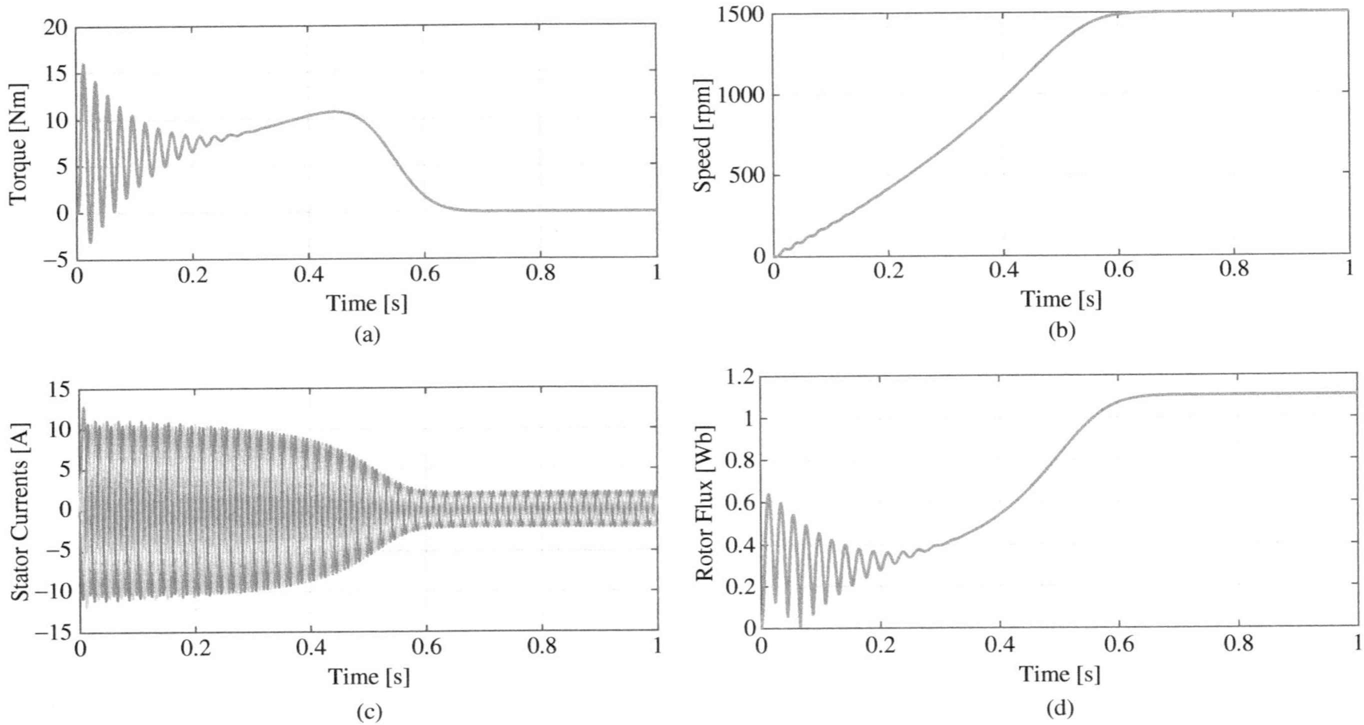

Figure 6.52 Dynamic response of three-phase induction motor.

N = Alternator speed is rpm
Synchronous speed of 3 ϕ induction motor
$N = 120 f / P_m$
$= 120 \times 50/8$
$= 750$ rpm
Speed of 3 ϕ induction motor $N = Ns\,(1-s)$
$= 750(1-0.03)$
$= 727.5$ rpm

Example 6.12 A motor generator set used for providing a variable frequency AC supply consists of a 3-ϕ synchronous and 24 pole 3 ϕ synchronous generator. The motor generated set is fed from 25 Hz, 3 ϕ ac supply. A six-pole 3 ϕ induction motor is electrically connected to the terminals of the synchronous generator and runs at a slip of 5%. Find

i) the frequency of generated voltage of the synchronous generator
ii) the speed at which the induction motor is running

Solution
Speed of motor generator set
$N_s = (120 \times f_1$ [supply frequency])/(number of poles on synchronous motor)
$= 120 \times 25/10$
$= 300$ rpm

(1) Frequency of generated voltage
f_z = speed of motor gen set voltage × no. of poles on syn gen/120
$= 300 \times 24/120$
$= 60$ Hz
(2) Speed of induction motor
$N_m = N_s\,(1-s)$
$= 120 f_z / P_m\,(1-s)$
$= 120 \times 60/6(1-0.05)$
$= 1140$ rpm

Example 6.13 A 3 ϕ 50 Hz induction motor has a full load speed of 960 rpm

(a) find slip
(b) Number of poles
(c) Frequency of rotor induced emf
(d) Speed of rotor field wrt rotor structure
(e) Speed of rotor field wrt Stator structure
(f) Speed of rotor field wrt stator field

Solution
Given $f = 50$ Hz (supply frequency)
$N = 960$ rpm
The number of poles will be six only (because at $P = 6$, $Ns = 1000$ which is nearer and greater than 960 rpm)

(a) Slip, $s = (N_s\text{-}N)/N_s \times 100\%$
$= (1000\text{–}960)/1000 \times 100\%$
$= 4\%$
(b) Number of poles = 6
(c) Frequency of rotor induced emf $= f_r = sf = 0.04 \times 50 = 2$ Hz
(d) Speed of rotor field wrt rotor structure $= 120f_r/P = 120 \times 2/6 = 40$ rpm
(e) Speed of rotor field wrt stator structure is actually the speed of stator filed wrt stator structure, $N_s = 1000$ rpm
(f) Speed of rotor field wrt stator field is zero

Example 6.14 A six pole 3 ϕ 50 Hz induction motor is running at full load with a slip of 4%. The rotor is star connected and its resistance and standstill reactance are 0.25 and 1.5 Ω per phase. The emf between slip ring is 100 V. Find the rotor current per phase and pf, assuming the slip rings are short circuited.

Solution

Rotor emf/phase at stand still $E_2 = 100\sqrt{3} = 57.7$ V
Rotor emf/phase at full load $= sE_2 = 0.04 \times 57.7 = 2.31$ V
Rotor reactance/phase at full Load $= SX_2 = 0.04 \times 1.5 = 0.06\ \Omega$
Rotor impedance/phase at full load $= \sqrt{((0.25)^2 + (0.06)^2)} = 0.257\ \Omega$
Full load Rotor current/phase = 2.31/0.257 = 9 A
Rotor pf = 0.25/0.257 = 0.97 lag

Example 6.15 The power input to a 500 V 50 Hz, six-pole, 3 ϕ squirrel case inductor motor running at 975 rpm is 40 kW. The stator losses are 1 kW and friction and windage losses are 2 kW. Calculate:

1) Slip in pu
2) Rotor Cu loss
3) Output Power HP

Solution

i) $N_s = 120f/P = 120 \times 50/6 = 1000$ rpm
$s = (N_s - N)/N_s = (1000\text{–}975)/1000 = 0.025$ pu
Power input to stator $P_1 = 40$ kW
Stator output power $= P_1$ – stator losses = 40–1 = 39 kW
Power input to rotor P_2 = Stator output power = 39 kW
ii) Rotor Cu loss $= sP_2 = 0.025 \times 39 = 0.975$ kW
$P_{mech} = P_2 - P_{cu} = 39\text{–}0.975 = 38.025$ kW
iii) Motor output $= P_{mech}$ – friction and windage loss
= 38.025–2
= 36.025 kW

Example 6.16 The power input to a three-phase induction motor is 60 kW. The stator losses total 1 kW. Find the total mechanical power developed and the rotor copper loss per phase if the motor is running with a slip of 3%.

Solution

Stator Input $P_{\text{in}} = 60$ kW, $s = 3\% = \frac{3}{100} = 0.03$ pu

Stator losses =1 kW

Stator output = 60–1 = 59 kW

Rotor input = Stator Output = 59 kW

Total rotor copper loss = s × rotor input = 0.03 × 1.77 kW

Rotor copper loss per phase $= \frac{1}{3} \times 1.77 = 0.59$ kW

Mechanical power developed = Rotor input–Rotor copper loss

= 59–1.77 = 57.23 kW

Example 6.17 A 25 hp, six-pole, 50 Hz, three-phase slip-ring induction motor runs at 960 revolutions per minute on full load with a rotor current per phase of 35 A. Allowing 250 W for the copper loss in the short-circuiting gear, and 1000 W for mechanical losses, find the resistance per phase of the three-phase rotor winding.

Solution

$f_1 = \frac{PN_s}{120}$

$$50 = \frac{6 \times N_s}{120}, N_s = \frac{120 \times 50}{6} = 1000\ \text{rpm}$$

$$s = \frac{N_s - N_r}{N_s} = \frac{1000 - 960}{1000} = 0.04\ \text{pu}$$

Rotor copper loss $= \frac{s}{1-s}$xmechanical power developed

$$3I_2^2R_2 + 250 = \frac{0.04}{1 - 0.04}(25 \times 746 + 1000)$$

$$3 \times I_2^2R_2 = 818.75 - 250$$

$$R_2 = \frac{568.75}{3 \times 35^2} = 0.15476\Omega$$

Example 6.18 A 220-V three-phase six-pole 50-Hz induction motor is running at a slip of 3.5%. Find:

(a) The speed of the magnetic fields in rpm
(b) The speed of the rotor in rpm
(c) The slip speed of the rotor
(d) The rotor frequency in Hz

Solution

(a) The speed of the magnetic fields is

$$N_{sunc} = \frac{120f_e}{P} = \frac{120 \times 50}{6} = 1000\text{r.p.m}$$

(b) The speed of the rotor is

$$n_m = (1 - s)\, n_{sync} = (1 - 0.035) \times 1000 = 965\text{r.p.m}$$

(c) The slip speed of the rotor is

$$n_{slip} = sn_{sync} 0.035 \times 1000 = 35\text{r.p.m}$$

(d) The rotor frequency is

$$f_r = \frac{n_{slip}P}{120} = \frac{35 \times 6}{120} = 1.75\text{Hz}$$

Example 6.19 A three-phase 60-Hz induction motor runs at 715 r/min at no load and at 670 r/min at full load.

(a) How many poles does this motor have?
(b) What is the slip at rated load?
(c) What is the speed at one-quarter of the rated load?
(d) What is the rotor's electrical frequency at one-quarter of the rated load?

Solution

(a) This machine has 10 poles, which produces a synchronous speed of

$$n_{sunc} = \frac{120 f_e}{P} = \frac{120 \times 60}{10} = 720\text{r.p.m}$$

(b) The slip at rated load is

$$s = \frac{n_{sync} - n_m}{n_{sync}} \times 100\% = \frac{720 - 670}{720} \times 100\% = 6.94\%$$

(c) The motor is operating in the linear region of its torque-speed curve, so the slip at one-quarter load will be

$$s = 0.25 \times 0.0694 = 0.0174$$

The resulting speed is

$$n_m = (1 - s)\, n_{sync} = (1 - 0.0174) \times 720 = 707\text{r.p.m}$$

(d) The electrical frequency at one-quarter load is

$$f_r = sf_e = 0.0174 \times 60 = 1.04\text{Hz}$$

Example 6.20 A 10 hp, 440 V (line-to-line), three-phase induction motor has synchronous speed of 1800 and runs at 1700 rpm at the full-load power output of 10 hp. The stator copper loss is 200 W and the rotational loss is 400 W. Assume the core losses are zero, i.e. $P_c = 0$. Determine

(a) Power developed, P_m (otherwise known as P_D)
(b) Slip
(c) Input power to the rotor, P_g
(d) Rotor copper losses, P_{rc}

(e) Total power input, P_{in}
(f) The magnitude of the current drawn by the motor, assuming the motor power factor is 0.80

Solution

$$P_m = P_{out} + P_{rotationalloss} = 7460 + 400 = \underline{7860W}$$
$$s = \frac{1800 - 1700}{1800} = \underline{0.0556}$$
$$P_g = \frac{1}{1-s}P_m = \underline{8322.7W}$$
$$P_{rc} = P_g - P_m = s{\cdot}P_g = \frac{s}{1-s}{\cdot}P_m = \underline{462.7W}$$
$$P_{in} = P_g + P_{sc} = \underline{8522.7W}$$
$$P = \sqrt{3}V{\cdot}I{\cdot}\cos\varphi$$
$$I = \frac{P}{\sqrt{3}V{\cdot}\cos\varphi} = \frac{8522.7}{\sqrt{3}\times 440\times 0.8} = \underline{13.98A}$$

Example 6.21 Tests are conducted on a 7 kW, 440 V, four-pole, 50 Hz, Star-connected three-phase induction motor with stator resistance per phase is 0.667 Ω.

No-Load Test:
Voltmeter Reading: 440 V
Ammeter Reading: 5.0 A
Wattmeter Reading: 400 W

Blocked-rotor Test:
Voltmeter Reading: 150 V
Ammeter Reading: 20 A
Frequency: 50 Hz
Wattmeter Reading: 1600 W
Calculate

(i) Parameters of the equivalent circuit
(ii) Torque produced by the machine
(iii) The input power and power factor
(iv) The Braking torque developed when two of its terminals are interchanged.

Solution

(i) **Using No-Load test data:**
Voltage per-phase:

$$V_{per-phase} = \frac{440}{\sqrt{3}} = 254.034V$$
$$\cos(\theta_o) = \frac{Pc}{\sqrt{3}V_1 I_0} = \frac{400}{\sqrt{3}\times 440\times 5} = 0.10497$$
$$\theta_o = 83.9744^0$$
$$R_c = \frac{V_{per-phase}}{|I_o|\cos(\theta_0)} = \frac{254.034}{0.52485} = 484.06\Omega$$

$$X_m = \frac{V_{per-phase}}{|I_o| \sin(\theta_0)} = \frac{254.034}{4.9423} = 51.0898\Omega$$

Using Blocked-rotor test data:

$$Z_{sc} = \frac{V_{sc}}{I_{sc}} = \frac{150/\sqrt{3}}{20} = 4.33012\Omega$$

$$\cos(\theta_{sc}) = \frac{P_{sc}}{\sqrt{3}V_{sc}I_{sc}} = \frac{1600}{\sqrt{3} \times 150 \times 20} = 0.3079$$

$$\theta_{sc} = 72.066^0$$

$$R_{sc} = R_1 + R_2' = Z_{sc} \times \cos(\theta_{sc}) = 4.33012 \times 0.3079 = 1.3332\Omega$$

$$R_2' = 1.3332 - R_1 = 1.3332 - 0.667 = 0.6665\Omega$$

$$X_{sc}^* = X_1 + X_2' = Z_{sc} * \sin(\theta_{sc}) = 4.33012 * 0.95141 = 4.1197\Omega$$

$$X_1 = X_2' = 4.1197/2 = 2.0598\Omega$$

The equivalent circuit obtained is shown in Figure E6.21:

(ii) Slip (s) = 0.04

$$N_s = 120 * 50/4 = 1500 \text{ rpm}$$

$$\omega_s = \frac{2\pi N_s}{60} = \frac{2 * \pi * 1500}{60} = 157.079$$

Torque:

$$T = \frac{3}{\omega_s} \frac{V^2 \frac{R_2'}{s}}{\left(R_1 + \frac{R_2'}{s}\right)^2 + (X_1 + X_2')^2}$$

$$T = \frac{3}{157.079} \frac{\left(\frac{440}{\sqrt{3}}\right)^2 \frac{0.6665}{0.04}}{\left(0.667 + \frac{0.6665}{0.04}\right)^2 + (4.1197)^2} = 64.73Nm$$

(iii) From the equivalent circuit, the rotor current (referred to the stator) is

$$V = \frac{440}{\sqrt{3}} = 254.034V$$

$$\bar{I}_2' = \frac{\bar{V}}{\left(R_1 + \frac{R_2'}{s}\right) + j(X_1 + X_2')} = \frac{254.03 < 0}{(0.667 + 0.6665/0.04) + j(4.1197)}$$

$$\bar{I}_2' = \frac{254.03 < 0}{17.3295 + j4.1197} = 13.87 - j3.295$$

From the shunt branch

$$I_c = \frac{V}{R_c} = \frac{254.03}{484.06} = 0.5247 < 0A$$

$$I_m = \frac{V}{jX_m} = -j\frac{254.03}{51.0898} = -j4.9722A$$

No-load current (shunt branch current)

$$\bar{I}_o = \bar{I}_c + \bar{I}_m = 0.5247 - j4.9722$$

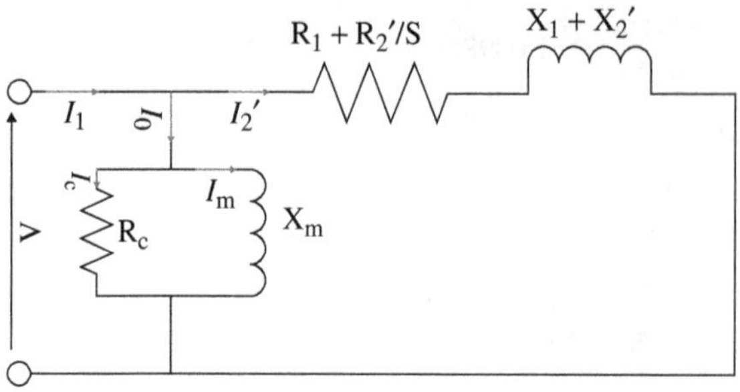

Figure E6.21 Model of a three-phase induction motor.

Input Current

$$\overline{I_1} = \overline{I_o} + \overline{I_2'} = 0.5247 - j4.9722 + 13.87 - j3.295$$

$$\overline{I_1} = 14.3947 - j8.2672 = 16.6 < -29.87$$

Input power factor

$$Pf = \cos(29.87) = 0.867\ (lagging)$$

Input Power

$$P_{in} = \sqrt{3}V_{line}I_{line}\cos(\theta) = \sqrt{3} \times 440 \times 16.6 \times 0.867 = 10.9683kW$$

(iv) Slip (s) = 0.04

$$N_s = 120 \times 50/4 = 1500 \text{ rpm}$$

$$N = (1 - S)N_s = 0.96 \times 1500 = 1440rpm$$

During Braking:

$$S_b = \frac{N_s - (-N)}{N_s} = \frac{1500 - (-1440)}{1500} = 1.96$$

$$T = \frac{3}{\omega_s} \frac{V^2 \frac{R_2'}{S}}{\left(R_1 + \frac{R_2'}{S}\right)^2 + (X_1 + X_2')^2}$$

$$T = \frac{3}{157.079} \frac{\left(\frac{440}{\sqrt{3}}\right)^2 \frac{0.6665}{1.96}}{\left(0.667 + \frac{0.6665}{1.96}\right)^2 + (4.1197)^2} = 23.3Nm$$

Example 6.22 A three-phase 20 kW, 440 V, four-pole, 1440 rpm, star-connected, 50 Hz induction motor develops a torque of 100 Nm at 5% slip. If the motor is supplied by 25 Hz source, calculate the voltage and speed for constant air-gap flux operation (v/f = constant).

Solution

Let the rated Voltage and frequency is V_o and f_o, respectively. For constant V/f, the following relation can be written

$$\frac{V_o}{f_o} = \frac{V_1}{f_1}$$
$$V_1 = f_1\frac{V_o}{f_o} = \left(\frac{25}{50}\right) \times 440 = 220V$$

Torque equation be simplified for low slip region as

$$\left(R_1 + \frac{R_2'}{s}\right)^2 + (X_1 + X_2')^2 \approx \left(\frac{R_2'}{s}\right)^2$$

Using the above approximation, torque equation is written as

$$T = \left(\frac{3V^2}{\omega_s R_2'}\right) s$$

At rated condition of V_o and f_o

$$T_o = \left(\frac{3V_o^2}{\omega_s R_2'}\right) s_o$$

Torque at any other voltage (V1) and frequency (f1)

$$T_1 = \left(\frac{3V_1^2}{\omega_s R_2'}\right) s_1$$

For constant torque (To = T1)

$$T_o = T_1 = \left(\frac{3V_o^2}{\omega_{so} R_2'}\right) s_o = \left(\frac{3V_1^2}{\omega_{s1} R_2'}\right) s_1$$
$$\left(\frac{3V_o^2}{\omega_{so}}\right) s_o = \left(\frac{3V_1^2}{\omega_{s1}}\right) s_1$$
$$\frac{s_o\left(\frac{f_1}{f_o}\right)V_o^2}{\omega_{s1}\left(\frac{f_1}{f_o}\right)\omega_{so}} = \frac{s_o V_o^2}{\omega_{so}}$$
$$s_1 f_1 = s_o f_o$$
$$s_1\left(\frac{120f_1}{P}\right) = s_o\left(\frac{120f_o}{P}\right)$$
$$s_1 N_{s1} = s_o N_{so}$$
$$N_{s1} - N_{r1} = N_{so} - N_{r0}$$

The synchronous speeds and Actual speed are

$$N_{s1} = \frac{120f_1}{P} = \frac{120 * 25}{4} = 750rpm$$
$$N_{so} = \frac{120f_o}{P} = \frac{120 * 50}{4} = 1500rpm$$

$$N_{ro} = 1440rpm$$
$$N_{r1} = ?$$

$$750 - N_{r1} = 1500 - 1440$$
$$N_{r1} = 690rpm$$

Example 6.23 A three-phase star-connected slip-ring type 50 Hz, 20 hp, 440 V, induction motor runs at 1460 rpm on full load. The per-phase equivalent circuit parameters are:

$$R_1 = 0.20\,\Omega, X_1 = 1.0\,\Omega, R_2' = 0.15\,\Omega, X_2' = 0.9\,\Omega, X_{\text{m}} = 40.0\,\Omega$$

(a) Calculate the number of poles of the stator winding of the machine.
(b) Calculate the starting torque developed by the machine.
(c) Calculate the value of the external resistance to be inserted in each phase of the rotor circuit such that the maximum torque occurs at starting. Use Thevenin's equivalent circuit.

Solution

(a) Since the motor is running at 1460 rpm, the next higher synchronous speed must be chosen as the field speed which is 1500 rpm.

$$N_s = (120 \times f)/P = 1500 \text{ rpm}$$

Therefore

$$P = (120 \times 50)/1500 = 4 \text{ Poles}$$

(b) The starting torque is given as

$$T_{st} = \frac{3|V_{TH}|^2 R_2'}{\omega_s\left[\left(R_{TH} + R_2'\right)^2 + \left(X_{TH} + X_2'\right)^2\right]}$$

Thevenin's equivalent impedance is

$$Z_{TH} = Z_1 \parallel jX_m = \frac{\left(R_1 + jX_1\right)jX_m}{R_1 + j\left(X_1 + X_m\right)}$$

$$Z_{TH} = \frac{\left(R_1 + jX_1\right)jX_m}{R_1 + j\left(X_1 + X_m\right)} = \frac{(0.2 + j1)j40}{0.2 + j(1 + 40)} = \frac{1.02\angle 78.7X40\angle 90}{41\angle 89.72} = 0.995\angle 78.98$$

Thevenin's equivalent voltage is obtained as shown in Figure 6.13a

$$|V_{TH}| = \frac{X_m}{\sqrt{R_1^2 + \left(X_1 + X_m\right)^2}}|V_1| = \frac{40}{\sqrt{(0.2)^2 + (41)^2}}\frac{440}{\sqrt{3}} = 247.835V$$

$$\omega_s = \frac{2\pi N_s}{60} = \frac{2\pi 1500}{60} = 157.08 rad/\sec$$

Starting torque

$$T_{st} = \frac{3|V_{TH}|^2 R_2'}{\omega_s\left[\left(R_{TH} + R_2'\right)^2 + \left(X_{TH} + X_2'\right)^2\right]} = \frac{3X(247.835)^2 X0.15}{157.08\left[(0.19 + 0.15)^2 + (0.976 + 0.9)^2\right]}$$

$$= \frac{27639.984}{570.982} = 48.4Nm$$

(c) Condition for maximum torque

$$S_{\max T} = \frac{R_2'}{\sqrt{R_{TH}^2 + (X_{TH} + X_2')^2}}$$

Since the maximum torque is required at starting, Put $S_{maxT} = 1$

$$S_{\max T} = \frac{R_2' + R_{ext}}{\sqrt{R_{TH}^2 + (X_{TH} + X_2')^2}} = \frac{0.15 + R_{ext}}{\sqrt{(0.19)^2 + (0.976 + 0.9)^2}} = 1$$

$$0.15 + R_{ext} = 1.8856$$

$$R_{ext} = 1.735\Omega/phase$$

Example 6.24 A three-phase squirrel-cage induction motor is to be started such that the maximum current drawn from a 440 V, three-phase supply is limited to 150 A. Compute the maximum permissible full-load kVA assuming the starting inrush current as five times its full load value for the following cases.

(a) For direct on-line start
(b) Autotransformer start with 60% tapping
(c) Star-Delta start

Solution

a. Maximum line current, $I_L = 150$ A,
Starting current $I_{st} = 5$ X Full-Load Current $= 5\ I_{fl}$
Since the maximum line current drawn from the supply is 150 A.
Therefore, $5I_{fl} = 150$ A

$$I_{fl} = 150/5 = 30\text{ A}$$

Maximum permissible rating of motor $KVA_{max} = \sqrt{3}V_L I_{fl} = \sqrt{3}X440X30 = 22.863$ kVA

b. Autotransformer start

$$I_{st} = x^2 I_{sc} = x^2\left(5I_{fl}\right)$$

$$150 = (0.6)^2\left(5I_{fl}\right)$$

$$I_{fl} = \frac{150}{5X(0.6)^2} = 83.33A$$

Maximum permissible rating of motor
$KVA_{max} = \sqrt{3}V_L I_{fl} = \sqrt{3}X440X83.33 = 63.505$ kVA

c. Star-Delta start

$$I_{st} = \frac{1}{3}\left(5I_{fl}\right)$$

$$I_{fl} = \frac{3}{5}I_{st} = \frac{3}{5}X150 = 90A$$

Maximum permissible rating of motor

$$KVA_{max} = \sqrt{3}V_L I_{fl} = \sqrt{3}X440X90 = 68.589\text{ kVA}$$

Example 6.25 A 3ϕ, 440 V, 150 kW, eight-pole wound-rotor induction motor controls the speed of a fan. The torque required for the fan varies as the square of the speed. At full load (150 kW) the motor slip is 0.03 with the slip rings short-circuited. The slip–torque relationship of the motor can be assumed to be linear from no load to full load. The resistance of each rotor phase is 0.02 Ω. Determine the value of resistance to be added to each rotor phase so that the fan runs at 600 rpm.

Solution

$$n_s = \frac{120 \times 50}{8} = 750rpm$$

Full-load speed $n_{FL} = 750 \times (1 - 0.03) = 727.5rpm$

$$T_{600rpm} = \left(\frac{600}{727.5}\right)^2 \times T_{FL} = T'$$

$$S_{600rpm} = \frac{750 - 600}{750} = \frac{1}{5}$$

with $\mathrm{R}_2 = 0.02\Omega$, slip for torque T′ is

$$S_{0.02\Omega} = 0.03 \times \frac{T'}{T_{FL}} = 0.03 \times \left(\frac{600}{727.5}\right)^2 = 0.0204$$

To develop the same torque at $s = \frac{1}{5}$,

$$\frac{R_2 + \mathrm{R}_{ext}}{1/5} = \frac{R_2}{0.0204}$$

$$5 \times \left(0.02 + \mathrm{R}_{ext}\right) = \frac{0.02}{0.0204}$$

$$\mathrm{R}_{ext} = 0.176\Omega$$

6.17 Summary

This chapter is dedicated to the theory and performance of a three-phase induction motor. Detailed analysis is included with simple explanations for greater understanding. Many solved and unsolved numerical examples are given that cover almost all aspects of three-phase induction motor principles and performance. Limiting operating condition are given for achieving maximum torque, maximum starting torque etc. Equivalent circuit and model is explained in the text. Speed control and starting methods are also elaborated. Finally, dynamic model is given which is used to develop Matlab/Simulink model of three-phase induction motor.

Problems

6.1 A 3Φ, four-pole, 50 Hz Induction Motor (I.M.) runs at 1460 rpm Find its percentage slip.

6.2 A two-pole, 50-Hz induction motor runs at 2890 rpm. Calculate (i) Synchronous speed of the motor, and (ii) slip of the motor.

6.3 A 12-pole 3Φ alternator driver at speed of 500 rpm supplies power to an eight-pole 3Φ induction motor. If the slip of motor is 0.03 pu, calculate the speed.

6.4 A four-pole induction motor has stator frequency of 50 Hz and rotor frequency of 2 Hz. Calculate the speed at which motor is running and the slip.

6.5 A 3Φ, four-pole induction motor is supplied from 3Φ 50 Hz AC supply. Find:

i) synchronous speed
ii) rotor speed when slip is 4%
iii) the rotor frequency when runs at 600 rpm

6.6 A motor generates set used for providing variable frequency AC supply consists of a 3Φ synchronous and 24 pole 3Φ synchronous generator. The motor generated set is fed from 25 Hz, 3Φ AC supply. A six-pole 3Φ induction motor is electrically connected to the terminals of the synchronous generator and runs at a slip of 5%. Find:

i) the frequency of generated voltage of synchronous generator
ii) the speed at which induction motor is running

6.7 A 12 pole 3Φ alternator is coupled to an engine running at 500 rpm If supplied a 3Φ induction motor having full speed of 1440 rpm Find the percentage slip, frequency of rotor current and number of poles of rotor.

6.8 The rotor of 3Φ induction motor rotates at 980 rpm when states are connected to 3Φ supply. Find the rotor frequency.

6.9 A 3Φ 50 Hz induction motor has a full load speed of 960 rpm
Find

i) slip
ii) Number of poles
iii) Frequency of rotor induced emf
iv) Speed of rotor field wrt rotor structure
v) Speed of rotor field wrt Stator structure
vi) Speed of rotor field wrt stator field

6.10 A 3Φ, 400 V wound rotor has delta connected stator winding and star connected rotor winding. The stator has 48 turns/phase while rotor has 24 turns per phase. Find the standstill or open-circuited voltage across the slip rings.

6.11 A six-pole 3Φ 50 Hz induction motor is running at full load with a slip of 4%. The rotor is star connected and its resistance and standstill reactance are 0.25 Ω and 1.5 Ω per phase. The emf between slip ring is 100 V. Find the rotor current per phase and pf, assuming the slip rings are short circuited.

6.12 A 50 Hz, eight-pole induction motor has full load slip of 4%. The rotor resistance and standstill reactance are 0.01 Ω and 0.1 Ω per phase respectively. Find:

i) The speed at which maximum torque occurs
ii) The ratio of maximum torque to full load torque

6.13 An eight-pole 3Φ, 50 Hz induction motor has rotor resistance of 0.025 Ω/phase and rotor standstill reactance of 0.1 Ω/phase. At what speed is the torque maximum? What proportion of maximum torque is the starting torque?

6.14 A 500 V, 3Φ, 50 Hz induction motor develops an output of 15 kW at 950 rpm. If the input pf is 0.86 lagging, Mechanical losses are 730 W and stator losses 1500 W Find

i) the slip
ii) the rotor Cu loss
iii) the motor input
iv) the line current

6.15 A six-pole 3Φ induction motor develops 30 HP including 2 HP mechanical losses at a speed of 950 rpm on 550 V, 50 Hz mains. The pf is 0.88 lagging
Find:

i) Slip
ii) Rotor Cu loss
iii) Total input if stator losses are 2 kW
iv) Efficiency (η)
v) Line current

6.16 A four-pole 50 Hz 3Φ induction motor running at full load, develops a torque of 160 N-m, when a rotor makes 120 complete cycles per minute, find out power output.

6.17 The rotor of a 3Φ, 50-Hz, six-pole induction motor takes 100 kW at 2 Hz. Determine (i) the rotor speed and (ii) the rotor copper losses.

6.18 A 50 Hz, 3Φ, eight-pole induction motor takes 46 kW at 720 rpm. The stator Cu-loss is 1.2 kW, stator core loss is 1.4 kW, and rotor mechanical losses are 1 kW. Find the motor efficiency.

6.19 The power input to a 500 V 50 Hz, six-pole, 3Φ squirrel case inductor motor running at 975 rpm is 40 kW. The stator losses are 1 kW and friction and windage losses are 2 kW.
Find:

i) Slip
ii) Rotor Cu loss
iii) Brake hp

6.20 A 50 kW, six-pole, 50 Hz, 3Φ slip-ring induction motor runs at 975 rpm on full-load with a rotor current per phase of 25 A. Assume 350 W for the copper loss and 1200 W for mechanical losses; find the resistance per phase of the three-phase rotor winding.

6.21 A 650 kW, 3Φ, 50 Hz, eight-pole induction motor has a rotor-impedance of (0.03 + j0.21) Ω at standstill. Full-load torque is obtained at 720 rpm. Calculate the speed at which maximum torque occurs and the external resistance per phase to be inserted in the rotor circuit to reach maximum torque at starting.

6.22 A 3Φ induction motor started using star-delta starter and it takes 175% of full-load line current and develops 30% of full-load torque at starting. Calculate the starting torque and current in term of full-load values, if an autotransformer with 75% tapping were employed.

6.23 A 12-pole, 3Φ, 50 Hz Induction motor runs at full load with a slip of 3%. Rotor standstill impedance per phase is (0.02 + j0.1) Ω, calculate the maximum available torque in terms of full-load torque. Also, determine the speed at which the maximum torque occurs.

6.24 The results of no-load and blocked rotor tests on a 3Φ Y-connected induction motor are:

No-load test	Blocked rotor test
Line-to-line voltage = 400 V	line-to-line voltage = 42 V
Input power = 1400 W	Input power = 2400 W
Input Current = 15.25 A	Input current = 54 A
Friction and windage loss = 550 W	

Determine the parameters of the approximate equivalent circuit.

6.25 The synchronous speed of a Y-connected induction motor is 1500 rpm. With blocked rotor, input power is 54 kW with 201.3 A. The rotor resistance per phase is 0.05 Ω and transformation ratio is 2. Calculate (i) Stator resistance per phase (ii) Starting torque.

6.26 The output of 3Φ IM is 25 kW and mechanical/rotational losses are 0.2 kW. Calculate the rotor copper losses at a slip of 5%.

6.27 A four-pole 3Φ IM produces maximum torque at 1200 rpm when supplied by a 50 Hz source. Given rotor resistance as 1 Ω/Phase. At what speed will it produce maximum torque, if supply frequency is 60 Hz. Neglect stator impedance.

6.28 A wound rotor type 3Φ IM develop Max torque at 1200 rpm when supplied from a 50 Hz source. Calculate the value of actual resistance to be added so that it develops maximum torque at starting. Given the rotor resistance as 1 Ω/phase. Neglect stator impedance.

6.29 A four-pole 3Φ, 50 Hz IM has rotor leakage reactance at standstill which is twice its rotor resistance. Calculate the frequency of supply so that maximum torque is obtained at starting.

6.30 A 3Φ 440 V, 50 Hz, four-pole IM takes 50 kW input power when running at 1440 rpm The stator losses are 1.4 kW and mechanical losses are 2.5 kW. Calculate the efficiency of the motor.

6.31 Prove that the approximate efficiency is given as $\eta = (1-s)/(1+s)$. Neglect motor and stator iron losses and assume stator copper loss. Given mechanical losses = 0 and stator iron losses = 0.

6.32 Calculate the reduction in maximum torque, if the supply Voltage is reduced by one-quarter of rated value.

6.33 Determine how the maximum torque and starting torque varies wrt frequency in a three-phase induction motor. Neglect stator impedance.

6.34 A 3Φ, 3.3 kV, four-pole IM takes 500 A starting current and starting torque of 600 Nm. Compute the tapping of autotransformer to reduce the starting current to 100 A. Also, calculate the corresponding starting torque.

6.35 A 15-kW, 230-V, three-phase, Y-connected, 60-Hz, four-pole squirrel-cage induction motor develops full-load internal torque at a slip of 3.5% when operated at rated voltage and frequency. For the purposes of this problem, rotational and core losses can be neglected. The following motor parameters, in ohms per phase, have been obtained

$$R_1 = 0.21\,\Omega\, X_1 = X_2 = 0.26\,\Omega\, X_m = 10.1\,\Omega$$

Determine the maximum internal torque at rated voltage and frequency, the slip at maximum torque, and the internal starting torque at rated voltage and frequency.

6.36 A three-phase induction motor, operating at rated voltage and frequency, has a starting torque of 135% and a maximum torque of 220%, both with respect to its rated-load torque. Neglecting the effects of stator resistance and rotational losses and assuming constant rotor resistance, determine

(a) the slip at maximum torque.
(b) the slip at rated load.
(c) the rotor current at starting (as a percentage of rotor current at rated load).

6.37 A squirrel-cage induction motor runs at a full-load slip of 3.7%. The rotor current at starting is 6.0 times the rotor current at full load. The rotor resistance and inductance are independent of rotor frequency and rotational losses, stray-load losses and stator resistance may be neglected. Expressing torque in per unit of the full-load torque, compute

(a) the starting torque.
(b) the maximum torque and the slip at which the maximum torque occurs.

6.38 A three-phase, Y-connected, 460-V (line-line), 25-kW, 60-Hz, four-pole induction motor has the following equivalent-circuit parameters in ohms per phase referred to the stator:

$$R_1 = 0.103\,\Omega\, R_2 = 0.225\,\Omega\, X_1 = 1.10\,\Omega\, X_2 = 1.13\,\Omega\, X_m = 59.4\,\Omega$$

The total friction and windage losses may be assumed constant at 265 W, and the core loss may be assumed to be equal to 220 W. With the motor connected directly to a 460-V source, compute the speed, output shaft torque and power, input power and power factor and efficiency for slips of 1%, 2%, and 3%. You may choose either to represent the core loss by a resistance connected directly across the motor terminals or by resistance R_c connected in parallel with the magnetizing reactance X_m.

6.39 A 3ϕ, 440 V, 50 Hz, six-pole induction motor has the following single-phase equivalent circuit parameters.

$$R_1 = 0.2\,\Omega, X_1 = 1.055\,\Omega,$$
$$R_2' = 0.28\,\Omega, X_2' = 1.055$$
$$X_m = 33.9\,\Omega$$

The induction motor is connected to a 3ϕ, 440 V, 50 Hz supply.

(a) Determine the starting torque.
(b) Determine the breakdown torque and the speed at which it occurs.
(c) The motor drives a load for which $T_L = 1.8$ N m. Determine the speed at which the motor will drive the load. Assume that near the synchronous speed the motor torque is proportional to slip. Neglect rotational losses.

6.40 A 3ϕ, 230 V, 50 Hz, six-pole induction motor has the following equivalent circuit parameters:

$$R_1 = 0.075\Omega, R_2' = 0.11\,\Omega,$$
$$L_1 = L_2' = 0.25mH$$
$$L_m = 15mH$$

The motor drives a fan. The torque required for the fan varies as the square of the speed and is given by

$$T_{fan} = 12.7 \times 10^{-3} \times \omega_m^2$$

Determine the speed, torque, and power of the fan when the motor is connected to a 3ϕ, 230 V, 50 Hz supply. Use the approximate equivalent circuit as shown in Figure

5.14b, and neglect rotational losses. For operation at low slip, the motor torque can be considered proportional to slip.

6.41 A 3ϕ, 25 hp, 440 V, 50 Hz, 1440 rpm, wound-rotor induction motor has the following equivalent circuit parameters:

$$R_1 = 0.25\ \Omega, X_1 = 1.2\Omega,$$
$$R_2' = 0.2\Omega, X_2' = 1.1$$
$$X_m = 35\ \Omega$$

The motor is connected to a 3ϕ, 440 V, 50 Hz, supply.
(a) Determine the number of poles of the machine.
(b) Determine the starting torque.
(c) Determine the value of the external resistance required in each phase of the rotor circuit such that the maximum torque occurs at starting. Use Thevenin's equivalent circuit.

6.42 A 3ϕ, squirrel-cage induction motor has a starting torque of 1.75 pu and a maximum torque of 2.5 pu when operated from rated voltage and frequency. The full-load torque is considered as 1 pu of torque. Neglect stator resistance.
(a) Determine the slip at maximum torque.
(b) Determine the slip at full-load torque.
(c) Determine the rotor current at starting in per unit – consider the full-load rotor current as 1 pu.
(d) Determine the rotor current at maximum torque in per unit of full-load rotor current.

6.43 A 3ϕ, 440 V, 50 Hz, four-pole wound-rotor induction motor develops full-load torque at a slip of 0.04 when the slip rings are short-circuited. The maximum torque it can develop is 2.5 pu. The stator leakage impedance is negligible. The rotor resistance measured between two slip rings is 0.5 Ω.
(a) Determine the speed of the motor at maximum torque.
(b) Determine the starting torque in per unit. (Full-load torque is one per-unit torque.)
(c) Determine the value of resistance to be added to each phase of the rotor circuit so that maximum torque is developed at the starting condition.
(d) Determine the speed at full-load torque with the added rotor resistance of part (c).

6.44 A 3ϕ, 200 hp, 440 V, 1440 rpm, 50 Hz induction motor has a power factor of 0.85 lagging and an efficiency of 90% at full load. If started with rated voltage, the starting current is six times greater than the rated current of the motor. An autotransformer is used to start the motor at reduced voltage.
(a) Determine the rated motor current.
(b) Determine the autotransformer output voltage to make the motor starting current twice the full-load current.

(c) Determine the ratio of the starting torque at the reduced voltage of part (b) to the torque at rated voltage.

6.45 A 3ϕ, 440 V, 50 Hz, 1440 rpm, 100 hp, four-pole squirrel-cage induction motor has negligible stator resistance and leakage inductance. The motor is to be operated from a 60 Hz supply.

(a) Determine the supply voltage if the air gap flux is to remain at the same value if it were operated from a 3ϕ, 440 V, 50 Hz supply.

(b) Determine the speed at full-load torque if the motor operates from the 60 Hz supply of part.

References

1. Boldea, I. and Nasar, S.A. (2002). *The Induction Machine Handbook*. CRC Press.
2. Lamine, A., Levi, E., and Cavagnino, A. (2006). Compensation of stray load loss induced detuning in rotor flux oriented induction machines. In: *Proceedings of the 41st International Universities Power Engineering Conference*, 685–689. Newcastle-upon-Tyne: IEEE.
3. Levi, E., Lamine, A., and Cavagnino, A. (2005). Detuned operation of vector controlled induction machines due to stray load losses. In: *Fourtieth IAS Annual Meeting. Conference Record of the 2005 Industry Applications Conference, 2005*, vol. 1, 500–507. Kowloon, Hong Kong: IEEE.
4. Levi, E., Lamine, A., and Cavagnino, A. (2006). Impact of stray load losses on vector control accuracy in current-fed induction motor drives. *IEEE Transactions on Energy Conversion* 21 (2): 442–450.
5. Levi, E., Lamine, A., and Cavagnino, A. (2005). Rotor-flux-oriented control of induction machines considering the stray load losses. In: *2005 IEEE 36th Power Electronics Specialists Conference*, 1325–1331. Recife: IEEE.
6. Krause, P., Wasynczuk, O., Sudhoff, S., and Pekarek, S. (2013). *Analysis of Electric Machinery and Drive Systems*, 3e, IEEE, Series, Print ISBN: 9781118024294. Wiley.
7. Krishnan, R. (2007). *Electric Motor Drives, Modeling, Analysis and Control*", ISBN-10: 8120321685. Prantice Hall of India.

7

Synchronous Machines

7.1 Preliminary Remarks

A synchronous machine is an AC rotating machine that rotates at a constant speed under the steady-state condition and the speed of rotation is proportional to the frequency of the current in its armature winding. The magnetic field produced by the current flowing in the armature winding (stator winding) rotates at the same speed as that of the field produced by the field winding current on the rotor. This causes the production of a steady torque. Synchronous generators are also called Alternators. The working principle is based on Faraday's law of electromagnetic induction. This machine rotates at a constant speed corresponding to frequency and number of poles, called the synchronous speed, hence known as the synchronous machine.

Synchronous machines are most commonly employed as generators in large power plants, such as thermal/gas-based power stations and hydroelectric power stations for a bulk power system. Because the rotor speed is proportional to the frequency of the supply current, synchronous motors are employed in a constant speed high-power drive system. Since the reactive power generated by synchronous machines can be adjusted by controlling the magnitude of the rotor field current, unloaded synchronous machines are also often installed in power systems solely for power-factor correction or control of reactive kVA flow. Such machines, known as *synchronous condensers*, may be more economical in large sizes than static capacitors. A synchronous machine either has a cylindrical rotor or a projected/salient pole rotor. The salient pole rotor is used for low-speed applications. This chapter provides the fundamental operation and characteristics for both the synchronous generator and the synchronous motor (both salient and non-salient pole structure). The equivalent circuit, armature reaction, determination of voltage regulation, testing of machines, operation with variable excitation for power-factor control, and operation of salient pole machines are elaborated on in this chapter. Various numerical examples are given for a clear understanding of the topics. Additionally, Matlab/Simulink models of the machine are given.

Electrical Machine Fundamentals with Numerical Simulation using MATLAB/SIMULINK, First Edition.
Atif Iqbal, Shaikh Moinoddin, and Bhimireddy Prathap Reddy.

Companion website: www.wiley.com/go/iqbal/electricalmachine

7.2 Synchronous Machine Structures

7.2.1 Stator and Rotor

The armature is a machine part in which alternating voltage is generated due to relative motion with respect to a magnetic flux. The armature winding of a conventional synchronous machine is almost invariably on the stator and is usually a three-phase winding. The stator has slots and armature winding is placed on these slots. The field winding is usually on the rotor and excited by DC supply or permanent magnets. The DC power supply required for excitation is usually supplied through a DC generator known as exciter, which is often mounted on the same shaft as the synchronous machine. Various excitation systems using AC exciter and power semiconductor-based AC-DC converters are used for high-power synchronous generators.

Rotor comes in two shapes: **round or cylindrical or non-salient rotor and salient pole rotor** as shown in Figure 7.1. Generally, a round rotor structure, also called non-salient poles, is used for high-speed synchronous machines, such as steam or gas turbine generators, while a salient pole structure is used for low-speed applications, such as hydroelectric generators. The salient pole structure is a dumbbell shape on which concentrated DC winding is wound or permanent magnets are placed. The pictures below (Figure 7.1) show the stator and rotor of a hydroelectric generator and the rotor of a turbine generator. The structure of cylindrical and salient pole synchronous machines are shown in Figure 7.2 and Figure 7.3, respectively.

The stator of the three-phase synchronous machine has a three-phase distributed winding similar to that of the three-phase induction machine. Unlike the DC machines, the stator winding, which is connected to the AC supply system, is sometimes called the *armature winding*. It is designed for high voltage and current.

The field winding is placed on the rotor that is supplied by means of an external DC source.

In a synchronous generator, a DC current is supplied to the rotor or field winding that produces a constant DC magnetic field. The rotor is rotated by an external prime mover (e.g.

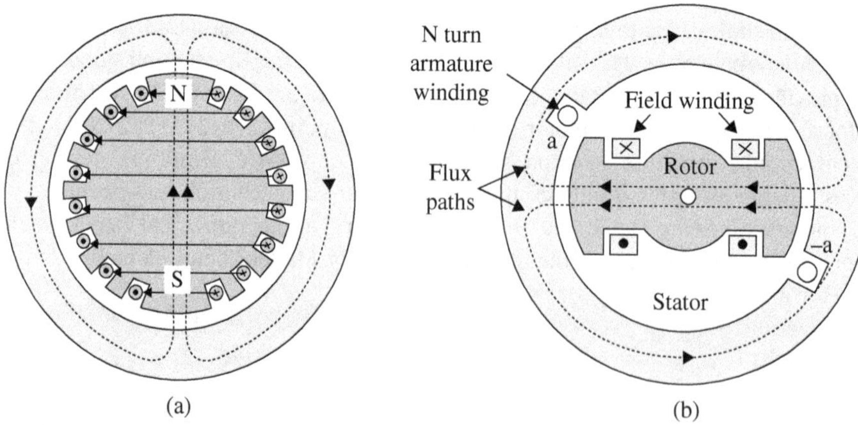

Figure 7.1 A cross-sectional view of the synchronous machines, (a) Round or cylindrical rotor and (b) salient-pole rotor structures.

turbines in large power stations). When the rotor is rotated, the magnetic field associated with the rotor also rotates at the same speed. Now, this rotating magnetic field induces a voltage on the armature (stator) due to Faraday's law of electromagnetic induction. If a load is connected to the armature of the synchronous generator, current flows and electrical power is fed to the load. The frequency of the armature depends upon the speed of the prime mover coupled to its rotor and the number of poles of the armature winding. This is the principle of operation of a synchronous generator.

The frequency of the armature induced voltage is given as

$$f = \frac{PN_s}{120} \tag{7.1}$$

Where

$$f \Rightarrow \text{Frequency of the armature voltage/current}$$
$$P \Rightarrow \text{Number of poles of armature winding}$$
$$N_s \Rightarrow \text{Speed of rotor/prime mover}$$

A three-phase synchronous motor is a doubly-fed motor, unlike a three-phase induction motor which is single-fed. In a synchronous motor, the main field is provided by the DC current supplied to its field winding placed in a rotor that produces a constant field. The stator (armature) is supplied by three-phase currents similar to a three-phase induction motor. The three-phase currents in the armature produce a rotating magnetic field. The rotor and stator fields have to be synchronized to produce torque. Since the rotor produces stationary field and stator produces a rotating field, no net torque is developed and the synchronous motors are **not self-starting**. Several means are used to start a synchronous motor. However, the principle is to synchronize the two fields as with that of a three-phase induction motor. Once both fields are synchronized the motor starts. One way is to start the synchronous motor by means of another small motor and once it reaches close to synchronous speed, the external means are withdrawn and the motor keeps on running. Detail of synchronous motor starting is given in a separate section.

Two common approaches are used to supply DC current to the field circuits on the rotating rotor

1. Supply the DC power from an external DC source to the rotor by means of slip rings and brushes
2. Supply the DC power from a special DC power source mounted on the shaft of the synchronous machine, called exciters.

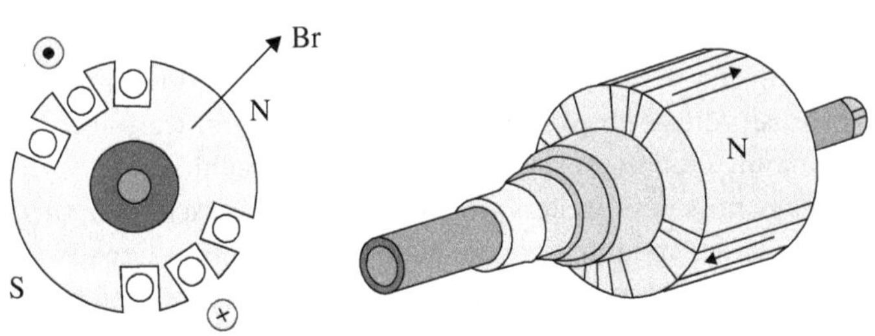

Figure 7.2 Cylindrical rotor machine.

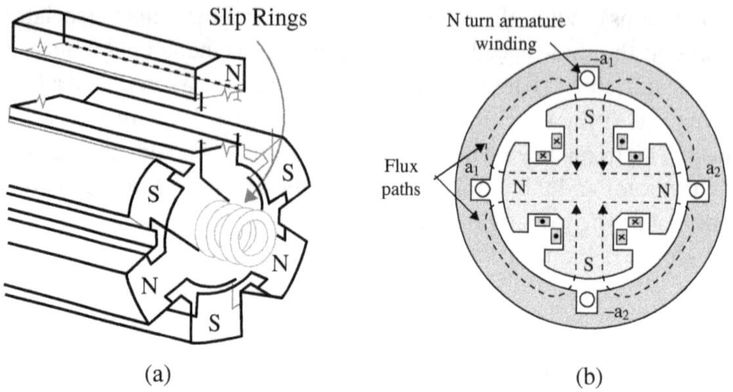

Figure 7.3 Salient-pole rotor, (a) six poles (b) four poles.

The field system is placed on the rotor and the armature is placed on the stator in a synchronous machine.

The field circuit is DC and low voltage and low power while the armature is a high voltage, high power, and AC. Using high voltage and high power in the rotating part poses serious technical challenges and it is mounted on the stator. On the rotor slip rings are carbon brushes and are used to connect to the external circuit. If AC is used on the rotor then four slip rings and four carbon brushes would be required while only two slip rings and brushes are needed for DC circuit connection. In a high-power machine, forced cooling is done using hydrogen. In such cases, ducts are provided which are convenient on the stator side.

The following are the advantages of placing a field on rotor and armature on stator:

- Reduced insulation requirements
- Reduced number of slip rings and brushes
- Machine size is reduced
- High speed of rotation is possible
- Ease in cooling of the machine
- Reduced mechanical vibrations, etc.

Slip rings are metal rings completely encircling the shaft of a machine but insulated from it. One end of a DC rotor winding is connected to each of the two slip rings on the machine's shaft. Graphite-like carbon brushes connected to DC terminals ride on each slip ring supplying DC voltage to field windings regardless of the position or speed of the rotor. The slip ring and brush arrangement are shown in Figure 7.4.

Slip rings and brushes have certain disadvantages: increased friction and wear (therefore, require maintenance), a brush voltage drop occurs that introduces significant electrical power losses. Still, this is commonly used in most small synchronous machines.

On large generators and motors, brushless exciters are used. A brushless exciter is a small AC generator whose field circuits are mounted on the stator and armature circuits are mounted on the rotor shaft. The exciter generator's three-phase output is rectified to DC

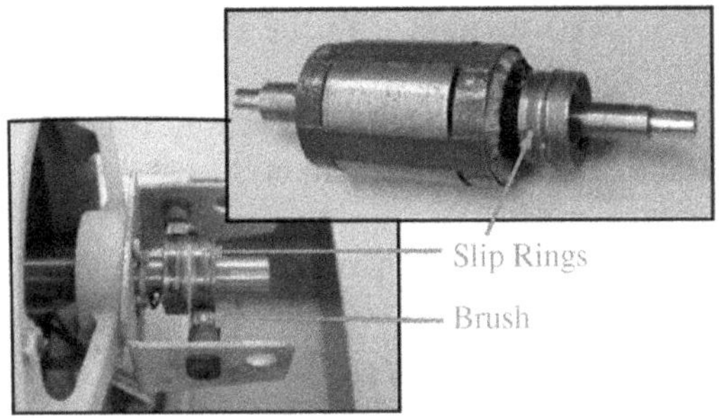

Figure 7.4 Brush and slip ring arrangement.

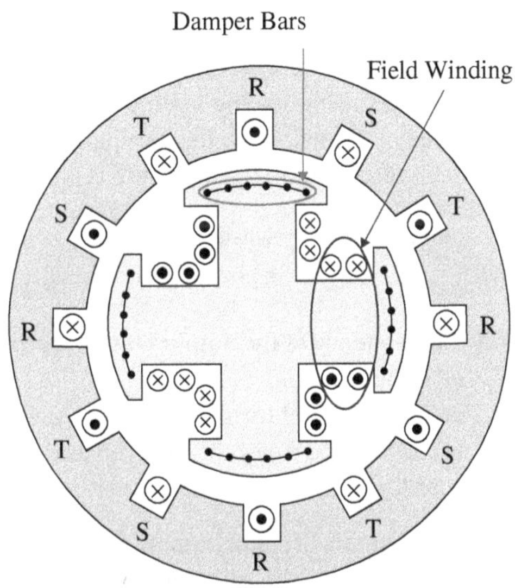

Figure 7.5 Damper winding in salient pole of the synchronous machine.

by an AC-DC power converter and supplies the main DC field circuit. It is possible to adjust the field current on the main machine by varying the small DC field current of the exciter generator (located on the stator). Since no mechanical contact occurs between the rotor and the stator, exciters of this type require far less maintenance.

The synchronous machines with salient poles (projected poles) have additional windings called damper or amortisseurs winding embedded on the rotor faces (Figure 7.5).

The main purpose of this winding is to start the synchronous motor as an induction motor or, in other words, to make it a self-starting machine. The short-circuited damper windings act as a squirrel cage induction motor rotor. In the beginning, an emf is induced in the damper winding due to the rotating magnetic field produced from the stator (armature), which causes current to flow in the short-circuited rotor. This current produces its flux. The interaction of these two fluxes produces a torque that makes the machine self-starting. Once the speed is close to the synchronous value, the machine is pulled into synchronism and starts rotating at synchronous speed. Once the machine speed is synchronous, the induced emf in the damper winding becomes zero and the induction effect is no longer there. Thus, the role of damper winding is finished at this instant.

7.3 Working Principle of the Synchronous Generator

A synchronous generator is a machine that converts mechanical power from a prime mover (turbines in the power station) to AC electric power at a specific voltage and frequency. A synchronous machine rotates at a constant speed called the synchronous speed that depends on the number of poles and speed of the machine.

The working principle of the synchronous generator is very simple. The DC excitation on the rotor produces a DC constant field. The rotor is rotated by the external prime mover at a constant speed corresponding to the required frequency of the output voltage and currents. When the rotor rotates, the associated field of the rotor also rotates at the same speed in the air gap. The rotating field cuts the stator/armature windings and hence emf is induced in the stator/armature windings. The induced emf in the armature depends upon the flux and speed of the machine. The main field or flux depends upon the excitation current and can be varied by changing the field excitation current.

The induced emf in the armature is given as

$$E_a = 4.44 f \phi_f T_{ph} K_W \tag{7.2}$$

Where

f is the frequency of the output voltage/current
ϕ_f is the flux per pole
T_{ph} is the number of turns per phase in the armature winding
K_W is the winding factor that is given as the product of the chording factor and distribution factor $K_W = K_c . K_d$ and the value is less than 1.

The schematic of stator/armature and rotor/field winding is presented in Figure 7.6.

The variation of the main field with the excitation current (DC) and induced emf with the excitation current is shown in Figure 7.7. When the excitation current is zero, there still exists a small emf $E_\phi$due to the residual magnetic field. The induced emf increases linearly with an increase in the excitation current up to the certain value of the excitation current. If excitation current is further increased, the emf does not increase due to the saturation of the magnetic circuit. This is called an open circuit or magnetization characteristics of the

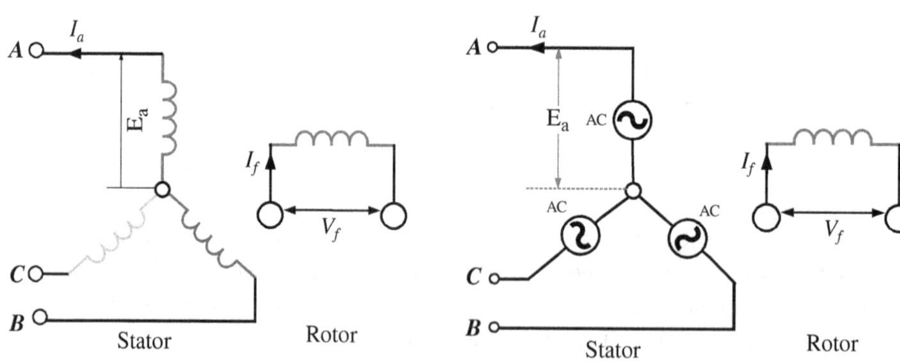

Figure 7.6 Stator and rotor windings in the synchronous generator.

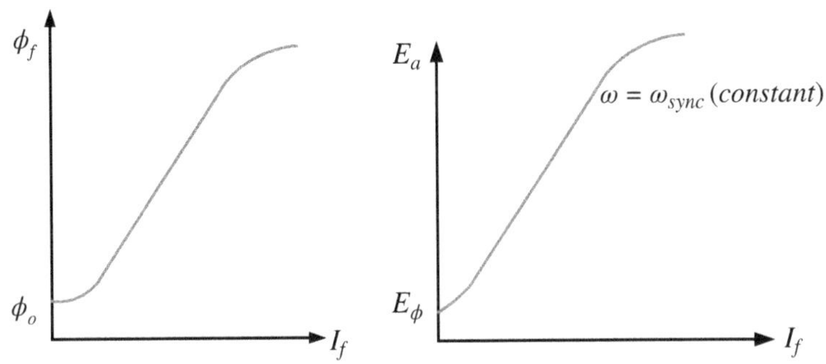

Figure 7.7 Open-circuit or magnetization characteristics.

synchronous machine. When no load is supplied by the machine, the terminal voltage and induced emf due to excitation are the same.

Example 7.1 A four-pole three-phase 50 Hz the synchronous machine has a flux of 0.05 Wb. The winding has 200 turns/phase with a winding factor (Kw) as 0.94. Compute the following;

(a) Speed in rpm and rad/sec
(b) Three-phase voltages as a function of time
(c) The root mean square (rms) phase voltage of the generator

Solution

(a) The speed of the generator is the synchronous

$$N_s = \frac{120f}{P} = \frac{120 \times 50}{4} = 1500\ rpm$$

$$\omega = \frac{2\pi N_s}{60} = \frac{2 \times 3.141 \times 1500}{60} = 157.08\ rad/\sec$$

(b) The peak-phase voltage is calculated as

$$E_{peak} = 4.44\sqrt{2}f\phi_f T_{ph}K_W$$

$$E_{peak} = 4.44\sqrt{2} \times 50 \times 0.05 \times 200 \times 0.94 = 2951.18V$$

Thus, the three-phase voltages are

$$e_a = 2951.18 \sin(157.08t)$$

$$e_b = 2951.18 \sin(157.08t - 120)$$

$$e_c = 2951.18 \sin(157.08t - 240)$$

(c) The rms phase voltage is

$$E_{rms} = \frac{E_{peak}}{\sqrt{2}} = \frac{2951.18}{\sqrt{2}} = 2086.8V$$

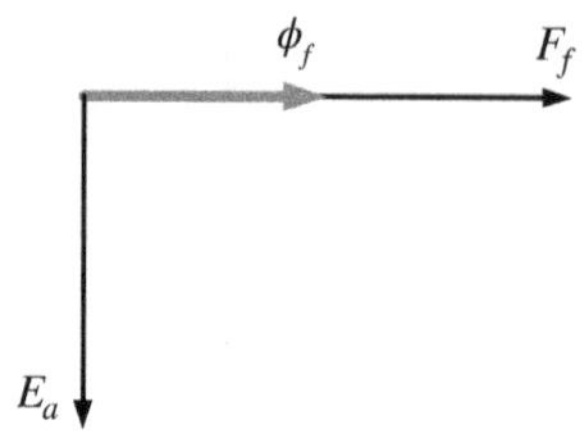

Figure 7.8 The synchronous generator under no-load condition.

7.3.1 The Synchronous Generator under No-Load

Under the no-load condition, emf is induced in armature windings. The emf lags the flux causing this emf by 90°. The phasor diagram under a no-load condition is shown in Figure 7.8. The main field flux is represented by ϕ_f and the main field mmf is F_f. The armature induced emf is E_a.

7.3.2 The Synchronous Generator under Load

When a three-phase electrical load is supplied by a three-phase synchronous generator, three-phase current flows through the armature winding. The current flowing through the armature windings produces its rotating magnetic field (ϕ_a). This is called *Armature reaction*. The net field/flux now in the air gap is the sum of the main field produced from the excitation winding and armature field produced from the armature winding.

$$\phi_r = \phi_f + \phi_a \tag{7.3}$$

Case 1:

Considering the Unity power factor load (resistive load), the current through armature winding is in phase with the induced emf (E_a). The armature field is along with the armature current. The resultant or net field is the phasor sum of the main field and armature field. The resultant emf is lagging behind the resultant field by 90°. The phasor diagram is shown in Figure 7.9.

The resultant emf is given as

$$E_r = 4.44 f \phi_r T_{ph} K_W \tag{7.4}$$

The emf under no-load condition is given by Eq. (7.3) and the emf under load condition is given by Eq. (7.4).

The armature reaction effect under the unity power factor load is purely cross magnetizing. The armature field is in quadrature to the main field. Under such a situation, the distribution of main field flux becomes non-uniform and distorted. The flux density under the trailing pole tips increase somewhat while under the leading pole tips it decreases. This is called the cross-magnetizing effect of the armature reaction.

Case 2:

Considering zero power factor (lagging) load (pure inductive), the current through armature winding lags 90° behind the induced emf (E_a). The armature field is along with the armature current. The resultant or net field is the phasor sum of the main field and armature field ($\phi_r = \phi_f - \phi_a$). The resultant emf is lagging behind the resultant field by 90°. The phasor diagram is shown in Figure 7.10.

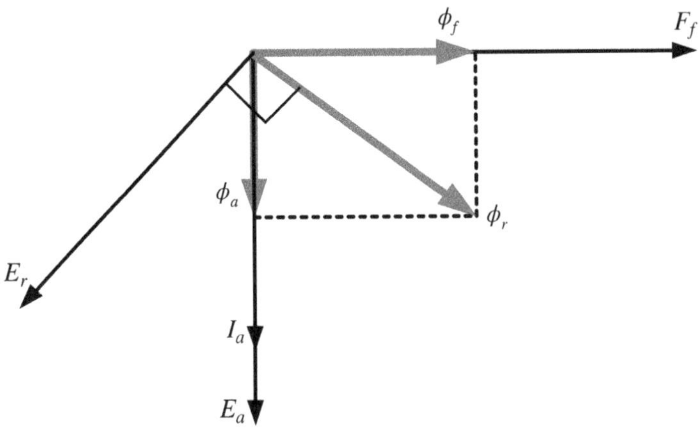

Figure 7.9 The synchronous generator under load condition (unity power factor).

Figure 7.10 The synchronous generator under load condition (zero power factor lagging).

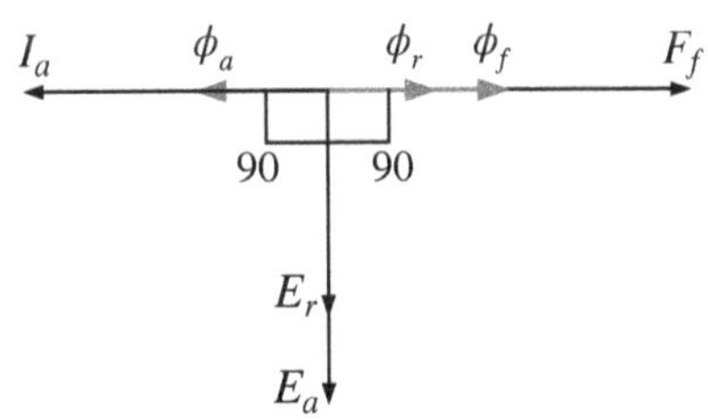

Under zero power factor lagging load condition, the armature reaction is purely demagnetizing (See Figure 7.10). Demagnetizing means the armature reaction causes a reduction in the main field.

Case 3:

Considering zero power factor (leading) load (pure capacitive), the current through armature winding leads 90° the induced emf (E_a). The armature field is along with the armature current. The resultant or net field is the phasor sum of the main field and armature field ($\phi_r = \phi_f + \phi_a$). The resultant emf is lagging behind the resultant field by 90°. The phasor diagram is shown in Figure 7.11.

Under zero power factor leading load condition, the armature reaction is purely magnetizing (See Figure 7.11). Magnetizing means the armature reaction causes an increase in the main field.

Case 4:

Considering lagging power factor load (resistive and inductive load), the current through armature winding lags by an angle say θ degree with the induced emf (E_a). The armature field is along with the armature current. The resultant or net field is the phasor sum of the main field and armature field ($\phi_r = \phi_f + \phi_a$). The resultant emf is lagging behind the resultant field by 90°. The phasor diagram is shown in Figure 7.12.

The armature flux in the previous cases had only one component, either horizontal or vertical. In the present case, the armature flux has two components (by resolving the armature flux along horizontal and vertical axes) along horizontal (along induced emf or quadrature

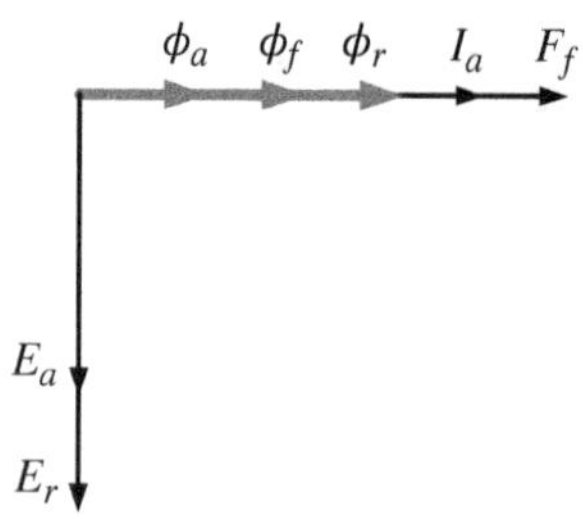

Figure 7.11 The synchronous generator under load condition (zero PF leading).

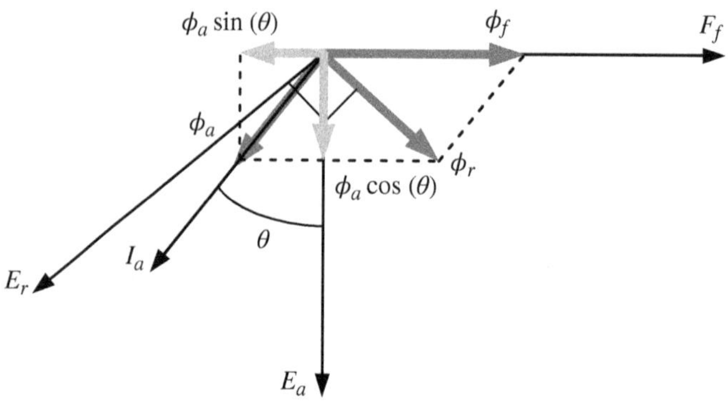

Figure 7.12 The synchronous generator under load condition (lagging power factor).

to the main field) and vertical (opposite to the main field) axes. Thus, it is seen that the armature reaction, in this case, has both a demagnetizing effect (due to the vertical axis component) and a cross-magnetizing effect (due to horizontal axis component).

The demagnetizing component of armature flux is

$$\phi_a \sin(\theta) \tag{7.5}$$

The cross-magnetizing component of armature flux is

$$\phi_a \cos(\theta) \tag{7.6}$$

Case 5:

Considering the leading power factor load (resistive and capacitive load), the current through armature winding leads by an angle say θ degree with the induced emf (E_a). The armature field is along with the armature current. The resultant or net field is the phasor sum of the main field and armature field ($\phi_r = \phi_f + \phi_a$). The resultant emf is lagging behind the resultant field by 90°. The phasor diagram is shown in Figure 7.13.

The armature flux has two components (by resolving the armature flux along horizontal and vertical axes) along horizontal (along induced emf or quadrature to the main field) and vertical (along the main field) axes. Thus, it is seen that the armature reaction, in this case, has both a magnetizing effect (due to vertical axis component or component along the main field) and cross magnetizing effect (due to horizontal axis component or quadrature to the main field).

The magnetizing component of armature flux is

$$\phi_a \sin(\theta) \tag{7.5}$$

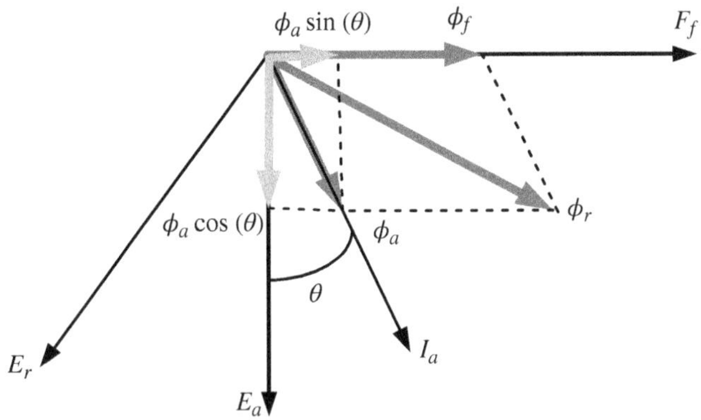

Figure 7.13 The synchronous generator under load condition (leading power factor).

The cross-magnetizing component of armature flux is

$$\phi_a \cos(\theta) \tag{7.6}$$

7.4 Working Principle of the Synchronous Motor

The synchronous machine is also used as a motor, converting electrical energy into mechanical energy. However, the synchronous motors are not self-starting contrary to induction motors which are self-starting. The synchronous motors are doubly-fed motors; rotors are supplied by DC power and armatures are fed by three-phase AC power. When both field and armature supply are turned on, the motor still does not produce torque. It is important to recall the general expression of electromagnetic torque in a round rotor machine which is given as

$$T_e = K\phi_s\phi_r \sin(\delta) \tag{6.2}$$

The speed of both the stator and rotor field should be the same to maintain the angle between them (δ). In the case of the synchronous motor, the main field produced by the rotor is stationary (DC) while the field produced from the stator rotates at the synchronous speed. Hence, the angle between the two fields continuously varies. On average in each cycle δ becomes 360° that makes $\sin(\delta) = \sin(360) = 0$, thus the average torque becomes zero and the motor does not start.

Non-self-starting of the synchronous motor can be further explained with the help of Figure 7.14.

When three-phase supply is given to three-phase stator windings, the rotating magnetic field is produced that rotates at the synchronous speed

$$N_s = \frac{120f}{P} \tag{7.7}$$

for example, for 50 Hz supply and two-pole winding, the field rotates at 3000 rpm, which is equivalent to a rotation of the magnetic poles. Consider the situation at a time instant say t = 0, where N1 and S1 are the poles created by the rotating magnetic field from the

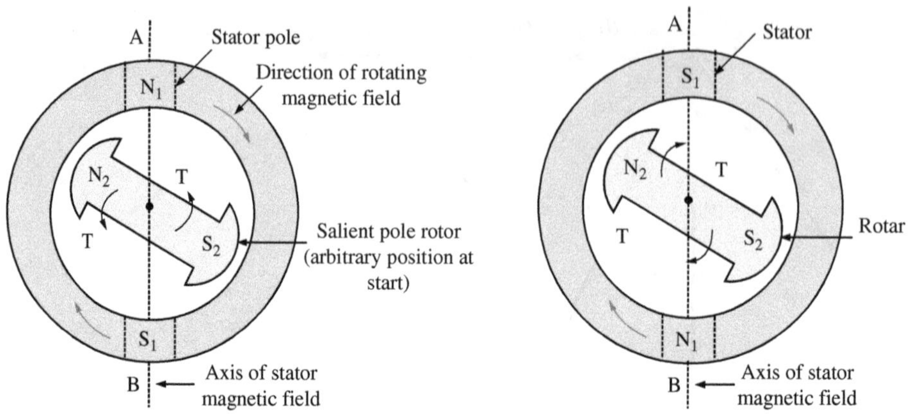

Figure 7.14 Starting principle of the synchronous Motor.

stator supply that rotates at high speed (3000 rpm in case of 50 Hz supply). The rotor field is created from DC supply to the rotor winding and these are stationary with N2 and S2 poles. At t = 0, north of the stator and north of the rotor repel each other trying to rotate the rotor in the anti-clockwise direction. Similarly, south of the stator and south of the rotor repel each other trying to rotate the rotor in the anti-clockwise direction. Due to the inertia of the rotor, the rotor does not rotate, rather it tries to move in the anti-clockwise direction. However, at instant t = t1, considering the stator poles have rotated for half a cycle. Under this situation, the south of the stator and north of the rotor face each other trying to rotate the rotor in a clockwise direction. Due to the inertia of the rotor, it does not rotate but rather just vibrates. Since the stator poles quickly change the position, the rotor is not able to rotate.

Considering the phenomena discussed above, it can be concluded that to start the rotor

- The rotor field should also rotate at the same speed as that of the stator field.
- The speed of the stator field should be slow in order to overcome the inertia of the rotor and allow the rotor to start moving in one direction.

7.5 Starting of the Synchronous Motor

7.5.1 Starting by External Motor

The basic principle of starting the synchronous motor is bringing it to the synchronous speed and then switching on the field supply. A small induction motor called a Pony motor is coupled to the shaft of the synchronous motor. The rated speed of the induction motor is higher than the synchronous speed of the synchronous machine. The first induction motor is energized which is a self-starting machine and it starts rotating which also runs the synchronous motor. As soon as the synchronous motor reaches the synchronous speed, the DC field supply is turned on and the induction motor is decoupled.

A DC motor is also used for starting the synchronous motor. A DC motor is coupled to the shaft of the synchronous motor. Once the speed of the DC motor reaches close to the synchronous speed, the synchronous motor is pulled into synchronism and kept running at a synchronous speed. The DC motor now runs as a DC generator supplying DC power to the field winding of the synchronous motor.

7.5.2 Starting by using Damper Winding

An additional winding called a damper winding in the form of copper bars similar to the squirrel cage rotor is provided on the pole faces of the salient poles of a synchronous motor. Stator supply is given while the rotor supply is switched off. The synchronous motor initially acts as an induction motor and, once the speed reaches close to the synchronous speed, the motor is pulled into synchronism and starts rotating at the synchronous speed. The DC supply to the field winding is switched on. As soon as the synchronous motor reaches the synchronous speed, the role of damper winding has vanished. There is no more relative speed between the rotor and the field and hence no induction effect takes place. Thus, there is no role of the damper winding once the synchronous motor starts running at the synchronous speed.

It is important to note that if there is any sudden load change or other transients and disturbances, the speed of the synchronous motor deviates from the synchronous speed, there will be an induced emf in the damper winding and restorative torque is produced by the damper winding and the speed of the synchronous motor is restored. This is the reason it is called 'damper winding' as it damps out the transient. Damper windings are also used in synchronous generators to take care of the transients.

7.5.3 Starting by Variable Frequency Stator Supply

In this method, a variable frequency power converter is used to supply the stator windings of the synchronous motor. The stator is supplied at a very low frequency so that the stator poles rotate slowly and the rotor poles can follow the stator poles. The low speed of the stator poles will cause locking of the stator and rotor poles and the rotor starts moving in one direction. The supply frequency can be increased gradually to the rated frequency value.

This is an expensive method since a full-rated power converter is required just to start the motor. However, if variable frequency speed is desired then this method is the most suitable.

7.6 Armature Reaction in Synchronous Motor

The stator or armature current direction in the synchronous motor is opposite to that of the synchronous generator. The armature reaction for all of the above cases (synchronous generator) can be understood by just considering the direction of the armature current.

Case–1:

Considering Unity power factor current being drawn from the mains by the synchronous motor armature when the motor is supplying a mechanical load. This current remains in phase with the induced emf (E_a). Thus, the armature field produced is along the armature current direction. The resultant or net field is the phasor sum of the main field and armature field. The resultant emf that is induced due to the resultant field is lagging by 90°. The phasor diagram under this condition is shown in Figure 7.15.

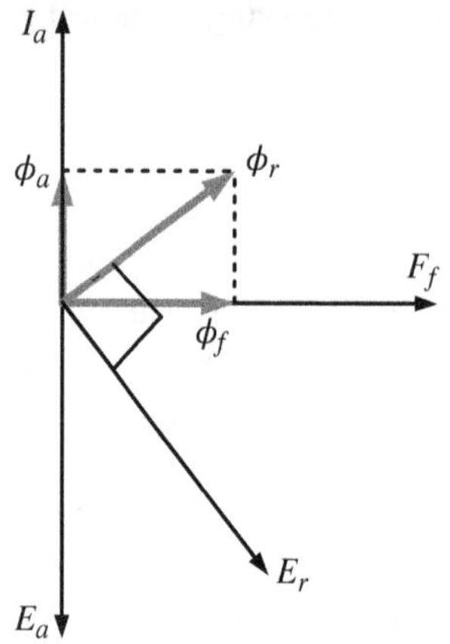

Figure 7.15 The synchronous motor drawing unity power factor current.

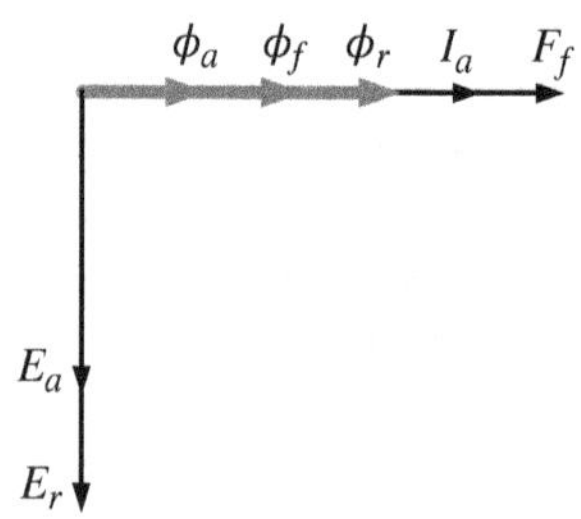

Figure 7.16 The synchronous motor drawing zero power factor lagging current.

Since the armature flux is in quadrature to the main flux, the armature reaction is cross-magnetizing in nature causing distortion in the resultant flux.

Case–2:

Considering the current drawn by the synchronous motor armature as zero power factor (lagging), the current through armature winding lags behind the induced emf (E_a) by 90°. However, the direction of current in the motor is opposite to the generator, it is shown leading by 90°. The armature field is along with the armature current. The resultant or net field is the phasor sum of the main field and armature field ($\phi_r = \phi_f + \phi_a$). The resultant emf is lagging behind the resultant field by 90°. The phasor diagram is shown in Figure 7.16.

Under zero power factor lagging current, the armature reaction is purely magnetizing (See Figure 7.16). Magnetizing means the armature reaction causes an increase in the main field. Thus, the resultant field is higher than the main field.

Case–3:

Considering the current drawn by the synchronous motor as zero power factor (leading), the current through armature winding leads by 90° ahead of the induced emf (E_a). However, the direction of current in the motor is opposite to the generator, it is shown lagging by 90°. The armature field is along with the armature current. The resultant or net field is the phasor sum of the main field and armature field ($\phi_r = \phi_f - \phi_a$). The resultant emf is lagging behind the resultant field by 90°. The phasor diagram is shown in Figure 7.17.

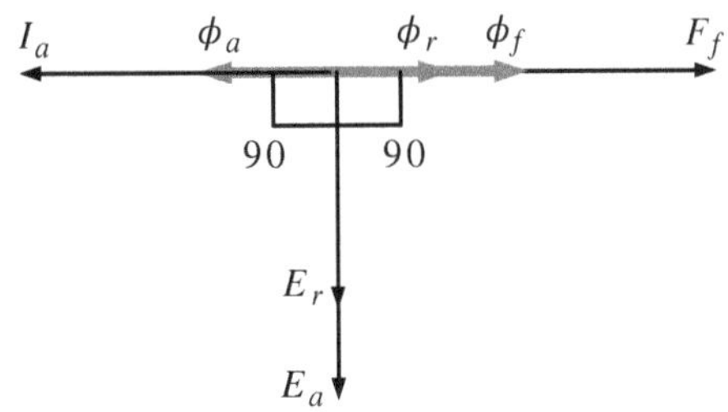

Figure 7.17 The synchronous motor drawing zero power factor leading current.

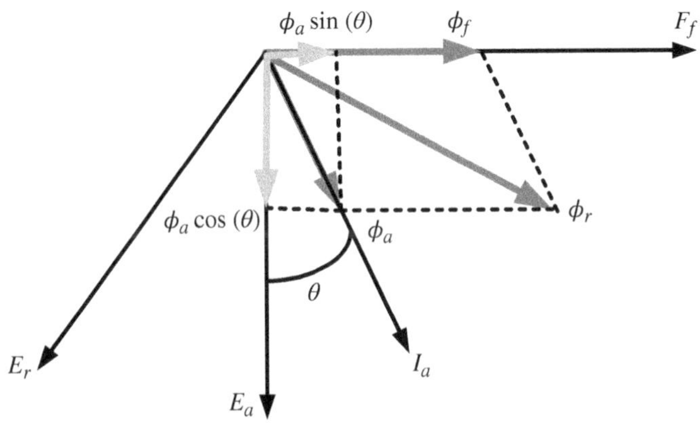

Figure 7.18 The synchronous motor drawing lagging power factor current.

Under zero power factor leading load condition, the armature reaction is purely demagnetizing (See Figure 7.17). Demagnetizing means the armature reaction causes a reduction in the main field.

Case 4:

Considering the current drawn by the synchronous motor is lagging power factor, the current through the armature winding lag by an angle say θ degree with the induced emf (E_a). However, the current is shown as leading because the motor current is opposite to that of the generator. The armature field is along with the armature current. The resultant or net field is the phasor sum of the main field and armature field ($\phi_r = \phi_f + \phi_a$). The resultant emf is lagging behind the resultant field by 90°. The phasor diagram is shown in Figure 7.18.

The armature flux has two components (by resolving the armature flux along horizontal and vertical axes) along horizontal (along induced emf or quadrature to the main field) and vertical (along the main field) axes. Thus, it is seen that the armature reaction, in this case, has both a magnetizing effect (due to a vertical axis component or component along the main field) and a cross-magnetizing effect (due to a horizontal axis component or quadrature to the main field).

The magnetizing component of armature flux is

$$\phi_a \sin(\theta) \tag{7.7}$$

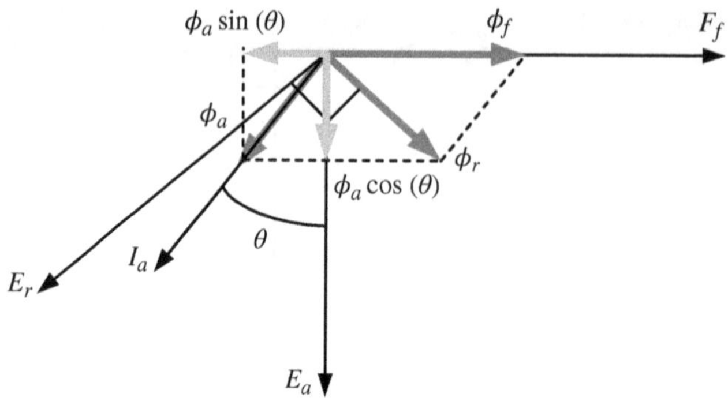

Figure 7.19 The synchronous motor drawing leading power factor current.

The cross-magnetizing component of armature flux is

$$\phi_a \cos(\theta) \tag{7.8}$$

Case 5:

Considering the synchronous motor is drawing leading power factor current through its armature winding. The current leads by an angle say θ degree with the induced emf (E_a). However, the current is shown as lagging since the motor current is opposite to the generator current. The armature field is along with the armature current. The resultant or net field is the phasor sum of the main field and armature field ($\phi_r = \phi_f + \phi_a$). The resultant emf is lagging behind the resultant field by 90°. The phasor diagram is shown in Figure 7.19.

The armature flux has two components (by resolving the armature flux along horizontal and vertical axes) along horizontal (along induced emf or quadrature to the main field) and vertical (opposite to the main field) axes. Thus, it is seen that the armature reaction, in this case, has both a demagnetizing effect (due to vertical axis component) and cross-magnetizing effect (due to horizontal axis component).

The demagnetizing component of armature flux is

$$\phi_a \sin(\theta) \tag{7.9}$$

The cross-magnetizing component of armature flux is;

$$\phi_a \cos(\theta) \tag{7.10}$$

A summary of the armature reaction effects in the synchronous generator and the synchronous motor are given in Table 7.1

7.7 Equivalent Circuit and Phasor Diagram of the Synchronous Machine

The equivalent circuit is derived on a single-phase basis assuming a balanced three-phase supply and steady-state condition. Time constants of damper winding and field winding are ignored.

Table 7.1 Summary of armature reaction in the synchronous generator and the synchronous motor.

Condition	The synchronous generator	The synchronous motor
I_a lags E_a by 90°	Demagnetizing	Magnetizing
I_a leads E_a by 90°	Magnetizing	Demagnetizing
Lagging pf; I_a lags E_a by a certain angle	Demagnetizing and Cross magnetizing	Magnetizing and cross magnetizing
Leading pf; I_a leads E_a by a certain angle	Magnetizing and Cross magnetizing	Demagnetizing and Cross magnetizing

When the field winding on the rotor supplied with DC current and the rotor is rotated at the synchronous speed by an external prime mover (generator), the rotating magnetic field is produced that induces emf in the armature winding. This is called excitation emf or internal emf, represented by E_a. When an electrical load is supplied by the synchronous generator, armature current flows, causing armature flux to be produced (current-carrying conductor produces its magnetic field). The armature flux has two components, namely, leakage flux and armature reaction flux.

$$\phi_a = \phi_{al} + \phi_{ar} \tag{7.11}$$

The leakage flux links the with the armature flux while the armature reaction flux links with both armature and field winding. These fluxes are in phase with the armature current. The induced emf due to these fluxes are lagging by 90°. Thus, the effect of the leakage flux and armature reaction flux is represented by leakage reactance and armature reaction reactance, respectively. The leakage flux and the armature reaction flux are combined and called the synchronous reactance. The armature winding has its own resistance that is included in the equivalent circuit representation.

$$X_s = X_{al} + X_{ar} \tag{7.12}$$

The combination of the synchronous reactance and armature resistance is called the synchronous impedance.

$$Z_s = R_a + jX_s \tag{7.13}$$

The per-phase equivalent circuit of the synchronous generator is shown in Figure 7.20 and that of the synchronous motor in Figure 7.21. The direction of armature current in the synchronous motor is opposite to that of the synchronous generator.

It is important to note that the iron/core loss is not shown in the equivalent circuit. However, the iron loss is measured from the no-load test and hence is a part of rotational losses.

The equivalent circuit of the synchronous generator is shown in Figure 7.20a. The field circuit is shown separately.

After combining the armature reaction reactance and leakage reactance the equivalent circuit is shown in Figure 7.20b (the field circuit is omitted).

The equivalent circuit of the synchronous motor is the same as that of the synchronous generator except for the direction of the current is reversed as shown in Figure 7.21.

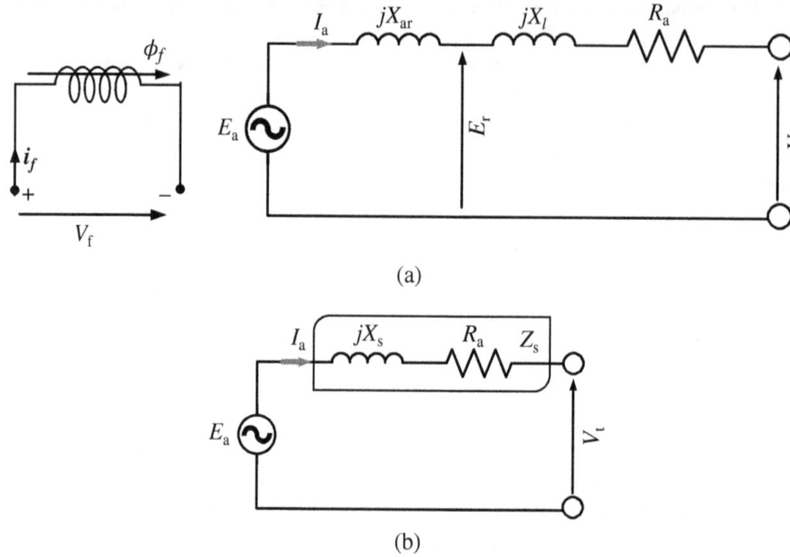

Figure 7.20 Equivalent circuit of the synchronous generator on a per-phase basis (a) and (b).

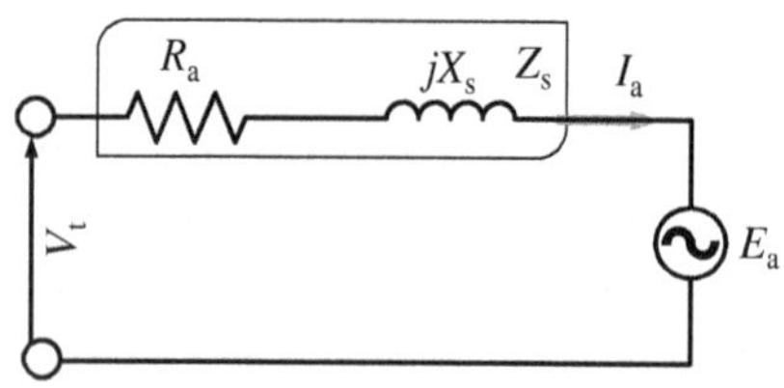

Figure 7.21 Equivalent circuit of the synchronous motor on a per-phase basis.

The phasor diagram can be drawn from the equivalent circuit equations for different power factor conditions.

7.7.1 Phasor Diagram of the Synchronous Generator

Case–1: Considering a lagging power load being supplied by the synchronous generator. The terminal voltage and the induced emf are linked by the following relationship (use Kirchhoff's Voltage Law (KVL) in Figure 7.20b)

$$E_a\angle\delta = V_t\angle 0 + I_aR_a + jI_aX_s \tag{7.14}$$

The phasor diagram of the synchronous machine is drawn considering Eq. (7.14). The armature terminal voltage ($V_t \angle 0$) is considered a reference. Armature current ($I_a \angle -\theta$) lags behind the armature terminal voltage by power factor angle. To the tip of the armature terminal voltage, resistive voltage drop and inductive voltage drop are added. Resistive voltage drop (I_aR_a) has the same phase as that of the armature current, while the reactive voltage drop (I_aX_s) leads the armature current by 90°. The sum of the armature terminal voltage, the resistive voltage drop and the inductive voltage drop is the armature induced emf. The angle between the induced emf and the terminal voltage is δ. This angle is also load angle. The resulting phasor diagram is shown in Figure 7.22.

Figure 7.22 Phasor diagram for lagging power factor load.

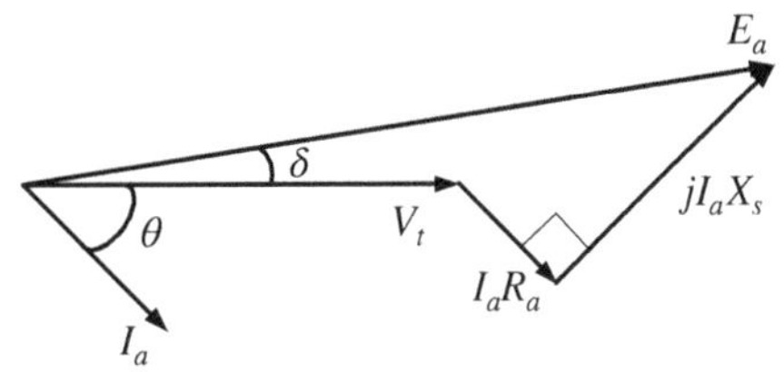

Figure 7.23 Phasor diagram for unity power factor load.

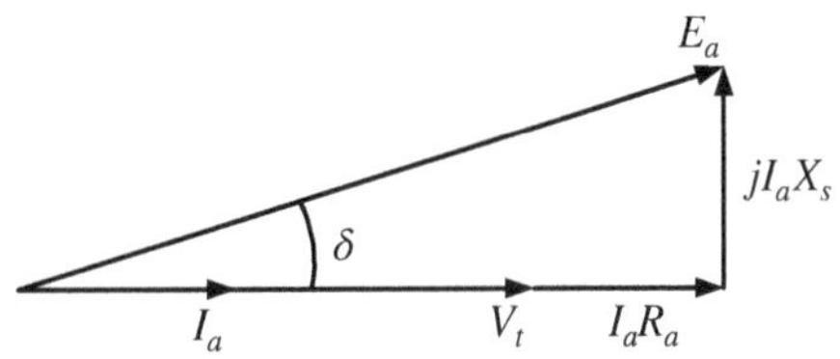

Case–2: Considering a unity power load being supplied by the synchronous generator. The terminal voltage and the induced emf are linked by the following relation (use KVL in Figure 7.20b)

$$E_a\angle\delta = V_t\angle 0 + I_aR_a + jI_aX_s \tag{7.14}$$

The phasor diagram of the synchronous machine is drawn considering Eq. (7.14). The armature terminal voltage ($V_t\angle 0$) is considered a reference. Armature current ($I_a\angle 0$) is in phase with the armature terminal voltage. To the tip of the armature terminal voltage, resistive voltage drop and inductive voltage drop are added. Resistive voltage drop (I_aR_a) has the same phase as that of the armature current, while the reactive voltage drop (I_aX_s) leads the armature current by 90°. The sum of the armature terminal voltage, the resistive voltage drop and the inductive voltage drop is the armature induced emf. The angle between the induced emf and the terminal voltage is δ. The resulting phasor diagram is shown in Figure 7.23.

Case–3: Considering a unity power load being supplied by the synchronous generator. The terminal voltage and the induced emf are related by the following relation (use KVL in Figure 7.20b)

$$E_a\angle\delta = V_t\angle 0 + I_aR_a + jI_aX_s \tag{7.14}$$

The phasor diagram of the synchronous machine is drawn considering Eq. (7.14). The armature terminal voltage ($V_t\angle 0$) is considered a reference. Armature current ($I_a\angle\theta$) is in phase with the armature terminal voltage. To the tip of the armature terminal voltage, resistive voltage drop and inductive voltage drop are added. Resistive voltage drop (I_aR_a) has the same phase as that of the armature current, while the reactive voltage drop (I_aX_s) leads the armature current by 90°. The sum of the armature terminal voltage, the resistive voltage drop and the inductive voltage drop is the armature induced emf. The angle between the induced emf and the terminal voltage is δ. The resulting phasor diagram is shown in Figure 7.24.

If the load is highly leading, the induced emf may become lower than the terminal voltage.

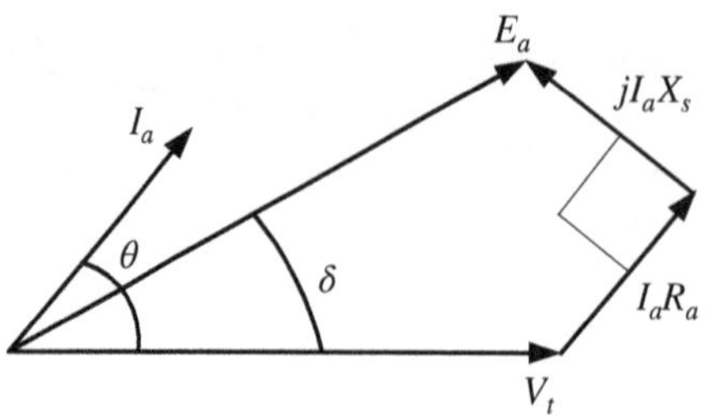

Figure 7.24 Phasor diagram for leading power factor load.

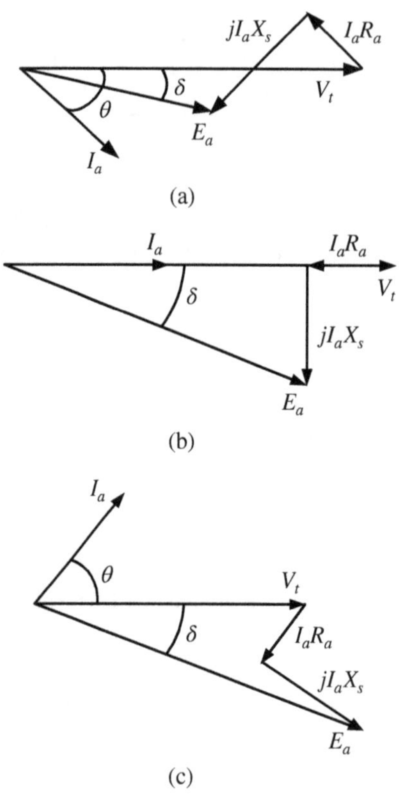

Figure 7.25 Phasor diagram of the synchronous motor, (a) lagging pf current, (b) unity pf current, and (c) leading pf current.

7.7.2 Phasor Diagram of the Synchronous Motor

Considering Figure 7.21, the equivalent circuit of the synchronous motor, the loop equation is written as

$$E_a\angle -\delta = V_t\angle 0 - I_aR_a - jI_aX_s \tag{7.15}$$

Assuming the terminal voltage as a reference, the phasor diagram is drawn for different power factor conditions and shown in Figure 7.25. The armature resistive and inductive voltage drops are subtracted from the terminal voltage to obtain the induced emf. The angle between the terminal voltage and induced emf, known as the power angle δ in the motoring case is negative.

Example 7.2 A 1 MVA three-phase star-connected 3.6 kV, 12-Pole, 50 Hz generator has the following parameters

Armature resistance $R_a = 0.15\,\Omega$/phase and the synchronous reactance $X_s = 3.5\,\Omega$/phase. Calculate the full-load generated voltage per phase at (a) 0.8 pf lagging, (b) unity power factor, (c) 0.8 leading power factor

Solution

The armature current under full-load condition

$$S = \sqrt{3}V_t I_a$$

$$I_a = \frac{S}{\sqrt{3}V_t} = \frac{1 \times 10^6}{\sqrt{3} \times 3500} = 198.1A$$

In Polar form the current is written as

$$I_a = 198.1\angle - 36.87$$

$$Z = R_a + jX_s = 0.15 + j3.5$$

(a) 0.8 pf lagging

$$E_a\angle\delta = V_t\angle 0 + I_a R_a + jI_a X_s$$

$$E_a\angle\delta = \frac{3500}{\sqrt{3}}\angle 0 + (198.1\angle - 36.87 * 0.15) + j\,(198.1\angle(-36.87 + 90) * 3.5)$$

$$E_a\angle\delta = 2020.72\angle 0 + 29.715\angle - 36.87 + j693.35\angle 53.13$$

$$E_a\angle\delta = 2020.72 + 23.772 - j17.83 + j416 - 554.68$$

$$E_a\angle\delta = 1489.812 + j398.17 = 1542.1\angle 14.96^0$$

(b) Unity pf

$$E_a\angle\delta = V_t\angle 0 + I_a R_a + jI_a X_s$$

$$E_a\angle\delta = \frac{3500}{\sqrt{3}}\angle 0 + (198.1\angle 0 * 0.15) + j\,(198.1\angle(0 + 90) * 3.5)$$

$$E_a\angle\delta = 2020.72\angle 0 + 29.715\angle 0 + j693.35\angle 90$$

$$E_a\angle\delta = 2020.72 + 29.715 - 693.35$$

$$E_a\angle\delta = 1357.1 = 1357.1\angle 0^0$$

(c) 0.8 Leading pf

$$E_a\angle\delta = V_t\angle 0 + I_a R_a + jI_a X_s$$

$$E_a\angle\delta = \frac{3500}{\sqrt{3}}\angle 0 + (198.1\angle 36.87 * 0.15) + j\,(198.1\angle(36.87 + 90) * 3.5)$$

$$E_a\angle\delta = 2020.72\angle 0 + 29.715\angle 36.87 + j693.35\angle 126.87$$

$$E_a\angle\delta = 2020.72 + 23.772 + j17.83 - j416 - 554.68$$

$$E_a\angle\delta = 1489.812 - j398.17 = 1542.1\angle - 14.96^0$$

Example 7.3 Consider the machine of Example of 7.2 operating as a motor. Compute the excitation voltage for (a) 0.8 pf lagging, (b) unity pf, and (c) 0.8 leading pf.

Solution

(a)

$$E_a\angle\delta = V_t\angle 0 - I_a R_a - jI_a X_s$$

$$E_a\angle\delta = \frac{3500}{\sqrt{3}}\angle 0 - (198.1\angle -36.87 * 0.15) - j(198.1\angle(-36.87+90) * 3.5)$$

$$E_a\angle\delta = 2020.72\angle 0 - 29.715\angle -36.87 - j693.35\angle 53.13$$

$$E_a\angle\delta = 2020.72 - 23.772 + j17.83 - j416 + 554.68$$

$$E_a\angle\delta = 2551.628 - j398.17 = 2582.51\angle -8.87^0$$

(b)

$$E_a\angle\delta = V_t\angle 0 - I_a R_a - jI_a X_s$$

$$E_a\angle\delta = \frac{3500}{\sqrt{3}}\angle 0 - (198.1\angle 0 * 0.15) - j(198.1\angle(0+90) * 3.5)$$

$$E_a\angle\delta = 2020.72\angle 0 - 29.715\angle 0 - j693.35\angle 90$$

$$E_a\angle\delta = 2020.72 - 29.715 + 693.35$$

$$E_a\angle\delta = 2684.36 = 2684.36\angle 0$$

(c)

$$E_a\angle\delta = V_t\angle 0 - I_a R_a - jI_a X_s$$

$$E_a\angle\delta = \frac{3500}{\sqrt{3}}\angle 0 - (198.1\angle 36.87 * 0.15) - j(198.1\angle(36.87+90) * 3.5)$$

$$E_a\angle\delta = 2020.72\angle 0 - 29.715\angle 36.87 - j693.35\angle 126.87$$

$$E_a\angle\delta = 2020.72 - 23.772 - j17.83 + j416 + 554.68$$

$$E_a\angle\delta = 2551.628 + j398.17 = 2582.51\angle 8.87^0$$

Example 7.4 A three-phase 50 Hz the synchronous motor is rated at 750 kW and 2.2 kV. The total losses in the motor including electrical and mechanical losses are 20 kW. The motor is supplied with 0.8 pf lagging current when fully loaded. Compute the excitation voltage per phase at full load, given the synchronous reactance as 1.5 Ω/phase. The armature resistance is negligible.

Solution

$$\text{Input power} = \text{Output power} + \text{Losses}$$

$$P_{in} = P_{out} + P_{losses}$$

$$P_{in} = 750 + 20 = 770\,\text{kW}$$

Thus, the input current is calculated as

$$I_a = \frac{P_{in}}{\sqrt{3}V_t\cos(\theta)} = \frac{780\times 10^3}{\sqrt{3}\times 2200\times 0.8} = 255.87A$$

The excitation voltage per-phase is

$$E_a = V_t - jI_aX_s = \frac{2200}{\sqrt{3}} - j255.87\angle - 36.87 \times 1.5$$

$$E_a = 1270.17\angle 0 - 383.8\angle 53.13 = 1270.17 - 230.28 - j307.04$$

$$E_a = 1039.89 - j307.04 = 1084.27\angle - 16.45$$

Example 7.5 A 220 V, three-phase Y-connected the synchronous motor has the synchronous impedance of (0.25 + j2.5) Ω/phase. The motor draws the rated current of 10 A at 0.8 pf leading. Calculate the rms value of line to line induced voltage.

Solution

$$V_t = \frac{220}{\sqrt{3}} = 127.01V$$

For motor, the generated emf for lagging power factor is given as

$$E_a = \sqrt{(V_t \cos\theta - I_aR_a)^2 + (V_t \sin\theta - I_aX_s)^2}$$

For the leading power factor, the sign of the power factor angle θ is changed. Therefore, the above equation modifies as

$$E_a = \sqrt{(V_t \cos\theta - I_aR_a)^2 + (V_t \sin\theta + I_aX_s)^2}$$

$$= \sqrt{(127.01 \times 0.8 - 10 \times 0.25)^2 + (127.01 \times 0.6 + 10 \times 2.5)^2}$$

$$= 141.65V$$

Therefore

$$E_a = \sqrt{3} \times 141.65 = 245V$$

Example 7.6 A three-phase 50 Hz star-connected the synchronous generator has R_a = 0.2 Ω/phase, and X_s = 1.5 Ω/phase. The generator supplies a current of 120 A at a power factor of 0.8 lagging to a feeder of resistance of 1.2 Ω/phase and line reactance of 2.5 Ω/phase. The voltage at the feeder end is 3.5 kV. Calculate the terminal voltage of the generator and the excitation/induced emf.

Solution

A generator supplying a load through a feeder is shown in Figure E7.6

Per-phase bus voltage:

$$V = \frac{3500}{\sqrt{3}} = 2020.72V$$

pf = 0.8 lagging

Hence per phase current is

$$I_a = 120\angle - 36.87$$

Feeder impedance :

$$Z = R + jX = 1.2 + j2.5$$

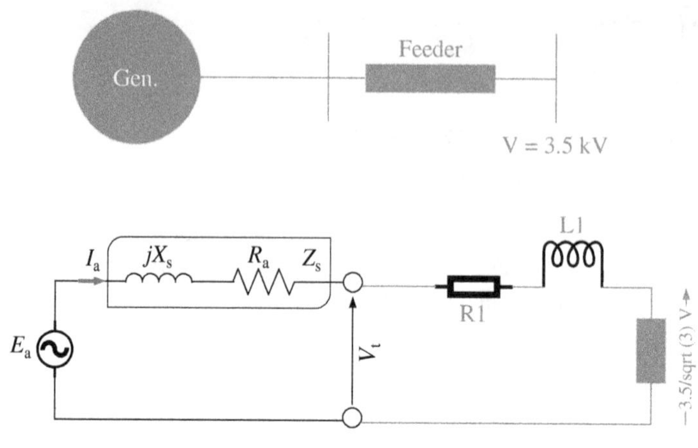

Figure E7.6 A generator supplying a load through a feeder.

Generator Terminal phase Voltage is

$$V_t = V + I_a(R + jX) = 2020.72\angle 0 + 120\angle - 36.87\,(1.2 + j2.5)$$

$$V_t = 2020.72\angle 0 + 120\angle - 36.87\,(2.773\angle 64.36) = 2020.72\angle 0 + 332.76\angle 27.48$$

$$V_t = 2020.72 + j0 + 295.215 + j153.55 = 2315.935 + j153.55 = 2321.02\angle 3.79$$

Generator Terminal line Voltage is

$$|V_{tL}| = \sqrt{3}\,(2321.02) = 4020.12V$$

Excitation per-voltage is

$$E_a\angle\delta = V_t + I_a\left(R_a + jX_s\right) = 2321.02\angle 3.79 + 120\angle - 36.87\,(0.2 + j1.5)$$

$$E_a\angle\delta = 2321.02\angle 3.79 + 120\angle - 36.87\,(1.513\angle 82.4)$$

$$E_a\angle\delta = 2321.02\angle 3.79 + 181.56\angle 45.53 = 2315.935 + j153.55 + 127.2 + j129.56$$

$$E_a\angle\delta = 2443.135 + j283.11 = 2459.48\angle 6.6$$

Excitation Line voltage is

$$|E_{aL}| = \sqrt{3}\,(2459.48) = 4259.94V$$

7.8 Open-Circuit and Short-Circuit Characteristics

7.8.1 Open-Circuit Curve

The relation between the induced emf in the armature and the supplied field current is called open-circuit characteristics. This curve is obtained by running a synchronous machine under no-load condition at the synchronous speed and supplying variable field current and measuring the terminal voltages. Since the machine is running under no-load condition, the induced emf and the terminal voltages are same. The circuit arrangement for the testing of the synchronous machine is shown in Figure 7.26. The open-circuit curve is in essence a relation between the fundamental component of the air-gap flux and the

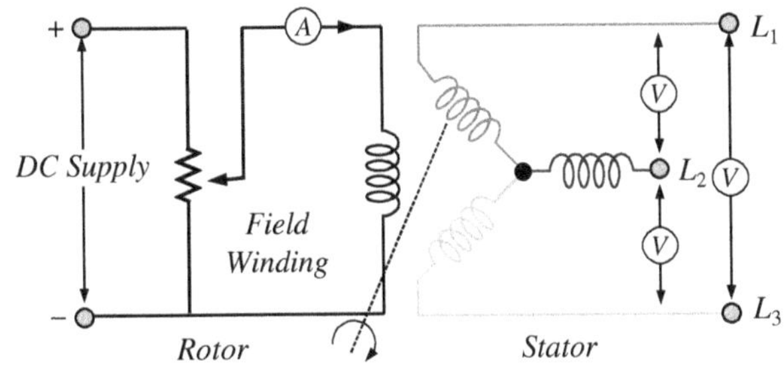

Figure 7.26 Open-circuit test of the synchronous machine.

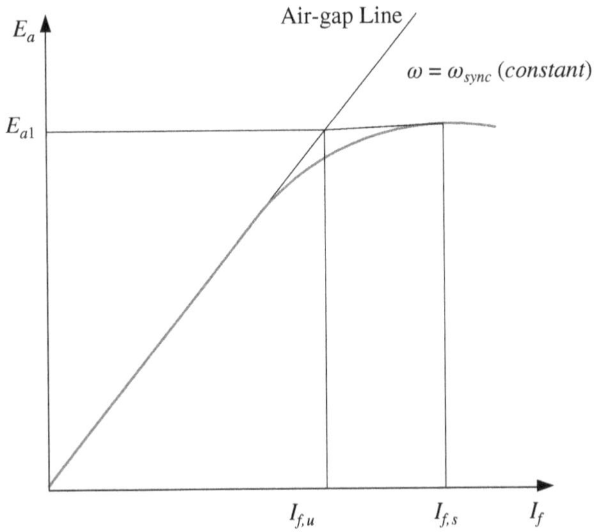

Figure 7.27 Open circuit characteristics of the synchronous machine.

field mmf acting on the magnetic circuit. The open-circuit curve gives a direct relation between the field and armature mutual inductance. The open-circuit characteristics are shown in Figure 7.7, is repeated here in Figure 7.27, considering zero residual magnetism. Due to the saturation effect, the reluctance of the flux path increases that weakens the effect of field current in producing the magnetic flux. The curve is linear in the initial part and saturates later. The linear part can be extended for higher values of the field current, which is called the air-gap line.

The air-gap line represents linear characteristics. This is possible when the machine does not saturate even for the high value of field current, called unsaturated operating conditions. In unsaturated operating conditions, the dominant reluctance is that of the air gap.

Consider the open-circuit curve of Figure 7.27, the field current required to produce the same amount of armature induced emf/terminal voltage in two cases (i) unsaturated operating condition and (ii) saturated operating condition. It is observed that the field current required under saturated condition is much higher than that in an unsaturated condition. The difference between these two field currents is the measure of the degree of saturation in the machine at that voltage.

It is important to note the wattmeter reading in the open-circuit test measures the no-load losses. No-load losses in the synchronous machine comprised of iron/core losses and windage and friction losses. Windage and friction losses depend upon the speed of the machine, which is usually constant at the synchronous speed and hence this loss component is constant. The core/iron loss depends upon the flux and, in turn, voltage. If the machine is not excited (zero field current) and running on no-load, the core loss will be zero and only windage and friction loss takes place. If the machine is excited (field current is supplied), both windage and friction and core losses take place. Thus, the core loss can be computed by taking the difference of the power consumed by machine with excitation and without excitation. It is common practice to consider core loss under load and no-load conditions the same.

$$\text{Power Measured under unexcited and no} - \text{load condition} = \text{P}_1$$

$$\text{Power Measured under excited and no} - \text{load condition} = \text{P}_2$$

$$\text{Core/Iron Loss} = \text{P}_2 - \text{P}_1$$

7.8.2 Short-Circuit Curve

The short-circuit characteristic of the synchronous machine is the plot between the armature current and field current under shorted armature conditions. The armature is short-circuited and field current is varied while running the machine at the synchronous speed obtain the short-circuit curve. It is a linear curve as shown in Figure 7.28.

The equivalent circuit considering the synchronous generator under short-circuit condition is shown in Figure 7.29. Under short-circuit condition, the voltage equation is given as

$$E_a = I_a\left(R_a + jX_s\right) \tag{7.16}$$

The armature resistance is small compared to the synchronous reactance. Neglecting the armature resistance, the current is purely inductive i.e. current lags the voltage by 90°. The phasor diagram under this condition is shown in Figure 7.30. The field mmf and the armature mmf are opposite to each other. The resultant mmf is small. Due to small resultant flux,

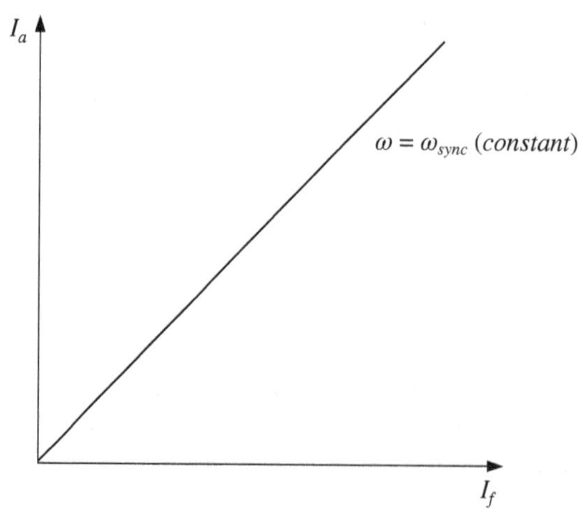

Figure 7.28 Short circuit characteristic of the synchronous machine.

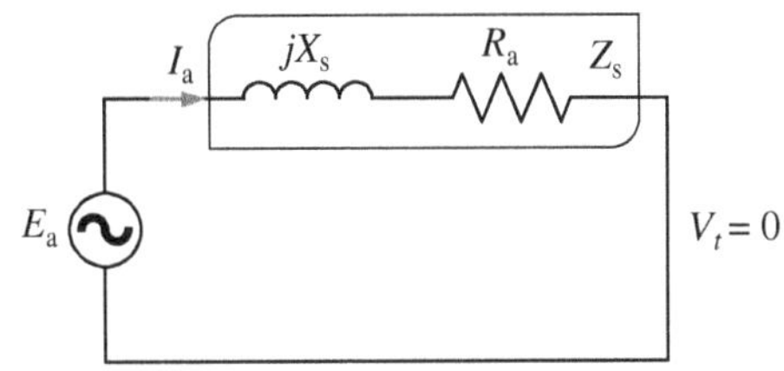

Figure 7.29 Equivalent circuit of the synchronous generator under short circuit condition.

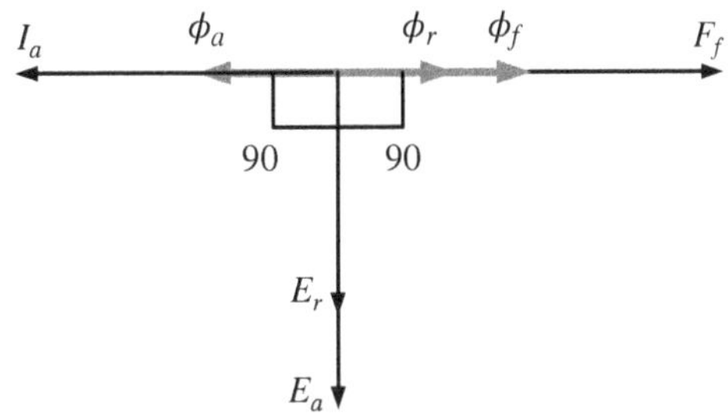

Figure 7.30 Phasor diagram of the synchronous generator.

the machine does not saturate and hence the short-circuit characteristic is linear in nature. The short-circuit curve remains linear for armature current well above the rated value (zero to about 125% of the rated). Generally, the reactance voltage drop is around 10%–20% of the rated value.

Under short-circuit condition, to flow the rated armature current, and only a small voltage is needed and hence the machine does not saturate. The machine generally saturates near the rated voltage.

7.8.3 The Unsaturated Synchronous Reactance

Unsaturated, the synchronous reactance can be computed using an open-circuit and short-circuit curves. Neglecting the armature resistance, the synchronous reactance is calculated directly from the curve of Figure 7.31. Using the air gap line, at any field current, the ratio of the voltage (from the air-gap line) and current (from SCC) is unsaturated the synchronous reactance.

$$X_{s,unsat} = \frac{ad}{dc} = \frac{E_{a1}}{I_{a1}} \tag{7.17}$$

Since the air gap line and short-circuit characteristic (SCC) are linear curves, the field current is not important. At any field current, the ratio will remain the same. Hence, the unsaturated synchronous reactance can be calculated at any field current value.

7.8.4 The Saturated Synchronous Reactance

When the machine is operating at rated voltage, it is common practice to assume that the machine is unsaturated. Under this condition, a modified air-gap line is drawn using two points (origin and corresponding to rated voltage) as shown in Figure 7.31. The saturated reactance is the ratio of the rated voltage under open circuit (measured from the modified

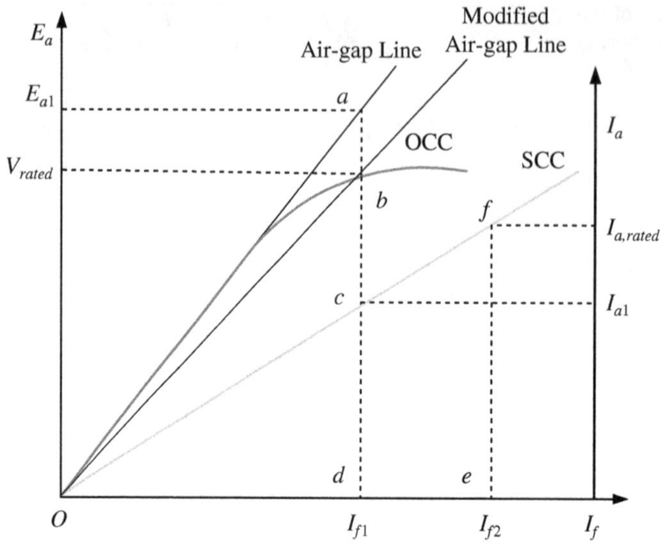

Figure 7.31 Open-circuit and short-circuit characteristics.

air-gap line) to the armature current on a short circuit.

$$X_{s,sat} = \frac{bd}{dc} = \frac{\text{Open Circuit Voltage}}{\text{Short Circuit Current}} = \frac{V_{rated}}{I_{a1}} \tag{7.18}$$

7.8.5 Short-Circuit Ratio

This is a figure of merit of the synchronous generator. It is defined as the ratio of field current required to produce rated voltage under open-circuit condition to the field current required to produce rated armature current under short-circuit condition. From Figure 7.31, the following relation can be written

$$SCR = \frac{od}{oe} = \frac{I_{f1}}{I_{f2}} \tag{7.19}$$

From Figure 7.31, the triangles ocd and ofe are similar and the following relation is written

$$SCR = \frac{od}{oe} = \frac{dc}{fe} \tag{7.20}$$

The synchronous reactance is obtained from Figure 7.31 as

$$\begin{aligned} X_s &= \frac{bd}{dc} \\ X_{s,base} &= \frac{V_{rated}}{I_{rated}} = \frac{bd}{fe} \\ X_{s,pu} &= \frac{X_s}{X_{s,base}} = \frac{\frac{bd}{dc}}{\frac{bd}{fe}} = \frac{fe}{dc} \end{aligned} \tag{7.21}$$

Substituting Eq. (7.21) in Eq. (7.20), the following is obtained

$$SCR = \frac{1}{X_{s,pu}} \tag{7.22}$$

The short-circuit ratio (SCR) is the inverse of the pu the synchronous reactance. Inductance or reactance is inversely proportional to the reluctance of magnetic circuit $\left(L = \frac{N^2}{\Re}\right)$. This means the SCR is directly proportional to the reluctance of the magnetic circuit. High SCR value means high value of reluctance. The synchronous machine with a high value of SCR means

- Better Voltage Regulation
- The improved steady-state stability limit
- High short circuit current under fault condition
- The larger size of the machine.

SCR can be raised by increasing the air-gap length of the machine. For greater air-gap length, the field magnetomotive force (mmf) is required to be increased for the same value of the excitation voltage. To increase the field mmf, either field current or number of turns in the winding is increased. This requires a longer length of the field poles and an overall diameter of the machine increases. Thus, SCR affects the size of the machine. A machine with a high value of SCR offers better voltage regulation and improved steady-state stability limits. Short-circuit fault current is high in the armature of the machine.

The synchronous machine with a low value of SCR means a high value of reactance and inductance. The high value of inductance means a low value of reluctance. Low reluctance means small air gap and smaller sized machines. Such a machine is tightly coupled to the grid and the dynamic response is fast. A machine with a low value of SCR has a low steady-state stability limit. A small SCR machine is less stable when operating in parallel with the other generators and in parallel with the infinite bus bar.

A typical value of SCR for a cylindrical rotor machine lies between 0.45 and 0.9 and for salient pole machine, this ranges from 1 to 1.5.

Example 7.7 The various tests are conducted on a 100 kVA, star-connected 480 V, 50 Hz the synchronous machine and the data is given below

Open Circuit Test

The line to line voltage: 480 V
Field Excitation current: 3.5 A

Short Circuit Test:

Armature Current:	90.5 A	120.2 A
Field Current:	3.5 A	4.3 A

Air-gap Line:

The line to line voltage: 480 V
Field Excitation Current: 2.95 A
Calculate the following

(a) Unsaturated synchronous reactance
(b) Saturated synchronous reactance at rated voltage
(c) The SCR

Solution

(a) The field current of 2.95 A is required for the rated line-to-line voltage of 480 V on the air-gap line produces a short-circuit armature current of
2.95 X[90.5/3.5] = 76.278 A

$$X_{s,unsat} = \frac{480}{\sqrt{3}\,(76.278)} = 3.63\Omega/\text{phase}$$

(b) The field current of 3.5 A produces the rated voltage on the open circuit and short-circuit armature current of 90.5 A. Therefore, the saturated synchronous reactance at the rated voltage is

$$X_{s,unsat} = \frac{480}{\sqrt{3}\,(90.5)} = 3.06\Omega/\text{phase}$$

(c) SCR = 3.5/4.3 = 0.814

7.9 Voltage Regulation

The terminal voltage of the synchronous generator varies with the load and load power factor as shown in Figure 7.32.

Voltage regulation is defined as the pu or percentage rise in the terminal voltage in the synchronous generator when a full load is removed, provided the excitation and speed are kept constant.

$$V.R. = \frac{V_{no-load} - V_{load}}{V_{load}} \quad p.u.$$

$$V.R. = \frac{V_{no-load} - V_{load}}{V_{load}} X100\% \quad percentage$$

$$V.R. = \frac{E_a - V_t}{V} X100\% \tag{7.23}$$

There are several methods to determine the voltage regulation of the synchronous machine. Some of the methods are listed here and described in the following section.

- Emf or synchronous impedance methods
- The ampere-turn or mmf method
- Zero power factor method or Potier Triangle method

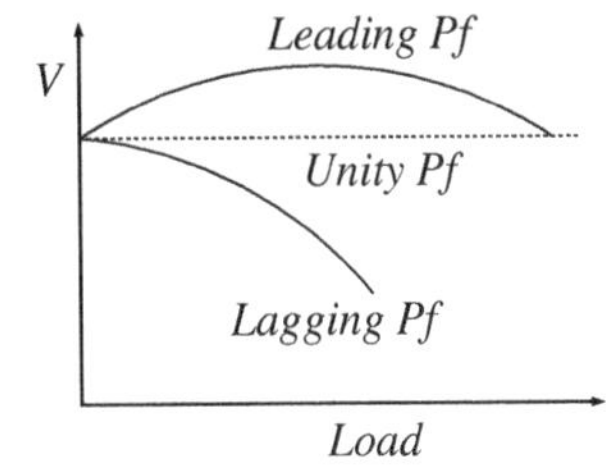

Figure 7.32 Terminal voltage vs load in the synchronous generator.

7.9.1 Emf or Synchronous Method

Voltage regulation is obtained by computing the emf using the phasor diagram. The phasor diagram is shown in Figures 7.22–7.24 for different power factors. To simplify the mathematical equations, the phasor diagrams are redrawn considering current as a reference instead of voltage.

The phasor diagrams for unity, lagging, and leading power factor loads are shown in Figures 7.33–7.35, respectively.

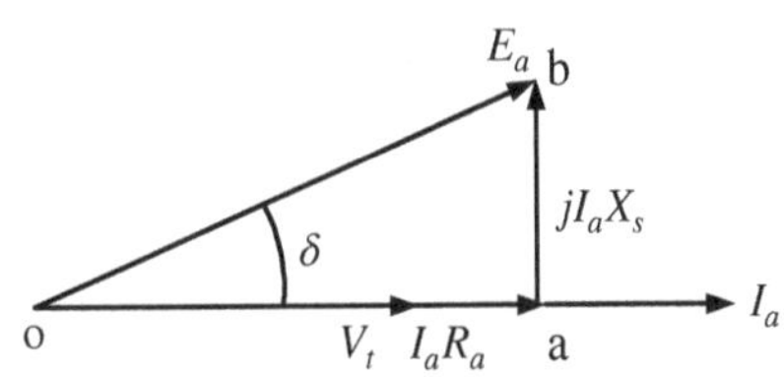

Figure 7.33 Phasor diagram of the synchronous generator for unity power factor.

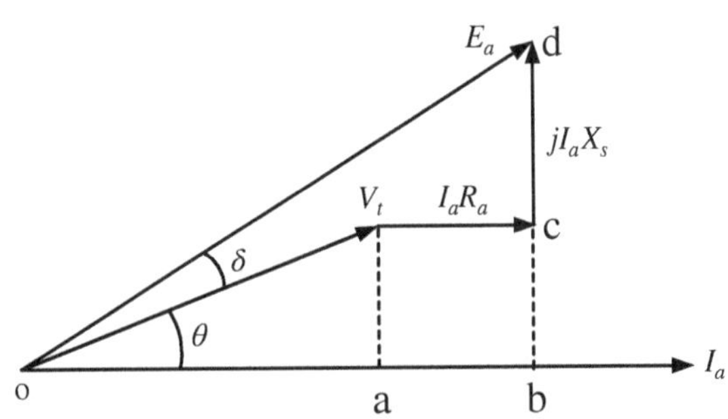

Figure 7.34 Phasor diagram of the synchronous generator for lagging power factor.

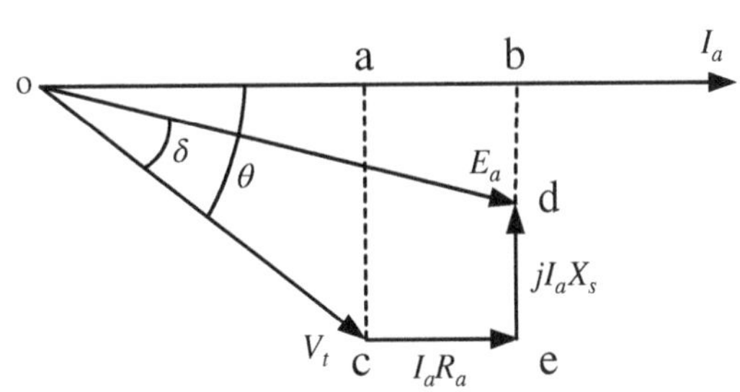

Figure 7.35 Phasor diagram of the synchronous generator for leading power factor.

Considering the right-angled triangle oab in Figure 7.33, the following equation is written

$$ob = \sqrt{(oa)^2 + (ab)^2}$$

$$E_a = \sqrt{(V_t + I_a R_a)^2 + (I_a X_s)^2}$$

$$V.R. = \frac{E_a - V_t}{V_t} \times 100\% \tag{7.24}$$

Considering the right-angled triangle obd in Figure 7.34, the following equation is written

$$od = \sqrt{(ob)^2 + (bd)^2}$$

$$od = \sqrt{(oa + ab)^2 + (bc + cd)^2}$$

$$E_a = \sqrt{(V_t \cos(\theta) + I_a R_a)^2 + (V_t \sin(\theta) + I_a X_s)^2}$$

$$V.R. = \frac{E_a - V_t}{V_t} \times 100\% \tag{7.25}$$

Considering the right-angled triangle obd in Figure 7.35, the following equation is written

$$od = \sqrt{(ob)^2 + (bd)^2}$$

$$od = \sqrt{(oa + ab)^2 + (ac - ed)^2}$$

$$E_a = \sqrt{(V_t \cos(\theta) + I_a R_a)^2 + (V_t \sin(\theta) - I_a X_s)^2}$$

$$V.R. = \frac{E_a - V_t}{V_t} \times 100\% \tag{7.26}$$

7.9.2 The Ampere-Turn or mmf Method

In the mmf method all three tests, the DC resistance test, open-circuit test, and short-circuit test are required to determine the voltage regulation by indirect mean. Unlike the emf method, where leakage reactance and armature reaction reactance are taken as fictitious reactance known as the synchronous reactance, in the mmf method, these reactances are considered separately. The equivalent mmfs required from the field current are taken to compensate for these reactances. Therefore, the flux from the field winding on the rotor will produce the rated terminal voltage at the output and, at the same time, it will compensate for the armature reaction. The field current in rotor winding will produce flux which is in phase with it. Now the rotating flux at the synchronous speed will induce a voltage in the armature, which will be lagging by 90° from it as shown in Eqs. (7.27) to (7.29).

$$\phi = \phi_m \sin(\omega t) \tag{7.27}$$

$$E = -N\frac{d\phi}{dt} = -N\phi_m\omega\cos(\omega t) \tag{7.28}$$

$$E = N\phi_m\omega\sin(\omega t - 90°) \tag{7.29}$$

From Eq. (7.29), the emf generated is lagging flux and thus field winding current (say I_{f1}) by an angle of 90°. Also, the field current (say I_{f2}) that is required for compensating the effect of armature reaction will have 180° phase shift from the load current or armature current. The resultant of these two field currents will be I_f and corresponding to this resultant field current, one can obtain the terminal voltage from the open-circuit characteristics.

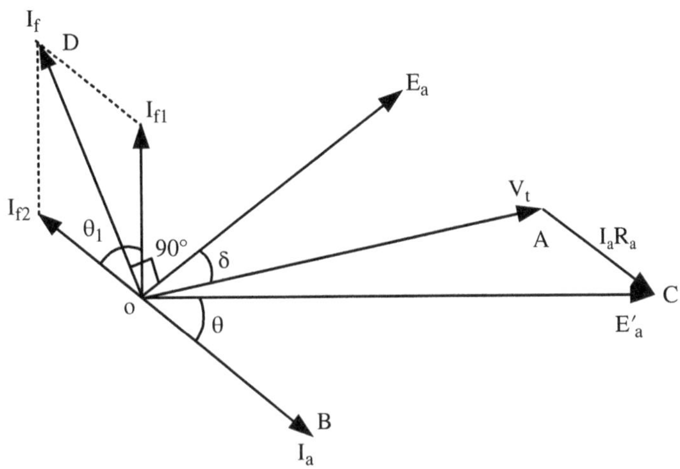

Figure 7.36 Phasor diagram for lagging power factor load in mmf method.

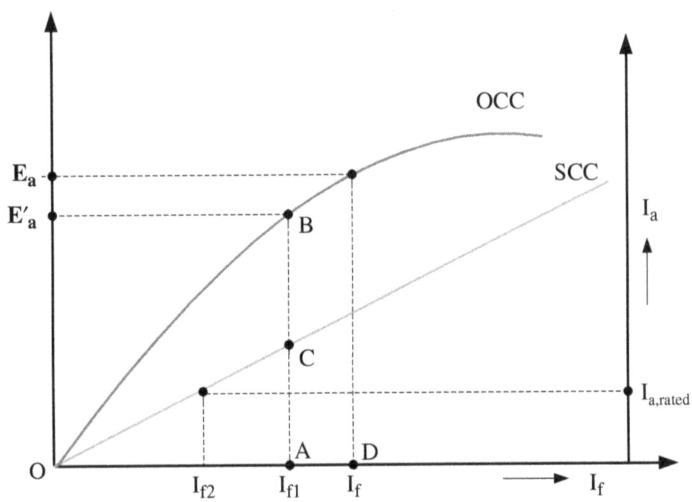

Figure 7.37 Phasor diagram showing open-circuit characteristics.

The phasor diagram is shown in Figure 7.36. The following steps are required to construct the phasor diagram for a lagging power factor load.

- The per-phase armature terminal voltage (V_t) along OA is taken as the reference phasor.
- The lagging armature current (I_a) is drawn along OB with a phase angle of (θ) from the reference armature terminal voltage.
- The armature resistance drop (I_aR_a) is drawn along AC and in phase with the armature current. Then join OC and it will represent the emf (E'_a).
- Find the field current (I_{f1}) corresponding to this emf (E'_a) from the open-circuit characteristics shown in Figure 7.37. The field current (I_{f1}) will lead by an angle of 90° from the emf (E'_a).

- From the short-circuit characteristics, find the field current (I_{f2}) corresponding to the rated armature current on the short-circuiting condition of the armature winding. This will correspond to the mmf required for compensating the armature reaction. The field current (I_{f2}) is in the opposite direction (180°) to armature current.
- Find the phasor sum of field currents (I_{f1}) and (I_{f2}), shown by OD in Figure 7.36 (say, I_f). Mathematically, it can be calculated by using Eq. (7.30)

$$I_f^2 = I_{f1}^2 + I_{f2}^2 + 2I_{f1}I_{f2}\cos\left(90^o - \phi\right) \tag{7.30}$$

- Corresponding to the resultant field current (I_f), obtain no-load voltage (E_a) from the open-circuit characteristics.
- Calculate the voltage regulation using Eq. (7.23)

$$V.R = \frac{E_a - V_t}{V_t} \times 100 \tag{7.23}$$

The mmf method is an optimistic method, i.e. it gives a lesser value of voltage regulation in comparison to the actual loading condition of the synchronous machine.

Example 7.8 For a three-phase, star connected, 1000- kVA, 2000 V, 50 Hz, the synchronous generator, the readings for open-circuit and short-circuit tests are as follows:

Field current (A)	:	**10**	**20**	**25**	**30**	**40**	**50**
Open-circuit terminal voltage (V)	:	790	1495	1762	1995	2355	2590
Short-circuit current (A)	:	–	198	248	305	–	–

The armatures per phase effective resistance is 0.2 Ω, determine the full-load voltage regulation at 0.8 pf (lagging).

Solution

The per-phase voltages

Per phase O.C (V)	**$790/\sqrt{3}$**	**$1495/\sqrt{3}$**	**$1762/\sqrt{3}$**	**$1995/\sqrt{3}$**	**$2355/\sqrt{3}$**	**$2590/\sqrt{3}$**
or	456.1	863.1	1017.3	1151.8	1359.7	1495.3

$$\text{Full load terminal voltage} = V_t = \frac{2000}{\sqrt{3}} = 1154.7V$$

$$\text{Rated armature current} = I_a = \frac{kVA \times 1000}{\sqrt{3} \times V_L} = \frac{100 \times 1000}{\sqrt{3} \times 2000} = 288.7A$$

At lagging power factor of 0.8, the terminal voltage $E'_a = V_t + I_a R_a = 1154.7 + (288.7\angle - \cos^{-1}(0.8)) \times 0.2$

$$E'_a = 1201.7\angle - 1.6^o V$$

To produce 1201.7 V at the terminal of alternator, the field current required is obtained from the open circuit test (OCC) and it is found to be 32 A. Therefore $I_{f1} = 32$ A.

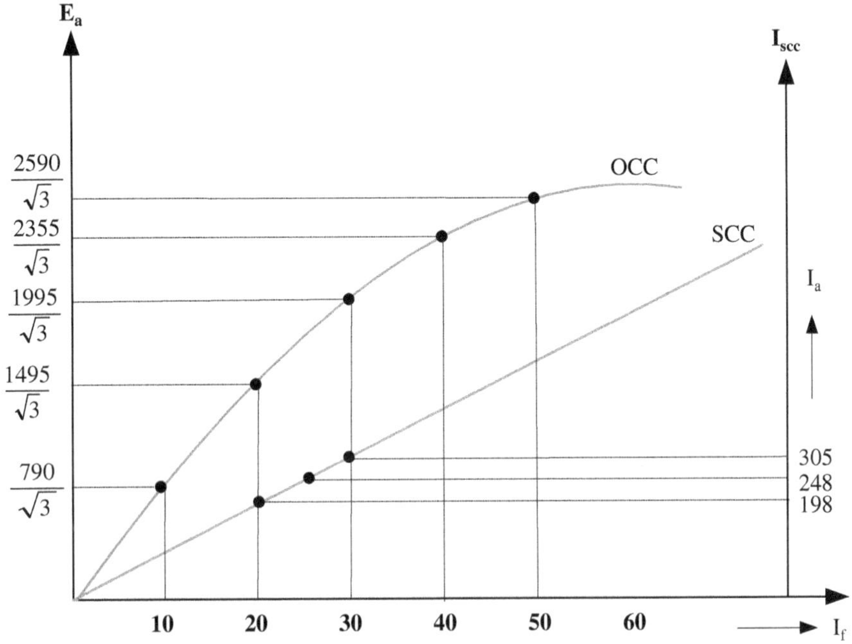

Figure E7.8 OCC and SCC curve of the alternator.

To produce a full-load armature current of 288.7, the field current required is found from the SCC and it is 29 A. Therefore $I_{f2} = 29$ A. Also, for $\cos(\theta) = 0.8$, $\theta = 36.87°$.

From the phasor diagram

$$I_{f2} = \left|I_{f2}\right| \angle\left(180^o - \theta\right) A$$
$$I_{f2} = 29\angle\left(180^o - 36.87^o\right) A$$
$$I_{f2} = 29\angle 143.13^o A$$

Similarly

$$I_{f1} = \left|I_{f1}\right| \angle\left(90^o - \delta\right) A$$
$$I_{f1} = 32\angle\left(90^o - 1.6\right) A$$
$$I_{f1} = 32\angle 88.4^o A$$

Therefore, the resultant field current

$$I_f = I_{f1} + I_{f2} = \left(32\angle 88.4^o + 29\angle 143.13^o\right) A$$
$$I_f = I_{f1} + I_{f2} = 54.18\angle 114.3^o A$$

The open-circuit voltage, corresponding to the field current of 54.18 A is found to be 1559 V. Therefore, $E_a = 1559$ V. Thus, the percentage voltage regulation is found as

$$V.R = \frac{E_a - V_t}{V_t} \times 100$$
$$V.R = \frac{1559 - 1154.7}{1154.7} \times 100 = 35.01\%$$

7.9.3 Zero-Power Factor Method or Potier Triangle Method

The zero-power factor method or Potier triangle method is superior to emf and mmf methods for calculating voltage regulation in alternators as a more accurate result is found. In this method, the effect of leakage reactance and armature resistance is treated like emf drop, whereas the armature reaction drop is dealt like mmf drop. In this method, the zero-power factor characteristic is drawn between the terminal voltage and field current. The field current is taken corresponding to the rated armature current value. The zero-power factor characteristic (ZPFC) is a curve of armature terminal voltage per phase plotted against the field current obtained by operating the machine with constant rated armature current at the synchronous speed and zero lagging power factor. The shape of ZPFC resembles the shape of open-circuit characteristics displaced downwards and to the right as shown in Figure 7.38a.

7.9.3.1 Steps for Drawing Potier Triangles

The mmf required to generate armature terminal voltage under open-circuit condition is F_l. If a purely inductive load is connected, the resultant mmf in the air gap is obtained by considering both the leakage reactance drop and armature reaction drop and it will be given by F_r. Now a line is drawn from F_r to open-circuit characteristics and it will cut it at C. The I_aX_{al} drop will shift point C to point B in the downwards direction. Now, a horizontal line through B will give rated terminal voltage from the open-circuit characteristic. Corresponding to armature reaction mmf mark point *A*. *Now* join points A, B, and C to obtain the Potier triangle. Similarly, points A_1, A_0 can be obtained and the locus these points will produce ZPSC at constant armature current. The Potier triangles thus obtained will be identical (Figure 7.38b).

7.9.3.2 Procedure to Obtain Voltage Regulation using the Potier Triangle Method

The phasor diagram to obtain voltage regulation is shown in Figure 7.39a. The rated terminal voltage (V_t) is drawn along OA. It is taken as the reference phasor. The rated armature

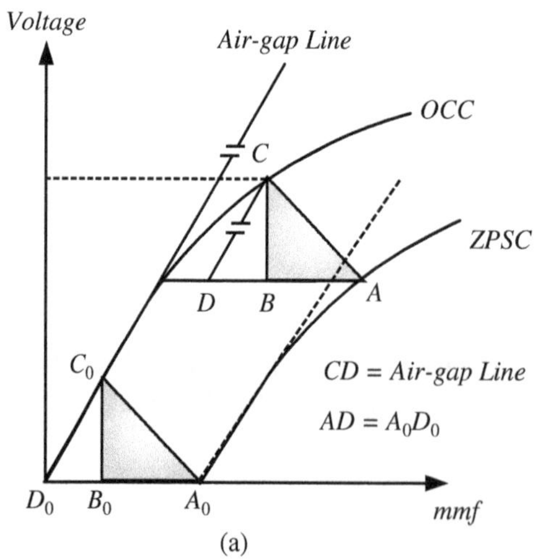

Figure 7.38a Phasor diagram for ZPSC.

Figure 7.38b Locus of Potier triangles for lagging power factor load.

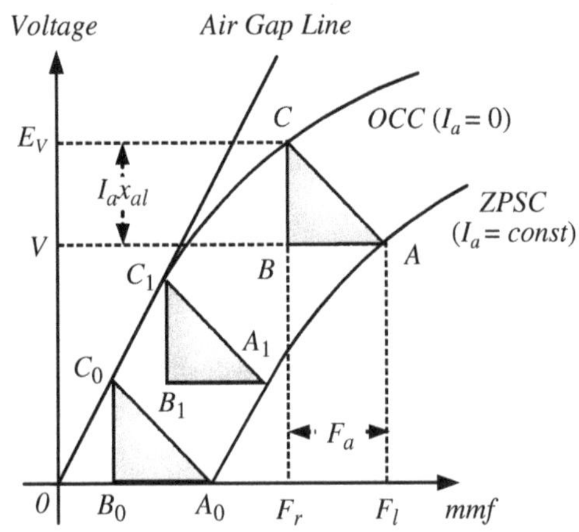

(b)

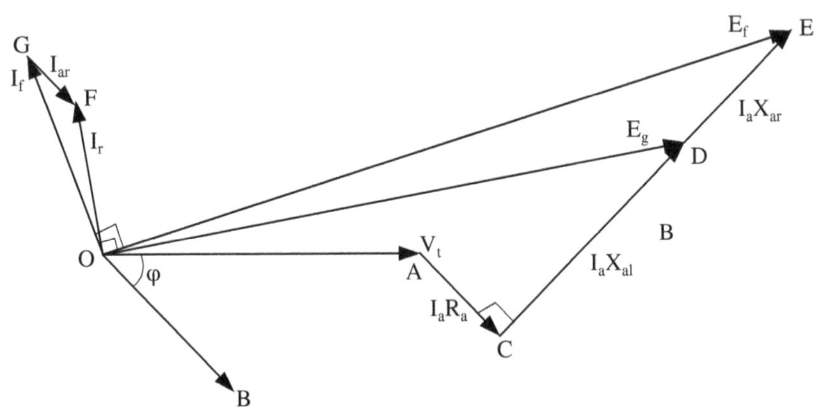

(a)

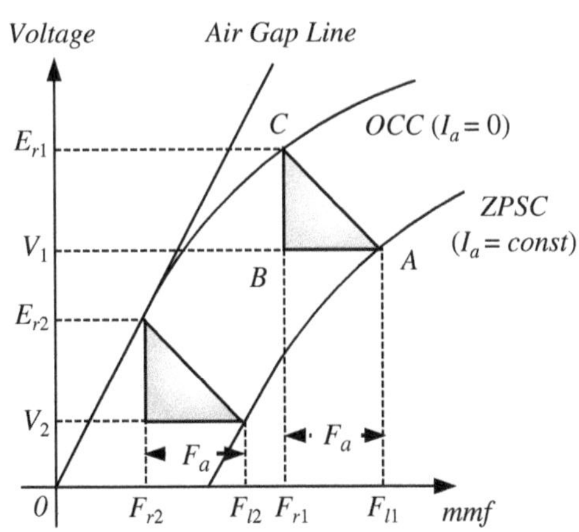

(b)

Figure 7.39 (a) Phasor diagram to obtain voltage regulation using Potier triangle method. (b) Phasor diagram for lagging power factor load in mmf method.

current (I_a) is drawn along OB at an angle of ϕ with the rated terminal voltage. The armature resistance drop is drawn in the direction of AC and phase with armature current. The leakage reactance drop is drawn in the direction of CD at an angle of 90° with the armature current. The phasor sum of rated terminal voltage, armature resistance drops, and leakage reactance drop will give E_g. The mmf or field the current required to generate E_g will be at an angle of 90° from it. It is drawn in the direction of OF. Let the required field current for generating E_g is I_r. Now, corresponding to armature reaction drop, the field current required which is 180° of phase with the armature current is drawn along GF. The resultant OF and GF will give the actual field current required to generate voltage E_f. Thus, the E_f will be at an angle of 90° from the desired field current if along with OE. The corresponding mmf values are shown in Figure 7.39b of the Potier triangles.

After finding the value of E_a, the voltage regulation is computed using equation (7.23)

$$V.R = \frac{E_a - V_t}{V_t} \times 100\% \tag{7.23}$$

Example 7.9 For a three-phase, star connected, 5000- kVA, 6600 V, 50 Hz, the synchronous generator, the readings for open-circuit and saturation curve (zero power factor curve) tests are as follows:

Field current (A)	:	32	50	75	100	140
Open-circuit terminal voltage (V)	:	3100	4900	6600	7500	8300
Saturation curve, zero pf (V)	:	0	1850	4250	5800	7000

The armatures per phase effective resistance are 0.75 Ω, determine the voltage regulation for a load of 500 A at 0.71 pf (lagging) using the Potier triangle method.

Solution

The OCC and ZPFC characteristics are first drawn, based on the data given in the problem, and are shown in Figure E7.9. At rated line voltage of 6600 V, a horizontal line is drawn and it meets the ZPFC at '*b*'. Now, at this line *bk* is taken which is equal to *Ob'* and its value is 32 A. A parallel line to AGL is drawn from *k* which cuts the OCC at point *c*. A vertical line to ca is drawn and the triangle *abc* thus formed is the Potier triangle. Here *ab* is the field current required to overcome the effect of armature reaction at rated-load condition and its value is found to be 25 A. Also, from the Potier triangle *ac* is found to be 900 V.

Therefore, leakage reactance drop $I_a x_{al}$ is $900/\sqrt{3}$ which is 579.6 per phase. Also, $I_a = 500$ (given) and thus the leakage reactance x_{al} is found to be 1.039 Ω.

Since cos(θ) = 0.71, therefore θ = 44.77°. If we consider armature current as reference phasor, then (Figure 7.42)

$$Ia = |Ia| \angle 0^o = 500\angle 0^o A$$

$$V_t = |V_t| \angle + \theta^o = 579.6\angle 44.77^o V$$

$$E_g = V_t + I_a R_a + jI_a x_{al}$$

$$E_g = 579.6\angle 44.77^o + 500\angle 0^o \times 0.75 + 500\angle 0^o \times 1.039 \times \angle 90^o$$

$$E_g = 4217\angle 49.4^o$$

$$E_{gl} = \sqrt{3} \times 4217 = 7303.8V$$

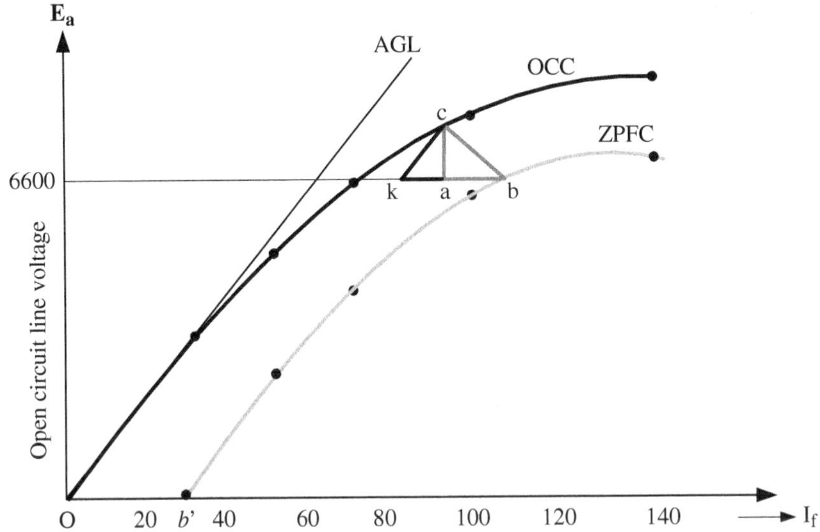

Figure E7.9 OCC and ZPFC for the problem (7.9).

Now, from the OCC, the field current corresponding to the line voltage of 7303.8*V* is found to be 95 A. Therefore, I_r is $95\angle(90^o + 49.4^o)$ which is $95\angle(139.4^o)$ A. Also, I_{ar} is equal to $25\angle 0^o$ A. The field current I_f is thus found to be

$$I_f = I_r - I_{ar} = 95\angle\left(139.4^o\right) - 25\angle 0^o = 115\angle\left(147.5^o\right) A$$

Corresponding to the field current of 115 A, the terminal voltage E_a is found from the OCC and its value is 7900 V. Therefore, the percentage voltage regulation as computed as

$$V.R = \frac{E_a - V_t}{V_t} \times 100$$

$$V.R = \frac{7900 - 6600}{6600} \times 100 = 19.7\%$$

7.10 Efficiency of the Synchronous Machine

The efficiency of the machine is defined as the ratio of power output in Watts to the power input in Watts. The input power is the sum of the output power and losses.

$$\eta = \frac{P_{out}}{P_{in}} \times 100\% = \frac{P_{out}}{P_{out} + P_{losses}} \times 100\% \tag{7.31}$$

Output power of the synchronous generator is given as

$$P_{out} = 3V_t I_a \cos(\theta) \tag{7.32}$$

Output power of the synchronous motor is given as

$$P_{out} = T\omega_s \tag{7.33}$$

Input power to a synchronous motor is given as

$$P_{in} = 3V_t I_a \cos(\theta) \tag{7.34}$$

Where T is the output torque and $\omega_s = \frac{2\pi N_s}{60}$ is the speed of machine in rad/sec.

To calculate efficiency, losses should be known. The losses in a machine are classified as

- Core Losses or Iron Losses
- Copper Losses
- Mechanical Losses, and
- Stray Losses

Core Losses: This loss takes place in the core of the machine. Core losses are comprised of hysteresis and eddy current losses. Hysteresis losses occur because of the magnetization and demagnetization of the core material. This loss depends upon the frequency and maximum flux density. This loss can be lessened by reducing the hysteresis loop area. Eddy current loss occurs because of the induced emf on the core that causes current to flow in the core material in the form of eddy. This loss is reduced by laminating the core that breaks the path of eddy current. Eddy current loss depends upon the frequency and maximum flux density.

Copper Losses: Copper loss takes place in the stator and rotor windings. This is also called I^2R loss because the loss occurs due to winding resistance. Considering the equivalent circuit of the synchronous machine, the copper loss is given by the following equation

$$P_{cu} = 3I_a^2R_a \tag{7.35}$$

The copper loss also takes place in the field winding of the machine.

Mechanical Losses: This loss takes place in the rotor side. This loss is also windage and friction losses. Windage loss occurs because the machine rotates against air friction. In a high-speed machine, windage loss is quite appreciable. Hence, for high power, high-speed machines, hydrogen is used for cooling in a closed-circuit form. Hydrogen has 14 times lower density compared to air, therefore, offering lower friction and drag. Friction loss takes place due to friction in bearing.

Stray Losses: The losses that occur because of unknown reasons are called stray losses. Some of the known reasons are leakage fluxes of the armature and variations of the flux distribution in the air gap. This loss is about 1% of the total power output.

The power-flow diagram for the synchronous generator and the synchronous motor is shown in Figure 7.40.

Example 7.10 A single-phase 220 kVA, 50 Hz, 1100 V synchronous generator has an armature resistance of 0.5 Ω. Under short-circuit condition, a field current of 25 A, gives the full-load current of 200 A. The emf generated on open-circuit with the field excitation is 800 V. Compute the synchronous impedance and reactance.

Solution

$$Z_s = \frac{\text{Open Circuit Voltage}}{\text{Short Circuit Current}} = \frac{800}{200} = 4\Omega$$

$$X_s = \sqrt{(Z_s)^2 - (R_a)^2} = \sqrt{(4)^2 - (0.5)^2} = 3.96\Omega$$

Example 7.11 Calculate the voltage regulation of the synchronous generator of Example 7.3 when delivering (a) unity power factor load, (b) lagging power factor of 0.8 and (c) leading power factor of 0.9

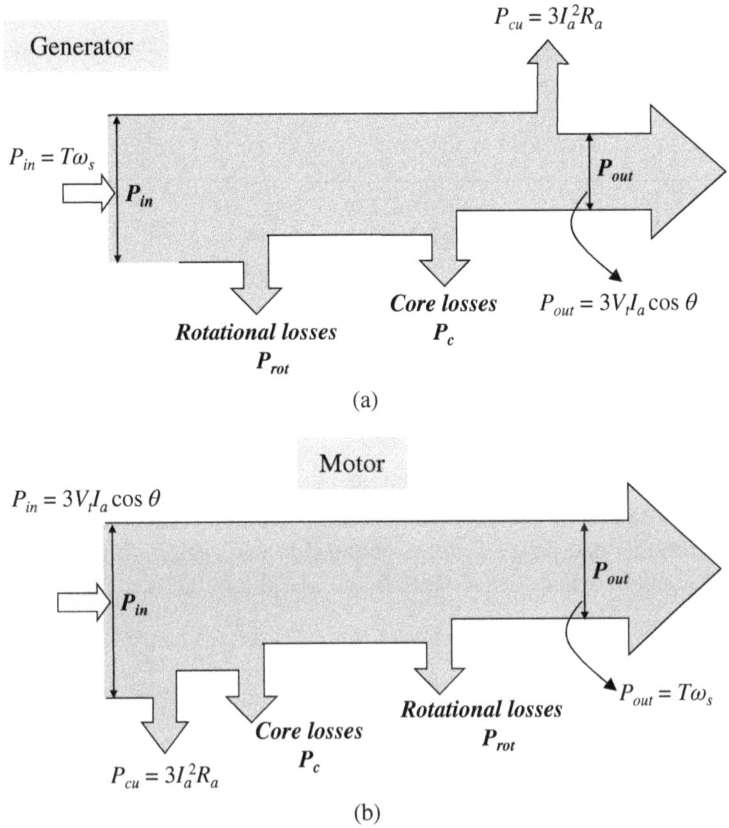

Figure 7.40 Power Flow diagram, (a) The synchronous generator, (b) The synchronous motor.

Solution

Full-load current $I_a = \frac{220 \times 10^3}{1100} = 200A$

$$R_a = 0.5\,\Omega; X_s = 3.96\,\Omega$$

(a) Using the emf equation

$$E_a = \sqrt{(V_t + I_aR_a)^2 + (I_aX_s)^2} = \sqrt{(1100 + 200 \times 0.5)^2 + (200 \times 3.96)^2}$$

$$E_a = \sqrt{(1440000) + (627264)} = 1437.8V$$

$$V.R. = \frac{E_a - V_t}{V_t} \times 100\%$$

$$V.R. = \frac{1437.8 - 1100}{1100} \times 100\% = 30.7\%$$

(b)

$$\theta = \cos^{-1}(0.8) = 36.87; \sin(36.87) = 0.6$$

$$E_a = \sqrt{(V_t\cos(\theta) + I_aR_a)^2 + (V_t\sin(\theta) + I_aX_s)^2}$$

$$E_a = \sqrt{(1100 \times 0.8 + 200 \times 0.5)^2 + (1100 \times 0.6 + 200 \times 3.96)^2}$$

$$E_a = \sqrt{(980)^2 + (1452)^2} = 1751.77$$

$$V.R. = \frac{E_a - V_t}{V_t} \times 100\% = \frac{1751.77 - 1100}{1100} \times 100\% = 59.25\%$$

(c)

$$E_a = \sqrt{\left(V_t \cos(\theta) + I_a R_a\right)^2 + \left(V_t \sin(\theta) - I_a X_s\right)^2}$$

$$E_a = \sqrt{(1100 \times 0.8 + 200 \times 0.5)^2 + (1100 \times 0.6 - 200 \times 3.96)^2}$$

$$E_a = \sqrt{(980)^2 + (-132)^2} = 988.85$$

$$V.R. = \frac{E_a - V_t}{V_t} \times 100\% = \frac{988.85 - 1100}{1100} \times 100\% = -10.1\%$$

Example 7.12 A three-phase star-connected 25 kVA, 440 V synchronous generator is supplying the rated load at 0.9 leading power factor. Calculate the excitation induced emf. Now with the same excitation, the load power factor is changed to 0.9 lagging. Compute the new terminal voltage. Given the armature resistance $R_a = 0.2\,\Omega$/phase and the synchronous reactance $X_s = 2.5\,\Omega$/phase.

Solution

Per-phase Voltage

$$V_t = \frac{440}{\sqrt{3}} = 254.03V$$

Rated current

$$I_a = \frac{S}{\sqrt{3}V} = \frac{25 \times 10^3}{\sqrt{3}(440)} = 32.8$$

$$I_a = 32.8\angle 25.84$$

The excitation voltage is

$$\begin{aligned} E_a\angle\delta &= V_t\angle 0 + I_a R_a + jI_a X_s = 254.03\angle 0 + 32.8\angle 25.84\,(0.2 + j2.5) \\ &= 254.03\angle 0 + 32.8\angle 25.84\,(2.508\angle 85.42) = 254.03\angle 0 + 82.26\angle 111.26 \\ &= 254.03 + j0 - 29.827 + j76.66 = 224.203 + j76.66 \\ &= 236.95\angle 18.87 \end{aligned}$$

Line voltage

$$E_{aline} = \sqrt{3}(236.95) = 410.4V$$

The excitation remains the same with a change in power factor

$$E_a = \sqrt{\left(V_t \cos(\theta) + I_a R_a\right)^2 + \left(V_t \sin(\theta) + I_a X_a\right)^2}$$

$$E_a^{\,2} = \left(V_t \cos(\theta) + I_a R_a\right)^2 + \left(V_t \sin(\theta) + I_a X_a\right)^2$$

$$(236.95)^2 = \left(0.9V_t + 32.8 \times 0.2\right)^2 + \left(0.435V_t + 32.8 \times 2.5\right)^2$$

$$56145.3 = V_t^2 + 83.1V_t + 6767$$
$$V_t^2 + 83.1V_t - 49378.3 = 0$$
$$V_t = (-267.61,\ \ 184.51)$$

The final choice is $V_t = 184.51$ V
Line voltage $V_{tLine} = \sqrt{3}(184.51) = 319.6V$

Example 7.13 A 440 V three-phase 50 Hz star-connected 40 kVA synchronous machine has a synchronous reactance of 3 Ω/phase and negligible armature resistance. Given the mechanical rotational losses as 0.5 kW and core losses as 1 kW. The connected load is 10 kW at a power factor of 0.8 leading. Calculate the current drawn by the machine.

Solution
Output power of the machine is $P_{out} = 10$ kW
Core loss or Iron Losses, $P_{core} = 1.0$ kW
Rotational Losses, $P_{W\&F} = 0.5$ kW
Copper Losses, $P_{cu} = 0$ (because $R_a = 0$)
Total Input power $P_{in} = P_{out} + P_{core} + P_{W\&F} + P_{cu} = 10 + 1 + 0.5 = 11.5$ kW
Line Voltage, $V_t = 440V$, pf = 0.8 leading

$$P_{in} = \sqrt{3}V_t I_a Cos(\theta)$$
$$I_a = \frac{P_{in}}{\sqrt{3}V_t \cos(\theta)} = \frac{11.5}{\sqrt{3}(440)(0.8)} = 18.86A$$

7.11 Torque and Power Curves

The synchronous generator is connected to an infinite bus bar. The terminal voltage and the speed of the machine are constant. This section derives the expression for the power transferred from the generator to an infinite bus. Also derived, is the expression for the torque that the synchronous motor can generate and deliver to the mechanical load, without losing synchronism. The power and torque expressions are derived in terms of voltages and machine parameters. The single-phase equivalent circuit of the synchronous generator supplying an infinite bus at a voltage $V_t < 0$ is shown in Figure 7.41. The bus voltage is considered as a reference phasor.

The voltages are written as

$$V_t = |V_t| \angle 0 \tag{7.36}$$
$$E_a = |E_a| \angle \delta \tag{7.37}$$
$$Z_s = R_a + jX_s = |Z_s| \angle \phi_s \tag{7.38}$$

The KVL equation in Figure 7.40 gives the following relation

$$E_a \angle \delta = V_t \angle 0 + I_a Z_s \tag{7.39}$$

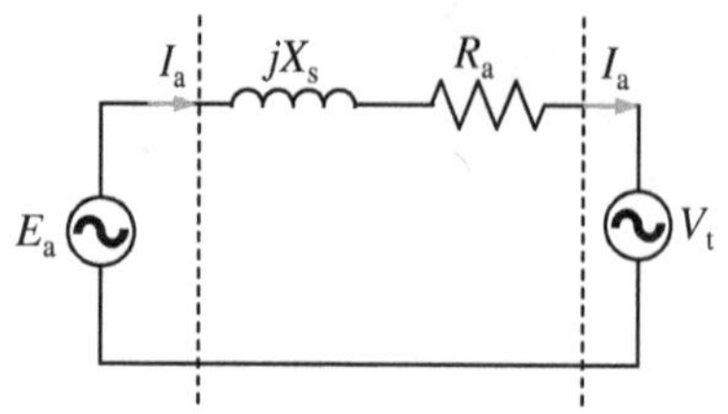

Figure 7.41 The synchronous generator supplying power to an infinite bus.

From above equation, current equation is obtained as

$$I_a = \frac{E_a\angle\delta - V_t\angle 0}{Z_s} \tag{7.40}$$

Complex Power Output from the generator is given as

$$S_{og} = P_{og} + jQ_{og} = 3V_t I_a^* \quad [I_a^* - \text{Current Conjugate}] \tag{7.41}$$

Substituting the expression of current from Eq. (7.40) to Eq. (7.41), the expression for the complex power is obtained as

$$S_{og} = 3V_t\angle 0\left(\frac{E_a\angle\delta - V_t\angle 0}{Z_s\angle\theta_s}\right)^* = 3V_t\angle 0\left(\frac{E_a}{Z_s}\angle(\delta-\theta_s) - \frac{V_t}{Z_s}\angle(-\theta_s)\right)^*$$

$$S_{og} = 3\frac{V_tE_a}{Z_s}\angle(\theta_s-\delta) - 3\frac{V_t^2}{Z_s}\angle\theta_s$$

$$S_{og} = 3\frac{V_tE_a}{Z_s}\cos(\theta_s-\delta) + j3\frac{V_tE_a}{Z_s}\sin(\theta_s-\delta) - 3\frac{V_t^2}{Z_s}\cos(\theta_s) - j3\frac{V_t^2}{Z_s}\sin(\theta_s)$$

$$S_{og} = 3\frac{V_tE_a}{Z_s}\cos(\theta_s-\delta) - 3\frac{V_t^2}{Z_s}\cos(\theta_s) + j\left(3\frac{V_tE_a}{Z_s}\sin(\theta_s-\delta) - 3\frac{V_t^2}{Z_s}\sin(\theta_s)\right) \tag{7.42}$$

7.11.1 Real/Active Output Power of the Synchronous Generator

The active power is the real part of the complex power while reactive power is the imaginary part of the complex power.

Output real power from the generator is

$$P_{og} = 3\frac{V_tE_a}{Z_s}\cos(\theta_s-\delta) - 3\frac{V_t^2}{Z_s}\cos(\theta_s) \tag{7.43}$$

Consider the impedance triangle as shown in Figure 7.42a.

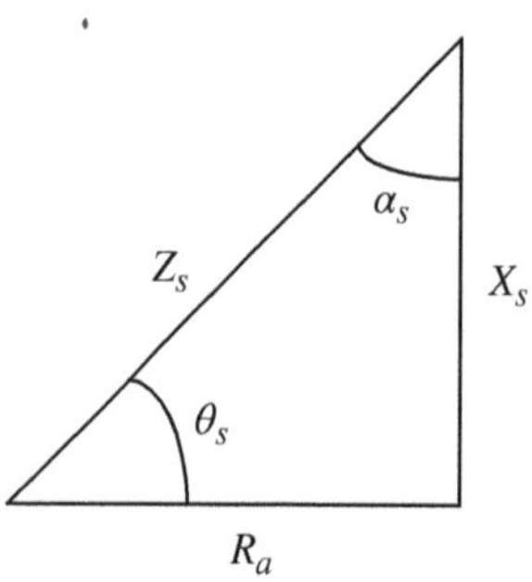

Figure 7.42a Impedance triangle.

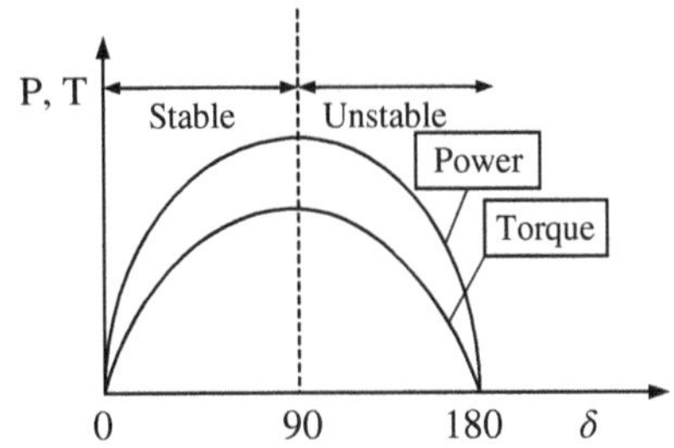

Figure 7.42b Power and torque vs torque angle for the synchronous generator.

From the impedance triangle of Figure 7.41, the angles are given are

$$\theta_s = \tan^{-1}\left(\frac{X_s}{R_a}\right); \alpha_s = 90^0 - \theta_s = 90^0 - \tan^{-1}\left(\frac{X_s}{R_a}\right); \cos\left(\theta_s\right) = \frac{R_a}{Z_s} \tag{7.44}$$

Put Eqs. (7.44) in (7.43), the following is obtained

$$P_{og} = 3\frac{V_t E_a}{Z_s}\cos\left(\theta_s - \delta\right) - 3\frac{V_t^2}{Z_s} \times \frac{R_a}{Z_s}$$

$$P_{og} = 3\frac{V_t E_a}{Z_s}\cos\left(90^0 - \left(\delta + \alpha_s\right)\right) - 3\frac{V_t^2}{Z_s^2}R_a$$

$$P_{og} = 3\frac{V_t E_a}{Z_s}\sin\left(\delta + \alpha_s\right) - 3\frac{V_t^2 R_a}{Z_a^2} \tag{7.45}$$

If the armature resistance $R_a = 0$, $\alpha_s = 0$, $|Z_s| = X_s$, the active power output of the synchronous generator is given as

$$P_{og} = 3\frac{V_t E_a}{X_s}\sin\left(\delta\right) = P_{max}\sin\left(\delta\right) \tag{7.46}$$

7.11.2 Reactive Output Power of the Synchronous Generator

The imaginary part of the complex power is the reactive power given as

$$Q_{og} = 3\frac{V_t E_a}{Z_s}\sin\left(\theta_s - \delta\right) - 3\frac{V_t^2}{Z_s}\sin\left(\theta_s\right) \tag{7.47}$$

From the impedance triangle, the following is written

$$Sin\left(\theta_s\right) = \frac{X_s}{Z_s}; \theta_s = 90^0 - \alpha_s \tag{7.48}$$

Substituting Eq. (7.48) in Eq. (7.47), the following is obtained

$$Q_{og} = 3\frac{V_t E_a}{Z_s}\sin\left(\theta_s - \delta\right) - 3\frac{V_t^2}{Z_s^2}X_s$$

$$Q_{og} = 3\frac{V_t E_a}{Z_s}\sin\left(90^0 - \left(\delta + \alpha_s\right)\right) - 3\frac{V_t^2}{Z_s^2}X_s$$

$$Q_{og} = 3\frac{V_t E_a}{Z_s}\cos\left(\delta + \alpha_s\right) - 3\frac{V_t^2}{Z_s^2}X_s \tag{7.49}$$

If the armature resistance $R_a = 0$, $\alpha_s = 0$, $|Z_s| = X_s$, the reactive power output of the synchronous generator is given as

$$Q_{og} = 3\frac{V_t E_a}{X_s}\cos(\delta) - 3\frac{V_t^2}{X_s} \tag{7.50a}$$

$$Q_{og} = 3\frac{V_t}{X_s}\left(E_a\cos(\delta) - V_t\right) \tag{7.50b}$$

$$Q_{og} = 0; E_a\cos(\delta) = V_t \Rightarrow \text{Normal Excitation} \Rightarrow \text{Unity pf operation}$$

$$Q_{og} > 0; E_a\cos(\delta) > V_t \Rightarrow \text{Over Excitation} \Rightarrow \text{Leading of operation}$$

$$Q_{og} < 0; E_a\cos(\delta) < V_t \Rightarrow \text{Under Excitation} \Rightarrow \text{Lagging pf operation}$$

7.11.3 Complex Input Power to the Synchronous Generator

Complex power input to the synchronous generator is given as

$$S_{ig} = P_{ig} + jQ_{ig} = E_a I_a^*$$

$$S_{ig} = 3E_a\angle\delta\left(\frac{E_a\angle\delta - V_t\angle 0}{Z_s\angle\theta_s}\right)^* = 3E_a\angle\delta\left(\frac{E_a}{Z_s}\angle(\delta-\theta_s) - \frac{V_t}{Z_s}\angle(-\theta_s)\right)^*$$

$$S_{ig} = 3\frac{E_a^2}{Z_s}\angle(\theta_s) - 3\frac{V_t E_a}{Z_s}\angle(\delta+\theta_s)$$

$$S_{ig} = 3\frac{E_a^2}{Z_s}\cos(\theta_s) + j3\frac{E_a^2}{Z_s}\sin(\theta_s) - 3\frac{V_t E_a}{Z_s}\cos(\delta+\theta_s) - j3\frac{V_t E_a}{Z_s}\sin(\delta+\theta_s)$$

$$S_{ig} = 3\frac{E_a^2}{Z_s}\cos(\theta_s) - 3\frac{V_t E_a}{Z_s}\cos(\delta+\theta_s) + j\left(3\frac{E_a^2}{Z_s}\sin(\theta_s) - 3\frac{V_t E_a}{Z_s}\sin(\delta+\theta_s)\right) \tag{7.51}$$

7.11.4 Real/Active Input Power to the Synchronous Generator

The real or active power is the real part of the complex power given in Eq. (7.51)

$$P_{ig} = 3\frac{E_a^2}{Z_s}\cos(\theta_s) - 3\frac{V_t E_a}{Z_s}\cos(\delta+\theta_s) \tag{7.52}$$

From the impedance triangle of Figure 7.42b, the following relations are written

$$\cos(\theta_s) = \frac{R_a}{Z_s}; \theta_s = 90^0 - \alpha_s \tag{7.53}$$

Putting Eq. (7.53) in Eq. (7.52), the following is obtained

$$P_{ig} = 3\frac{E_a^2}{Z_s}\frac{R_a}{Z_s} - 3\frac{V_t E_a}{Z_s}\cos(90^0 + \delta - \alpha_s)$$

$$P_{ig} = 3\frac{E_a^2 R_a}{Z_s^2} + 3\frac{V_t E_a}{Z_s}\sin(\delta - \alpha_s) \tag{7.54}$$

If the armature resistance $R_a = 0$, $\alpha_s = 0$, $|Z_s| = X_s$, the active power input in a synchronous generator is given as

$$P_{ig} = 3\frac{V_t E_a}{X_s}\sin(\delta) \tag{7.55}$$

7.11.5 Reactive Input Power to the Synchronous Generator

The imaginary part of the complex power is reactive power.

$$Q_{ig} = 3\frac{E_a^2}{Z_s}\sin(\theta_s) - 3\frac{V_t E_a}{Z_s}\sin(\delta + \theta_s) \tag{7.56}$$

From the impedance triangle, the following is written

$$Sin(\theta_s) = \frac{X_s}{Z_s}; \theta_s = 90^0 - \alpha_s \tag{7.48}$$

Substituting Eq. (7.48) in Eq. (7.56), the following is obtained

$$Q_{ig} = 3\frac{E_a^2}{Z_s} * \frac{X_s}{Z_s} - 3\frac{V_t E_a}{Z_s}\sin(\delta + 90^0 - \alpha_s)$$

$$Q_{ig} = 3\frac{E_a^2 X_s}{Z_s^2} - 3\frac{V_t E_a}{Z_s}\cos(\delta - \alpha_s) \tag{7.57}$$

If the armature resistance $R_a = 0$, $\alpha_s = 0$, $|Z_s| = X_s$, the reactive power output of the synchronous generator is given as

$$Q_{ig} = 3\frac{E_a^2}{X_s} - 3\frac{V_t E_a}{X_s}\cos(\delta) \tag{7.58}$$

Mechanical power input to the generator = $P_{ig} + P_{rotational}$
Torque equation is derived for the synchronous generator/motor as

$$T_e = \frac{P}{\omega_s} = \frac{3}{\omega_s}\frac{V_t E_a}{X_s}\sin(\delta) = T_{\max}\sin(\delta) \tag{7.59}$$

Power and torque vary as a sinusoidal function of δ which is called torque or power angle as shown in Figure 7.42. The synchronous generator operating in $0 \le \delta \le 90°$ is called stable region and for $\delta \ge 90°$, the region is called an unstable region.

7.12 Maximum Power Output of the Synchronous Generator

For maximum power output from the generator, the following condition is required

$$\frac{\partial P_{og}}{\partial \delta} = 0; \frac{\partial^2 P_{og}}{\partial \delta^2} < 0$$

Differentiating Eq. (7.45) wrt to δ and equating to zero

$$\frac{\partial P_{og}}{\partial \delta} = \frac{\partial}{\partial \delta}\left(3\frac{V_t E_a}{Z_s}\sin(\delta + \alpha_s) - 3\frac{V_t^2 R_a}{Z_a^2}\right) = 0$$

$$3\frac{V_t E_a}{Z_s}\cos(\delta+\alpha_s)=0$$
$$\cos(\delta+\alpha_s)=0$$
$$\delta+\alpha_s=90^\circ$$
$$\delta=90^\circ-\alpha_s=\theta_s \tag{7.60}$$

Hence, for the maximum output power from the synchronous generator, the load angle δ should be equal to the impedance angle θ_s.

For the special cases with armature resistance $R_a=0$, the power output is given by Eq. (7.46).

$$P_{og}=3\frac{V_t E_a}{X_s}\sin(\delta)=P_{max}\sin(\delta) \tag{7.46}$$

The maximum output power is achieved at $\delta=90^\circ$ and is given as

$$P_{max}=3\frac{V_t E_a}{X_s} \tag{7.61}$$

This is called steady-state power limit or static stability limit.

The terminal voltage is generally constant because the generator is connected to the infinite bus. The synchronous reactance is also remained constant, thus the maximum power can be varied by changing the value of armature induced emf E_a. The armature induced emf is varied by varying the field current. However, there is a limit to the value of the field current and the induced emf E_a.

The maximum torque, also called Pull out torque, is given as

$$T_{max}=\frac{P_{max}}{\omega_s}=3\frac{V_t E_a}{\omega_s X_s} \tag{7.62}$$

The power and torque characteristics of the synchronous generator and motor are shown in Figure 7.43.

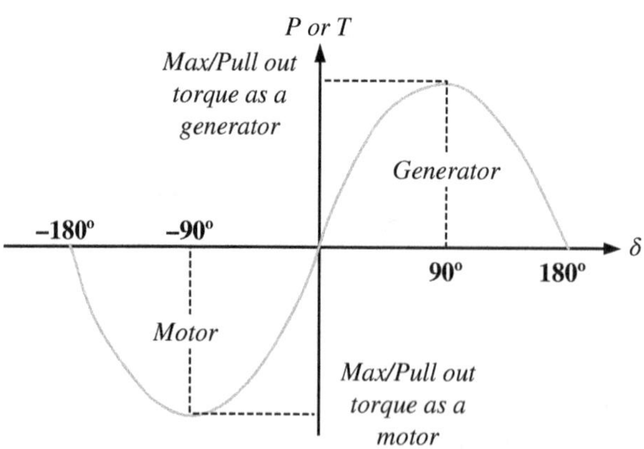

Figure 7.43 Power and torque curve for the synchronous generator and motor.

Example 7.14 A three-phase 50 Hz star-connected the synchronous motor is operated from a 6.6 kV grid with a unity power factor. The synchronous reactance of the motor is 30 Ω/phase. With a load angle of 30°, calculate the power delivered to the motor.

Solution

As the synchronous motor is operated at unity power factor, the reactive power observed by the motor is zero

$$Q = \frac{V_t^2}{X_s} - \frac{E_a V_t}{X_s} \cos \delta = 0$$

$$\frac{V_t^2}{X_s} = \frac{E_a V_t}{X_s} \cos \delta$$

$$E_a = \frac{V_t}{\cos \delta} = \frac{6.6}{\sqrt{3}X \cos 30°} = 4.4kV$$

The power delivered to motor is given by

$$P_{in} = \frac{3E_a V_t}{X_s} \sin \delta = \frac{3 \times 4.4 \times 10^3 \times 6.6 \times 10^3}{\sqrt{3} \times 30} \sin 30°$$

$$P_{in} = 838.2kW$$

Example 7.15 A three-phase star-connected 50 Hz, 440 V, four-pole the synchronous motor draws full-load current at 0.8 pf lagging. Under this condition, the ratio of the developed torque to the maximum torque is 0.5. Compute the load angle δ.

Solution

The relation between the maximum torque and developed torque is given as

$$T_e = \frac{P}{\omega_s} = \frac{3}{\omega_s} \frac{V_t E_a}{X_s} \sin(\delta) = T_{\max} \sin(\delta)$$

$$\frac{T_e}{T_{\max}} = \sin(\delta) = 0.5$$

$$\delta = 60$$

Example 7.16 A three-phase star-connected 50 Hz, 440 V, four-pole the synchronous motor is operating at a load angle $\delta = 30^0$. If the motor load is increased to two times, keeping other parameters constant, compute the new load angle.

Solution

The power is given as

$$P = P_{\max} \sin(\delta) \quad (1)$$

$$P_1 = P_{\max} \sin(\delta_1) \quad (2)$$

Dividing (2) by (1)

$$\frac{P_1}{P} = \frac{P_{\max} \sin(\delta_1)}{P_{\max} \sin(\delta)} = 2$$

$$\sin(\delta_1) = 2 \sin(\delta) = 2 \sin(24) = 0.8134$$

$$\delta_1 = 54.4^0$$

Example 7.17 A three-phase synchronous generator is connected to an infinite bus with an excitation voltage of 1.2 pu. The generator has a synchronous reactance of 1.15 pu with negligible resistance. The generator is supplying a real power of 0.7 pu to the bus. Assuming the infinite bus voltage to be 1 pu. Calculate the reactive power supplied by the generator.

Solution

Power delivered to the infinite bus

$$P = \frac{V_t E_a}{X_s} \sin(\delta)$$

$$0.7 = \frac{1 \times 1.2}{1.15} \sin(\delta)$$

$$\sin(\delta) = 0.6708$$

$$\delta = 42.13^0$$

Reactive Power Delivered

$$Q = \frac{V_t E_a}{X_s} \cos(\delta) - \frac{V_t^2}{X_s} = \frac{1 \times 1.2}{1.15} \cos(42.13) - \frac{(1)^2}{1.15} = 0.7738 - 0.8695 = -0.957 p.u$$

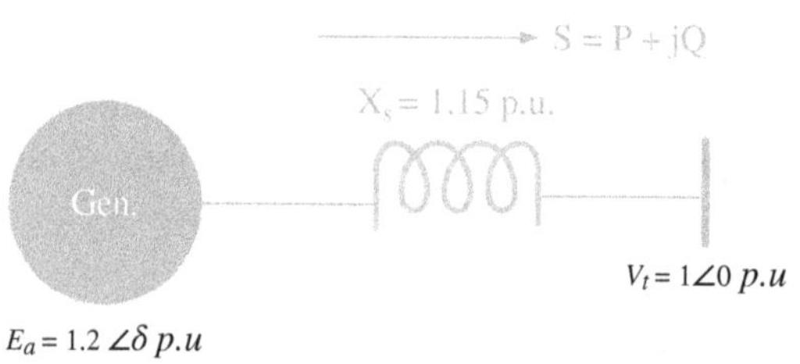

Figure E7.17 Generator supplying a load through the feeder to an infinite bus.

Example 7.18 The synchronous generator is connected to an infinite bus and is delivering half the maximum power. Compute the load angle.

Solution

$$P = P_{\max} \sin(\delta) = 0.5 P_{\max}$$

$$\sin(\delta) = 0.5$$

$$\Rightarrow \delta = 30^0$$

Example 7.19 The synchronous motor is connected to an infinite bus at 1 pu voltage and takes 0.8 pu current at unity power factor. Its synchronous reactance is 0.9 pu and resistance is negligible.

(a) Calculate the excitation voltage and load angle.
(b) Keeping the excitation constant, if the load is reduced such that the motor takes 90% of the previous current, calculate the new power factor.

Solution

(a)

$$R_a = 0; X_s = 0.9 p.u.; V_t = 1\angle 0 p.u.; I_a = 0.8\angle 0 p.u.$$

$$E_a = V_t - jI_a X_s = 1\angle 0 - j0.8\angle 0 (0.9) = 1.23\angle - 35.8$$

(b)

$$E'_a = 1.23; I'_a = 0.9I_a = 0.72p.u.$$
$$E_a^2 = \left(V_t \cos(\theta) - I_a R_a\right)^2 + \left(V_t \sin(\theta) + I_a X_s\right)^2$$
$$(1.23)^2 = (1 \times \cos(\theta))^2 + (1 \times \sin(\theta) + 0.72 \times 0.9)^2$$
$$= \cos^2(\theta) + \sin^2(\theta) + 1.3\sin(\theta) + 0.42$$
$$\Rightarrow \theta = 4.1$$
$$pf = \cos(\theta) = 0.998 lead$$

7.13 Capability Curve of the Synchronous Machine

The operation of the synchronous generator is restricted by the following factors

- Heating limit of the armature winding
- Heating limit of the field winding
- Maximum Power limit

The capability curve is a graphic representation of the limit of the operating condition. The curve is plotted in a complex power plane. The curve is derived from the phasor diagram (with $R_a = 0$). The phasor diagram with zero armature resistance is shown in Figure 7.44.

Consider the tip of V_t as the origin, the x and y axes are voltages. The curve should be drawn in a power plane; hence, the voltage needs to be changed to power by multiplying with a suitable factor.

Consider the triangle OAB, OB is the reactive voltage drop component jI_aX_s, resolving this in horizontal and vertical components

$$OA = I_a X_s \cos\left(90^0 - \theta\right) = I_a X_s \sin(\theta) --- \text{Horizontal component (voltage)}$$

$$AB = I_a X_s \sin\left(90^0 - \theta\right) = I_a X_s \cos(\theta) --- \text{Vertical component (voltage)}$$

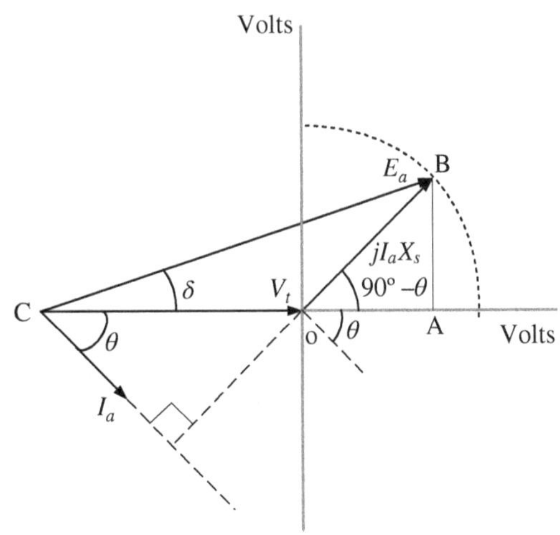

Figure 7.44 Phasor diagram of the synchronous generator with $R_a = 0$.

The origin is point O with the coordinate (0, 0) and the point C is $(-V_t, 0)$. To convert the voltage axes to power axes, use $3\frac{V_t}{X_s}$as a multiplying factor. This yield power component as

$$O'A' = I_a X_s \sin(\theta) * 3\frac{V_t}{X_s} = 3V_t I_a \sin(\theta) = Q$$

---Horizontal component (Reactive Power)

$$A'B' = I_a X_s \cos(\theta) * 3\frac{V_t}{X_s} = 3V_t I_a \cos(\theta) = P$$

---Vertical component (Active Power)

The origin is point O with the coordinate (0, 0) and the point C is $(-V_t, 0)$. Converting the voltage components to power components using the multiplying factor $3\frac{V_t}{X_s}$.

$$C' = \left(-V_t * 3\frac{V_t}{X_s}, 0\right) = \left(-3\frac{V_t^2}{X_s}, 0\right)$$

$$C'B' = E_a * 3\frac{V_t}{X_s} = 3\frac{V_t E_a}{X_s} = \varepsilon$$

$$O'B' = I_a X_s * 3\frac{V_t}{X_s} = 3V_t I_a = S --- \text{Apparent Power}$$

The voltage diagram of Figure 7.44 is translated into the power curve as shown in Figure 7.45.

The operation limit is imposed due to armature current limit (I_a) and field current (I_f) limit. The length of the phasor O'B′ representing apparent power decides the limit of armature current.

Since the terminal voltage (bus voltage) is constant hence the phasor

$$O'B' = KI_a$$

The length of the phasor C′B′ representing maximum power decides the limit of field current.

Since the terminal voltage (bus voltage) is constant and the induced emf Ea is proportional to the flux which is directly proportional to the field current and hence the phasor

$$C'B' = K'E_a = K_1 I_f$$

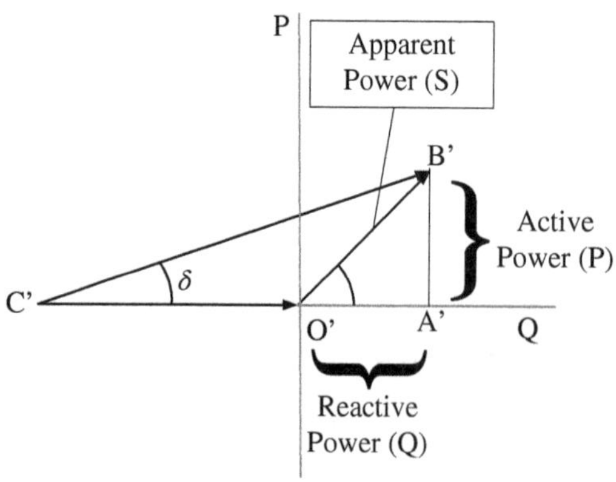

Figure 7.45 Power diagram.

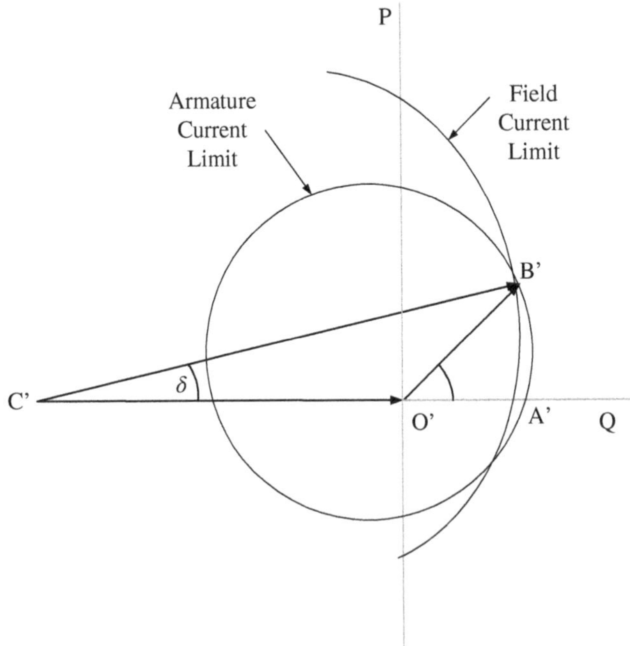

Figure 7.46 Armature and field currents locus.

Assuming the rated current and rated induced emf condition, two circular loci are drawn. The armature current cannot exceed the circular locus and the field current cannot exceed the circular locus as shown in Figure 7.46.

The operating region of the synchronous generator lies within the common region between the two circular loci or, in other words, the synchronous generator capability curve is the overlapping region of the two limiting circular loci as shown in Figure 7.47.

It is also possible to show the active power limit of the generator or maximum prime mover power or the static stability limit. This can be shown by a constant line at the limiting active power value as shown in Figure 7.48.

Example 7.20 A three-phase the synchronous machine has $R_a = 0.2\,\Omega$/phase and $X_s = 1.5\,\Omega$/phase. Given the excitation voltage $E_a = 420\angle -12$ and the terminal voltage $V_t = 440\angle 0$, is this machine operating as a generator or motor?

Calculate the amount of active and reactive power exchanged with the grid system.

Solution

Given the following data

$$V_t = 440\angle 0$$

$$E_a = 420\angle -12$$

$R_a = 0.2\,\Omega$/phase and $X_s = 1.5\,\Omega$/phase

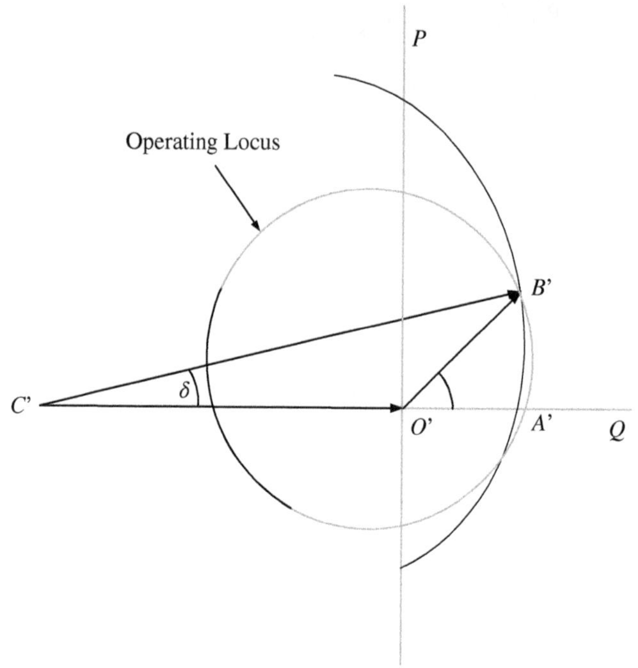

Figure 7.47 Operating locus of the synchronous generator considering rated armature and field current limits.

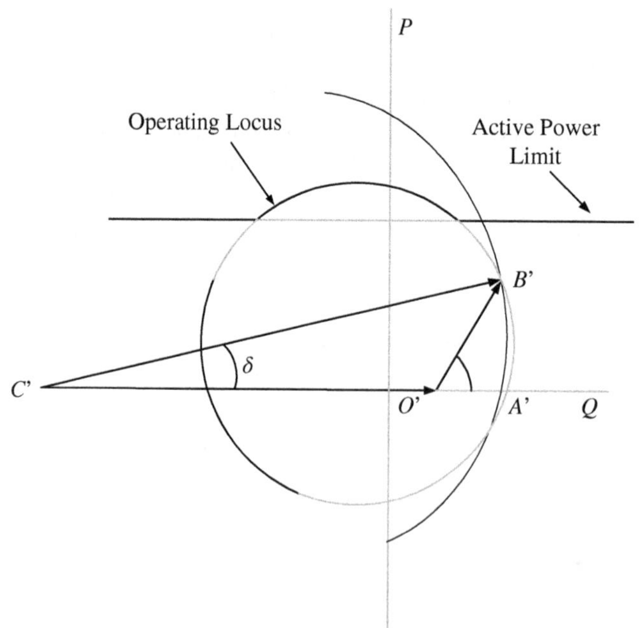

Figure 7.48 Operating locus of the synchronous generator considering rated armature, rated field current, and active power limits.

Since the load angle is $\delta = -12$, means the E_a lags V_t by 12°, hence the machine operates as a motor.

The current drawn by the motor is

$$E_a = V_t - I_a\left(R_a + jX_s\right)$$

$$I_a = \frac{V_t - E_a}{\left(R_a + jX_s\right)} = \frac{440\angle 0 - 420\angle -10}{0.2 + j1.5} = \frac{440 - 413.62 + j72.93}{0.2 + j1.5} = \frac{26.38 + j72.93}{0.2 + j1.5}$$

$$I_a = \frac{77.55\angle 70.11}{1.513\angle 82.4} = 51.25\angle -12.3$$

pf

$$\cos(\theta) = \cos(12.3) = 0.97 lag$$

Active Power

$$P = \sqrt{3}V_t I_a \cos(\theta) = \sqrt{3}(440)(51.25)0.97 = 37.886 kW$$

Reactive Power

$$\sin(12.3) = 0.213$$

$$Q = \sqrt{3}V_t I_a \sin(\theta) = \sqrt{3}(440)\,51.25\,(0.213) = 8.32 kVAr$$

7.14 Salient Pole Machine

The rotor construction of a synchronous machine are of two types; (i) cylindrical or non-salient pole and (ii) salient pole (projected pole) of dumbbell shape. The synchronous machine with a projected pole or salient pole is treated differently. A salient pole synchronous machine is illustrated in Figure 7.49.

The air-gap length is non-uniform in this type of the synchronous machine. The axis along the pole face is known as the direct axis or D-axis while the inter-polar axis is called the quadrature axis or Q-axis. The direct and quadrature axes in a two-pole and four-pole salient pole synchronous machine is shown in Figure 7.50.

The length of the air-gap along the D-axis is shorter than the length of the air-gap along the Q-axis. Since the reluctance is proportional to the length of the air-gap $\left(\mathfrak{R}_g = \frac{l_g}{\mu_0 A_g}\right)$, it is different for the two axes. The reluctance along D-axis is smaller than the reluctance along Q-axis. Due to different reluctances, the machine analysis is carried out as discussed in the subsequent section.

The inductance is defined as $\left(L = \frac{N^2}{\mathfrak{R}}\right)$ and the inductive reactance is defined as ($X_L = 2\pi fL$). This indicates high-inductive reactance for lower reluctance. Hence, the reactance along D-axis is higher than the reactance along Q-axis. The synchronous reactances along D-axis and Q-axis is defined as

$$X_{ds} = X_{ad} + X_{al} \tag{7.63}$$

$$X_{qs} = X_{aq} + X_{al} \tag{7.64}$$

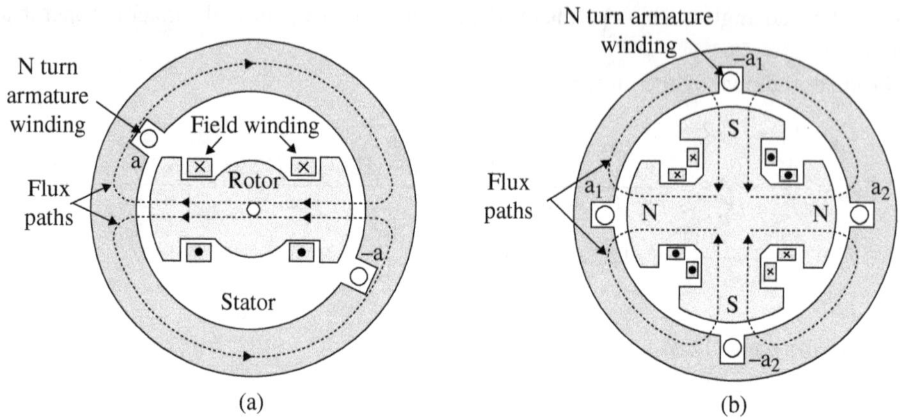

Figure 7.49 Salient pole the synchronous machine, (a) Two-pole structure, (b) Four-pole structure.

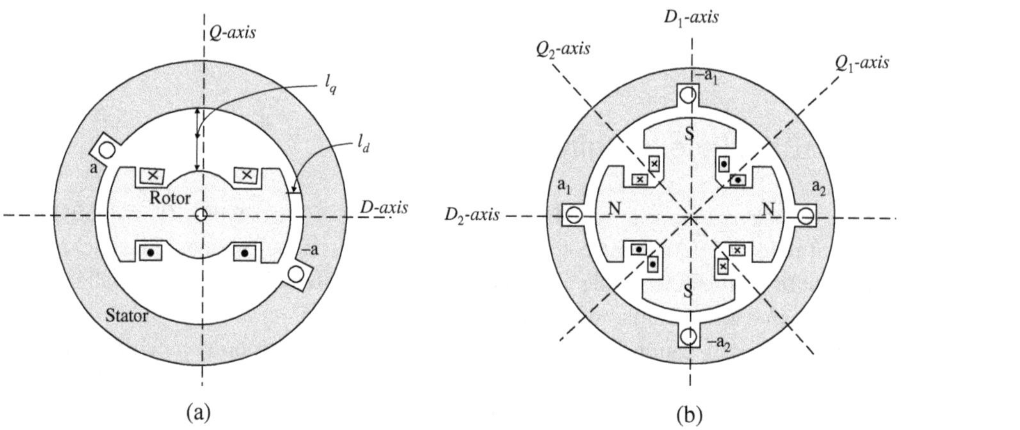

Figure 7.50 Salient pole of the synchronous machine, showing the direct and quadrature axes, (a) Two-pole structure, (b) Four-pole structure.

Where

X_{ad} is the D-axis armature reaction reactance
X_{al} is the leakage reactance
X_{aq} is the Q-axis armature reaction reactance

The leakage reactance is independent of rotor axis position since it is caused by leakage flux linking only the stator winding.

The armature current and terminal voltage are resolved into the D-axis and Q-axis

$$I_a = I_d + jI_q \tag{7.65}$$

$$V_t = V_d + jV_q \tag{7.66}$$

7.14.1 Phasor Diagram of a Salient Pole Synchronous Generator

To draw the phasor diagram, let us first write down the equation from the equivalent circuit of the cylindrical rotor machine

$$E_a\angle\delta = V_t\angle 0 + I_a Z_s \tag{7.39}$$

In this equation, the armature current is resolved into the D and Q axes and the above equation is modified as

$$E_a\angle\delta = V_t\angle 0 + I_a r_a + jI_d X_d + jI_q X_q \tag{7.67}$$

It is to be noted that the resistance part is the same as that of the cylindrical rotor machine and the reactance drops are now written along with D and Q axes. This is obvious as the reactances are only affected due to rotor saliency.

To draw the phasor diagram, first of all, the terminal voltage is drawn as a reference. The no-load induced emf E_a is located somewhere above the terminal voltage at an angle δ and along the Q-axis. This is always the case in a generator where E_a leads the terminal voltage V_t. However, the angle δ is not known but will be computed from the known values of terminal voltage, armature current, and reactances. The computation of angle δ is shown after drawing the phasor diagram (Figure 7.51).

The angle between the armature current and no-load induced emf is called the internal power factor angle and is given as

$$\phi = \theta + \delta \tag{7.68}$$

(see Figure 7.52)

From the phasor diagram, the components of currents and voltage are written as

$$\begin{aligned} I_d &= I_a \sin(\phi) = I_a \sin(\delta+\theta) \\ I_q &= I_a \cos(\phi) = I_a \cos(\delta+\theta) \end{aligned} \tag{7.69}$$

$$\begin{aligned} V_d &= V_t \sin(\phi) = V_t \sin(\delta+\theta) \\ V_q &= V_t \cos(\phi) = V_t \cos(\delta+\theta) \end{aligned} \tag{7.70}$$

From the phasor diagram, the following expression is written

$$V_q = V_t \sin(\delta) = I_q X_q - I_a R_a Sin(\phi)$$

$$V_t \sin(\delta) = I_a \cos(\delta+\theta) X_q - I_a R_a \sin(\delta+\theta)$$

$$\begin{aligned} V_t \sin(\delta) = {} & I_a X_q (\cos(\theta)\cos(\delta) - \sin(\theta)\sin(\delta)) \\ & - I_a R_a (\cos(\theta)\sin(\delta) + \sin(\theta)\cos(\delta)) \end{aligned}$$

$$\begin{aligned} & V_t \sin(\delta) + I_a X_q (\sin(\theta)\sin(\delta)) + I_a R_a (\cos(\theta)\sin(\delta)) = \\ & \quad I_a X_q (\cos(\theta)\cos(\delta)) - I_a R_a (\sin(\theta)\cos(\delta)) \end{aligned}$$

$$\sin(\delta)\left(V_t + I_a X_q \sin(\theta) + I_a R_a \cos(\theta)\right) = \cos(\delta)\left(I_a X_q (\cos(\theta)) - I_a R_a (\sin(\theta))\right)$$

$$\tan(\delta) = \frac{I_a X_q (\cos(\theta)) - I_a R_a (\sin(\theta))}{V_t + I_a X_q \sin(\theta) + I_a R_a \cos(\theta)}$$

$$\delta = \tan^{-1}\left[\frac{I_a X_q (\cos(\theta)) - I_a R_a (\sin(\theta))}{V_t + I_a X_q \sin(\theta) + I_a R_a \cos(\theta)}\right] \tag{7.71}$$

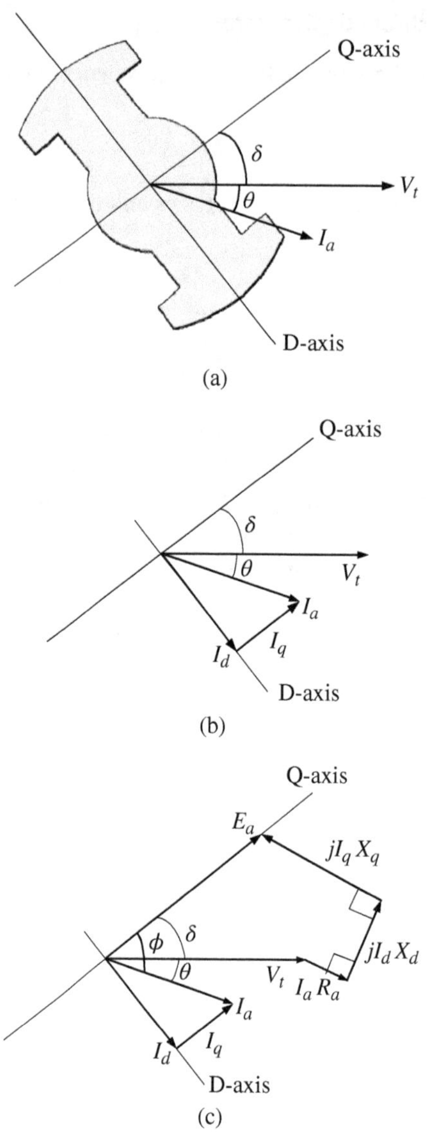

Figure 7.51 Phasor diagram of a salient pole of the synchronous generator.

If armature resistance is neglected

$$\tan(\delta) = \frac{I_a X_q (\cos(\theta))}{V_t + I_a X_q \sin(\theta)}$$

$$\delta = \tan^{-1}\left[\frac{I_a X_q (\cos(\theta))}{V_t + I_a X_q \sin(\theta)}\right] \tag{7.72}$$

The angle δ can thus be computed from the knowledge of the terminal voltage, the armature current, and the power factor angle, and then the phasor diagram can be drawn.

Phasor diagrams can be redrawn in the same fashion if D and Q axes are considered first and then location of the position of voltages and currents.

Example 7.21 A 1 MVA star-connected 2.2 kV three-phase salient-pole synchronous generator has d-axis reactance $X_d = 2\,\Omega$/phase, and $X_q = 1.5\,\Omega$/phase. Neglecting the armature resistance, calculate the excitation voltage for rated-load operation at a power factor of 0.8 lagging. Neglect all losses.

Solution

Let us redraw the phasor diagram of Figure 7.52. Let OC = E_a' which is along Q-axis and Ea. This gives the direction of excitation voltage Ea i.e. its angle is δ.

From triangle ABC, the relation can be written

$$\cos(\phi) = \frac{BC}{AC} = \frac{jI_qX_q}{AC}$$

$$AC = \frac{jI_qX_q}{\cos(\phi)}$$

$$I_q = I_a\cos(\phi)$$

$$AC = \frac{jI_qX_q}{\cos(\phi)} = \frac{jI_a\cos(\phi)X_q}{\cos(\phi)} = jI_aX_q$$

Thus, the following phasor relation can be written

OC = OA + AC

$$E_a'\angle\delta = V_t\angle 0 + jI_a\angle\theta X_q$$

Using the above relation, the load angle δ is calculated.

The per-phase terminal voltage

$$V_t = \frac{2200}{\sqrt{3}} = 1270.17V$$

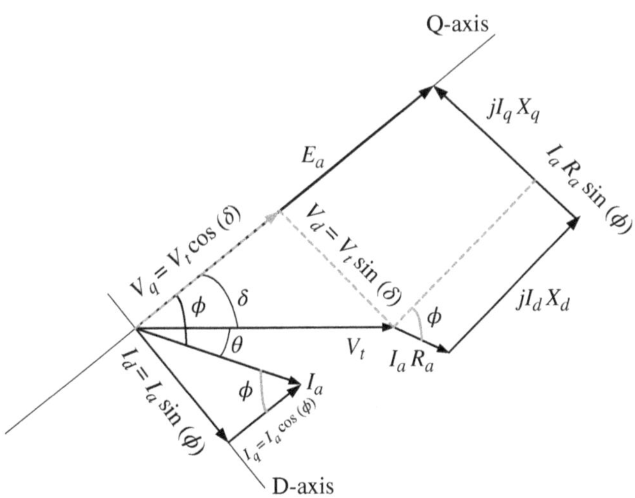

Figure 7.52 Phasor diagram of a salient pole of the synchronous generator showing all D-Q components.

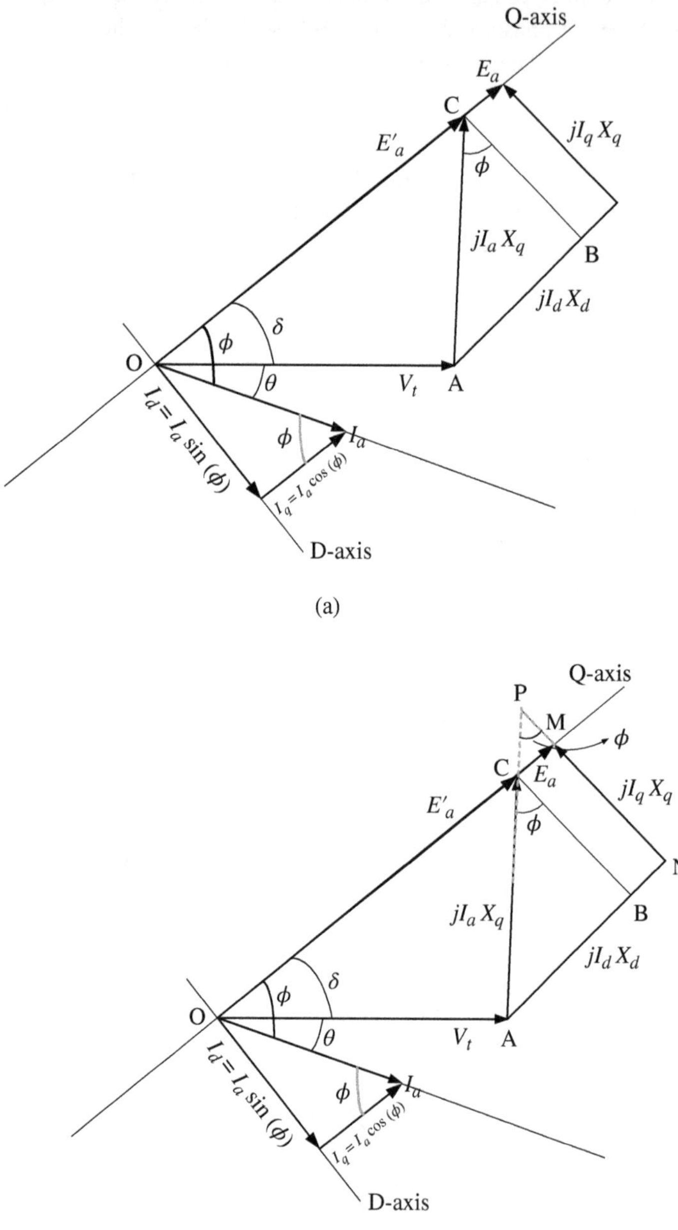

Figure E7.21 Phasor diagram of a salient pole synchronous generator for Method I (a) and (b).

The phase current is

$$I_a = \frac{P}{\sqrt{3}V_{tL}\cos(\theta)} = \frac{1\times 10^6}{\sqrt{3}(2200)(0.8)} = 328.04A$$

$$\overline{I_a} = 328.04\angle - 36.87A$$

$$E'_a\angle\delta = V_t\angle 0 + jI_a\angle\theta X_q = 1270.17\angle 0 + 328.04\angle - 36.87\,(1.5)\angle 90$$

$$E'_a\angle\delta = 1270.17 + j0 + 492.06\angle 53.13 = 1270.17 + 295.236 + j393.65$$

$$E'_a\angle\delta = 1565.406 + j393.65 = 1614.14\angle 14.11$$

The angle between I_a and E_a is thus obtained as

$$\phi = \theta + \delta = 36.87 + 14.11 = 50.98$$

The two components of the armature current are calculated as

$$I_d = I_a \sin(\phi) = 328.04 \sin(50.98) = 254.86A$$

$$I_q = I_a \cos(\phi) = 328.04 \cos(50.98) = 206.53A$$

In phasor form:

The angle of I_d is the angle between V_t and I_d
The angle of I_q is the angle between V_t and I_q

$$I_d = 254.86\angle(-90 + \delta) = 254.86\angle(-90 + 14.11) = 254.86\angle - 75.89A$$

$$I_q = 206.53\angle\delta = 206.53\angle 14.11A$$

$$E_a\angle\delta = V_t\angle 0 + jI_dX_d + jI_qX_q = 1270.17 + j254.86\angle - 75.89\,(2) + j206.53\angle 14.11\,(1.5)$$

$$E_a\angle\delta = 1270.17 + 509.72\angle 14.11 + 309.795\angle 104.11$$

$$E_a\angle\delta = 1270.17 + 494.34 + j124.26 - 75.52 + j300.45$$

$$E_a\angle\delta = 1688.99 + j424.71 = 1741.57\angle 14.11$$

Alternative Method I:
Let us first compute the load angle

$$\delta = \tan^{-1}\left[\frac{I_aX_q(\cos(\theta))}{V_t + I_aX_q\sin(\theta)}\right] = \tan^{-1}\left[\frac{328.04\,(1.5)\,0.8}{1270.17 + 328.04\,(1.5)\,0.6}\right]$$

$$= \tan^{-1}\left[\frac{393.648}{1565.406}\right] = 14.11$$

$$I_d = 254.86\angle(-90 + \delta) = 254.86\angle(-90 + 14.11) = 254.86\angle - 75.89A$$

$$I_q = 206.53\angle\delta = 206.53\angle 14.11A$$

$$E_a\angle\delta = V_t\angle 0 + jI_dX_d + jI_qX_q = 1270.17 + j254.86\angle - 75.89\,(2) + j206.53\angle 14.11\,(1.5)$$

$$E_a\angle\delta = 1270.17 + 509.72\angle 14.11 + 309.795\angle 104.11$$

$$E_a\angle\delta = 1270.17 + 494.34 + j124.26 - 75.52 + j300.45$$

$$E_a\angle\delta = 1688.99 + j424.71 = 1741.57\angle 14.11$$

Alternative Method II:

Extend the line AC to point P and form the triangle CPM with angle P equal to ϕ. Now from triangle APN, the following relation is written

$$\sin(\phi) = \frac{jX_d I_d}{AP}$$

$$AP = \frac{jX_d I_d}{\sin(\phi)} = \frac{jX_d I_a \sin(\phi)}{\sin(\phi)} = jX_d I_a$$

Now using triangle CPM, the following relation holds good

$$\sin(\phi) = \frac{CM}{CP}$$

$$CM = CP\sin(\phi) = (AP - AC)\sin(\phi) = \left(jI_a X_d - jI_a X_q\right)\sin(\phi)$$

$$CM = jI_a \sin(\phi)\left(X_d - X_q\right) = jI_d\left(X_d - X_q\right)$$

Thus, the excitation voltage is obtained as

OM = OC + CM

$$|E_a| = |E_a'| + \left|\left(X_d - X_q\right) I_d\right|$$

$$|E_a| = 1614.14 + (2 - 1.5) \times 254.86 = 1741.57V$$

7.14.2 Power Delivered by a Salient Pole Synchronous Generator

To obtain the equation for power delivered by the synchronous generators, assuming $R_a = 0$, the phasor diagram is redrawn as shown in Figure 7.53.

The terminal voltage is resolved into D and Q axes as

$$V_t = V_d + jV_q$$

$$I_a = I_d + jI_q \tag{7.73}$$

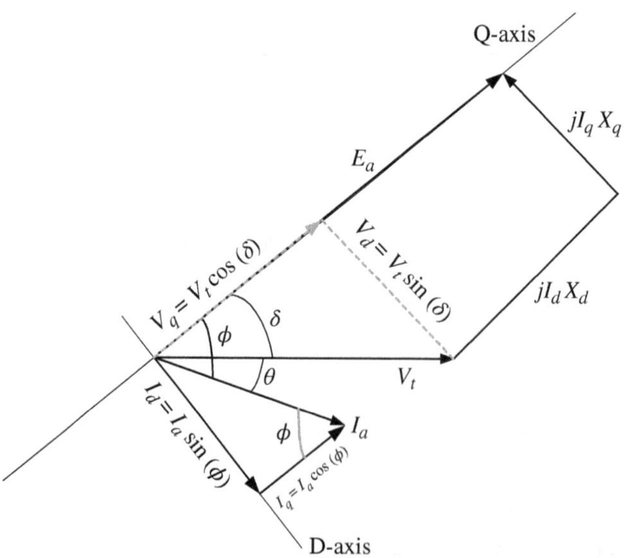

Figure 7.53 Phasor diagram of a salient pole of the synchronous generator with $R_a = 0$.

The apparent power delivered by the synchronous generator is given as

$$S = V_t I_a^*$$

$$S = \left(V_d + jV_q\right)\left(I_d - jI_q\right) = V_d I_d - jV_d I_q + jV_q I_d + V_q I_q$$

$$S = \left(V_d I_d + V_q I_q\right) + j\left(V_q I_d - V_d I_q\right) = P + jQ$$

$$P = V_d I_d + V_q I_q$$

$$Q = V_q I_d - V_d I_q$$

Active/real power delivered is further obtained as

$$P = V_d I_d + V_q I_q \tag{7.74}$$

Substituting V_{d} and V_{q} from phasor diagram

$$P = V_d I_d + V_q I_q = V_t \sin(\delta) I_d + V_t \cos(\delta) I_q \tag{7.75}$$

Further looking at the phasor diagram, the following relations are written

$$V_t \sin(\delta) = I_q X_q$$

$$V_t \cos(\delta) = E_a - I_d X_d$$

The expression for I_{d} and I_{q} is found from the above equations as

$$I_d = \frac{E_a - V_t \cos(\delta)}{X_d}$$

$$I_q = \frac{V_t \sin(\delta)}{X_q}$$

Substituting the expression of I_{d} and I_{q} in Eq. (7.75)

$$P = V_t \sin(\delta)\left[\frac{E_a - V_t \cos(\delta)}{X_d}\right] + V_t \cos(\delta)\frac{V_t \sin(\delta)}{X_q}$$

$$P = \frac{E_a V_t}{X_d}\sin(\delta) + \frac{V_t^2}{2}\left[\frac{1}{X_q} - \frac{1}{X_d}\right]\sin(2\delta) \tag{7.76}$$

For a three-phase generator, this is multiplied by a factor 3.

$$P = 3\frac{E_a V_t}{X_d}\sin(\delta) + 3\frac{V_t^2}{2}\left[\frac{1}{X_q} - \frac{1}{X_d}\right]\sin(2\delta) \tag{7.77a}$$

$$P = S_1 \sin(\delta) + S_2 \sin(2\delta) \tag{7.77b}$$

It is seen that the real power has two components; one varies as angle sin(δ) while the second term varies as sin(2δ). The second term called reluctance power has a smaller peak value. This appear due to the saliency in the machine. The graphic representation of the total active or real power delivered by a salient pole the synchronous generator is shown in Figure 7.54.

The peak value of the active power occurs at $\delta < \frac{\pi}{2}$.

Similarly, the reactive power delivered by a three-phase salient pole the synchronous generator equation is obtained as

$$Q = 3\frac{E_a V_t}{X_d}\cos(\delta) - 3\frac{V_t^2}{2X_d X_q}\left[\left(X_d + X_q\right) - \left(X_d - X_q\right)\cos(2\delta)\right] \tag{7.78}$$

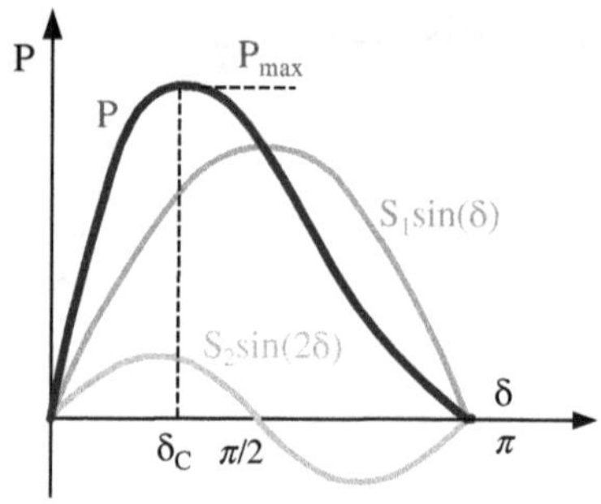

Figure 7.54 Active power vs delta curve.

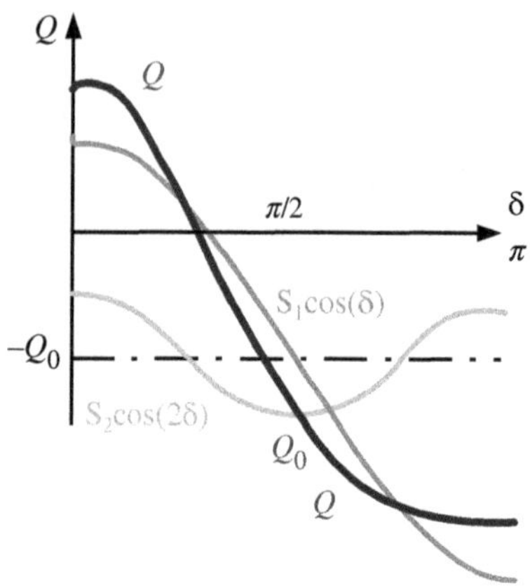

Figure 7.55 Reactive power vs delta curve.

A graphical view of the reactive power delivered is shown in Figure 7.55.

Torque equation is obtained as

$$T_e = \frac{P}{\omega_m} = \frac{P}{2\pi N_s} = \frac{1}{2\pi N_s}\left[3\frac{E_a V_t}{X_d}\sin(\delta) + 3\frac{V_t^2}{2}\left[\frac{1}{X_q} - \frac{1}{X_d}\right]\sin(2\delta)\right]$$

$$T_e = \frac{1}{2\pi N_s}\left[3\frac{E_a V_t}{X_d}\sin(\delta) + 3\frac{V_t^2}{2}\left[\frac{X_d - X_q}{2X_d X_q}\right]\sin(2\delta)\right] \tag{7.79}$$

The generated torque has two components one due to excitation and other due to reluctance.

$$T_{eE} = \frac{3E_a V_t}{2\pi N_s X_d}\sin(\delta) \tag{7.80}$$

$$T_{eX} = \frac{3V_t^2}{2\pi N_s}\left[\frac{X_d - X_q}{2X_d X_q}\right]\sin(2\delta) \tag{7.81}$$

The reluctance torque is independent of the excitation.

7.14.3 Maximum Active and Reactive Power Delivered by a Salient Pole Synchronous Generator

7.14.3.1 Active Power

Total Power for 3Φ is given as

$$P_{3\Phi} = \frac{3V_tE_a}{X_d}\sin\delta + \frac{3V_t^2}{2}\left(\frac{1}{X_q} - \frac{1}{X_d}\right)\sin 2\delta \qquad (7.77a)$$

For maximum power

$$\frac{dP_{3\Phi}}{d\delta} = 0 \qquad (7.82)$$

$$\frac{d}{d\delta}\left[P_{3\Phi}\right] = \frac{d}{d\delta}\left[\frac{3V_tE_a}{X_d}\sin\delta + \frac{3V_t^2}{2}\left(\frac{1}{X_q} - \frac{1}{X_d}\right)\sin 2\delta\right] = 0 \qquad (7.83)$$

$$\frac{3V_tE_a}{X_d}\times\cos\delta + \frac{3V_t^2}{2}\times\left(\frac{X_d - X_q}{X_qX_d}\right)\times 2\times\cos 2\delta = 0$$

$$\frac{3V_tE_a}{X_d}\times\cos\delta = -\left(\frac{3V_t^2}{2}\times\left(\frac{X_d - X_q}{X_qX_d}\right)\times 2\times\cos 2\delta\right)$$

$$E_a\cos\delta = \frac{X_q - X_d}{X_q}\cos 2\delta \qquad (7.84)$$

Putting $\cos 2\delta = 2\cos^2\delta - 1$ to above equation, the following is obtained

$$E_a\cos\delta = \frac{X_q - X_d}{X_q}\left(2\cos^2 2\delta - 1\right) \qquad (7.85)$$

Simplifying the above equation:

$$\frac{2V_t\left(X_d - X_q\right)}{X_q}\cos^2\delta - 3V_t\cos\delta - \frac{2V_t\left(X_d - X_q\right)}{X_q} = 0 \qquad (7.86)$$

The quadratic Eq. (7.86) can be solved for δ considering $a = \frac{2V_t(X_d - X_q)}{X_q}$, $b = -3V_t, and$ $c = -\frac{2V_t(X_d - X_q)}{X_q}$, the value of δ_c for maximum power is obtained as

$$\delta_c = \cos^{-1}\left[\frac{-b \pm \sqrt{b^2 - 4ac}}{2a}\right] \qquad (7.87)$$

7.14.3.2 Reactive Power

Maximum power delivered by the machine is obtained by differentiating the reactive power wrt angle δ.

$$Q = 3\frac{E_a V_t}{X_d}\cos(\delta) - 3\frac{V_t^2}{2X_d X_q}\left[\left(X_d + X_q\right) - \left(X_d - X_q\right)\cos(2\delta)\right]$$

$$\frac{dQ}{d\delta} = \frac{d}{d\delta}\left(3\frac{E_a V_t}{X_d}\cos(\delta) - 3\frac{V_t^2}{2X_d X_q}\left[\left(X_d + X_q\right) - \left(X_d - X_q\right)\cos(2\delta)\right]\right) = 0$$

$$\frac{dQ}{d\delta} = -3\frac{E_a V_t}{X_d}\sin(\delta) - 3\frac{2V_t^2}{2X_d X_q}\left[\left(X_d - X_q\right)\sin(2\delta)\right] = 0$$

$$\cos(\delta) = -\frac{E_a X_q}{2V_t\left(X_d - X_q\right)}$$

Substituting back the value of cos(δ) in the reactive power equation, the maximum reactive power delivered is obtained as

$$Q_{\max} = -3\frac{V_t^2}{X_d} - \frac{E_a^2 X_q}{4X_d\left(X_d - X_q\right)} \tag{7.88}$$

The power curve for motoring and generating modes are shown in Figure 7.56.

Example 7.22 A lossless salient pole synchronous motor has reluctance along the q-axis twice that of along d-axis and operating in such a manner as excitation voltage is equal to the terminal voltage. Calculate the ratio of reluctance torque to the electromagnetic torque developed when it is delivering a maximum load.

Solution

$$P = \frac{3V_t E_a}{X_d}\sin\delta + \frac{3V_t^2}{2}\left(\frac{1}{X_q} - \frac{1}{X_d}\right)\sin 2\delta$$

$$E_a = V_t;\ \mathfrak{R}_q = 2\mathfrak{R}_d$$

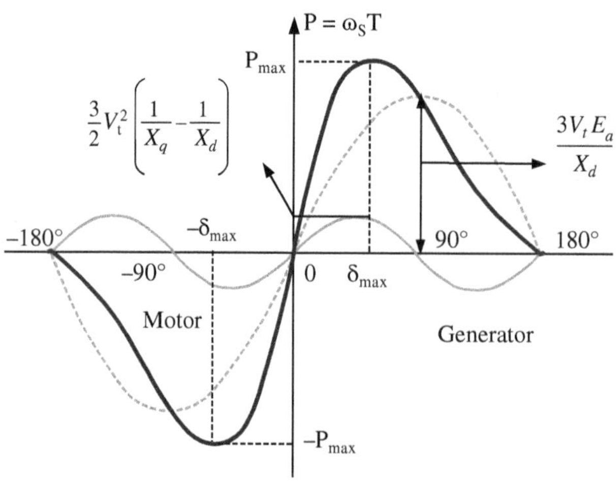

Figure 7.56 Power vs delta for motoring and generating modes.

$$X_d = \frac{\omega N^2}{\Re_d}; X_q = \frac{\omega N^2}{\Re_q}$$

$$X_q = \frac{X_d}{2}$$

$$P = \frac{V_t^2}{X_d}\sin(\delta) + \frac{V_t^2}{2X_d}\sin(2\delta)$$

Torque developed:

$$T = \frac{P}{\omega_s}$$

$$T = \frac{V_t^2}{\omega_s X_d}\sin(\delta) + \frac{V_t^2}{2\omega_s X_d}\sin(2\delta)$$

Electromagnetic Torque developed: $T_{em} = \frac{V_t^2}{\omega_s X_d}\sin(\delta)$

Reluctance Torque developed: $T_{rel} = \frac{V_t^2}{2\omega_s X_d}\sin(2\delta)$

Ratio

$$\frac{T_{rel}}{T_{em}} = \frac{\sin(2\delta)}{2\sin(\delta)} = \cos(\delta)$$

For maximum load

$$\frac{dP}{d\delta} = 0 = \frac{V_t^2}{X_s}(\cos(\delta) + \cos(2\delta))$$

$$(\cos(\delta) + \cos(2\delta)) = 0$$

$$2\cos^2(\delta) + \cos(\delta) - 1 = 0$$

$$\cos(\delta) = \frac{-1 \pm \sqrt{1+8}}{4} = \frac{3-1}{4} = \frac{1}{2}$$

$$\frac{T_{rel}}{T_{em}} = \frac{1}{2}$$

Example 7.23 A three-phase salient pole synchronous machine has $X_d = 0.95$ pu and $X_q = 0.75$ pu. What maximum percentage power this machine can deliver without loss of synchronism if the excitation voltage is zero.

Solution

The power supplied by a three-phase salient pole synchronous machine is given as

$$P_{3\Phi} = \frac{3V_t E_a}{X_d}\sin\delta + \frac{3V_t^2}{2}\left(\frac{1}{X_q} - \frac{1}{X_d}\right)\sin 2\delta$$

Since $E_a = 0$

$$P_{3\Phi} = \frac{3V_t^2}{2}\left(\frac{1}{X_q} - \frac{1}{X_d}\right)\sin 2\delta$$

For maximum power

Put $V_t = 1, X_d = 0.95, X_q = 0.75, \delta = 45$

$$P_{max} = \frac{3V_t^2}{2}\left(\frac{1}{X_q} - \frac{1}{X_d}\right) = \frac{3(1)^2}{2}\left(\frac{1}{0.75} - \frac{1}{0.95}\right) = 1.5\,(1.333 - 1.052) = 0.4215p.u.$$

Maximum power the machine will deliver is 42.15% of its rated value.

7.15 Synchronization of an Alternator with a Bus-Bar

Before discussing the parallel operation, it is important to understand the characteristics of a large power system.

The large power system behaves like a large generator having virtually zero internal impedance and infinite rotational inertia. Such a system of constant voltage and frequency, regardless of the load, is called infinite bus-bar system. The infinite bus is a power system so large that its voltage and frequency remain constant regardless of how much real and reactive power is drawn from it or supplied.

- The voltage of the infinite bus-bar is always constant because the incoming machine is too small to have any impact.
- The frequency remains constant because the rotational inertia of the connected generator is too large to have an impact on the speed of the system.

Consider a large power system comprised of a high number of synchronous generators connected in parallel and feeding the bus-bar as shown in Figure 7.57. Each generator is assumed to have synchronous impedance of Z_s. Since there are n number of generators, the equivalent impedance is $Z_{eq} = Z_s/n$, n being a high number $Z_{eq} \approx 0$, thus the voltage equation is written as

$$E_a = V_t + IZ_{eq} \tag{7.89}$$

$$E_a = V_t \tag{7.90}$$

This shows that the voltage remains constant irrespective of the load.

The moment of inertia of one generator is J, and since there is n number of generators, the total inertia is $J_{eq} = nJ$.

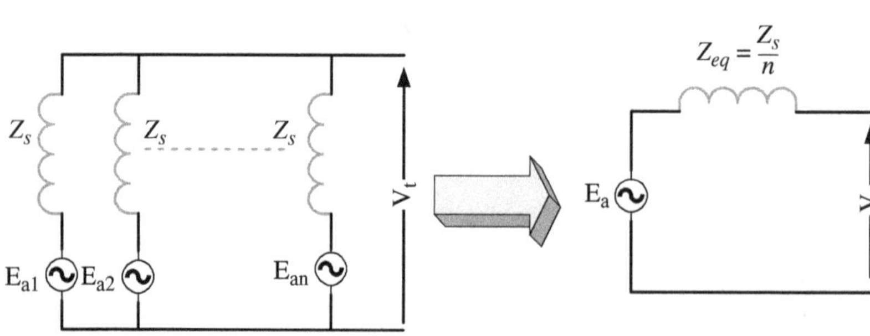

Figure 7.57 The large number of generators connected in parallel to form an infinite bus.

Thus, the torque equation is written as

$$T = J_{eq}\frac{d\omega_r}{dt} = nJ\frac{d\omega_r}{dt} \tag{7.91}$$

$$\frac{d\omega_r}{dt} = \frac{T}{nJ} \approx 0 \tag{7.92}$$

$$\omega_r = \text{constant} \tag{7.93}$$

Since n is high, the acceleration is zero and thus the speed of the generator system is constant, which indicates constant frequency.

Synchronization is the process of connecting two synchronous generators in parallel to supply a common load or connecting a synchronous generator with an infinite bus-bar. If two generators are connected in parallel, the machine already supplying the load is called, 'running machine' and the new machine to be connected is called 'incoming machine' as shown in Figure 7.58a. Parallel operation is used when the load exceeds the capacity of one generator.

For synchronizing or paralleling two generators or a generator with infinite bus-bar the following conditions require to be met

1. The voltages of the incoming machine and running machine or infinite bus-bar should be the same.
2. The frequency of the incoming machine and running machine or infinite bus-bar should be the same.
3. The phase sequence of the incoming machine and running machine or infinite bus-bar should be the same.
4. The phase shifts of the incoming machine and running machine or infinite bus-bar should be the same.

Once these conditions are met, the incoming machine is connected by means of the paralleling switches. To check if these conditions are met the following equipment is used, (i) voltmeter to check the voltages of the bus-bar and incoming machine, (ii) three lamps are connected across the lines as shown in Figure 7.58b, (iii) synchroscope to check the relative speed of both machines or incoming machine and bus-bar (in other words to check the frequency)

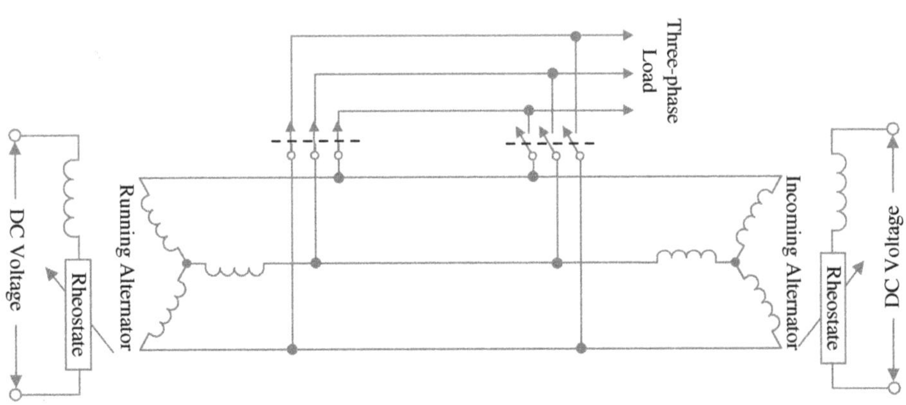

Figure 7.58 (a) Two alternators supplying a common load. (b) Synchronization procedure.

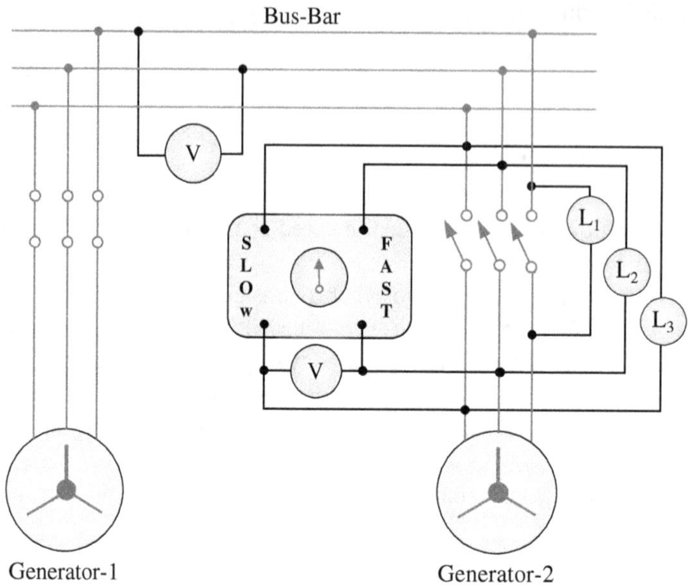

Figure 7.58 *(Continued)*

7.15.1 Process of Synchronization

To meet the above-stated conditions, the incoming machine is connected to the infinite bus-bar via a switch. Three lamps are connected across the line as shown in Figure 7.58b. This connection is used in all dark methods of synchronization.

- Voltages are measured by voltmeters; if the voltage produced by the incoming machine is lower or higher, the field excitation of this generator is changed.
- The speed of the incoming machine wrt the bus-bar is shown by the synchroscope. If the synchroscope shows slow or fast, the prime mover input is changed to fix the frequency.
- If all the lamps are dark at one instant and the synchroscope needle is at the centre, and the voltmeters are showing equal voltages, this ensures all conditions are met. The switch is closed at this moment.

Consider the following situations:

- Voltages are not the same, but frequency and phase sequence are the same.
  Note that the voltage phasors of the infinite bus are E_A, E_B, and E_C while the phasor of the incoming machine is E_a, E_b, and E_c. The voltages across the three bulbs will be E_{Aa}, E_{Bb}, E_{Cc}. As the voltages are not equal, while frequency and phase sequences are the same, the phasors will overlap each other but a small voltage difference appears across each lamp. This will cause all lamps to glow constantly with equal brightness (Figure 7.59).
- Frequencies are not the same, but voltage and phase sequence are the same.
  In this situation, at one instant of time, the two phasors are overlapping each other. However, as time passes, one phasor moves at a faster speed, causing a continuously increasing voltage across the lamps. Thus, the brightness of the lamp keeps on increasing and after

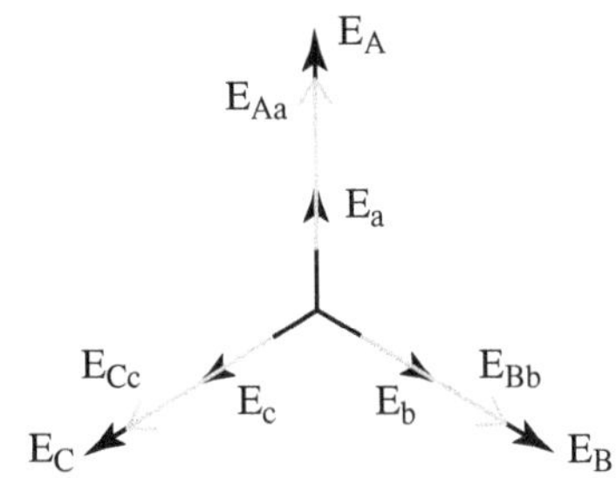

Figure 7.59 Two phasors when voltages are different.

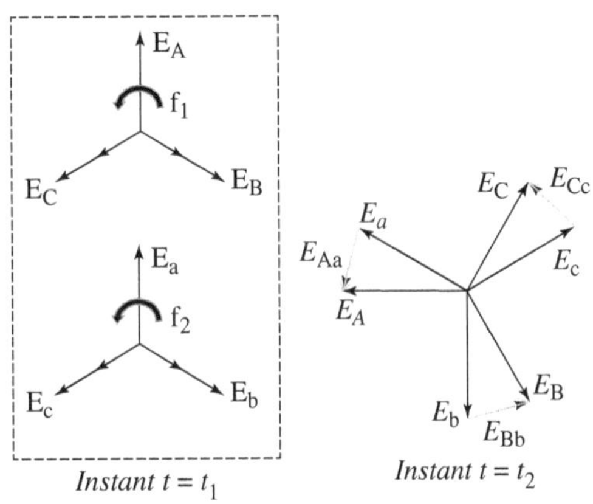

Figure 7.60 Different frequencies.

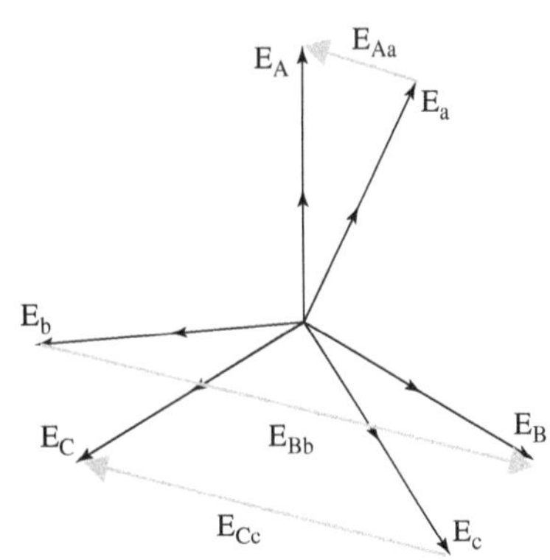

Figure 7.61 Different phase sequence connection.

reaching maximum (which can be double the rated voltage of each when both phasors are opposite each other), it starts dimming together. Thus, the rating of the bulb used should be of double voltage. In the case of high voltage, the potential transformer is used to connect the bulbs. This is illustrated in Figure 7.60.

- Phase sequence are not the same, but voltage and frequencies are the same.
  In this situation, different voltages appear across each bulb. Thus, they glow with different brightness but that remains constant. The phasors are shown in Figure 7.61.

7.16 Operation of a Synchronous Machine Connected to an Infinite Bus-Bar (Constant V_t and f)

7.16.1 Motor Operation of Change in Excitation at Fixed Shaft Power

When the synchronous machine is connected to an infinite bus-bar, the terminal voltage and frequency remain fixed and the shaft power is assumed constant. Shaft power constant means, in the case of a generator, the input prime mover power is constant and, in the case of the synchronous motor, the load torque is constant. Thus, the following conditions are written from Eq. (7.46), repeated below

$$P = 3\frac{V_t E_a}{X_s}\sin(\delta) \tag{7.46}$$

Since the terminal voltage and power are constants the power equation is written as

$$E_a \sin(\delta) = K_1 \tag{7.94}$$

Also, power is given as

$$P = 3V_t I_a \cos(\phi) \tag{7.95}$$

As the terminal voltage and power are constants, the following condition is obtained

$$I_a \cos(\phi) = K_2 \tag{7.96}$$

Based on the two conditions Eq. (7.94) and Eq. (7.96) the phasor diagram is for varying Excitation Voltage E_a. The phasor diagram for motoring operation is shown in Figure 7.62a and for generating mode is shown in Figure 7.62b. The excitation current I_f is varied to vary the Excitation Voltage E_a. The excitation current is the DC current supplied to the rotor field winding. The excitation voltage phasor slides along the locus shown in the phasor diagram such that condition Eq. (7.94) is met. Similarly, the current phasor slides along the vertical locus such that condition Eq. (7.96) is satisfied. In other words, it is said that by varying the machine excitation keeping the power constant, excitation voltage varies such that $E_a \sin(\delta)$ remains constant, and armature current varies such that $I_a \cos(\phi)$ remains constant.

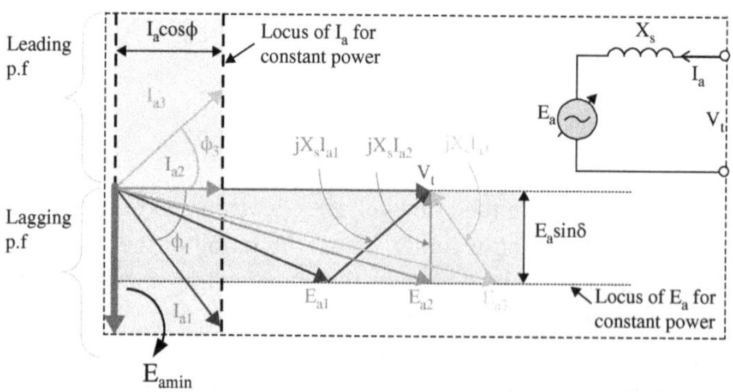

Figure 7.62 (a) Phasor diagram of the synchronous motor for varying excitation and constant power. (b) Phasor diagram of the generating mode.

Figure 7.62 (*Continued*)

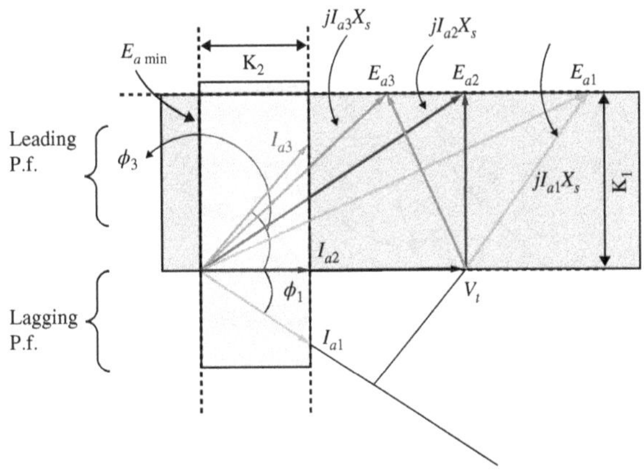

For the case of the motor

$$E_a\angle\delta = V_t\angle 0 - I_aX_s\angle(\pm\phi + 90)$$

Let us define three different excitations conditions

Normal Excitation: The excitation when the power factor is unity
Under Excitation: Amount of excitation less than the normal
Over Excitation: Amount of excitation more than the normal

For the case of generator

$$E_a\angle\delta = V_t\angle 0 + I_aX_s\angle(\pm\phi + 90)$$

Observations from the phasor diagrams (Figure 7.62a,b)
Motoring Operation:

- The motor draws a unity power factor current under normal excitation condition ($E_a = E_{a2}$)
- Motor draws lagging power factor current when under excited ($E_a = E_{a1}$)
- Motor draws leading power factor current when over-excited ($E_a = E_{a3}$)

Generator Operation:

- Generator supplies unity power factor current under normal excitation condition ($E_a = E_{a2}$)
- Generator supplies leading power factor when under excited ($E_a = E_{a3}$)
- Generator supplies lagging power factor when over-excited ($E_a = E_{a1}$)

V-Curve:

Looking at Figure 7.62, the armature current decreases as the excitation or field current is gradually increased, reaching a minimum when at normal excitation. The armature current rises further with an increase in the field excitation. The curve between armature current (I_a) and field current (I_f) at constant shaft power, is called the V-curve as shown in Figure 7.63.

- The curve between armature current and field current gives V-curves of the synchronous motor.

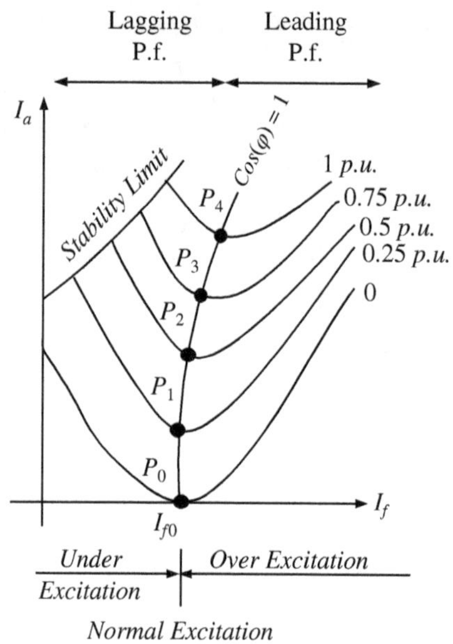

Figure 7.63 V-curve the synchronous motor.

- V-curves show that armature current is minimum at the unity power factor.
- At leading or lagging power factors, armature current is more than its value at a unity power factor.
- Increasing the field current, above the level for minimum armature current, causes the motor to operate at leading pf
- The motor is now overexcited and delivers reactive power to the system.

V-Curve can be physically explained from the fact, that the terminal voltage V_t remains fixed, and hence the flux should remain fixed. The flux has two components; main field flux (due to field excitation I_f) and armature flux (due to load current I_a). Under normal excitation conditions, the field current is sufficient to provide the total flux demanded by the terminal voltage, and hence no magnetizing current is drawn by the armature, and the machine operates at unity power factor.

When field excitation is reduced from normal, the flux produced by the field current is reduced, and to make up the deficiency, the armature draws magnetizing current and hence the machine operates at lagging power factor.

In the over-excited condition, the field current produces a higher amount of flux. To reduce the flux, the armature draws demagnetizing current, and thus the machine operates at leading power factor.

The characteristic curve between power factor and excitation current is called inverted V-curve as shown in Figure 7.64. It is seen from Figure 7.62 that the power factor angle decreases when excitation is increased, reaches zero, and then increases further. This causes the power factor to increase, reach unity, and then decreases as shown in Figure 7.64.

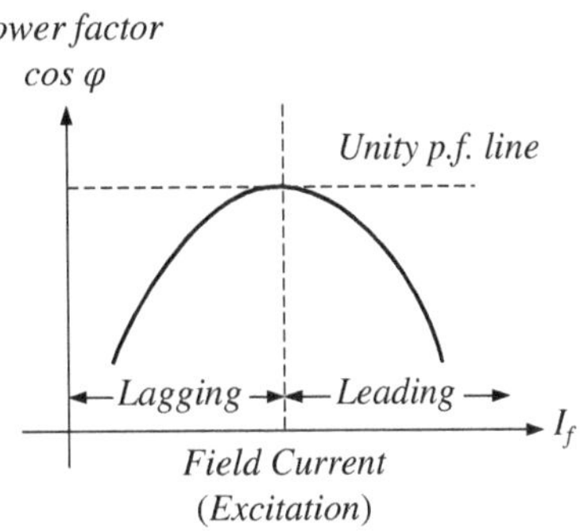

Figure 7.64 Inverted V-curve.

The synchronous Condenser:

The synchronous motor running under the no-load condition with variable excitation is called the synchronous condenser. The power factor of the current drawn by the motor is controlled simply by changing the field excitation. Reactive power is injected into the grid when there is a deficiency by running the synchronous motor in an over-excited condition. Reactive power is absorbed from the grid by running the motor in under an excited condition. Thus, the smooth variation of reactive power is obtained.

7.16.2 Generator Operation for Change in Output Power at Fixed Excitation

Consider the synchronous generator supplying a load as shown in Figure 7.65. Assuming excitation is constant (hence no change in induced emf E_a) and there is an increase in the load at the same power factor. The voltage relationship in the synchronous generator is given as

$$E_a \angle \delta = V_t \angle 0 + I_a X_s \angle (\pm\phi + 90)$$

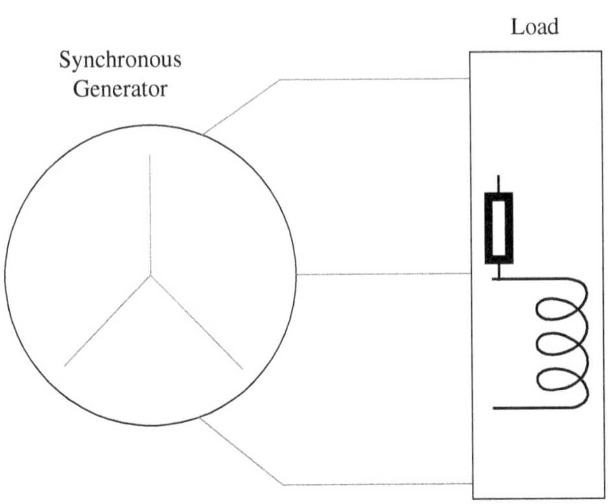

Figure 7.65 A generator supplying a load.

When the load is increased, the magnitude of induced emf does not change because excitation remains constant. The induced emf slides along the circular periphery with a shift in the load angle (increased load angle δ'). The reactance voltage drop increases without change in the power factor angle. The current is increased at the same power factor angle. The phasor diagrams for lagging, unity, and leading power factor conditions are shown in Figure 7.66.

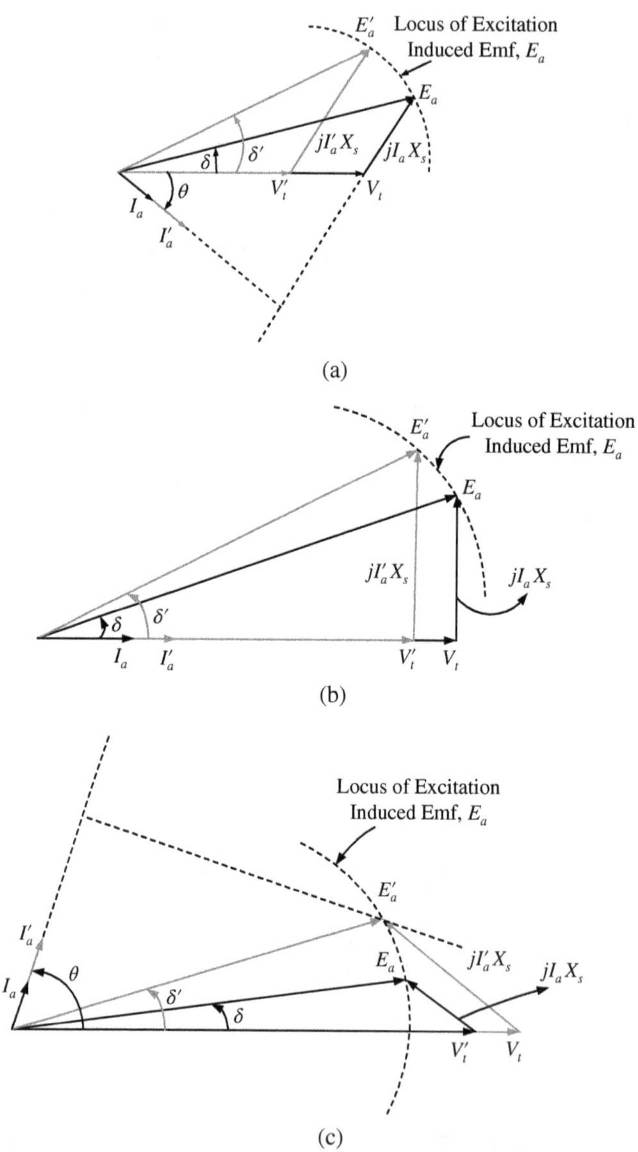

Figure 7.66 Phasor diagram of the synchronous generator supplying an isolated load, (a) lagging power factor, (b) unity power factor, and (c) leading power factor.

The following conclusion is drawn from the phasor diagrams

- When lagging load is added and increased (higher inductive reactive power demand, +Q), terminal voltage reduces significantly.
- When the unity power factor load is increased, there is a slight drop in the terminal voltage.
- When the leading load is increased (higher capacitive reactive power demand, −Q), the terminal voltage increases.

Example 7.24 A three-phase, 400 V, 5 kW, star connected the synchronous motor having a synchronous reactance of 10 Ω/phase is operating at 50% load, unity pf. Now, the excitation is increased by 1%. What will the new load be in percent, if the power factor is to be kept the same? Neglect all losses and consider linear magnetic circuit.

Solution

Full-load current

$$I_a = \frac{P}{\sqrt{3} \times V_t \times pf} = \frac{5 \times 10^3}{\sqrt{3} \times 400 \times 1} = 7.22A$$

$$E_a = \sqrt{\left(V_t \cos\theta - I_a R_a\right)^2 + \left(V_t \sin\theta + I_a X_s\right)^2} \quad (1)$$

With unity power factor and negligible power loss, (1) simplifies as

$$E_a = \sqrt{(V_t)^2 + (I_a X_s)^2} \quad (2)$$

$$= \sqrt{\left(\frac{400}{\sqrt{3}}\right)^2 + (10 \times 3.6)^2}$$

$$= 233.73V$$

Excitation will increase 1% then E_{a2}

$$E_{a2} = 233.73 \times 1.01 = 236V$$

From (2)

$$I_a = \frac{\sqrt{E_{2a}^2 - V_t^2}}{X_s} = \frac{\sqrt{(236)^2 + \left(\frac{400}{\sqrt{3}}\right)^2}}{10} = 4.8932A$$

Therefore, the new load % is given by

$$Load\,(\%) = \frac{4.8932}{7.22} \times 100 = 67.83\%$$

Example 7.25 A three-phase 440 V, 10 kW, the delta-connected synchronous motor has a synchronous reactance of 10 Ω/phase and is operating at full-load unity power factor. If the excitation is increased by 1%, what will the new load current be if the power factor remains the same? Neglect all losses.

Solution

$$P = \sqrt{3}V_t I_a \cos(\theta)$$

$$I_a = \frac{P}{\sqrt{3}V_t \cos(\theta)} = \frac{10 \times 10^3}{\sqrt{3}(440)(1)} = 13.12A$$

The excitation voltage for this condition

$$E_a^2 = V_t^2 + (I_a X_s)^2 = (440)^2 + (13.12 \times 10)^2 = 210817.63$$

$$E_a = 459.15V$$

The excitation is increased by 1%, keeping the same power factor, the excitation voltage is

$$E_a' = 1.01E_a = 1.01\,(459.15) = 463.74V$$

$$(E_a')^2 = (V_t)^2 + (I_a' X_s)^2$$

$$(463.74)^2 = (440)^2 + (10I_a')^2$$

$$215054.78 = 193600 + 100(I_a')^2$$

$$I_a' = 14.64A$$

Example 7.26 A 440 V three-phase star-connected synchronous motor draws 40 A at a unity power factor from a 440 V power grid. The field excitation current is 3 A and the synchronous reactance is 2 Ω/phase. Assuming linear open-circuit characteristics

(a) Calculate the per-phase excitation induced emf.
(b) What is the torque angle?
(c) What is the stability limit under this condition?
(d) How much field current is required to operate the motor at 0.8 pf leading?
(e) Compute the new torque angle in part (d).
(f) How much field current is required to operate the motor at 0.8 pf lagging?

Solution

(a) The terminal voltage per-phase is

$$V_t = \frac{440}{\sqrt{3}} = 254.03V$$

$$E_a = V_t - I_a\left(R_a + jX_s\right) = 254.03 - 40\angle 0\,(2\angle 90) = 254.03 - 80\angle 90$$

$$E_a = 254.03 - j80 = 266.33\angle -17.48$$

Per-phase excitation induced emf = 266.33 V

(b) Load angle $\delta = -17.48$
(c) The stability limit is the maximum power developed by the machine

$$P_{\max} = \frac{3V_t E_a}{X_s} = \frac{3\,(254.03)\,(266.33)}{2} = 101.483kW$$

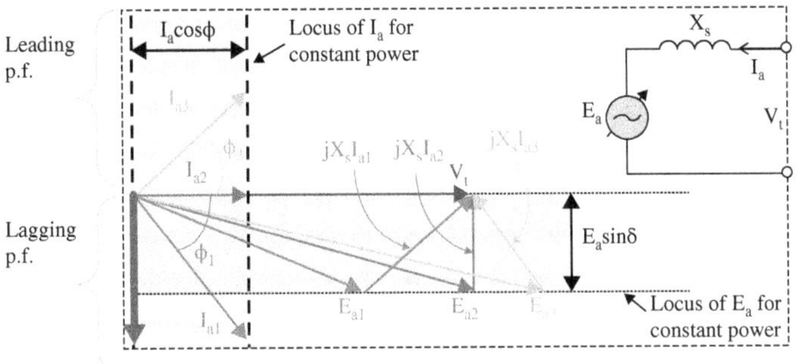

Figure E7.26 Phasor diagram of a synchronous motor for variable excitation at constant power.

(d) Consider the phasor diagram when the motor is drawing a unity power factor and then start drawing leading current with constant power

Since the power supplied to the motor is constant, $I_a \cos(\theta)$ it remains constant.

$$I_{a2}\cos(\theta_2) = I_{a1}\cos(\theta_1)$$
$$I_{a2}(0.8) = 40(1)$$
$$I_{a2} = \frac{40}{0.8} = 50\angle 36.87$$

The new terminal voltage is

$$E_{a2} = V_t - I_{a2}(jX_s) = 254.03 - 50\angle 36.87(2\angle 90) = 254.03 - 100\angle 126.87$$
$$E_{a2} = 254.03 + 60 - j80 = 314.03 - j80 = 324.06\angle - 14.3$$

The excitation voltage is proportional to the field current and the field flux is proportional to the field current.

$$E_{a2} \infty \psi_{f1};\ \psi_{f1} \infty I_{f1}$$
$$E_{a2} \infty \psi_{f2};\ \psi_{f2} \infty I_{f2}$$
$$\frac{I_{f2}}{I_{f1}} = \frac{E_{a2}}{E_{a1}}$$
$$I_{f2} = \left(\frac{E_{a2}}{E_{a1}}\right) I_{f1} = \left(\frac{324.06}{266.33}\right) \times 2 = 2.43A$$

Thus, the new field current required is 2.43 A.

(e) The new torque angle is –14.3°
(f) The new field current is computed in the same way as that of part (c)

$$I_{a2}\cos(\theta_2) = I_{a1}\cos(\theta_1)$$
$$I_{a2}(0.8) = 40(1)$$
$$I_{a2} = \frac{40}{0.8} = 50\angle - 36.87$$
$$E_{a2} = V_t - I_{a2}(jX_s) = 254.03 - 50\angle - 36.87(2\angle 90) = 254.03 - 100\angle 53.13$$
$$E_{a2} = 254.03 - 60 - j80 = 194.03 - j80 = 209.87\angle - 22.4$$

The new field current

$$\frac{I_{f2}}{I_{f1}} = \frac{E_{a2}}{E_{a1}}$$

$$I_{f2} = \left(\frac{E_{a2}}{E_{a1}}\right) I_{f1} = \left(\frac{209.87}{266.33}\right) \times 2 = 1.57A$$

7.17 Hunting in the Synchronous Motor

Hunting is the phenomenon of oscillation of the synchronous motor rotor about its equilibrium point when a sudden load is applied to the motor.

Considering the no-load condition, where stator and rotor are locked to each other and continue to rotate. The load angle $\delta = 0$ under the no-load condition as shown in Figure 7.67. Under this condition, electromagnetic torque produced by the motor is zero (note, a small torque will be produced by the motor to overcome the no-load losses, but for simplicity of explanation, it is assumed zero). The electromechanical equation of motor-mechanical load combination is given as

$$T_e - T_L = J\frac{d\omega_r}{dt} + B\omega_m \tag{7.97}$$

Where

T_e is the electromagnetic torque produced by the motor
T_L is the load torque applied to the motor in the opposite direction to T_e
J is the inertia of the motor load combination
B is the friction and damping constant of the load.

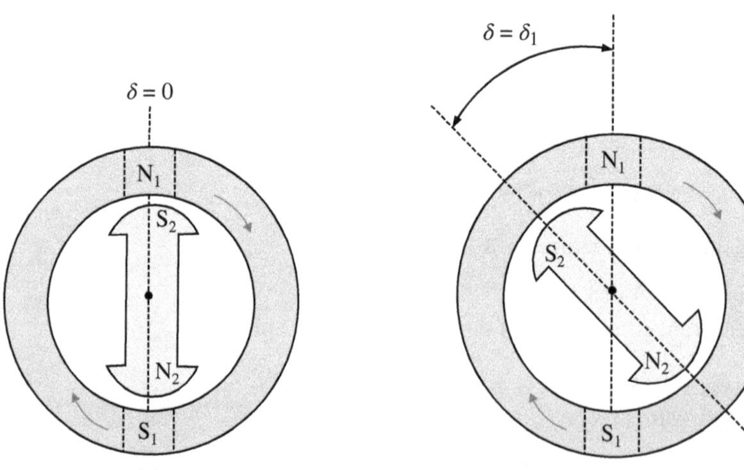

Figure 7.67 The synchronous motor operation, (a) no-load condition, $\delta = 0$, (b) loaded condition $\delta = \delta_1$.

If $B = 0$, the above equation is written as

$$T_e - T_L = J\frac{d\omega_r}{dt} \tag{7.98}$$

Three different situations are possible

$T_e = T_L$; in this condition, the acceleration $\frac{d\omega_r}{dt} = 0$ and speed ω_r is constant. This is called the equilibrium steady-state condition.
$T_e < T_L$; The motor decelerates and speed reduces, $\frac{d\omega_r}{dt} < 0$
$T_e > T_L$; The motor accelerates and the speed increases, $\frac{d\omega_r}{dt} > 0$

Consider a sudden-load torque is applied to the motor, the equilibrium situation is disturbed, T_e becomes less than T_L and the motor speed drops below the synchronous speed, the rotor falls back and the load angle increases as shown in Figure 7.67.

As soon as the load angle increases, the electromagnetic torque increases since $T_e = \frac{V_t E_a}{\omega_r X_s}\sin(\delta) = K\sin(\delta)$ and as δ increases, the torque increases. Thus, the increased torque once again balances the load torque demand. However, due to inertia of the motor-load combination, the rotor does not stop at the required position δ_1, rather it overshoots. Due to overshooting of the rotor position from δ_1 (the load angle becomes more than the required value), the electromagnetic torque produced is more than the required value. Hence, the motor accelerates and it reaches the synchronous speed. However, once again, it does not stop at this point but it overshoots. This phenomenon can be understood from Figure 7.68. Thus, there is a constant oscillation of rotor about its equilibrium position, called hunting. The oscillation will be sustained if there is no damping. However, if there is suitable damping, these oscillations damp out.

Impact of Hunting

- It causes severe mechanical stress on the motor shaft
- There is huge surge in current and power
- There is increased machine losses and increased temperature rise
- The machine may lose synchronism

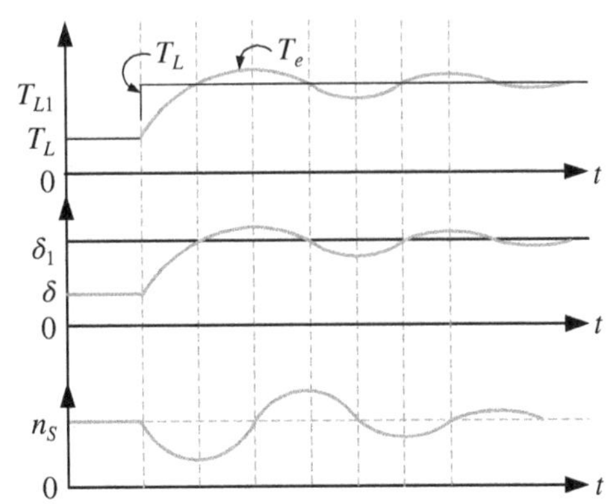

Figure 7.68 Hunting in the synchronous motor.

Reduction in Hunting

- By damper winding or amortisseur winding
- By designing a motor with appropriate synchronizing power coefficient or stiffness
- By using flywheel (flywheel is a mechanical energy device)

7.17.1 Role of the Damper Winding

When the motor rotates at the synchronous speed under steady-state condition, there is no torque produced by the damped winding. As soon as the speed of the motor is reduced due to the sudden application of load, there will be a relative motion between the stator field (running at the synchronous speed) and the rotor as the rotor speed becomes lower than the synchronous speed. Thus, this relative speed causes an induction effect in the damper winding. It produces the torque in the motor that causes the motor to reach the synchronous speed quickly. Thus, the oscillation is reduced.

7.18 Parallel Operation of Synchronous Generators

Two or more generators connected in parallel to supply a common load is called the parallel operation of the synchronous generators. Parallel operation is needed when the load demand increases and one generator is not enough to supply the load demand. The power share among the two or more generators is controlled using the adjustment in their frequency-power curve no-load set-up point.

The frequency-power curve of a generator is made from a linear curve using Governor system installed with the generator. A typical frequency-power curve of a synchronous generator is shown in Figure 7.69. The characteristic is called a drooping curve. At no-load condition, the frequency is f_{nl} which drops to a few percent (2–5%) to f_{fl} at full-load shown in Figure 7.69a. The same curve in terms of speed of generators is shown in Figure 7.69b. The no-load set point f_{nl} or n_{nl} is adjusted using the governor in order to obtain the desired frequency under loaded condition.

The speed droop (SD) is defined by the equation:

$$SD = \frac{n_{nl} - n_{fl}}{n_{fl}} \times 100\%$$

$$SD = \frac{f_{nl} - f_{fl}}{f_{fl}} \times 100\% \tag{7.99}$$

Since the rotor speed is related to the electrical frequency of the generator, the power output of the synchronous generator in terms of frequency is given as

$$P = m_p \left(f_{nl} - f\right) \tag{7.100}$$

Where

m_{p} is the slope of the frequency-power curve given in MW/Hz or kW/Hz.
f_{nl} is the no-load frequency
f is the system operating frequency at any load

Figure 7.69 (a) Frequency-power curve of a synchronous generator. (b) Speed-power curve of a synchronous generator.

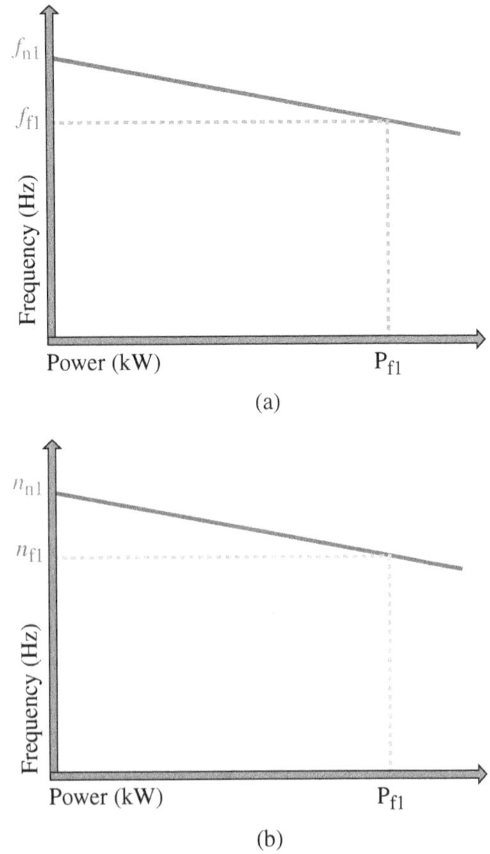

Similarly, a relationship exists between the terminal voltage of the synchronous generator and reactive power, called voltage-reactive power curve as shown in Figure 7.70. In the diagram, negative means leading reactive power and positive is the lagging reactive power. When lagging load is increased, the voltage decreases. When the leading load is added, the terminal voltage increases.

- This characteristic is not essentially linear however, voltage regulators are used to making this curve linear.
- The characteristic curve can be moved up and down by changing the no-load terminal voltage set point from the voltage regulator.
- As with the frequency-power characteristic, this curve plays an important role in the parallel operation of the synchronous generators.
- It is important to note that when a single generator is supplying a load, the real power P and reactive power Q supplied by the generator will be the amount demanded by the load attached to the generator – the P and Q supplied cannot be controlled by the generator's controls.
- Therefore, for any given real power, the governor set points to control the generator's operating frequency f and for any given reactive power, the field current controls the generator's terminal voltage V_t.

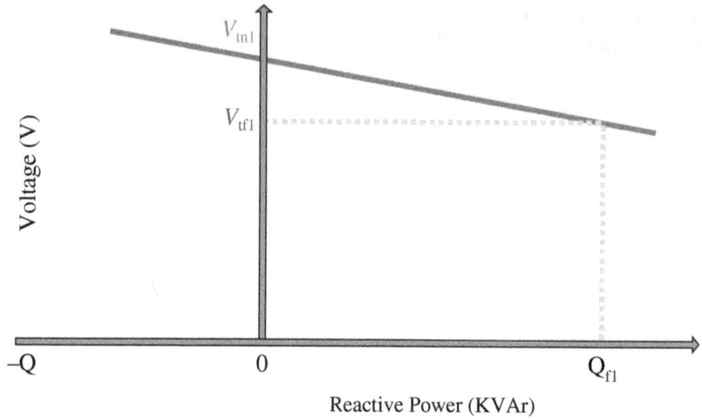

Figure 7.70 Voltage-reactive power curve.

7.18.1 The Synchronous Generator Operating in Parallel with the Infinite Bus Bar

Consider a synchronous generator is connected to an infinite bus and is supplying a load shown in Figure 7.71.

- When a generator is connected in parallel with another generator or an infinite bus, *the frequency and terminal voltage of all the machines must be the same. The frequency and voltage of the generators are dictated by the infinite bus.*
- The frequency-power characteristics of two machines (connected in parallel) or machine and infinite bus (connected in parallel) are plotted back to back, with a common vertical axis.
- The voltage-reactive power characteristics of two machines (connected in parallel) or machine and infinite bus (connected in parallel) are plotted back to back, with a common vertical axis.
- Such a sketch is called a *house diagram*.

The house diagram is shown in Figure 7.72 for a generator connected in parallel with an infinite bus. The generator curve is drooping in nature (the drop-in frequency with an

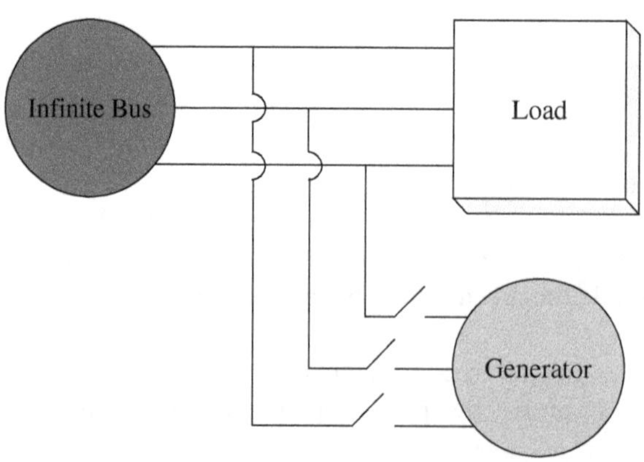

Figure 7.71 A synchronous generator connected in parallel with an infinite bus.

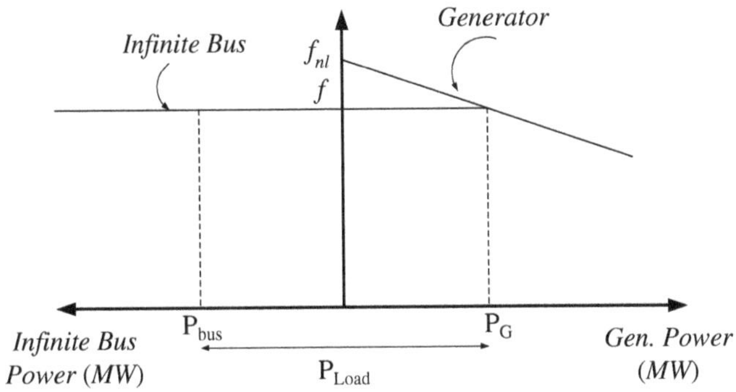

Figure 7.72 Generator and infinite bus are connected in parallel.

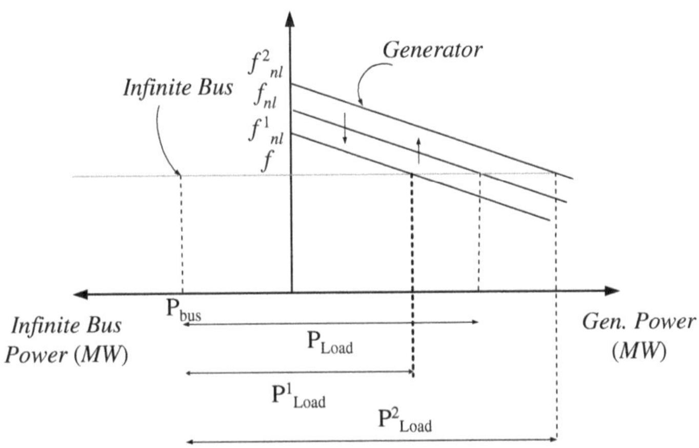

Figure 7.73 Change in power share of the generator operating in parallel with an infinite bus.

increase in the output power) while the infinite bus is a straight line (no change in frequency with a change in output power).

If the power share of the generator needs to be increased, the curve is shifted upward by increasing the no-load frequency value. Similarly, if the power share of the generator needs to be decreased, the curve is shifted downward by reducing the no-load frequency set point. This is explained in Figure 7.73.

If the two generators are connected in parallel and supplying a common load, their frequency-power characteristics are drawn back to back against a common vertical axis, called a house diagram as shown in Figure 7.74. The no-load frequency set points are f_{nlG1} and f_{nlG2} for Generator 1 and Generator 2, respectively. The operating frequency of the system is f, where Generator 1 is supplying P_{G1} and Generator 2 is supplying P_{G2} with a total load of $P_{load} = P_{G1} + P_{G2}$.

The power share by two generators and the operating frequency of the system is adjustable by changing the set points of both generators as shown in Figure 7.75. Consider a situation where the power share of Generator 1 is increased with a simultaneous reduction in power share of Generator 2, without changing the operating frequency. To

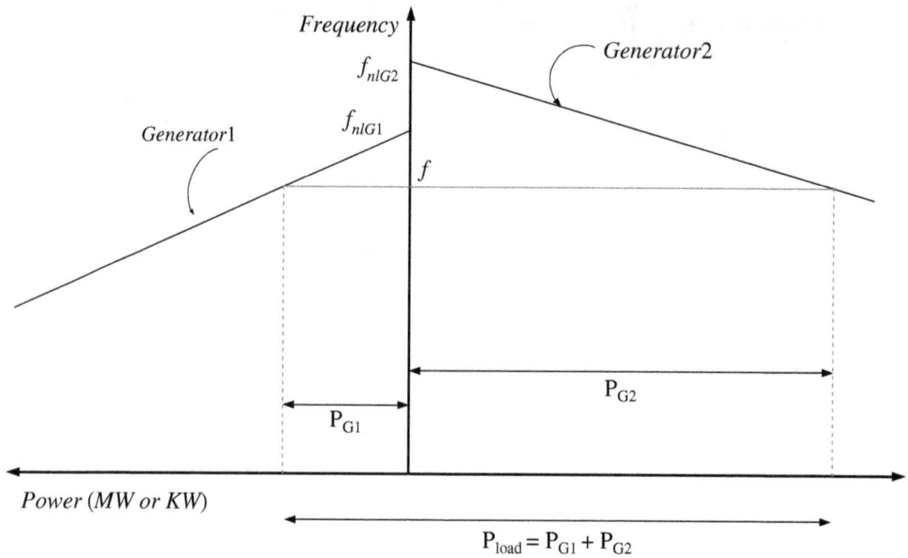

Figure 7.74 House diagram of two parallel-connected generators.

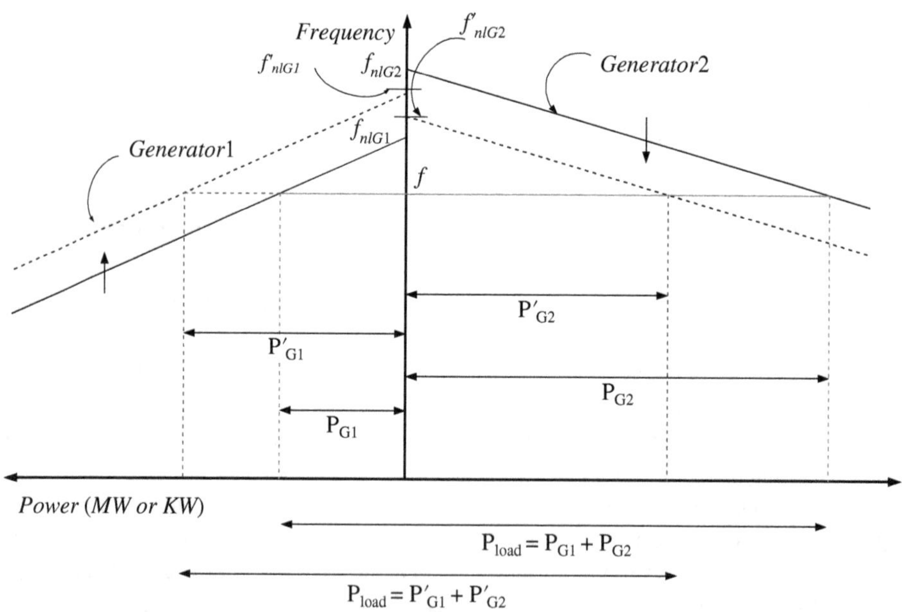

Figure 7.75 Varying power share among two parallel-connected generators.

achieve the operating condition, the no-load frequency set point of Generator 1 is increased from f_{nlG1} to f'_{nlG1}. At the same time, the no-load frequency set point of Generator 2 is decreased from f_{nlG2} to f'_{nlG2}. In this way, the power share of Generator 1 is increased while the power share of Generator 2 is reduced, without changing the system operating frequency f.

In case the operating frequency needs to increase, the no-load set point of both generators should be moved up. Similarly, to decrease the operating frequency, both set points should be simultaneously moved down.

The voltage-reactive power characteristic is also adjusted accordingly in order to change the reactive power share among two generators or terminal voltage of the load.

- The frequency and terminal voltage of the generators are controlled by the system to which it is connected.
- The governor set points of the generator control the real power supplied by the generator to the system.
- The field current in the generator controls the reactive power supplied by the generator to the system.

Example 7.27 Two 1 MVA three-phase synchronous generators are operating in parallel and delivering a load of 1600 KVA at 0.9 pf lagging. If one machine is operating at 0.5 pf lagging and supplying 900 KVA. Calculate the power delivered by the second machine and the power factor at which it is operating.

Solution

$$S_{Load} = 1600kVA = P_{Load} + jQ_{Load}$$
$$P_{Load} = S_{Load}\cos(\theta) = 1600 \times 0.9 = 1440KW$$
$$Q_{Load} = S_{Load}\sin(\theta) = -1600 \times 0.436 = -697.42kVAr$$

-ve reactive power since it is lagging
Power supplied by generator 1:

$$P_1 = S_1\cos(\theta) = 900 \times 0.5 = 450KW$$
$$Q_1 = S_1\sin(\theta) = -900 \times 0.866 = -779.422kVAr$$

Power supplied by generator 2:

$$S_2 = S_{Load} - S_1 = P_{Load} + jQ_{Load} - (P_1 + jQ_1)$$
$$S_2 = 1440 - j697.42 - (450 - j779.422) = 990 + j82 = 993.4\angle 4.73$$

KVA supplied by generator 2 is 993.4 KVA at a power factor of 0.996 leading

Example 7.28 Two generating units rated 300 and 400 MW have governor speed regulations of 6% and 4% respectively from no load to full load. Both generating units operating in parallel share a load of 600 MW. Assuming the free governor action, calculate the load share by the larger unit. The no-load set point is given as 50 Hz.

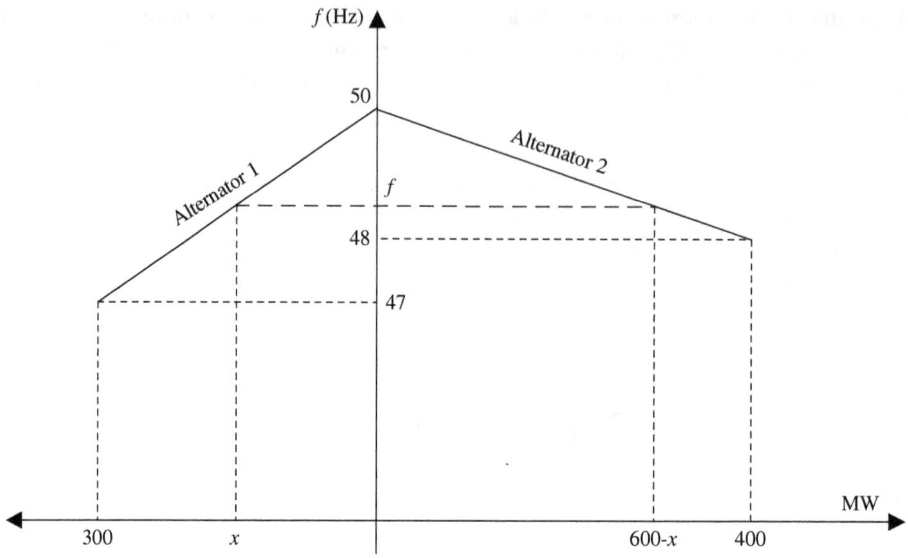

Figure E7.28 Power sharing among two parallel-connected generators.

Solution

$$\%of\,droop = \frac{f_{nl} - f_{fl}}{f_{fl}} \times 100$$

$$f_{fl} = \frac{f_{nl}}{1 + \frac{\%droop}{100}}$$

Alternator 1:

$$f_{fl1} = \frac{50}{1 + \frac{6}{100}} = 47$$

Alternator 2:

The frequency vs. load characteristic is shown below:

With f as a common frequency of operation, applying a similar triangle property

$$\frac{50 - f}{3} = \frac{\times}{300}$$

$$15000 - 300f = 3x$$

$$3x + 300f = 15000 \qquad (1)$$

And

$$\frac{50 - f}{2} = \frac{600 - x}{400}$$

$$20000 - 400f = 1200 - 2x$$

$$2x - 400f = -18800 \qquad (2)$$

From (1) and (2):

$$x = 200, f = 48$$

Load share by alternator 2 = x = 200 MW
Load share by alternator 1 (larger unit) = 600–x = 400 MW

Example 7.29 Two identical 2 MVA, 1 kV, 0.8 pf lagging, 50 Hz, three-phase alternators are connected in parallel and supplying a load. The frequency-power curve of the two alternators has a different droop. When the field excitation of the two alternators is equal, generator 1 delivers 1 kA at 0.9 pg lagging and generator 2 delivers 0.8 kA at 0.75 pf lagging.

(a) Calculate the real power and reactive power supplied by each alternator
(b) What is the power factor of the load?

Solution

(a) Generator 1 supplying 1000 A at 0.9 pf, the active power supplied is thus

$$P_{G1} = \sqrt{3}V_tI_a\cos(\theta) = \sqrt{3}(1000)(1000)(0.9) = 1.558MW$$

Reactive power supplied

$$\cos(\theta) = 0.9$$
$$\Rightarrow \theta = 25.84$$
$$\sin(\theta) = 0.4358$$
$$Q_{G1} = \sqrt{3}V_tI_a\sin(\theta) = \sqrt{3}(1000)(1000)(0.4358) = 0.755MVAr$$

For generator 2

$$P_{G2} = \sqrt{3}V_tI_a\cos(\theta) = \sqrt{3}(1000)(800)(0.75) = 1.039MW$$

Reactive power supplied

$$\cos(\theta) = 0.75$$
$$\Rightarrow \theta = 41.40$$
$$\sin(\theta) = 0.6614$$
$$Q_{G2} = \sqrt{3}V_tI_a\sin(\theta) = \sqrt{3}(1000)(800)(0.6614) = 0.916MVAr$$

(b) Total active power supplied by the two generators combined

$P_{total} = P_{G1} + P_{G2} = 1.558 + 1.039 = 2.597$ MW
Total reactive power supplied by the two generators
$Q_{total} = Q_{G1} + Q_{G2} = 0.755 + 0.916 = 1.671$ MVAr
Load power factor

$$pf = \cos\left(\tan^{-1}\left(\frac{Q_{total}}{P_{total}}\right)\right) = \cos\left(\tan^{-1}\left(\frac{1.671}{2.597}\right)\right) = 0.84lag$$

Example 7.30 A power plant has four synchronous generators each of 250 MVA, 13.6 KV, 0.9 pf lagging with identical speed droop curve. They are operating in parallel to supply the load. The governors are adjusted to produce a 3 Hz drop from no-load to full-load. Three of these available generators are supplying 180 MW at a frequency of 50 Hz, and the fourth generator is for the reserve to supply all incremental load changes in order to keep the system frequency at 50 Hz.

(a) At one instant of time, the total system load is 580 MW at a frequency of 50 Hz. Calculate the no-load frequency of each generator.
(b) If the system load increased to 680 MW, what will be the new system frequency if the no-load set points are not changed?
(c) What should the no-load set point be of the generator 4 to restore the system frequency to 50 Hz?
(d) For the operating condition of part (c) if the fourth generator is tripped, what will the system frequency be?

Solution

(a)

The slope of the frequency-power curve m_p is given as
m_p = (Full load power)/(droop from no-load to full load)
Full load power in MW = Full load MVA * power factor = 250 * 0.9 = 225 MW
Therefore

$$m_p = (225)/(3) = 75\ \text{MW/Hz}$$

The total load is 580 MW, thus the load share among different generators are
P_{G1} = 180 MW, P_{G2} = 180 MW, P_{G3} = 180 MW and P_{G4} = 580–180*3 = 40 MW
Thus, the no-load frequency set point is calculated as

$$P_{G1} = m_p\left(f_{nl1} - f\right)$$
$$180 = 75\left(f_{nl1} - 50\right)$$
$$f_{nl1} = 50 + \frac{180}{75} = 50 + 2.4 = 52.4Hz$$
$$f_{nl1} = f_{nl2} = f_{nl3} = 52.4Hz$$

For generator 4 the no-load set point is

$$P_{G4} = m_p\left(f_{nl4} - f\right)$$
$$40 = 75\left(f_{nl4} - 50\right)$$
$$f_{nl4} = 50 + \frac{40}{75} = 50 + 0.533 = 50.533Hz$$

(b) The new load is 680 MW

Thus, the power and frequency are related as

$$P_{G1} = m_p\left(f_{nl1} - f\right); P_{G2} = m_p\left(f_{nl2} - f\right); P_{G3} = m_p\left(f_{nl3} - f\right); P_{G4} = m_p\left(f_{nl4} - f\right)$$
$$P_{load} = P_{G1} + P_{G2} + P_{G3} + P_{G4}$$
$$P_{load} = m_p\left(f_{nl1} - f\right) + m_p\left(f_{nl2} - f\right) + m_p\left(f_{nl3} - f\right) + m_p\left(f_{nl4} - f\right)$$
$$680 = 75\left[3\left(f_{nl1} - f\right) + \left(f_{nl4} - f\right)\right] = 75\left[3f_{nl1} + f_{nl4} - 4f\right]$$
$$\frac{680}{75} = \left[3 \times 52.4 + 50.533 - 4f\right]$$
$$\frac{680}{75} = 207.733 - 4f$$
$$4f = 207.733 - \frac{680}{75} = 198.666$$
$$f = 49.66Hz$$

Thus, the system frequency drops below 50 Hz and is equal to 49.66 Hz

(c) The load power is 680 MW, the three generators supply 180 × 3 = 540 MW and hence the power supplied by the fourth generator is 680–540 = 140 MW

Thus, the new no-load set point is calculated as

$$P_{G4} = m_p\left(f_{nl4} - f\right)$$
$$140 = 75\left(f_{nl4} - 50\right)$$
$$f_{nl4} = 50 + \frac{140}{75} = 50 + 1.866 = 51.866Hz$$

(d) Generator 4 is out of service so only generators 1, 2 and 3 are supplying the total load, thus the new system frequency will be

$$P_{G1} = m_p\left(f_{nl1} - f\right); P_{G2} = m_p\left(f_{nl2} - f\right); P_{G3} = m_p\left(f_{nl3} - f\right);$$
$$P_{load} = P_{G1} + P_{G2} + P_{G3}$$
$$P_{load} = m_p\left(f_{nl1} - f\right) + m_p\left(f_{nl2} - f\right) + m_p\left(f_{nl3} - f\right)$$
$$680 = 75\left[3\left(f_{nl1} - f\right)\right] = 75\left[3f_{nl1} - f\right]$$
$$\frac{680}{75} = \left[3 \times 52.4 - 3f\right]$$
$$\frac{680}{75} = 157.2 - 3f$$
$$f = \frac{1}{3}\left(157.2 - \frac{680}{75}\right) = 49.37Hz$$

7.19 Matlab/Simulink Model of a Salient Pole Synchronous Machine

The simulation model of a salient pole synchronous machine is modelled in Matlab/Simulink by using basic blocks, as shown in Figure 7.76. The synchronous Machine (SM) model is designed, based on the mathematical equations in the rotor reference frame, where the machine operates in generator or motor modes. The operating mode of the machine is dependent on the sign of the mechanical power (positive for generator mode, negative for motor mode).

A two-pole, three-phase star connected salient pole type SM is modelled. The rotor of the machine is modelled by considering three damper windings and one field winding. Here, the field winding (*fd* windings) and one damper winding (*kd* winding) has the same magnetic axis with resistance of R_{fd} and R_{kd} respectively. The other two damper windings are displaced by 90^0 ahead of field windings which are *kq1* and *kq2* windings with a resistance of R_{kq1} and R_{kq2} respectively. To make the self-stating of the SM, the three damper windings are shorted and one field winding is excited with DC supply. Detailed modelling and explanation are given in [1].

Parameters of the synchronous motor

V_m = peak value of the stator phase voltages = 230*sqrt(2)V
V_f = field winding voltage = 30 V
F = frequency of the supply = 50 Hz
R_s = stator winding resistance = 1.62 Ω

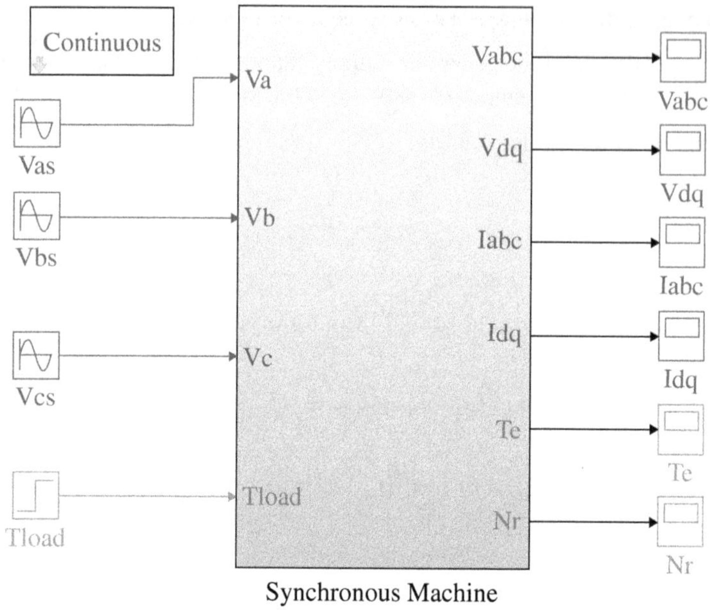

Figure 7.76 Matlab/Simulink model of the synchronous machine.

R_{kq1} = q-axis damper winding 1 resistance = 4.772 Ω
R_{kq2} = q-axis damper winding 2 resistance = 4.772 Ω
R_{kd} = field axis aligned damper winding resistance = 3.142 Ω
R_{fd} = field winding resistance 1.208 Ω
L_{ls} = leakage inductance = 0.004527 H
L_{md} = d-axis magnetizing inductance = 0.1086 H
L_{mq} = q-axis magnetizing inductance = 0.05175 H
L_{lfd} = field winding leakage inductance = 0.01132 H
L_{lkd} = field axis aligned damped winding leakage inductance = 0.0074 H
L_{lkq1} = damper winding 1 leakage inductance = 0.0074 H
L_{lkq2} = damper winding 2 leakage inductance = 0.0074 H
P = number of poles = 2
T_{l} = load torque = 10 Nm
J = moment of inertia = 0.01 kg-m^2
ψ = Flux linkage

The mathematical equations used in the modelling of the SM are given as [1];

1. ***Electrical equations***

$$v^r_{qdos} = R_s i^r_{qdos} + p\psi^r_{qdos} + \omega_r \psi^r_{qdos}$$
$$v^r_{qdr} = R_r i^r_{qdr} + p\psi^r_{qdr} \tag{7.101}$$

Where

$$\left[\psi^r_{qdos}\right] = \left[\psi^r_{qs} - \psi^r_{ds} \;\; 0\right]$$

2. ***Flux linkage equations***

$$\begin{bmatrix} \psi^r_{qdos} \\ \psi^r_{qdr} \end{bmatrix}_{7*1} = [L]_{7*7} \begin{bmatrix} i^r_{qdos} \\ i^r_{qdr} \end{bmatrix}_{7*1}$$

Where

$$[L]_{7*7} = \begin{bmatrix} (L_{ls}+L_{mq}) & 0 & 0 & L_{mq} & L_{mq} & 0 & 0 \\ 0 & (L_{ls}+L_{md}) & 0 & 0 & 0 & L_{md} & L_{md} \\ 0 & 0 & L_{ls} & 0 & 0 & 0 & 0 \\ L_{mq} & 0 & 0 & (L_{lkq1}+L_{mq}) & L_{mq} & 0 & 0 \\ L_{mq} & 0 & 0 & L_{mq} & (L_{lkq2}+L_{mq}) & 0 & 0 \\ 0 & L_{md} & 0 & 0 & 0 & (L_{lfd}+L_{md}) & L_{md} \\ 0 & L_{md} & 0 & 0 & 0 & L_{md} & (L_{lkd}+L_{md}) \end{bmatrix} \tag{7.102}$$

3. ***Mechanical equations***

Torque equation of the machine is

$$T_e = \frac{3}{2}\frac{P}{2}\left(\psi^r_{ds} i^r_{qs} - \psi^r_{qs} i^r_{ds}\right) \tag{7.103}$$

Mechanical equation relating the speed and torque is given as

$$T_e = J\frac{d\omega_r}{dt} + B_m\omega_r + T_L \tag{7.104}$$

Where J is the inertia, B_m is the viscous friction and T_L is the load torque.

The park transformation matrix is

For changing three-phase stationary variables to arbitrary reference frame can expressed as

$$f_{qdo} = K_s * f_{abcs} \tag{7.105}$$

Where

$$\underline{K}_s = \sqrt{\frac{2}{3}} \begin{bmatrix} \cos\theta & \cos(\theta-120) & \cos(\theta-240) \\ \sin\theta & \sin(\theta-120) & \sin(\theta-240) \\ \sqrt{\frac{1}{2}} & \sqrt{\frac{1}{2}} & \sqrt{\frac{1}{2}} \end{bmatrix}$$

And for converting back arbitrary reference frame to three-phase stationary frame,

$$f_{abcs} = (K_s)^{-1} f_{qdos} \tag{7.106}$$

$$\underline{K}^{-1}_s = \begin{bmatrix} \cos\theta & \sin\theta & 1 \\ \cos(\theta-120) & \sin(\theta-120) & 1 \\ \cos(\theta-240) & \sin(\theta-240) & 1 \end{bmatrix}$$

For transforming one reference frame to other reference frame

$$f^{y}_{qdos} = K^{y}_{x} f^{x}_{qdos} \tag{7.107}$$

$$\underline{K}^{y}_{x} = \begin{bmatrix} \cos(\theta_y - \theta_x) & -\sin(\theta_y - \theta_x) & 0 \\ \sin(\theta_y - \theta_x) & \cos(\theta_y - \theta_x) & 0 \\ 0 & 0 & 1 \end{bmatrix}$$

With the help of Eqs. (7.101) to (7.107) the SM is modelled in the MATAB/Simulink as shown in Figure 7.76. The inputs are three-phase balanced supply and torque reference (T_{load}). The three-phase balanced supply is given to the machine model, and these voltages are converted as V_{abc} *to* V_{qd} *to* V^{r}_{qd}. Stator and rotor flux linkage estimation are done with the help of Eq. (7.101), by considering the input voltage and resistive drop of each phase in *qd* domain, as shown in Figure 7.77. After that, currents in the *qd* domain are estimated by using flux linkages to current transformation by using Eq. (7.102). From the *qd* currents and *qd* flux linkages, torque and speed equations of the machine are calculated, i.e. with Eqs. (7.103) and (7.104), as shown in Figures 7.77 and 7.78.

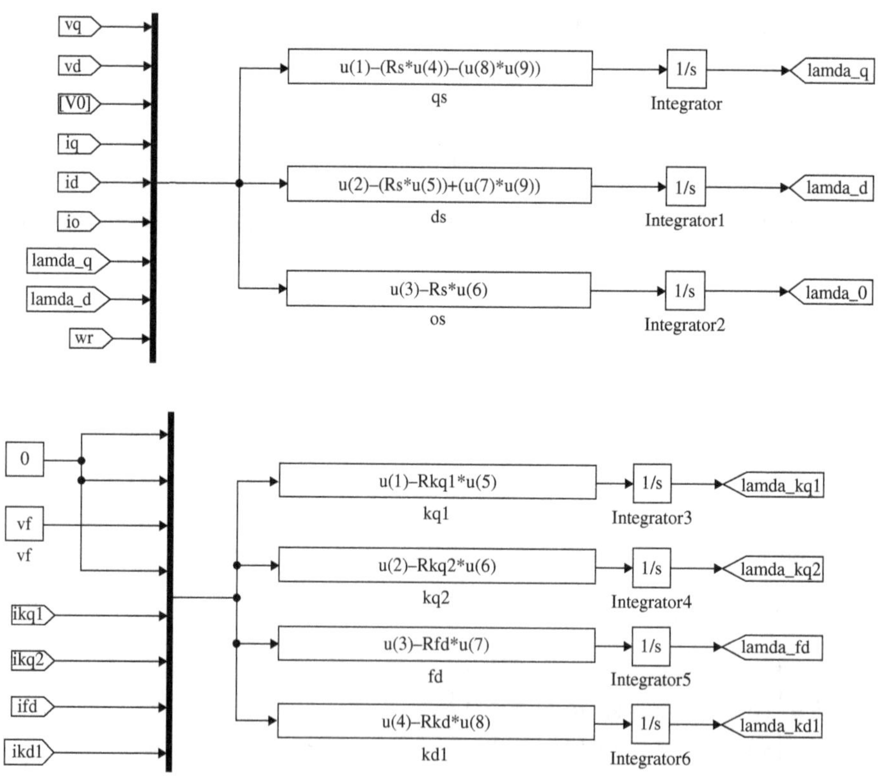

Figure 7.77 Stator and Rotor flux linkage blocks in *qdo*.

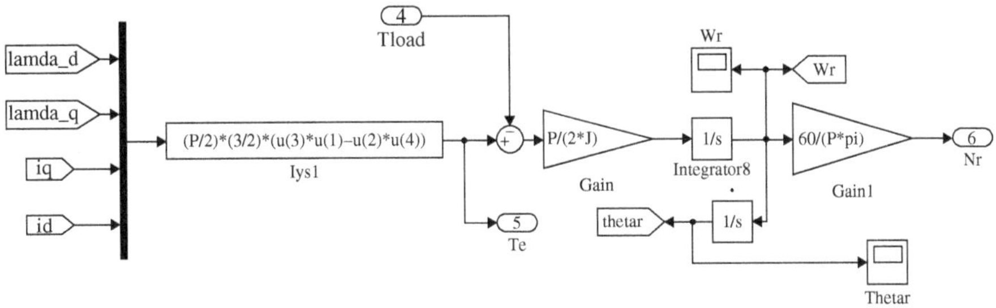

Figure 7.78 Mechanical torque and speed conversion block.

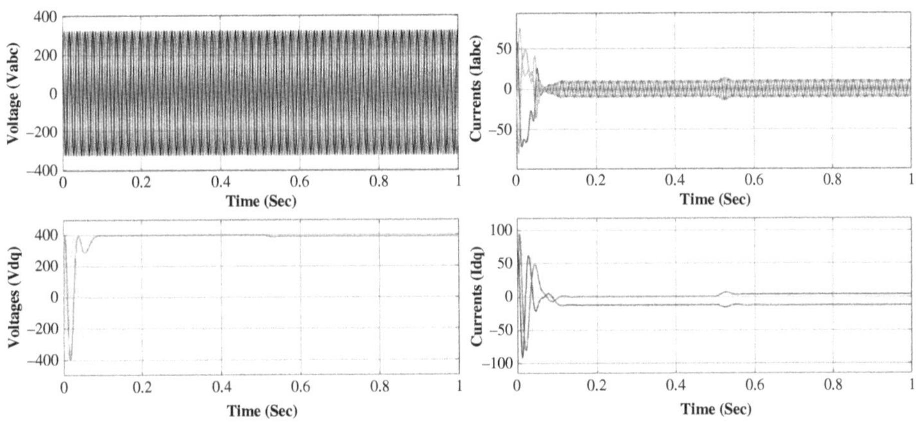

Figure 7.79 Stator phase voltage and currents in *abc* and *qd* domains for motor.

7.19.1 Results Motoring Mode

Figures 7.79 and 7.80 show the results for the synchronous machine operating as a motor. At the time of starting the machine, the stator, as well as rotor poles, are not locked magnetically which results in the transients in phase currents as well as in torque. As soon as the poles are magnetically locked, the machine reaches steady-state and the transients dies down. From the torque and speed plot, it is observed that load torque is applied at time instant $t = 0.5$ s, i.e., 10 Nm. Even although the load applied on the motor, speed is constant i.e., 3000 rpm with a fundamental frequency 50 Hz and two Poles.

7.19.2 Results Generator Mode

In the modelling of the salient pole synchronous generator, the mathematical equations, as well as parameters and other values, is the same as the motor, i.e., Eqs. (7.101) to (7.107). To operate the synchronous motor as a generator, the sign of load torque is reversed (−ve). Results for the synchronous generator are shown in Figures 7.81 and 7.82, where the currents, voltages, torque and speed plots are shown. From the torque and speed plot it is observed that the load torque sign is -ve and it is applied at time instant $t = 0.5$ s, i.e., −10 Nm. Also, in generator mode, the speed of the SM is constant i.e., 3000 rpm with a fundamental frequency 50 Hz and two Poles.

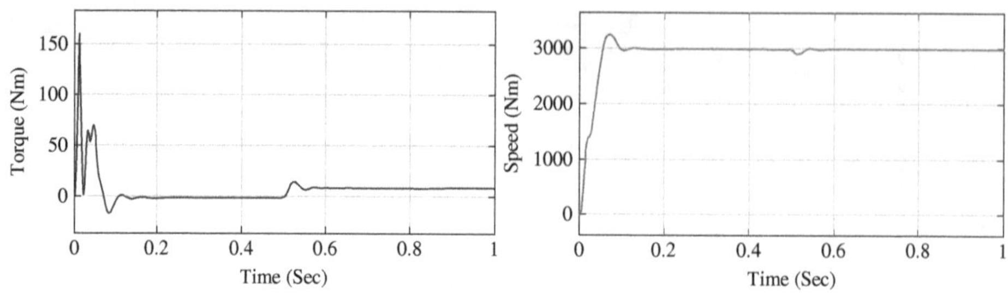

Figure 7.80 Torque and speed plots of the synchronous machine operating as a motor (load torque = +ve).

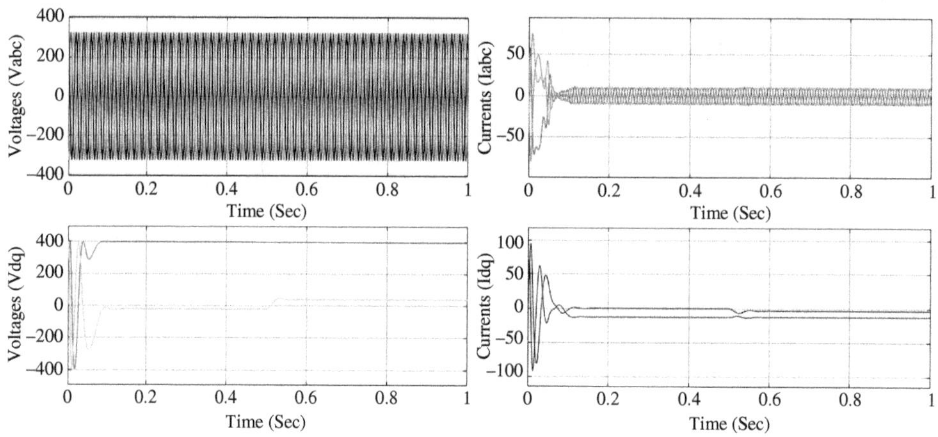

Figure 7.81 Stator phase voltage and currents in *abc* and *qd* domains for generator.

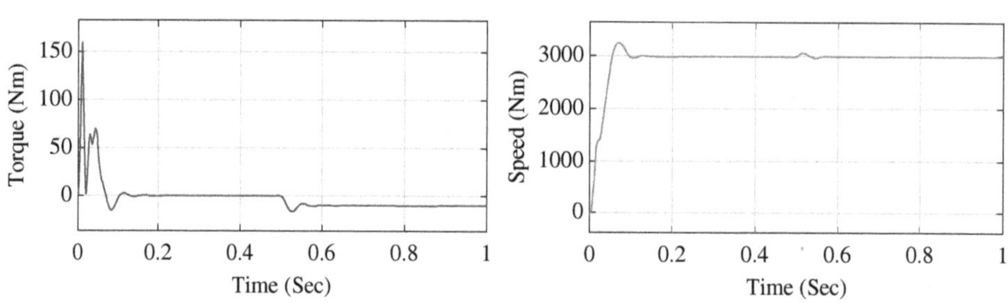

Figure 7.82 Torque and speed plots of the synchronous machine operating as generator (load torque = −ve).

7.20 Summary

This chapter described the principle of operation, theory, analysis and modelling of a three-phase synchronous machine. Both salient pole and non-salient pole machines are explained in this chapter. Simple and detailed explanation is given for understanding the fundamental concepts. Synchronization of the synchronous generator is described as well as the impact of different parameter changes. Limiting conditions are given based on fundamental laws. Numerous solved and unsolved examples are given for clear understanding of the chapter. Finally, a dynamic model of the machine is given which is used for developing Matlab/Simulink model. Students can change variables and as well as the structure of the machine to analyse the characteristics and behaviour of the machine.

Problems

7.1 At a location in Europe, it is necessary to supply 200 kW of 60-Hz power. The only power sources available operate at 50 Hz. It has been decided to generate the power by means of a motor-generator set consisting of the synchronous motor driving a synchronous generator. How many poles should each of the two machines have in order to convert 50-Hz power to 60-Hz power?

7.2 A 60-Hz, three-phase the synchronous motor is observed to have a terminal voltage of 460 V (line-line) and a terminal current of 120 A at a power factor of 0.95 lagging. The field current under this operating condition is 47 A. The machine synchronous reactance is equal to 1.68 Ω (0.794 per unit on a 460-V, 100-kVA, three-phase base) Assume the armature resistance to be negligible.
Calculate

(a) The generated voltage E_a in volts
(b) The magnitude of the field-to- armature mutual inductance L_{af}, and
(c) The electrical power input to the motor in kW and in horsepower.

7.3 A three-phase synchronous generator is operated at 460 V, 60 Hz supply. The synchronous reactance of the machine is 1.68 Ω and field to armature inductance is 20 mH. Calculate the field current required to supply a load of 85 kW, 0.95 pf leading.

7.4 The following data are taken from the open- and short-circuit characteristics of a 45-kVA, three-phase, Y-connected, 220-V (line-to-line), six-pole, 60-Hz synchronous machine. From the open-circuit characteristic: Line-to-line voltage = 220 V field current = 2.84 A.
From the short-circuit characteristic

Armature current (A)	118	152
Field current (A)	2.20	2.84

From the air-gap line
Field current = 2.20 A
Line-to-line voltage = 202 V
Compute the unsaturated value of the synchronous reactance, its saturated value at rated voltage in accordance with Eq. (5.29), and the short-circuit ratio. Express the synchronous reactance in ohms per phase and per unit using the machine rating as a base.

7.5 The manufacturer's data sheet for a 26-kV, 750-MVA, 60-Hz, three-phase the synchronous generator indicates that it has a synchronous reactance $X_s = 2.04$ and a leakage reactance $X_{al} = 0.18$, both in per unit on the generator base. Calculate
(a) the synchronous inductance in mH
(b) the armature leakage inductance in mH, and
(c) the armature phase inductance L_{aa} in mH and per unit.

7.6 A 25-MVA, 11.5 kV the synchronous machine is operating as a synchronous condenser. The generator short-circuit ratio is 1.68 and the field current at rated voltage, no load is 420 A. Assume the generator to be connected directly to an 11.5 kV source.
a) What is the saturated synchronous reactance of the generator per unit and in ohms per phase? The generator field current is adjusted to 150 A.
b) Draw a phasor diagram, indicating the terminal voltage, internal voltage, and armature current.
c) Calculate the armature current magnitude (per unit and amperes) and its relative phase angle with respect to the terminal voltage.
d) Under these conditions, does the synchronous condenser appear inductive or capacitive to the 11.5 kV system?
e) Repeat parts (b) through (d) for a field current of 700 A.

7.7 A 3.6 kV, 1.5 MVA, salient-pole synchronous generator with $X_d = 1.2\,\Omega$/phase and $X_q = 0.8\,\Omega$/phase is supplying 50% load at rated voltage and 0.9 pf lagging. Calculate the excitation voltage.

7.8 A salient-pole synchronous generator with saturated the synchronous reactances $X_s = 1.57$ per unit and $X_q = 1.34$ per unit is connected to an infinite bus of rated voltage through an external impedance $X_{bus} = 0.11$ per unit. The generator is supplying its rated MVA at 0.95 power factor lagging, as measured at the generator terminals.
a) Draw a phasor diagram, indicating the infinite-bus voltage, the armature current, the generator terminal voltage, the excitation voltage, and the rotor angle (measured with respect to the infinite bus).
b) Calculate the per-unit terminal and excitation voltages, and the rotor angle in degrees.

7.9 A three-phase salient pole synchronous machine has $X_d = 0.9$ pu and $X_q = 0.7$ pu. What maximum per unit power can this machine deliver without loss of synchronism if the excitation voltage is zero? Also, calculate per unit armature current and per unit reactive power delivered by the machine.

7.10 In a synchronous machine, the excitation voltage is given as $E_a = 390\angle 15$ and the terminal voltage is $V_t = 400\angle 0$. The armature resistance is 0.2 and the synchronous reactance is 3 Ω/phase. The machine is delta connected.
a) Is this machine operating as a generator or motor?
b) Compute the real and reactive power delivered or consumed by the machine.

c) If the excitation voltage is $E_a = 450\angle -20$ and the terminal voltage is $V_t = 420\angle 0$, determine whether the machine is supplying real and reactive power or consuming.

d) Compute the real and reactive power delivered or consumed by the machine in part (c).

7.11 A 2.2 kV, 500 hp., 50 Hz, six-pole, star-connected synchronous motor has an efficiency of 90% at full load and pf is 0.9 leading. The armature resistance is 0.5 Ω/phase and the synchronous reactance of 5 Ω/phase. Calculate the following when motor is running at full load condition

(a) Output rated torque
(b) Input electrical power
(c) Current drawn by the motor
(d) Excitation phase and line voltages

7.12 A 440 V, 50 Hz, three-phase star-connected synchronous motor has negligible armature resistance and the synchronous reactance is 1.2 Ω/phase. The output power by the motor is 50 kW when the load angle $\delta = 20^0$.

a) Calculate the excitation voltage
b) Calculate the armature current drawn by the machine
c) Calculate the maximum power supplied by the machine if the field current is kept constant.

7.13 A 440 V, 400 kVA, 0.9 pf lagging, three-phase 50 Hz, star-connected synchronous generator with $X_s = 0.3\ \Omega$/phase is supplying the synchronous motor. The synchronous motor is 440 V, 60 kW, star-connected with $X_s = 1.5\ \Omega$/phase. The terminal voltage of the synchronous generator is adjusted to 440 V when the motor is drawing rated power at unity power factor. What will the magnitude and angle of the excitation induced voltage Ea for generator and motor be?

7.14 In a factory, the following loads are used
Induction motor: 1000 HP, 0.7 average power factor, 0.85 average efficiency
Lightening and heating load: 100 kW
A three-phase synchronous motor is installed to provide 300 hp to a new process. The synchronous motor operates at 92% efficiency. Determine the kVA rating of the synchronous motor if the overall factory power factor is to be raised to 0.95. Determine the power factor of the synchronous motor.

7.15 A three-phase 60 Hz supply and two three-phase synchronous machines are available. Determine the speed, and a suitable number of poles for each synchronous machine to provide.

(a) A three-phase, 180 Hz supply
(b) A three-phase 500 Hz supply

7.16 The following test data are obtained for a three-phase 195 MVA, 15 kV, 60 Hz star connected the synchronous machine
Open-Circuit test:

I_f (A)	150	300	450	600	750	900	1200
V_{LL} (kV)	3.75	7.5	11.2	13.6	15	15.8	16.5

Short-circuit test
$I_f = 750$ A, $I_a = 7000$ A
The armature resistance is very small.

a) Draw the open-circuit characteristic, the short-circuit characteristic, air gap line, and the modified air gap line.
b) Determine the unsaturated and saturated values of the synchronous reactance in ohms and per unit.
c) Find the field current required if the synchronous machine is to deliver 100 MVA at rated voltage, and 0.8 leading power factor.

7.17 A three-phase star-connected the synchronous generator having the synchronous impedance of $0 + j\ 10\ \Omega$/phase supplying 1.732 kW at 0.5 leading power factor. Calculate its voltage regulation.

7.18 The following test results are obtained for a three-phase 750 MVA, 25 kV, 60 Hz, 3600 rpm, star-connected synchronous machine at rated speed.

I_f (A)	V_{LL}(kV) Open-circuit test	I_a (A) Short-circuit test	V_{LL}(kV) Air-gap test
1500	25	10 000	30

a) Determine the number of poles of the synchronous machine
b) Determine the unsaturated and saturated values of the synchronous reactance in ohm and per unit
c) The short-circuit test is performed at constant field current (1500 A) but at different speeds 1000, 2000, 3000, and 3600 rpm. Determine the short-circuit current at these speeds.
d) Determine the field current if the synchronous machine delivers rated MVA to an infinite bus at 0.9 lagging power factor.

7.19 A three-phase 500 MVA, 25 kV, 60 Hz alternator has a synchronous reactance of 1.5 pu. The alternator is connected to an infinite bus of 25 kV through a feeder of reactance 0.25 pu. The terminal voltage as the alternator is maintained to 25 kV for any loading, by means of a voltage regulator that adjust the field current.

a) Draw the phasor diagram.
b) Determine the current and power factor of the alternator.
c) Determine the excitation voltage of the alternator.

7.20 A 400 V, 40 kVA, 0.8 pf leading delta-connected 50 Hz synchronous machine has negligible armature resistance while the synchronous reactance is 2.2 Ω. The friction and windage losses are 1.8 kW and core losses are 0.7 kW. The shaft supplies 10 kW load at a pf of 0.8 leading. Calculate the supply current.

7.21 The nameplate of star-connected synchronous motor has the following information. 20 000 HP, 1800 rpm, 1.0 PF, 6.6 kV, 1350 A, three-phase 60 Hz, Excitation voltage 120 A, 5.5 A.
The per unit the synchronous reactance is 0.95 and the per unit resistance is 0.012.
a) Determine the number of poles of the synchronous motor.
b) Determine the synchronous reactance and resistance in ohm
c) For the rated condition
   i) Determine output torque in newton-meter.
   ii) Determine efficiency.
   iii) Determine rotational loss.
   iv) Determine power loss in the field circuit.
   v) Determine excitation-induced emf E_a

7.22 A three-phase 1 MVA, 2300 V, 60 Hz synchronous machine has negligible stator resistance and a saturated synchronous reactance $X_s = 1.25\ \Omega$/phase at rated terminal voltage. The efficiency of the machine is 0.95 at rated speed. The machine is connected to an infinite bus.
a) Determine the X_s in per unit.
b) Determine the excitation voltage and power angle when the machine is operated as a synchronous motor at 0.85 lagging power factor and delivers 500 hp.
c) The field current is now reduced by 40% keeping the power output the same as in (b). Find the stator current and power factor. Will the motor lose synchronism?

7.23 A three-phase 10 MVA, 2300 V, 60 Hz the synchronous machine. $X_s = 0.9$ pu and negligible stator resistance. The machine is connected to an infinite bus. If $V_t = 2300\angle 0°$ V and $E_a = 3450\angle 120°$ V,
a) Is the machine operating as a generator or motor?
b) Determine the power transfer (MW) and power factor of the machine.

7.24 A synchronous generator with synchronous reactance of 0.8 pu is supplying rated power at a unity power factor at a terminal voltage of 1.1 pu. Neglecting the armature resistance, calculate the angle behind the synchronous reactance with respect to the angle of the terminal voltage.

Reference

1 Krause, P.C., Wasynczuk, O., Sudhoff, S.D., and Pekarek, S.D. (2002). *Analysis of Electric Machinery and Drive Systems*. Wiley-IEEE Press.

8

Single-Phase and Special Machines

8.1 Preliminary Remarks

A high number of motors are manufactured as single-phase motors for use in homes, offices, hospitals, airports, factories, and other commercial and residential applications. These are small motors with fractional horsepower ratings. They are cheap motors made in variety of design and sizes. Single-phase motors are made as induction motors and synchronous motors.

This chapter gives the fundamental concepts related to different types of single-phase motors including induction and synchronous machines. Construction, principle of operation, equivalent circuit and testing of machines are elaborated. Single-phase induction motors are inherently non-self-starting machines. Additional winding, known as auxiliary winding, is included to make them self-starting. Once the machine is under running condition, auxiliary winding is cut-off. However, in capacitor run motors, the capacitor remains in the circuit. Analysis of single-phase induction machines is complex and is based on double revolving-field theory and cross-field theory. The book only describes double revolving-field theory.

8.2 Single-phase Induction Machine

The single-phase induction motor is structurally similar to the polyphase induction motor, except that (i) the stator has single-phase winding and (ii) a centrifugal switch to remove the auxiliary winding used for starting purposes. Similar to a DC motor, a single-phase induction motor has two principal parts, one rotating and the other fixed. In single-phase induction motors, the stationary part is Stator and the rotary part is Rotor.

The constructional descriptions of the single-phase induction motor are shown in Figure 8.1. It consists of a stator. The stator consists of a laminated structure consisting of stampings. The stampings are placed on its periphery so that it carries the winding known as the stator winding. A single-phase AC supply is exciting. The laminated construction minimizes iron losses. The stampings are made from silicone steel material, which minimizes the loss of hysteresis. The winding of the stator is a wound for a certain number of poles. When excited by a single-phase AC supply, the stator winding is wound for a certain

Electrical Machine Fundamentals with Numerical Simulation using MATLAB/SIMULINK, First Edition.
Atif Iqbal, Shaikh Moinoddin, and Bhimireddy Prathap Reddy.

Companion website: www.wiley.com/go/iqbal/electricalmachine

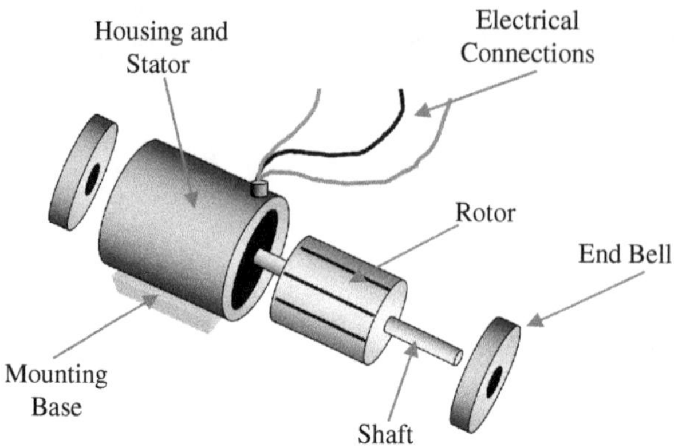

Figure 8.1 Single-phase induction motor.

number of poles and means that the stator produces the magnetic field which produces the effect of a certain number of poles.

The number of poles for which the winding of the stator is wound determines the synchronous speed of the motor. The synchronous speed is defined as N_s and has a fixed relationship to the supply frequency f and the number of poles P. The relationship is given by

$$N_s = \frac{120f}{P} \tag{8.1}$$

The construction of the rotor of the single-phase induction motor is similar to the squirrel cage of the three-phase induction motor. The rotor is cylindrical in shape and has slots all over its periphery. Slots do not operate in parallel but are slightly skewed, since the skew prevents the stator and rotor tooth locking magnetically and makes the motor movement slower and smoother. The squirrel cage rotor is made from brass, bronze, or copper plates. The rotor bars are positioned on the periphery of the rotor slot. Copper or aluminium rings are called end rings which permanently shorten the rotator conductors. These rotary conductors are fixed to the end ring to provide the mechanical strength, creating a complete closed circuit like a cage. Hence, they have been called 'Squirrel cage induction motors'. Since the bars are shortened permanently by the rings, resistance is very low and additional resistance cannot be applied. The lack of slip ring and brushes makes single-phase induction motor very straightforward and robust.

8.2.1 Field System in a Single-phase Machine

Usually, the winding used in a single-phase induction motor (IM) stator is distributed all around the periphery of the stator slots. The rotor is a squirrel cage, because the rating for this type of motor is low, as compared to a three-phase IM. As the stator winding is fed by a single-phase supply, the flow of flux in the air gap only alternates (pulsates), rather than a rotating one produced by a poly-phase winding (maybe two or three) in an IM stator. This type of field cannot produce a torque if the rotor is at a standstill. A single-phase IM, unlike a three-phase IM, is not self-starting. However, as shown below, when the rotor is initially

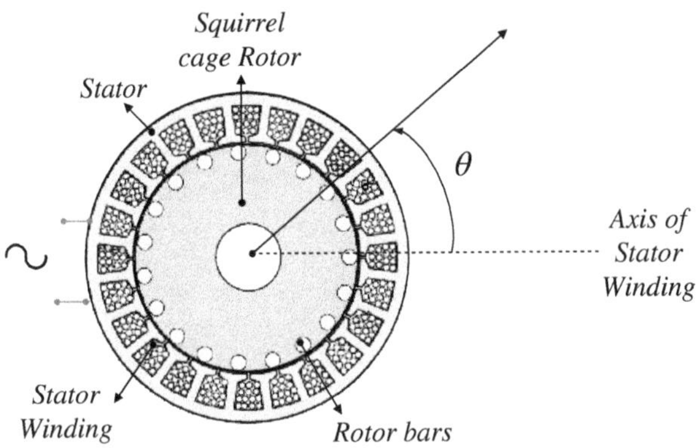

Figure 8.2 Cross-sectional view of a single-phase induction motor.

given a certain torque in either direction, a torque is immediately generated in the motor. The motor then reaches its final speed, which is always below its synchronous speed. This is now explained by the theory of double-field revolution.

Assuming the current fed to stator winding is given as

$$i_m = I_m \cos(\omega t) \tag{8.2}$$

This current produces sinusoidal mmf along the axis of stator (as shown in Figure 8.2) given as

$$\mathfrak{F} = Ni_m = NI_m \cos(\omega t) = F_m \cos(\omega t) \tag{8.3}$$

where N is the effective number of turns in the stator winding and ω is the angular frequency of the stator supply current.

The mmf along the rotor position θ is given as

$$F(\theta, t) = \mathfrak{F} \cos(\theta) = F_m \cos(\omega t) \cos(\theta) \tag{8.4}$$

The Eq. (8.3) can be written as

$$\begin{aligned} F(\theta, t) &= \frac{F_m}{2} \cos(\omega t - \theta) + \frac{F_m}{2} \cos(\omega t + \theta) \\ F(\theta, t) &= F_f + F_b \\ F_f &= \frac{F_m}{2} \cos(\omega t - \theta) \\ F_b &= \frac{F_m}{2} \cos(\omega t + \theta) \end{aligned} \tag{8.5}$$

The net mmf is resolved into two revolving mmfs called F_f and F_b. The two mmfs are rotating in nature and rotate at angular frequency of ω although in opposite directions. Thus, the net field is not revolving but pulsating in nature. The amplitude varies but the position remains fixed. The mmf produces the flux. The resultant flux has two components; one forward rotating and one backward rotating. The position of fluxes at different instants of time is shown in Figure 8.3. It is seen that the flux changes the amplitude at a fixed position. This type of flux is called a pulsating flux. Figure 8.4a shows the resultant flux with

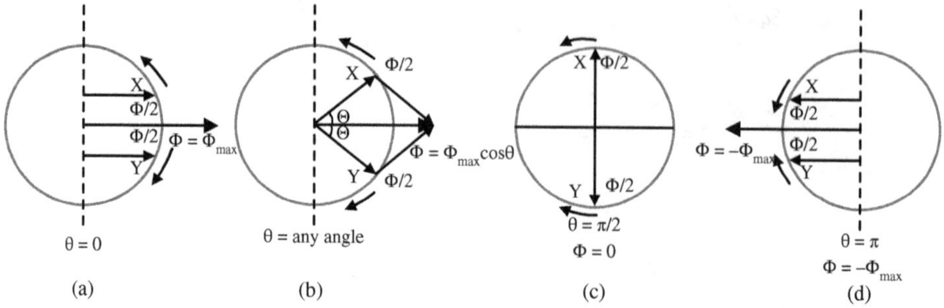

Figure 8.3 Resultant flux with respect to angle.

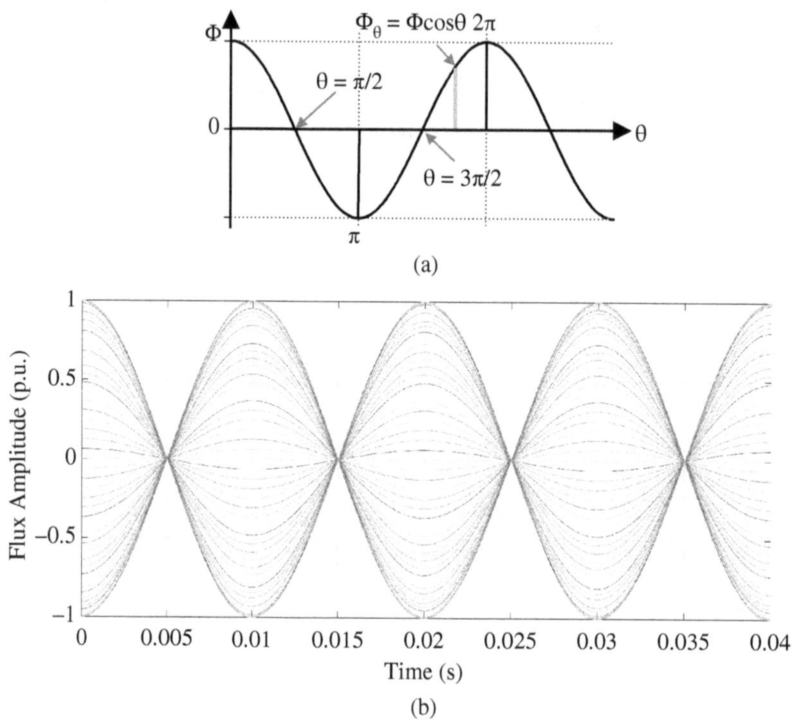

Figure 8.4 (a) Variation of flux with angle. (b) Flux waveform for different rotor positions.

respect to time. Flux waveform for two cycles (assuming 50 Hz supply frequency- 20 ms) is also depicted in Figure 8.4b.

Let us assume that the clockwise direction of the rotating flux corresponds to the positive or forward direction and anti-clockwise corresponds to the negative or backward direction. As both fluxes are rotating in opposite direction to each other, the emf is induced in the rotor bars which results in the rotor current, as the rotor bars are short-circuited by end rings. Both currents produce two different torques of equal magnitude but in opposite direction. The resultant torque is the sum of these two torques (forward and backward) and is shown in Figure 8.5.

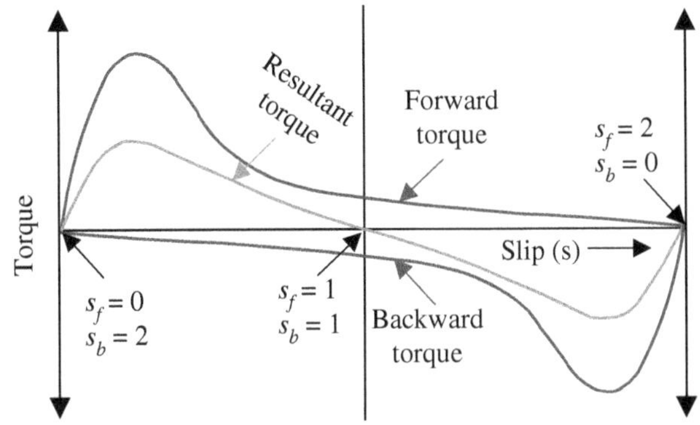

Figure 8.5 Speed-torque characteristic of single-phase IM.

For the forward direction, the forward slip of rotor s_f with respect to forward-rotating field is defined as

$$s_f = s = \frac{N_s - N}{N_s} \tag{8.6}$$

Similarly, for the backward rotation, the backward slip s_b is defined as

$$s_b = \frac{-N_s - N}{-N_s} = 2 - s \tag{8.7}$$

If the motor is stationary, the slip due to both forward and backward field will be 1, i.e. $s_f = s_b = 1$. These two torques are equal and opposite at this instant and the resultant torque is zero. Therefore, the resultant starting torque for the IM is zero, thus the IM is not self-starting. But if the motor (rotor) is started or rotated by some auxiliary means in any direction, say the anti-clockwise direction (Forward) direction, the forward torque exceeds the backward torque, which results in a positive torque. The motor speeds forward; the forward torque being greater than the backward torque. As the motor rotates in the forward direction, the resulting torque is positive.

8.3 Equivalent Circuit of Single-phase Machines

In this section, the steady-state model for a single winding of a single-phase IM is given. The steady-state circuit model does not include effects such as pulsating torques. Also, this is not the case if a split-phase winding (with or without capacitors) is used. In that case, the backward-rotating field can be totally or partially eliminated. Let us define the following terms

R1 = Stator winding resistance
X1 = Stator winding leakage reactance
Xm = Magnetizing reactance
R2′ = Rotor resistance referred to the stator side
X2′ = Rotor leakage reactance referred to the stator side
V1 = Stator supply voltage

Vf = Forward induced emf
Vb = Backward induced emf

The equivalent circuit of a single-phase induction motor is shown in Figure 8.6a for standstill condition when the slip s = 1.

According to double-revolving field theory, this equivalent circuit is divided into two halves as shown in Figure 8.6b. The first part shows the effect of rotating-forward flux and the second part shows the effect of rotating-backward flux.

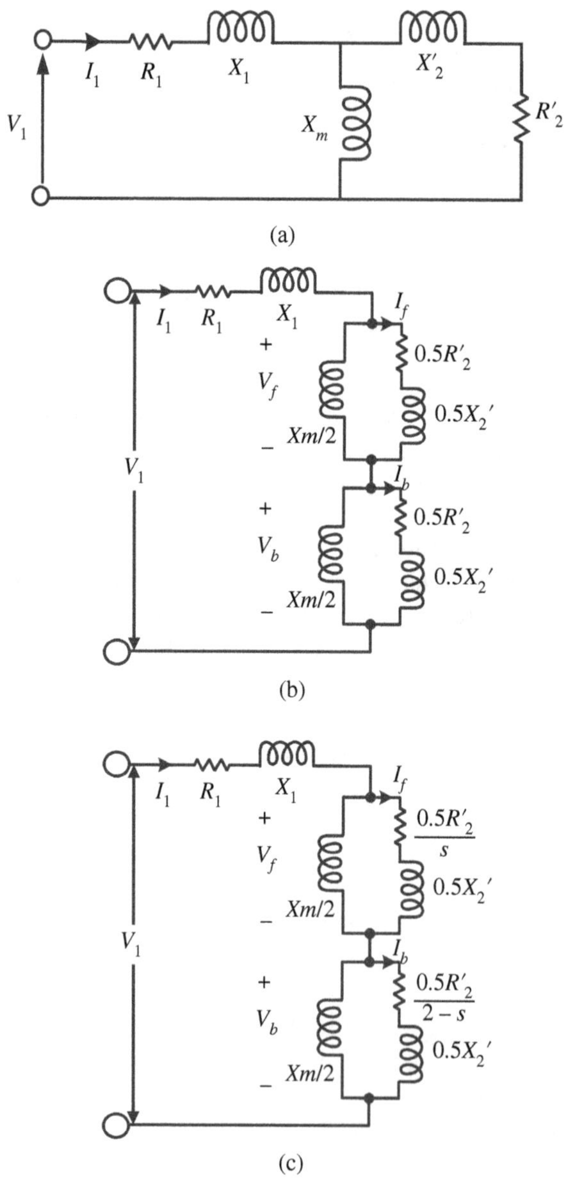

Figure 8.6 Equivalent circuit of a single-phase induction motor, (a,b) for standstill condition, and (c) for running condition.

Under running condition, the equivalent circuit is modified considering the forward and backward slips as shown in Figure 8.6c. The effective rotor resistance in relation to the forward-rotating flux is $\frac{0.5R_2'}{s}$ and $\frac{0.5R_2'}{2-s}$ in relation to the backward-rotating flux.

8.3.1 Equivalent Circuit Analysis

The analysis of equivalent circuit is carried out considering Figure 8.6c.

Equivalent impedance corresponding to the forward-rotating field

$$0.5Z_f = 0.5X_m \parallel \left(\frac{0.5R_2'}{s} + j\frac{0.5X_2'}{s}\right) = 0.5R_f + j0.5X_f \tag{8.8}$$

Equivalent impedance corresponding to the backward-rotating field

$$0.5Z_b = 0.5X_m \parallel \left(\frac{0.5R_2'}{(2-s)} + j\frac{0.5X_2'}{(2-s)}\right) = 0.5R_b + j0.5X_b \tag{8.9}$$

The equivalent circuit is shown in Figure 8.7 considering the parallel equivalent of the shunt branch components.

If magnetizing reactance is neglected $X_m \to \infty$, the impedance is

$$\begin{aligned} 0.5Z_f &= \frac{0.5R_2'}{s} + j\frac{0.5X_2'}{s} \\ \frac{R_f}{2} &= \frac{0.5R_2'}{s}; R_f = \frac{R_2'}{s} \\ \frac{X_f}{2} &= \frac{0.5X_2'}{s}; X_f = \frac{X_2'}{s} \end{aligned} \tag{8.10}$$

$$\begin{aligned} 0.5Z_b &= \frac{0.5R_2'}{2-s} + j\frac{0.5X_2'}{2-s} = 0.5R_b + j0.5X_b \\ \frac{R_b}{2} &= \frac{0.5R_2'}{2-s}; R_b = \frac{R_2'}{2-s} \\ \frac{X_b}{2} &= \frac{0.5X_2'}{2-s}; X_b = \frac{X_2'}{2-s} \end{aligned} \tag{8.11}$$

Air-gap power corresponding to the forward-rotating field

$$P_{gf} = (I_1)^2(0.5R_f) = \frac{1}{2}(I_1)^2\left(\frac{R_2'}{s}\right) \tag{8.12}$$

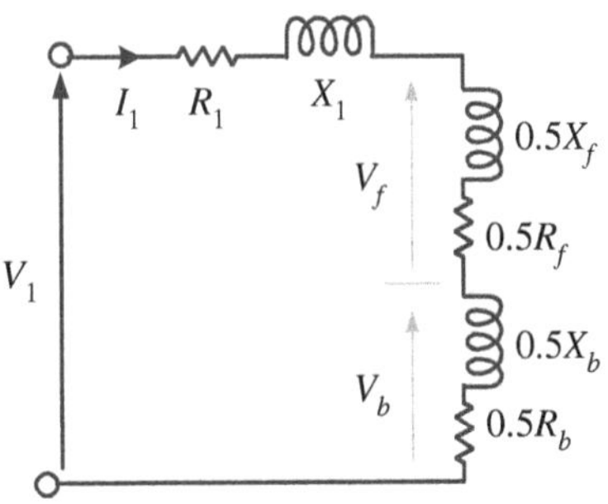

Figure 8.7 Equivalent circuit of a single-phase induction motor.

Air-gap power corresponding to the backward-rotating field

$$P_{gb} = (I_1)^2 (0.5R_b) = \frac{1}{2}(I_1)^2 \left(\frac{R_2'}{2-s}\right) \tag{8.13}$$

Torque developed corresponding to the forward-rotating field

$$T_f = \frac{P_{gf}}{\omega_s} \tag{8.14}$$

Torque developed corresponding to the backward rotating field

$$T_b = \frac{P_{gb}}{\omega_s} \tag{8.15}$$

Net developed torque

$$T_e = T_f - T_b = \frac{1}{2\omega_s}(I_1)^2 (R_f - R_b)$$

$$T_e = \frac{1}{2\omega_s}(I_1)^2 \left(\frac{R_2'}{2} - \frac{R_2'}{(2-s)}\right) \tag{8.16}$$

Output power from single-phase induction motor

$$P_{out} = T_e(\omega) \tag{8.17}$$

Rotor copper loss

$$P_{rc} = s_f P_{gf} + s_b P_{gb} = sP_{gf} + (2-s) P_{gb} \tag{8.18}$$

8.3.1.1 Approximate Equivalent Circuit

Approximate equivalent circuit is used for analysis and performance evaluation of a single-phase induction motor. Slip under full-load condition is between 5% and 10%. Thus, the backward-rotating field resistance $\frac{0.5R_2'}{2-s}$ varies from $\frac{R_2'}{3.9}$ to $\frac{R_2'}{3.8}$. Thus, it can be approximated equal to $\frac{R_2'}{4}$. The approximate equivalent circuit is shown in Figure 8.8a. Looking at the parallel branch of backward field; $0.5X_m \gg \sqrt{(0.25R_2')^2 + (0.5X_2')^2}$, the magnetizing branch of backward field is neglected and the resultant approximate equivalent circuit is shown in Figure 8.8b. This is further redrawn as Figure 8.8c in a more convenient form by combining the series branch of stator impedance and backward field impedance where

$$R = R_1 + \frac{R_2'}{4}$$

and

$$X = X_1 + 0.5X_2'$$

The approximate equivalent circuit of Figure 8.8c is similar to that of a three-phase induction motor.

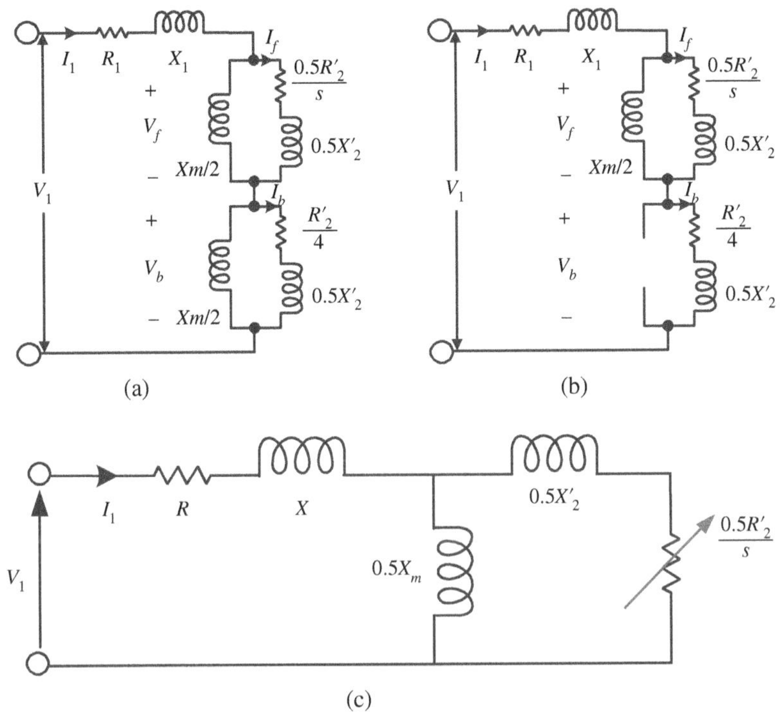

Figure 8.8 Approximate equivalent circuit of a single-phase induction motor.

8.3.1.2 Thevenin's Equivalent Circuit

Thevenin's equivalent circuit is drawn for the single-phase induction motor, considering the load point a to b as shown in Figure 8.9. Thevenin's equivalent impedance is obtained by looking from a to b and is parallel, equivalent of the magnetizing branch and stator impedance.

$$Z_{th} = R_{th} + jX_{th} = j\left(0.5X_m\right) \parallel (R + jX) = \frac{j\left(0.5X_m\right)(R + jX)}{\left(R + jX + j0.5X_m\right)} \tag{8.19}$$

Thevenin's voltage is the open-circuit voltage across point a to b.

$$V_{th} = \frac{j\left(0.5X_m\right)}{\left(R + j\left(0.5X_m + X\right)\right)} V_1 \tag{8.20}$$

Maximum Torque

Condition for maximum torque is obtained as

$$\frac{0.5R_2'}{s_{mT}} = \sqrt{\left(R_{th}\right)^2 + \left(X_{th} + 0.5X_m\right)^2}$$

$$s_{mT} = \frac{R_2'}{2\sqrt{\left(R_{th}\right)^2 + \left(X_{th} + 0.5X_m\right)^2}} \tag{8.21}$$

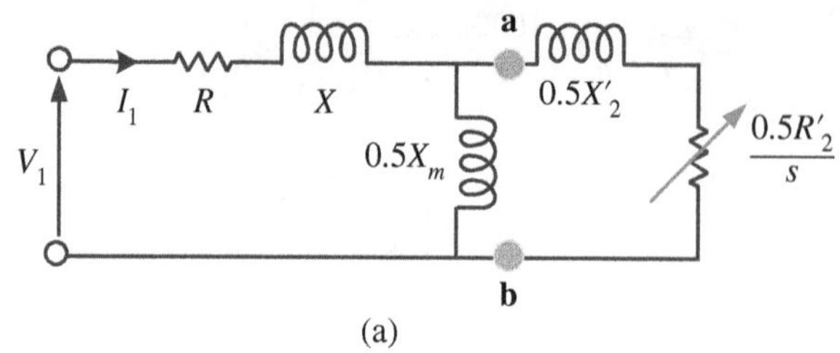

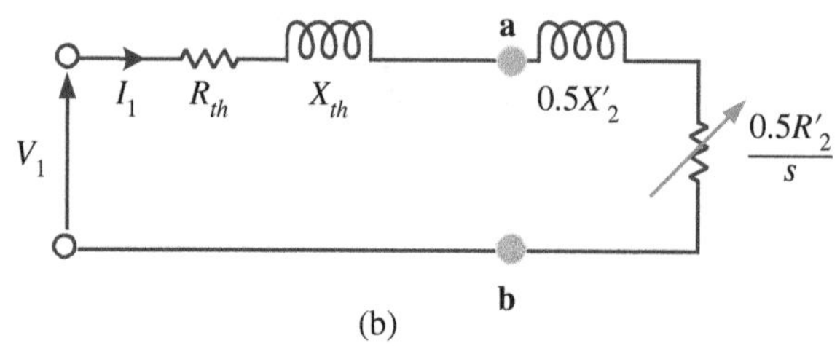

Figure 8.9 Thevenin's equivalent circuit of a single-phase induction motor.

Maximum torque is given as

$$T_{e\max} = \frac{V_{th}^2}{\omega_s} \frac{1}{2\left[R_{th} + \sqrt{(R_{th})^2 + (X_{th} + 0.5X_m)^2}\right]} \tag{8.22}$$

slip at which maximum torque occurs depends upon the rotor resistance. The slip increases with a rise in the rotor resistance.

Maximum torque also depends upon the rotor resistance and it decreases with a rise in the rotor-resistance value. It is of note that the maximum torque is independent of rotor resistance in a three-phase induction motor.

8.4 How to Make a Single-phase Induction Motor Self Starting

A single-phase motor can be made self-starting, if a rotating magnetic field can be produced. To produce rotating magnetic field, two windings are required with 90° phase shift. An additional winding known as auxiliary winding can be placed in addition to the main winding inside the stator with 90° space displaced as shown in Figure 8.10. The current supplied to the auxiliary winding is 90° phase shifted in time.

The current through the main winding is given by Eq. (8.2). The current through auxiliary winding is given by

$$i_{aux} = I_m \cos(\omega t + 90) \tag{8.23}$$

mmf produced by auxiliary current along rotor position θ is given as

$$\Im_{aux} = Ni_{aux} \cos(\theta + 90)$$

$$\Im_{aux} = NI_m \cos(\omega t + 90) \cos(\theta + 90)$$

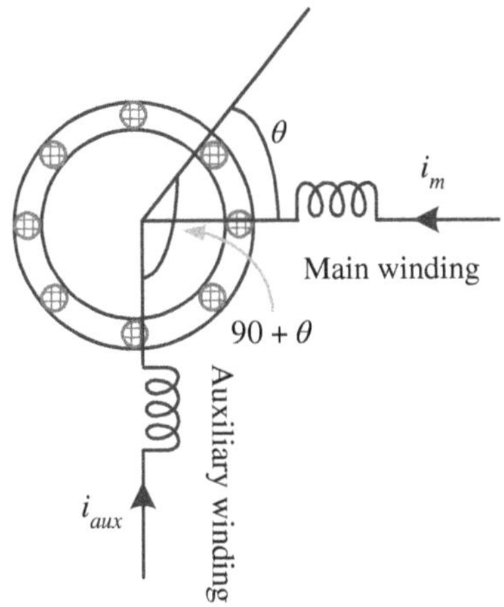

Figure 8.10 Single-phase induction motor with main and auxiliary winding.

$$\Im_{aux} = F_m \sin(\omega t)\sin(\theta)$$

$$\Im_{aux} = \frac{F_m}{2}\cos(\omega t - \theta) - \frac{F_m}{2}\cos(\omega t + \theta) = F_f - F_b \tag{8.24}$$

The mmf produced due to the current in auxiliary winding is given by Eq. (8.24). This also has two components, one rotating in the forward direction and one rotating in the backward direction.

The resultant mmf is the sum of mmf taking into account main and auxiliary winding.

$$\Im_{res} = \Im_m + \Im_{aux} = F_f + F_b + F_f - F_b = 2F_f$$

$$\Im_{res} = F_m \cos(\omega t - \theta) \tag{8.25}$$

The resultant mmf is forward rotating with amplitude $F_m = NI_m$ and speed is synchronous. Therefore, the single-phase induction motor is started as a two-phase motor (two identical windings displaced by 90° in space and supplied by two currents 90 time displaced in time).

To create a phase difference of 90° between the main and auxiliary winding currents, an appropriate value of capacitor is used in series with auxiliary winding. The following section discusses the design of capacitor for obtaining 90° phase shift between the main and auxiliary windings currents.

Let us consider a capacitor C is connected in a series with auxiliary winding to create a phase shift of 90° between main and auxiliary winding's currents as shown in Figure 8.11.

Rm = Resistance of main winding
Xm = Leakage reactance of main winding
Raux = Resistance of auxiliary winding
Xaux = Leakage reactance of auxiliary winding
I1 = Total supply current
Im = Main winding current
Ia = Auxiliary winding current

Phasor diagram showing main winding and auxiliary winding currents and applied voltage is shown in Figure 8.12.

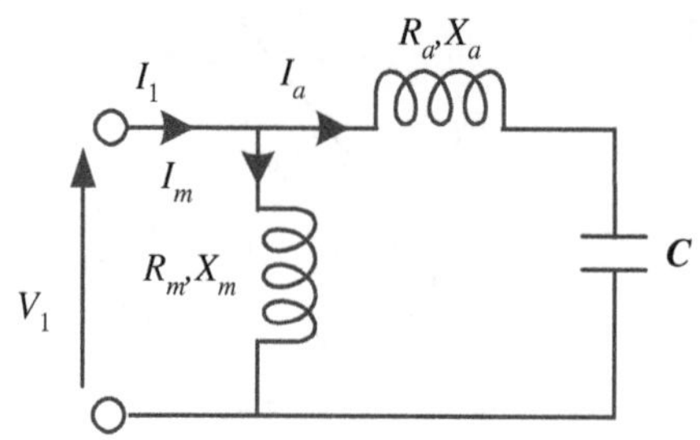

Figure 8.11 Main and auxiliary winding with series-connected capacitor to create phase shift.

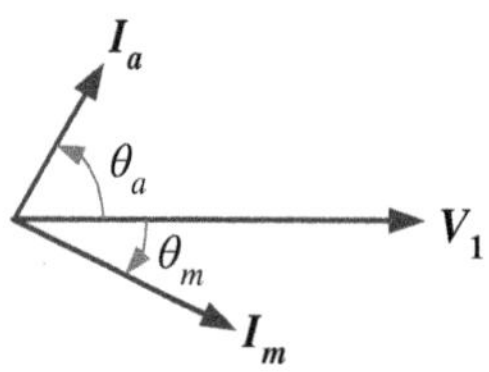

Figure 8.12 Phasor diagram of main and auxiliary winding currents.

Impedance of main winding

$$Z_m = R_m + jX_m = |Z_m| \angle \theta_m \tag{8.26}$$

Applied voltage

$$V_1 = |V_1| \angle 0$$

Current through main winding

$$I_m = \frac{V_1}{Z_m} = \frac{|V_1|}{|Z_m|} \angle - \theta_m \tag{8.27}$$

Main winding current lags the applied voltage by an angle θ_m, where

$$\theta_m = \tan^{-1}\left(\frac{X_m}{R_m}\right) \tag{8.28}$$

Impedance of auxiliary winding

$$Z_a = R_a + jX_a - jX_c = R_a - j\left(X_c - X_a\right) = |Z_a| \angle - \theta_a \tag{8.29}$$

For a phase shift of 90° between the two currents, the current through auxiliary winding should be leading in nature (Figure 8.12). Hence, the impedance angle of auxiliary winding is considered as negative.

Current through auxiliary winding

$$I_m = \frac{V_1}{Z_a} = \frac{|V_1|}{|Z_a|} \angle \theta_a \tag{8.30}$$

Auxiliary winding current leads the applied voltage by an angle θ_a, where

$$\theta_a = \tan^{-1}\left(\frac{X_c - X_a}{R_a}\right) \tag{8.31}$$

$$\theta_m + \theta_a = 90$$

Equation (8.31) is used to determine the value of capacitance.

Example 8.1 The impedance of main winding of a single-phase induction motor is 3.5 + j 3 Ω and impedance of auxiliary winding is 7.5 + j 4 Ω. Find the value of capacitor needed to connect in series with auxiliary winding to achieve a phase shift of 90° between two currents. The currents are supplied at 50 Hz.

Solution

Given, main winding impedance as

$$Z_m = 3.5 + j\,3\,\Omega = 4.6\angle 40.6^0$$
$$\theta_m = 40.6^0$$

Main winding current I_m lags the applied voltage by 40.6°

Given the impedance of auxiliary winding as

$$Z_a = 7.5 + j4 = 7.5 - j\,(X_c - 4) = |Z_a| \angle -\theta_a$$
$$\theta_a = \tan^{-1}\left(\frac{X_c - 4}{7.5}\right)$$
$$\theta_a + \theta_m = 90$$
$$\Rightarrow \tan^{-1}\left(\frac{X_c - 4}{7.5}\right) + 40.6 = 90$$
$$\Rightarrow \left(\frac{X_c - 4}{7.5}\right) = \tan(40.6 + 90) = 1.166$$
$$\Rightarrow X_c = 12.75\Omega$$

$$X_c = \frac{1}{2\pi fC}$$
$$C = \frac{1}{2\pi f X_c} = \frac{1}{2\pi\,(50)\,(12.75)} = 249.6\mu F$$

Example 8.2 A single-phase induction motor with only the main winding excited would exhibit the following response when running at synchronous speed

(a) Rotor current is zero
(b) Rotor current is non-zero and at slip frequency
(c) Forward and backward rotating fields are equal
(d) Forward field is more than the backward field

Solution

(b)

$$\text{Forward slip } s_f = \frac{N_s - N}{N_s} = s$$

$$\text{Backward slip } s_b = \frac{N_s + N}{N_s} = 2 - s$$

When rotor rotates at synchronous speed; $N_s = N$

$$\text{Forward slip } s_f = s = 0$$

$$\text{Backward slip } s_b = 2 - 0 = 2$$

Hence, the relative speed between forward field and rotor winding is zero. However, there is a relative motion between backward field and rotor so there is induced current in rotor corresponding to backward field at a frequency $f_2 = s_b f = 2f$

Example 8.3 A 220 V, four-pole, 50 Hz, single-phase induction motor has the following parameters

$$R1 = 2.3\,\Omega, X1 = X2' = 3.2\,\Omega, R2' = 4.2\,\Omega, Xm = 74\,\Omega$$

Iron losses is given as 80 W.

Windage and Friction loss is 20 W

If this motor is running at 1425 rpm at rated voltage and frequency, determine the following

(a) Stator current
(b) Input Power factor
(c) Power output
(d) Output Torque
(e) Efficiency

Assuming auxiliary winding as open circuited.

Solution

Given the equivalent circuit shown in Figure E8.3

$$\text{Synchronous speed } N_s = \frac{120f}{P} = \frac{120\,(50)}{4} = 1500\ rpm$$

Slip is calculated as $s = \frac{N_s - N}{N_s} = \frac{1500 - 1425}{1500} = 0.05$

Let us first calculate the impedance

Forward impedance

$$Z_f = \frac{j0.5X_m\left[\frac{0.5R_2'}{s} + j0.5X_2'\right]}{\frac{0.5R_2'}{s} + j0.5X_2' + j0.5X_m} = 17.66 + j20.75\Omega$$

Backward impedance

$$Z_b = \frac{j0.5X_m\left[\frac{0.5R_2'}{2-s} + j0.5X_2'\right]}{\frac{0.5R_2'}{2-s} + j0.5X_2' + j0.5X_m} = 0.99 + j1.562\Omega$$

Total impedance

$$Z = Z_f + Z_b = 20.95 + j25.5 = 33.1\angle 50.4$$

(a) Stator current drawn by motor

$$I_1 = \frac{V_1}{Z} = \frac{220\angle 0}{33.1\angle 50.4} = 6.64\angle -50.4$$

(b) Power Factor Cos(50.4) = 0.637 lagging

Figure E8.3 Equivalent circuit of induction motor.

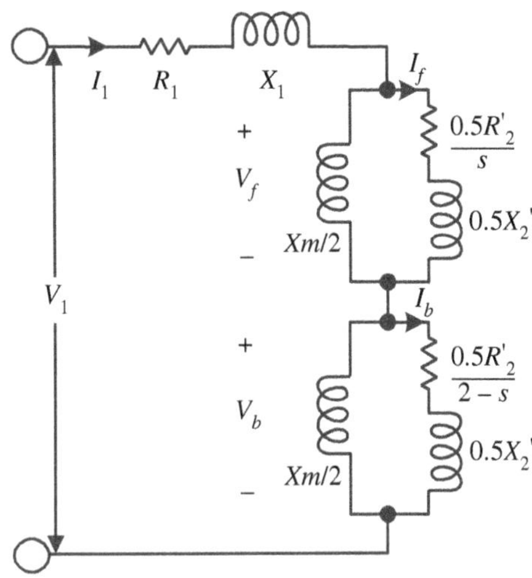

(c) Power output is calculated by computing the total power input to the rotor side (air gap power)

$$\text{Air-gap power due to forward field } P_{gf} = (I_1)^2 R_f = (6.64)^2\,(17.66) = 778.62W$$

$$\text{Air-gap power due to backward field } P_{gb} = (I_1)^2 R_b = (6.64)^2\,(0.99) = 43.64W$$

Mechanical Power Developed by machine

$$P_m = (1 - s)\left(P_{gf} - P_{gb}\right) = (1 - 0.05)\,(778.62 - 43.64) = 698.22W$$

Output power = Mechanical power – Rotational Losses

Rotational Losses = Windage and Friction Losses + Iron Losses = 80 + 20 = 100 W

$$P_{out} = P_m - R.Losses = 698.22 - 100 = 598.22W$$

(d) Output Torque

$$\text{Angular speed } \omega_r = \frac{2\pi N}{60} = \frac{2\pi\,(1425)}{60} = 149.22 rad/s$$

$$T_e = \frac{P_{out}}{\omega_r} = \frac{598.22}{149.22} = 4Nm$$

(e) Efficiency is calculated as

$$\text{Input power } P_{in} = V_1 I_1 \cos(\theta) = 220\,(6.64)\,(0.637) = 930.53W$$

$$\text{Efficiency } \eta = \frac{P_{out}}{P_{in}} X100\% = \frac{598.22}{930.53} X100\% = 64.3\%$$

Example 8.4 A 220 V, 50 Hz, four-pole single-phase induction motor draws a line current of 50 < −75 at standstill condition with main winding excited with rated voltage. For a slip of 0.04 compute the following

(a) Current drawn under running condition
(b) Power factor
(c) Mechanical power developed
(d) Torque developed
(e) Output torque

Neglect the stator impedance, magnetizing current and rotational losses.

Solution

Stator impedance and magnetizing branch is neglected.

Impedance under standstill condition (s = 1) is

$$Z_2 = R_2' + jX_2' = \frac{V_1}{I_1} = \frac{220\angle 0}{50\angle -75} = 4.4\angle 75 = 1.138 + j4.25$$

$$R_2' = 1.138\Omega; X_2' = 4.25\Omega$$

Under running condition for a slip of 0.04

Forward impedance

$$Z_f = \frac{0.5R_2'}{s} + j\frac{X_2'}{2} = \frac{0.5\,(1.138)}{0.04} + j\frac{(4.25)}{2} = 14.225 + j2.125$$

Backward impedance

$$Z_b = \frac{0.5R_2'}{2-s} + j\frac{X_2'}{2} = \frac{0.5\,(1.138)}{1.96} + j\frac{(4.25)}{2} = 0.29 + j2.125$$

Total input impedance under running condition

$$Z = Z_f + Z_b = 14.225 + j2.125 + 0.29 + j2.125 = 14.515 + j4.25 = 15.124\angle 16.32$$

(a) Current drawn by motor under running condition

$$I_1 = \frac{V_1}{Z} = \frac{220\angle 0}{15.124\angle 16.32} = 14.54\angle -16.32$$

(b) Power facto cos (16.32) = 0.96

(c) Mechanical power developed

$$P_m = (1-s)\left(I_1\right)^2\left(R_f - R_b\right) = (1-0.04)\,(14.54)^2\,(14.225-0.29) = 2.881kW$$

(d) Torque developed $T_e = \frac{P_m}{\omega_r} = \frac{P_m}{(1-s)\omega_s} = \frac{2.88kW}{(1-0.04)\frac{2\pi(1500)}{60}} = 19Nm$

(e) Output torque is the same as developed torque as the friction and core losses are neglected.

8.5 Testing of an Induction Machine

To determine the equivalent circuit parameters, three different tests are performed, namely

- DC test
- No-load test
- Blocked rotor test

8.5.1 DC Test

This test is performed to determine the stator winding resistance. A suitable DC voltage is applied and DC current is measured. The resistance is obtained as

$$R_{1dc} = \frac{V_{dc}}{I_{dc}} \tag{8.32}$$

The actual resistance is 10–20% higher due to skin effect.

8.5.2 No-load Test

To conduct a no-load test, the motor is allowed to run without load and rated voltage is applied at rated frequency.

Measured quantities

Voltmeter reading: Rated Voltage (V_1)
Ammeter Reading: No-load current (I_{nl})
Wattmeter reading: No-load losses (P_{nl})

Under no-load condition, the slip is almost zero $s \approx 0$. For zero slip, the forward resistance is infinite $\left(\frac{R_2'}{s} = \infty\right)$.

The backward resistance is $\left(\frac{0.5R_2'}{2-s} = \frac{R_2'}{4}\right)$

Furthermore, this backward resistance is much smaller than the magnetizing reactance $\left(\frac{X_m}{2} \gg \sqrt{\left(\frac{R_2'}{4}\right)^2 + \left(\frac{X_2'}{2}\right)^2}\right)$, hence the magnetizing branch in the backward circuit is neglected. The equivalent circuit become as shown in Figure 8.13.

Parameter Calculation:

$$\text{No-load power loss } P_{nl} = V_1 I_{nl} \cos\left(\theta_{nl}\right)$$

$$\cos\left(\theta_{nl}\right) = \frac{P_{nl}}{V_1 I_{nl}} \tag{8.33}$$

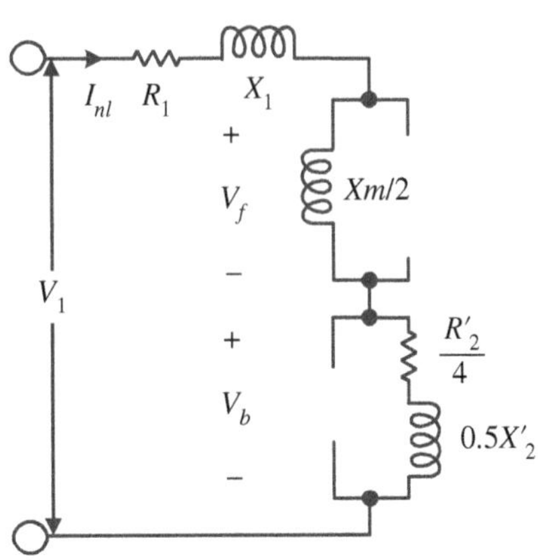

Figure 8.13 Equivalent circuit under no-load condition.

No-load equivalent impedance

$$Z_{nl} = \frac{V_1}{I_{nl}} \tag{8.34}$$

No-load leakage reactance

$$X_{nl} = Z_{nl} \sin\left(\theta_{nl}\right) \tag{8.35}$$

From the circuit of Figure 8.13, the no-load impedance is given as

$$Z_{nl} = R_1 + \frac{R_2'}{4} + j\left(X_1 + 0.5X_2' + 0.5X_m\right) \tag{8.36}$$

The no-load reactance component is given as

$$X_{nl} = \left(X_1 + 0.5X_2' + 0.5X_m\right) \tag{8.37}$$

If X_1 and $0.5X_2'$ are known (obtained from blocked-rotor test), X_m can be calculated from Eq. (8.37).

8.5.3 Blocked-Rotor Test

To conduct a blocked-rotor test, the motor is mechanically blocked from rotation. Reduced voltage is applied to the main winding (keeping the auxiliary winding open) so that rated current flows.

Measured quantities

Voltmeter reading: Reduced Voltage (V_{sc})
Ammeter Reading: Rated current (I_{sc})
Wattmeter reading: Copper losses (P_{sc})

Due to s = 1, the voltage needed to flow rated current is small. As the applied voltage is low, the flux produced is small thus the magnetizing current is also low and the magnetizing branch can be neglected, leading to the equivalent circuit as shown in Figure 8.14.

Parameter Calculation

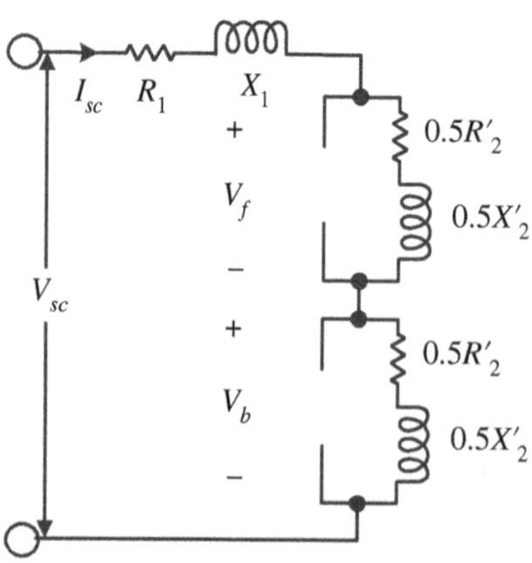

Figure 8.14 Equivalent circuit under blocked-rotor condition.

Impedance under blocked-rotor condition

$$Z_{sc} = \frac{V_{sc}}{I_{sc}} \tag{8.38}$$

The equivalent resistance is given as

$$R_{sc} = R_1 + R_2' = \frac{P_{sc}}{\left(I_{sc}\right)^2} \tag{8.39}$$

Stator resistance R_1 is obtained by DC test.

The equivalent leakage reactance is calculated as

$$X_{sc} = X_1 + X_2' = \sqrt{\left(Z_{sc}\right)^2 - \left(R_{sc}\right)^2} \tag{8.40}$$

Leakage reactance X_1 and X_2' cannot be separated. It is a common practice to assume them equal.

Example 8.5 A 220 V, four-pole, 50 Hz, 300 W single-phase induction motor is tested whose results are given below.

No-load test data: 220 V, 80 W, 3.2 A
Blocked Rotor test data: 90 V, 200 W, 9 A
DC test data: 40 V, 16 A

Compute the equivalent circuit parameters.

Solution

DC test: 40 V, 16 A

$$R_{1dc} = \frac{V_{dc}}{I_{dc}} = \frac{40}{16} = 2.5\Omega$$

AC resistance is assumed to be 15% higher than the DC value;

$$R_1 = 1.15R_{dc} = 1.15\,(2.5) = 2.875\Omega$$

Calculation from blocked-rotor test data

$$Z_{sc} = \frac{V_{sc}}{I_{sc}} = \frac{90}{8} = 11.25\Omega$$

$$R_{sc} = R_1 + R_2' = \frac{P_{sc}}{\left(I_{sc}\right)^2} = \frac{250}{(8)^2} = 3.9\Omega$$

$$R_2' = R_{sc} - R_1 = 3.9 - 2.875 = 1.025\Omega$$

$$X_{sc} = X_1 + X_2' = \sqrt{\left(Z_{sc}\right)^2 - \left(R_{sc}\right)^2} = \sqrt{(11.25)^2 - (3.9)^2} = 10.55\Omega$$

Thus, the leakage reactance

$$X_1 = X_2' = \frac{X_{sc}}{2} = 5.27\Omega$$

Calculation from no-load test data

$$\cos\left(\theta_{nl}\right) = \frac{P_{nl}}{V_1 I_{nl}} = \frac{80}{220\,(3.2)} = 83.47^\circ$$

$$\sin\left(\theta_{nl}\right) = 0.993$$

$$Z_{nl} = \frac{V_1}{I_{nl}} = \frac{220}{3.2} = 68.75\Omega$$

$$X_{nl} = Z_{nl}\sin\left(\theta_{nl}\right) = 68.75\,(0.993) = 68.26\Omega$$

$$X_{nl} = \left(X_1 + 0.5X_2' + 0.5X_m\right)$$

$$X_m = X_{nl} - \left(X_1 + 0.5X_2'\right) = 68.26 - (5.27 + 0.5\,(5.27)) = 60.355\Omega$$

8.6 Types of Single-Phase Induction Motors

As mentioned above, the induction motor does not self-start. There are many ways to start a single-phase induction motor by itself. Single-phase induction motors are classified into the following types using the starting method.

8.6.1 Split-Phase Induction Motor

This is one of the most commonly used single-phase induction motors. The main components of the split-phase motor include main winding, auxiliary winding and a centrifugal switch. This is the simplest way to establish a rotating magnetic field with two windings in the same stator core as shown in the Figure 8.15a and the respective phasor diagram shown in Figure 8.15b. The auxiliary or starting winding has a series resistance that makes its impedance highly resistant. It is not the same winding as the main winding, but has fewer turns of a much smaller diameter than the main winding. This will reduce the voltage lagging of the starting current. The main winding is inductive in nature to lag the voltage by a certain angle. This winding is designed to operate at synchronous speeds of 75% and above. These two windings are connected to the input AC supply in parallel. Because of the inductive nature, the supply voltage lags at a large angle, with the current induced by the starting winding almost being in phase with the voltage due to its resistive nature. Thus, there is a phase difference between these currents. The resulting current produces a rotating magnetic field and thus the starting torque. The centrifugal switch is connected to the starting winding in series. When the motor is 75–80% of the synchronous speed, the centrifugal switch is opened mechanically, thereby removing the auxiliary winding. The motor, therefore, only works with the main winding. Split-phase motors give a poor start torque because of the small difference in phases between main and auxiliary currents. The power factor of these motors is also poor. These are mainly used for loads that are easily started like blowers, fans, washing machines, grinders, etc.

8.6.2 Capacitor-Start Induction Motor

This motor is similar to the split-phase motor, but the capacitor is also connected to auxiliary winding in series. This is a modified version of the split-phase motor. The use of a capacitor increases the phase angle between the two currents (main and auxiliary) and hence the

Figure 8.15 (a) Split-phase IM and (b) phase diagram.

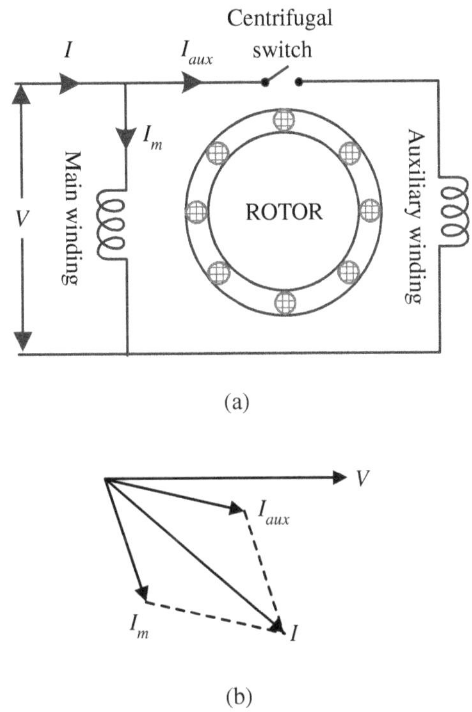

starting torque, as the capacitor draws a leading current. This is the main reason why a capacitor is used in single-phase induction motors. The capacitor is electrolytic and is only designed for alternating current usage. Due to the lower cost of capacitors, these motors are becoming more popular in various applications. The capacitors are designed for a certain service cycle, but not for continuous use. The scheme of the starting capacitor motor is shown in Figure 8.16. The operation of this motor is similar to the split-phase motor, where additional winding provides the starting torque. Once the speed has been picked up, the additional winding and capacitor are removed by the centrifugal switch. The difference is that the torque produced by the motor is higher than the split-phase motor because of the use of the capacitor. Due to the presence of a capacitor, the current by means of an auxiliary winding causes the voltage at an angle to be greater than that of a split case. The phase difference between main and auxiliary currents is thus increased and it also increases the starting torque. The motor's performance is the same as the split-phase motor when it runs at full speed. The inrush currents in this motor are reduced due to the capacitor. These motors have a very high starting torque of up to 300% of the full-load torque. At rated load and rate of speed, however, the power factor is low. These motors are used for domestic as well as industrial applications such as water pumps, grinders, machine lathes, compressors, etc.

8.6.3 Capacitor-Start Capacitor-Run Induction Motor (Two-Value Capacitor Method)

These motors are also referred to as two-value capacitor motors. They combine the advantages of the start capacitor type and induction motors of a permanent capacitor

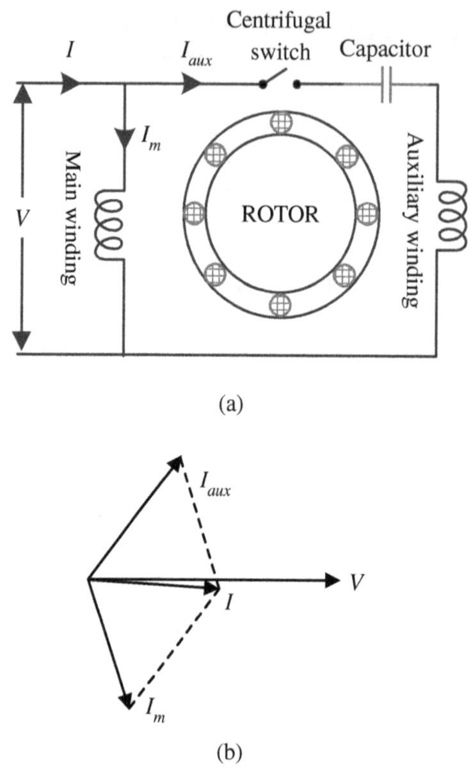

Figure 8.16 (a) Capacitor start IM and (b) phase diagram.

type. This motor consists of two capacitors with a different capacitance value for start and run. A high-value capacitor is used at starting conditions while a low value is used to run conditions. It is important to note that this motor uses the same winding arrangement as the capacitor-start motor during start-up and working conditions. The schematic configuration of this motor is shown in Figure 8.17. At first, both start and run capacitors are connected to the auxiliary winding in series. The motor-starting torque is therefore more comparable with other motor types. The centrifugal switch disconnects the starting capacitor after the motor reaches some speed and releases the running capacitor in series with the auxiliary winding. Both running and auxiliary windings remain in operating condition thus improving the motor's power factor and efficiency. These are the most common single-phase motors due to the high start torque and improved power factor. They are applied in compressors, fridges, air conditioners, conveyors, ceiling fans, air blowers and so on.

8.7 Single-Phase Induction Motor Winding Design

The principal objective of the starting winding (auxiliary) is to develop the starting torque. The starting winding can, however, be designed for maximum starting torque or optimizing starting torque per ampere of the starting current. A design procedure is described in this section to achieve these goals. First, an expression is derived for the

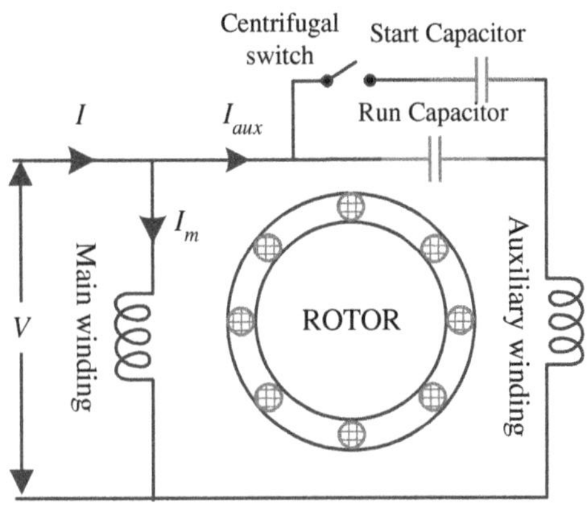

Figure 8.17 Capacitor start capacitor run induction motor.

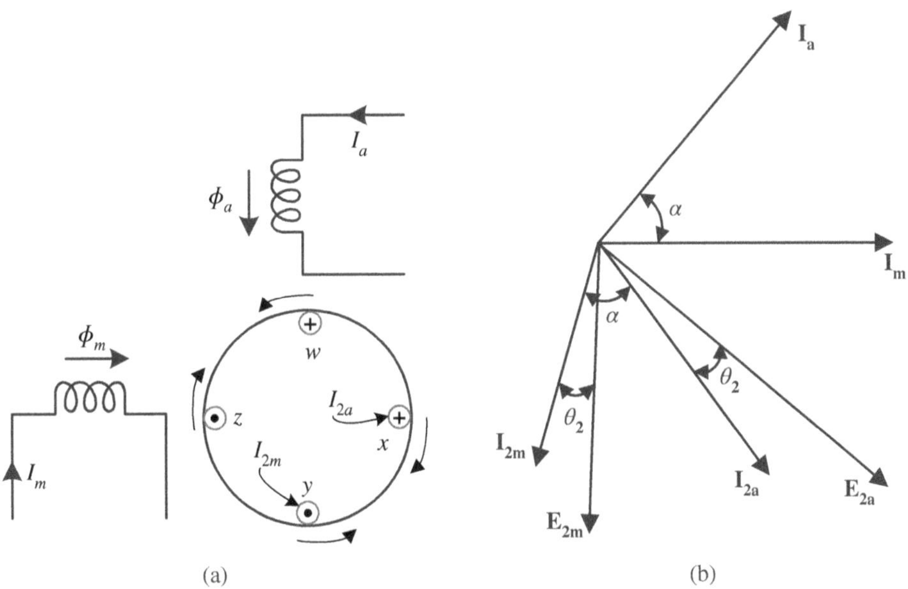

Figure 8.18 Phasor diagram of the single-phase induction motor (IM).

starting torque. Figure 8.18a shows the main and auxiliary windings along with the cage rotor which can be represented by two-phase equivalent windings having coil *w-y* and *x-z* as two coils. Figure 8.18b shows the phasor diagram for the single-phase IM. The cage rotor is equivalent to the two-phase winding. Let N_2, R_2, and X_2 be the number of turns, resistance and inductance of these two windings. The current flowing through the main and auxiliary winding generates the flux which causes the induced voltage which is given as

$$\left.\begin{aligned} E_{2m} &= 4.44 N_2 f \phi_m \\ E_{2a} &= 4.44 N_2 f \phi_a \end{aligned}\right\} \tag{8.41}$$

As both windings are inductive in nature their respective current lags the induced voltage. Therefore, the phase angle between voltage and current is given as

$$\theta_2 = \cos^{-1}\frac{R_2}{\sqrt{R_2^2+X_2^2}} = \cos^{-1}\frac{R_2}{Z_2} \tag{8.42}$$

The torque is developed due to the interaction of the fluxes generated by the main and auxiliary windings. The torque generated by the main winding flux, T_m is given as

$$T_m \propto \phi_m I_{2a}\cos\left(90^\circ + \theta_2 - \alpha\right) \tag{8.43}$$

and the torque generated by the auxiliary winding flux, T_a is given as

$$T_a \propto \phi_a I_{2m}\cos\left(90^\circ + \theta_2 + \alpha\right) \tag{8.44}$$

The total starting toque T_s is given as

$$T_s \propto \left[\phi_m I_{2a}\cos\left(90^\circ + \theta_2 - \alpha\right) - \phi_a I_{2m}\cos\left(90^\circ + \theta_2 + \alpha\right)\right] \tag{8.45}$$

However, as the flux generated by the windings is proportional to their respective current, the following equations relate to both fluxes and currents.

$$\phi_m I_{2a} = \phi_a I_{2m} \tag{8.46}$$

Therefore

$$T_a \propto \phi_a I_{2m}\cos\theta_2 \sin\alpha \tag{8.47}$$

From (8.41)

$$\phi_a = \frac{E_{2a}}{4.44N_2 f} = \frac{I_{2m}Z_2}{4.44N_2 f} \tag{8.48}$$

From (8.42), (8.47), and (8.48)

$$T_s \propto \frac{I_{2m}I_{2a}R_2\sin\alpha}{4.44N_2 f} \tag{8.49}$$

As the current I_{2m} and I_{2a} are propositional to I_m and I_a, therefore

$$T_s \propto I_a I_m \sin\alpha$$

$$T_s = K I_a I_m \sin\alpha \tag{8.50}$$

From the above equation, it is clear that the starting torque is dependent on the main and auxiliary winding currents and the phase difference between these two currents.

The main winding in split-phase motors is designed to satisfy the running motor operation, while the auxiliary winding is designed to produce the desired starting torque in line with the main winding, without excessive starting current. A convenient approach is to take several turns for the starting winding and calculate the value for the desired starting torque for the starting winding resistance. If the optimal torque and current starting design is required, a range of values can be calculated for the starting winding turns until optimum design is achieved. The resistance in the auxiliary winding can be determined to maximize the starting torque if the number of turns N_a for the starting winding is specified. Let Z_m

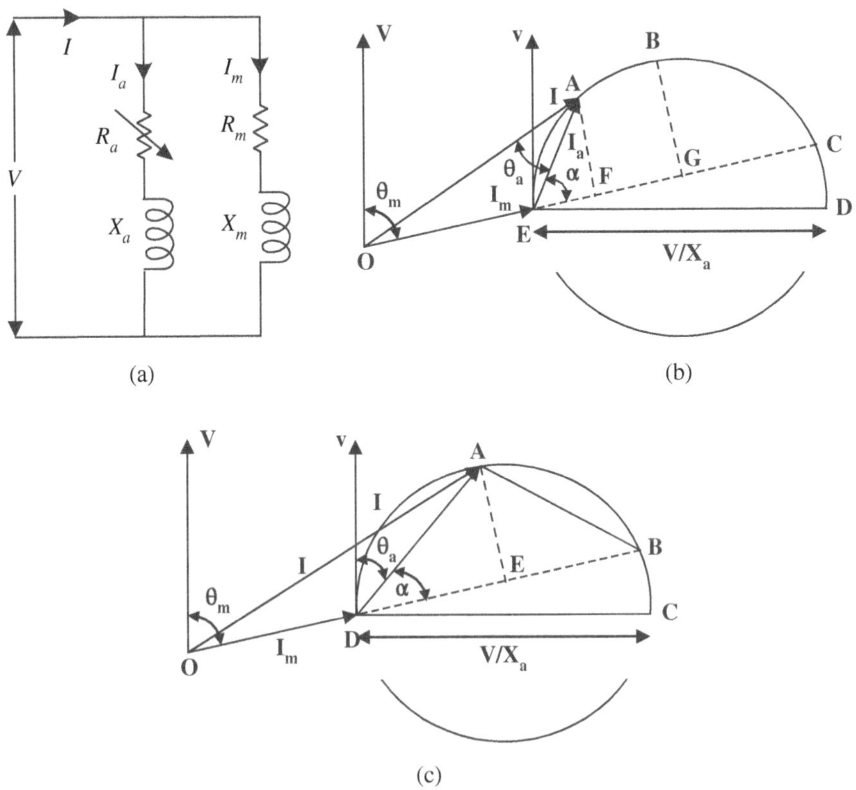

Figure 8.19 Split-phase induction motor.

and Z_a represents the impedence of the main and auxiliary winding and can be represented as

$$\left.\begin{aligned} Z_m &= R_m + jX_m \\ Z_a &= R_a + jX_a \end{aligned}\right\} \tag{8.51}$$

8.7.1 Split-Phase Induction Motor

Figures 8.19a and 8.19b show the equivalent circuit and phasor diagram of the split-phase induction motor respectively. As the nature of winding is inductive, the current phasor lags the voltage by an angle θ_m. The source current is the sum of the main and auxiliary winding currents and is shown as I=OA. As the magnitude of the main winding current I_m has fixed magnitude, the magnitude of I depends on the variation of auxiliary winding current I_a, thus resulting in the variation of the angle α.

For a very high value of R_a, the magnitude of I_a will be very low. Thus, I will be very close to I_m. Similarly, for a very low value of R_a, the current I_a can be represented by phasor ED which lags V by an angle of 90°. Therefore, the locus of input current I and current I_a forms a semicircle with a diameter of phasor ED. As the current I_m has a fixed value, from (10), the equation for starting torque can be represented as

$$T_s \propto I_a \sin\alpha = \text{Length of } AF \tag{8.52}$$

To achieve maximum starting torque, i.e. phasor AF, the operating point must be point B which is mid-point between C and E, thus the length of phasor AF changes to BG as shown in Figure 8.19c. As the phasor AE represents the radius of the semicircle and DC represents the diameter of the semicircle

$$\theta_m = 2\theta_a = \alpha \tag{8.53}$$

Therefore

$$\theta_a = \cot^{-1}\frac{1+\cos\theta_m}{\sin\theta_m} \tag{8.54}$$

Similarly, the resistance of the starting current is given as

$$\frac{R_a}{X_a} = \frac{1+\frac{R_m}{|Z_m|}}{\frac{X_m}{|Z_m|}} = \frac{R_m+|Z_m|}{X_m}$$

$$\Rightarrow R_a = \frac{X_a}{X_m}\left(R_m+|Z_m|\right) = \left|\frac{N_a}{N_m}\right|^2\left(R_m+|Z_m|\right) \tag{8.55}$$

The current of the starting winding is calculated as

$$I_a = \frac{V}{Z_a} = \frac{V}{\sqrt{R_a^2+X_a^2}} \tag{8.56}$$

8.7.2 Capacitor-Start Motors

Figure 8.20a shows the schematic circuit diagram for the capacitor start motor with its phasor diagram shown in Figure 8.20b. Due to being inductive in nature, the current I_m lags the voltage V by an angle θ_m. As a capacitor is integrated into the auxiliary winding, the current I_a leads the voltage by an angle θ_a. If the value of C is very high, the capacitive reactance will be very low, thus the magnitude of current I_a will be very low and $I = I_m$. Similarly, if the value of C is selected such that, $X_C = X_a$, the current I_a will have its maximum value as shown by phasor DB and will be in phase with voltage V. As the magnitude of I_m is fixed, the starting torque is given by

$$T_s \propto I_a \sin\alpha = \text{Length of } DB \tag{8.57}$$

The phasor of DB is maximum when it is equal to the diameter of the circle. Therefore, from the geometry of the phasor diagram

$$\theta_a = \frac{90^\circ - \theta_m}{2} \tag{8.58}$$

$$\tan\theta_a = \sqrt{\left(\frac{1-\cos 2\theta_a}{1-\cos 2\theta_a}\right)} = \frac{R_m}{|Z_m|+X_m} \tag{8.59}$$

Also

$$\tan\theta_a = \frac{X_c - X_a}{R_a} \tag{8.60}$$

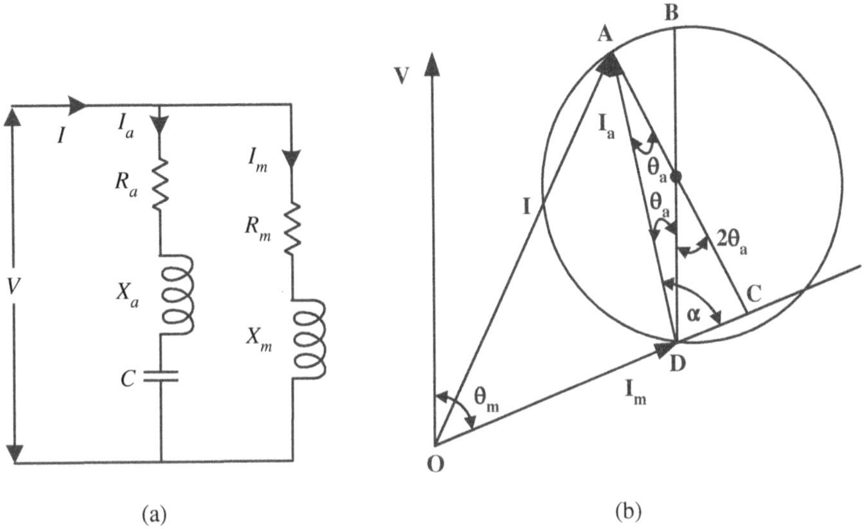

Figure 8.20 Capacitor start motor.

From (8.58) and (8.59)

$$X_C = X_a + \frac{R_m}{|Z_m| + X_m}$$

$$\Rightarrow C = \frac{1}{\left[X_a + \frac{R_m}{|Z_m| + X_m}\right] \times \omega} \tag{8.61}$$

Equation (8.61) gives the value of the capacitor for maximum starting torque.

Example 8.6 A single-phase, split-phase induction motor has main and auxiliary impedances which are 2 + j4 and 4 + j3 respectively. The input voltage rating of this motor is 150 V, 50 Hz. Calculate the capacitance value of the series connected capacitor to produce a pure-forward mmf wave.

Solution

$$\text{phase angle of the main winding} = \phi_m = \tan^{-1}\left(\frac{X_m}{R_m}\right) = \tan^{-1}\left(\frac{4}{2}\right) = 63.43°$$

For a pure-forward mmf, the phase angle between the main and auxiliary winding must be 90°. The required angle which needs to be achieved through the capacitor in auxiliary winding must be 63.43° − 90° = −26.57°

$$\phi_a = \tan^{-1}\left(\frac{X_a - X_C}{R_a}\right)$$

$$-26.57 = \tan^{-1}\left(\frac{4 - X_C}{3}\right)$$

$$\Rightarrow X_C = 5.5\Omega$$

$$C = \frac{1}{2\pi f \times X_C} = \frac{1}{2\pi \times 50 \times 5.5} = 578.7\mu F$$

Example 8.7 A single-phase capacitor-start induction motor, four-pole, 240 V, 50 Hz, has the following impedances at standstill conditions.

$$Z_m = 3.5 + j6.2$$
$$Z_a = 7.2 + j4.8$$

(a) To produce a 90° phase shift between main and auxiliary current, calculate the value of the starting capacitor.
(b) Compare the starting torques and starting current with and without the starting capacitor.

Solution

(a)

$$\phi_m = \tan^{-1}\left(\frac{X_m}{R_m}\right) = \tan^{-1}\left(\frac{6.2}{3.5}\right) = 60.55^\circ$$

$$\phi_a = \tan^{-1}\left(\frac{X_a - X_C}{R_a}\right) = 60.55^\circ - 90^\circ$$

$$-29.45 = \tan^{-1}\left(\frac{4.8 - X_C}{7.2}\right)$$

$$\Rightarrow X_C = 8.86\Omega$$

$$C = \frac{1}{2\pi f \times X_C} = \frac{1}{2\pi \times 50 \times 8.86} = 359.3\mu F$$

(b) Without starting capacitor

$$I_m = \frac{240}{3.5 + j6.2} = 33.71\angle -60.55A$$

$$I_a = \frac{240}{7.2 + j4.8} = 27.74\angle -33.69A$$

Starting current

$$I_{st} = I_m + I_a = 33.71\angle -60.55 + 27.74\angle -33.69$$

$$I_{st} = 59.78\angle -48.45A$$

Starting Torque

$$T_{st} \propto I_m I_a \sin\alpha$$

$$\alpha = \phi_a - \phi_m = -33.69 - (-60.55) = 26.86^\circ$$

With starting capacitor

$$Z_{a,total} = 7.2 + j4.8 - j8.86 = 7.2 - j4.06$$

$$I_a = \frac{240}{7.2 - j4.06} = 29.04\angle 29.42A$$

Starting current

$$I_{st} = I_m + I_a = 33.71\angle -60.55 + 29.04\angle 29.42$$

$$I_{st} = 44.50\angle -19.82A$$

Starting Torque

$$T_{st} \propto I_m I_a \sin \alpha$$

$$\alpha = \phi_a - \phi_m = 29.42 - (-60.55) = 89.97^\circ$$

Ratio of starting current with and without stating capacitors

$$\frac{I_{st|with}}{I_{st|without}} = \frac{44.5}{59.78} = 0.744$$

Ratio of starting torque with and without stating capacitors

$$\frac{T_{st|with}}{T_{st|without}} = \frac{I_m I_a \sin \alpha|_{with}}{I_m I_a \sin \alpha|_{without}} = \frac{29.04 \times \sin 89.97}{27.74 \times \sin 26.86} = 2.32$$

8.8 Permanent Split-Capacitor (PSC) Motor

The Permanent Split-Capacitor (PSC) motor also has a cage rotor and two coils called the main and auxiliary coils similar to the capacitor start and start capacitor start motor. It only has a capacitor connected to the starting winding in series. The capacitor C is permanently connected both in initial and running conditions in the circuit. The connection diagram of a PSC motor is shown in Figure 8.21. It is also called a motor with a single value. As the capacitor is always in the circuit, there is no centrifugal switch for this type of motor. There is always the auxiliary winding on the circuit. Thus, the motor functions as a balanced two-phase motor. The motor produces a uniform torque and is noise-free.

PSC motor advantages

- ➢ There is no centrifugal switch needed.
- ➢ Efficiency is high.
- ➢ As the capacitor is permanently connected to the circuit, the power factor is high.

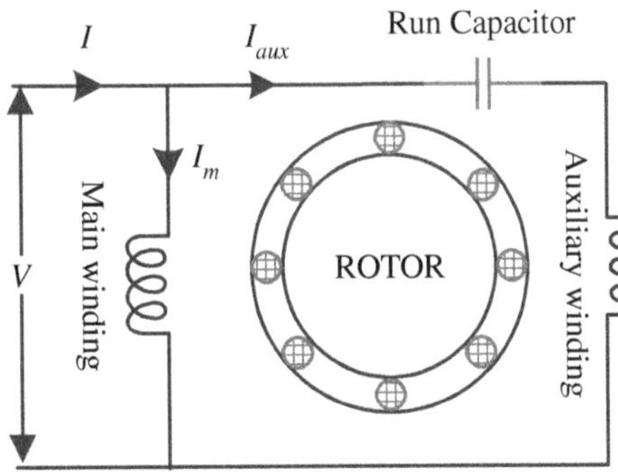

Figure 8.21 Permanent-split capacitor motor.

- It has a higher torque pull-out.

The motor limitations are as follows

- The paper capacitor cannot be used for continuous operation in the motor as an electrolyte capacitor. The cost of the paper capacitor is higher and the size is also large compared to the same electrolytic capacitor.
- It has a small starting torque and a load torque less than full.

PSC Motor applications

- Used in heaters and air conditioners in fans and blowers.
- Used in compressors for refrigeration.
- Used in office machinery.

8.9 Shaded-Pole Induction Motor

This motor employs entirely different techniques to start the motor compared to other motors we have discussed so far. This motor uses no auxiliary winding or even a rotating field, but a shading coil that sweeps across the poles is sufficient to drive the motor. The field is therefore moving from one side of the pole to the other. While these motors are small, inefficient and have low-starting torque, they are used in several applications because of their outstanding characteristics, such as roughness, low initial cost, small size and simple design. A shadowed-pole motor consists of a stator with salient pole(s) and a squirrel-type rotor. In this way, the stator is specially built to produce moving magnetic fields. Every pole is divided into two parts: shaded and unshaded. A shading portion is a slot, which is cut across the laminations at about one-third from one corner, with a heavy copper ring around it (also referred to as a shading coil or a copper-shade band). This part of the shading coil is usually called the shaded part of the pole and the rest of the part is referred to as the unshaded part as illustrated in Figure 8.22.

8.10 Universal Motor

A universal motor is a special motor type designed for operation with either a DC or a single-phase AC. These motors are generally series wound (series of reinforcement and field windings) and therefore produce high starting torque. Therefore, universal motors are usually integrated into the device they are intended to drive. Most universal motors operate at higher speeds, which exceed 3500 rpm. The reactance voltage drop in AC is higher than the DC supply with the same voltage, thus AC provides a lower speed.

A universal motor construction is very similar to that of a DC machine. It consists of a stator with field poles. Field coils on the field poles are wound. The entire magnetic track (stator field circuit and armature) is, however, laminated. Lamination is necessary to minimize the currents induced by eddy during operation on AC. The rotary frame of the wound type is equipped with straight or skewed slots. When current flows in the winding field, an electromagnetic field is produced. The same current is also generated by the reinforcement

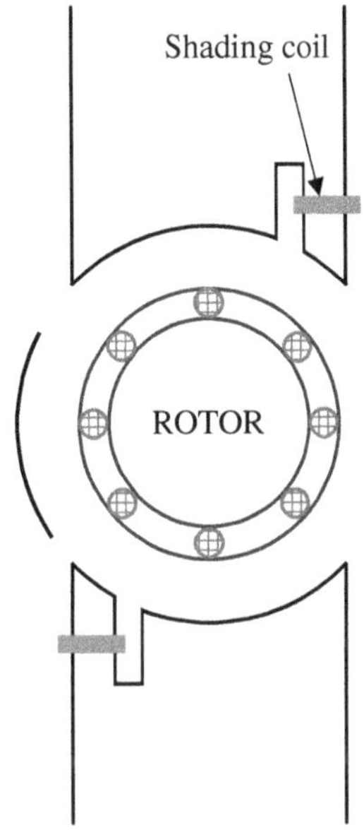

Figure 8.22 Shaded-pole induction motor.

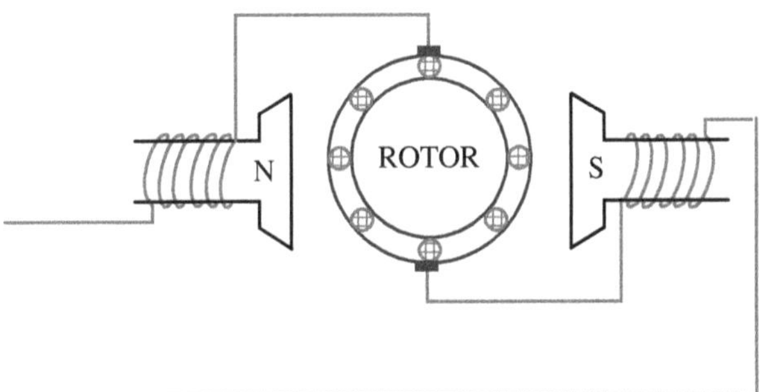

Figure 8.23 Universal motor.

conductors. When a conductor is placed in an electromagnetic field, a mechanical force is experienced. The rotor begins to rotate due to this mechanical force, or torque. Fleming's left-hand rule is the direction of this force. It still produces unidirectional torque when fed with AC supply. Because of the series connection of armature winding and field winding, they are in the same phase. Therefore, with AC polarity changing periodically, the current direction in armature and winding in field reverses simultaneously. Therefore, the direction of the magnetic field and direction of the bracket current reverts to the same direction of force as the armature conductors. So, the universal motor operates on the same principle as the DC series motor, regardless of AC or DC supply. Universal motors are used in a variety of home appliances such as vacuum cleaners, food and beverage mixers, domestic sewing machines, drills, mixers etc. (Figure 8.23)

8.11 Switched-Reluctance Motor (SRM)

SR motors are a special type of continuous operation variable reluctance motor. The switched reluctance motor (SRM) is an electric motor that works by torque reluctance. In contrast to standard DC brushed-motor types, power is delivered to windings in the stator instead of the rotor. This simplifies mechanical design greatly, as power does not need to be supplied to a moving part, but complicates the electrical design because a certain type of switching system has to be used to supply power to various windings. Electronic devices can be used precisely, making SRM configurations easier. Its main disadvantage is the ripple of torque. Properly designed SR motors are capable of low-weight high-torque motors that are attractive to vehicles and aerospace applications. The machines are structurally simple, but relatively advanced controls can be required for optimal performance. Figure 8.24 shows a diagram of a switched-reticence motor. The SRM has wound field coils for stator windings similar to a DC motor. However, no magnets are attached to the rotor. It is a solid, salient-polar rotor (with magnetic poles projected) composed of soft magnetic material (often laminated steel). The magnetic reluctance of the rotor creates a force which aims to align the rotor pole with the nearest stator pole when power is applied to the stator. To maintain a rotation, an electronic control system switches the windings of successive stator poles sequentially to 'lead' the rotor pole, pulling it forward. The switched-reluctance motor uses an electronic position sensor to determine the angle of the rotor shaft and the solid-state electronic to switch the stator windings, which allows the dynamic control of pulse timing and shaping. This differs from the similar induction motor, which also energizes windings in a rotating phase. In an SRM, magnetization of the rotor is static. The absence of a slip by SRM enables the rotor position to be accurately known and the motor can be stepped arbitrarily slowly.

8.12 Permanent Magnet Synchronous Machines

Like any rotating electric motor, a permanent magnet synchronous motor consists of a rotor and a stator. The stator is the stationary part. The rotor is the rotating component. The

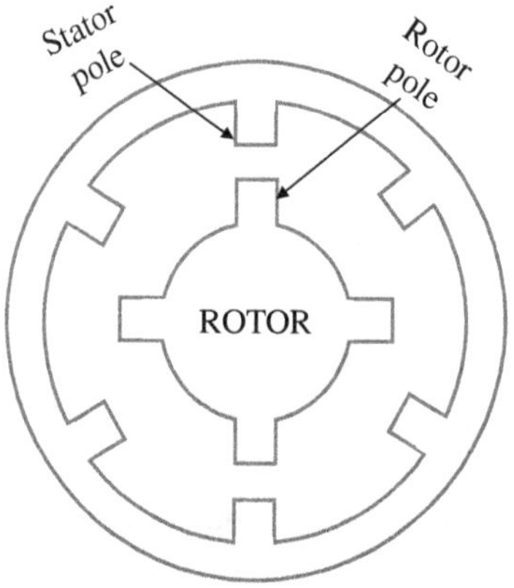

Figure 8.24 Switched reluctance motor.

rotor is made up from permanent magnets. As these are permanent magnets, materials with high coercive force are used. The principle of a permanent magnet synchronous motor is based on the interaction of the stator rotating magnetic field and the magnetic field of the rotor. The concept of the synchronous motor's rotating magnetic field is the same as that of a three-phase induction motor. In line with the Ampere Law, the magnetic field of the rotor, which interacts with the stator's synchronous alternating current, generates a torque, forcing the rotor to rotate. A constant magnetic field is created by permanent magnets on the PMSM rotor. The rotor pole interlocks with the rotating magnetic field of the stator at a synchronous speed of turning of the rotor with the stator field. In that respect, when the PMSM is connected to the current three-phase system, it cannot start itself.

8.13 Brushless DC Motor

Brushless DC motors are common in industrial applications worldwide. The brushless DC motors do not contain brushes and use DC current. It often helps to explain how a brushed DC motor works first as it was used before brushless DC motors were available. A brushed DC motor has permanent magnets with a rotating armature inside it. The permanent magnets are called stators, which are stationary on the outside. The rotating armature, which contains an electromagnet, is called the rotor. When the electric current is flowing to the armature, the rotor spins 180° in a brushed DC motor. In order to go any further, the electromagnet pole needs to flip. As the rotor rotates, the brushes come into contact with the stator, flip the magnetic field and allow the rotor to rotate a full 360° (Figure 8.25).

A brushless DC motor is essentially pushed inside, thereby eliminating the need for brushes to push the electromagnetic field. The permanent magnets are on the rotor and the electromagnet is on the stator in brushless DC motors, as shown in Figure 8.26. A computer then loads the electromagnets in the stator to turn a full 360° of the rotor. Brushless DC motors typically have efficiency of 85–90%, while brushed motors generally have efficiency of only 75–80%. Eventually, the brushes become wearable, sometimes causing dangerous sparking that limits a brushed motor's life. Brushless DC motors are quiet, lighter and long-life. Because the electrical current is controlled by computers, brushless motors can be far more accurate. Brushless DC motors offer a number of advantages over other types

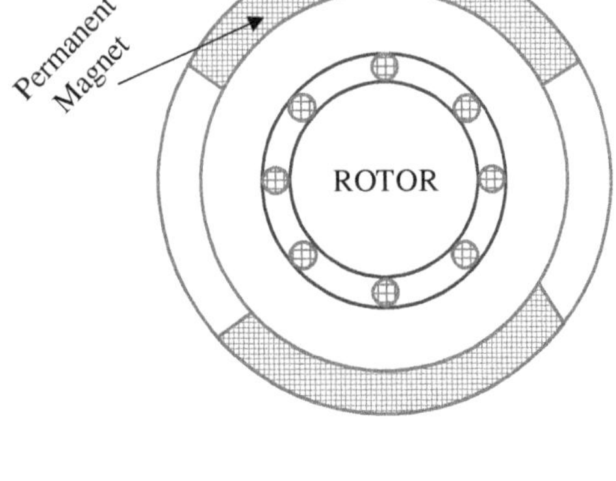

Figure 8.25 Permanent magnet synchronous motor.

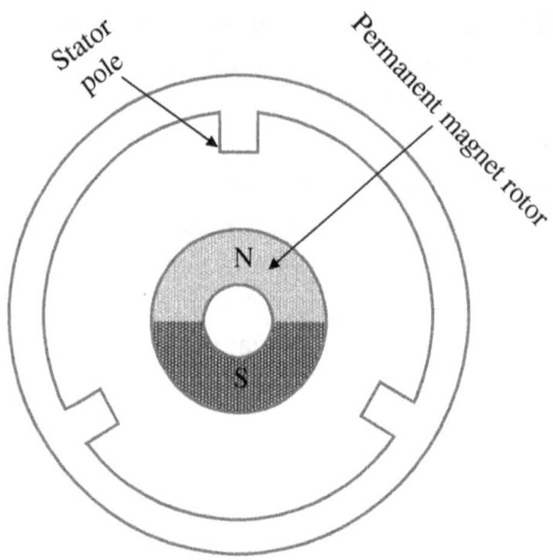

Figure 8.26 Brushless DC motor.

of electric motors, which is why they have become part of many household products and may be a major growth factor for service robots both within and outwit industry. Because of all these advantages, brushless DC motors often work in modern devices that require low noise and heat, especially in continuously-running devices. This might include washing machines, air-conditioning systems and other electronics. They could even be the main source of power for service robots, which require very careful force control for reasons of safety (Figure 8.26).

8.14 Mathematical Model of the Single-phase Induction Motor

The two-phase realization of the single-phase induction motor is possible in the d-q model. To build a motor model, the stator and rotor voltage equations are required. From the three-phase induction motor modelling, the voltage and torque equations can be obtained. Here, the complexity of the voltage equations exists because of the time-varying parameters between stator and rotor circuits. In order to get rid of time-dependent parameters, a transformation is required. The general transformation refers to the motor variables to an arbitrary angular speed reference frame [1]. The voltage equations are given as:

$$\left.\begin{aligned} R_s i_{qs} + \frac{d\phi_{qs}}{dt} &= V_{qs} \\ R_S i_{ds} + \frac{d\phi_{ds}}{dt} &= V_{ds} \\ R'_r i'_{qr} + \frac{d\phi'_{qr}}{dt} - \frac{N_s}{N_S}\omega_r \phi'_{dr} &= V'_{qr} \\ R'_R i'_{dr} + \frac{d\phi'_{dr}}{dt} + \frac{N_s}{N_S}\omega_r \phi'_{qr} &= V'_{dr} \end{aligned}\right\} \tag{8.62}$$

where R_s, R_S, R'_r, and R'_R represents the main winding, and auxiliary winding resistances of stator and rotor respectively. i_{qs}, i_{ds}, i'_{qr}, and i'_{dr} represents the q and d axis currents of stator and rotor respectively. ϕ_{qs}, ϕ_{ds}, ϕ'_{qr}, and ϕ'_{dr} represents the q and d axis flux of stator and rotor respectively. V_{qs}, V_{ds}, V'_{qr}, and V'_{dr} represents the q and d axis voltages of stator and

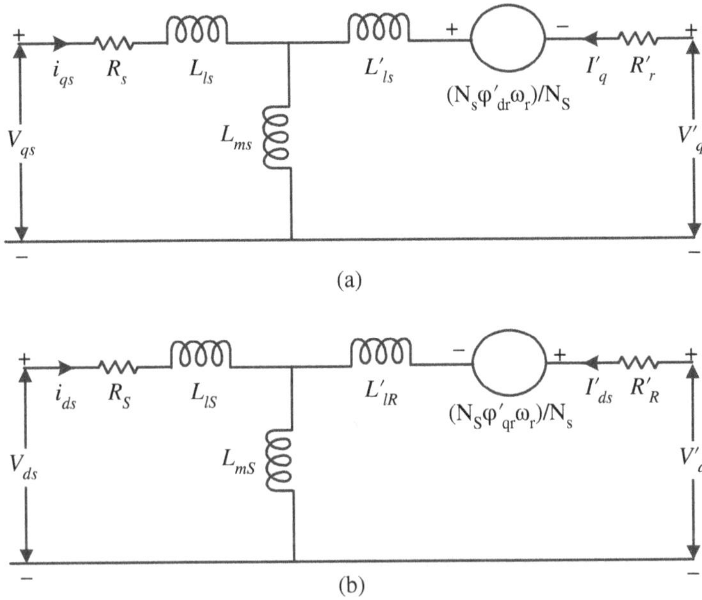

Figure 8.27 Equivalent circuit of the single-phase induction motor.

rotor respectively. N_s and N_S denotes the effective numbers of main and auxiliary winding turns. ω_r is the angular velocity of the rotor. The torque equations are given as

$$T_e = \left[\left(N_S/N_s\right)\varphi'_{qr}i'_{dr} - \left(N_s/N_S\right)\varphi'_{dr}i'_{qr}\right] \times P \tag{8.63}$$

where P is the number of poles.

For the mechanical system, the equation of toque is given as

$$T_e = T_L + J\frac{d\omega_r}{dt} + B_m\omega_r \tag{8.64}$$

Based on the above formulation, Figure 8.27 shows the equivalent circuit diagrams in the q- and the d-axes. The terms L_{ms} and L_{mS} represent the main and auxiliary winding magnetizing inductance. L_{ls} and L_{lS} represent the main and auxiliary winding leakage inductance of stator. L'_{ls} and L'_{lS} represent the main and auxiliary winding leakage inductance of rotor.

8.15 Simulink Model of a Single-Phase Induction Motor

As the two windings of the stator are displaced 90° in space. The single-phase induction motor has a non-linear system. Figure 8.28 shows the MATLAB model of the split-phase induction motor. The input voltages are transformed to the d-q axis as shown in transformation block and given in (8.65) and these voltages are used to simulate the induction motor.

$$\begin{bmatrix} V_q \\ V_d \end{bmatrix} = \begin{bmatrix} \cos\theta_r & -\sin\theta_r \\ -\sin\theta_r & -\cos\theta_r \end{bmatrix} \begin{bmatrix} V_a \\ V_b \end{bmatrix} \tag{8.65}$$

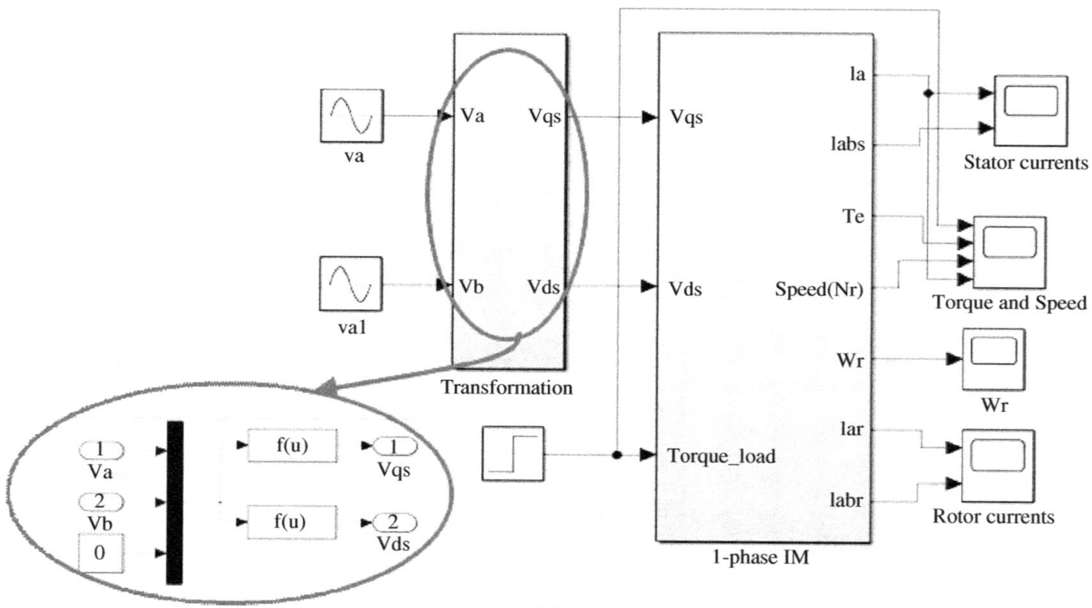

(a)

Figure 8.28 Split-phase induction motor model.

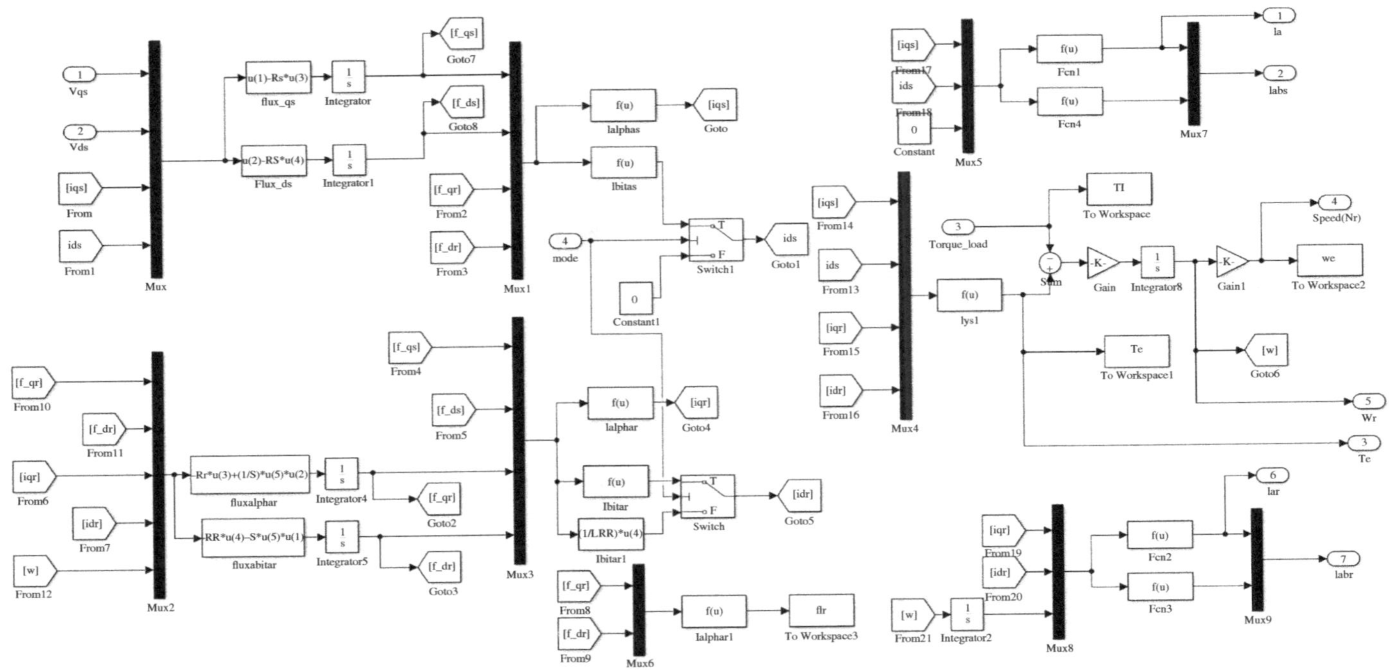

(b)

Figure 8.28 (*Continued*)

The input voltage is fixed to 110 V with 50 Hz and is designed to operate at 1500 rpm. The load torque has a step variation with a magnitude of 2 N m as shown in Figure 8.29. At a time, t = 2.0 seconds, the load is applied. Before that, the motor is running at its rated speed of 1500 rpm. The magnitude of starting current of the motor is 27 A. As the motor reaches its rated speed, the magnitude of current decreases as depicted in Figure 8.29. As the load is applied to motor, the speed decreases from 1500 rpm and settles at 1230 rpm. The peak magnitude of load current is 24 A.

Figure 8.30 shows the MATLAB model of the single-phase capacitor-start induction motor. The difference from the split-phase IM is the subtraction of the d-axis voltage due to the presence of the capacitor. Figure 8.31 shows the different waveforms for the capacitor start motor. For the same operating conditions, peak-load current decreases to 10 A from 24 A with split-phase motor. The presence of the capacitor reduces the starting current and this also reduces the starting torque ripples.

Furthermore, the SIMULINK model of the third type of single-phase IM i.e. capacitor start capacitor run motor has been depicted in Figure 8.32. In this, the additional capacitor voltage is included in the SIMULINK model. Figure 8.33 illustrates the simulated waveforms for the change of load torque of 2 N m at a time of two seconds. The inclusion

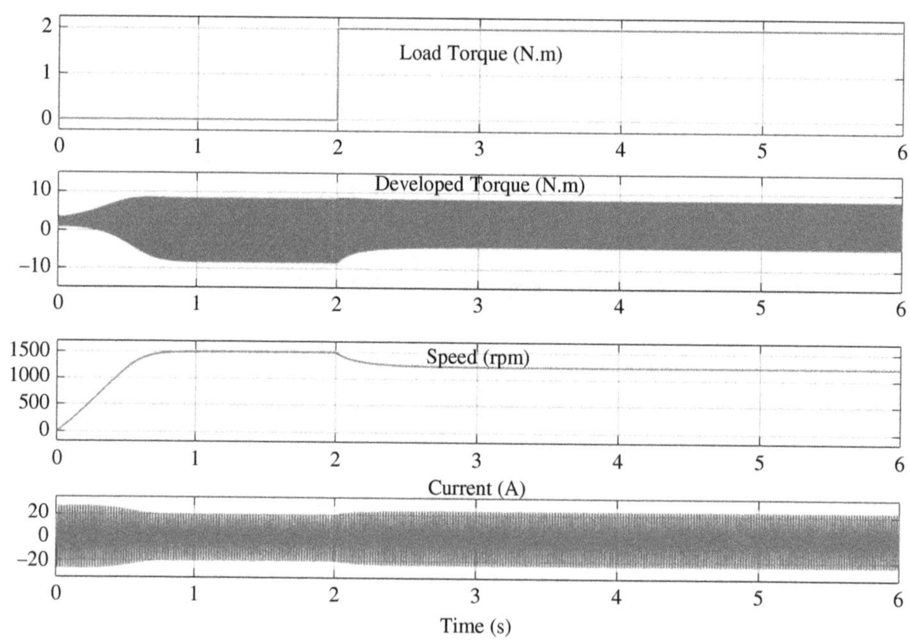

Figure 8.29 Simulated waveforms for the split-phase IM.

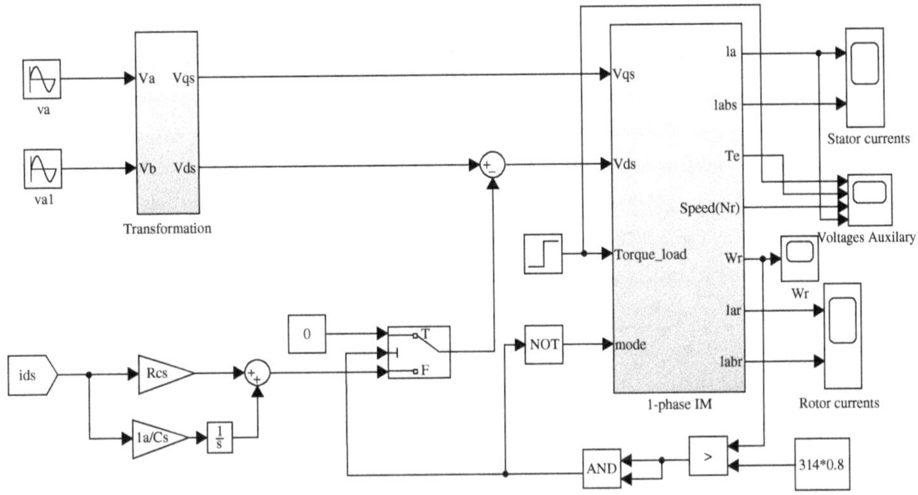

Figure 8.30 Capacitor start single-phase induction motor.

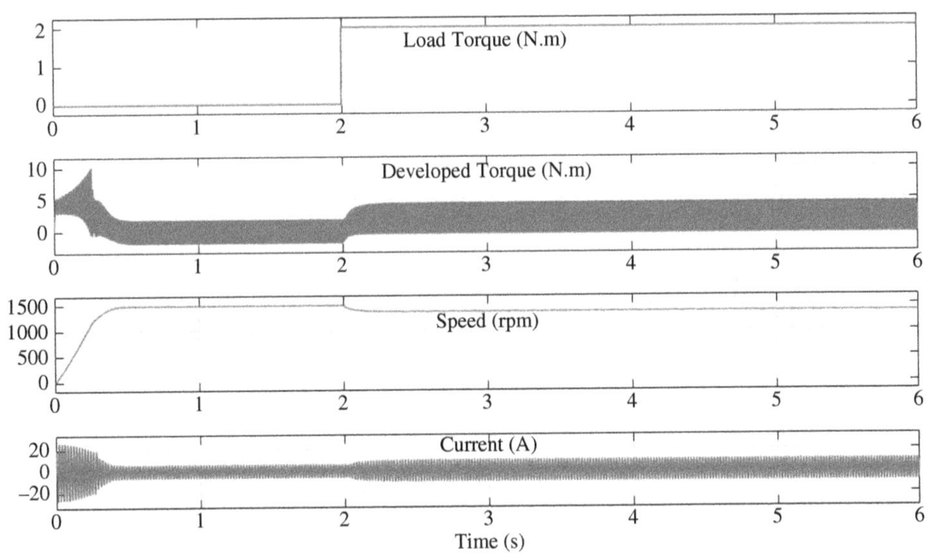

Figure 8.31 Simulated waveforms for the capacitor start IM.

of the capacitor during the running conditions reduces the torque ripple. The speed is higher than the split-phase motor and it reaches 1400 rpm after loading conditions. Furthermore, the load current is 7 A which is one-third of the current in split-phase motor.

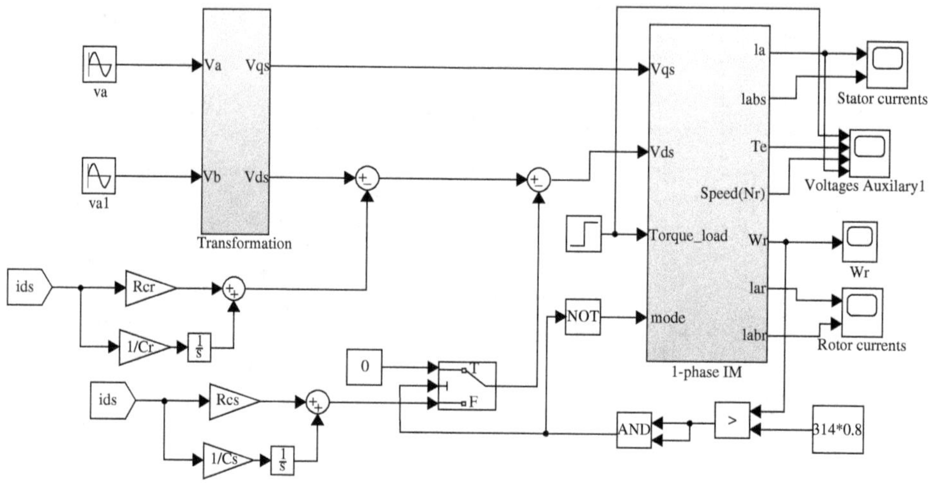

Figure 8.32 SIMULINK model of capacitor-start capacitor-run motor.

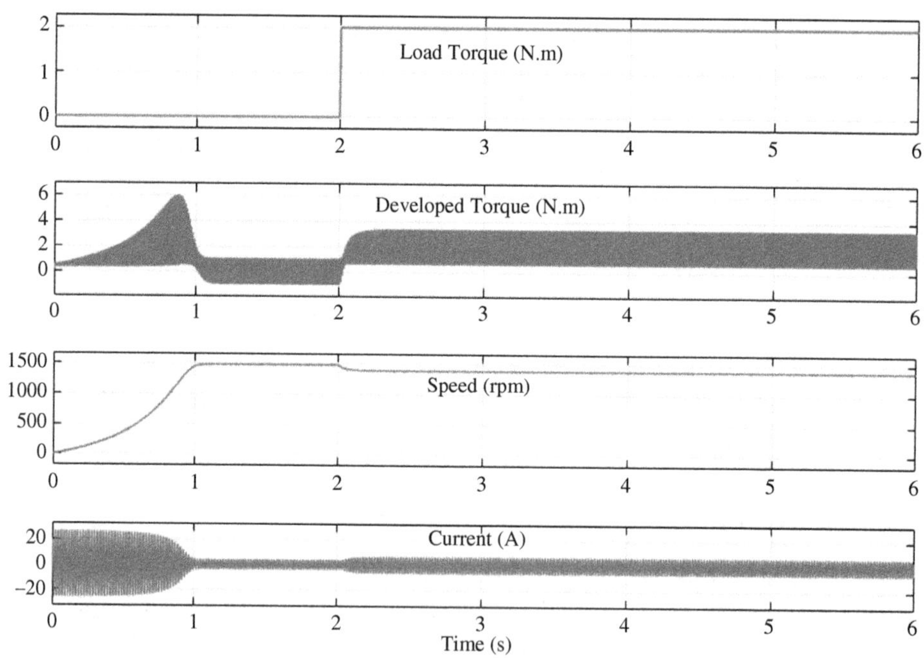

Figure 8.33 Simulated waveforms for the capacitor-start capacitor-run IM.

8.16 Summary

This chapter is dedicated to the discussion on single-phase induction motors. Types and principles of operation are explained. Equivalent circuit and performance of single-phase induction motor is given, along with the testing and methods to determine the parameters of machine.

Problems

8.1 Tests were conducted on a 1-hp 50 Hz, single-phase induction motor, the test data are given as follows

$$\text{No-Load Test}: V_{nl} = 220\,\text{V}, I_{nl} = 3.5\,\text{A}, P_{nl} = 180\,\text{W}$$

$$\text{Blocked Rotor Test}: V_{sc} = 80\,\text{V}, I_{sc} = 9\,\text{A}, P_{sc} = 400\,\text{W}$$

Given stator winding resistance Rs = 1.5 Ω
Calculate the equivalent circuit parameters and show the equivalent circuit.

8.2 Compute the torque developed and efficiency of the single-phase induction of Problem 8.1 when the motor is running at 2880 rpm.

8.3 A 1 hp, four-pole, 50 Hz, 220 V single-phase induction motor is running at a slip of 10%. The equivalent circuit of the machine is given in the following Figure P8.3

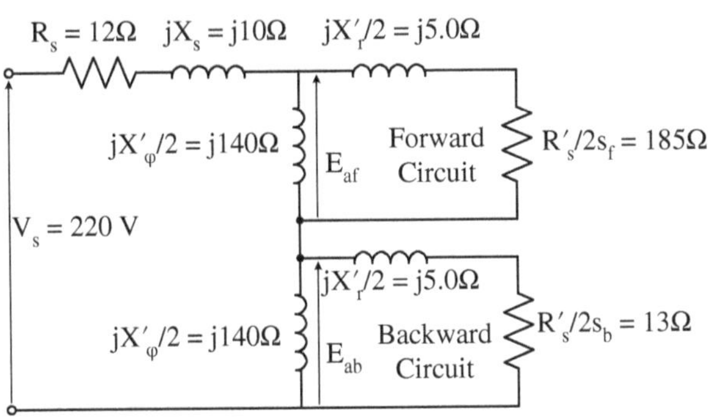

Figure P8.3 Equivalent circuit for a 1-hp, single-phase induction motor.

Compute the following
(a) Input Current
(b) Power factor
(c) Developed power
(d) Developed torque

8.4 A 1 hp, four-pole, 50 Hz, 220 V single-phase induction motor is running at a slip of 12%. The equivalent circuit of the machine is given in the following Figure P8.4

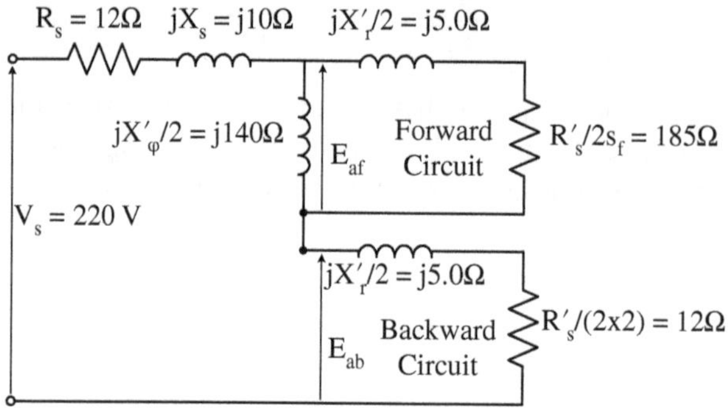

Figure P8.4 Approximate equivalent circuit for a 1-hp, single-phase induction motor.

Compute the following;
(a) Input Current
(b) Power factor
(c) Developed power
(d) Developed torque

8.5 For Problems 8.3 and 8.4, compute the losses and efficiency of the motor considering the core losses = 25 W and mechanical rotational losses = 20 W

8.6 A 0.5 hp, 110 V, 50 Hz, four-pole, capacitor start single-phase induction motor has the following parameter, number of poles = four, Stator main-winding resistance = 2 Ω, stator auxiliary winding resistance = 7 Ω, stator main winding leakage reactance = 2.5 Ω, stator auxiliary winding leakage reactance = 2.8 Ω, reactance of capacitor (connected in series with the auxiliary winding) = 60 Ω, rotor circuit resistance referred to the stator side = 3.5 Ω, rotor circuit leakage reactance referred to stator side = 2 Ω, magnetizing reactance referred to the stator = 70 Ω. Compute the starting torque developed by the machine given the effective turn ratio of auxiliary winding to main winding is 1.2. See Figure P8.6.

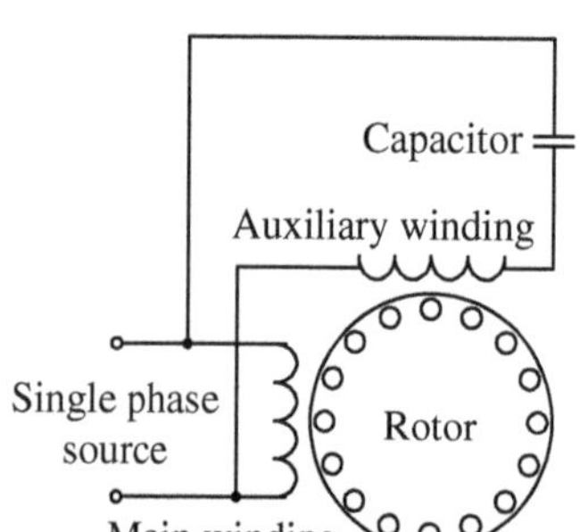

Figure P8.6 Single-phase, capacitor induction motor.

8.7 A 0.5 hp, 110 V, resistance start split-phase induction motor draws starting current in the main winding $I_m = 12$ A at a lagging power factor angle of 50° and the starting current in the auxiliary winding Ia = 8 A at a lagging power factor angle of 12°. Compute the following

(a) the total starting current drawn the machine

(b) Starting torque

Given the machine constant as 0.19 Vs/A.

8.8 A 110 V, 50 Hz, 4 mhp four-pole, draws a current of 0.5 A under-rated load condition and input power is 15 W. The full-load speed of the motor is 1300 rpm and no-load speed 1460 rpm. No-load power is 6 W and no-load input current is 250 mA. The stator resistance under DC condition is 25 Ω. Compute the efficiency under full-load condition (Figure P8.8).

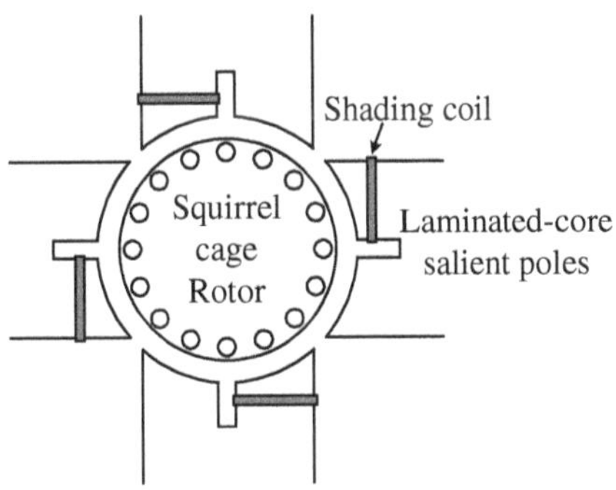

Figure P8.8 Four-pole, shaded-pole motor with a squirrel-cage rotor.

8.9 A single-phase reluctance motor with eight rotor poles has sinusoidal reluctance variation as shown in Figure P8.9. The supply is 110 V 50 Hz with maximum reluctance of 2.5×10^7 AT/Wb and minimum reluctance of 0.9×10^7 AT/Wb. Compute the developed torque and speed of the machine.

8.10 A single-phase reluctance motor is shown in the Figure P8.10. The dimensions of the motor are; cross-sectional area is 2.5×2.5 cm^2, mean length of magnetic flux path is 6 cm. The length of each air-gap is 5 mm. The variation of reluctance is sinusoidal. The q-axis reluctance is four times d-axis reluctance. Compute the maximum value of average mechanical power developed.

8.11 The maximum torque of a shaded-pole motor with four-pole, 50 Hz, 15 mhp, is 70 mNm. The speed-torque curve of the shaded-pole motor is given in Figure P8.11. Compute the following

(a) Speed and power at breakdown condition

(b) Speed when motor is operating at rated condition

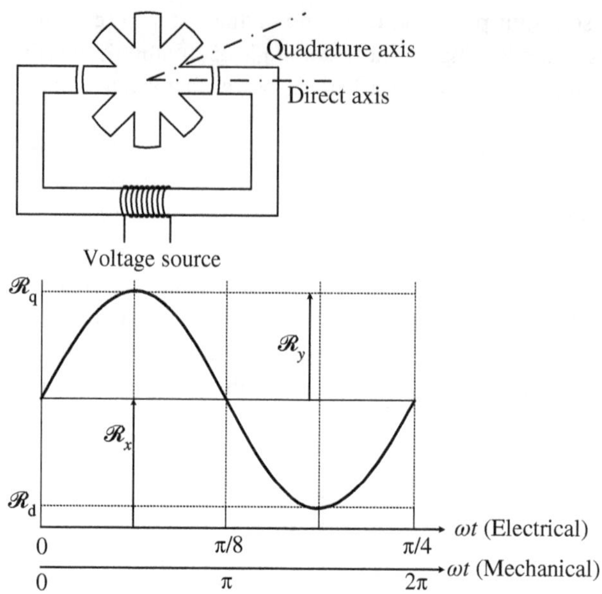

Figure P8.9 (a) Reluctance motor with a singly fed eight-pole rotor. (b) Sinusoidal variation of reluctance for an eight-pole reluctance motor.

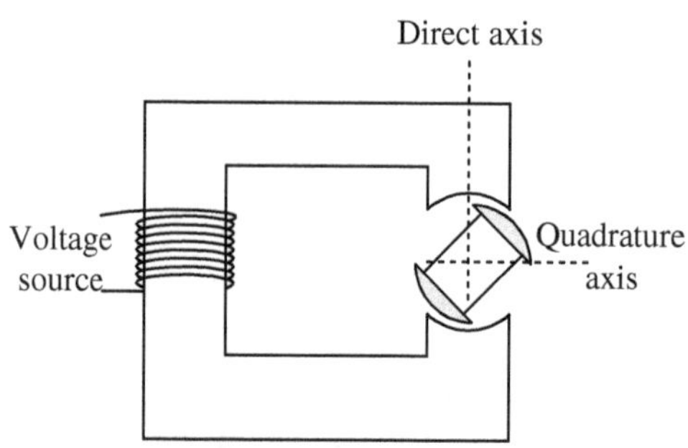

Figure P8.10 Singly fed reluctance motor.

8.12 A universal motor is supplied by DC source draws starting line current of 4 A and generate a starting torque of 3 Nm. The supply is changed to AC with a voltage of 220 V, 50 Hz. Compute for the AC supply;

(a) The starting torque

(b) The mechanical power developed at 4 A.

(c) Operating power factor

Given the total resistance of the motor circuit as 2.5 Ω, the inductance as 30 mH. The rotational losses are neglected and assuming linear magnetic circuit.

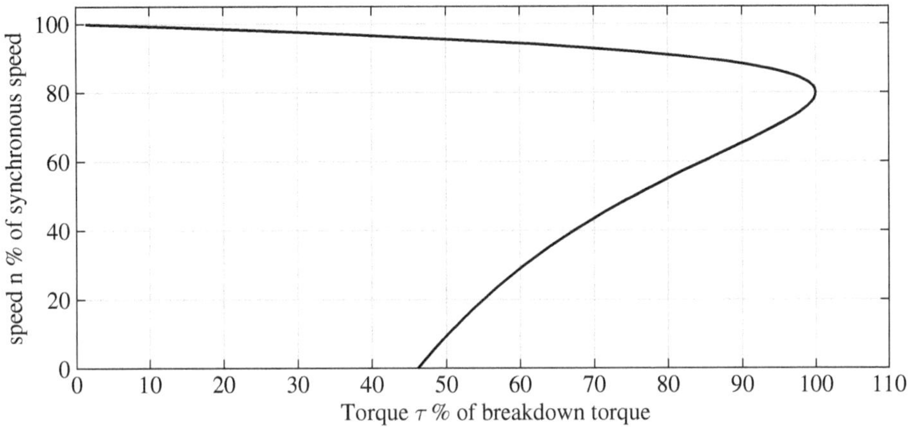

Figure P8.11 Shaded-pole speed-torque curve for a fractional horsepower motor.

Reference

1 Krause, P.C. (2002). *Analysis of Electric Machinery and Drive Systems*. IEEE Press.

9

Motors for Electric Vehicles and Renewable Energy Systems

9.1 Introduction

In the present day, transportation is a key factor for country advancement. The rapid urbanization and increased population have led to a sudden rise in travel demand, which results in an increase of gasoline-powered vehicles. These vehicles are causing environmental pollution and global warming by emitting hazardous gases. Moreover, fast enervation, the high price of gasolines and the necessity of ecological systems to augment research towards green transportation. Green transportation is nothing more than an efficient and effective way to use renewable energy resources (e.g. solar, wind, and fuel cells) for running vehicles with suitable structural modification and which do not affect the environment. Electric vehicles are the best alternative for replacing the IC Engine dependent conventional vehicles due to their benefits such as high efficiency, high reliability, zero air pollution, wider speed-torque range, smooth and less noisy operations [1–10]. These vehicles play a major role in urban transportation, which helps in the reduction of hazardous air pollutants, and hence will save nature.

The basic overview of the Electric vehicle is shown in Figure 9.1 [6, 10]. The key enabling parts are Battery, Power Converters, Electric Motor, and Gear Box. The battery pumps the required energy to drive the vehicle. Amongst all parts, the electric drive (motor along with power converter) plays a key role in controlling as well as meeting the required speed and torque. The preferable machines for electric traction are induction machines (IMs), switched reluctance machines, permanent-magnet synchronous machines (PMSMs) and brushless machines (BLDC). In these machines, BLDC motors have been used for low-power applications. However, the PMSM and IM are preferable for high-power applications [2, 10–14].

On the other side, to meet the global energy demand the renewable energy systems (RESs) are penetrating at a massive rate due to the scarcity of gasoline resources. In these renewable energy generators, power converters, electrical machines, gearbox (in wind power) are the key components. Amongst all of this, electrical machines are a significant component in all RES applications for attaining the required power level, as in wind energy systems, hydropower plants, Flywheel energy storage system, hydrokinetic systems, tidal power systems, etc. The induction and synchronous machines are major contributors for faster penetration of RESs.

Electrical Machine Fundamentals with Numerical Simulation using MATLAB/SIMULINK, First Edition.
Atif Iqbal, Shaikh Moinoddin, and Bhimireddy Prathap Reddy.

Companion website: www.wiley.com/go/iqbal/electricalmachine

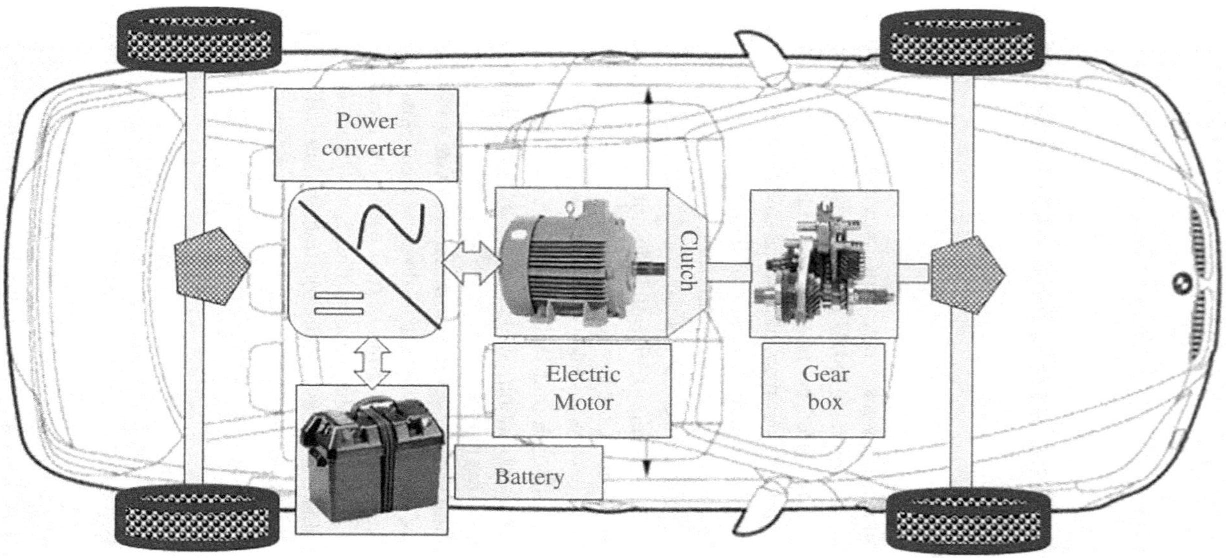

Figure 9.1 Basic overview of the electric vehicle with conventional three-phase IM [10].

In this chapter, the importance and applicability of different types of electrical machines for electric vehicles and RESs are presented. The basic components of different EVs (battery-based, hybrid and fuel-cell type) are also discussed, where the battery, electric motors, power converters, transmission system and other components are explained in detail. The charging technologies like wired and wireless charging are explained. The basic charging methods, for example, constant current (CC), constant voltage (CV), constant current and constant voltage (CCCV) etc., has been presented. Present challenges and requirements of existing electrical machines for the EVs have been discussed. Later, a variety of electrical vehicles manufactured by different EV companies with respect to the type of commercially available motors and transmission system are given. In addition to this, the requirements and challenges involved in electrical machines for RESs are also discussed. Finally, existing electrical machines for RES applications are given.

9.2 Components of Electric Vehicles

The electric vehicle's structure is quite simple compared with conventional gasoline-based vehicles. In EVs, the electric motor propels the vehicle through a transmission system. The primary footsteps towards electric vehicles have been taken in the early years of the eighteenth century by Ányos Jedlik (in 1827) and Robert Anderson (in 1839) [1]. After that in 1881, the French scientist Gustave Trouvé developed the first EV with rechargeable batteries [1]. Even although the mass production of EVs started in the early nineteenth century by 'Studebaker Automobile Company', USA, their market was restricted due to the lack of battery technology and cheap costs of gasoline-based vehicles. In the present century, the necessity of eco-friendly systems, scarcity of gasolines, significant advancements in battery technology has boosted the market as well as a focus on electric vehicles over conventional gasoline-based vehicles.

9.2.1 Types of EVs

Electric vehicles are classified depending on the source of energy [5–9, 14], for example

1) Battery Electric Vehicles (BEVs): Battery based electric drive
2) Hybrid Electric Vehicles (HEVs): Combination of Battery based electric drive and gasoline-based IC engine
3) Fuel Cell Electric Vehicles (FCEVs): Hydrogen is transformed into electrical energy by chemical reactions in a fuel cell.

The basic block diagrams of all types of vehicles are shown in Figure 9.2. The conventional IC engine-based vehicle is shown in Figure 9.2a, where fossil fuels (like petrol, diesel etc.,) are used to drive the vehicle. This usage of fossil fuel will result in hazardous emissions as well as global warming. The hybrid-type EVs such as non-plug-in HEV, parallel plug-in HEV and series plug-in HEV are shown in Figure 9.2b–d respectively. The block diagram of battery-based EV is shown in Figure 9.2e. Detailed discussion of each type of EV is presented in the following sections.

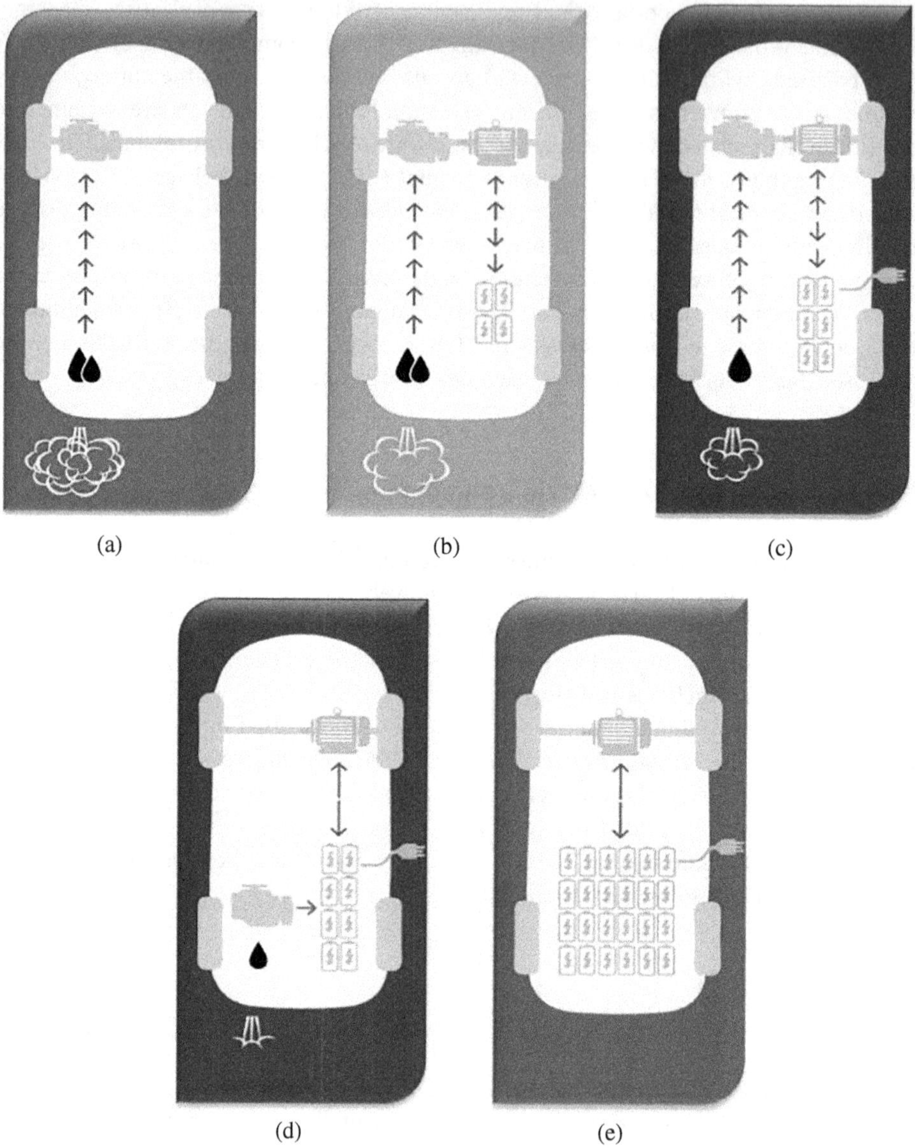

(a) (b) (c) (d) (e)

Figure 9.2 Types of existed vehicles; (a) Conventional IC engine drive, (b) Non plug-in hybrid drive, (c) Parallel plug-in hybrid drive, (d) Series plug-in hybrid drive, (e) Battery-based electric drive [14].

9.2.1.1 Battery-Based EVs

The battery-based electric vehicles are called pure electric vehicles or all-electric vehicles. In this type of EV, the source of energy is a battery, as shown in Figure 9.2e. The batteries are charged either through the utility grid or renewable energy sources (such as solar, wind or fuel cells etc.,). In solar PV-based EVs, the panels are installed on the vehicle itself. The energy stored in the battery is regulated by DC-DC converters and DC to AC inverters to produce the required power to the propulsion electric motor. The electric motor converts electrical energy into mechanical energy in the form of torque and speed. The developed torque of the machine is delivered to the wheels through the gearbox or transmission system. Depending on the pressure applied on the accelerator by the user, the electric motor obtains regulated power from the battery bank through power converters and delivers the required speed and torque to propel the vehicle. When a brake applied by the user, the motor decelerates and acts as a generator to return energy to the battery.

The components of the battery-based Electric vehicles are shown in Figure 9.3, and listed below.

- Electric motor
- Power inverter
- DC-DC converter
- Battery bank
- Controller module
- Transmission or gearbox
- Cooling system
- Charging circuit.

Examples of Battery based EVs include

Tesla Model S, X and 3, Nissan Leaf, Toyota Rav4.B, MW i3, Hyundai Ioniq, Volkswagen e-Golf, Ford Focus Electric, Mitsubishi i-MiEV and many more.

9.2.1.2 Hybrid EVs

In the HEVs the energy sources are gasoline-based IC engine as well as battery-based electric drive. The IC engine work as a normal gasoline-based vehicle, where the fuel is pumped into the engine to generate the required speed and torque for driving the vehicle. The electric motor gets the energy from the battery and delivers the required speed and torque to the wheels through the transmission system. In this type of EV, internal combustion (IC) engine and electric motor are connected to the same shaft of the wheels, and simultaneously drive the vehicle depending on the user requirement. These HEVs are further classified into, (i) Standard HEV, (ii) Plug-in HEV.

9.2.1.2.1 Standard Hybrid EVs

In a standard Hybrid Electric Vehicle (SHEV), the batteries are charged through ICE and regenerative braking during the deceleration of the vehicle; the block diagram is shown in Figure 9.2b. The charging system is on-board internally within the vehicles. The external charging through charging port from the utility grid or renewable power sources is not present in the SHEV. In the SHEV initially, the vehicle runs based on the gasoline-based IC engine when speed goes above a certain value or during uphill driving. The power from the battery is used when a vehicle is running under the steady condition at low speed. A combination of gasoline and battery are used to run the vehicle optimally. The objective is to obtain an efficient run cycle. In addition to this, the battery also supplies the power for the auxiliary loads. The key components of the SHEV are shown in Figure 9.4.

Examples of SHEV:

Toyota Hybrid Prius and Camry, Honda Civic Hybrid.

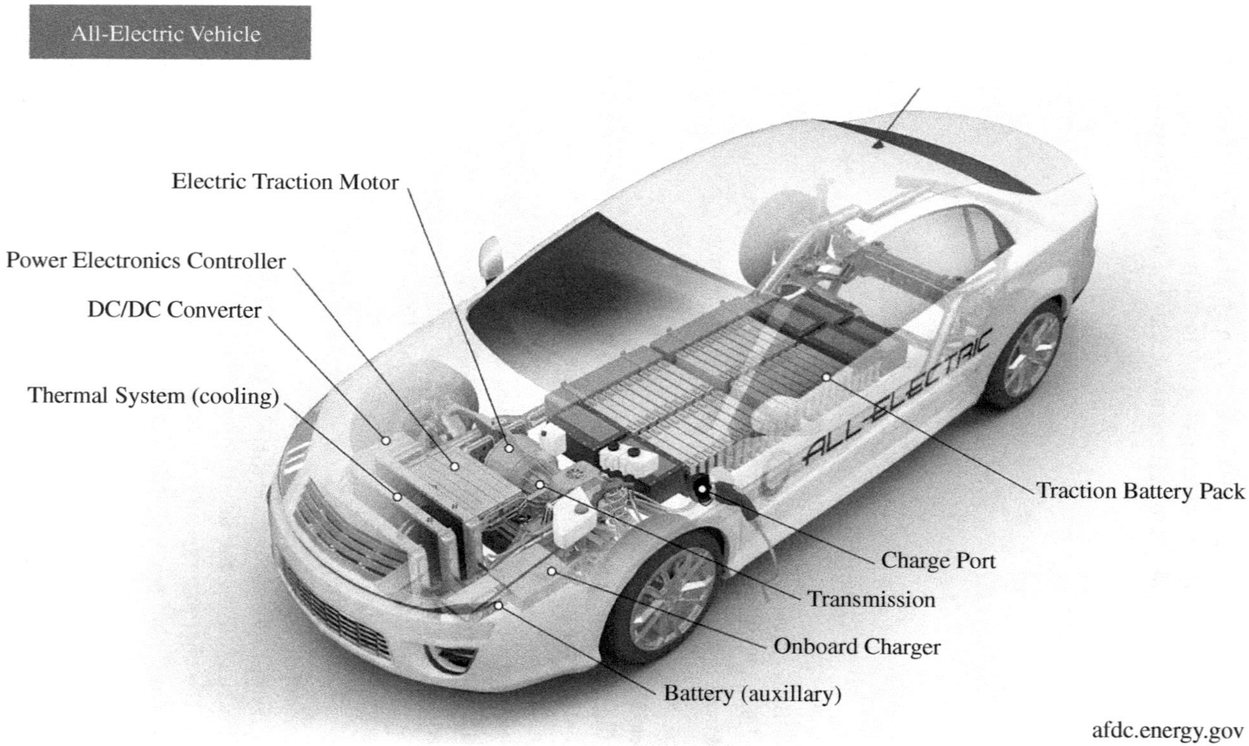

Figure 9.3 Components of a battery-based EV [6].

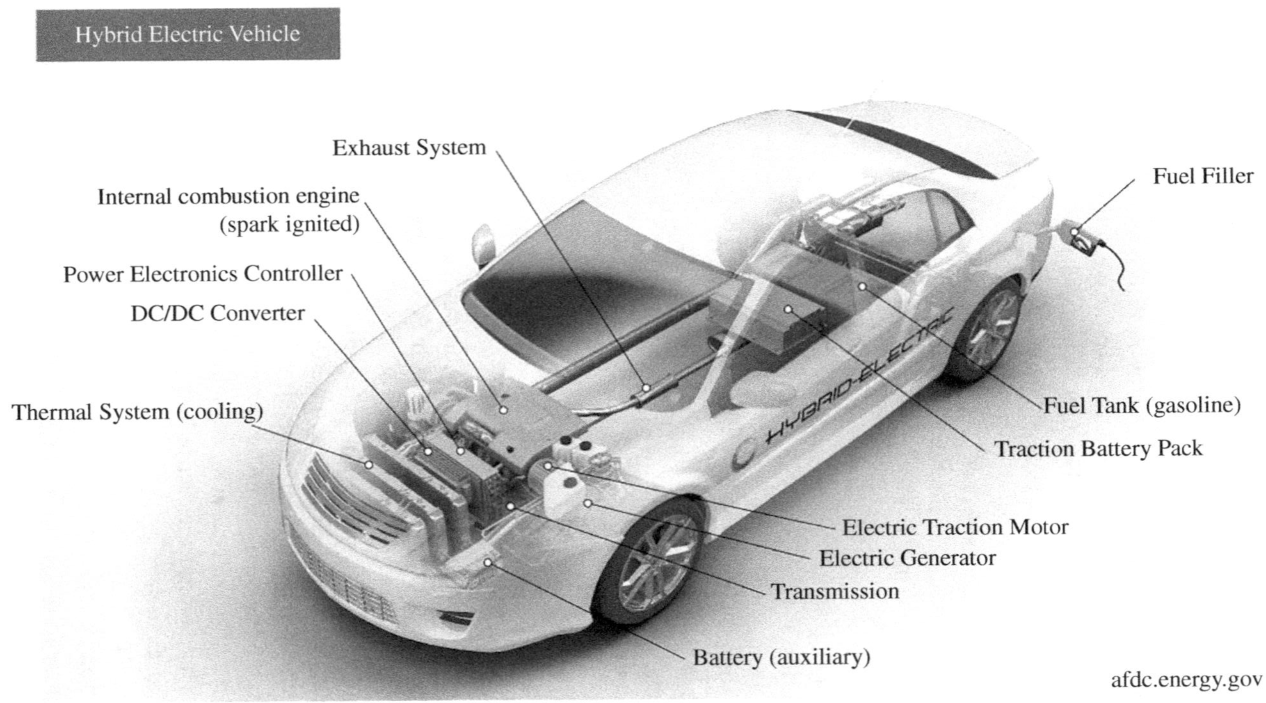

Figure 9.4 Components of a hybrid EV. [7].

9.2.1.2.2 Plug-in Hybrid EVs

In the Plug-in Hybrid Electric Vehicle (PHEV), an external charging port is available to charge the battery from the utility grid, charging stations or renewable energy resources. This PHEV works in All-Electric mode similar to BEV and Hybrid mode similar to HEV. In general, the PHEVs start in all-electric mode, where the power is taken from the batteries only. Once the battery is discharged fully, the mode of operation is changed to the IC engine-based drive.

In addition to external charging, the batteries in PHEVs can also charge with IC engine and regenerative braking mode, which depend on the powertrain configuration, i.e. series PHEV block diagram is shown in Figure 9.2d, parallel PHEV is shown in Figure 9.3c, series-parallel configurations. The key components of the PHEV are shown in Figure 9.5.

Examples of PHEV

Porsche Cayenne and Panamera S E-hybrid, Kia Optima, Audi A3 E-Tron, Hyundai Sonata, BMW 330e, Ford C-Max Energi.

The major components in HEVs are

- Electric motor
- Power Inverter
- DC-DC converter
- Battery bank
- Controller module
- Transmission or Gear box
- Cooling system
- Charging circuits (only for PHEV)
- IC Engine
- Fuel Tank

9.2.1.3 Fuel-Cell EVs

The fuel cell electric vehicles (FCEVs) work the same as battery-based EVs, where the battery is replaced by a hydrogen-based fuel cell. In FCEVs, the hydrogen (chemical energy) is directly converted into the electrical energy by using the fuel cell, which is fed to the electric motor to drive the vehicle. Also, in FCEVs, the speed/torque of an electric motor is regulated by using the power electronic converters. The FCEVs consist of a battery bank to store the energy during regenerative braking, which supplies the power to meet torque demand during acceleration time and for auxiliary loads. In some FCEVs, the plug-in electric charging port will be available along with the higher ratings of the battery banks to meet the higher acceleration torques as well as long range. The mileage of FCEV depends on the tank size of hydrogen fuel.

Examples of FCEV

Hyundai Nexo and Tucson, Honda Clarity Fuel Cell, Toyota Mirai

The key components of the FCEV are shown in Figure 9.6, which are given below

- Fuel-cell stack
- Hydrogen storage tank
- Electric motor
- Power Inverter

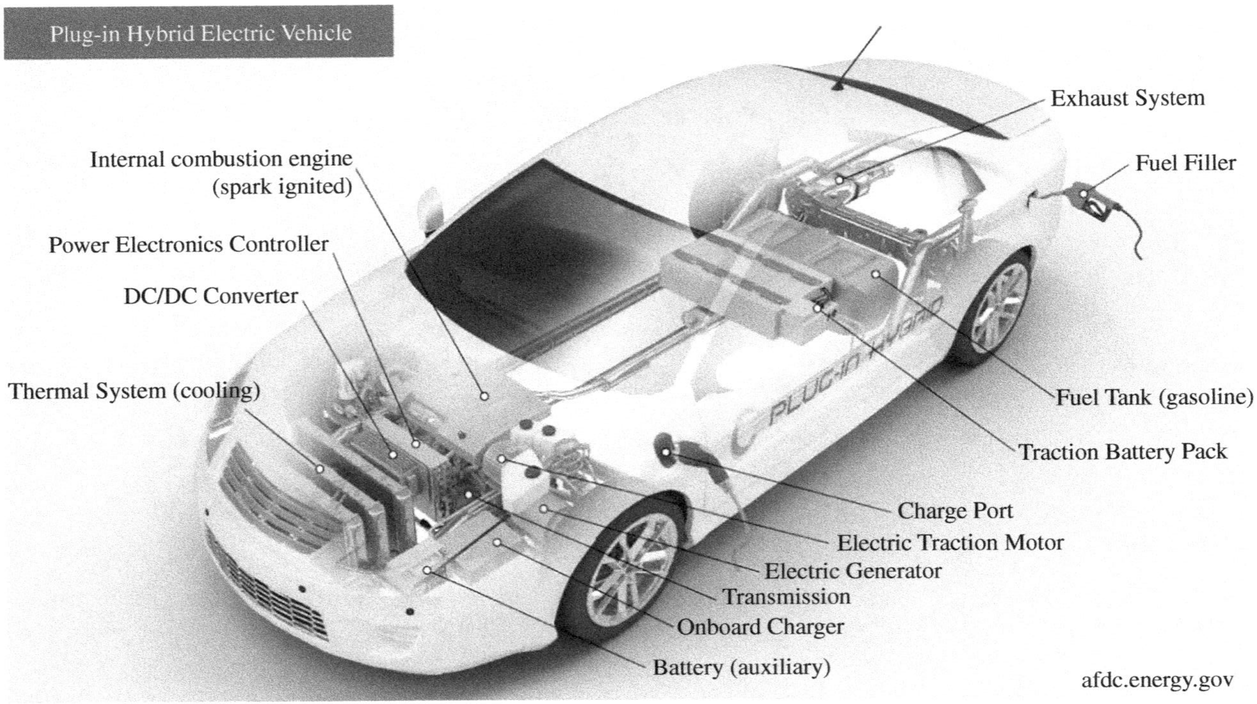

Figure 9.5 Components of a plug-in hybrid EV. [8].

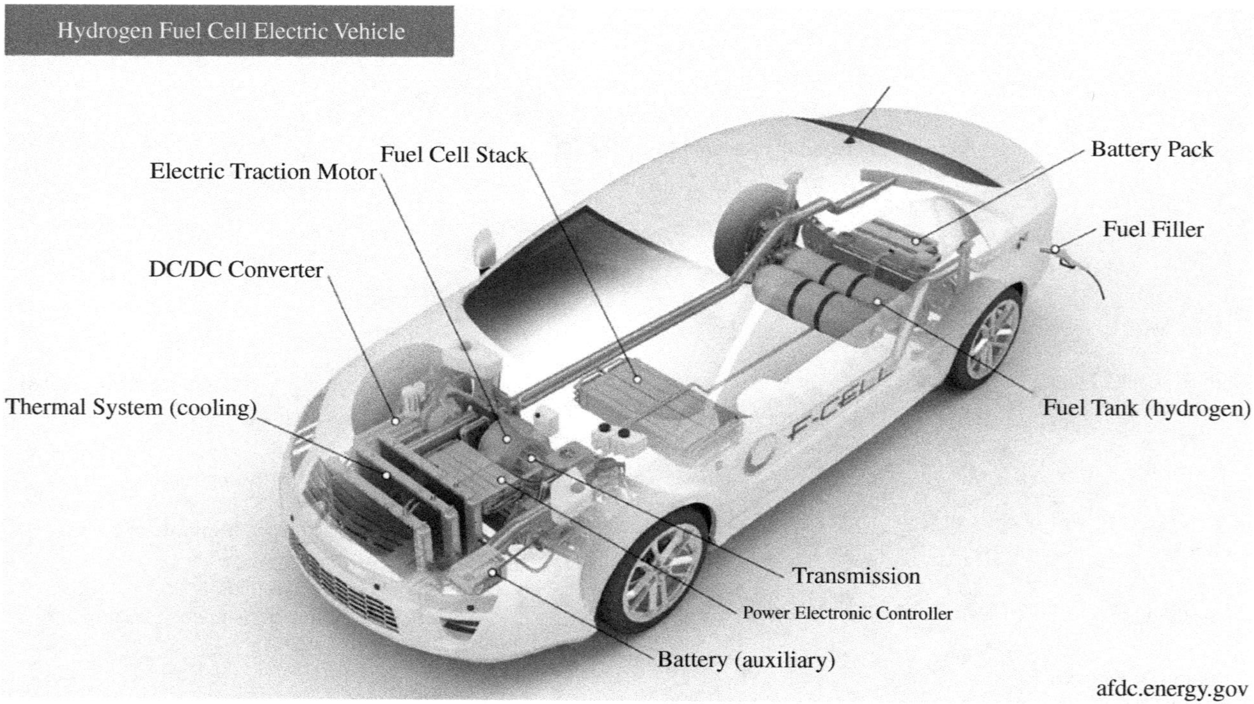

Figure 9.6 Components of a fuel-cell based EV. [9].

- DC-DC converter
- Battery bank
- Controller module
- Transmission or Gearbox
- Cooling system

9.2.2 Significant Components of EVs

The components of each type of EV are given in the previous section. In this section, detailed discussion on the significant components of EVs (i.e. battery, electric motors, power converters, transmission system) are presented.

9.2.2.1 Battery Bank

The battery bank is a DC type of electrical energy storage system that stores energy in chemical form and releases it in electrical form and has the same role as that of gasoline in conventional vehicles. The battery consists of several electro-chemical cells which convert the stored chemical energy into electrical energy. The DC energy supplied by the battery is regulated by a DC-DC converter as well as a power inverter to deliver the power to an electric motor. The power flow from the battery is regulated with the user-defined control. Batteries used in EVs are of rechargeable type which are arranged in series and parallel to meet the required voltage and current ratings [5–8, 15].

To select a battery for any application, capacity will play a crucial role, in terms of ampere-hours (Ah) and terminal voltage. The discharge current of a battery is represented in terms of C-rate, i.e. 1C means battery discharge time is one hour. The transformation rate of an electric vehicle is regulated by battery technology. In the present century, massive advancements in battery technology have boosted the new generation of electric vehicles, i.e. Li-ion, Metal-Air batteries, Zebra batteries etc., Presently available batteries in the market are, Lead-acid, Nickel, Zebra, Lithium and Metal-air batteries [15].

1) Lead-acid batteries are the most common type used in households and conventional gasoline-based vehicles. This type of battery is the cheapest among all battery types, but the weight and volume per kWh are high. In addition to this, these types of battery hamper ecological systems during the production as well as disposal process.
2) Nickel batteries are another type of commonly used battery, which are sub-classified into Nickel-zinc, Nickel-iron, Nickel-cadmium (Ni-Cd) and Nickel-metal hydride (Ni-MH). The Nickel-Zinc batteries are eco-friendly but the ratings and life of the battery are less. The Nickel-iron battery suffers from heavier weight, maintenance cost and self-discharge rate. Ni-Cd and Ni-MH batteries can be used in EVs but the charging times and charging control complexity are higher.
3) Zero emission batteries research activity (ZEBRA), are basically two types, i.e. sodium-sulphur and sodium nickel chloride (NaNiCl). These batteries have higher efficiency at high temperatures. The advantages are, they have a similar or higher energy density to lithium batteries, lower cost and longer life. The drawbacks include lesser efficiency compared to Li-ion batteries because of the self-discharge losses.
4) Lithium-ion batteries are a promising candidate for modern EV applications, due to their eco-friendly nature, low weight/kWh, higher efficiency, higher specific power

and longer life. The limitations of lithium batteries are higher production costs and the requirement of protection circuits for the safe operation of every pack of batteries. Lithium batteries are sub-classified as, Lithium–iron sulphide (FeS), Lithium–iron phosphate ($LIFePO_4$), Lithium-ion Polymer (LiPo), Lithium-ion, Lithium-titanate ($LiTio/NiMno_2$). For a high-power density of EVs, lithium iron phosphate will give better characteristics. A Lithium-titanate battery has faster charging capability compared to the lithium-ion batteries.

5) Zinc-air batteries are another promising battery type, which have a higher energy density as well as higher specific energy over the lithium battery [15]. The major limitations are limited cycle life, they are bulky in size with low specific power. This type of battery is still in the research stage and is commercially not available.

9.2.2.1.1 Ultra-Capacitors

Ultra-capacitor (UC) or super capacitor has much higher capacitance, the charging as well discharging rates are very high, but the structure is same as the conventional capacitor. However, the storage of energy in an ultra-capacitor is limited. Its energy storage density is less than 20% of the lead acid battery. The ultra-capacitor has less maintenance, longer life and doesn't depend on temperature variation. At present, three types of UC are commonly using in EVs, i.e. double-layer carbon/carbon electric capacitors (DLEC), pseudo-capacitors and hybrid capacitors, which are classified depending on energy storage mechanism and electrode materials. UCs are mainly used in the EVs for quick or transient discharge of energy to meet faster acceleration and climbing gradient roots. The specific power density of UCs will be 1000–2000 kW/kg with a 95% efficiency [15]. The UC can offer a longer lifetime. The ultra-capacitor is still at the beginning of its development. The cost is expected to decrease and the energy density will increase rapidly over the next few years.

9.2.2.1.2 Comparison of Batteries and Ultra Capacitors

In the application of EVs, the super capacitors (ultra-capacitors) are used when fast discharge/charge of energy is required for short-term power requirements (such as faster acceleration or hill climbing). Whereas batteries are considered as the best candidate for long-term discharge/charge requirements. The combination of the battery, as well as super capacitors, will result in improved performance of the EV, i.e. (i) reduce the stress on the battery, (ii) longer service life of a battery, (iii) faster discharge to meet the quick transient power, (iv) during regenerative braking, faster charging of the battery will happen with UC. The unique difference between the battery as well as ultra-capacitors is shown in Figure 9.7. The chemistry (electrochemical reactions) involved in the battery technology determines the operating voltage, charge/discharge capabilities. Whereas in the ultra-capacitors, the type of dielectric material defines the maximum allowable voltage. The detailed comparison in terms of specific energy, specific power, cost, lifetime and other details are given in Table 9.1 [16].

9.2.2.1.3 Various Capacities of EVs

EVs are classified as three types, based on the capacity, as well as purpose of usage, i.e. low-duty EVs, medium-duty EVs, heavy-duty EVs. The low-duty vehicles are mainly used for city transportation, short distance, individual transportation (two people), for example, two-wheelers. The rating of batteries used for low-duty EVs in terms of a range of battery,

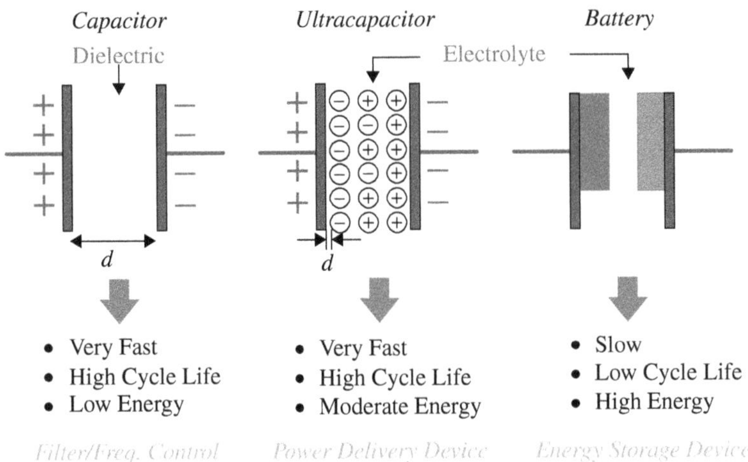

Figure 9.7 Comparison between an ultra-capacitor and a battery.

Table 9.1 Comparative analysis of the super-capacitor and battery

Performance comparison	Super capacitor	Battery
Charge time	1–10 s	10–60 min
Cycle life	1 million or 30 000 h	500 and higher
Cell voltage	2.3–2.75 V	3.6–3.7 V
Specific energy (Wh/kg)	5	100–200
Specific power (W/kg)	Up to 10 000	1000–3000
Cost per Wh	20$	0.5$–1$ (large system)
Lifetime (in vehicle)	10–15 yr	5–10 yr
Charge temperature	–40 °C to 65 °C (–40 °F to 149 °F)	0 °C to 45 °C (32 °F to 113 °F)
Discharge temperature	–40 °C to 65 °C (–40 °F to 149 °F)	–20 °C to 60 °C (–4 °F to 140 °F)

kWh, voltage and current etc., are shown in Figure 9.8a. Medium-duty EVs come in sharing-based transportation, longer transportation like three-wheelers, four-wheelers and the respective ratings of the battery are shown in Figure 9.8b. Examples of medium-duty EVs are shown in Table 9.2. The heavy-duty EVs are mainly used for goods carriers, passenger travel purposes, like buses, trucks, cranes etc., the ratings of batteries used in this type of EVs are shown in Figure 9.2c.

9.2.2.1.4 EV-Charging Technologies

At present, the penetration of EV is directly influenced by the charging infrastructure of the vehicle. The charging systems of the EVs have been classified based on the power transfer medium between the vehicle and charging unit, i.e. wired and wireless power transfer. The block diagram for different charging technologies available for EVs is shown in Figure 9.9 and the comparison between wired and wireless charging is shown in Figure 9.10 [17–20].

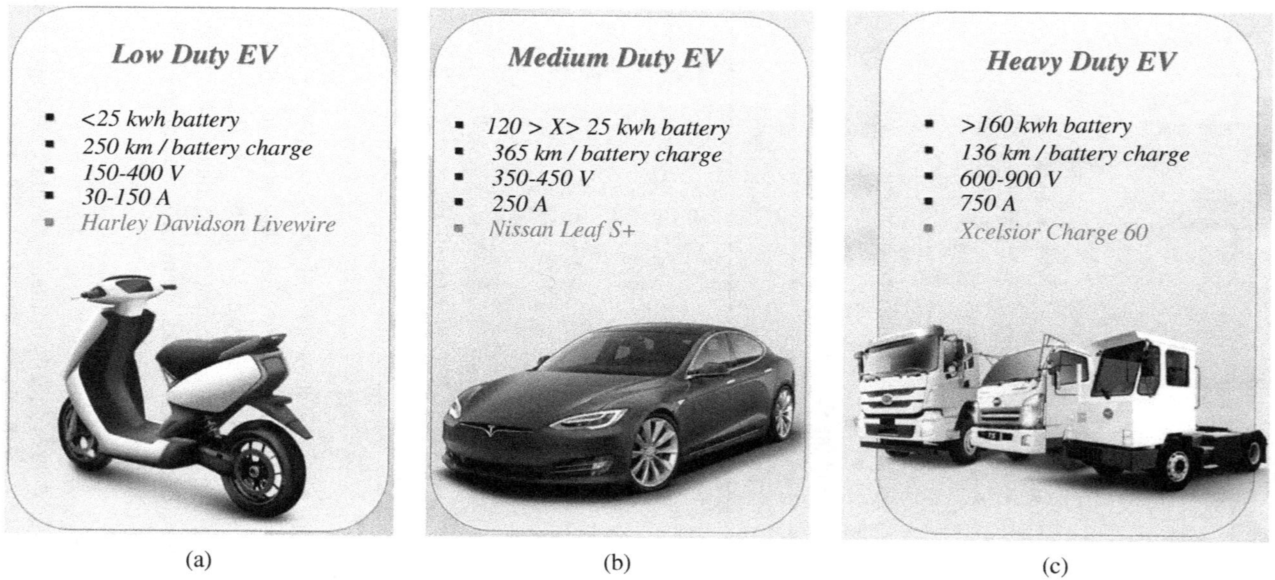

Figure 9.8 Rating of batteries for various EVs.

Table 9.2 Different types of electric motors used by various automobile companies [5].

Manufacturer	Model	Top speed	Acceleration (seconds)	Year	Type of motor	Transmission
Audi	Audi e-tron 55(2018)	200 km/h (124 mph)	5.7	2018	three-phase PMSM	Single speed with fixed ratio
BMW	i3	150 km/h (93 mph)	8	2013–2014	three-phase PMSM	Single speed with fixed ratio
BYD	BYD e6	100 mph (160 km/h)	8	2009	three-phase PMSM	Single speed with fixed ratio
Chevrolet	Bolt EV	150 km/h (93 mph) (speed limited)	6.5	2016	three-phase PMSM	Electronic Precision Shift, single speed
Citroën	C-Zero	130 km/h (81 mph)	15.9	2010	three-phase PMSM	Single speed reduction gear
Honda	Clarity Electric		7.7	2017–2020	Fuel cell-powered three-phase PMSM	Single speed
Hyundai	Ioniq Electric	165 km/h (103 mph)	9.9	2016	three-phase PMSM	Single speed constant ratio
	Kona Electric	167 km/h (104 mph)	6.4	2018	three-phase PMSM	Single-speed reduction gear
Jaguar Land Rover	Jaguar I-Pace	200 km/h (124 mph), electronically limited	4.8	2018		
Kia	Soul EV	145 km/h (90 mph)	11.2	2014	three-phase PMSM	Direct drive reduction gear
	e-Niro	167 km/h (104 mph) [13]	7.8	2018	three-phase PMSM	Direct drive reduction gear

(continued)

Table 9.2 (Continued)

Manufacturer	Model	Top speed	Acceleration (seconds)	Year	Type of motor	Transmission
Mahindra	e2o plus	82 km/h (51 mph)		2016	three Phase IM	
	e-Verito	85 km/h (53 mph)		2017	3 Phase IM	
Mercedes-Benz	B-Class Electric Drive	160 km/h (99 mph)	7.9	2015	synchronous motor	Single speed fixed ratio
	EQC	180 km/h (112 mph)	4.9	2019	three-phase IM	Single speed fixed ratio
Mitsubishi	i-MiEV	130 km/h (81 mph)		2009	three-phase PMSM	Direct drive reduction gear
MW Motors	Luka EV	148 km/h (92 mph)	9.3	July 2016	Four-wheel drive, wheel hub motors (BLDC)	Single gear
Nissan	Leaf	150 km/h (93 mph)		2010	three-Phase PMSM	Single speed constant ratio
Porsche Taycan	Porsche			2019	2 AC synchronous electric motors	1-speed direct-drive And 2-speed automatic
Peugeot	i0n	130 km/h (81 mph)	15.9	2010	three-phase PMSM	Single speed reduction gear
Rayttle	E28	80 km/h (50 mph)		2014	three-Phase PMSM	Single speed constant ratio

(continued)

Table 9.2 (Continued)

Manufacturer	Model	Top speed	Acceleration (seconds)	Year	Type of motor	Transmission
Renault	Fluence Z.E. / SM3 Z.E.	135 km/h (84 mph), electronically limited		2010	synchronous motor	Single speed constant ratio
	Zoe	135 km/h (84 mph), electronically limited	13.5 11.4	2012	synchronous motor	Single speed constant ratio
	Twizy	80 km/h (50 mph) (Urban 80 version)		2012	synchronous motor	Single speed constant ratio
SEAT	Mii Electric	130 km/h (81 mph)	12.3	2019		
Skoda	CITIGOe IV	130 km/h (81 mph)	12.3	2019		
Sono Motors	Sion	140 km/h (87 mph)	9	2019	three-phase IM	
Stevens	ZeCar	90 km/h (56 mph)		March 2008	Induction motor	
Tesla	Model S	261 km/h (162 mph)	2.5	2012	three-phase IM	Single gear Transmission
	Model X	250 km/h (160 mph)	2.8	2015	three-phase IM	Single-speed transaxle gearboxes
	Model 3	261 km/h (162 mph)	3.4	2017	three-phase IM	Single gear Transmission
	Model Y	241 km/h (150 mph)	3.7	2020	three-phase IM	Single gear Transmission
Volkswagen	e-Golf	150 km/h (93 mph)	9.6	2014	three-phase synchronous motor	
Volkswagen	e-up!	130 km/h (81 mph)	11.9	Updated in 2019	three-phase synchronous motor	

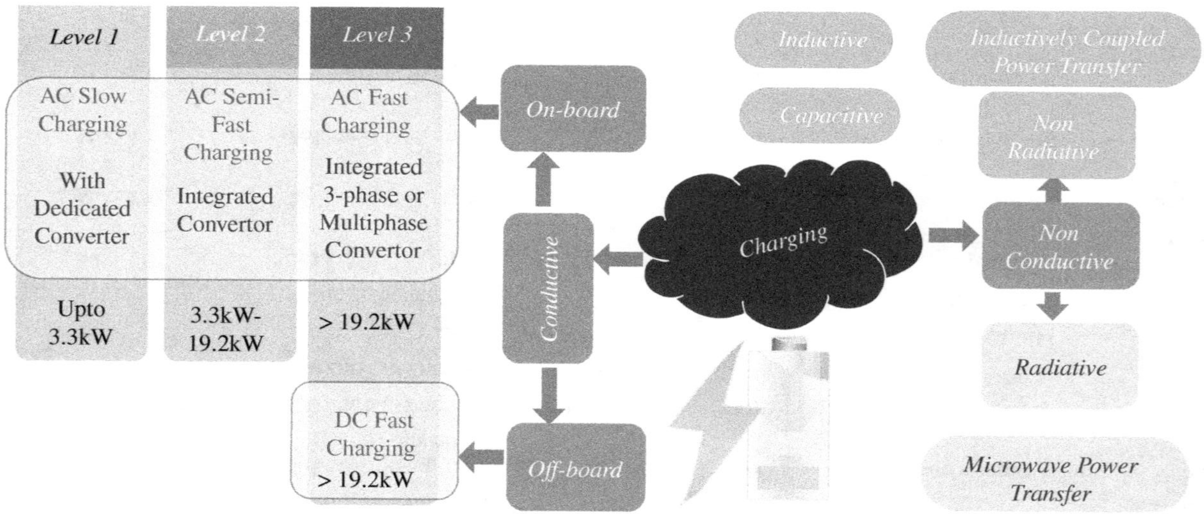

Figure 9.9 Charging technologies for EV batteries.

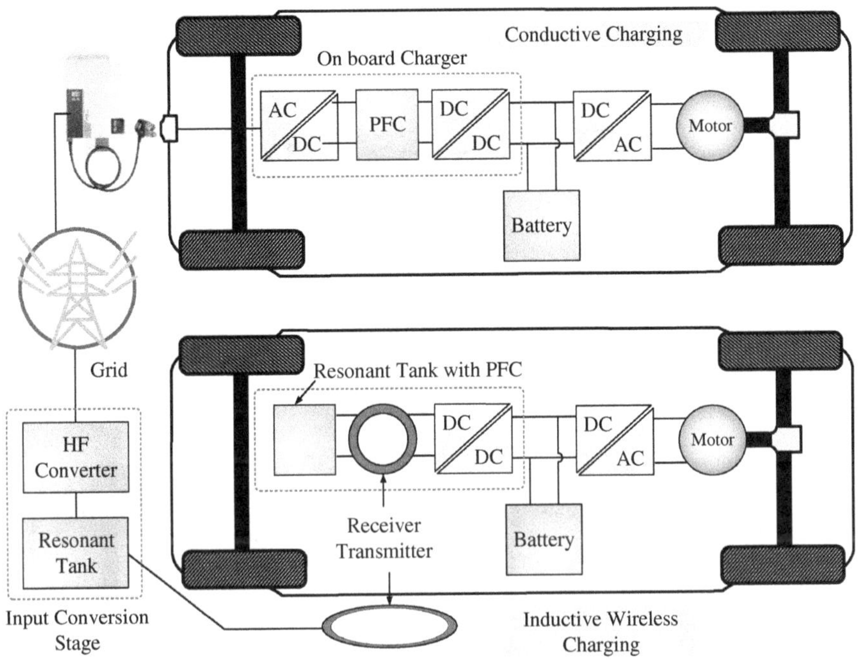

Figure 9.10 Comparison between conductive and nonconductive charging schemes of EVs.

Conductive Charging for EVs In this type of charging, a physical connection between the vehicle and charging unit has been used to charge the battery of the EV called conductive or wired charging. The key components of wired-charging infrastructure in EVs are AC-DC converter, DC-DC converter and power factor correction (PFC) circuits. Depending on the power level, as well as the infrastructure used for charging in the EV, conductive charging has been classified as on-board (AC charging) and off-board (DC charging). A detailed explanation of conductive charging technologies has been presented in [17, 18].

1) **On board chargers:** The charging infrastructure (AC-DC converter, DC-DC converter, PFCs) are installed inside the vehicle. This on-board charging system is further classified into Levels 1, 2, and 3. In Level 1, charging the power level is up to 3.3 kW and charging is done with dedicated converters (separate DC-AC and PFC circuits) which are placed in the EV, as shown in Figure 9.11a. This Level 1 charging comes under AC slow charging, i.e. home charging. In Level 2 charging, the power level is 3.3 kW to 19.2 kW, where the battery is charging by using the available AC-DC converter and machine of the drive system, i.e. called an integrated on-board charger. In Level 2 charging, the machine along with AC-DC converter will act as an inverter and motoring mode while running mode of the vehicle, whereas when the vehicle is in charging mode, the same AC-DC converter and machine has realized as rectifier along with the PFC circuits, as shown in Figure 9.11b. The Level two charger uses semi-fast charging, similar to office charging. This integrated charger has benefits in terms of greater efficiency, saving on weight as well as volume of the vehicle. However, with this integrated charger, both charging and driving modes of the vehicle cannot be done at the same time, but driving and regeneration are possible. In the Level 3 charging operation, the charging infrastructure is capable of carrying a high current for faster charging, but the operation is the same as

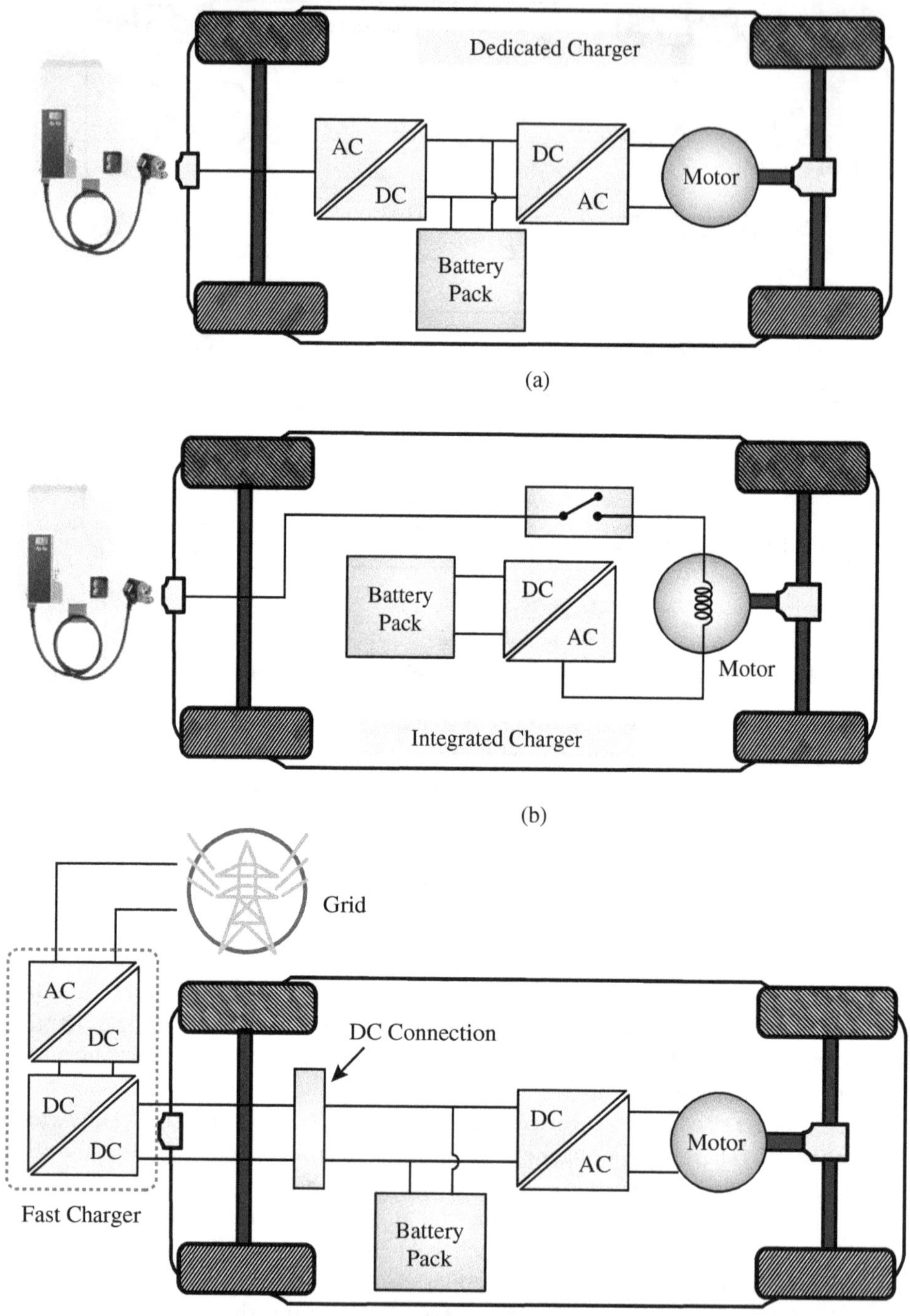

Figure 9.11 Conductive (wired) charging technologies for EVs, (a) Level 1 charger with dedicated converter, (b) Level 2 charger with integrated converter, (c) off-board DC fast charging.

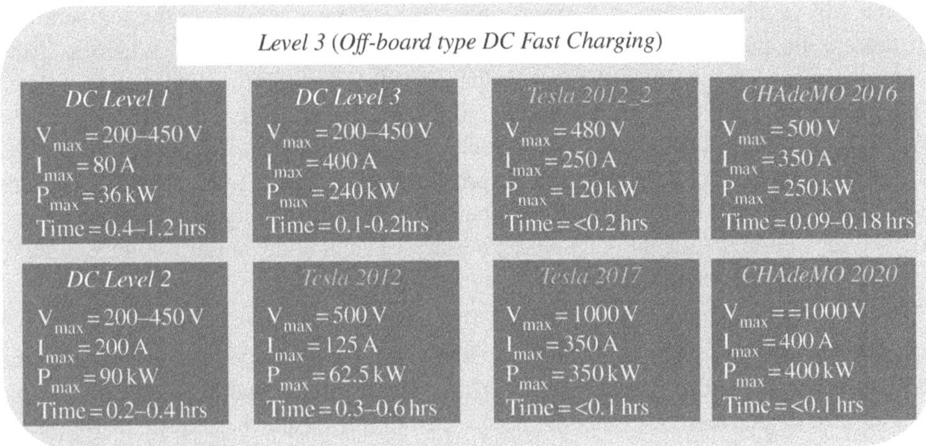

Figure 9.12 Available off-board type of DC fast-charging levels.

Level 2 charging. In Level 3 charging, the power level is >19.2 kW, where the interleaved structures of PFCs or multiphase machines have been used to handle the high currents.

2) **Off-board charger:** The charging infrastructure (AC-DC converter, DC-DC converter, PFCs) are installed outside the vehicle, as shown in Figure 9.11c. This off-board charger, directly connected to the battery of EV, helps with fast charging by pumping high currents. Since the DC is connected to the battery directly this charging is also called DC fast charging and the power level is >19.2 kW. Different types of the available off-board chargers are shown in Figure 9.12.

Non-conductive Charging for EVs In this type of charging, there is a physical connection between the vehicle and charging unit. The power is transferred through a medium, i.e. magnetic fields established between the inductive coils called non-conductive or wireless or inductive charging, as shown in Figure 9.10 (bottom). The principle will be the same as a transformer with air as core and with a large air gap. In wireless charging technologies, the distance between the transmitter and receiver inductive coils, as well as coil designs, define the efficiency as well as the effectiveness of the system. With a short distance, the low range of power can be efficiently transferred easily, but transferring high power over long distances will be a challenging task where significant research has been going on in academia as well as in industry [19]. The non-conductive (wireless) charging has been classified based on the range of frequencies, i.e. (i) non-radiative (inductive, capacitive and other), in the order of kHz, and (ii) radiative in the order of MHz. Detailed studies and advancements related to wireless power transfer charging technologies have been given in [19, 20].

9.2.2.1.5 Battery Charging Methods

When the battery is stated as fully discharged, this means the battery voltage is lower than the cut-off voltage (10–20% drop of rated voltage) or the state of charge of a battery is lower than 20%. At this set point or condition, the battery has to be charged. The over-charging and over-discharging of a battery will affect its lifetime as well as the performance of the battery. To charge the battery, the current has to be pumped according to ratings of battery. If the charging current is high, the chemical reaction of battery is not sufficient which will eventually raise the battery internal resistance as well as the temperature. If there is no proper control of current, the battery will be damaged. Whereas, if the charging current is too small, the battery life, as well as temperature, will not be affected but longer times will be required for charging the battery. The temperature will be the deciding parameter of battery life as well as charging performance. Therefore, to obtain the optimized performance of battery with respect to lesser charging times, charging methods will play a significant role. The basic charging methods of the battery are, Constant Voltage (CV) type, Constant Current (CC) type, Constant Current and Constant Voltage (CCCV) type, Multistage Constant Current (MCC) type [21].

Constant Voltage (CV) Type In this type of charging, the constant voltage is applied to the battery during the entire charging period. The constant voltage charging method is simple among all charging methods, and is realized with a simple DC power supply. The DC power supply can be realized with a step-down transformer followed by a rectifier along with a small filter to charge the battery. The output voltage of the rectifier is regulated, according to the predefined constant voltage for charging the battery. This CV method offers, no over-voltages, longer life of the battery and less chance of damage. This method is CV charging, which regulates a predefined constant voltage to charge batteries.

The charging current is inversely proportional to the battery charging level (SOC), i.e. if the battery charges up, the charging current will come down. During recharging of a fully discharged battery, or at the low SOC condition of the battery, a high current value is required to charge the battery, which will affect the battery life due to the increased temperature of the battery.

Constant Current (CC) Type In this type of charging, the constant current is pumped into the battery during the entire charging period. This method is mainly used for nickel-cadmium and nickel-metal hydride batteries. Compared to the CV technique, the CC method will charge the battery in a short time. Using this CC method, the charging current rate influences the battery performance with respect to capacity utilization, charging current, battery life with respect to temperature rise. Once the battery is fully charged, or >95% of SOC conditions, if we stop the charging method or not reduce the current rate, this results in overcharging, damage, and a reduction in a battery lifetime.

Constant Current and Constant Voltage (CCCV) Type The CCCV type is a combination of CC and CV methods, where the CC method is used for initial charging and after reaching the maximum safe threshold voltage, the CV method is used, as shown in Figure 9.13. This method overcomes the challenges of both CC and CV modes. In this method, starting with the CC method helps in faster charging and the CV method helps in self-regulation of the charging current at higher SOC levels or, greater than the maximum safe threshold voltage. The CC method defines the charging time of the battery and the CV method defines the capacity utilization. This CCCV method is mainly used for Lead-acid, Li-ion batteries. This method helps to improve battery performance with respect to charging times, temperature rise and lifetime. But the key challenge of the CCCV method is selecting the appropriate constant voltage and current in the respective modes of charging.

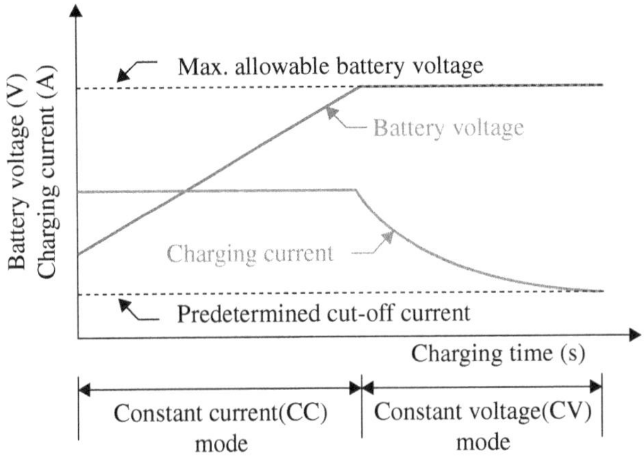

Figure 9.13 Constant current constant voltage charging method.

Multistage Constant Current (MCC) Type In this MCC charging method, the charging of battery has done in several CC stages, where the magnitude of CC at each stage is regulated according to the voltage of the battery. The current magnitude is gradually decreased as the terminal voltage reaches a default voltage threshold, as shown in Figure 9.14. This method is suitable for charging the lead-acid, NiMH, and Li-ion batteries.

There are several advanced battery charging methods, like pulsed current charging, burp or reflex charging, constant power and constant voltage charging, Trickle charge, Sinusoidal ripple current, Constant Temperature constant voltage, Float Charge or continuous charging method as well as other methods given in [21].

9.2.2.2 DC-DC Converters

The DC-DC converters regulate the DC power of the battery. In between the power inverter and battery bank, a high-power DC-DC converter is placed which acts as a boost converter

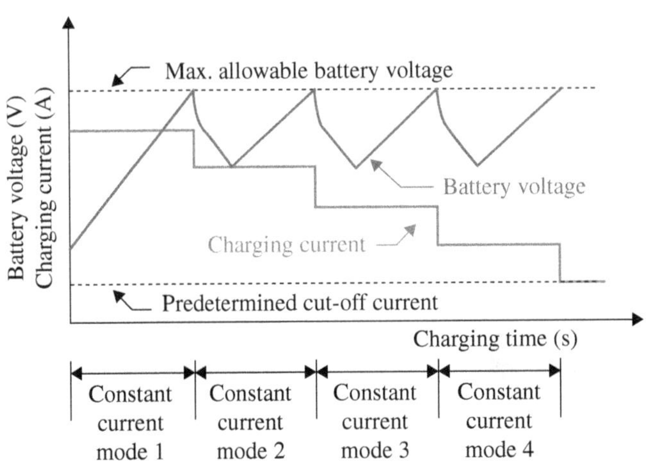

Figure 9.14 Multistage constant current charging method.

while the vehicle is in driving mode (motoring mode of the electric machine), and acts like a buck converter while the vehicle is in regenerative braking mode (generating mode of the electric machine). In addition to this, few other buck converters are used within the EV to run vehicle accessories and recharge some auxiliary batteries. The most commonly DC-DC converters are classified into two types

1) Non-isolated type: Buck, Boost, Buck Boost, Cuk, SEPIC and some advanced converters.
2) Isolated: Buck, Boost, Forward, Fly back, Push pull, Half bridge, Full bridge and others.

For details on converters, readers are recommended to look into a book on Power Electronics.

9.2.2.3 Power Inverter

The power inverter converts the DC output power of the DC-DC converter into AC power. The inverter frequency and modulation index (output voltage magnitude) are controlled with respect to the required speed of the drive. The most common type of inverter is a three-phase two-level inverter for driving the three-phase electrical machine, as shown in Figure 9.15. Depending on the number of phases of the machine, the number of lags in the power inverter has to be increased. With the conventional two-level inverter, the possible voltage levels are either $+V_{dc}$ or 0 by operating the top switch and bottom switch, respectively. With the advent of fast-acting switches, the inverter can be controlled by using PWM schemes [10]. The usage of high-switching frequencies in PWM, modulated inverters help to shift all lower-order harmonics near the fundamental frequency to a higher-order of frequencies, which will reduce the filter size (and it depends on the carrier frequency). Nowadays, wide-band gap devices such as SiC and GAN switches are increasingly used for EV applications.

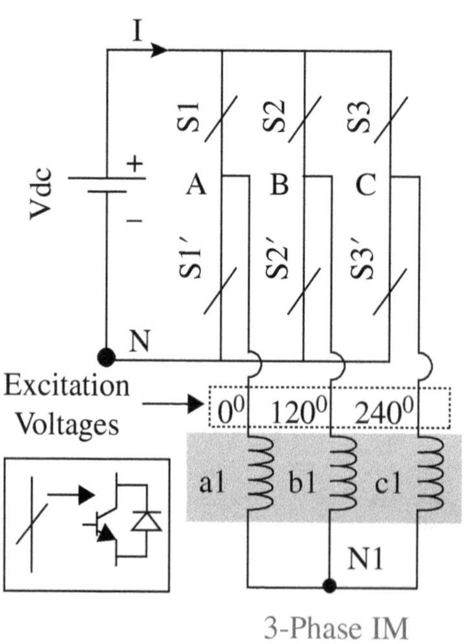

Figure 9.15 Conventional 2-level three-phase inverter for three-phase induction motor.

9.2.2.4 Electric Motor

The electric motor converts electrical energy into mechanical energy (in terms of speed and torque). The shaft of a motor is connected to the wheels through a transmission system. The vehicle speed and torque are controlled with respect to the motor control. In driving mode, the electrical machine works as a motoring action, and the same machine works in generating action for regenerating braking. Preferable machines for electric traction are Induction machines (IM), switched reluctance machines (SRMs), permanent-magnet synchronous machines (PMSM) and brushless machines (BLDC) [10–13]. In these machines, SRM and BLDC motors have been used for low-power applications. The PMSM and IM are preferable for high-power applications like EVs, traction and propulsion etc., The machine for EVs has to be selected based on constraints like enriched torque-speed range with high efficiency, high-power handling capability, volume and high-fault tolerant capability.

9.2.2.5 Transmission System or Gear Box

To meet the required torque-speed characteristics of the vehicle, the torque converters (gearbox) have to be used, which will convert the rated torque and speed of the electric machine in a certain gear ratio. In low-gear operation, the gear ratio has to be high in order to meet the high-torque requirement for starting acceleration and hill climbing. In high-gear operation, the gear ratio has to be low to meet the high-speed requirement of the vehicle. The renowned powertrains of the electric vehicles (EV) like Audi e-tron, Tesla Model S, Nissan Leaf, and Hyundai Ioniq models have been designed with a single-speed gearbox system to attain wider speed range.

9.2.2.6 Other Components

The IC engine, fuel tank, cooling systems, working and operating are the same as conventional IC engine-based vehicles. The hydrogen tank in an FCEV is similar to the petrol or diesel tanks in conventional vehicles.

The charging port in the BEV and PHEV allow the refilling or charging of the battery presented in the vehicle with an external power source like a utility grid or renewable energy sources. In the on-board type of charging circuits, the machine inductances along with the inverter legs act as PFC circuits (similar to interleaved boost converter).

The Cooling mechanism is different for conventional IC engine-based vehicles and EVs. In EVs, the air-cooling mechanism is used to maintain the temperature of the motor, power converters and other accessories within the limits.

9.3 Challenges and Requirements of Electric Machines for EVs

The development of eco-friendly systems as well meeting the global transport demand requires technological advancements in EVs research and development. As discussed earlier, in the electric vehicle system the electrical machine converters turn the stored electrical energy of the battery into mechanical energy. The entire design of the EV mainly depends on the electric machine because it is a key factor for the vehicle, power density, torque speed characteristics, accurate control, efficiency and etc., Because of this, the major challenges and the requirements of electric machines for EVs are explained in this section.

9.3.1 Challenges of Electric Machines for EVs

Nowadays, most of the EV manufacturing automobile companies are designing EVs with PMSMs due to the high-power density and higher efficiency due to less copper loss in rotor. A few other automobile companies are using induction motors (IM), brushless DC (BLDC) machines and switched reluctance machines (SRM) for designing the EVs. There is no universal rule for choosing a specific motor for EV application. The choice depends upon many factors such as cost, application, weight, torque-speed range etc. The major challenges associated with electrical machines for EVs are [10–13],

1) Permanent magnet dependency in PMSM and BLDC motors
2) Ageing effects of permanent magnets in PMSMs
3) Reliability issues in all electrical machines
4) High-power density with reduced Volume for IMs
5) Efficiency improvement in IMs
6) Noise and torque-ripple issues in SRMs
7) High-power handling issues of all electrical machines
8) Complex control structures in SRMs and BLDC machines
9) Temperature-dependent permanent magnet characteristics in PMSM, BLDC motors
10) Type of cooling requirement
11) Installation size and space constraints in EVs
12) Less sensitive to environmental variations, etc.

9.3.2 Requirements of Electric Machines for EVs

In order to select a drive for electric traction applications, the key constraints are as follows [10–13]

- Enhanced torque-speed range with high efficiency
- High-power handling capability
- High torque for starting and hill climbing and high power for high-speed cruising
- Acceptable cost
- Low acoustic noise and low torque ripple
- The volume of a machine
- Fast dynamic response
- Low maintenance
- Compact in size
- High overload capability
- High reliability and robustness

The operating machine speed torque characteristics of an ideal machine for electric traction are shown in Figure 9.16. The machine torque should be high and constant in the start-up process from a standstill, to allow a fast and smooth acceleration. The machine should deliver constant power when the maximum rated power is reached until the maximum continuous speed of the vehicle is achieved without interruptive shifts of gear. Ultimately, for safety reasons, high speed or low-power region should be accessible with high-speed driving and part mechanics. An internal combustion motor does not have such an operational feature as a low-revolution engine does not produce a proper torque and the constant power zone is very narrow, as a peak, meaning that, as a supporting transmission

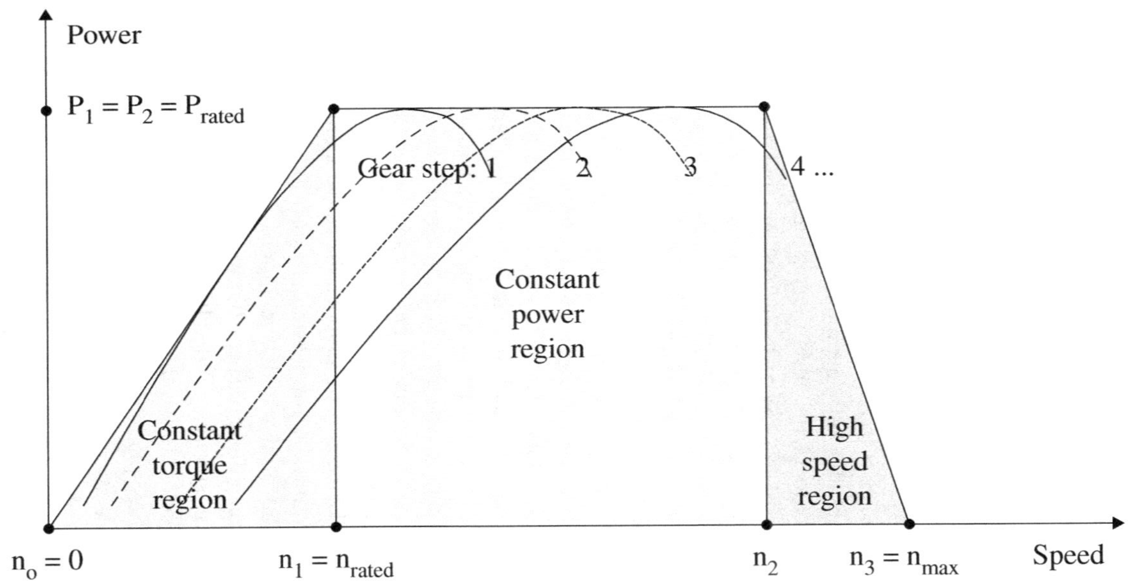

Figure 9.16 Different operating conditions for an ideal machine [12].

component, a multi-stage gearbox is required as shown in Figure 9.16. Nevertheless, an electric machine with a gearless or a simple single-phase transmission can be equipped for the same performance characteristics given in Figure 9.16. The solution for the gearless direct drive can be supported by distributed in-wheel machines, whereas the single-speed transmission is preferred for weight reasons in classical configurations.

For electrical machines, the width of a constant power region or field-weakening zone depends on the output necessary for the corresponding vehicle and on used transmission used. The wider constant power region, usually with a speed ratio of n_2/n_1 shown in Figure 9.16 is up to four, and is necessary for electrical vehicles with single-step transmissions. Electrical equipment specifications in electric and hybrid-electric vehicles include competitive prices, low weight and volume, high default tolerance, wider field weakening, high-energy efficiency and a high-power factor.

The specifications should allow for an automotive with qualities such as reasonable price, lightweight, durable, easy transmission, long-driving ranges and low inverter ratings. Tesla EV models, Mahindra Reva and others use the induction machines as traction motors. The induction motors of the Tesla models are powered by Li-ion batteries and for a nominal power range of 185 KW at 4500–8500 rpm and a speed limit of 14 000 rpm with a constant power range ratio of nearly two followed by a maximum speed ratio of approximately three [12].

In all the IM, PMSM and BLDC and other types of machines, the average torque in the high-speed region is inversely proportional to the speed square. At a rated rpm, the maximum torque density is 7.2 Nm/kg for the air-cooled motor in the Tesla Roadster [12]. The continuous torque density is not defined but is usually around half the maximum value. The continuous torque density of industrial motors with the same scale are about 1.0 Nm/kg for comparison but industrial motors usually operate at lower temperatures, have a long lifespan and are regulated by a broad variety of industrial standards. A permanent magnet is the cutting-edge electric machine type used in electric and hybrid electric vehicles. Today, the machine category dominates this market entirely and is used in the most common commercial hybrid electric vehicles.

For meeting the above discussed requirements, the conventional three-phase machines have to be oversized. However, the power/torque is distributed over three phases which means it has less reliability under faulty conditions [10]. Moreover, the three-phase machines have a higher magnitude of space harmonics, which results in higher torque ripples, high noise and lower efficiency. With all these concerns, the three-phase machines may not be the best solution for electric traction applications in the near future.

Over the conventional three-phase solutions, the multiphase induction motors (MIMs) are finding more and more attention progressively in the industry as well as academia [10], because of the improved efficiency with lower space harmonics, a wider range of speed and torque, the ability to handle high power with reduced voltage ratings of power electronic devices, better torque/power distribution, high-fault tolerant capability etc. The electric machines typically driven with power converters, i.e. decoupled from the utility grid, which means the number of phases of the machine need not be same as utility grid (which is three phases). This advantage motivates the researchers in academia and industry for further research on multiphase machines. Moreover, the power and torque are shared by m number of phases that help to handle the high power with improved fault tolerant capability.

9.4 Commercially Available Electric Machines for EVs

Electric machines are the heart of present modern automobile industry market as well as the transformation of Electric vehicles. The force or torque required to propel a vehicle is produced by the electrical machine, where the torque/force is produced due to the interaction of the current and magnetic field. Electrical machines are broadly classified into AC and DC machines. In earlier days the DC machines were used for electric propulsion applications, whereas in modern EVs, AC machines are gaining their interest due to their high-power density, commutators-free operation, higher reliability, higher efficiency, lower cost and less maintenance. The principle of operations for all rotating machine is similar, which is dependent on the basic relationships

The electromagnetic torque developed by a machine

$$T = kI_a \tag{9.1}$$

where k is the machine constant (a function of magnetic and physical properties), and I_a is the machine phase current.

The relationship between the back emf induced and angular speed of the machine is

$$E = k\omega_r \tag{9.2}$$

At present, most of the EV manufacturing companies are using permanent magnet three-phase machines due to the high-power density, the fact that they are compact in size and have greater efficiency.

The induction machines (IM), switched reluctance machines (SRMs), permanent-magnet synchronous machines (PMSM) and brushless machines (BLDC) are the majority of traction motors which are commercially available and used by automobile companies. Among these machines, SRM and BLDC motors have been used for low- and medium-power applications. The PMSM and IM are preferable for high-power electric traction applications. Different types of electric vehicles manufactured by various automobile companies with the associated type of motor, transmission, year of manufactured, and other details are presented in Table 9.2.

9.4.1 DC Motors

DC machines are suitable for electric propulsion because of their wider torque-speed characteristics and simple speed control. Among all DC machines, the high starting torque capability of the DC series motor makes the best suitable candidate for electric traction applications. The initial powertrain design of electrical vehicles in the 1900s was done with DC series motors because of the easy speed control, and their ability to withstand capability of sudden variations in load. Moreover, the present electric traction (like locomotives in India) use DC series motors. However, the major limitations of DC motors that restricts the progress rate in EVs are, bulky in construction and size, commutators, brushes, lower efficiency, higher maintenance, lower reliability and a rotor that consists of windings.

9.4.2 Induction Motor

Looking back over 150 years, Nikolas Tesla invented the Induction Motor (IM) which works on the Mutual Induction Principle [10]. The three-phase IMs are popular in the power industry because of the ruggedness, lower maintenance and economical, however, efficiency is slightly less than the synchronous motors. The wider range of speed-torque characteristics of IM drive can be achieved by using the V/f, field-oriented controls and other vector control techniques. With appropriate vector-control techniques, the maximum torque at starting can be attained which will be most suitable for electric traction applications. Compared to slip ring IMs, the squirrel cage IMs have gained more interest due to their simplicity in control, less maintenance, and a longer life. The flux density (B) of IM can be adjusted by maintaining the voltage and frequency ratio, which helps in controlling the desired torque of the machine. The major advantages to adapt for EVs are, low cost, magnet-free design, ruggedness, insensitive to temperature variation and are simple to control. However, induction motors suffer from the lower power factor as well as efficiency at light loads because reactive power is required to magnetize the rotor, i.e. is drawn from the battery source.

9.4.3 Permanent Magnet Synchronous Motors (PMSM)

Synchronous machines (SM) are another type of AC machine, which works on the magnetic-locking phenomenon. If an electromagnet is placed inside this rotating magnetic field, it is magnetically locked with the rotating magnetic field and the former rotates with the rotating magnetic field at the same speed that is at synchronous speed. In the conventional SM, the rotor is made by electromagnets and field winding is excited with DC supply. To increase the power density, as well as eliminating the field windings on rotor, permanent magnet SM (PMSM) is implemented. Because of the higher power density, compact size and higher efficiency compared to IM, the PMSM has seen tremendous growth and the majority of automobile companies have adopted this for EV applications, as shown in Table 9.2.

In the PMSM, the permanent magnets are placed on the rotor to attain the sinusoidal back EMF or trapezoidal back EMF. The materials with high coercive force are used as permanent magnets (PM) materials. Because of this, the cost of the machine is high. The major limitations of the PMSM machines are, high dependence of PMs, ageing effects of PMs and magnetic properties are temperature sensitive. In addition to this, PMSM is more sensitive to overheating and, operating at overloads can lead to permanent loss of magnetic strength of the magnets, due to high temperature.

9.4.4 Brushless DC Motors

Brushless motor is another permanent magnet motor designed for attaining the trapezoidal back EMF. This motor is similar to the DC motor category with the brushless operation. Unlike conventional DC machines, in the brushless DC motor (BLDC) the commutation is done with power electronic converters along with hall sensors installed on the stator. Because of the permanent magnets, the BLDC motors also give high-power density, high-starting torque, higher efficiency as compared with IMs and brushed DC machines. The BLDC motor is mostly used in two-wheeler bikes and three-wheelers.

The BLDC motors are classified as outer rotor and inner rotor types. If the rotor is present the outside and stator is present inside, then that motor is called a Hub motor. These hub motors are directly placed in the wheel and connect to the exterior rotor. Because of this, the overall system efficiency increases, system weight comes down, saves the space in the vehicle and doesn't require an additional gear system. This motor is widely preferred by electric cycle manufacturers like Hullikal, Tronx, Spero, light speed bicycles, etc.; and bike manufacturers like NDS Eco Motors, etc.

If the rotor of a BLDC machine is placed inside the rotor and stator is placed outside similarly to conventional machines, then the machine is called an inner BLDC motor. These inner-type BLDC motors require a transmission system to drive the wheels. As compared with the outer rotor type, this inner rotor type of BLDC motor is bulky. Many three-wheeler manufacturers such as Goenka Electric Motors, Speego Vehicles, Kinetic Green, Volta Automotive use BLDC motors. In addition to this, low- and medium-performance scooter manufacturers also use BLDC motors for propulsion.

The major limitations of the BLDC motors are, high dependency of PMs, ageing effects, temperature-dependent PM characteristics, and higher cost etc.

9.4.5 Switched Reluctance Motors (SRMs)

The Switched Reluctance Motor (SRM) is a type of variable reluctance machine which is designed as a doubly-salient structure with an unequal number of stator and rotor poles. The SRMs are robust and simple in construction. Due to the saliency of a machine, the torque production is entirely due to reluctance. The stator has concentrated phase windings independently arranged on the stator poles. The rotor also has a simple structure with a stack of steel laminations without any windings or permanent magnets. The SRM offers high- power density, greater efficiency, no permanent magnets, low inertia, easier to cool due to the absence of rotor windings and high torque-to-inertia ratio which is easier to adopt for EVs. However, the higher torque ripple, noise, vibrations, no-linear variation of inductances, complex control as well as switching circuits, the requirement of position sensors and associated converter have restricted the progress of SRMs.

9.5 Challenges and Requirements of Electric Machines for RES

In 1821, Michael Faraday first demonstrated the principle of electromechanical energy conversion. William Sturgeon invented the first electric machine, which was a DC motor, in 1832. Nikola Tesla designed the rotating magnetic field based first AC motor, invented in 1882, was in fact, an induction motor. Whereas AC machines, including synchronous generators and induction generators, have been widely used to work well for power generation using gasolines. Why have huge research activities been underway on electric machinery in recent years? The answer is the desire for new machine technologies to enable increased use of efficient and effective renewable energy [22].

All electric machines for the collection of renewable energy are currently expanded from fossil-fuel power plants. Traditional generators are relatively ineffective in converting renewable energy to electricity, and are particularly inefficient at exploiting wind or wave power. For example, in 2010, hydro-power generation was approximately 3427 TWh, so only one single day would lose 94 TWh/yr due to unanticipated machine maintenance costing around 1050 million dollars/yr, while in 2010 wind generation was approximately 430TWh, allowing an extra energy efficiency of approximately 43 TWh/yr only to improve 1% in machine efficiency [22].

Renewable energy sources such as wind or waves are intermittent and slow to move, creating challenges in how these movements are effectively captured and converted to electricity. The intermittence not only refers to a large difference in magnitudes but also to low predictability of change. This intermittent operation causes a sudden power failure or a time difference in power and frequency. It is also a difficult task to capture slow motion. For example, wind speed is generally around 5–20 m/s, which is equivalent to a rotational wind turbine speed of approximately 5–20 rpm. This low speed of rotation requires a low-speed wind generator, which leads to a bulky size and heavy weight. In other cases, a mechanical gearbox is necessary to increase the wind turbine speed by approximately 100 times which, in turn, drives the wind-generator rotor. This system can allow a high-speed wind-generator design, thus reducing its size and weight, but it needs the mechanical gearbox that incurs additional cost and loss of transmission along with the need for lubrication and maintenance. In fact, the downtime for repairing gearboxes is the longest of all components of repair wind power repair. It takes about one week even if the necessary replacement gearbox is available. Similarly, the wave generator's low-speed design problem and the mechanical gearbox's wear and tear problem are present in wave-power generation.

In general, for RESs, both induction and synchronous machines are derived from the previous century. To meet the present global electricity demand novel machines and advancements have been provided to attain more efficient operation. The challenges of the electrical machines for EVs are also applicable for RESs. However, the key challenges associated with electrical machines for RESs are

1) Efficiency improvement in induction machines
2) Permanent magnet dependency in PMSMs
3) Reliability issues in all electrical machines
4) High-power handling issues of all electrical machines
5) High-power density with reduced volume for wind applications
6) Less sensitivity to environmental variations
7) Installation size and space constraints for wind power plans.

Electrical machines for the collection of renewable energy should not be regarded as a subset of industrial electrical machinery as they have a fundamental difference in requirements. The renewable energy machine should form an individual class of electrical machines with the following characteristics

- High efficiency over wide ranges of torque and speed to increase energy utilization
- High-power density to reduce overall weight and size
- Wide range of speeds to harness energy at various speeds
- High reliability to minimize operational failure or other failures
- Rugged to eliminate maintenance costs and possible maintenance outages
- High strength to withstand tough operating conditions and natural conditions
- Good voltage control to maintain voltage of the system
- High-power factor to improve transmission of power

- Low cost to reduce the cost of the system
- Fast dynamic response
- Compact in size for wind power plant
- High-overload capability.

9.6 Commercially Available Electric Machines for RES

The electrical machines are classified into two groups; DC and AC machines. This classification becomes inappropriate with the advent of new machine types. Figure 9.17 illustrates the classification of electric machines applied to the production of renewable power; however, branches which cannot be made viable for the production of renewable. They are divided into two main groups; commutator and commutator-less. The present research trend focuses on developing new types of commutator-less machines. There are three main types of electrical machines used in the generation of renewable energy: DC, induction and synchronous. They have fundamentally different topologies of machines as shown in Figure 9.18.

9.6.1 DC Machine

DC machines can be grouped as the self-excited DC and separately excited DC types which are based on the methods of field excitation. Based on the source of field excitation, the wound field DC and PM DC can also be grouped. The interconnection between the field windings and the armature winding or the use of PM excitation determines that the entire group consists of the separately excited DC, shunt DC, DC series and PM DC types. The field and armature circuits are independent of each other for the separately excited DC machine. The field and armature circuits are connected in parallel for the shunt DC machine. The field and armature circuits are connected in series for the DC series machine. The PM field is uncontrollable for the PM DC machine. The PM DC machine provides greater power density and efficiency than its counterparts on the wound field, thanks to the space-efficient benefit offered by PMs and the lack of field losses. All DC machines are affected by the same problem of commutators and brushes. Commutators cause ripples of power and reduce the rotor speed, while brushes cause friction and radio-interference in frequency. In addition, the periodic maintenance of commutators and brushes is always required due to wear and tear. These disadvantages make them less reliable and unsuitable for an operation that is maintenance-free. DC machinery's main advantages are its maturity and simplicity. DC machines are unattractive to commercially renewable power generation, due to their relatively low efficiency and maintenance requirements. Due to their simplicity, the separately excited DC and PM DC types are, however, attractive for educational demonstration of small-scale wind power generation typically under 1 kW.

9.6.2 Induction Machines

The most mature technology among various commutator-less machines is the induction machine. Induction machines of two types are available: the wound rotor and the cage rotor. The slip-ring induction machine consists of slip rings and carbon brushes to enable its

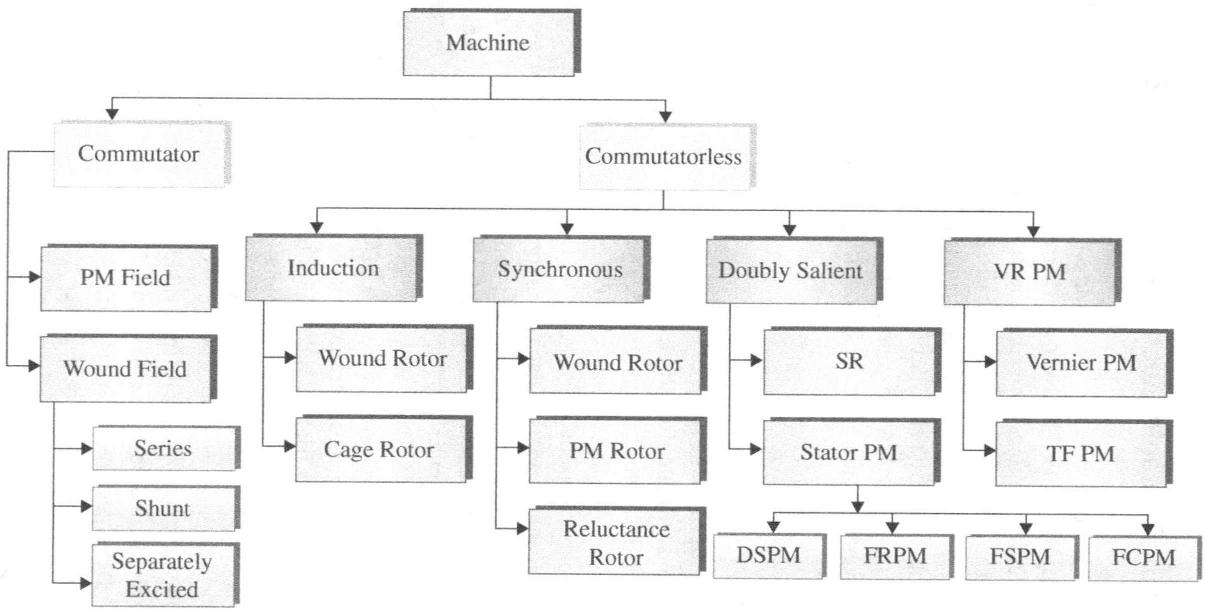

Figure 9.17 Classification of electric machines.

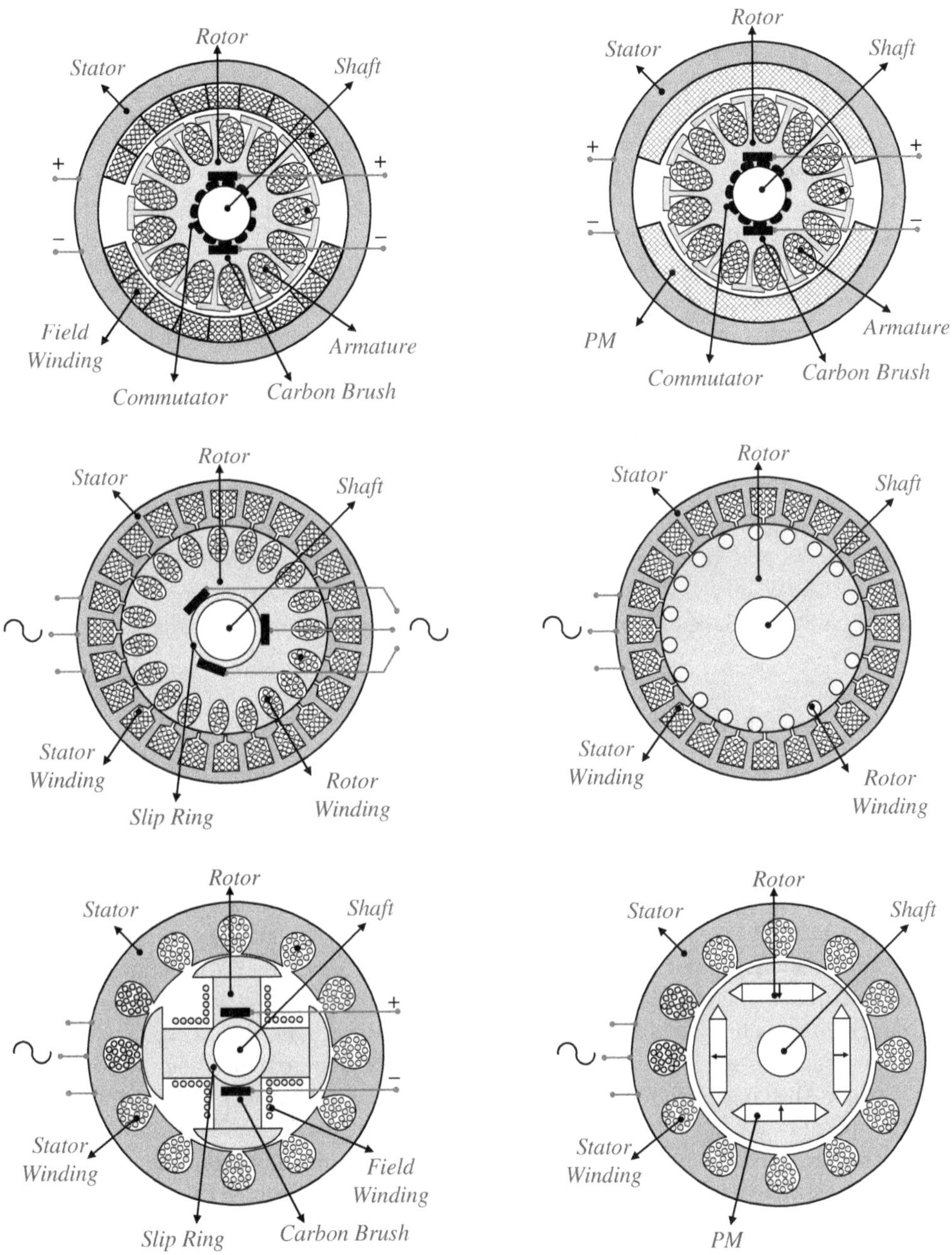

Figure 9.18 Existing machine types for renewable energy [22].

rotor sliding to connect to the external circuit. When a frequency converter is used for the rotor winding to control sliding power and the stator windings are attached to the grid, the wound rotor induction machine is commonly referred to as the doubly-fed induction generator (DFIG) that is widely used for the generation of wind energy. Acciona, Alstom-Ecotia, Gamesa, General Electric, Vestas, Mitsubishi, Nordex and Repower are the main players in the DFIG wind-turbine industry. For example, the Alta Wind Energy Center is currently the largest wind farm in the United States with an installed capacity of 1020 MW and 390 Vestas DFIGs with 1.5 or 3 MW of each. The main reason for the DFIG's popularity in wind farms is that the frequency of output can remain synchronized with the grid at various wind speeds. The wound-rotor induction machine or the derived DFIG is, however, less attractive to new wind farms because of the requirement for maintenance and lack of robustness. The DFIG is unattractive for other renewable energy collecting except for the use of wind turbine technology for wave power harvest. The induction cage-rotor machine is more efficient and more powerful than its counterpart to the wound-rotor by using the squirrel cage to replace the winding distributed rotor and removal of carbon brushes and slip rings. It also benefits from high robustness and low cost. In general, this cage-rotor induction machine is freely referred to as the IG used with various renewable energy sources, such as wind, wave, hydropower, tidal streams, solar-thermal energy, geothermal power, and biomass thermal energy. The Walney Wind Farm in the UK, for example, is the world's largest offshore wind farm with an installed capacity of 367 MW, which includes 102 Siemens IGs each with 3.6 MW [22].

9.6.3 Synchronous Machines

Synchronous machines are used almost exclusively in power production using gasolines. They also are prevalent in thermal power plants and hydro plants which adopt steam turbines and hydro turbines with nearly fixed speeds, respectively. Two main types of synchronous machines exist, namely the wound rotor and the PM rotor. The wound-rotor synchronous machine incorporates the DC field winding in the rotor, generally regenerated by the external DC provision through slip rings and carbon brushes. This synchronous wound-rotor machine is generally referred to as the synchronous generator (SG) that attracts renewable electricity generation systems with high-power ratings. For example, Gorges Dam is the world's largest installed hydro-power plant with a capacity of 21GW, which has 30 main SGs, each 700 MW [22]. These SGs are produced by the companies: Alstom, ABB and Harbin Motor, Voith, General Electric, Siemens and Oriental Motor. Compared to IG, this SG uses greater efficiency for the production of electricity and can operate at, or near, a unit power factor. The use of slip rings and carbon brushes leads to maintenance requirements that are especially disadvantageous for small-scale wind, wave, small hydro and tidal systems' energy harvesting streams.

The synchronous permanent magnet (PM) machine has the definite advantages of high efficiency utilizing magnets that replace the DC field winding in the rotor due to the absence of a copper loss, a high-power density caused by the elimination of DC field winding and a fast response due to a less electrostatic time-constant of the rotor. The demagnetization features of viable PM materials, including ferrite, aluminium nickel-cobalt (Al-Ni-co), samarium-cobalt (Sm-co) and neodymium-iron-boron materials (Nd-Fe-B) are shown in Figure 9.19. The Nd-Fe-B may be preferable because it can achieve a high residual

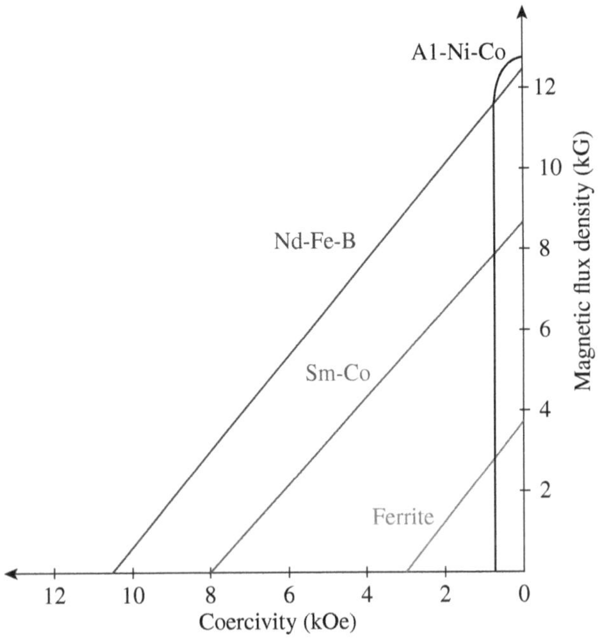

Figure 9.19 PM characteristics [22].

magnetism measuring the magnetic field's strength, high coercivity, which denotes the resistance to demagnetization and large-scale energy (BH_{max}), which represents the magnetic energy density.

The synchronous PM machine is categorized as the PM brushless AC (BLAC) and PM brushless DC (BLDC) machines, based on the induced emf waveforms. The PM-based BLAC machine is characterized by sinusoidal back electromotive force (emf) with sinusoidal AC current while the trapezoidal emf with rectangular AC current characterizes the PM-based BLDC machine. In fact, the PM-based BLAC is also referred to as a PMSM. The PMSM can be classified as surface-assembled, surface-inserted, interior-radial and interior-circumferential topologies as illustrated in Figure 9.20, according to the position of PMs in the rotor. The PM synchronous machine is increasingly attractive for renewable energy applications with medium-power ratings because of its high efficiency and power density, although it is suffering from the drawbacks of relatively high PM material costs and temperature-sensitive PM characteristics. It is generally known as the PMSG and is especially attractive for the latest wind, wave, small hydro and tidal-storm generation systems. For example, the Siemens PMSG in Hovsore, Denmark, can deliver up to 6 MW for offshore wind-power generation

9.6.4 Advanced Machines for Renewable Energy

The latest research and development of renewable energy machines have focused on solving the challenges mentioned in Section 9.5. The major focus is on

1) Higher fault tolerant machines
2) Magnet-free high-power density electrical machines

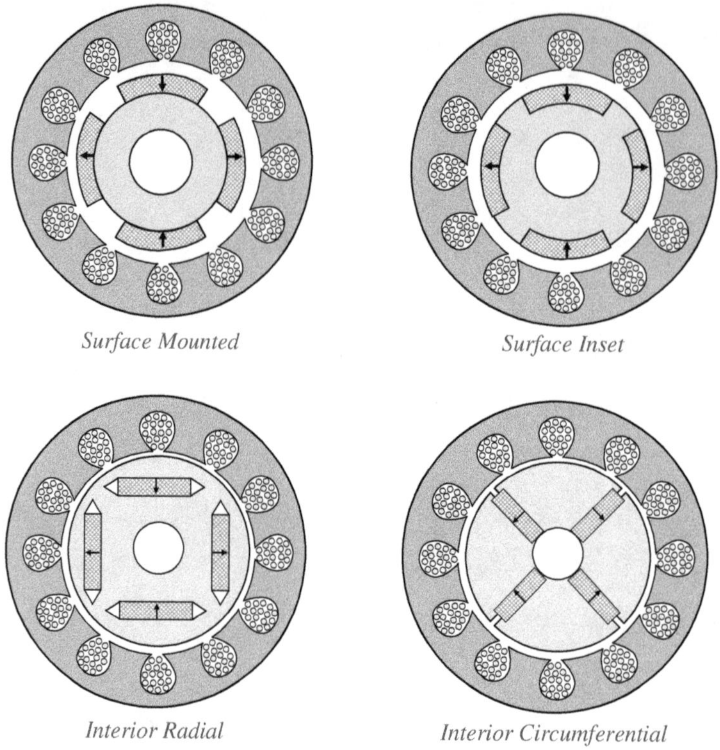

Figure 9.20 PM synchronous machine topologies [22].

3) Multiphase machines wrt IM and PMSMs
4) Electromagnet-based advanced synchronous machines
5) Doubly salient magnet-less machine topologies.

9.7 Summary

In this chapter, the electrical machines for electrical vehicles and RESs are discussed. The major components of EVs such as the battery, electric motor, power inverter, transmission system and others are discussed in detail. The challenges and requirements for selecting an electric machine for EVs as well as renewable energy applications have been presented in detail. The major electrical machines used in various EV models have also been presented in detail. The advantages and disadvantages of DC machines, induction machines, PM synchronous machines, SRM, BLDC machines for EV have been presented. The classifications of charging technologies for EVs, for example, wired and wireless charging have been discussed, where the power levels of each type of charging technology have also been presented. The charging methods of a battery such Constant Voltage (CV), Constant Current (CC), Constant Current and Constant Voltage (CCCV), Multistage Constant Current (MCC) are all explained. For renewable energy applications, induction generators, PM synchronous generators and other machines are explained.

References

1 Wikipedia (2020). Electric vehicle. https://en.wikipedia.org/wiki/Electric_vehicle (accessed 30 September 2020).
2 Husain (2003). *Electric and Hybrid Vehicles Design Fundamentals*. Boca Raton, FL: CRC.
3 Chan, C.C. (2002). The state of the art of electric and hybrid vehicles. *Proceedings of the IEEE* 90 (2): 247–275.
4 Mandela, K.C. (2018). Future of Electric Cars in India. https://electricvehicles.in/future-of-electric-cars-in-india (accessed 30 September 2020).
5 Wikipedia (2020). List of electric cars currently available.https://en.wikipedia.org/wiki/List_of_electric_cars_currently_available (accessed 30 September 2020).
6 US Department of Energy (2018). How Do All-Electric Cars Work? https://afdc.energy.gov/vehicles/how-do-all-electric-cars-work (accessed 30 September 2020).
7 US Department of Energy (2018). How Do Hybrid Electric Cars Work? https://afdc.energy.gov/vehicles/how-do-hybrid-electric-cars-work (accessed 30 September 2020).
8 US Department of Energy (2018). How Do Plug-In Hybrid Electric Cars Work? https://afdc.energy.gov/vehicles/how-do-plug-in-hybrid-electric-cars-work (accessed 30 September 2020).
9 US Department of Energy (2018). How Do Fuel Cell Electric Vehicles Work Using Hydrogen? https://afdc.energy.gov/vehicles/how-do-fuel-cell-electric-cars-work (accessed 30 September 2020).
10 Reddy, B.P. (2019). Analysis and Performance Improvement of Pole Phase Modulated Induction Motor Drives for Traction Applications. PhD thesis. Indian Institute of Technology.
11 Zhu, Z.Q. and Howe, D. (2007). Electrical Machines and Drives for Electric, Hybrid, and Fuel Cell Vehicles. *Proceedings of the IEEE* 95 (4): 746–765.
12 Zelaya De La Parra, H., Magnussen, F., and Bosga, S. (2009). Challenges for electric machines and power electronics in automotive applications. In: *International Conference on Ecological Vehicles and Renewable Energies, (EVER'09)*, 1–9. EVER.
13 Bostanci, E., Moallem, M., Parsapour, A., and Fahimi, B. (2017). Opportunities and challenges of switched reluctance motor drives for electric propulsion: a comparative study. *IEEE Transactions on Transportation Electrification* 3 (1): 58–75.
14 Magnusson, S. (2020). NZ Electric Car Guide. www.leadingthecharge.org.nz/nz_electric_car_guide (accessed 30 September 2020).
15 Tie, S.F. and Tan, C.W. (2013). A review of energy sources and energy management system in electric vehicles. *Renewable and Sustainable Energy Reviews* 20: 82–102.
16 Kerns, J. (2015). What's the Difference Between Batteries and Capacitors? https://www.machinedesign.com/automation-iiot/batteries-power-supplies/article/21831866/whats-the-difference-between-batteries-and-capacitors (accessed 30 September 2020).
17 Rubino, L., Capasso, C., and Veneri, O. (2017). Review on plug-in electric vehicle charging architectures integrated with distributed energy sources for sustainable mobility. *Applied Energy* 207: 438–464.
18 Veneri, O. (2017). *Technologies and Applications for Smart Charging of Electric and Plug-in Hybrid Vehicles*. Springer.

19 Ahmad, A., Alam, M.S., and Chabaan, R. (2018). A comprehensive review of wireless charging technologies for electric vehicles. *IEEE Transactions on Transportation Electrification* 4 (1): 38–63.

20 Patil, D., McDonough, M.K., Miller, J.M. et al. (2018). Wireless power transfer for vehicular applications: overview and challenges. *IEEE Transactions on Transportation Electrification* 4 (1): 3–37.

21 Electropaedia (2005). Battery Chargers and Charging Methods. https://www.mpoweruk.com/chargers.htm (accessed 30 September 2020).

22 Chau, K.-T., Li, W., and Lee, C.H.T. (2012). Challenges and opportunities of electric machines for renewable energy (Invited Paper). *Progress in Electromagnetics Research B* 42: 45–74.

10

Multiphase (More than Three-Phase) Machines Concepts and Characteristics

10.1 Preliminary Remarks

Electrical machines are at the heart of various day-to-day appliances such as washing machines, air conditioners (ACs), chillers, refrigerators, pumps, electric vehicles (EVs), and all industrial applications which enable mankind to function. High-power electrical machines include, (i) synchronous machines; where rated torque is produced only at synchronous speed (slip speed = 0) and works on the principle of magnetic locking, (ii) induction machines where rated torque is produced in all ways other than the synchronous speeds (slip speed $\neq 0$) and works on the principle of mutual induction. The detailed construction, working principles and control of the conventional three-phase induction machines and synchronous machines are presented in Chapters 6 and 7, respectively. In high-power applications like traction, naval ship propulsion, aircraft, and heavy cranes for lifting etc., the traditional three-phase machines need to be oversized to meet the required range of speed and torque, require higher per-phase voltage ratings of power electronic devices for handling the high power and also the output power deteriorates drastically under fault conditions. To overcome these, the machine with a higher number of phases has been reported for high-power applications, where power/torque of the multiphase machine is split over m-phases.

This chapter describes the advantages of the multiphase machines ($m > 3$, where m = number of phases) over the conventional three-phase machines. The working principles, winding design details of the multiphase machines as well as three-phase machines are presented. Moreover, the mathematical modelling equations for generalized m-phase machines with different reference frames are included in this chapter. The basic vector control schemes are also given. At the end of this chapter, the Matlab/Simulink modelling of different multiphase machines with associated results and discussion are elaborated on to visualize the performance characteristics of multiphase machines.

10.2 Necessity of Multiphase Machines

Nowadays, with the advancement in power-electronics technology, to obtain variable speed and torque, electric machines are typically driven by power converters, i.e. decoupled from

Electrical Machine Fundamentals with Numerical Simulation using MATLAB/SIMULINK, First Edition.
Atif Iqbal, Shaikh Moinoddin, and Bhimireddy Prathap Reddy.

Companion website: www.wiley.com/go/iqbal/electricalmachine

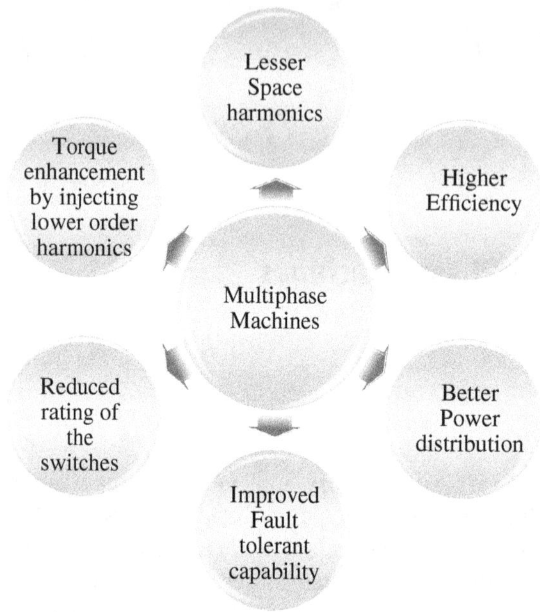

Figure 10.1 Advantages of the multiphase machines over three-phase machines.

the utility grid, which means the number of phases of the machine need not be the same as the utility grid (which is three phases). Moreover, the speed/torque can be controlled by two 90° (independently) displaced controllable current components irrespective of phase order. Apart from this, the key advantages of multiphase machines are presented in Figure 10.1.

10.2.1 Evolution of Multiphase Machines

Looking back over 150 years, Nikolas Tesla invented the Induction Motor (IM) which works on the Mutual Induction Principle [1]. Three-phase IMs are popular in the power industry because of high ruggedness, lower maintenance and are economical but efficiency is slightly less than with synchronous motors. In 1968, 'E. A. Klingshirn' analysed the performance of the poly-phase IM's with respect to losses and time harmonics present in the input voltage [2]. In the late 1960s, induction motors were driven with silicone-controlled rectifier (SCR) based inverters, in which the switching frequency was in the order of few Hz's [3, 4]. Because of this, the six-step output voltage of SCR based inverter had dominant 5th and 7th lower-order harmonics, that resulted in higher torque ripple, as well as the higher losses in machine.

The first work towards the Multiphase Induction Motors (MIM) had been done by E. E. Ward and colleagues in 1969 [3, 5], where authors explored the advantage of the five-phase inverter-fed MIM. This work provided the details of mathematical modelling, the study of torque and current ripple of the five-phase inverter-fed MIM drive wrt three-phase induction motor (IM) drive. The main conclusion from this work was the torque ripple percentage of the five-phase inverter-fed MIM was reduced by 3 times compared to traditional three-phase inverter-fed IM drive. In 1974, R. H. Nelson and P. C. Krause presented a six-phase IM for reducing torque pulsations caused by the switching of the stator voltages [6, 7]. The six-phase two-pole machine winding was realized as two three-phase windings

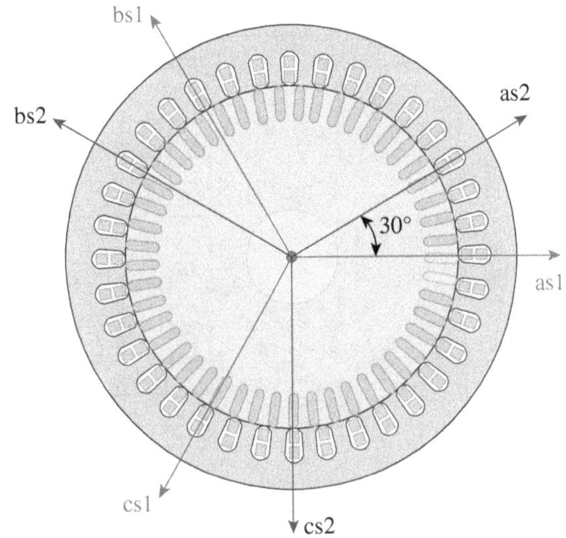

Figure 10.2 A two-pole six-phase asymmetrical IM diagram with two three-phase winding sets.

displaced by an angle of 30° as shown in Figure 10.2. The neutrals of these three-phase winding sets were isolated for arresting the circulating currents. Moreover, the authors analysed the torque ripple profile for both symmetrical (60° phase shift between all phase windings) and asymmetrical six-phase machines (30° phase shift between the two three-phase winding sets, as given in Figure 10.2) in comparison with three-phase IM's [6]. The conclusions of this study were; (i) 30° phase shifted windings with isolated neutrals will give better torque ripple profile over 0° and 60° phase shifted windings, (ii) the torque ripple frequency is shifted to 12 times that of the modulating frequency wrt 6 times that of conventional three-phase IM's. Later, T.M. Jahns proposed a phase-redundancy technique for improving the reliability of the solid-state AC drives [8], where the performance analysis of six-phase IM was presented for various open-circuit and short-circuit faults. The m-phase machine is able to operate with (m – 3) phases, however, the windings have to be designed for resisting the transient currents if the number of phases is minimal. In 1983, E. A. Klingshirn analysed the performance characteristics of the MIM with multiple three-phase winding sets [9]. In these papers, authors stated that the order of time harmonics will increase wrt to the number of phases, where the magnitude of time harmonics is inversely proportional to its order. The time harmonics present in the output voltage of the inverter increase the I^2R loss of stator, but the I^2R loss associated with the rotor is minimal because some of these time harmonics do not induce the currents in the rotor circuit. Moreover, the increase in the order of harmonics will result in lower torque pulsations. This work explains about the performance description and theoretical analysis of the third multiples of the phase windings such as six-phase, nine-phase, 12-phase and 18-phase, etc. These machines are able to operate with open-circuit faults, where the performance of the machine following the fault is directly related to the number of healthy phases.

Gopakumar and colleagues proposed a modified current source fed six-phase IM for reducing torque pulsations with reduced ratings of the SCR's [10], as shown in Figure 10.3. In this proposed current source inverter (CSI) configuration, a constant current source

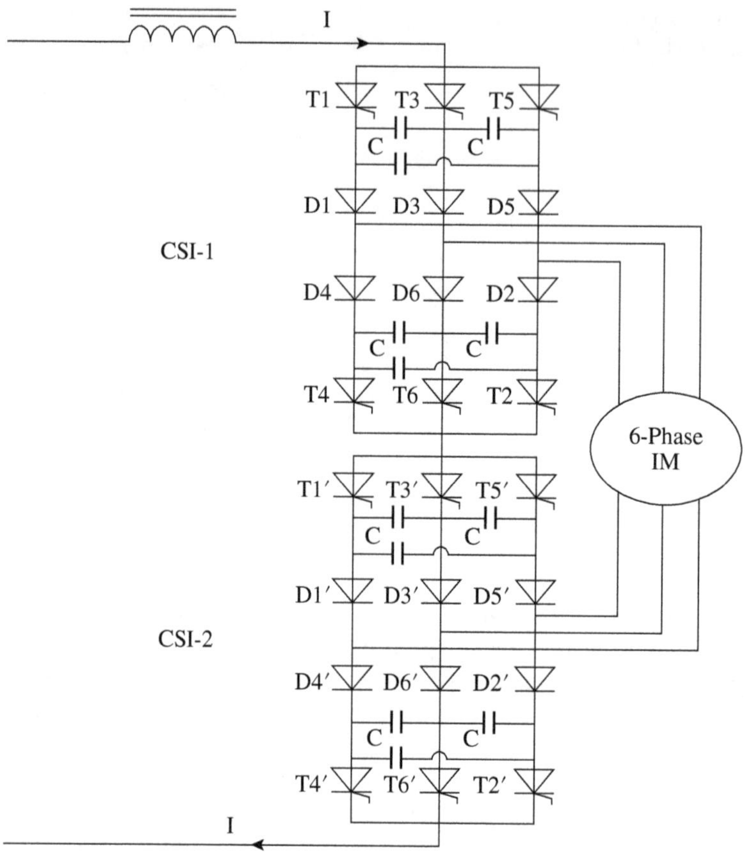

Figure 10.3 The series connection of dual CSI fed six-phase IM circuit.

is used for feeding two three-phase CSIs. These two CSIs are connected in series and the magnitude of a current flowing through each CSI is the same. Because the winding disposition and series-connected CSIs will enhance the harmonic profile of the air gap magneto-motive force (mmf), this results in lower torque ripple.

In 1993, Gopakumar and colleagues also presented a pulse width modulated (PWM) voltage source inverter-fed split-phase IM drive [11]. Here, the split-phase means two three-phase winding sets are displaced by an angle of 30°, as shown in Figure 10.4. In this work, a dodecagonal (12-sided) space vector PWM has been reported, which gives the maximum possible output voltage of the inverter is 0.643*Vdc as compared to 0.577*V_{dc} with a three-phase system. Moreover, the possible switching states have increased to 48 wrt the eight switching states with a conventional Hexagon PWM. The significant advantage of this technique is that the harmonics of the order $6n \pm 1$ (where $n = 1, 3, 5, 7\ldots$) will present in the pole-voltages but these harmonics will not appear in air gap flux due to the winding disposition. However, the remaining $6n \pm 1$ (where $n = 2, 4, 6, 8\ldots$) harmonics of the order will not appear in the voltage waveform itself, because of the split-phase concept. This elimination of the lower-order harmonics from the air gap flux results in better torque ripple compared to three-phase inverter-fed IM drives. The field-oriented control of the

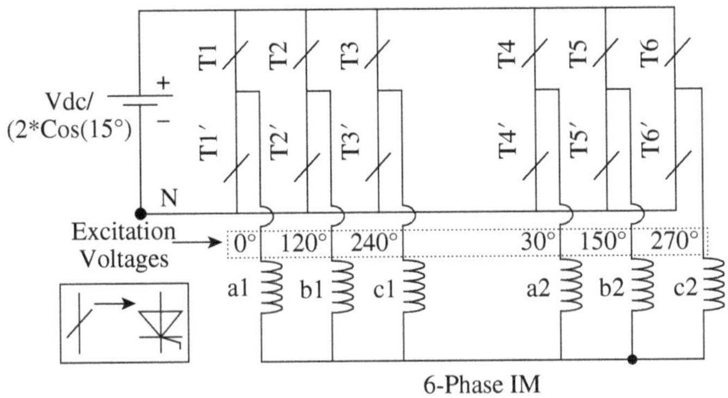

Figure 10.4 Voltage-source inverter (VSI) fed spilt-phase IM drive [11].

split-phase IM is presented in [12, 13], where voltage and current measurements are used to find the speed of the motor.

In the present century, advancement in power electronic semiconductor technology has augmented the research towards the MIM drives because of the lower cost and high-switching frequency capability of the switches (such as the metal oxide silicone field effect transistor [MOSFET] and insulated-gate bipolar transistor [IGBT]) as compared with SCRs [3, 4]. G. K Singh is the person who reported on the first survey paper on MIM drives, where the author presented the potential benefits in terms of torque ripple, efficiency and space harmonics wrt time harmonics and fault-tolerant operation over three-phase IMs [12]. In 2003, Steve Williamson proposed a mathematical analysis for recognizing the harmonic fields produced by the time harmonics present in the supply voltage and stator joule losses of the MIM drive. Later, many survey papers on MIM drives were reported in the twentieth century and are discussed [3, 9, 12, 14–16], where the history and evolution of multiphase machines with their characteristic advantages are presented.

10.2.2 Advantages of Multiphase Machines

10.2.2.1 Better Space Harmonics Profile

The fifth and seventh harmonics are dominant in three-phase machines, where these harmonics will not appear in the air-gap flux because of the split-phase IM concept (i.e. with a six-phase machine) [3, 9, 12, 14–16], which reduces the torque ripple. The current equation of the xth phase of the m-phase IM is as follows [15, 17],

$$i_x(t) = \Re e \sum_q \sqrt{2} I_q e^{jq[wt-((x-1)\pi/m)]} \tag{10.1}$$

Here, m is the number of phases, I_q is the qth harmonic current, and w is related to the inverter output frequency. The rated current of the m-phase machine is I, number of stator conductors/phase are Z and d is the diameter of the stator bore, the electric loading is [17],

$$K = \frac{mIZ}{\pi d} \tag{10.2}$$

Table 10.1 Harmonic fields produced by the different time-harmonic components in three-phase IM.

Order of time harmonics	N = 0	N = 1	N = −1	N = 2	N = −2	N = 3	N = −3	N = 4	N = −4
q = 1	**1p**	**−5p**	**7p**	**−11p**	**13p**	**−17p**	**19p**	**−23p**	**25p**
q = 5	**5p**	**−1p**	**11p**	**−7p**	**17p**	**−13p**	**23p**	**−19p**	**29p**
q = 7	7p	1p	13p	−5p	19p	−11p	25p	−17p	31p
q = 11	11p	5p	17p	−1p	23p	−7p	29p	−13p	35p
q = 13	13p	7p	19p	1p	25p	−5p	31p	−11p	37p

The current distribution of the *2p*-pole *m*-phase excited machine is

$$J_s(\theta, t) = \Re e \sum_{v} \sum_{q} \sqrt{2} K_q k_{dv} k_{pv} e^{j[qwt - v\theta]} \tag{10.3}$$

In Eq. (10.3), the stator current has both fundamental frequency component w and associated time-harmonic components which are integer multiples of the fundamental frequency. k_{dv} and k_{pv} are the distribution and pitch factors of the m-phase machine. The rms value of the stator current density for a particular harmonic component is [17]

$$J^s_{qv} = K_q k_{dv} k_{pv} \tag{10.4}$$

Here K_q is the electric loading related to Iq current and v is the order of space harmonic component [15, 17], i.e.

$$v = p(q - 2Nm) \tag{10.5}$$

Where N is a positive integer, it equals to 0, ±1, ±2, ±3, ±4… and p is the pole pairs of the machine.

Based on Eq. (10.5), spatial harmonics fields produced by different time-harmonic components (e.g. $q = 1, 3, 5, 7\ldots$) in a three-phase $2p$-pole machine are presented in Table 10.1. From this table, it can be seen that the fifth and seventh are dominant lower-order spatial harmonics associated with fundamental frequency time-harmonic components $(q = 1)$. As compared to the fundamental component $(q = 1)$, the magnitude of the time harmonics $(q = 3,5,7\ldots)$ present in the supply voltage is less, therefore, the magnitude of space harmonics associated with these harmonics is also less. Similarly, space-harmonic fields associated with different time harmonics for five-phase $2p$-pole and nine-phase $2p$-pole IM are presented in Tables 10.2 and 10.3, respectively.

From Tables 10.1 to 10.3, it is noticed that the order of spatial harmonics is increasing with respect to the higher number of phases, which means the magnitude of these harmonics will reduce. Based on this concept, many six-phase IM drives were presented in [8–11, 15], where the authors eliminated the fifth and seventh harmonics present in the three-phase machine with an appropriate phase shift between the two three-phase winding sets.

10.2.2.2 Better Torque Ripple Profile

The interaction between two opposite rotating air-gap fields with the same pole number will result in the torque pulsations of a machine [17]. The magnitude of these torque pulsations

Table 10.2 Harmonic fields produced by the different time-harmonic components in five-phase IM.

Order of time harmonics	N = 0	N = 1	N = −1	N = 2	N = −2	N = 3	N = −3	N = 4	N = −4
q = 1	**1p**	**−9p**	**11p**	**−19p**	**21p**	**−29p**	**31p**	**−39p**	**41p**
q = 3	3p	−7p	13p	−17p	23p	−27p	33p	−37p	43p
q = 7	7p	−3p	17p	−13p	27p	−23p	37p	−33p	47p
q = 9	**9p**	**−1p**	**19p**	**−11p**	**29p**	**−21p**	**39p**	**−31p**	**49p**
q = 11	11p	1p	21p	−9p	31p	−19p	41p	−29p	51p

Table 10.3 Harmonic fields produced by the different time-harmonic components in nine-phase IM.

Order of time harmonics	N = 0	N = 1	N = −1	N = 2	N = −2	N = 3	N = −3	N = 4	N = −4
q = 1	**1p**	**−17p**	**19p**	**−35p**	**37p**	**−53p**	**55p**	**−71p**	**73p**
q = 5	5p	−13p	23p	−31p	41p	−49p	59p	−67p	77p
q = 7	7p	−11p	25p	−29p	43p	−47p	61p	−65p	79p
q = 11	11p	−7p	29p	−25p	47p	−43p	65p	−61p	83p
q = 13	13p	−5p	31p	−23p	49p	−41p	67p	−59p	85p
q = 17	**17p**	**−1p**	**35p**	**−19p**	**53p**	**−37p**	**71p**	**−55p**	**89p**

rely on; (i) The difference in speed of these two interacting fields (in terms of frequency), (ii) The product of these two magnitudes of interacting fields. Based on the data given in Table 10.1, it is clear that for three-phase machines, space harmonic fields associated with the time harmonics of order $q = 1$ and $q = 5$ are rotating in opposite directions with the same number of poles, which results in pulsating torque with a frequency of $6w$. Similarly, from Tables 10.2 and 10.3 it is seen that the frequency of the pulsating torques for five- and nine-phase MIM are $10w$ and $18w$, respectively.

The generalized expression for calculating the time harmonics, which causes the high torque pulsations in the m-phase machine is as follows [14, 15, 17],

$$q = 2Nm \pm 1 \tag{10.6}$$

In the above equation N = 1,2,3,4.... The frequency of the torque pulsations is given as

$$w_{pulsating\ Torque} = 2Nmw \tag{10.7}$$

From Eqs. (10.6) and (10.7), it is can be seen that the fifth and seventh time harmonics caused for torque pulsations in the three-phase machine and 17th and 19th time harmonics caused for the torque pulsations in a nine-phase machine, which means the nine-phase machine has better torque ripple over the three-phase machine. In order to obtain better torque ripple profile in the m-phase machine, $2Nm \pm 1$ time harmonics have to be eliminated from the inverter supply voltage.

10.2.2.3 Improved Efficiency

Where two machines are designed with the same core material and the same number of poles for achieving the same output power/torque at the same speed with $m1$ and $m2$ number of phases respectively, at this condition, both machines must have the same stator-current density value, then from Eqs. (10.1) and (10.8) the relationship for two machines can be derived as [17]

$$K_1^{m1} k_{d1}^{m1} k_{p1}^{m1} = K_1^{m2} k_{d1}^{m2} k_{p1}^{m2} \tag{10.8}$$

Where k_{dx} and k_{px} are the distribution and pitch factors of the m_x-phase machine $(x = 1,2)$, the expression distribution and pitch factors are as follows [17]

$$k_{dv} = \frac{\sin\left(\frac{v\pi}{2mp}\right)}{\frac{v\pi}{2mp}}, \quad k_{pv} = \cos\frac{v\alpha}{2} \tag{10.9}$$

If both machines have the same number of conductors/slot and the same amount of copper then the stator copper loss can be expressed as

$$\frac{P_{S_{cu}}^{m1}}{P_{S_{cu}}^{m2}} = \left[\frac{K^{m1}}{K^{m2}}\right]^2 = \left[\frac{k_{d1}^{m2} k_{p1}^{m2}}{k_{d1}^{m1} k_{p1}^{m1}}\right]^2 \tag{10.10}$$

From Eq. (10.10), the percentage loss reduction in the multiphase machine over three-phase machine can be calculated, which is given in Table 10.4. As compared to a three-phase machine, the stator copper loss is reduced by 7.88% for a nine-phase machine, which results in greater efficiency.

10.2.2.4 Fault Tolerant Capability

In the case of a three-phase inverter fed three-phase IM, if a fault occurs in either one phase machine winding or switch fault inverter, the machine has to operate in the open delta with some external arrangements (to drive the motor from a standstill). The performance of the three-phase machine will deteriorate massively (i.e. power will reduce by 1.732 times) and it will lose self-starting capability. If the fault occurs in more than one phase or more than one switch of the three-phase IM drive, the system has to be shut down as there is no other alternative to maintain the operation [3, 12–16, 18, 19]. However, if a fault occurs in an m-phase machine either one or two phases the derating will be minimum and the machine does not lose its self-starting capability, which means a machine with a greater number phases will have better fault tolerance compared with the three-phase machine, this phenomenon can be observed from the Figure 10.5.

To maintain the minimum derating in output power/torque after fault condition (one phase open-circuited) of m-phase machine, the current in the healthy phases must be controlled. The control schemes of the healthy phase currents are as follows [15, 18, 19],

Table 10.4 Stator copper-loss reduction in MIM with respect to three-phase IM [17].

Number of phases	4	5	6	9	15
Stator I^2R loss reduction (%)	3.98	5.75	6.7	7.88	8.48

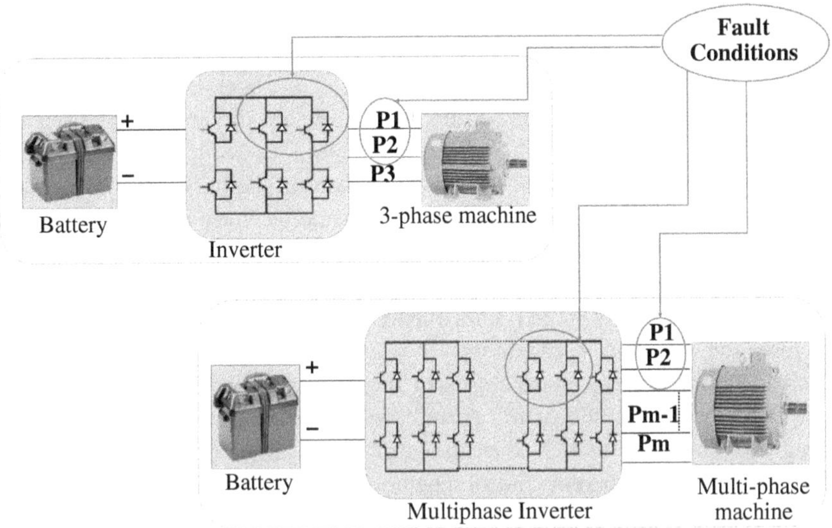

Figure 10.5 Fault conditions of the three-phase machine and multiphase machine.

Scheme-1: The magnitude and phase of the currents in healthy phases must be maintained in the same way, compared to pre-fault condition, which will reduce the torque/power (0.78% change in output for the 9-phase machine) accordingly. This scheme will reduce the stator joule loss by a magnitude of $(m - 1)/m$, but the slip, as well as rotor copper loss, will increases.

Scheme-2: If the healthy phase currents are increased by a factor $\sqrt{m/(m-1)}$, the derating in output is half of Scheme-1. The copper joule loss with this scheme will increase slightly as compared to Scheme-1.

Scheme-3: If the healthy phase currents are increased by a factor $m/(m - 1)$, then the derating in output is zero, which means the machine is capable of delivering the same output in post-fault condition as compared to pre-fault condition. However, the stator copper loss will increase compared to Scheme-1 and 2.

If the m-phase machine has a single neutral point, then the above-mentioned schemes will provide better performance after the fault. However, control complexity and increase in stator/rotor copper loss are limitations of the Schemes 1–3.

Scheme-4: If the machine has $m = xk$ number of phases, then the machine windings can be arranged as the k windings groups ($k = 2,3,4,5\ldots$) with x number of phases per group ($x = 3,4,5,6\ldots.$). For example, the nine-phase machine can be arranged as three three-phase winding groups ($m = 3*3$). If the neutrals of these three-phase winding groups are isolated and are excited with isolated three-phase inverters, a simplest post-fault control strategy can be applied.

If a fault occurs in one of the phases in the first group of the nine-phase machine, then isolate the faulty group from the system irrespective of the number of faulty phases within the group. This scheme is simple in control compared to the other three schemes but the output power/torque will reduce accordingly. The stator joule loss will not increase in this scheme. In post-fault operation, taking out the one three-phase winding group from the nine-phase machine, the output power/torque will reduce by (1/3) times as compared to the pre-fault condition.

As discussed earlier, the first article related to a fault-tolerant operation of the six-phase machine is explored by the author T.M. Jahns in 1980 where the author has used the phase redundancy technique [8]. In 1994, Jen-Ren Fu and T. A. Lipo have reported a novel control technique for MIMs, which is capable of controlling the drive continuously under fault condition of either complete inverter leg or motor phase [19]. Based on this control technique, the performance of the three-phase, four-phase, six-phase, five-phase, and seven-phase IMs under the complete loss of one phase winding is analysed [18]. For example, if an open-circuit phase fault occurs due to the inverter leg fault in a three-phase machine it loses the self-starting capability. However, rated torque can be maintained as a pre-fault condition, where phase angles of the healthy phases have to be shifted appropriately and the magnitude of the current/voltage will increase by 1.732 times. Whereas, in a five-phase machine, in order to maintain the same rated torque under the open circuit phase fault, two current-control schemes were presented in [19], for example

i) Simply add the zero-sequence component to the remaining healthy phases which will produce the same air-gap flux compared to the pre-fault condition, but the currents in healthy phases are unequal, i.e. current magnitudes in two phases are increased by 1.17 and two other phases by 1.9 times.
ii) Similar to the three-phase machine control, a five-phase system also provides an appropriate phase shift between the four healthy phase currents, which will increase the all phase current magnitudes by only 1.38 times.

Many fault-tolerant control schemes are proposed for multiphase machines to operate the machine under different faults, such as a single-phase fault or multiple-phase faults or short-circuit phase faults or switch or leg faults of the inverter. From this, it can be observed that the machine with the greater number of phases will have a higher fault tolerance with minimum derating in output compared to the conventional three-phase machines.

10.2.2.5 Reduced Ratings of Semiconductor Switches and Better Power/Torque Distribution

In general, in three-phase machines, the output power/torque is distributed over three phases. Whereas in multiphase machines output power/torque is distributed over m-phases [3, 12, 14–16], which can be observed from Figure 10.6. As compared to the three-phase machine, if the input power to the drive, volt per turn and the current flowing through each phase is the same then the required DC-link voltage requirement and blocking voltage ratings of the semiconductor switches will come down by 2 times for a six-phase machine, i.e. shown in Figure 10.6 [9–11]. Based on the reduced voltage/power per phase advantage, a nine-phase machine is reported in [20] for wind-energy conversion systems.

10.2.2.6 Torque Enhancement by Injecting Lower-Order Harmonics into Stator Currents

The torque enhancement of multiphase machines is possible only with concentrated winding because the winding function has higher third-harmonic component. In the literature [3, 12. 14–16, 21, 22], many control schemes are presented for enhancing the torque range of

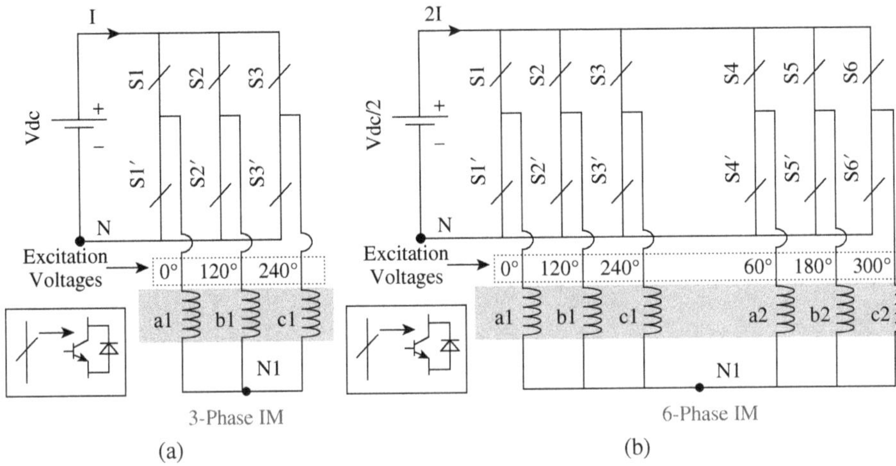

Figure 10.6 Power circuit configuration (a) three-phase IM, (b) six-phase IM.

Table 10.5 The enhancement in torque of 11-Phase IM with different harmonic injection [22].

Harmonic injection	Fundamental	Up to third	Up to fifth	Up to seventh	Up to ninth
Enhancement in torque	1	1.137	1.229	1.259	1.274

five-phase, seven-phase nine-phase and 11-phase IMs by injecting lower-order harmonics into stator currents. In 2002, Thomas A. Lipo has presented a control technique for enhancing the torque of six-phase IM by injecting the zero-sequence 3rd harmonic component in the stator currents [21]. If the saturation of the stator teeth is depending on the flux density of the core (high peak value of the flux density), then the increase in torque production of six-phase IM by injecting the third harmonic is 40% over the three-phase machine. However, if the flux density of the core and stator teeth are fixed, then the possible enhancement in torque is only 14%. In 2010 [22], A. Abdelkhalik investigated the torque enhancement of 11-phase IM by injecting the third, fifth, seventh, and ninth harmonics into the stator currents. The enhancement of the 11-phase machine with different harmonics are injected into stator currents is given in Table 10.5. From this table, it is seen that by injecting the harmonics up to the ninth (including third, fifth, seventh, and ninth) the torque of 11-phase IM is enhanced by 27.4% as compared with three-phase IM.

10.2.3 Applications of Multiphase Machines

Multiphase machines are gaining more attention in the high-power applications like ship propulsion, traction, more electric aircraft and electric vehicles due to the advancement in power semiconductor technology, the ability to handle high power with reduced power/current per phase and high-fault tolerance, which are shown in Figure 10.7 [3, 12, 14–16]. Moreover, the key advantages of the multiphase machines to make the substitute for traditional three-phase systems are mentioned in Table 10.6.

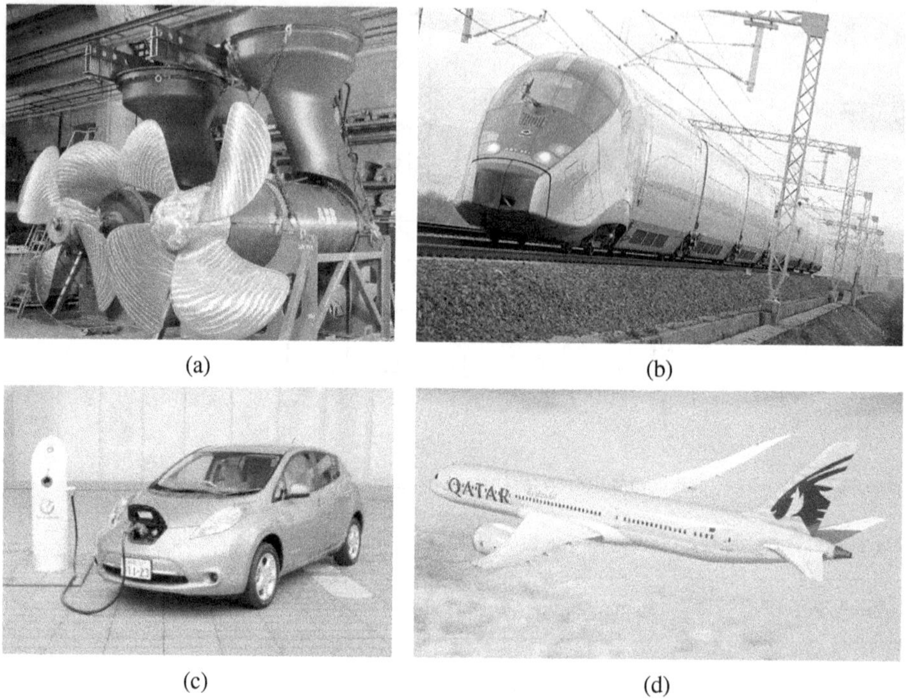

Figure 10.7 Application of multiphase machines; (a) ship propulsion, (b) electric traction, (c) EV, (d) more electric aircraft.

Table 10.6 Comparison of three-phase machines with multiphase machines.

Attributes	**Traditional three-phase machine**	**Multiphase machine (m-phase)**
Frequency of torque pulsations (f = fundamental frequency)	6f	2mf (>three-phases)
Dominant lower-order space harmonics	5 and 7	2 m ± 1
Power/Torque distribution per phase (Rated Power/Torque = P/Te)	P/3 (Te/3)	P/m (Te/m)
Fault condition either switch or phase	NA (has to operate in open delta configuration)	Yes
Torque enhancement by injecting harmonics into stator currents	NA	Yes (Concentrated Winding)

10.3 Working Principle

The basic working principle of rotating electrical machines is 'when a current-carrying conductor placed in a magnetic field it produces a force' or "interaction of two magnetic fields (stator and rotor) produces torque". Induction and synchronous machines are used for high-power applications. In both induction and synchronous machines, the stator winding, as well as constructional arrangement, are same. The basic principles are

1) Induction motor works on the electromagnetic induction
2) The synchronous motor works on magnetic locking phenomenon

The operation and working of multiphase induction and synchronous machines are the same as the three-phase machines. The detailed working and operation of the three-phase induction motor and synchronous motor are discussed in Sections 6.3 and 7.3, respectively. In these machines, the stator windings are fed with AC power which creates the magnetic field in the air gap of the machine. This magnetic field is rotating at a synchronous speed according to the frequency and poles of the machine. In this section, a brief discussion of the operation of the multiphase machine is presented.

10.3.1 Multiphase Induction Machine

The three-phase induction motors are fed with three-phase AC supply or three-phase inverters to generate a rotating magnetic field in the airgap. This rotating magnetic field is linked with rotor bars (in squirrel-cage rotor) or conductors (in slip-ring rotor type), which will induce the electromotive force (emf) in the rotor bars, according to the electromagnetic induction principle. The short-circuited type of rotor (with end rings in squirrel cage or external resistance in slip ring), a current will induce in the rotor due to the induced emf. This flow of currents will produce the respective magnetic field in the rotor. The interaction of stator and rotor magnetic fields will induce the force according to the direction of the rotating magnetic field of the stator. The difference in speed of two magnetic fields will give the slip speed, where, if the slip is zero, the force/torque of the machine will zero.

Similar to the three-phase machines in multiphase machines, the multiphase supply is created by power converters or phase-shift transformers (generally not recommended due to control and bulky in volume). The multiphase supply fed to the stator creates the multiphase rotating magnetic field with lesser space harmonics (discussed in Section 10.2.2), i.e. the phase shift is $2\pi/m$ (m = number of phases). The rest of the operation and analysis of multiphase induction motor is the same as the three-phase machine, which is presented in Section 6.3. The squirrel-cage type rotor structure will be the same for three-phase as well as multiphase machines. In a squirrel-cage rotor, the number of phases induced are same as that of the stator. In the case of the slip-ring rotor type, the number of windings on a rotor should also be the same as the number stator windings as associated resistance circuit.

10.3.2 Multiphase Synchronous Machine

The synchronous machine is a doubly exciting machine, where the stator is supplied with AC supply and the rotor is supplied with DC supply. Similar to the stator of the induction machine, feeding AC supply to the stator of the synchronous machine will produce the rotating magnetic field in the air gap of the machine, which will rotate at synchronous speed. Because of the DC supply in rotor, constant field will induce due to the unidirectional current. The rotor structure can either be of salient pole or non-salient pole type. Here, the stator magnetic field poles will vary in every half cycle and the rotor field poles will remain the same due to the DC supply. If a magnetic locking between two magnetic fields exists, the machine will produce torque and rotates at a synchronous speed. The magnetic locking is a phenomenon of the magnetic poles i.e. attractive forces between two different poles and repulsive forces between two similar poles. However, if a rotor is at standstill position, due to inertia, the rotor is unable to rotate or magnetic lock in any direction according to stator rotating magnetic field, this prevent the motor from self-starting. To bring the rotor into synchronism or magnetic locking, the external arrangement has to be added to the motor, for example, damper windings and coupling of the external motor to a prime mover (using DC or induction machines). Detailed operation and analysis of the three-phase synchronous machine are presented in Section 7.3.

In multiphase synchronous machines, the stator consists of multiple phases and is excited with multiphase power converters. The rotor of the multiphase machine will be the same as a three-phase machine, i.e. with one field winding on the rotor. The working operation and analysis of multiphase synchronous machines are the same as a three-phase synchronous machine. However, depending on the application and user requirement, multiple damper windings can be used in the case of multiphase machines, but it is not mandatory.

10.4 Stator-Winding Design

The armature winding of the rotating machines is a group of conductors, which are responsible for building the magnetic field when that is excited. In an m-phase machine, the windings are symmetrically placed on stator slots with a phase displacement of $2\pi/m$ to achieve the equal-induced emf in all phase windings. The detailed explanation of the different type of windings of electrical machines is presented in Section 4.2. In this chapter, the design and analysis of the multiphase machines are discussed. The main terminologies used for winding analysis are

Conductor: A piece of copper wire presented in stator slots, which is responsible for the production of emf and flux when it is exciting, shown in Figure 10.8a.

Turn: The series connection of two conductors with additional polarity of their emfs, shown in Figure 10.8b.

Coil: The series connection of multiple turns on the same magnetic poles, shown in Figure 10.8b.

Coil Side: Coil has two sides which are placed on two different poles.

Coil Span: The distance between two coil sides. The distance between these two coil sides depends on the type of winding, for example, for full-pitch winding, it is equal pole pitch (180°), for short-pitch winding it is <180° and for over-pitch winding it is >180°. The coil span will be represented in terms of slots, pole pitch, teeth and electrical degrees. This coil span is also called a coil pitch, as shown in Figures 10.8 and 10.9.

Pole Pitch: The distance between two adjacent poles, i.e. between the centre points of the North pole to South Pole, which always equals 180°, shown in Figure 10.8c.

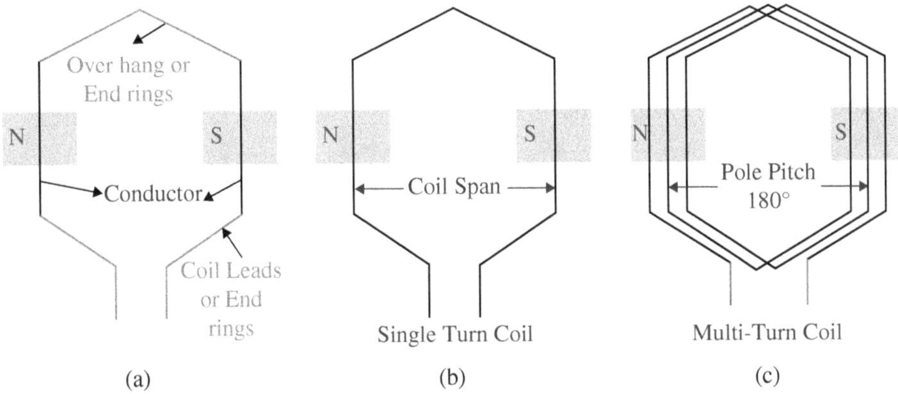

Figure 10.8 Coil description, (a) conductor's arrangement, (b) single-turn coil, (c) multi-turn coil.

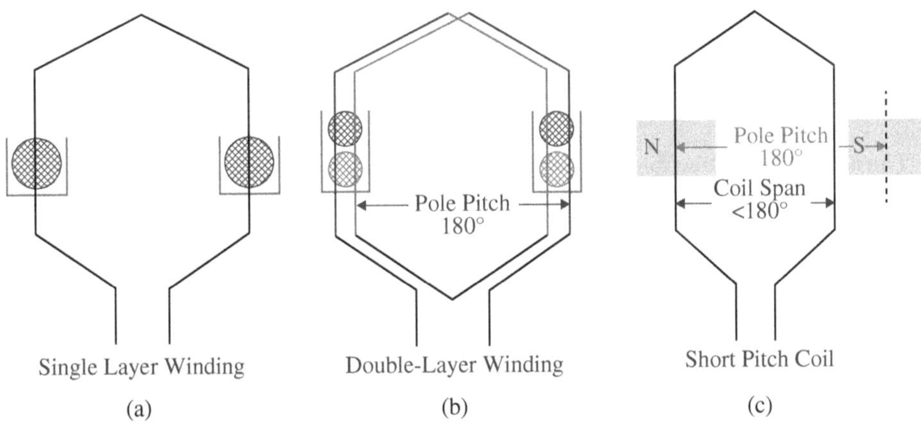

Figure 10.9 (a) single-layer winding, (b) double-layer winding, (c) short-pitch coil.

Phase Spread: Slots per pole per phase, also called the phase span or phase belt.

The armature windings of AC machines are classified depending on supply (single-phase and three-phase), coil pitch (short pitch and full pitch), the number of layers/slot (single layer and double/multi-layer), type of placing (concentrated or distributed) and end connections of coils (lap and wave). Generally, in synchronous and induction machines, the distributed type of single-layer or double-layer windings are used.

Single-Layer Windings: The stator slot of a machine is filled with only one coil side. With the single-layer winding, only integral-slot full-pitch winding is possible, as shown in Figure 10.9a.

Double-Layer Windings: The stator slot of a machine is filled with two coil sides, as shown in Figure 10.9b. If the slot is filled with n number of coil sides, then it is called an n-layer winding. With the double-layer winding, both integral slot winding (slots per pole per phase is an integer number) and fractional slot winding (slots per pole per phase is a fractional number) are possible.

The Nomenclature for this winding analysis is

- ➢ The number of phases = m
- ➢ Number of poles = $P = 2p$, where p = pole pairs
- ➢ Pole pitch = Q/P
- ➢ Number of stator slots = $Q = 2mpq$,
- ➢ q = slots/pole/phase = phase spread = phase belt = phase span
- ➢ If the slots/pole/phase
- ➢ $q = 1$ or any integer; integral slot full-pitch winding if coil pitch = pole pitch
- ➢ $q = 1$ or any integer; integral slot short-pitch winding if coil pitch < pole pitch, shown in Figure 10.9c.
- ➢ q = fraction like 3/2 or 5/2: fractional slot full- pitch winding if coil pitch = pole pitch
- ➢ q = fraction like 3/2 or 5/2: fractional slot short-pitch winding coil pitch < pole pitch.

The advantages of the short-pitch windings over full-pitch windings are

- Saving of copper due to end-ring connections
- Reduction of harmonics in emf and mmf waveforms

The advantages of fractional slot windings over integral slot windings are

- Flexibility to select any number of slots as well as coil span
- Reduction of harmonics in emf and mmf waveforms

In the design of winding diagrams, the multi-turn coil shown in Figure 10.10a is represented as a bunch of turns, shown in Figure 10.10b. In this multi-turn coil, all turns of the winding are connected in series to attain the additive polarity of emf. The practical multi-turn coils placed in different slots are shown in Figure 10.10c, where these coils are connected in series by using the leads of each multi-turn coil. In the winding diagrams presented in the following subsections, the multi-turn coil shown in Figure 10.10b is used.

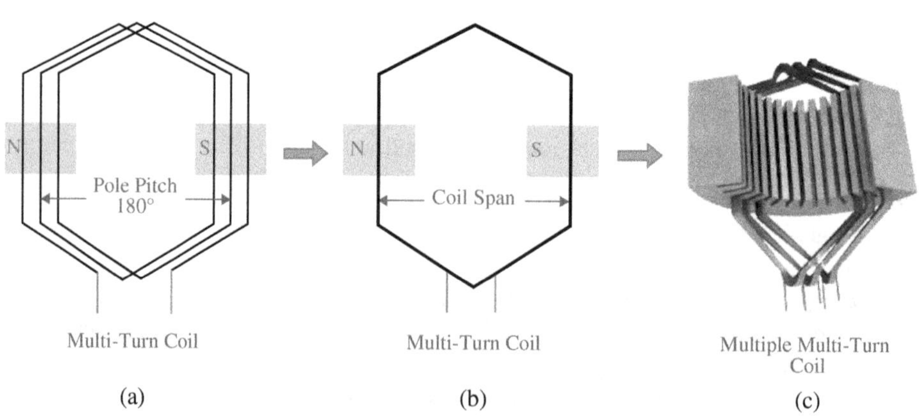

Figure 10.10 Presentation of the multi-turn coil.

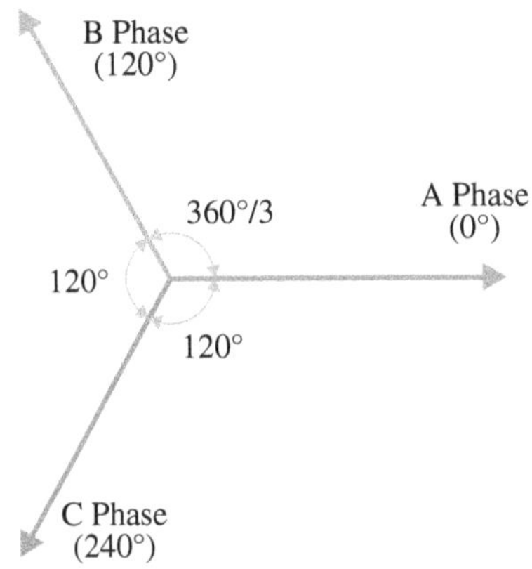

Figure 10.11 Three-phase machine vector distribution.

10.4.1 Three-Phase Windings

In a three-phase machine, the conductors associated with the three phases are distributed over a stator circumference symmetrically with a phase shift of 120°, as shown in Figure 10.11. To understand the three-phase windings, different possible types of windings such as single-layer winding and double-layer winding (fractional slot and short pitch windings) are illustrated in this section.

Let us consider an example of a 24 slot, three-phase, four-pole machine. The general nomenclature required for the design of the winding diagram is presented below

Slots/pole/phase=$\frac{Q}{2mp}$ $= 24/(4*3) = 2 =$ integer so only an integral-slot winding is possible, (single-layer, double-layer and short-pitch winding).

Phase spread $=2$ slots

Pole pitch $= 24/4 = 6$

Coil span = pole pitch $=6$ slots

With this nomenclature for the example 24 slot, three-phase, four-pole machine, the possible windings of the machine are

a) Single-layer full-pitch winding
b) Single-layer short-pitch winding
c) Double-layer full-pitch winding
d) Double-layer short-pitch winding

10.4.1.1 Single-Layer Full-Pitch Winding

The single-layer full-pitch winding for the 24 slot, three-phase, four-pole machine is shown in Figure 10.12 and the winding details in terms of slot angle, slot number and phase excitation is presented in Table 10.7. The coil side lying under the North pole is connected to the coil side lying on the South pole with a coil span of 180° (which is a span of one pole pitch). To attain the additive polarities of the induced emf, all coils are connected in series, i.e. the second end of one coil is connected to the first end adjacent coil of the same phase winding. This series connection of the coils is used for high voltage and low current purpose.

Detailed winding connections of the three phases are

A-phase: Starts at 0°

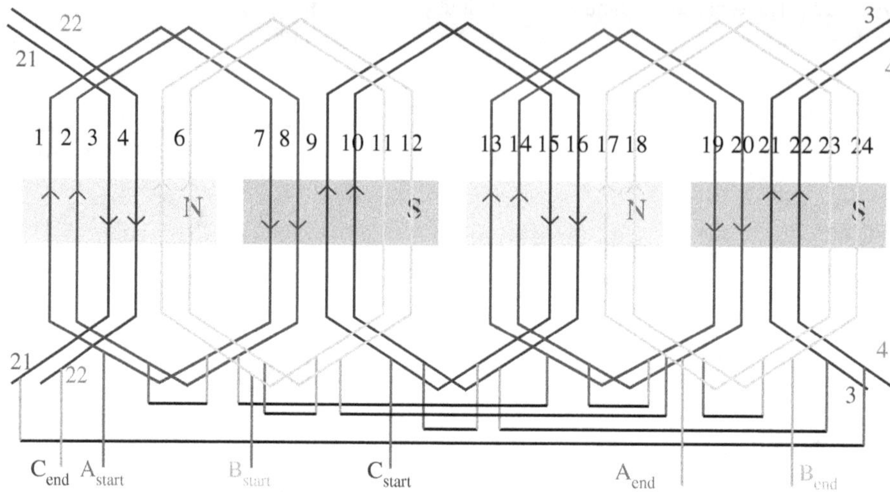

Figure 10.12 Single-layer full-pitch winding diagram of 24-slot three-phase four-pole machine.

Associated Slots: 1, 2, 7, 8, 13,14, 19, 20

Connections: 1 => 7 (1 + 6 = 7), 2 => 8 (2 + 6 = 8), 13 => 19 (13 + 6 = 19), 14 => 20 (14 + 6 = 20) are the four coils associated with A-phase and the coil span is six slots.

The connections required to attain the additive polarity of the emf in all four coils are, 7 = > 2, 8= > 13, 19= > 14, clearly shown in Figure 10.12.

Starting and ending terminals are 1 and 20 respectively.

B-phase: Starts at 120°

Slots: 5,6, 11,12, 17,18, 23,24

Connections: 5 => 11, 6 => 12, 17 => 23, 18 => 24, are the four coils associated with B-phase and the coil span is six slots.

The connections required to attain the additive polarity of the emf in all four coils are, 11 => 6, 12 => 17, 23 => 18, clearly shown in Figure 10.12.

Starting and ending terminals are, 5 and 24 respectively

C-phase: Starts at 240°

Slots: 3, 4, 9, 10, 15, 16, 21, 22

Connections: 9 => 15, 10 => 16, 21 => 3 (21 + 6 = 27–24 = 3), 22 => 4 (22 + 6 = 28–24 = 4), are the four coils associated with C-phase and the coil span is six slots.

The connection of the 21st coil side is 21 + 6 = 27; here, the coil side number exceeds the slot number i.e. 24. As there is no 27th coil side on the stator winding, to attain symmetry, the number slots 24 (coil sides) must be subtracted from 27, i.e. 27–24 = 3. So, the connection is 21 => 3.

The connections required to attain the additive polarity of the emf in all four coils are, 15 => 10, 16 => 21, 3 => 22, clearly shown in Figure 10.12.

Starting and ending terminals are, 9 and 22 respectively.

Table 10.7 Winding details of single-layer winding diagram of 24-slot three-phase four-pole machine.

i	a	a	*c	*c	b	b	*a	*a	c	c	*b	*b	a	a	*c	*c	b	b	*a	*a	c	c	*b	*b
ii	1	2	3	4	5	6	7	8	9	10	11	12	13	14	15	16	17	18	19	20	21	22	23	24
iii	0	30	60	90	120	150	180	210	240	270	310	340	0	30	60	90	120	150	180	210	240	270	310	340
iv	N						S						N						S					

i) Phases (*indicates the negative direction of current), ii) Slots, iii) Slot Angle in Electrical Degrees, iv) Poles

(i) Phases (* indicates the negative direction of current), (ii) Slots, (iii) Slot Angle in Electrical Degrees, (iv) Poles

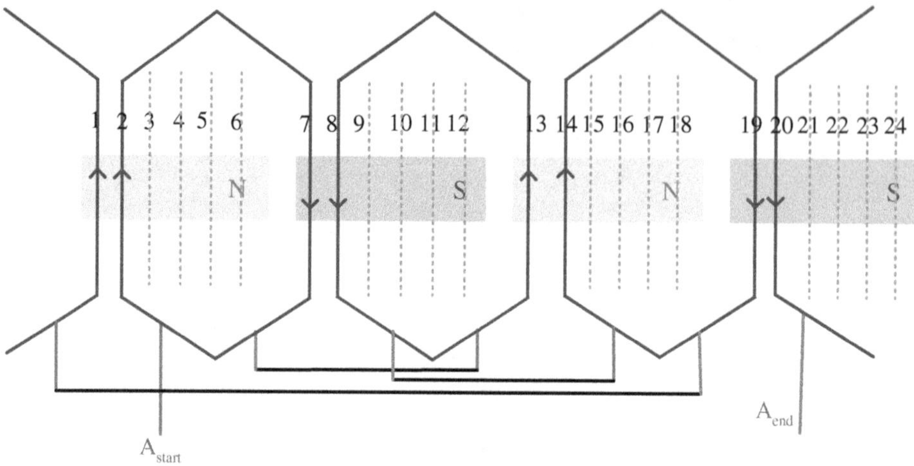

Figure 10.13 Single-layer short pitch winding diagram of 24-slot three-phase four-pole machine.

10.4.1.2 Single-Layer Short-Pitch Winding

The single-layer short-pitch winding for the 24 slot, three-phase, four-pole machine is shown in Figure 10.13, with a coil span of five stator slots (short-pitched by one stator slot). The winding details are the same as Table 10.7, but the only difference is the coil span.

The detailed winding connections of the three phases with a Coil span of five slots or 150° electrically (short-pitched by one slot or 30°) are

A-phase: Starts at 0°

Associated Slots: 1, 2, 7, 8, 13, 14, 19, 20

Connections: 2=>7 (2+5 = 7), 13=>8 (13–5 = 8), 14=>19 (14+5 = 19), 1=>20 (1–5 = –4+24 = 20), are the four coils associated with A-phase and the coil span is six slots.

The connections required to attain additive polarity of the emf in all four coils are, 7=>13, 8=>14, 19=>1, clearly shown in Figure 10.13.

Starting and ending terminals are two and 20 respectively.

B-phase: Starts at 120°

Slots: 5, 6, 11, 12, 17, 18, 23, 24

Connections: 6=>11, 17=>12, 18=>23, 5=>24 are the four coils associated with B-phase and the coil span is six slots.

The connections required to attain additive polarity of the emf in all four coils are, 11=>17, 12=>18, 23=>5, clearly shown in Figure 10.12.

Starting and ending terminals are, six and 24 respectively

C-phase: Starts at 240°

Slots: 3, 4, 9, 10, 15,16, 21, 22

Connections: 10=>15, 21=>16, 22=>3, 9=>4 are the four coils associated with C-phase and the coil span is six slots.

The connections required to attain additive polarity of the emf in all four coils are, 15=>21, 16=>22, 3=>9, clearly shown in Figure 10.12.

Starting and ending terminals are, 10 and 4 respectively

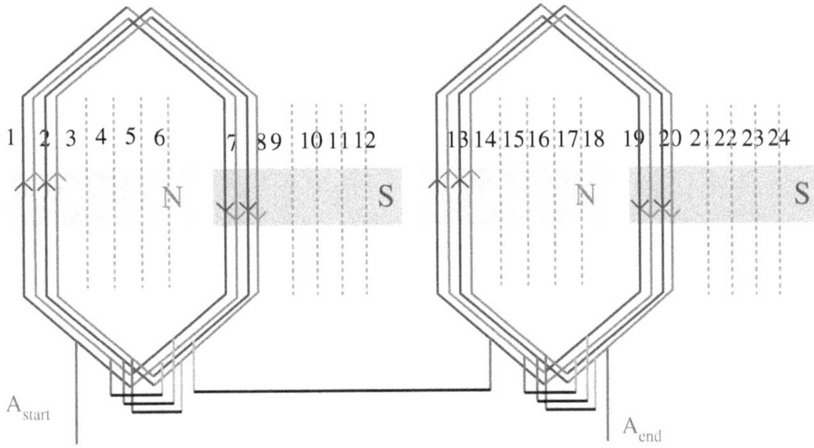

Figure 10.14 Double layer full pitch winding diagram of 24-slot three-phase four-pole machine.

10.4.1.3 Double-Layer Full-Pitch Winding

The double-layer winding for the 24 slot, three-phase, four-pole machine is shown in Figure 10.14 and the winding details in terms of slot angle, slot number and phase excitation are presented in Table 10.8. In this double-layer winding each slot has two coil sides, i.e. upper coil side (brown colour) and bottom coil side (green colour). The upper side of the coil side lying under North pole is connected to the bottom coil side lying on the South pole with a coil span of 180° (which is a span of one pole pitch or six slots) and vice versa. This arrangement of coil sides gives the lower leakage inductance as well as the enhanced harmonic profile of the emf which results in improved performance. Similar to the single-layer winding, in this double-layer winding all coil sides are also connected in series with an additive polarity.

Detailed winding connections of the three phases with a coil span of six slots or 180° electrically are

A-phase: Starts at 0°

Associated Slots: 1, 2, 7, 8, 13, 14, 19, 20

Connections: 1 top => 7 bottom (vice versa, i.e. 1 bottom => 7 top), 2 top => 8 bottom (vice versa), 13 top => 19 bottom (vice versa), 14 top => 20 bottom, (vice versa), are the eight coils associated with A-phase and the coil span is six slots. The end connections of coils have connected in series to attain the additive polarity of the emf in all eight coils are, 7 bottom => 1 bottom, 7 top => 2 top, 8 top => 13 top, 19 top => 14, clearly shown in Figure 10.14.

Starting and ending terminals are, 1 top and 20 top, respectively.

Similar to the A-phase, the B-phase and C-phase windings are arranged in respective slots with a displacement of 120°, as presented in Table 10.8

10.4.1.4 Double-Layer Short-Pitch Winding

The double-layer short-pitch winding for the 24 slot, three-phase, four-pole machine is shown in Figure 10.15, with a coil span of five stator slots (short-pitched by one stator slot). The winding details are the same as the Table 10.8, but the only difference is coil span, i.e.

Table 10.8 Winding details of double-layer full-pitch winding diagram of 24-slot three-phase four-pole machine.

i	a	a	*c	*c	b	b	*a	*a	c	c	*b	*b	a	a	*c	*c	b	b	*a	*a	c	c	*b	*b
	a	a	*c	*c	b	b	*a	*a	c	c	*b	*b	a	a	*c	*c	b	b	*a	*a	c	c	*b	*b
ii	1	2	3	4	5	6	7	8	9	10	11	12	13	14	15	16	17	18	19	20	21	22	23	24
iii	0	30	60	90	120	150	180	210	240	270	310	340	0	30	60	90	120	150	180	210	240	270	310	340
iv	N						S						N						S					

i) Phases (*indicates the negative direction of current), ii) Slots, iii) Slot Angle in Electrical Degrees, iv) Poles

(i) Phases (* indicates the negative direction of current), (ii) Slots, (iii) Slot Angle in Electrical Degrees, (iv) Poles

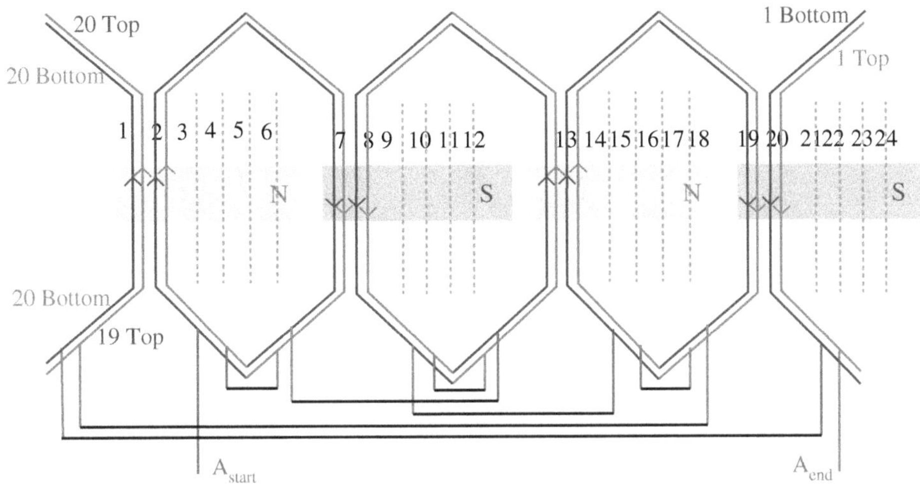

Figure 10.15 Double layer short pitch winding diagram of 24-slot three-phase four-pole machine.

short-pitched by one slot or 30°. In this short-pitch winding, upper coil side associated with the North pole is also connected to the bottom coil side associated with the South pole.

10.4.1.5 Fractional-Slot Winding

The fractional-slot winding for 18 slot, three-phase, four-pole machines are shown in Figure 10.16. The design details of the fractional-slot winding are as follows

- ➢ Number of phases = *3*
- ➢ Number of poles = $P = 4$
- ➢ Number of stator slots = $Q = 18$
- ➢ Coil span = Pole pitch = $\frac{18}{4} = 4\frac{1}{2}$ = four or five slots
- ➢ q = slots/pole/phase= $\frac{18}{4*3} = \frac{3}{2} = 1\frac{1}{2}$ slots

So, the conductors of each phase placed in 1.5 slots under each pole, each phase totally occupies six slots of the machine, and the detailed winding details, slot information are presented in Table 10.9.

The detailed winding connections of the A-phase with a Coil span of 4 slots or 180° electrically are,

A-phase: Starts at 0°

Associated Slots: 1, 2, 5, 6, 10, 11, 14, 15

Connections: 1 top => 5 bottom, 2 top => 6 bottom, 6 top => 10 bottom, 10 top => 14 bottom, 11 top => 15 bottom, 15 top => 1 bottom are the associated coils of A-phase and the coil span is four slots and the end connections are shown in the Figure 10.16.

Similar to the A-phase, the B-phase, and C-phase windings are arranged in respective slots with a displacement of 120°, as presented in Table 10.9

10.4.2 Five-Phase Windings

In a symmetrical five-phase machine, the coils associated with five phases are distributed over a stator circumference symmetrically with a phase shift of 360°/5 = 72°, as shown

Table 10.9 Winding details of fractional-slot winding for 18 slot, three-phase, four-pole machine.

i	a	a	*c	b	b	*a	c	c	*b	a	a	*c	b	b	*a	c	c	*b
	a	*c	*c	b	*a	*a	c	*b	*b	a	*c	*c	b	*a	*a	c	*b	*b
ii	1	2	3	4	5	6	7	8	9	10	11	12	13	14	15	16	17	18
iii	0	40	80	120	160	200	240	280	320	0	40	80	120	160	200	240	280	320
iv			N				S					N				S		
i) Phases (*indicates the negative direction of current), ii) Slots, iii) Slot Angle in Electrical Degrees, iv) Poles																		

(i) Phases (* indicates the negative direction of current), (ii) Slots, (iii) Slot Angle in Electrical Degrees, (iv) Poles

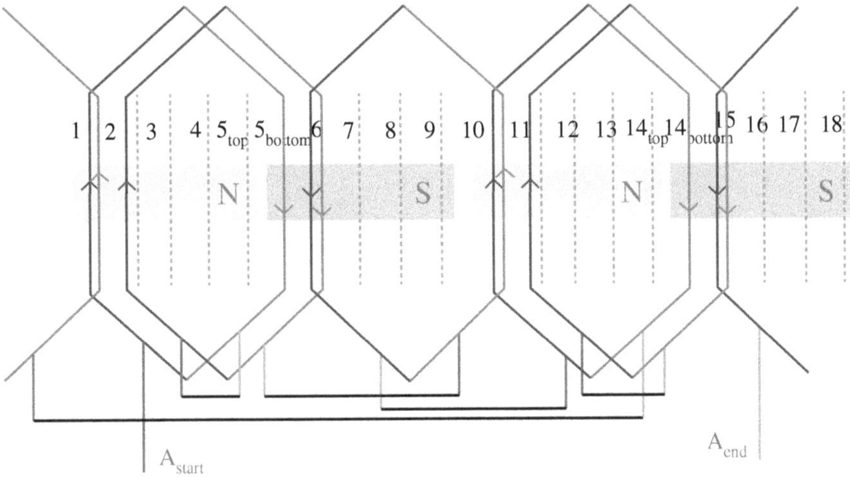

Figure 10.16 Fractional slot winding for an 18 slot, three-phase, four-pole machine.

Figure 10.17 Five-phase machine vector distribution.

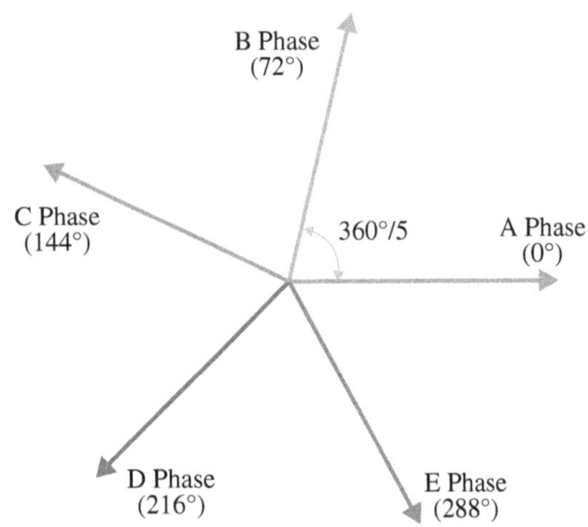

in Figure 10.17. For clear understanding, the five-phase machine windings, consider an example with a 20 slot, five-phase, four-pole machine. The general nomenclature required for the design of winding diagram is presented below

Slots/pole/phase $= \frac{Q}{2mp} = \frac{20}{4*5} = 1$

Phase spread = 1 slot

Pole pitch = 20/4 = 5

Coil span = pole pitch = five slots for full pitch winding

The single-layer winding diagram of the A-phase of a five-phase machine has shown in Figure 10.18 and the winding details of all phases with respect to slot angle, current direction and poles have presented in Table 10.10.

The detailed winding connections of the three phases are

A-phase: Starts at 0°

Associated Slots: 1, 6, 11, 16

Table 10.10 Winding details of single-layer full-pitch winding for 20 slot, five-phase, four-pole machine.

i	a	*d	b	*e	c	*a	d	*b	e	*c	a	*d	b	*e	c	*a	d	*b	e	*c
ii	1	2	3	4	5	6	7	8	9	10	11	12	13	14	15	16	17	18	19	20
iii	0	36	72	108	144	180	216	252	288	324	0	36	72	108	144	180	216	252	288	324
iv			N					S					N					S		
i) Phases (*indicates the negative direction of current), ii) Slots, iii) Slot Angle in Electrical Degrees, iv) Poles																				

(i) Phases (* indicates the negative direction of current), (ii) Slots, (iii) Slot Angle in Electrical Degrees, (iv) Poles

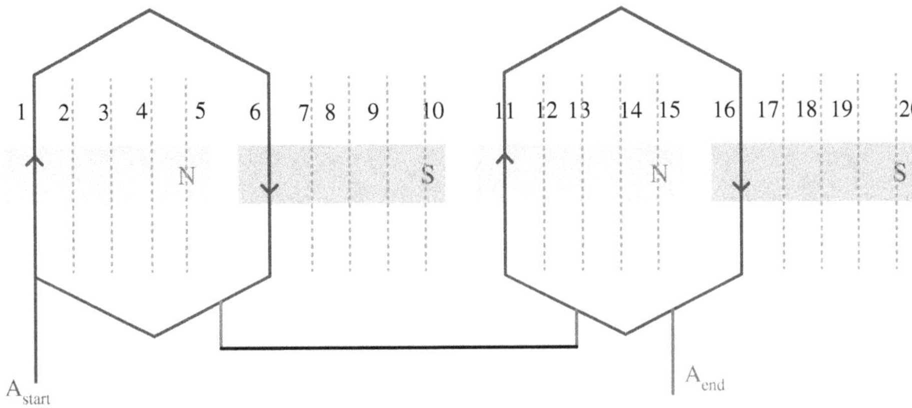

Figure 10.18 Single-layer full-pitch winding for 20 slot, five-phase, four-pole machine.

Connections: 1 => 6 (1 + 5 = 6), 11 => 16 (11 + 5 = 16) are the two coils associated with A-phase and the coil span is five slots. These two coils are connected in series to get additive polarity, 6 => 11 is connected.

Starting and ending terminals are 1 and 16 respectively.

B-phase: Starts at 72°

Associated Slots: 3, 8, 13, 18

Connections: 3 => 8 (3 + 5 = 8), 13 => 18 (13 + 5 = 18) are the two coils associated with A-phase and the coil span is five slots. These two coils are connected in series to get additive polarity, 8 => 13 is connected.

Starting and ending terminals are 3 and 18, respectively.

C-phase: Starts at 144°

Associated Slots: 5, 10, 15, 20

Connections: 5 => 10 (5 + 5 = 10), 15 => 20 (15 + 5 = 20) are the two coils associated with A-phase and the coil span is five slots. These two coils are connected in series to get additive polarity, 10 => 15 is connected.

Starting and ending terminals are 5 and 20, respectively.

D-phase: Starts at 216°

Associated Slots: 7, 12, 17, 2

Connections: 7 => 12 (7 + 5 = 12), 17 => 2 (17 + 5 = 22–20 = 2) are the two coils associated with A-phase and the coil span is five slots. These two coils are connected in series to get additive polarity, 12 => 17 is connected.

Starting and ending terminals are 7 and 2, respectively.

E-phase: Starts at 288°

Associated Slots: 9, 14, 19, 4

Connections: 9 => 14 (9 + 5 = 14), 19 => 4 (19 + 5 = 24–20 = 4) are the two coils associated with A-phase and the coil span is five slots. These two coils are connected in series to get additive polarity, 15 => 19 is connected.

Starting and ending terminals are 9 and 4, respectively.

The double-layer winding configuration, as well as fractional slot windings of five-phase machines, will be the same as the three-phase winding diagrams presented in Section 10.4.1.

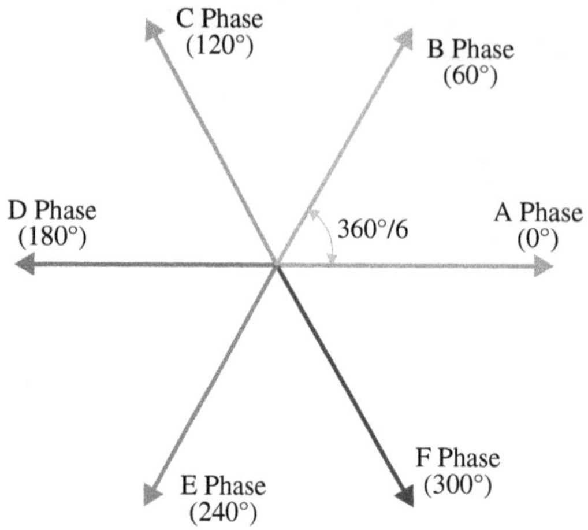

Figure 10.19 Symmetrical winding based six-phase machine vector distribution.

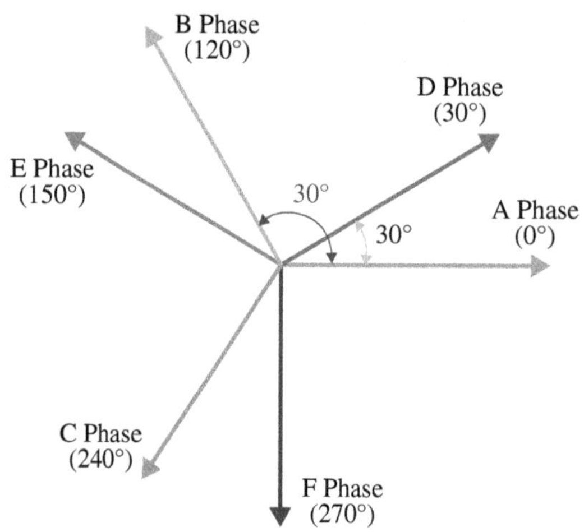

Figure 10.20 Asymmetrical winding based six-phase machine vector distribution.

10.4.3 Six-Phase Windings

The winding design of a six-phase machine can be designed symmetrically or asymmetrically, i.e.

1) Symmetrical winding arrangement of all phases are displaced by an angle of $360°/6 = 60°$, presented in Figure 10.19.
2) Asymmetrical winding arrangement, the six-phase windings of a machine are grouped into two three-phase winding groups. These two groups have been displaced by an angle of 30° and each phase three-phase group consists of three 120° displaced symmetrical windings, presented in Figure 10.20.

The generalization for a machine has $m = xk$ number of phases, then the machine windings can be arranged as the k windings groups ($k = 2,3,4,5...$) with x number of phases per group ($x = 3,4,5,6...$). Here, phase displacement between the k number of winding groups is equal to $\frac{360^0}{m*2}$.

10.4.3.1 Symmetrical Winding of Six-Phase Machine

The phase distribution of symmetrical six-phase machine has shown in Figure 10.19, where the phases are displaced by an angle of 360°/6 = 60°. A 24 slot, six-phase, four-pole symmetrical machine has considered and the general nomenclature is

Slots/pole/phase= $\frac{Q}{2mp} = \frac{24}{4*6} = 1$

Phase spread =1 slot

Pole pitch = 24/4 = 6

Coil span = pole pitch = six slots for full pitch winding.

The winding details of all six phase windings with respect to slot angle, current direction and poles have presented in Table 10.11. From Figure 10.19 and Table 10.11, it can be observed that the A-phase and D-phase windings are complementary to each other (i.e. 180° apart each other) which will reduce the useful flux as well as the torque of the machine. With the single-layer winding, the winding of symmetrical six-phase machine is not possible which can be observed in Table 10.11, where the overlapping of complementary phases is happening. For the design, the winding of a six-phase symmetrical machine, double-layer winding is mandatory and the same is presented in Figure 10.21 and Table 10.12.

With the winding details of symmetrical six-phase 24-slot four-pole machine shown in the Table 10.12, both full-pitch winding as well as short-pitch windings are possible. The detailed winding connections of A-phase with full-pitch winding are

A-phase: Starts at 0°

Associated Slots: 1, 6, 7, 12, 13, 18, 19, 24.

Connections: 1 => 7, 12 => 16, 13 => 19, 24 => 6, are the four coils associated with A-phase and the coil span is six slots, and the end connections to achieve the additive polarity of induced emf in all coils are shown in Figure 10.21.

Starting and ending terminals are, 6 and 7, respectively,

Similarly, the other phase winding coils are connected according to the winding details given in Table 10.12.

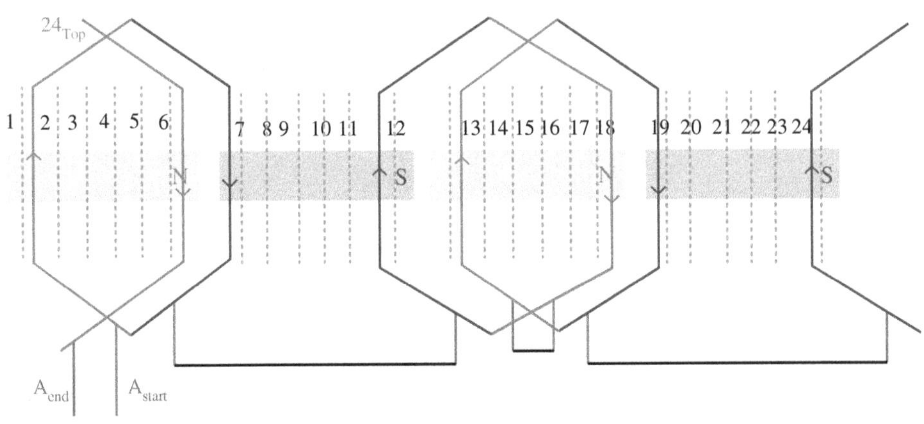

Figure 10.21 Double-layer winding diagram of symmetrical 24 slot, six-phase, four-pole machine.

Table 10.11 Winding details of symmetrical 24 slot, six-phase, four-pole machine.

i	*d		*e		*f		*a		*b		*c		*d		*e		*f		*a		*b		*c	
	a		b		c		d		e		f		a		b		c		d		e		f	
ii	1	2	3	4	5	6	7	8	9	10	11	12	13	14	15	16	17	18	19	20	21	22	23	24
iii	0	30	60	90	120	150	180	210	240	270	300	330	0	30	60	90	120	150	180	210	240	270	300	330
iv	N						S						N						S					
i) Phases (*indicates the negative direction of current), ii) Slots, iii) Slot Angle in Electrical Degrees, iv) Poles																								

(i) Phases (* indicates the negative direction of current), (ii) Slots, (iii) Slot Angle in Electrical Degrees, (iv) Poles

Table 10.12 Double-layer winding details of symmetrical 24 slot, six-phase, four-pole machine.

i	*d	b	*e	c	*f	d	*a	e	*b	f	*c	a	*d	b	*e	c	*f	d	*a	e	*b	f	*c	a
	a	*e	b	*f	c	*a	d	*b	e	*c	f	*d	a	*e	b	*f	c	*a	d	*b	e	*c	f	*d
ii	1	2	3	4	5	6	7	8	9	10	11	12	13	14	15	16	17	18	19	20	21	22	23	24
iii	0	30	60	90	120	150	180	210	240	270	300	330	0	30	60	90	120	150	180	210	240	270	300	330
iv	N						S						N						S					
i) Phases (*indicates the negative direction of current), ii) Slots, iii) Slot Angle in Electrical Degrees, iv) Poles																								

(i) Phases (* indicates the negative direction of current), (ii) Slots, (iii) Slot Angle in Electrical Degrees, (iv) Poles

10.4.3.2 Asymmetrical Winding

The phase distribution of an asymmetrical six-phase machine has shown in Figure 10.20, where the phases are displaced by an angle of 360°/6*2 = 30°. A 24 slot, six-phase, four-pole asymmetrical machine has considered the general nomenclature will be same as symmetrical winding like Slots/pole/phase=$\frac{Q}{2mp} = \frac{24}{4*6} = 1$, phase spread =1 slots, Pole pitch = 24/4 = 6, Coil span = pole pitch = six slots for full pitch winding.

The winding details of the all six phase windings with respect to slot angle, current direction and poles for asymmetrical machines are given in Table 10.13. In this asymmetrical winding, the A-phase winding and D-phase winding are displaced by an angle of 30°, whereas in symmetrical winding, same phases are complementary to each other (i.e. 180° apart each other). With this asymmetrical winding, both single-layer, as well as double-layer winding are possible for six-phase machines. With the winding details given in Table 10.13, different winding arrangements presented in Section 10.4.1 can be implemented for six-phase machine asymmetrical machines.

10.4.4 Nine-Phase Windings

The nine-phase machine can be grouped as three three-phase winding groups and the winding can be designed in a symmetrical or asymmetrical manner, i.e.

1) Symmetrical winding arrangement all phases are displaced by an angle of 360°/9 = 40°, as shown in Figure 10.22.
2) Asymmetrical winding arrangement, the windings are displaced by angle of $\frac{360^0}{m*2}$=20°, as shown in Figure 10.23.

Similar to the six-phase machine windings discussed in the previous section, the winding analysis of the nine-phase machine can be analysed. A nine-phase 36-slot, four-pole machine is considered for this both symmetrical as well as asymmetrical winding details are presented in Table 10.14 and Table 10.15, respectively. The general nomenclature is

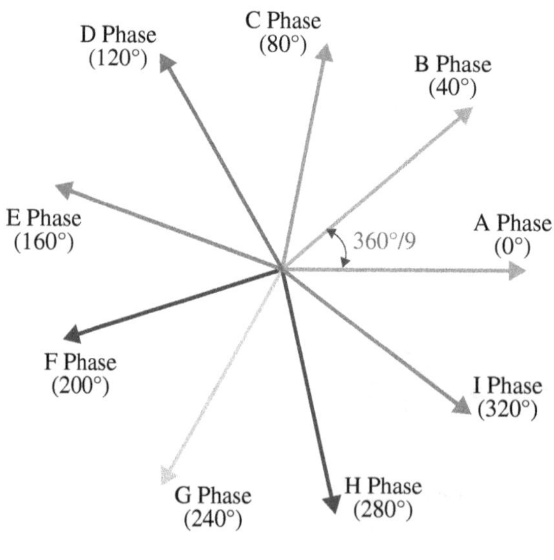

Figure 10.22 Symmetrical winding based nine-phase machine vector distribution.

Table 10.13 Single-layer winding details of asymmetrical 24 slot, six-phase, four-pole machine.

i	a	d	*c	*f	b	e	*a	*d	c	f	*b	*e	a	d	*c	*f	b	e	*a	*d	c	f	*b	*e
ii	1	2	3	4	5	6	7	8	9	10	11	12	13	14	15	16	17	18	19	20	21	22	23	24
iii	0	30	60	90	120	150	180	210	240	270	300	330	0	30	60	90	120	150	180	210	240	270	300	330
iv	N						S						N						S					
i) Phases (*indicates the negative direction of current), ii) Slots, iii) Slot Angle in Electrical Degrees, iv) Poles																								

(i) Phases (* indicates the negative direction of current), (ii) Slots, (iii) Slot Angle in Electrical Degrees, (iv) Poles

Table 10.14 Single-layer winding details of symmetrical 36-slot, nine-phase, two-pole machine.

i	a	*f	b	*g	c	*h	d	*i	e	*a	f	*b	g	*c	h	*d	i	*e
ii	1	2	3	4	5	6	7	8	9	10	11	12	13	14	15	16	17	18
iii	0	20	40	60	80	100	120	140	160	180	200	220	240	260	280	300	320	340
iv					N									S				

i	a	*f	b	*g	c	*h	d	*i	e	*a	f	*b	g	*c	h	*d	i	*e
ii	19	20	21	22	23	24	25	26	27	28	29	30	31	32	33	34	35	36
iii	0	20	40	60	80	100	120	140	160	180	200	220	240	260	280	300	320	340
iv	N									S								

i) Phases (*indicates the negative direction of current), ii) Slots, iii) Slot Angle in Electrical Degrees, iv) Poles

(i) Phases (* indicates the negative direction of current), (ii) Slots, (iii) Slot Angle in Electrical Degrees, (iv) Poles

Table 10.15 Single-layer winding details of asymmetrical 36-slot, nine-phase, two-pole machine.

i	a	d	g	*c	*f	*i	b	e	h	*a	*d	*g	c	f	i	*b	*e	*h
ii	1	2	3	4	5	6	7	8	9	10	11	12	13	14	15	16	17	18
iii	0	20	40	60	80	100	120	140	160	180	200	220	240	260	280	300	320	340
iv	N									S								

i	a	d	g	*c	*f	*i	b	e	h	*a	*d	*g	c	f	i	*b	*e	*h
ii	19	20	21	22	23	24	25	26	27	28	29	30	31	32	33	34	35	36
iii	0	20	40	60	80	100	120	140	160	180	200	220	240	260	280	300	320	340
iv	N									S								
i) Phases (*indicates the negative direction of current), ii) Slots, iii) Slot Angle in Electrical Degrees, iv) Poles																		

(i) Phases (* indicates the negative direction of current), (ii) Slots, (iii) Slot Angle in Electrical Degrees, (iv) Poles

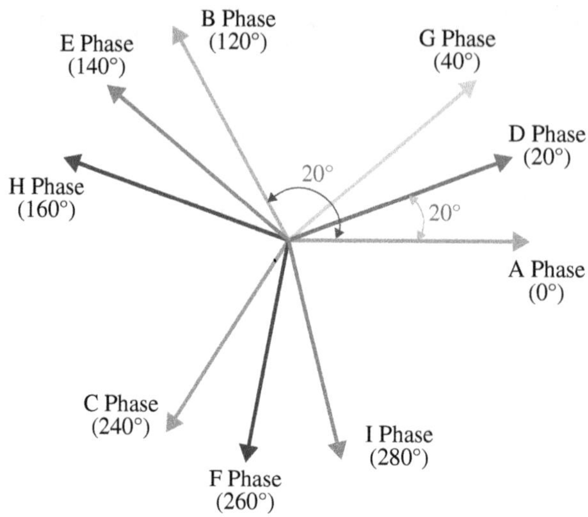

Figure 10.23 Asymmetrical winding based nine-phase machine vector distribution.

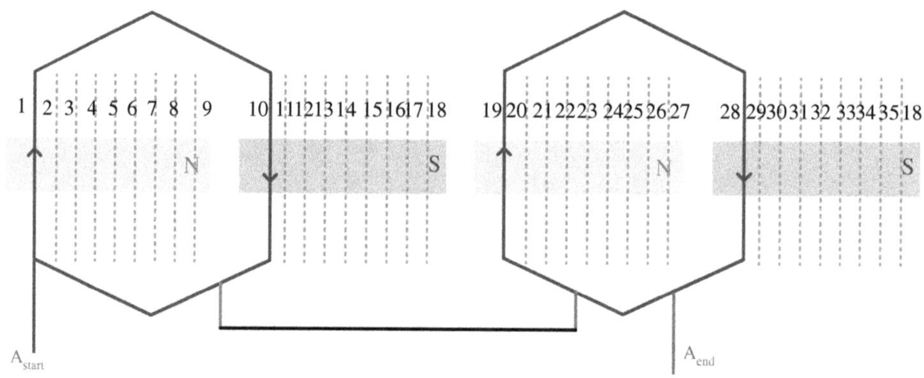

Figure 10.24 Single-layer winding details of symmetrical 36 slot, nine-phase, four-pole machine.

Slots = Q = 36, Poles = 4, phases = 9,

Slots/pole/phase = phase span = phase spread= $\frac{Q}{2mp} = \frac{24}{4*9} = 1$

Pole pitch = 36/4 = 9

coil span = 9 slots for full pitch winding.

The winding arrangement for symmetrical nine-phase 36-slot four-pole machine is shown in Figure 10.24. The details of the A-phase winding are

A-phase: Starts at 0°

Associated Slots: 1, 10, 19, 28

Connections: 1 => 10, 19 => 28, are the two coils associated with A-phase and the coil span is nine slots.

Starting and ending terminals are 1 and 28, respectively.

Similarly, the other phases winding coils are connected according to the winding details given in Table 10.14.

The asymmetrical nine-phase 36-slot four-pole machine has presented in Table 10.15, where the slot angle, current direction and poles are shown. The 120° displaced windings (Phases A, D, and G) in symmetrical winding are displaced by an angle of 20° $\frac{360^0}{(9*2)}$ in the asymmetrical winding. With the design details given in Table 10.15, various winding arrangements like single- layer, double-layer and short-pitch windings can be designed, similar to the analysis given in Section 10.4.1.

10.5 Mathematical Modelling of Multiphase Machines

The mathematical modelling of multiphase machines has been extensively reported in several research publications as well as textbooks with the reference of modelling equations of three-phase machines [3, 7, 12, 14–16]. The modelling of the multiphase machines involved some assumptions, such as

1) All phase windings are symmetrically distributed over a stator circumference with an angle of $2\pi/m$ degrees, where m = number of phases. The mmf, as well as flux distribution in the air gap of the machine, is assumed to be purely sinusoidal without any harmonics
2) Machine has a uniform air gap.
3) Slotting effect is neglected.
4) The resistances of stator/rotor windings have no variation wrt temperature and frequency (skin effects) variations.
5) The hysteresis, eddy current and other losses are neglected.
6) Nonlinearity of the magnetic material and saturation effects are neglected.
7) Machine windings are star-connected and the neutral isolated.

10.5.1 Mathematical Modelling of Multiphase Induction Machines in Original Phase-Variable Domain

Consider an m-phase machine with symmetrical winding to form a P number of poles on a stator circumference, as shown in Figure 10.25a. The stator windings are denoted as i_s, where $i_s = 1,2,3....m$ and s represent the stator winding and that are distributed symmetrically with an angle of $2\pi/m$ electrical degrees. Similar to the stator, the rotor of an induction machine has symmetrical windings (for squirrel cage bars has shored i.e. voltages of all three windings are equal to zero) and the rotor windings are represented by i_r, where $i = 1,2,3....m$ and r represent the rotor winding. The mathematical modelling of any rotating machine involves the electrical equations, magnetic flux linkage equations and mechanical equations. In general, all electrical machines are in a resistive and inductive nature.

The electrical equations are

$$\left[V_{sm}\right] = \left[R_{sm}\right]\left[i_{sm}\right] + \frac{d}{dt}\left[\lambda_{sm}\right] = \left[R_{sm}\right]\left[i_{sm}\right] + p\left[\lambda_{sm}\right] \tag{10.11}$$

$$\left[V_{rm}\right] = \left[R_{rm}\right]\left[i_{rm}\right] + \frac{d}{dt}\left[\lambda_{rm}\right] = \left[R_{rm}\right]\left[i_{rm}\right] + p\left[\lambda_{rm}\right] \tag{10.12}$$

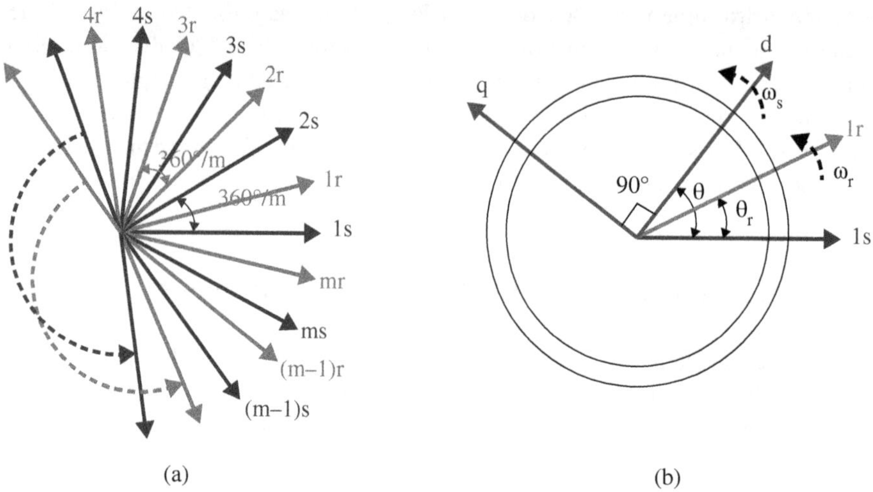

Figure 10.25 (a) m-phase machine stator and rotor axis representation (phase shift is $2\pi/m$ electrical degrees), (b) dq axis representation with respect to stator and rotor axis.

In Eqs. (10.11) and (10.12), the resistance of stator and rotor are the diagonal matrices, i.e.

$$\left[R_{sm}\right] = R_s \left[I_m\right]$$

where the R_s is the stator resistance (i.e. constant) and the $[I_m]$ is an identity matrix with a dimension of $m*m$

$$\left[R_{rm}\right] = R_r \left[I_m\right]$$

where the R_r is the rotor resistance (i.e. constant) and the $[I_m]$ is an identity matrix with a dimension of $m*m$

And the single-column matrices of stator and rotor, i.e.

$$\left[V_{sm}\right] = \left[V_{s1}\,V_{s2}\,V_{s3}\,V_{s4}\cdots\cdots\cdots\cdots V_{sm}\right]^T \text{ and } \left[V_{rm}\right] = \left[V_{r1}\,V_{r2}\,V_{r3}\,V_{r4}\cdots\cdots\cdots\cdots V_{rm}\right]^T$$

$$\left[i_{sm}\right] = \left[i_{s1}\,i_{s2}\,i_{s3}\,i_{s4}\cdots\cdots\cdots\cdots i_{sm}\right]^T \text{ and } \left[i_{rm}\right] = \left[i_{r1}\,i_{r2}\,i_{r3}\,i_{r4}\cdots\cdots\cdots\cdots i_{rm}\right]^T$$

$$\left[\lambda_{sm}\right] = \left[\lambda_{s1}\,\lambda_{s2}\,\lambda_{s3}\,\lambda_{s4}\cdots\cdots\cdots\cdots \lambda_{sm}\right]^T \text{ and } \left[\lambda_{rm}\right] = \left[\lambda_{r1}\,\lambda_{r2}\,\lambda_{r3}\,\lambda_{r4}\cdots\cdots\cdots\cdots \lambda_{rm}\right]^T$$

Under the assumption of the linear magnetic system, the flux linkage equations are

$$\begin{bmatrix} \lambda_{sm} \\ \lambda_{rm} \end{bmatrix} = \begin{bmatrix} \left[L_s\right]_{m*m} & \left[L_{sr}\right]_{m*m} \\ \left[L_{sr}\right]^T_{m*m} & \left[L_r\right]_{m*m} \end{bmatrix} \begin{bmatrix} i_{sm} \\ i_{rm} \end{bmatrix} \tag{10.13}$$

Since the windings are symmetrically distributed, the inductance matrices are

$$\left[L_s\right]_{m*m} = \begin{bmatrix} L_{s11} & L_{ms}\cos\left(\frac{2\pi}{m}\right) & L_{ms}\cos\left(\frac{4\pi}{m}\right) & \cdots\cdots\cdots & L_{ms}\cos\left(\frac{2(m-1)\pi}{m}\right) \\ L_{ms}\cos\left(\frac{2(m-1)\pi}{m}\right) & L_{s22} & L_{ms}\cos\left(\frac{2\pi}{m}\right) & \cdots\cdots\cdots & L_{ms}\cos\left(\frac{2(m-2)\pi}{m}\right) \\ L_{ms}\cos\left(\frac{2(m-2)\pi}{m}\right) & L_{ms}\cos\left(\frac{2(m-1)\pi}{m}\right) & L_{s33} & \cdots\cdots\cdots & L_{ms}\cos\left(\frac{2(m-3)\pi}{m}\right) \\ \cdots\cdots & \cdots\cdots & \cdots\cdots & \cdots\cdots\cdots & \cdots\cdots \\ \cdots\cdots & \cdots\cdots & \cdots\cdots & \cdots\cdots\cdots & \cdots\cdots \\ L_{ms}\cos\left(\frac{2\pi}{m}\right) & L_{ms}\cos\left(\frac{4\pi}{m}\right) & L_{ms}\cos\left(\frac{6\pi}{m}\right) & \cdots\cdots\cdots & L_{smm} \end{bmatrix} \tag{10.14}$$

$$[L_r]_{m*m} = \begin{bmatrix} L_{r11} & L_{mr}\cos\left(\frac{2\pi}{m}\right) & L_{mr}\cos\left(\frac{4\pi}{m}\right) & \ldots\ldots\ldots & L_{mr}\cos\left(\frac{2(m-1)\pi}{m}\right) \\ L_{mr}\cos\left(\frac{2(m-1)\pi}{m}\right) & L_{r22} & L_{mr}\cos\left(\frac{2\pi}{m}\right) & \ldots\ldots\ldots & L_{mr}\cos\left(\frac{2(m-2)\pi}{m}\right) \\ L_{mr}\cos\left(\frac{2(m-2)\pi}{m}\right) & L_{mr}\cos\left(\frac{2(m-1)\pi}{m}\right) & L_{r33} & \ldots\ldots\ldots & L_{mr}\cos\left(\frac{2(m-3)\pi}{m}\right) \\ \ldots\ldots & \ldots\ldots & \ldots\ldots & \ldots\ldots\ldots & \ldots\ldots \\ \ldots\ldots & \ldots\ldots & \ldots\ldots & \ldots\ldots\ldots & \ldots\ldots \\ L_{mr}\cos\left(\frac{2\pi}{m}\right) & L_{mr}\cos\left(\frac{4\pi}{m}\right) & L_{mr}\cos\left(\frac{6\pi}{m}\right) & \ldots\ldots\ldots & L_{rmm} \end{bmatrix} \tag{10.15}$$

The computation of self-inductance (L_{sxx}or L_{rxx}) and the mutual inductances are (L_{sxy}or L_{syx}or L_{rxy}or L_{ryx}) depending on the geometry, as well as the nature of machine windings. For a three-phase machine [6], $L_{sxx} = L_{ls} + L_{ms}$ (i.e. $L_{s11} = L_{s22} = L_{s33} = L_{ls} + L_{ms}$) and the mutual inductance are, $L_{s12} = L_{s23} = L_{s31} = L_{ms}\cos(2\pi/3)$ and $L_{s21} = L_{s32} = L_{s13} = L_{ms}\cos(4\pi/3)$. Similarly, for any phase machine, self and mutual inductance between the two phases windings can be calculated by using the matrices presented in Eq. (10.14). The mutual inductance between the stator and rotor winding depends on the angle between the two windings, which is directly proportional to the cosine function of the angle (θ). The angle θ is defined as the angle between the stator phase-*s1* and rotor phase-*r1* in the direction of rotation, as shown in Figure 10.25b. The position of the rotor at time t is represented as

$$\theta = \int \omega \; dt + \theta(0),$$

Where ω is the speed of the machine over a period of time. So, the mutual inductance between stator and rotor is as follows

$$[L_{sr}]_{m*m} = [L_{sr}]^T_{m*m} = L_{ms} \begin{bmatrix} \cos\theta & \cos\left(\theta - \frac{2\pi}{m}\right) & \cos\left(\theta - \frac{4\pi}{m}\right) & \ldots\ldots\ldots & \cos\left(\theta - \frac{2(m-1)\pi}{m}\right) \\ \cos\left(\theta - \frac{2(m-1)\pi}{m}\right) & \cos\theta & \cos\left(\frac{2\pi}{m}\right) & \ldots\ldots\ldots & \cos\left(\theta - \frac{2(m-2)\pi}{m}\right) \\ \cos\left(\theta - \frac{2(m-2)\pi}{m}\right) & \cos\left(\frac{2(m-1)\pi}{m}\right) & \cos\theta & \ldots\ldots\ldots & \cos\left(\theta - \frac{2(m-3)\pi}{m}\right) \\ \ldots\ldots & \ldots\ldots & \ldots\ldots & \ldots\ldots\ldots & \ldots\ldots \\ \ldots\ldots & \ldots\ldots & \ldots\ldots & \ldots\ldots\ldots & \ldots\ldots \\ \cos\left(\theta - \frac{2\pi}{m}\right) & \cos\left(\theta - \frac{4\pi}{m}\right) & \cos\left(\theta - \frac{6\pi}{m}\right) & \ldots\ldots\ldots & \cos\theta \end{bmatrix} \tag{10.16}$$

Eqs. (10.11)–(10.15) describe the electrical and magnetic flux linkages of the *m*-phase machine. The resultant voltage equation of the multiphase machine is

$$\begin{bmatrix} V_{sm} \\ V_{sm} \end{bmatrix} = \begin{bmatrix} [R_{sm}] \\ [R_{rm}] \end{bmatrix} \begin{bmatrix} i_{sm} \\ i_{sm} \end{bmatrix} + p \begin{bmatrix} [L_s]_{m*m} & [L_{sr}]_{m*m} \\ [L_{sr}]^T_{m*m} & [L_r]_{m*m} \end{bmatrix} \begin{bmatrix} i_{sm} \\ i_{rm} \end{bmatrix} \tag{10.17}$$

The stator and rotor voltage, as well as flux linkage equations, can be referred to either rotor side or stator side by using proper turns ratio. The change of mechanical energy in a rotating system with one mechanical input is [3, 7, 12, 14–16]

$$dW_m = -T_e\left(\frac{2}{P}\right)d\theta_e \tag{10.18}$$

Where T_e represents the electromagnetic torque and $\theta_e$represents the electrical angular displacement of the rotor, i.e. $\theta_e = (P/2)\theta_m$, the number of poles of a machine is P and θ_m is the actual displacement of rotor. In Eq. (10.18), the electromagnetic torque is positive for motoring action. In a linear magnetic system, the co-energy, as well as field energy, will be the same so that the actual torque equation of the machine is as follows

$$T_e = \frac{P}{2}\frac{dW_m}{d\theta} \tag{10.19}$$

In the above equation, the L_s and L_r matrices are not dependant on the rotor position, so the resultant torque equation of a machine is

$$T_e = \frac{P}{2}\left[i_{sm}\right]^T \frac{d}{d\theta}\left[L_{sr}\right]\left[i_{rm}\right] \tag{10.20}$$

The mechanical equation of machine with respect to rotor movement (only one-degree freedom) is,

$$T_e - T_L = J\left(\frac{2}{P}\right)\frac{d\omega}{dt} + B\left(\frac{2}{P}\right)\omega \tag{10.21}$$

Where, T_e is the electric torque generated by the motor, T_L load torque applied on motor J is the inertia of moving parts (units are kilogramme meter^2 [Kg/m^2] and joules second^2 [J.s^2]), B is the friction coefficient (pound mass.feet^2 [lbm.ft^2]) and ω the angular speed of rotation.

10.5.2 Transformation Matrix for Multiphase Machines

The original phase variable model voltage equations of the multiphase induction machine (Eq. 10.17) involve the time-dependent inductances which results in high-computation time as well as a slow response. These issues can be addressed by transforming the m-phase machine equations to dq domain. This two-dimensional model of the machine will generate the same mmf of the m-phase machine variable model. Park's transformation matrix gives the relation between the phase variables as well as dqo variables [3, 7, 12, 14–16], i.e. given as

$$\left[f_{d,q,x1,x2...0}\right] = \left[T_s\right]\left[f_{1,2,3,..m}\right] \tag{10.22}$$

Where $[T_s]$ is the Park's transformation matrix, and $[f_{d,q,x1,x2...0}]$ is the two-dimensional representation of the voltage, flux and currents of the stator or rotor-phase variable equations are the column matrices i.e. $[f_{d,q,x1,x2...0}] = [f_d\, f_q\, f_x\, f_y\, f_{x1}\, f_{y1} \cdot\cdot\cdot f_{xz}\, f_{yz}\, f_{0+}\, f_{0-}]^T$, $[f_{1,2,3,..m}]$ represents original phase variables of the machine, i.e. $[f_{1,2,3,..m}] = [f_1\, f_2\, f_3\, f_4 \cdot\cdot\cdot f_{m-1}\, f_m]^T$. For a three-phase machine, the Park's transformation matrix is [7]

$$\left[T_s\right] = \sqrt{\frac{2}{3}}\begin{bmatrix} Cos\theta & Cos\left(\theta - \frac{2\pi}{3}\right) & Cos\left(\theta - \frac{4\pi}{3}\right) \\ Sin\theta & Sin\left(\theta - \frac{2\pi}{3}\right) & Sin\left(\theta - \frac{4\pi}{3}\right) \\ \frac{1}{\sqrt{2}} & \frac{1}{\sqrt{2}} & \frac{1}{\sqrt{2}} \end{bmatrix} \tag{10.23}$$

Similar to the three-phase machine, the Park's transformation matrix for a m-phase machine is

$$[T_s] = \sqrt{\frac{2}{m}} \begin{bmatrix} \cos\theta & \cos\left(\theta - \frac{2\pi}{m}\right) & \cos\left(\theta - \frac{4\pi}{m}\right) & \cdots\cdots & \cos\left(\theta - \frac{2(m-1)\pi}{m}\right) \\ \sin\theta & \sin\left(\theta - \frac{2\pi}{m}\right) & \sin\left(\theta - \frac{4\pi}{m}\right) & \cdots\cdots & \sin\left(\theta - \frac{2(m-1)\pi}{m}\right) \\ \cos 2\theta & \cos 2\left(\theta - \frac{2\pi}{m}\right) & \cos 2\left(\theta - \frac{4\pi}{m}\right) & \cdots\cdots & \cos 2\left(\theta - \frac{2(m-1)\pi}{m}\right) \\ \sin 2\theta & \sin 2\left(\theta - \frac{2\pi}{m}\right) & \sin 2\left(\theta - \frac{4\pi}{m}\right) & \cdots\cdots & \sin 2\left(\theta - \frac{2(m-1)\pi}{m}\right) \\ \cos 3\theta & \cos 3\left(\theta - \frac{2\pi}{m}\right) & \cos 3\left(\theta - \frac{4\pi}{m}\right) & \cdots\cdots & \cos 3\left(\theta - \frac{2(m-1)\pi}{m}\right) \\ \sin 3\theta & \sin 3\left(\theta - \frac{2\pi}{m}\right) & \sin 3\left(\theta - \frac{4\pi}{m}\right) & \cdots\cdots & \sin 3\left(\theta - \frac{2(m-1)\pi}{m}\right) \\ \cdots\cdots & \cdots\cdots & \cdots\cdots & \cdots\cdots & \cdots\cdots \\ \cdots\cdots & \cdots\cdots & \cdots\cdots & \cdots\cdots & \cdots\cdots \\ \cos z\theta & \cos\left[z\left(\theta - \frac{2\pi}{m}\right)\right] & \cos\left[z\left(\theta - \frac{4\pi}{m}\right)\right] & \cdots\cdots & \cos\left[z\left(\theta - \frac{2(m-1)\pi}{m}\right)\right] \\ \sin z\theta & \sin\left[z\left(\theta - \frac{2\pi}{m}\right)\right] & \sin\left[z\left(\theta - \frac{4\pi}{m}\right)\right] & \cdots\cdots & \sin\left[z\left(\theta - \frac{2(m-1)\pi}{m}\right)\right] \\ \frac{1}{\sqrt{2}} & \frac{1}{\sqrt{2}} & \frac{1}{\sqrt{2}} & \cdots\cdots & \frac{1}{\sqrt{2}} \\ \frac{1}{\sqrt{2}} & -\frac{1}{\sqrt{2}} & \frac{1}{\sqrt{2}} & \cdots\cdots & -\frac{1}{\sqrt{2}} \end{bmatrix} \tag{10.24}$$

In the above transmission matrix, the position of dq reference frame depends on the stationary reference frame, where the angular displacement of the d-axis from the stationary reference is θ. The d and q axes are perpendicular to each other, with no coupling between them (similarly all two-dimensional vectors in each sub plane). The $dq0$ reference frame is rotating at a synchronous speed of ω_s over a period of time. The coefficient of a transformation matrix will be $\sqrt{2/m}$ which replicates the original machine in two-dimensional domain. To achieve the symmetry of two-dimensional as well as the original machine, the zero-sequence components (last two rows) has added to the transformation matrix. Since the machine is balanced and star connected isolated neutral the sum of all phase currents is equal to zero. If the coefficient of the matrix is $2/m$ then the last two rows of the transf2ormation matrix should be $1/2$ instead of $1/\sqrt{2}$. In a multiphase machine, if the phase number is an ODD integer

- The machine phase variables are converted into $(m - 1)/2$ number of two-dimensional sub planes
- The last row of transformation matrix will not exist, i.e. only one zero-sequence component will exist
- The multiplication factor of θ for the last two-dimensional sub plane is $z = (m - 1)/2$

If the phase number is an EVEN integer,

- The machine phase variables are converted into $(m - 2)/2$ number of two-dimensional sub planes
- The last row of transformation matrix will exist, i.e. two zero-sequence components will exist
- The multiplication factor of θ for the last two-dimensional sub plane is $z = (m - 2)/2$

The a-phase rotor axis is displaced by an angle of θ_r with respect to the stationary reference frame, as shown in the Figure 10.25b. So, the displacement between the $dq0$ with respect the rotor-reference axis is $\beta = \theta - \theta_r$. The inverse matrix of Park's transformation will give the original phase variables of the machine from the *dq0* two-dimensional variables, i.e., $[f_{1,2,3,..m}] = [T_s]^{-1}[f_{d,q,x1,x2...0}]$. Here, the $[T_s]^{-1} = [T_s]^T$ and the coefficient of the inverse transformation matrix will be unity.

10.5.3 Modelling of Multiphase Induction Machines in Arbitrary Reference Frames

Considering the assumptions mentioned at the start of Section 10.5 and Park's transformation matrix, the phase variable model of the multiphase machine can be transformed into the arbitrary reference frame. The electrical equations (voltage and flux linkage) of the machine model in the arbitrary reference frame are

$$\left[f_{d,q,x1,x2...0}\right] = \left[T_r\right]\left[f_{1,2,3,..m}\right]$$

Here, the transformation matrix $[T_r]$ in the arbitrary reference frame (rotor reference axis) is the same as Eq. (10.24), the only difference is instead of θ substitute β. According to the analysis of the three-phase machines [3, 7], the equations of the machines in the arbitrary reference frame are derived from Eqs. (10.17) and (10.24),

$$\left[V_{sm}\right] = \left[R_{sm}\right]\left[i_{sm}\right] + p\left\{\left[L_s\right]\left[i_{sm}\right] + \left[L_{sr}\right]\left[i_{rm}\right]\right\}$$

To convert the phase variable voltage equation into *dq0* domain, i.e., as follows

$$\begin{aligned}\left[T_s\right]\left[V_{sm}\right] &= \left[T_s\right]\left[R_{sm}\right]\left[T_s\right]^{-1}\left[T_s\right]\left[i_{sm}\right]\\ &+p\left\{\left[T_s\right]\left[L_s\right]\left[T_s\right]^{-1}\left[T_s\right]\left[i_{sm}\right] + \left[T_s\right]\left[L_{sr}\right]\left[T_r\right]^{-1}\left[T_r\right]\left[i_{rm}\right]\right\}\end{aligned} \tag{10.25}$$

The resultant stator voltage equation of multiphase machine after solving Eq. (10.25),

$$\left[V_{s(d,q,x1,x2...0)}\right] = R_s\left[I_m\right]\left[i_{s(d,q,x1,x2...0)}\right] + p\left[\lambda_{s(d,q,x1,x2...0)}\right] - \omega\left[\lambda_{s(q,d)}\right] \tag{10.26}$$

The rotor voltage equation is

$$\left[V_{r(d,q,x1,x2...0)}\right] = R_r\left[I_m\right]\left[i_{r(d,q,x1,x2...0)}\right] + p\left[\lambda_{r(d,q,x1,x2...0)}\right] - \left(\omega - \omega_r\right)\left[\lambda_{r(q,d)}\right] \tag{10.27}$$

In Eqs. (10.26) and (10.27), $[I_m]$ is an identity matrix and voltage, current and flux linkage matrices are column matrices, i.e.

$$\left[V_{s(d,q,x1,x2...0)}\right] = \left[V_{sd}\,V_{sq}\,V_{sx1}\,V_{sy1}\,V_{sx2}\,V_{sy2}\cdots\cdots\cdots\cdots V_{s0}\right]^T$$

$$\left[i_{s(d,q,x1,x2...0)}\right] = \left[i_{sd}\,i_{sq}\,i_{sx1}\,i_{sy1}\,i_{sx2}\,i_{sy2}\cdots\cdots\cdots\cdots i_{s0}\right]^T$$

$$\left[\lambda_{s(d,q,x1,x2...0)}\right] = \left[\lambda_{sd}\,\lambda_{sq}\,\lambda_{sx1}\,\lambda_{sy1}\,\lambda_{sx2}\,\lambda_{sy2}\cdots\cdots\cdots\cdots \lambda_{s0}\right]^T$$

Similarly, the rotor equation has the column matrices. The matrices $[\lambda_{s(q,d)}][\lambda_{r(q,d)}]$ are also the column matrices i.e. $[\lambda_{s(q,d)}] = [\lambda_{sq}\text{ - }\lambda_{sd}]$ and $[\lambda_{r(q,d)}] = [\lambda_{rq}\text{ - }\lambda_{rd}]$ where the other components will be zero (since the torque is only produced due to the fundamental component).

The flux equations of multiphase induction machine in the arbitrary reference frame are

$$\begin{bmatrix}\left[\lambda_{s(d,q,x1,x2...0)}\right]\\ \left[\lambda_{r(d,q,x1,x2...0)}\right]\end{bmatrix} = \begin{bmatrix}\left[T_s\right]\left[L_s\right]\left[T_s\right]^{-1} & \left[T_s\right]\left[L_{sr}\right]\left[T_r\right]^{-1}\\ \left[T_r\right]\left[L_{sr}\right]^T\left[T_s\right]^{-1} & \left[T_r\right]\left[L_r\right]\left[T_r\right]^{-1}\end{bmatrix}\begin{bmatrix}\left[i_{s(d,q,x1,x2...0)}\right]\\ \left[i_{r(d,q,x1,x2...0)}\right]\end{bmatrix} \tag{10.28}$$

Where

$$[T_s]\,[L_s]\,[T_s]^{-1} = \begin{bmatrix} L_{s11} & 0 & 0 & 0 & \dots\dots\dots & 0 \\ 0 & L_{s22} & 0 & 0 & \dots\dots\dots & 0 \\ 0 & 0 & L_{s33} & 0 & \dots\dots\dots & 0 \\ .. & .. & .. & .. & \dots\dots\dots & .. \\ .. & .. & .. & .. & \dots\dots\dots & .. \\ 0 & 0 & 0 & 0 & \dots\dots\dots & L_{smm} \end{bmatrix} = diag\,[L_{s11}\ L_{s22}\ L_{s33}\ \dots\dots\dots\ L_{smm}] \tag{10.29}$$

$$[T_r]\,[L_r]\,[T_r]^{-1} = \begin{bmatrix} L_{r11} & 0 & 0 & 0 & \dots\dots\dots & 0 \\ 0 & L_{r22} & 0 & 0 & \dots\dots\dots & 0 \\ 0 & 0 & L_{r33} & 0 & \dots\dots\dots & 0 \\ .. & .. & .. & .. & \dots\dots\dots & .. \\ .. & .. & .. & .. & \dots\dots\dots & .. \\ 0 & 0 & 0 & 0 & \dots\dots\dots & L_{rmm} \end{bmatrix} = diag\,[L_{r11}\ L_{r22}\ L_{r33}\dots\dots\dots\ L_{rmm}] \tag{10.30}$$

$$[T_s]\,[L_{sr}]\,[T_r]^{-1} = [T_r]\,[L_{sr}]^T\,[T_s]^{-1} = \begin{bmatrix} L_M & 0 & 0 & 0 & \dots\dots\dots & 0 \\ 0 & L_M & 0 & 0 & \dots\dots\dots & 0 \\ 0 & 0 & 0 & 0 & \dots\dots\dots & 0 \\ .. & .. & .. & .. & \dots\dots\dots & .. \\ .. & .. & .. & .. & \dots\dots\dots & .. \\ 0 & 0 & 0 & 0 & \dots\dots\dots & 0 \end{bmatrix} \tag{10.31}$$

In the stator inductance matrix (Eq. 10.29), only the first two diagonal elements consist of mutual inductance terms, i.e. $L_{s11} = L_{s22} = L_{ls} + L_M$ and the remaining elements $L_{s33} = L_{s44} = \dots..\ L_{smm} = L_{ls}$. Similarly, for the rotor inductance matrix $L_{r11} = L_{r22} = L_{lr} + L_M$ and the remaining elements $L_{r33} = L_{r44} = \dots.\ L_{rmm} = L_{lr}$,. the mutual inductance of the machine is $L_M = (m/2)L_{ms}$, where L_{ms} = magnetizing inductance.

From Eqs. (10.26) to (10.31), stator voltage and flux linkage equations are

$$V_{sd} = R_s i_{sd} + p\lambda_{sd} - \omega\lambda_{sq} \tag{10.32}$$

$$V_{sq} = R_s i_{sq} + p\lambda_{sq} + \omega\lambda_{sd} \tag{10.33}$$

$$V_{sx1} = R_s i_{sx1} + p\lambda_{sx1} \tag{10.34}$$

$$V_{sy1} = R_s i_{sy1} + p\lambda_{sy1}$$

similarly, other two-dimensional sub plane equations and final zero sequence is

$$V_{s0} = R_s i_{s0} + p\lambda_{s0} \tag{10.35}$$

$$\lambda_{sd} = L_{s11} i_{sd} + L_M i_{rd} \tag{10.36}$$

$$\lambda_{sq} = L_{s22} i_{sq} + L_M i_{rq} \tag{10.37}$$

$$\lambda_{sx1} = L_{s33} i_{sx1}$$

similarly, other two-dimensional sub-plane equations and final zero-sequence equation

$$\lambda_{s0} = L_{smm} i_{s0} \tag{10.38}$$

The rotor voltage and flux linkage equations are

$$V_{rd} = R_r i_{rd} + p\lambda_{rd} - \omega\lambda_{rq} \tag{10.39}$$

$$V_{rq} = R_r i_{rq} + p\lambda_{rq} + \omega\lambda_{rd} \tag{10.40}$$

$$V_{rx1} = R_r i_{rx1} + p\lambda_{rx1} \tag{10.41}$$

$$V_{ry1} = R_r i_{ry1} + p\lambda_{ry1}$$

similarly, other two-dimensional sub-plane equations and final zero-sequence equation

$$V_{r0} = R_r i_{r0} + p\lambda_{r0} \tag{10.42}$$

In the above rotor-voltage equations, the voltage will be zero, as the rotor is the squirrel-cage type where the bars are short circuited.

$$\lambda_{rd} = L_{r11} i_{rd} + L_M i_{sd} \tag{10.43}$$

$$\lambda_{rq} = L_{r22} i_{rq} + L_M i_{sq} \tag{10.44}$$

$$\lambda_{rx1} = L_{r33} i_{rx1}$$

similarly, other two-dimensional sub-plane equations and final zero-sequence equation

$$\lambda_{r0} = L_{rmm} i_{r0} \tag{10.45}$$

The electromagnetic torque equation in the arbitrary reference frame can be derived from the equation

$$T_e = \frac{P}{2}\left[\left[T_s\right]^{-1}\left[i_{s(d,q,x1,x2...0)}\right]\right]^T \frac{d}{d\theta}\left[L_{sr}\right]\left[T_r\right]^{-1}\left[i_{r(d,q,x1,x2...0)}\right]$$

$$T_e = \frac{P}{2} L_M\left[i_{qs} i_{dr} - i_{ds} i_{qr}\right] \tag{10.46}$$

The mechanical equation of machine is the same as Eq. (10.21),

$$T_e - T_L = J\left(\frac{2}{P}\right)\frac{d\omega}{dt} + B\left(\frac{2}{P}\right)\omega \tag{10.47}$$

10.5.4 Commonly used Reference Frames

The mathematical modelling equations of the multiphase induction machine in an arbitrary rotating reference frame is the most general form of representation. From arbitrary reference frame modelling equations, any type of reference frame equations can be derived by substituting the appropriate angular speed of reference frame, [3, 7, 14]

1) **Stationary reference frame,** $\omega = 0$
   The modelling of the machine in stationary reference frame is simple and easy to implement. This reference frame is more suitable when the stator voltages are unbalanced or discontinued but the rotor voltages are balanced or short-circuited (zero).
2) **Synchronous reference frame,** $\omega = \omega_s$
   The most commonly used reference frame is the synchronous reference frame where the *dqo* reference frame is rotating at synchronous speed ($\omega_s = d\theta/dt$). This reference frame is most accurate for analysing the transient and dynamic response of the machine and also for variable frequency applications.

3) **Rotor reference frame** $\omega = \omega_r$
   The rotor reference frame is most accurate under unbalanced conditions of the rotor side either with supply voltage variations or faults. Whereas the stationary and synchronous reference frames are only useful under balanced conditions of the rotor.

For transforming one variable from one reference to another reference frame without involving original phase variables, the rotational transformation matrix is

$$\left[f_{\alpha,\beta,x1,x2\ldots0}\right] = [C]\left[f_{d,q,x1,x2\ldots0}\right] \tag{10.48}$$

Where

$$[C] = \begin{bmatrix} \cos\theta & \sin\theta & 0 & 0 & 0 & \ldots\ldots & 0 \\ -\sin\theta & \cos\theta & 0 & 0 & 0 & \ldots\ldots & 0 \\ 0 & 0 & 1 & 0 & 0 & \ldots\ldots & 0 \\ 0 & 0 & 0 & 1 & 0 & \ldots\ldots & 0 \\ \ldots & \ldots & \ldots & \ldots & \ldots & \ldots\ldots & \ldots \\ \ldots & \ldots & \ldots & \ldots & \ldots & \ldots\ldots & \ldots \\ 0 & 0 & 0 & 0 & 0 & \ldots\ldots & 1 \end{bmatrix} \text{ and } [C]^{-1} = [C]^T \tag{10.49}$$

For easy understanding, in order to convert stationary reference frame to rotating reference frame

$$V_{\alpha\beta} = V_{dq}\, e^{-j\theta}$$

where

$$V_{\alpha\beta} = V_\alpha + jV_\beta, V_{dq} = V_d + jV_q \text{and } e^{-j\theta} = \cos\theta - j\sin\theta \tag{10.50}$$

For converting rotating reference frame to stationary reference frame

$$V_{\alpha\beta} = V_{dq}\, e^{j\theta} \text{ where } e^{j\theta} = \cos\theta + j\sin\theta \tag{10.51}$$

10.5.5 Modelling of a Multiphase Synchronous Machine

The mathematical modelling voltage and flux linkage equations and transformation matrices presented in the previous sections are also valid for the synchronous machines. The differences and points to consider in the modelling of a synchronous machine as compared with the induction machine [3, 7, 14] are

1) The stator-winding arrangement is the same in synchronous and induction machines, so the voltage and flux linkage equations are the same in both machines.
2) Transformation matrices are same.
3) Rotor structure will be different in the synchronous machines (damper windings in an induction machine but here, the rotor structure may either be cylindrical or salient).
4) Rotor always rotates at a synchronous speed.
5) For eliminating the time-varying inductances, the rotor should be fixed to the rotor reference frame.

Consider an m-phase synchronous machine with the symmetrical winding to form a P number of poles on a stator circumference. The stator windings are distributed symmetrically with a phase angle of $2\pi/m$ electrical degrees. The rotor consists of one field winding and y damper windings ($y = 1,2,3\ldots.$) on d-axis and k damper windings on q-axis of the rotor ($k = 1,2,3\ldots.$) and r represents the rotor winding.

The modelling equations of synchronous machines in the rotor reference frame are as follows

$$V_{sd} = R_s i_{sd} + p\lambda_{sd} - \omega\lambda_{sq} \tag{10.52}$$

$$V_{sq} = R_s i_{sq} + p\lambda_{sq} + \omega\lambda_{sd} \tag{10.53}$$

$$V_{sx1} = R_s i_{sx1} + p\lambda_{sx1} \tag{10.54}$$

$$V_{sy1} = R_s i_{sy1} + p\lambda_{sy1}$$

similarly, other two-dimensional sub-plane equations and the final zero-sequence equation is

$$V_{s0} = R_s i_{s0} + p\lambda_{s0} \tag{10.55}$$

$$\lambda_{sd} = L_{sd} i_{sd} + L_{md} i_{fd} + L_{md} i_{ry1d} + L_{md} i_{ry2d} + L_{md} i_{ry3d} + \ldots\ldots\ldots\ldots\ldots + L_{md} i_{ryd} \tag{10.56}$$

$$\lambda_{sq} = L_{sq} i_{sq} + L_{mq} i_{rk1q} + L_{mq} i_{rk2q} + L_{mq} i_{rk3q} + \ldots\ldots\ldots + L_{mq} i_{rkq} \tag{10.57}$$

$$\lambda_{sx1} = L_{ls} i_{sx1}$$

similarly, other two-dimensional sub-plane equations and final zero-sequence equation is

$$\lambda_{s0} = L_{ls} i_{s0} \tag{10.58}$$

Since the flux linkages of the various windings are different, the mutual inductances of d-axis and q-axis interms of magnetizing inductances (M_d, M_q) are $L_{md} = (m/2)\,M_d$ and $L_{mq} = (m/2)\,M_q$. In the above equations, $L_{sd} = L_{ls} + L_{md}$ and $L_{sq} = L_{ls} + L_{mq}$.

The rotor-voltage equations of windings present on the d-axis are

$$V_{rf} = R_{rf} i_{rf} + p\lambda_{rf} \tag{10.59}$$

$$V_{ry1d} = R_{ry1d} i_{ry1d} + p\lambda_{ry1d} \tag{10.60}$$

$$V_{ry2d} = R_{ry2d} i_{ry2d} + p\lambda_{ry2d} \tag{10.61}$$

Finally, the yth number of damper winding on d-axis voltage equation is

$$V_{ryd} = R_{kr} i_{ryd} + p\lambda_{ryd} \tag{10.62}$$

and voltage equations of damper windings on q-axis are

$$V_{rk1q} = R_{rk1q} i_{rk1q} + p\lambda_{rk1q} \tag{10.63}$$

$$V_{rk2q} = R_{rk2q} i_{rk2q} + p\lambda_{rk2q} \tag{10.64}$$

Finally, the kth number of damper winding on q-axis voltage equation is

$$V_{rkq} = R_{rkq} i_{rkq} + p\lambda_{rkq} \tag{10.65}$$

Here, the damper winding voltages will be zero, since they are circuited similar to the induction motor.

The flux linkage equations of field winding and damper windings on d-axis are

$$\lambda_{rfd} = L_{rfd} i_{rfd} + L_{md} i_{sd} + L_{md} i_{ry1d} + L_{md} i_{ry2d} + L_{md} i_{ry3d} \cdots \cdots \cdots + L_{md} i_{ryd} \tag{10.66}$$

$$\lambda_{ry1d} = L_{ry1d} i_{ry1d} + L_{md} i_{sd} + L_{md} i_{rfd} + L_{md} i_{ry2d} + L_{md} i_{ry3d} + \cdots \cdots \cdots + L_{md} i_{ryd} \tag{10.67}$$

$$\lambda_{ry2d} = L_{ry2d} i_{ry2d} + L_{md} i_{sd} + L_{md} i_{rfd} + L_{md} i_{ry1d} + L_{md} i_{ry3d} + \cdots \cdots \cdots + L_{md} i_{ryd} \tag{10.68}$$

and the yth number of damper winding on d-axis flux linkage equation is

$$\lambda_{ryd} = L_{ryd} i_{ryd} + L_{md} i_{sd} + L_{md} i_{rfd} + L_{md} i_{ry1d} + L_{md} i_{ry2d} + \cdots \cdots \cdots + L_{md} i_{r(y-1)d} \tag{10.69}$$

The flux linkage equations of the damper windings on q-axis

$$\lambda_{rk1q} = L_{rk1q} i_{rk1q} + L_{mq} i_{sq} + L_{mq} i_{rk2q} + L_{mq} i_{rk3q} + \cdots \cdots \cdots + L_{mq} i_{rkq} \tag{10.70}$$

$$\lambda_{rk2q} = L_{rk2q} i_{rk2q} + L_{mq} i_{sq} + L_{mq} i_{rk1q} + L_{mq} i_{rk3q} + \cdots \cdots \cdots + L_{mq} i_{rkq} \tag{10.71}$$

and the kth number of damper winding on q-axis flux linkage equation is

$$\lambda_{rkq} = L_{rkq} i_{rkq} + L_{mq} i_{sq} + L_{mq} i_{rk1q} + L_{mq} i_{rk2q} + \cdots \cdots \cdots + L_{mq} i_{r(k-1)q} \tag{10.72}$$

Since the damper windings are short circuited, the voltage is equal to zero. In the above equations, $L_{rfd} = L_{lrf} + L_{mf}$, $L_{ry1d} = L_{lry1d} + L_{md}$, $L_{ry2d} = L_{lry2d} + L_{md}$, ….. $L_{ryd} = L_{lryd} + L_{md}$ and $L_{rk1q} = L_{lrk1q} + L_{mq}$, $L_{rk2q} = L_{lrk2q} + L_{mq}$, ….. $L_{rkq} = L_{lrkq} + L_{mq}$, the d-axis and q-axis damper-winding resistance and their leakage inductances are different and the same is considered in the modelling equations. In general, to make the self-starting of synchronous machines, will require a minimum of one damper winding on the d-axis and one damper winding on the q-axis, i.e. in the above equation $y = k = 1$ can be used for simple modelling of machine [14].

The electromagnetic torque equation in the rotor reference frame can be

$$T_e = \frac{m}{2}\frac{P}{2}\left[\lambda_{sd} i_{sq} - \lambda_{sq} i_{sd}\right] \tag{10.73}$$

The mechanical equation of machine is the same as Eq. (10.21),

$$T_e - T_L = J\left(\frac{2}{P}\right)\frac{d\omega}{dt} + B\left(\frac{2}{P}\right)\omega \tag{10.74}$$

10.6 Vector Control Techniques for Multiphase Machines

The scalar control techniques of any machine will give the accurate speed and torque response. However, the torque and flux have a coupling effect and is directly influenced by the voltage or current and frequency which may result in a slower response as well as the instability of the system. For this vector, control techniques are introduced, where the decoupled concept of torque and flux components is implemented similar to the fully compensated separately excited DC machines. The construction of fully compensated separately excited DC machines is done in such a way that the field and armature currents are perpendicular to each other, resulting in independent control of the armature or field currents. This means the armature current control does not affect the field flux

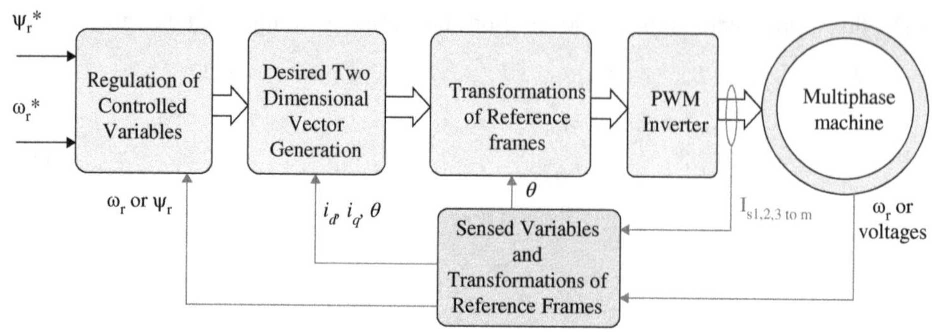

Figure 10.26 General block diagram of vector control of multiphase machine.

produced by the field current. The same characteristic is emulated in the AC machine by implementing vector controls. As the name suggests, it is control of both magnitudes and the position of current to decouple flux and torque. Based on this, many vector control techniques are implemented for three-phase machines. If multiphase machine windings are symmetrically distributed on stator circumference to attain sinusoidal mmf in the air gap, all vector control techniques of three-phase machines are applicable for multiphase machines but the only difference is transformation. In the vector control, the flux component related to the *d*-axis current (i_d) and torque component is related to the *q*-axis current (i_q). The transformation matrix and detailed modelling equations for multiphase machines is presented in Sections 10.5.2 and 10.5.3. The basic control block diagram of multiphase machines is presented in Figure 10.26. In this control diagram, the actual machine control variables such as speed and currents/voltages are sensed and processed to sense variables and transformation of the reference frame block. This block generates the actual vector control parameters, for example, flux, rotor position, currents and voltages in two-dimensional domains (DC components). The regulation of controlled variable blocks also regulates the machine torque as well as flux parameters to achieve the desired operating conditions. The transformation of reference frame block converts the DC components of currents/voltages to actual machine currents/voltage reference waves for PWM logic implementation. The PWM inverter switches according to the reference waves generated by the transformation of the reference frame block. In this book, the basic vector control techniques are discussed for multiphase machines.

10.6.1 Indirect Field-Oriented Control or Vector-Control Techniques for Multiphase Induction Machines

Vector control is also called field-oriented control. The field is aligned along *d*-axis to achieve the decoupled torque and flux control. To align field along the *d*-axis, the position of the field should be known. If the position of the field is measured, it is called 'Direct field-oriented control', and if the position is estimated using a mathematical formulation, it is called 'Indirect field-oriented control'. If the rotor field is used to align along the *d*-axis, it is called rotor field-oriented control. If the stator field is aligned along the *d*-axis to achieve vector control, it is known as 'stator field-oriented control'. The most popular among these types are 'Indirect rotor field-oriented control'.

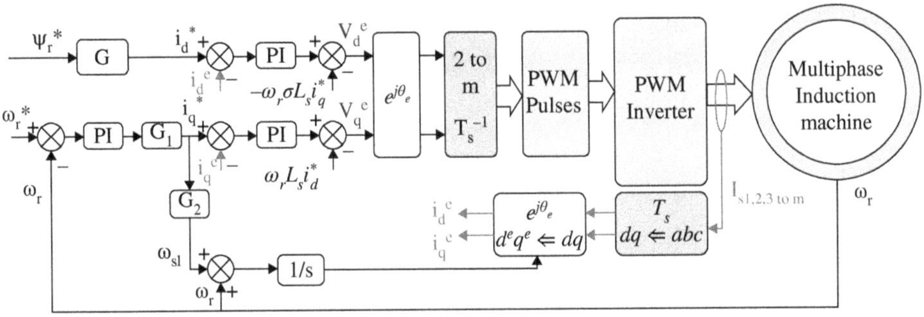

Figure 10.27 Indirect field-oriented vector control of multiphase induction machine.

Figure 10.28 Two-dimensional vector representation of the multiphase machine.

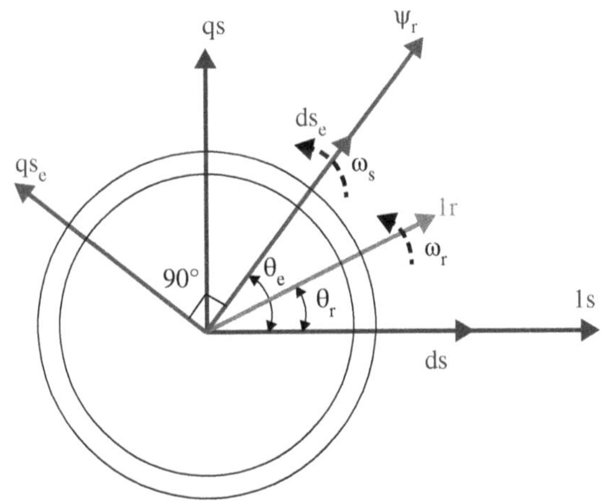

The Indirect field-oriented control (IFOC) is basically two types; (i) current-based control and, (ii) voltage-based control. The IFOC of multiphase induction motor is shown in Figure 10.27. The d-axis of stator is fixed with the first phase of stator ($1\,s$) and it is behind the rotor flux vector by an angle of θ_e. The first phase of rotor is behind the rotor flux vector by an angle of θ_r, shown in Figure 10.28. The rotor flux space vector is kept aligned at all times with the real axis (d-axis) of the synchronous reference frame, while the q-axis is perpendicular to it, i.e. the speed of rotor flux vector is the same as the synchronous reference frame speed. The relative speed of the rotor first phase axis to rotor flux vector is ω_{sl}, i.e. the rotor first phase is behind an angle of $\theta_{sl} = \theta_e - \theta_r$. Since the rotor flux space vector is completely aligned with the real axis its imaginary component always remains equal to zero, i.e.

$$\psi^e_{rd} = \psi_r$$
$$\psi^e_{rq} = 0 \tag{10.75}$$

The modelling equations of the multiphase induction motor are presented in Section 10.5.3. The d-axis rotor equations of the multiphase induction machine in synchronous rotating reference frame are

$$V^e_{rd} = 0 = R_r i^e_{rd} + p\psi^e_{rd} - \omega_{sl}\psi^e_{rq} \tag{10.76}$$

$$\psi_{rd}^e = L_{r11} i_{rd}^e + L_M i_{sd}^e \tag{10.77}$$

By substituting Eq. (10.75) (where the $\psi_{rq}^e = 0$ and ψ_{rd}^e = constant) in Eq. (10.76, 10.77) results in

$$i_{rd}^e = 0$$

$$\psi_{rd}^e = L_M i_{sd}^e \tag{10.78}$$

The q-axis rotor equations of the multiphase induction machine in synchronous rotating reference frame are

$$V_{rq}^e = 0 = R_r i_{rq}^e + p\psi_{rq}^e + \omega_{sl}\psi_{rd}^e \tag{10.79}$$

$$\psi_{rq}^e = L_{r11} i_{rq}^e + L_M i_{sq}^e \tag{10.80}$$

By substituting Eq. (10.75) (where the $\psi_{rq}^e = 0$ and ψ_{rd}^e = constant) in Eq. (10.79, 10.80) results in

$$R_r i_{rq}^e + \omega_{sl}\psi_{rd}^e = 0 \Rightarrow \omega_{sl} = \frac{-R_r i_{rq}^e}{\psi_{rd}^e} \tag{10.81}$$

$$\psi_{rq}^e = 0 = L_{r11} i_{rq}^e + L_M i_{sq}^e$$

$$i_{rq}^e = \frac{-L_M i_{sq}^e}{L_{r11}} \tag{10.82}$$

The resultant slip speed equation from Eqs. (10.78), (10.81), and (10.82) is

$$\omega_{sl} = \frac{R_r}{L_{r11}} \frac{i_{sq}^e}{i_{sd}^e} = \frac{1}{\tau_r} \frac{i_{sq}^e}{i_{sd}^e} \text{ or } \omega_{sl} = \frac{L_m}{\tau_r} \frac{i_{sq}^e}{\psi_{rd}^e} \tag{10.83}$$

Where $\tau_r = \frac{L_{r11}}{R_r}$ is a rotor time constant.

The remaining rotor voltage and flux linkage equations are

$$V_{rx1}^e = R_r i_{rx1}^e + p\psi_{rx1}^e \tag{10.84}$$

$$V_{ry1}^e = R_r i_{ry1}^e + p\psi_{ry1}^e \tag{10.85}$$

Similarly, other two-dimensional sub-planes can be written and the zero-sequence equation is

$$V_{r0}^e = R_r i_{r0}^e + p\psi_{r0}^e \tag{10.86}$$

$$\psi_{rx1}^e = L_{r33} i_{rx1}^e \tag{10.87}$$

$$\psi_{ry1}^e = L_{r44} i_{ry1}^e \tag{10.88}$$

Similarly, other two-dimensional sub-planes can be written and the zero-sequence equation is

$$\psi_{r0}^e = L_{rmm} i_{r0}^e \tag{10.89}$$

The electromagnetic torque equation is

$$T_e = \frac{m}{2}\frac{P}{2} L_M \left[i_{sq} i_{rd} - i_{sd} i_{rq} \right] \tag{10.90}$$

The resultant torque component of the machine form Eqs. (10.77), (10.80), and (10.90) gives the

$$T_e = \frac{m}{2}\frac{P}{2}\frac{L_M}{L_r}\left[i^e_{sq}\psi^e_{rd}\right] \tag{10.91}$$

In the above equation, i^e_{sq}, ψ^e_{rd}represent the torque and flux components of the machine, respectively. From Eqs. (10.32) to (10.46) (clearly presented in Section 10.5.3), it is concluded that, in the multiphase machines, the torque is produced by only dq components (first two-dimensional plane).

The voltage-based control of IFOC vector control scheme is presented in Figure 10.27. The phase currents of multiphase machine are transformed into dq domain by using the transformation matrix $[T_s]$ in the dq stationary reference frame. With the help of the rotational transformation matrix given in Section 10.5.4, the components are transformed into$d^e q^e$ domain wrt the angular position of the rotor θ_e. As compared to IFOC of three-phase machines given in [3, 7, 12, 14–16], the control of multiphase machine is different in terms of transformation matrix conversion only since the windings are symmetrically distributed in a sinusoidal manner. The V^e_d and V^e_q vectors are attained from the flux component and speed error, respectively, which are transformed into actual modulating vectors of multiphase machine by taking the inverse transformation of the $[T_s]$ matrix.

The current control-based IFOC diagram is presented in Figure 10.29, where the control is similar to the voltage-based control shown in Figure 10.27 except for voltage loops and PWM generation. After generating the flux ($i_d{}^*$) and torque references ($i_q{}^*$), the rotation block is used to convert one reference frame to another. From these two-dimensional reference currents, the reference phase variable currents are generated by using the inverse transformation matrix. The reference phase variable currents are compared with the actual phase currents and the error is fed to the hysteresis band control. The hysteresis band control will generate the variable frequency PWM pulses according to the desired current control. The major shortcoming of the hysteresis current control is the variable inverter switching frequency over a period of output voltage. According to Eqs. (10.76) to (10.91), the gain blocks G, G1, G2 in the IFOC are calculated.

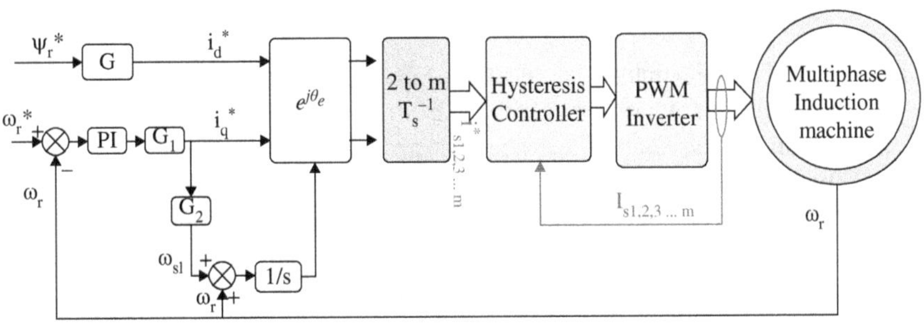

Figure 10.29 Current control based IFOC vector control diagram of multiphase-induction motor.

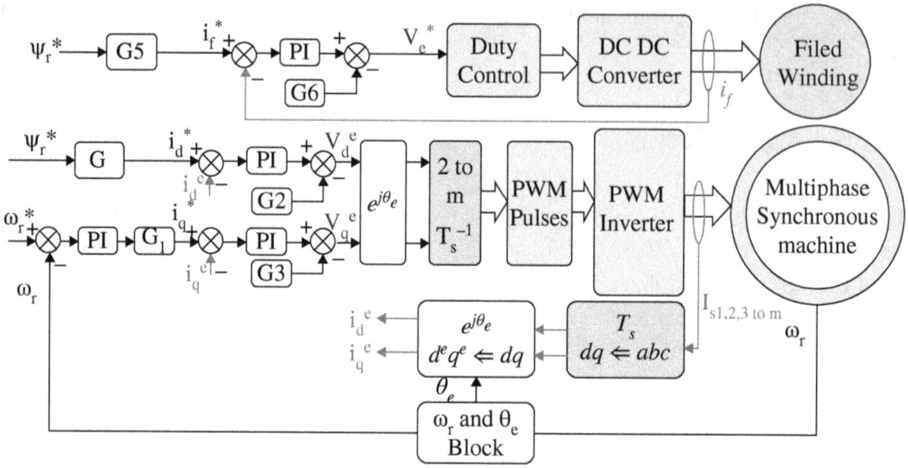

Figure 10.30 Indirect field-oriented vector control of multiphase synchronous machine.

10.6.2 Vector Control for Multiphase Synchronous Machines

The basic indirect field-oriented vector control of synchronous machine is presented in Figure 10.30. The detailed mathematical modelling and transformation matrices are presented in Sections 10.5.5 and 10.5.2, respectively. Similar to the multiphase induction machine control, the IFOC of synchronous motor also has voltage-based control and current-based control. The procedure of the vector control involves in

- Sensed phase voltages or currents have to transform into dq domain by using transformation matrices.
- From the position sensor of the machine, speed and position are calculated.
- The dq domain variables are transformed from one reference frame to another (synchronous rotating frame).
- The torque (i_q) reference is generated from the speed comparator and respective voltage reference is generated accordingly.
- The flux (i_d) component is generated with appropriate gain and the respective voltage reference is generated. If field weakening is considered, the flux components have to be generated from the field weakening lookup table.
- The field winding (i_f) component is controlled by generating the appropriate duty cycle to the DC-DC converter.

In the vector control of the synchronous machine, G, G1, G5 are gains to attain the desired reference-vector magnitudes, which are calculated from the modelling equations similar to the induction-machine analysis. The G3, G4, and G6 are feed-forward gains to obtain accurate control of the machine. The current-based vector control of the synchronous motor is similar to voltage-based vector control as shown in Figure 10.30, except voltage reference generation loops. In this current-based control, the reference currents are compared with actual currents and the error is fed to the hysteresis band control, which generates the required pulses for driving the inverter. In the permanent magnet synchronous machines (PMSM), the field winding on the rotor is absent and the rotor is made with permanent magnets. The control of the multiphase PMSM is the same as the conventional synchronous machine vector control shown in Figure 10.30; the only difference is the field winding and associated control has to be removed.

10.7 Matlab/Simulink Model of Multiphase Machines

In the previous section, detailed modelling and analysis of multiphase machines, are given. The modelling of the multiphase machines can be done in a reference frame, according to user convenience. In this book, detailed modelling of nine-phase induction motor, as well as nine-phase synchronous motor is presented.

10.7.1 Dynamic Model of the Nine-Phase Induction Machine

In this book, the dynamic model of four-pole nine-phase symmetrical induction motor is modelled in Matlab/Simulink, as shown in Figure 10.31. The rotor of a machine is a squirrel-cage type and the machine is modelled in the stationary reference frame. The mathematical equations of the machine are presented in Section 10.5.3. The parameters considered for the modelling of nine-phase induction motor are

V_m = peak value of the stator phase voltages = $230*sqrt(2)V$
Magnetizing inductance $M = 0.168H$
Mutual inductance $L_m = M*9/2\ H$

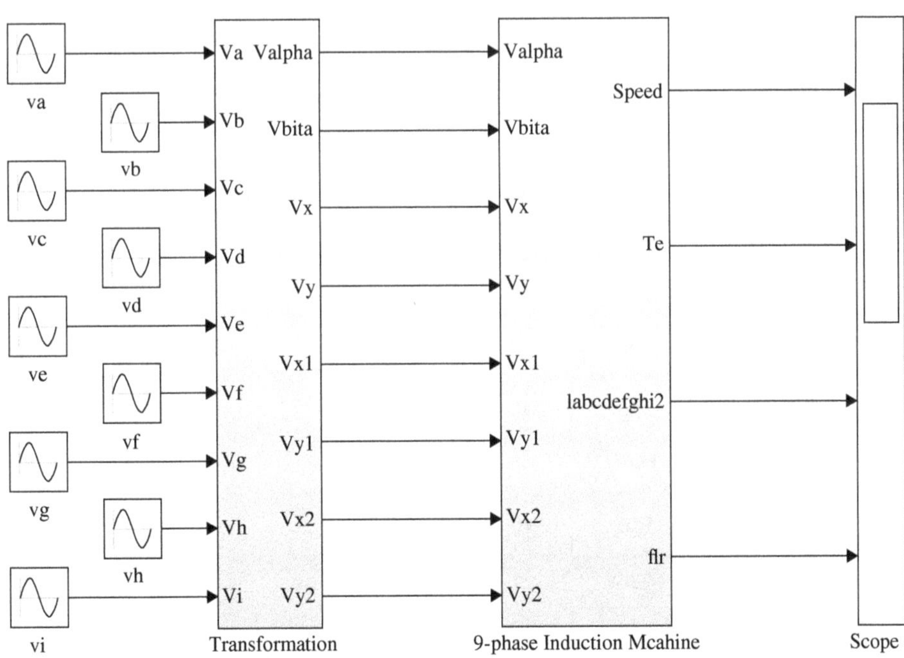

Figure 10.31 Matlab/Simulink dynamic model of nine-phase four-pole induction machine.

Leakage inductance of stator $L_{ls} = 0.04\,H$
Leakage inductance of rotor $L_{lr} = 0.04\,H$
Self-inductance of the stator $L_s = L_{ls} + L_m\,H$
Self-inductance of rotor $L_r = L_{ls} + L_m\,H$
Poles $P = 4$
J = moment of inertia = $0.03\,kg\text{-}m^2$
Stator Resistance $R_s = 14\,\Omega$
Rotor Resistance $R_r = 8.8\,\Omega$
Load torque $T_l = 15\,Nm.$

The transformation of voltages to two-dimensional domain (dq) in the stationary reference frame ($\omega_a = 0$) is shown in Figure 10.32. In this block, the transformation matrix shown in Section 10.5.2 is written in the form of equations. The stator voltage, as well as flux-linkage equations, are shown in Figure 10.33, where the mathematical modelling equations present in Section 10.5.3 are used. The flux linkages are derived from the dq voltages and resistive drops; the dq currents are derived from the flux linkages and inductance matrices. The rotor voltage and flux linkage equations are shown in Figure 10.34 (presented

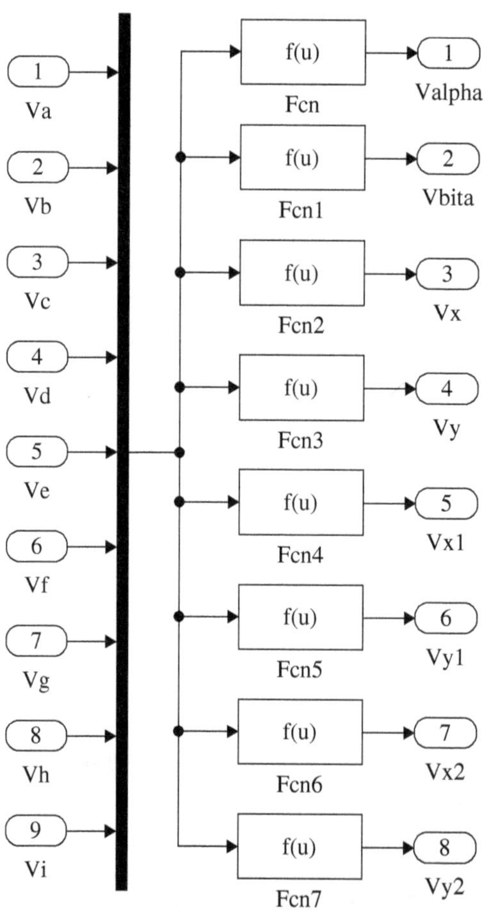

Figure 10.32 Equations of the transformation matrix for nine-phase to two-phase conversion.

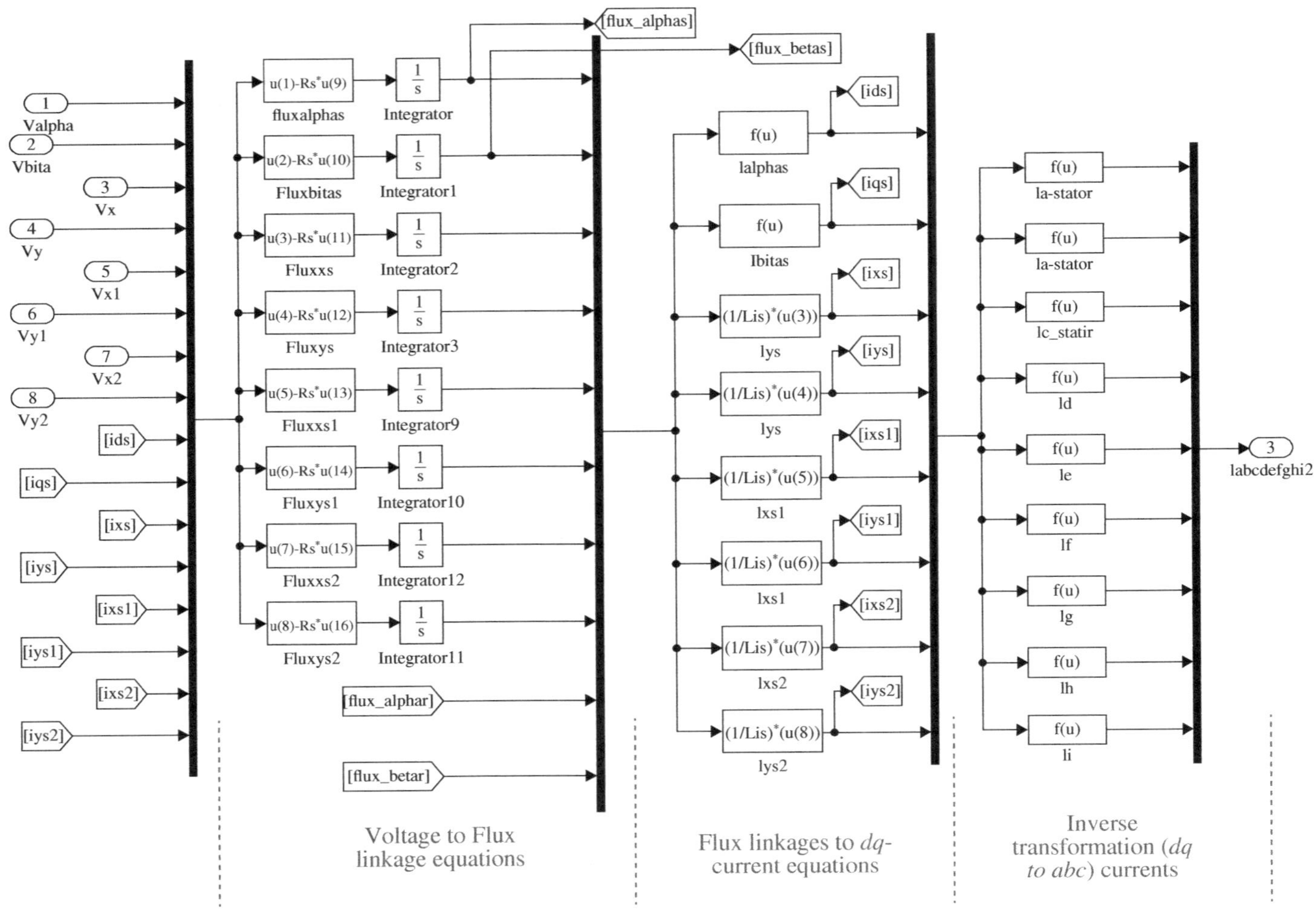

Figure 10.33 Stator voltage and flux linkage equations of nine-phase induction machine.

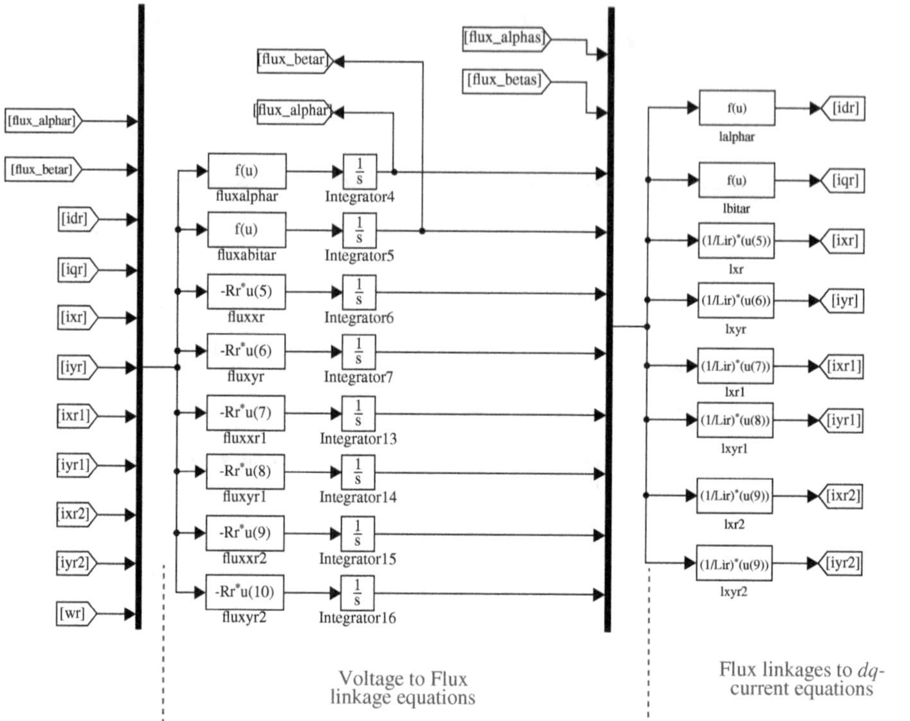

Figure 10.34 Rotor voltage and flux linkage equations of nine-phase induction machine.

in Section 10.5.3). The mechanical and rotor flux equations are presented in Figure 10.35, where the mechanical equations and load torque equations are implemented.

The results of the nine-phase symmetrical IM model are shown in Figure 10.36, where the machine starting transient response as well as steady-state response are demonstrated. In starting the machine under no-load condition, the transient in torque and flux are happening due to the high inrush currents to establish the magnetic field in the air gap (high magnetizing currents) as well as poor-power factor. The machine achieves the steady-state speed at 0.2 s, i.e. 1500 rpm with four-pole and 50 Hz frequency, where the machine is operating at no load (slip is zero) and the losses of the machine are neglected. The respective torque during steady state under no-load condition is 0 Nm. At the time, 0.5 s the load torque (15 Nm) is applied on the machine, this results in speed reduction to 1415 rpm and the currents of machine are increased accordingly. In the implemented model of the nine-phase IM, the mechanical power is positive because of the machine is operating as a motor. To attain the generator operation, the mechanical power (torque) of the machine has to be considered as negative, but the entire modelling and analysis will be same.

10.7.2 Dynamic Model of the Nine-Phase Synchronous Machine

The dynamic model of four-pole nine-phase symmetrical salient pole, a synchronous machine is modelled in Matlab/Simulink, as shown in Figure 10.37, where the machine is

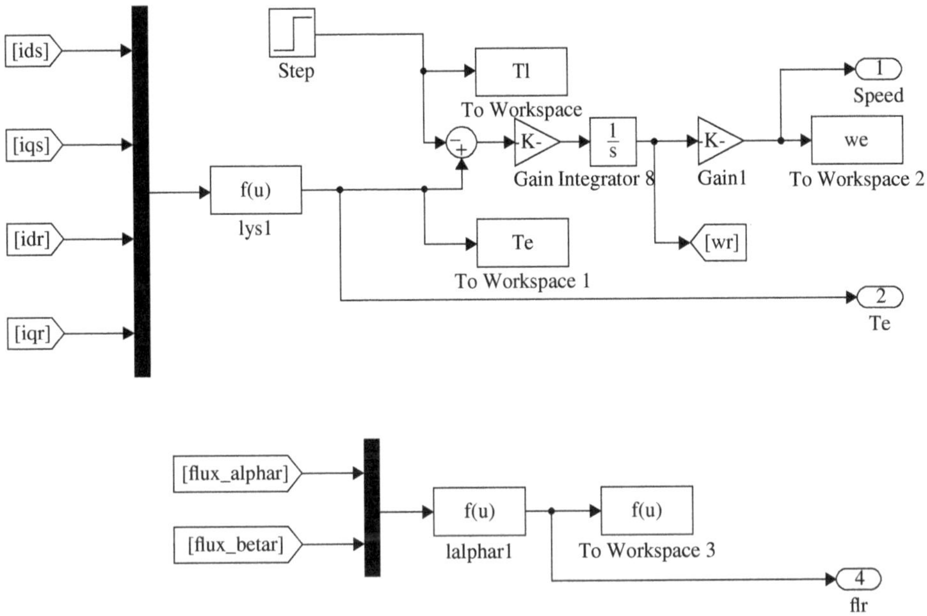

Figure 10.35 Mechanical torque equations and rotor flux equations of nine-phase induction machine.

modelled in the rotating reference frame. The rotor of a machine is salient-pole round-rotor type with one field winding on the *d*-axis. Two damper windings are also placed on the *d*-axis (*kd*) and *q*-axis (*kq*) of the rotor respectively with an angle of 90°. The detailed mathematical modelling equations and transformation matrices of the machine are presented in Section 10.5.5 and 10.5.2, respectively. The parameters considered for modelling of nine-phase salient pole synchronous machines are

V_m = peak value of the stator phase voltages = *230*sqrt(2)V*
V_f = field winding voltage = *60 V*
f = frequency of the supply = *50 Hz*
R_s = stator winding resistance = *1.62 Ω*
R_{kq} = q-axis damper winding 1 resistance = *5.8 Ω*
R_{kd} = field axis aligned damper winding resistance = *3.8 Ω*
R_{fd} = field winding resistance *1.45 Ω*
L_{ls} = leakage inductance = *0.009 H*
L_{md} = d-axis mutual inductance = *0.3258 H*
L_{mq} = q-axis mutual inductance = *0.1553 H*
L_{lfd} = field winding leakage inductance = *0.02264 H*
L_{lkd} = field axis aligned damped winding leakage inductance = *0.0074 H*
L_{lkq} = damper winding 1 leakage inductance = *0.0074 H*
P = number of poles = *2*
T_l = load torque = *30 Nm*,
J = moment of inertia = *0.01 kg-m²*.

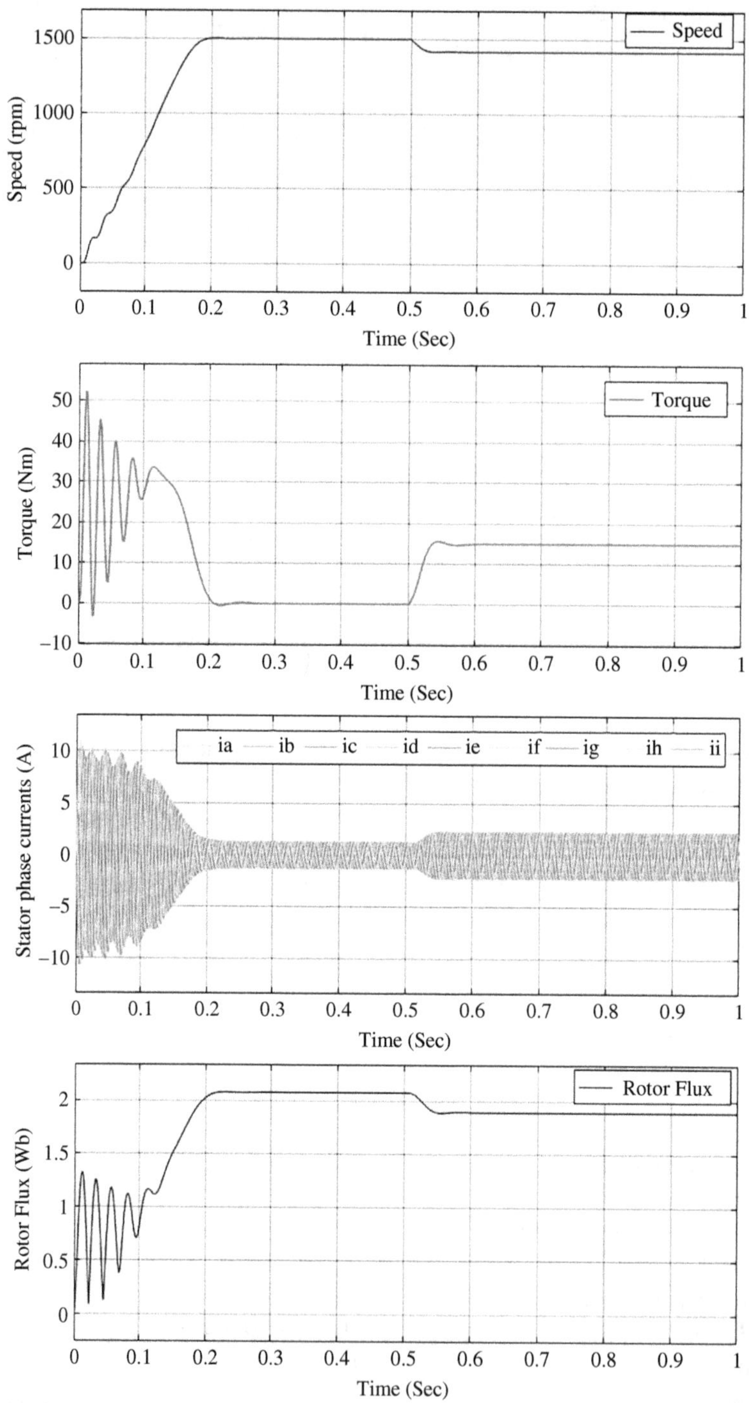

Figure 10.36 Matlab/Simulink results of the nine-phase induction machine.

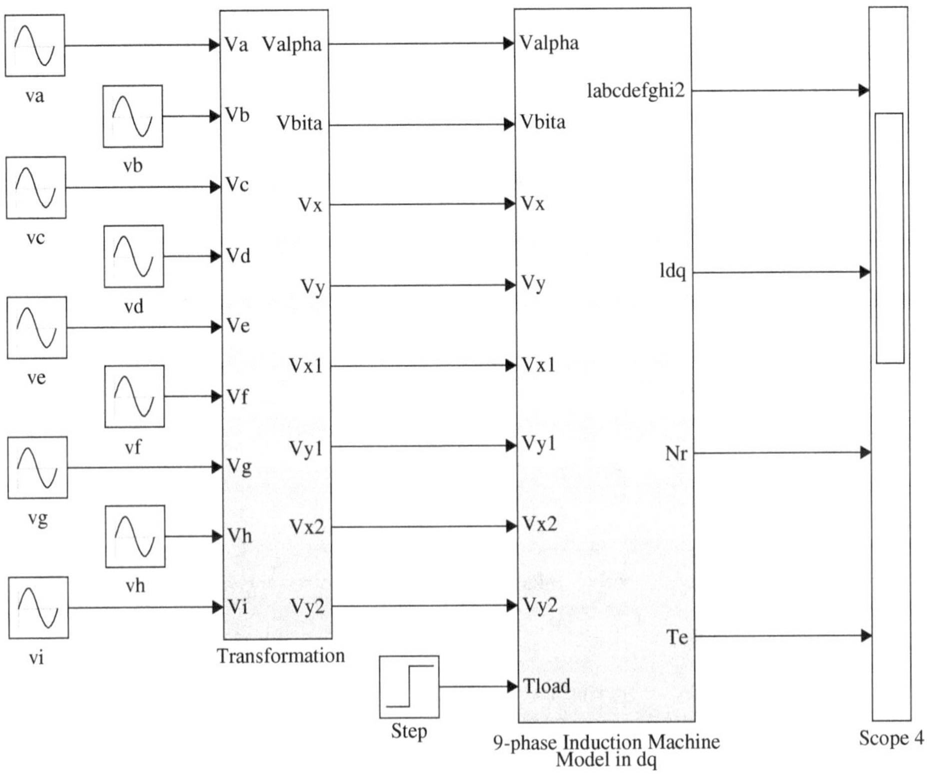

Figure 10.37 Matlab/Simulink dynamic model of nine-phase two-pole synchronous machine.

The self-inductances of stator and rotor are

$$L_{sd} = L_{ls} + L_{md}$$
$$L_{sq} = L_{ls} + L_{mq}$$
$$L_{fd} = L_{lfd} + L_{md}$$
$$L_{kd} = L_{lkd} + L_{md}$$
$$L_{kq} = L_{lkq} + L_{mq}.$$

Similar to the nine-phase induction motor, the transformation of voltages to two-dimensional domain (*dq*) is shown in Figure 10.32, which is also the same for the synchronous machine. Since the model is implemented in the rotor reference frame, the *dq* voltages are transformed into a rotating reference frame by using the rotation matrix presented in Section 10.5.4. The stator voltage to flux linkage equations are shown in Figure 10.38 and Figure 10.39, respectively, where the mathematical modelling equations present in Section 10.5.5 are used. The rotor voltages to flux-linkage equations are presented in the Figure 10.40. From the flux linkages to *dq* currents and then *abc* currents, inverse transformation, has derived from the flux linkages and inductance matrices shown in Figure 10.41, where the inductance matrices given in Section 10.5.5 are used. The mechanical load-torque equations are presented in Figure 10.42.

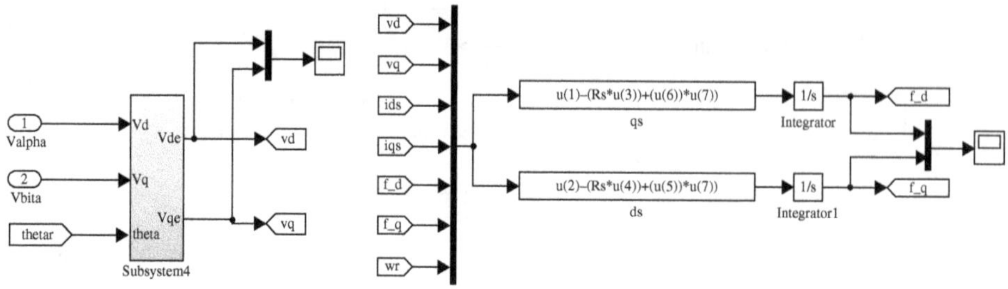

Figure 10.38 Stator *dq* voltage and flux linkage equations of nine-phase synchronous machine.

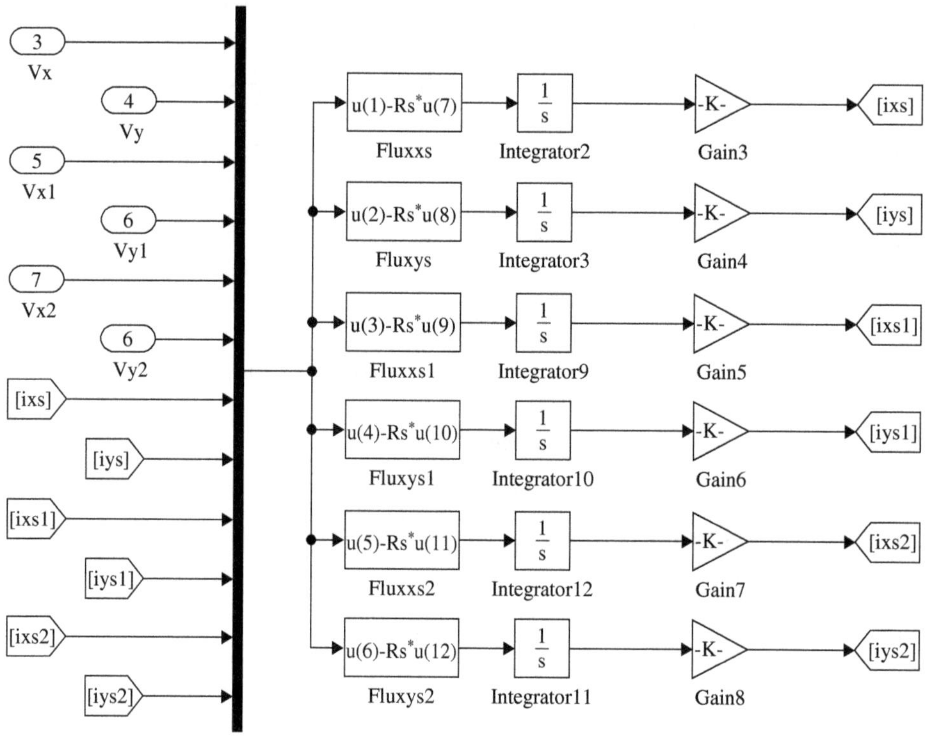

Figure 10.39 Stator voltage (x,y,x1...) and flux linkage equations of nine-phase synchronous machine.

The results of the symmetrical nine-phase salient-pole synchronous machine model are shown in Figure 10.43, where the machine starting transient response as well as steady-state response are presented. During the start of the machine under no-load condition, the transient in torque and currents are happening due to the un-locking effect of the magnetic poles. The machine achieves the steady-state speed at 0.15 s, i.e. 3000 rpm with two-pole and 50 Hz frequency. The machine is operating at no load up to 0.5 s, after that, a load of 30 Nm is applied on the machine, i.e. positive torque results motor operation. As the machine is of synchronous type, the speed of the machine is also in constant under-load condition, i.e.

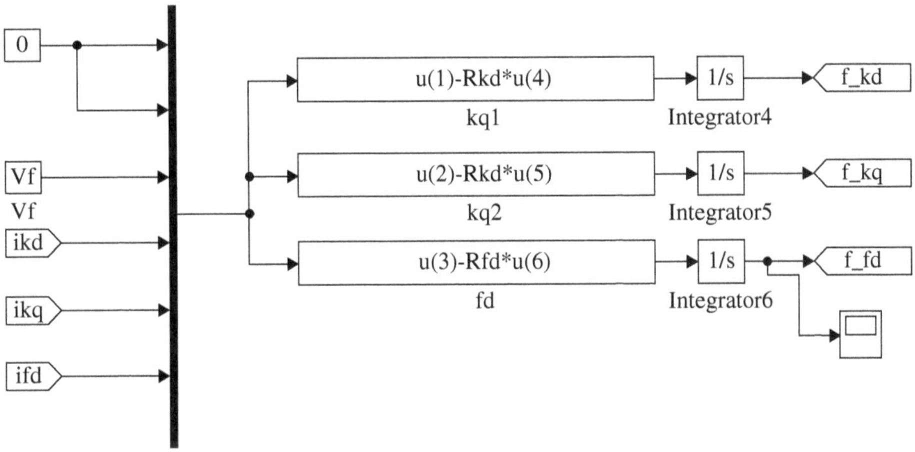

Figure 10.40 Rotor voltage and flux linkage equations of nine-phase synchronous machine.

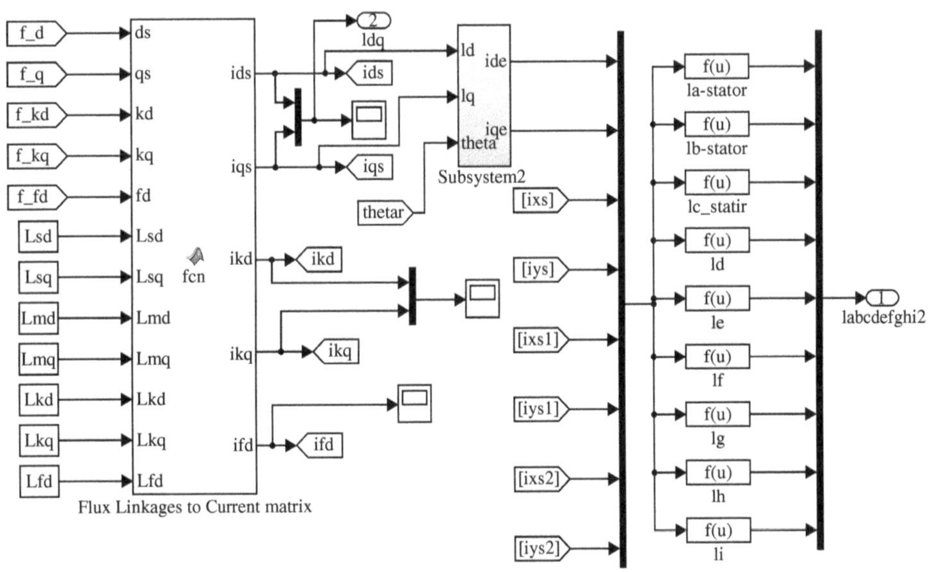

Figure 10.41 Flux linkages to current conversion equations of nine-phase synchronous machine.

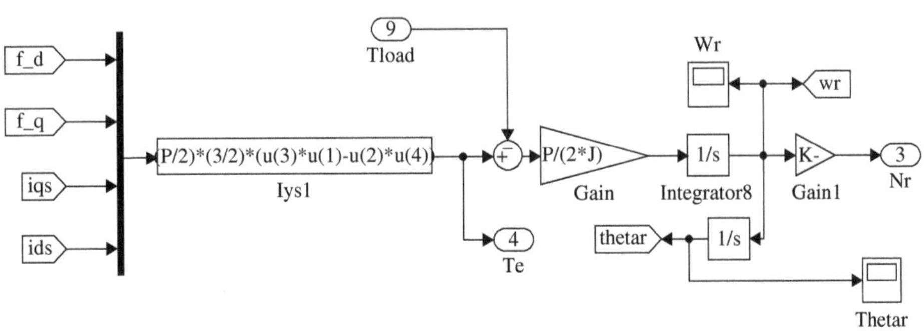

Figure 10.42 Mechanical torque equation of the nine-phase synchronous machine.

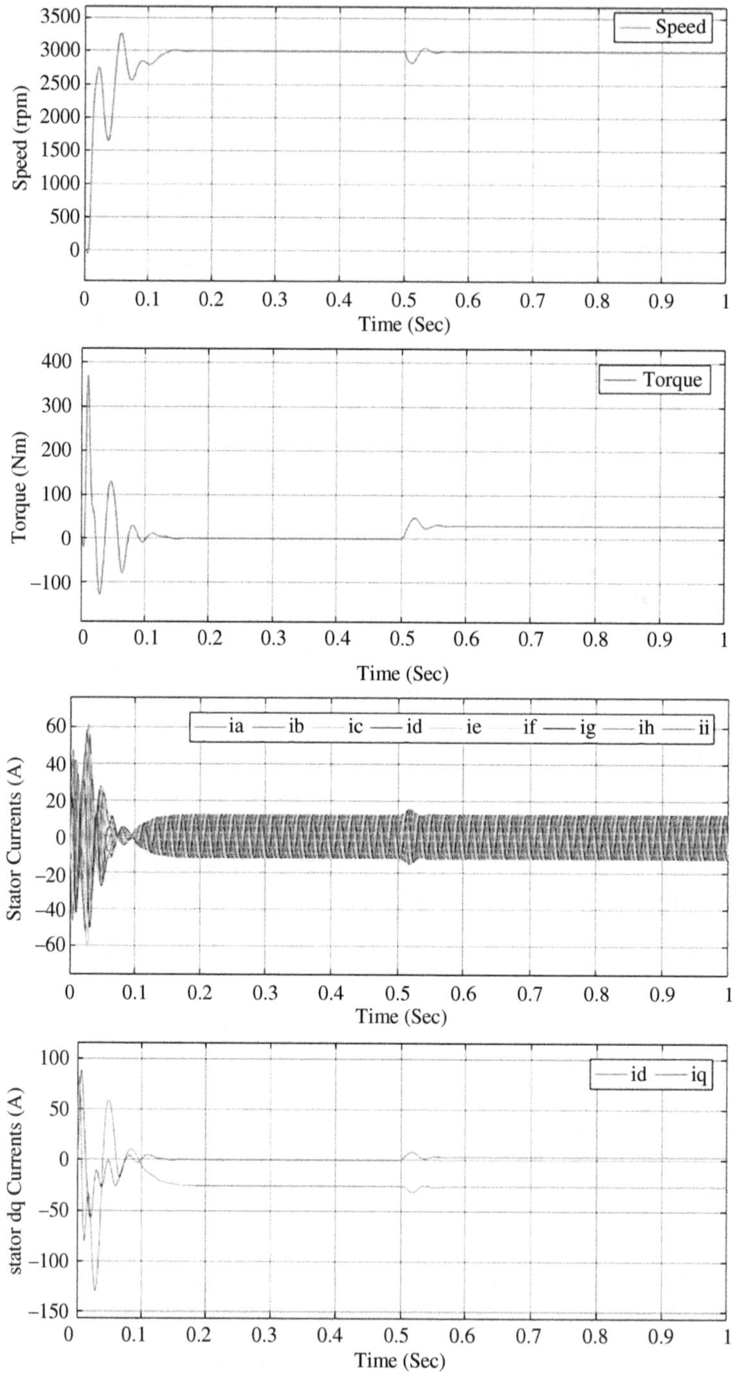

Figure 10.43 Matlab/Simulink results of the nine-phase synchronous machine.

3000 rpm. To operate, the same model as a generator, the polarity of the torque applied on the machine has to be reversed, where the modelling and analysis will be same.

10.8 Summary

In this chapter, the overview of multiphase machines (induction machines and synchronous machines) are discussed. The evolution of multiphase machines, and its applications, as well as detailed advantages compared to traditional three-phase solutions, is discussed. The winding diagrams of the three-phase, five-phase, six-phase, and nine-phase machines are presented in detail, and this analysis will help in the understanding of any number of phase machines. Moreover, the single-layer and double-layer, short-pitch and full-pitch, fractional and integral slot windings are presented. To control the power/torque of machine precisely, the mathematical modelling will play a crucial role. So, in this chapter, detailed mathematical modelling of m-phase machines (induction and synchronous machines) is discussed, where the commonly used reference frames, transformation matrices and modelling equations of the machine in different reference frames are presented. The basic vector-control techniques of the multiphase machines are also discussed in this chapter. Finally, the mathematical equations of the multiphase machines are implemented in the Matlab/Simulink and the associated blocks, as well as results, are discussed in detail.

Problems

10.1 Consider a three-phase, 12-pole machine with 36 slots. And design the following winding diagrams
(i) Single-layer full-pitch winding
(ii) Double-layer short-pitch winding with a coil pitch 2

10.2 Construct the winding table and key winding design details (such as coil pitch, number slots per phase and other) for a three-phase, six-pole machine with 60 slots.

10.3 Consider a five-phase four-pole machine with 60 slots. Design winding tables and design details for the following
(i) Double-layer full-pitch winding
(ii) Double-layer short-pitch winding with a short pitch by 12° or 1 slot.

10.4 Consider a six-phase, six-pole machine with 36 slots. And design the following winding diagrams
(i) Consider the machine with symmetrically distributed windings (symmetrical six-phase machine). Construct the winding tables for single-layer and double-layer windings. Give the appropriate observations and draw the possible winding diagram.

(ii) Consider the machine with asymmetrically distributed windings (Asymmetrical six-phase machine). Construct the winding tables for single-layer and double-layer full-pitch windings.

Draw the single-layer short-pitch winding with a short pitch of one stator slot.

10.5 Consider a nine-phase, four-pole machine with 36 slots. And design the following winding diagrams

(i) Single-layer full-pitch winding

(ii) Double-layer short-pitch winding with a coil pitch 2

10.6 Analyse the winding diagrams designed in Problems 1 and 5 and make observations and comparisons. Design a reconfigurable winding for a machine with 36 slots, which can be operated as nine-phase four-pole and three-phase 12-pole. Hint: use the pole-phase modulation methods discussed in Chapter 6.

10.7 Consider a 15-phase machine with four-pole and 120 slots. Construct a winding table for single-layer winding and double-layer short-pitch winding with a short pitch of one slot.

10.8 Consider a nine-phase machine with 36 slots and six poles. Draw the possible winding diagrams.
Hint: Design the multi-layer winding.

References

1 Lamme, B.G. (1921). The story of the induction motor. *Journal of the American Institute of Electrical Engineers* 40 (3): 203–223.

2 Klingshirn, E.A. and Jordan, H.E. (1968). Polyphase induction motor performance and losses on nonsinusoidal voltage sources. *IEEE Transactions on Power Apparatus and Systems* PAS-87 (3): 624–631.

3 White, C. and Woodson, H.H. (1959). *Electromechanical Energy Conversion*. New York: Wiley.

4 Sawhney, A.K. (1984). *A Course in Electrical Machine Design*. New Delhi: Dhanpat Rai and Sons.

5 Ward, E.E. and Harer, H. (1969). Preliminary investigation of an inverter fed 5-phase induction motor. *Proceedings of the Institution of Electrical Engineers* 116 (6): 980–984.

6 Nelson, R.H. and Krause, P.C. (1974). Induction machine analysis for arbitrary displacement between multiple winding sets. *IEEE Transactions on Power Apparatus and Systems* PAS-93 (3): 841–848.

7 Krause, P., Wasynczuk, O., Sudhoff, S., and Pekarek, S. (2013). *Analysis of Electric Machinery and Drive Systems*. IEEE Press.

8 Jahns, T.M. (1980). Improved reliability in solid-state ac drives by means of multiple independent phase-drive units. *IEEE Transactions* IAS-16: 321–331.

9 Klingshirn, E.A. (1983). High phase order induction motors—Part I—Description and theoretical considerations. *IEEE Transactions on Power Apparatus and System* 102 (1): 47–53.

10 Gopakumar, K., Sathiakumar, S., Biswas, S.K., and Vithayathil, J. (1984). Modified current source inverter fed induction motor drive with reduced torque pulsations. *Proceedings of the Institution of Electrical Engineers* 131 (4), pt. B: 159–164.

11 Gopakumar, K., Ranganthan, V.T., and Bhat, S.R. (1993). Split-phase induction motor operation from PWM voltage source inverter. *IEEE Transactions on Industry Applications* 29 (5): 927–932.

12 Singh, G.K. (2002). Multi-phase induction machine drive research – a survey. *Electric Power Systems Research* 61: 139–147.

13 Gopakumar, K., Ranganathan, V.T., and Bhat, S.R. (1994). Vector control of induction motor with split phase stator windings. *European Power Electronics and Drives Journal* 7 (1–2): 569–574.

14 Wilamowski, B.M. and Irwin, J.D. (2011). Power electronics and motor drives. In: *The Industrial Electronics Handbook* (ed. J.D. Irwin), 185–194. CRC Press/Taylor & Francis.

15 Levi, E., Bojoi, R., Profumo, F. et al. (2007). Multiphase induction motor drives - a technology status review. *IET Electric Power Applications* 1 (4): 489–516.

16 Levi, E. (May 2008). Multiphase electric machines for variable-speed applications. *IEEE Transactions on Industrial Electronics* 55 (5): 1893–1909.

17 Williamson, S. and Smith, A.C. (2003). Pulsating torque and losses in multiphase induction machines. *IEEE Transactions on Industry Applications* 39 (4): 986–993.

18 Apsley, J. and Williamson, S. (2006). Analysis of multiphase induction machines with winding faults. *IEEE Transactions on Industry Applications* 42 (2): 465–472.

19 Jen-Ren, F. and Lipo, T.A. (1994). Disturbance-free operation of a multiphase current-regulated motor drive with an opened phase. *IEEE Transactions on Industry Applications* 30 (5): 1267–1274.

20 Zoric, I., Jones, M., and Levi, E. (2018). Arbitrary power sharing among three-phase winding sets of multiphase machines. *IEEE Transactions on Industrial Electronics* 65 (2): 1128–1139.

21 Lyra, R.O.C. and Lipo, T.A. (2002). Torque density improvement in a six phase induction machine with third harmonic current injection. *IEEE Transactions on Industry Applications* 38 (5): 1351–1360.

22 Abdelkhalik, M.M. and Barry, W. (2010). Eleven-phase induction machine: steady-state analysis and performance evaluation with harmonic injection. *IET Electric Power Applications* 4 (8): 670–685.

11

Numerical Simulation of Electrical Machines using the Finite Element Method

11.1 Introduction

The Finite Element Method (FEM) is a numerical-based technique, which can be used for solving the partial differential equations (PDEs) with various space variables such two-space variables (2D analysis) and three-space variable (3D analysis) [1]. In this FEM, to attain the solution for a complicated problem, the targeted system is divided into finite elements or small regions, and finding the solution for each finite element (region) with simple equations. The roots of the FEM were established at the start of the nineteenth century by Courant, Turner and Martin, where the authors analysed the system with algebraic equations and numerical analysis [1, 2]. In the 1960s Clough, implemented the FEM by using Laplace and Poisson's equations. Later in the 1970s, Cheung, Zienkiewicz and his team implemented an open source-based FEM simulation software, after which NASA sponsored for software NASTRAN, and UC Berkeley implemented the FEM program [1, 2].

FEM is used for solving the problems (such as ordinary and elliptic partial differential equations [EPDEs]) involved in various engineering fields and industrial products, for example

1) Electrical engineering: Rotating machines, thermal as well as electrical analysis of power electronics, sensors, magnetic elements, electric actuators, thermal analysis of machines, and other electromagnetics (EMs)
2) Mechanical engineering: Acoustics, noise analysis, vibrations, heat transfer, fluid flows, metal forming, thermodynamics, and many more.
3) Structural engineering: Structural strength and design, structural interaction of fluid flows, mass diffusions, material design etc.
4) Industry-related products: Design and analysis (including thermal and mechanical design) of refrigerators, AC, coolers, fans, blowers, flights, electric vehicles, and many more.

The process involved in the finite element analysis (FEA) is presented in Figure 11.1. The implementation process of FEM for various problems is called FEA. The engineering problem consists of the actual structure or component of desired object (loads like motors). The realization problem involves the mathematical model, which consists of ordinary

Electrical Machine Fundamentals with Numerical Simulation using MATLAB/SIMULINK, First Edition.
Atif Iqbal, Shaikh Moinoddin, and Bhimireddy Prathap Reddy.

Companion website: www.wiley.com/go/iqbal/electricalmachine

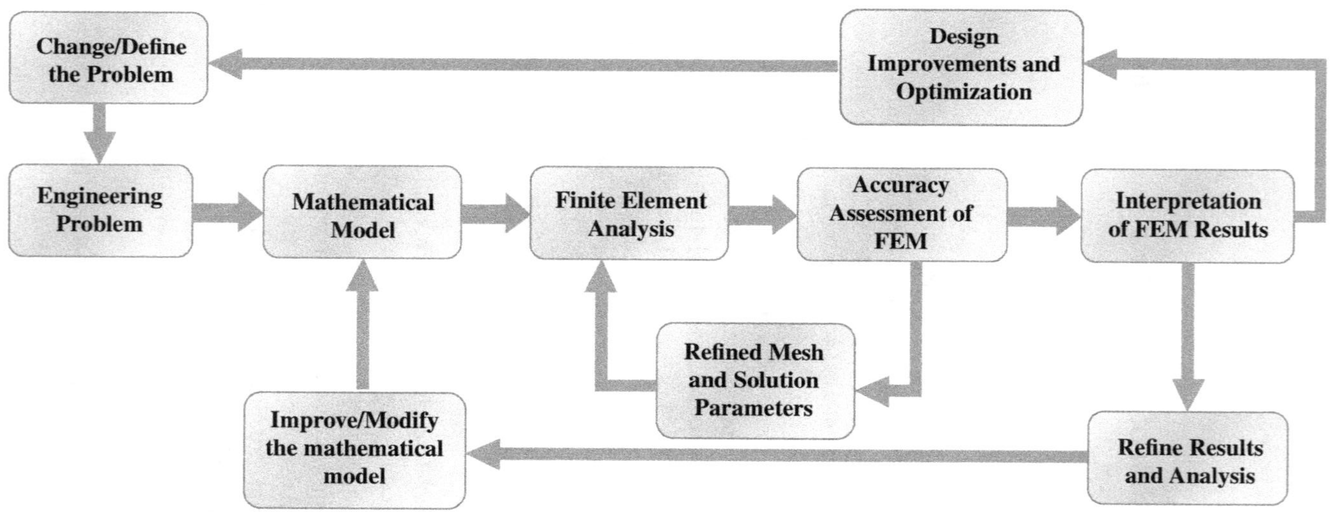

Figure 11.1 The process of finite element analysis.

or differential equations. This mathematical model is realized with some assumptions including geometry, kinematics, material law, loading, boundary conditions etc. [3]. The mathematical model is solved with FEA by dividing the system into smaller parts which are called finite elements. To attain greater accuracy, FEA is done with a higher number of meshing elements by selecting the optimal solution parameters. The equations of finite elements are simple and implemented with a differential form of Maxwell equations, and are then all assembled to attain the resultant system equations to solve the actual problem. In FEA, the error function is minimized by using the variation principles and adaptive meshing, which help find the approximate solution of the problem.

This chapter mainly focuses on the basic equations helpful for FEM analysis and EM analysis of electrical machines by using Ansys Maxwell 2D. The methods to solve EM analysis, detailed Maxwell equations, principles of the FEM are described. The electrical machine implementation in Ansys Maxwell and its inverter excitation in the Simplorer environment of Ansys are given in detail. The example models of three-phase machines and the associated results are explained to visualize the magnetic behaviour (in terms of flux distribution, flux density), mechanical torque, speed, electrical currents, and voltage characteristics.

11.2 Methods of Solving EM Analysis

From the electrical engineering perspective, the FEA is a significant tool to analyse and study the electric characteristics, magnetic behaviour, thermal behaviour, mechanical torque or speed and associated noise as well as vibrations in rotating machines. The basic types of analysis in the FEM are, for example, EM, magnetostatic, eddy current/current flow, Thermal analysis [4–6].

EM analysis is used to study the trainset as well as steady-state behaviour of the rotating machines, transformers and all EM devices. For analysing the behaviour of solenoids, permanent magnets, and magnetic shields etc., magnetostatic analysis can be adopted. In these EM and magnetostatic analyses, the key parameters to study are magnetic flux lines, flux density, various forces, time-varying inductances, torques, current densities, etc. The eddy current/current flow analysis is mainly used for analysing the various losses of the system such as eddy current loss, hysteresis loss, other power loss, currents densities etc. To analyse the temperature distribution, heat loss wrt temperature variation, thermal gradients etc. of the system are analysed with thermal analysis. The capacitive system behaviour wrt variation of capacitances, electric forces and fields are analysed with electrostatic analysis.

EM analysis is solved by using analytical and numerical techniques. The analytical techniques will solve the problems with the closed-form equations by using standard theorems. These analytical techniques give the exact solutions for a given problem. However, these techniques are time-consuming and sometimes solutions are impossible for some problems. Whereas, numerical techniques approximate solutions with minimal error by using computer-based simulations. These techniques require less time and solutions for most problems are possible. Classification of EM analysis is shown in Figure 11.2.

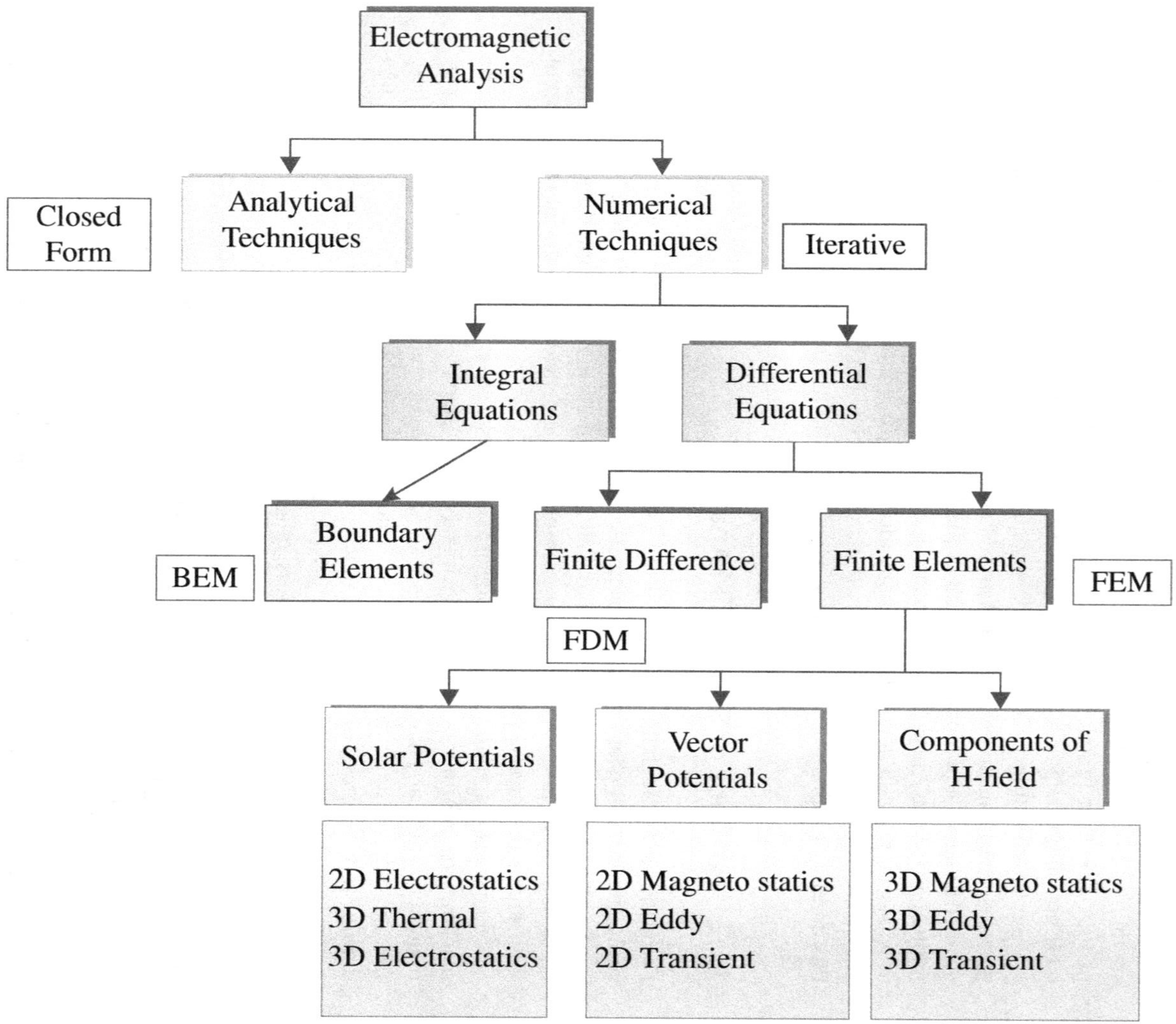

Figure 11.2 Methods of solving EM analysis.

11.2.1 Analytical Techniques

Basic analytical methods for solving the EM analysis are,

1) Separation of variables
2) Series expansion
3) Conformal mapping
4) Integral methods.

Of all methods, separation of variables is the most common and simple type of method to solve EM analysis. The separation of a variable (SV) is also called a Fourier method [2, 7–12]. The SV method is a simple method for solving the partial differential equation but the boundary conditions of the equations will be linear and homogeneous. In this SV, the solution for a given problem is separated into a product of various functions in which each product involves only one variable. Let us consider a simple equation of the heat conduction of a rod in a 2D domain $(0 \leq s \leq L)$, for a constant thermal coefficient. The Partial Different Equation (PDE) is

$$\frac{\partial u}{\partial j} = k\frac{\partial^2 u}{\partial s^2} + \frac{Q(j,s)}{c\rho}, \quad j > 0 \ and \ 0 < s < L \tag{11.1}$$

where the initial condition is

$$u(s,0) = f(s), \qquad 0 < s < L \tag{11.2}$$

If the temperature at the both ends of a rod is constant, then the boundary conditions are

$$\begin{aligned} u(0,j) &= T_1(j) \\ u(L,j) &= T_2(j) \qquad t > 0 \end{aligned} \tag{11.3}$$

The PDE in Eq. (11.1) is linear and homogenous for $Q(s, j) = 0$ and the boundary conditions $T_1(j) = T_2(j) = 0$.

$$\frac{\partial u}{\partial j} = k\frac{\partial^2 u}{\partial s^2} \quad 0 < s < L, j > 0 \tag{11.4}$$

$$u(0,j) = 0, \ u(L,j) = 0 \quad and \quad u(s,0) = f(s) \tag{11.5}$$

In the SV method, consider an example to determine a solution for the given product form

$$u(s,j) = x(s)y(j) \tag{11.6}$$

In the above equation $x(s)$ is the function of s and $y(j)$ is the function of j. From Eq. (11.6)

$$\frac{\partial u}{\partial j} = x(s)\frac{\partial y(j)}{\partial j} \tag{11.7}$$

$$\frac{\partial^2 u}{\partial s^2} = \frac{\partial^2 x(s)}{\partial s^2}y(j) \tag{11.8}$$

From Eqs. (11.4, 11.7, 11.8), the resultant heat equation in (11.1) is

$$x(s)\frac{\partial y(j)}{\partial j} = k\frac{\partial^2 x(s)}{\partial s^2}y(j) \tag{11.9}$$

Dividing the Eq. (11.9) with *k*x(s)y(j)* on both sides gives the

$$\frac{1}{ky(j)}\frac{\partial y(j)}{\partial j} = \frac{1}{x(s)}\frac{\partial^2 x(s)}{\partial s^2} \tag{11.10}$$

Equation (11.10) gives the solution for the heat Eq. (11.1), where the variables have separated, i.e. left-hand side only in terms of j variables and right-hand side in terms of s variables. According to the basic equations of the SV method, the PDEs of rectangular, cylindrical and spherical coordinate systems can be calculated by using the Laplace, heat and wave equations. The major steps involved in solving any PDE with the SV method are [12]

(1) Independent variables must be separated
(2) Solves the solution for separate equations of the independent variables, according to the boundary conditions
(3) Combines all solutions of separate equations to satisfy the other boundary conditions of particular PDEs.

In analytical techniques, the series expansion is mainly used for solving the PDEs, under the following conditions

1) The separation of independent variables is not possible
2) If variables are separated but boundary conditions are not satisfied with the solutions

Let us consider an example of Poisson's Equation in a Cube

$$\nabla^2 V = \frac{\partial^2 V}{\partial x^2} + \frac{\partial^2 V}{\partial y^2} + \frac{\partial^2 V}{\partial z^2} = -f(x,y,z) \tag{11.11}$$

The boundary conditions are

$$V(0,y,z) = V(a,y,z) = V(x,0,z) = 0 \tag{11.12}$$

$$V(x,b,z) = V(x,y,0) = V(x,y,c) = 0 \tag{11.13}$$

In the above equations, $f(x, y, z)$ is a given function, where the independent variables, x, y, and z are not separable. However, Eqs. (11.11–11.13) can be solved by using the Poisson's equation with series expansion [13].

11.2.2 Numerical Techniques

Numerical techniques are used to attain approximate solutions for given PDEs in an iterative manner by defining the set of discrete points (mesh elements) in computer-based simulations [2, 7–9]. The solution of PDE with discrete points is also called discretization. This discretization of a given PDE will play a crucial role in finding an accurate solution. The mesh size or discrete element size is nothing but the distance between two adjacent discrete elements. The problems related to EM analysis of Electrical Engineering involves the second-order differential or integral of both equations. Let us consider the general second-order PDE,

$$A\ddot{U}_{xx} + 2B\ddot{U}_{xy} + C\ddot{U}_{yy} + \ldots.\ (\textit{lower order terms}) = g \tag{11.14}$$

where the A, B, C are coefficients of the second-order PDE, which may depend on x and y. The $\ddot{U}_{xx} = \frac{\partial^2 U}{\partial x^2}$, $\ddot{U}_{xy} = \ddot{U}_{yx} = \frac{\partial^2 U}{\partial x \partial y}$ and $\ddot{U}_{yy} = \frac{\partial^2 U}{\partial y^2}$. In Eq. (11.14), for a particular region of the xy-plane, if the coefficients relation is $A^2 + B^2 + C^2 > 0$ then the PDE is second order in that particular region. In Eq. (11.14), $g = 0$ for homogeneous, $g \neq 0$ for non-homogeneous, and the equation is analogous to the conic- section equation (i.e. the intersection of the surface of a cone with a plane). To visualize the conic sections and quadratic forms as elliptic, hyperbolic and parabolic based on the discriminant $B^2 - 4AC$ [7–9]. Similar to this, for discriminant of the second-order PDE given in Eq. (11.14) is classified into various planes is $(2B)^2 - 4AC = 4(B^2 - AC)$, i.e.

1) *Elliptic partial differential equations (EPDEs):* $B^2 - AC < 0$
   The EPDE solutions are simple to analyse due to the lower magnitudes of coefficients and the defined boundaries (i.e. equation and solutions are defined). This EPDE is mainly used for steady-state solutions, i.e. for analysing the boundary value problems. The examples of the EPDE are

$$\text{Poisson's equation}: \nabla^2 A = -\mu J \tag{11.15}$$

$$\text{Laplace's equation}: \nabla^2 A = 0 \tag{11.16}$$

2) *Hyperbolic partial differential equations (HPDE):* $B^2 - AC > 0$
   The hyperbolic partial differential equation (HPDE) is used to solve propagation problems (discontinuities of derivatives in the initial data). An example of HPDE is a wave equation, i.e.

$$\nabla^2 V - \mu\varepsilon \frac{\partial^2 V}{\partial t^2} = -\frac{\rho}{\varepsilon} \tag{11.17}$$

3) *Parabolic partial differential equations (PPDE):* $B^2 - AC = 0$
   The parabolic partial differential equation (PPDE) which is parabolic at every point, can be realized as an analogous form of heat equation by changing the independent variables. The example for a PPDE is the heat and diffusion equation

$$\nabla^2 T - \frac{C}{\lambda}\frac{\partial T}{\partial t} = -\frac{Q}{\lambda} \tag{11.18}$$

$$\nabla^2 A - \mu\sigma \frac{\partial A}{\partial t} = -\mu J \tag{11.19}$$

The discretization approaches used in the numerical techniques are

- Finite difference method
- Finite element method
- Boundary element method
- Moment's method
- Monte Carlo method.

With the help of these discretization methods used in numerical techniques, the approximate solution for a given problem can be attained. However, to attain an accurate solution, the technique should be stable, convergent and consistent, [2] i.e.

Stable: The solution of a technique will be within boundaries or limits while solving the problem

Convergent: If the solution of a technique results in a real solution as the time-step and mesh size tends to zero

Consistent: If the error in a solution tends to zero as the time-step and mesh size tends to zero.

11.2.2.1 Finite Difference Method

In the 1920s, to solve the nonlinear hydrodynamic equations, A. Thom implemented the finite difference method (FDM) [7]. The FDM is mainly used for solving nonlinear PDEs and different field problems, where the solution is realized by using the finite differential equation of a given problem. The steps involved in solving a PDE by using the FDM are

(1) The solution region of a given problem has to divide into an appropriate number of grid nodes or vertices.
(2) Approximate the given differential equation by FDM that are related to the dependent variable at a point in the solution region to its values at neighbouring points.
(3) Solve all differential equations with the given initial, as well as boundary conditions.

Let us consider a function $f(x)$, as shown in Figure 11.3. The derivative of $f(x)$ can be analysed by the following equations

The forward-difference formula for the slope of arc PB

$$f'(x_0) = \frac{f(x_0 + \Delta x) - f(x_0)}{\Delta x} \tag{11.20}$$

The backward-difference formula for the slope of arc AP

$$f'(x_0) = \frac{f(x_0) - f(x_0 - \Delta x)}{\Delta x} \tag{11.21}$$

The central-difference formula for the slope of arc AB

$$f'(x_0) = \frac{f(x_0 + \Delta x) - f(x_0 - \Delta x)}{2\Delta x} \tag{11.22}$$

The second-order derivative of function $f(x)$ at P is

$$f''(x_0) = \frac{f'(x_0 + \Delta x/2) - f'(x_0 - \Delta x/2)}{\Delta x}$$
$$= \frac{1}{\Delta x}\left[\frac{f(x_0 + \Delta x) - f(x_0)}{\Delta x} - \frac{f(x_0) - f(x_0 - \Delta x)}{\Delta x}\right]$$
$$f''(x_0) = \frac{f(x_0 + \Delta x) - 2f(x_0) + f(x_0 - \Delta x)}{[\Delta x]^2} \tag{11.23}$$

With the help of FDM, the EPDE, HPDE, and PPDEs can be solved by implementing the finite difference equations for the discretized vertices or nodes of the given problem.

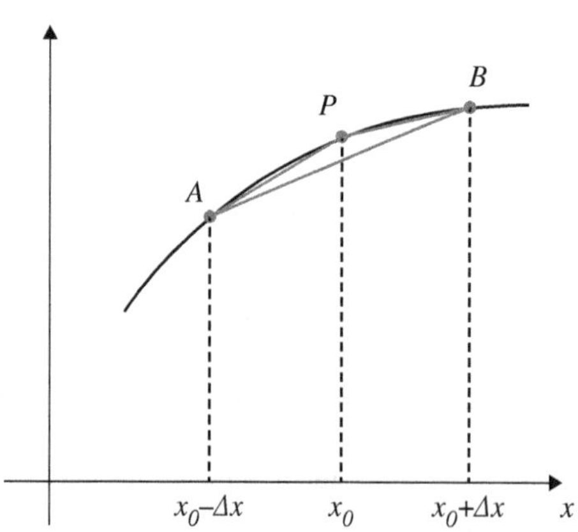

Figure 11.3 The derivative of $f(x)$ at point P using different FDM methods (like forward, backward and central difference).

11.2.2.2 Finite Element Method

The FEM is a significant and powerful numerical technique for solving the complex geometries and nonhomogeneous equations, especially EM problems [7–14]. The solution of any EM problem with FEM involves the following steps

- Discretize the solution region into small elements or meshing elements, called finite elements
- Analyse the governing equations of every finite element
- Integrate the solutions of all finite elements of the solution region
- Solve the complete system equation from the finite element equations and its solutions.

Typical elements for 1D, 2D and 3D problems are shown in Figure 11.4.

11.2.2.3 Solution of Laplace Equation Using the Finite Element Method

For a clear understanding of steps involved in the FEM analysis of the 2D solution region for the Laplace's equation, $\nabla^2 V = 0$ is explained in the following section.

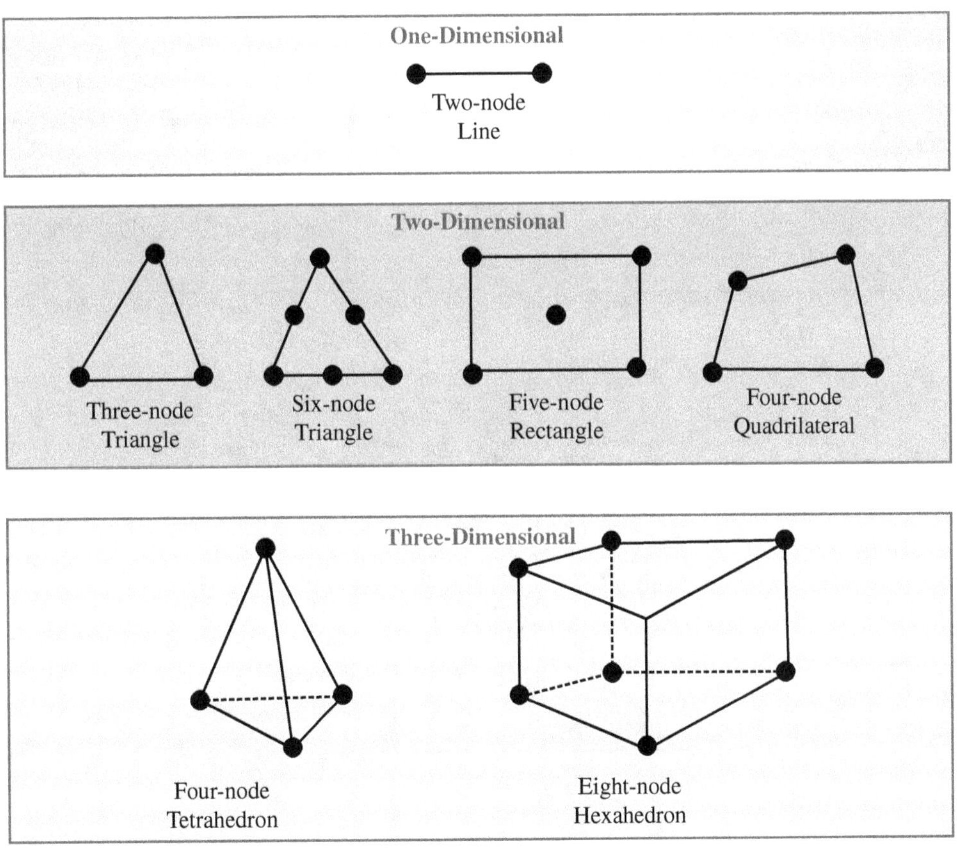

Figure 11.4 Different types of finite elements.

11.2.2.3.1 Discretization of the Finite Elements

Let us consider a 2D solution region for finding the potential distribution $V(x, y)$, presented in Figure 11.5. The finite elements of the solution region are presented in Figure 11.5, which consists of both triangle (three nodes, i.e. 1, 2, 3, 4, 5, and 7) as well as quadrilaterals (four nodes, i.e. 6, 8, and 9). In general, for practical problems, the solution region must be divided with equal nodes that are either triangle or quadrilateral for simplicity and fast computation. Because of this, the quadrilaterals in Figure 11.5b are rearranged as the triangles, i.e. the solution region consists of 12 triangles, as shown in Figure 11.5c.

The approximate solution of a given 2D region is nothing but the sum of potential distributions of all triangles, i.e.

$$V(x,y) \approx \sum_{e=1}^{N} V_e(x,y) \tag{11.24}$$

where N is the number of finite elements (triangles) in the solution region, V_e is the potential distribution with the element e. The generalized representation of potential distribution

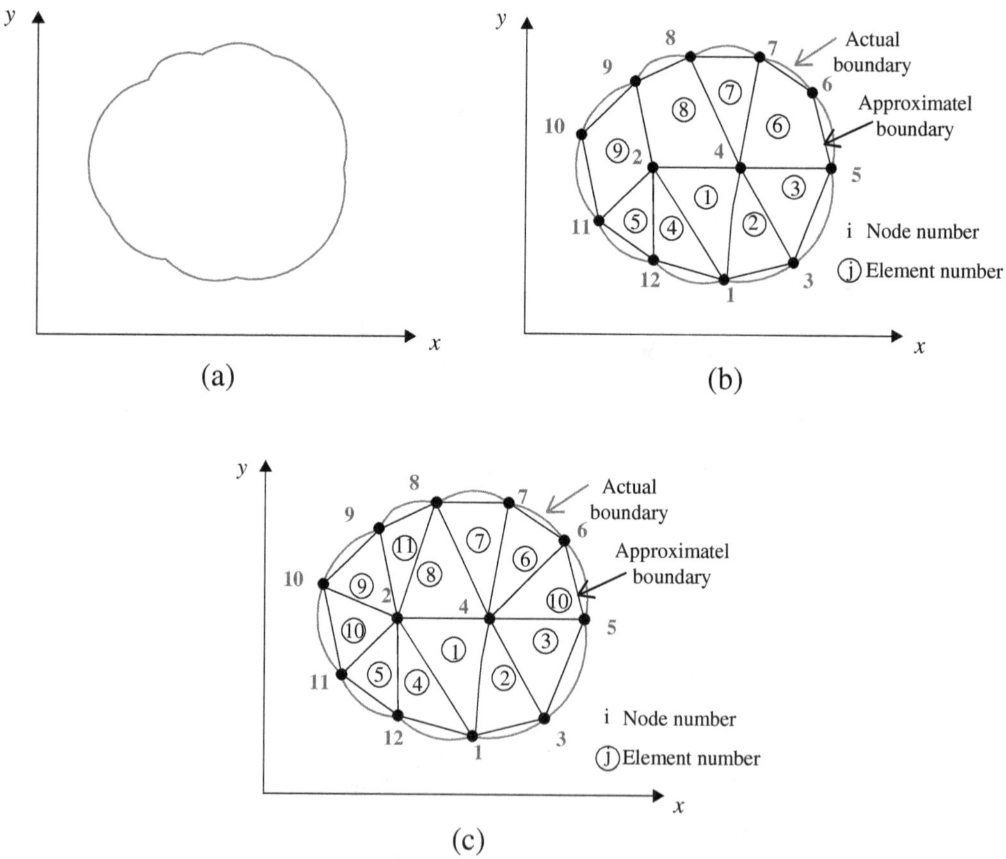

Figure 11.5 (a) Solution region, finite element discretization; (b) with triangular and rectangular elements, (c) with only triangular elements.

V_e for a triangular element is

$$V_e(x, y) = a + bx + cy \tag{11.25}$$

Similarly, the quadrilateral element is

$$V_e(x, y) = a + bx + cy + dxy \tag{11.26}$$

The constants *a, b, c, d* have to be calculated to attain potential distribution $V_e\ (x,y)$ of a finite element in the solution region. The electric field within the element is assumed to be similar to the linear variation of potential inside the triangular element given in Eq. (11.25), i.e.

$$E_e = -\nabla V_e = -ba_x - ca_y \tag{11.27}$$

11.2.2.3.2 Finite Element Equations

The triangular elements 1, 2, and 3 shown in the Figure 11.5c are considered and the respective vector potentials are V_{e1}, V_{e2}, and V_{e3}, the constants *a, b*, and *c* can be calculated by taking inverse transformations.

$$\begin{bmatrix} V_{e1} \\ V_{e2} \\ V_{e3} \end{bmatrix} = \begin{bmatrix} 1 & x_1 & y_1 \\ 1 & x_2 & y_2 \\ 1 & x_3 & y_3 \end{bmatrix} \begin{bmatrix} a \\ b \\ c \end{bmatrix} = [S] \begin{bmatrix} a \\ b \\ c \end{bmatrix} \tag{11.28}$$

$$\begin{bmatrix} a \\ b \\ c \end{bmatrix} = \begin{bmatrix} 1 & x_1 & y_1 \\ 1 & x_2 & y_2 \\ 1 & x_3 & y_3 \end{bmatrix}^{-1} \begin{bmatrix} V_{e1} \\ V_{e2} \\ V_{e3} \end{bmatrix} = [S]^{-1} \begin{bmatrix} V_{e1} \\ V_{e2} \\ V_{e3} \end{bmatrix} \tag{11.29}$$

The overall vector potential (V_e) of the triangular elements 1, 2, 3 derived from Eqs. (11.25, 11.29) is,

$$V_e = [1 \ \ x \ \ y] \frac{1}{2A} \begin{bmatrix} (x_2y_3 - x_3y_2) & (x_3y_1 - x_1y_3) & (x_1y_2 - x_2y_1) \\ (y_2 - y_3) & (y_3 - y_1) & (y_1 - y_2) \\ (x_3 - x_2) & (x_1 - x_3) & (x_2 - x_1) \end{bmatrix} \begin{bmatrix} V_{e1} \\ V_{e2} \\ V_{e3} \end{bmatrix} \tag{11.30}$$

$$V_e = \sum_{i=1}^{3} \alpha_i(x, y) V_{ei} \tag{11.31}$$

In the above equations, A is the area of triangular element or det of the matrix S, i.e.

$$\begin{aligned} A &= \frac{1}{2}[(x_1y_2 - x_2y_1) + (x_3y_1 - x_1y_3) + (x_2y_3 - x_3y_2)] \\ &= \frac{1}{2}[(x_2 - x_1)(y_3 - y_1) + (x_3 - x_1)(y_2 - y_1)] \end{aligned} \tag{11.32}$$

The α_i are linear interpolation functions of the triangular elements 1, 2, and 3, which are also named element shape functions, i.e.

$$\alpha_1 = \frac{1}{2A}[(x_2y_3 - x_3y_2) + x(y_2 - y_3) + y(x_3 - x_2)] \tag{11.33}$$

Similarly, the α_2, α_3 are calculated by using Eqs. (11.32, 11.33). Let us consider the energy per unit length is W_e for the element *e*. The W_e wrt the Laplace equation $\nabla^2 V = 0$ is

$$W_e = \frac{1}{2}\int \varepsilon |E_e|^2 dS = \frac{1}{2}\int \varepsilon |\nabla V_e|^2 dS \tag{11.34}$$

From Eqs. (11.31) and (11.34) [7]

$$\nabla V_e = \sum_{i=1}^{3} V_{ei} \nabla \alpha_i \tag{11.35}$$

$$W_e = \frac{1}{2} \sum_{i=1}^{3} \sum_{j=1}^{3} \varepsilon V_{ei} \left[\int \nabla \alpha_i \nabla \alpha_j dS \right] V_{ej} \tag{11.36}$$

In the above equation, the coefficient matrix is defined as

$$C_{ij}^{(e)} = \int \nabla \alpha_i \nabla \alpha_j dS \tag{11.37}$$

$$W_e = \frac{1}{2} \varepsilon [V_e]^T [C^{(e)}][V_e] \tag{11.38}$$

where the potential matrix and coefficient matrix

$$[V_e]^T = [V_{e1} \quad V_{e2} \quad V_{e3}] \ \ and \ \ [C^{(e)}] = \begin{bmatrix} C_{11}^{(e)} & C_{12}^{(e)} & C_{13}^{(e)} \\ C_{21}^{(e)} & C_{22}^{(e)} & C_{23}^{(e)} \\ C_{31}^{(e)} & C_{32}^{(e)} & C_{33}^{(e)} \end{bmatrix} \tag{11.39}$$

where the coefficient matrix elements can be calculated from Eqs. (11.32, 11.33, 11.37) as follows

$$\begin{aligned} C_{12}^{(e)} &= \int \nabla \alpha_1 . \nabla \alpha_2 ds \\ &= \frac{1}{4A^2} [(y_2 - y_3) + (y_3 - y_1) + (x_3 - x_2)(x_1 - x_3)] \int dS \\ &= \frac{1}{4A} [(y_2 - y_3) + (y_3 - y_1) + (x_3 - x_2)(x_1 - x_3)] \end{aligned} \tag{11.40}$$

Similarly, other elements of the coefficient matrix can be calculated. The non-diagonal elements will be, $C_{rs}^{(e)} = C_{sr}^{(e)}$, $r = 1, 2, 3$ and $s = 1, 2, 3$.

11.2.2.3.3 Finite Elements Assembling

In the previous section, a solution for the triangular elements 1, 2, 3 was calculated. Let us consider when n number of triangular elements (finite elements, 12 for the example shown in Figure 11.5c), the energy associated with all elements after assembling is

$$W = \sum_{e=1}^{N} W_e = \frac{1}{2} \varepsilon [V_e]^T [C^{(e)}][V_e] \tag{11.41}$$

Similar to Eq. (11.39), the potential and coefficient matrices consist of n number of nodes and N number of elements, the respective matrices size are (n*1) and (n*n).

Consider the three triangular elements 1, 2, and 3 in Figure 11.5, i.e. presented in Figure 11.6 for clear understanding. The resultant coefficient matrix with the 1, 2, and 3 triangular elements is

$$[C] = \begin{bmatrix} C_{11} & C_{12} & C_{13} & C_{14} & C_{15} \\ C_{21} & C_{22} & C_{23} & C_{24} & C_{25} \\ C_{31} & C_{32} & C_{33} & C_{34} & C_{35} \\ C_{41} & C_{42} & C_{43} & C_{44} & C_{45} \\ C_{51} & C_{52} & C_{53} & C_{54} & C_{55} \end{bmatrix} \tag{11.42}$$

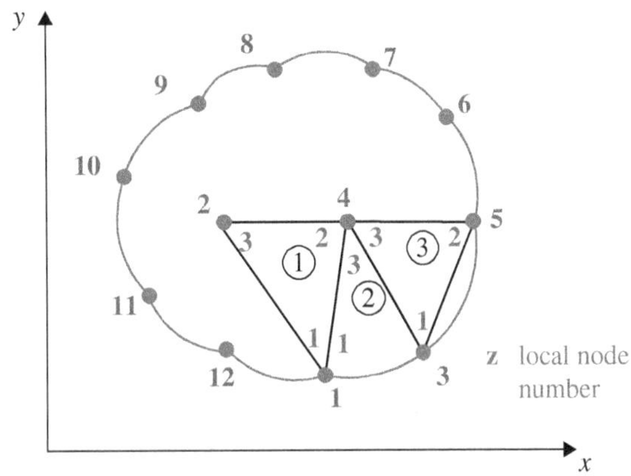

Figure 11.6 Assembling of three triangular elements, with the local node *(z)* numbering *(1, 2, and 3)* of each element.

The elements of a matrix [C] are calculated according to the combination from all finite elements containing nodes i and j. The node 1 is common for elements 1 and 2 in the Figure 11.6, i.e.

$$C_{11} = C_{11}^{(1)} + C_{11}^{(2)} \tag{11.43}$$

Node 2 only belongs to element 1, i.e.

$$C_{22} = C_{33}^{(1)} \tag{11.44}$$

Node 4 consists of the elements 1, 2, 3 i.e.

$$C_{44} = C_{22}^{(1)} + C_{33}^{(2)} + C_{33}^{(3)} \tag{11.45}$$

Nodes 1 and 4 simultaneously belong to elements 1 and 2, i.e.

$$C_{14} = C_{41} = C_{12}^{(1)} + C_{13}^{(2)} \tag{11.46}$$

Nodes 2 and 3 do not consist of a direct link or common element, so

$$C_{23} = C_{32} = 0 \tag{11.47}$$

Similarly, the elements of the matrix [C] can be calculated. If the [C] matrix consists of n number of nodes and N elements, the procedure will be same as above. The basic properties of the matrix [C] are

- The matrix is symmetric, $C_{rs}^{(e)} = C_{sr}^{(e)}$
- If there is no coupling or direct link between the i, j (local nodes) the $Cij = 0$.

11.2.2.3.4 Resulting Equations of Solution Region

In general, the energy of solution region should be minimum to satisfy Laplace's equation. Because of this, the derivative of energy wrt the potentials at each node is zero, i.e.

$$\frac{\partial W}{\partial V_1} = \frac{\partial W}{\partial V_2} = \frac{\partial W}{\partial V_3} = \dots\dots\dots = \frac{\partial W}{\partial V_n} = 0$$

$$\text{or} \qquad \frac{\partial W}{\partial V_k} = 0, \quad k = 1, 2, 3, \dots. n \tag{11.48}$$

From Eqs. (11.41, 11.42)

$$0 = \frac{\partial W}{\partial V_1} = 2V_1C_{11} + V_2C_{12} + V_3C_{13} + V_4C_{14} + V_5C_{15}$$
$$+ V_2C_{21} + V_3C_{31} + V_4C_{41} + V_5C_{51}$$

or

$$0 = V_1C_{11} + V_2C_{12} + V_3C_{13} + V_4C_{14} + V_5C_{15} \tag{11.49}$$

$$0 = \frac{\partial W}{\partial V_1} = \sum_{i=1}^{n} V_i C_{ik} \tag{11.50}$$

In the above equations, n is the number of nodes associated with each finite element (triangular element). The equations of each element with associated nodes are analysed with Eq. (11.50), where the $k = 1, 2, 3 \ldots .n$, with this, the resultant solution of a system can be derived, i.e.

$$[V]^T = [V_1 \quad V_2 \quad V_3 \ldots \ldots \ldots V_n] \tag{11.51}$$

The equation of a potential to all elements can be calculated with the iteration or band matrix methods [7].

11.3 Formulation of 2-Dimensional and 3-Dimensional Analysis

Maxwell 2D and Maxwell 3D are FEA-based software methods/interactives for solving the engineering problems (static, frequency-domain and time-varying EM and electric fields), i.e. the EM, eddy current and transient problems related to the electrical engineering. Here, the Maxwell 2D and 3D will solve the EM problems by using the Maxwell equations over a solution region or space, with appropriate design materials, initial and boundary conditions.

A) ***2D Simulation***

The Maxwell 2D is used to solve the EM field problems for the considered motor model with suitable boundary conditions, appropriate materials and source conditions applying Maxwell's equations over a finite region of space. In Maxwell 2D, two geometry models such as Cartesian and Axisymmetric models are available. Also, six different solvers available such as for *Electric fields*: Electrostatic, AC Conduction, DC Conduction, for *Magnetic Fields:* Magnetostatic, Eddy Current, Transient Magnetic.

These types of simulation are used when the geometry of the machine or structure to be analysed is assumed to be large in 2D compared to 3D. Also, if the solution parameters are interested in 2D only. In this method, the geometry of a problem is discretized into small 2D elements such as triangles and quads. The set of algebraic equations for solving the problem are obtained using the discretized elements. The required field in each element is approximated with a second-order quadratic polynomial.

B) ***3D Simulation***

The Maxwell 3D is a high performance interactive finite element software to solve electric, eddy current, magnetostatic and transient problems. Similar to 2D, there are also six different solvers available for solving any problem. These simulations consider complete machines/systems for the analysis and these are more realistic compared to 2D as they capture the gradients along three directions. In these simulations, the exact boundary conditions can be captured and asymmetries can be captured as it considers geometries in all three directions. The computational time for these simulations is more and even storage space is high.

In this method, the geometry of the problem is discretized into small solid elements (3D) such as tetrahedral and hexagonal. The set of algebraic equations for solving the problem are obtained using the discretized elements.

It must be noted that while performing either 2D or 3D simulations, a mesh convergence study has to be performed for the better results. Here, convergence means the mesh refinement has to perform to such an extent that the simulation results should not be altered for any change in mesh density. The basic comparison between 2D and 3D FEM simulations are presented in Table 11.1.

11.3.1 Maxwell Equations

In the seventeenth and beginning of the eighteenth century, the scientist's Ampere, Faraday, Gauss, Lenz, Coulomb, Lorentz, Laplace, and others have reported the EM phenomenon for electrical as well as other engineering problems [2, 8]. In the 1860s, the great scientist Maxwell established a set of PDEs that are a combination of Gauss, Ampere, Lorentz and Faraday's laws to solve the all complex EM problems (such as complicated geometry and non-linearity of magnetic materials and other fields). These equations mainly provide a mathematical model for electrical, radio, optical, structural and engineering technologies. Based on the frequency system the Maxwell equations are of two types, i.e. high frequency (radio or microwaves) and low frequency, as shown in Figure 11.7.

11.3.1.1 Gauss Law

Gauss's law gives the relationship between the static electric field and its cause of source (electric charges). The static electric field points move towards the negative charges and

Table 11.1 The comparison between 2D and 3D simulations.

Parameters	2D simulations	3D simulations
Geometry	Considers only two major directions of the geometry	Considers all three directions of the geometry
Meshing/modelling	Simple, as it considers only planar dimensions of the machine	Complex, as it considers all dimensions of the machine
Computational cost	Now, as the time required to solve the 2D models is less due to simplification of the geometry.	High, as the time required to solve the 3D models is more due to consideration of complete geometry.
Accuracy	Relatively less accurate with respect to realistic condition	More accurate as it resembles the realistic condition
Element type	Triangular and Quadrilateral	Tetrahedral and Hexagonal

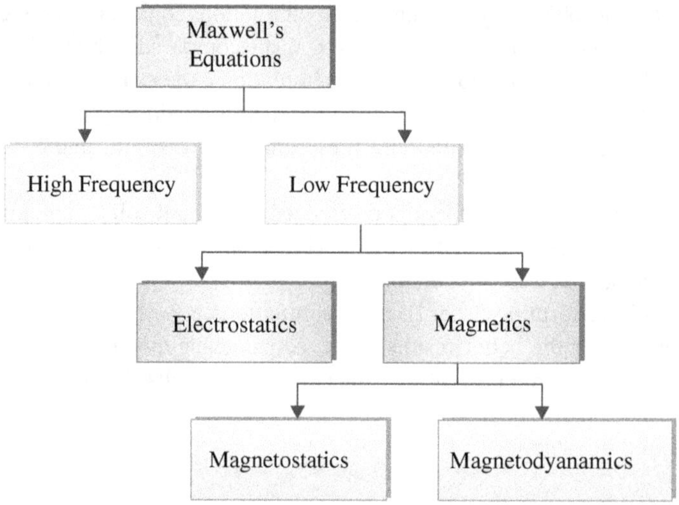

Figure 11.7 Classification of Maxwell equations.

away from the positive charges, the resultant electric field flow in a closed region will be directly proportional to the charge enclosed by the surface. The net charge for a volume V, which is enclosed by a surface S, related to the net electric flux by that surface is given by Eq. (11.52).

$$\oint\!\!\oint_S \varepsilon.E.dS = \iiint_V \rho.dV \Rightarrow D = \varepsilon.E$$

$$\oint\!\!\oint_S D.dS = \iiint_V \rho.dV \tag{11.52}$$

where electrical displacement flux density *(D)*, charge density *(ρ)*. The Gauss integral law for a problem with different materials is

$$n.(\varepsilon_a.E_a - \varepsilon_b.E_a) = \sigma_s \tag{11.53}$$

where surface charge density is σ_s and n represents normal component of E.

11.3.1.2 Gauss Law of Magnetism

Gauss's law for magnetism defines that, the net magnetic field through a closed surface (or Gaussian surface) is zero. Moreover, this law states that monopoles of a magnetic field (single pole) will not exist.

$$\oint\!\!\oint_S B.dS = 0 \tag{11.54}$$

Gauss's integral law of magnetic flux for different material is

$$n.(B_a - B_b) = 0 \tag{11.55}$$

11.3.1.3 Ampere's Integral Law

The Maxwell integral form of Amperes law defines the induced magnetic field around any closed loop, depending on the electrical current (Ampere's law) and displacement current (Maxwell's equation) flowing in that enclosed surface.

$$\oint_C H.dl = \iint_S J.dS + \frac{d}{dt}\iint_S D.dS \tag{11.56}$$

where H is the magnetic field intensity, J is the current density. Ampere's law for discontinuous magnetic field intensities is

$$n.(H_a - H_b) = K \tag{11.57}$$

where K and n are the surface current density and normal component of magnetic field intensity respectively.

11.3.1.4 Faraday's Integral Law

Maxwell Faraday's law is similar to Faraday's law of induction principle. This law states that the required work per unit charge to move the charge present in an enclosed surface is directly related to the rate of change in magnetic flux through that surface.

$$\oint_C E.dl = -\frac{d}{dt}\oiint_S B.dS \tag{11.58}$$

The Faraday's law of continuity is

$$n.(E_a - E_b) = 0 \tag{11.59}$$

11.3.1.5 Differential Form of Maxwell Equations

The summary of Maxwell equations in the form of integrals has presented in Table 11.2. The differential form of the Maxwell equations can be derived by using the divergence and stokes theorems.

Table 11.2 Integral form of Maxwell equations for FEM analysis.

Name	Integral law
Gauss's law	$\oiint_S \varepsilon.E.dS = \iiint_V \rho.dV \Rightarrow D = \varepsilon.E$
	$\oiint_S D.dS = \iiint_V \rho.dV$
Gauss's law of magnetism	$\oiint_S B.dS = 0$
Ampere's law	$\oint_C H.dl = \iint_S J.dS + \frac{d}{dt}\iint_S D.dS$
Faraday's law	$\oint_C E.dl = -\frac{d}{dt}\oiint_S B.dS$

A) **Gauss's law**

$$\oiint_S \varepsilon.E.dS = \iiint_V \rho.dV$$

$$\oiint_S \varepsilon.E.dS = \iiint_V \nabla \bullet (\varepsilon.E).dV$$

$$\iiint_V \rho.dV = \iiint_V \nabla \bullet (\varepsilon.E).dV$$

$$\nabla \bullet (\varepsilon.E) = \rho \tag{11.60}$$

B) **Gauss's Law of Magnetism**

$$\oiint_S B.dS = 0$$

$$\oiint_S B.dS = \iiint_V \nabla \bullet (B).dV$$

$$\iiint_V \nabla \bullet (B).dV = 0$$

$$\nabla \bullet (B) = 0 \tag{11.61}$$

C) **Ampere's Law**

$$\oint_C H.dl = \iint_S J.dS + \frac{d}{dt}\iint_S D.dS$$

$$\oint_C H.dl = \iint_S \nabla \times H.dS$$

$$\iint_S \nabla \times H.dS = \iint_S J.dS + \frac{d}{dt}\iint_S D.dS$$

$$\nabla \times H = J + \frac{dD}{dt} \tag{11.62}$$

D) **Faraday's Law**

$$\oint_C E.dl = -\frac{d}{dt}\iint_S B.dS$$

$$\oint_C E.dl = \iint_S \nabla \times E.dS$$

$$\iint_S \nabla \times E.dS = -\frac{d}{dt}\iint_S B.dS$$

$$\nabla \times E = -\frac{dB}{dt} \tag{11.63}$$

The summary of differential forms of Maxwell equations presented in Table 11.3.

Table 11.3 Differential form of Maxwell equations for FEM analysis.

Name	Integral law
Gauss's law	$\nabla \bullet (\varepsilon . E) = \rho$
Gauss's law of magnetism	$\nabla \bullet (B) = 0$
Ampere's law	$\nabla \times H = J + \frac{dD}{dt}$
Faraday's law	$\nabla \times E = -\frac{dB}{dt}$

11.3.2 FEM Adaptive Meshing

The flow chart of FEM adaptive meshing is presented in Figure 11.8. As discussed in Section 11.2.2.2, to attain the algebraic equations of solution region, the complete geometry of problem has to be divided into small finite elements or meshing elements (discretization, i.e. triangles in Maxwell 2D and tetrahedral in 3D). The models can be meshed manually or automatically the mesher. Finally, the assembly of all finite elements is called a Mesh of the solution region shown in Figure 11.9, where meshing over the surface area of three-phase induction motor (IM) is given.

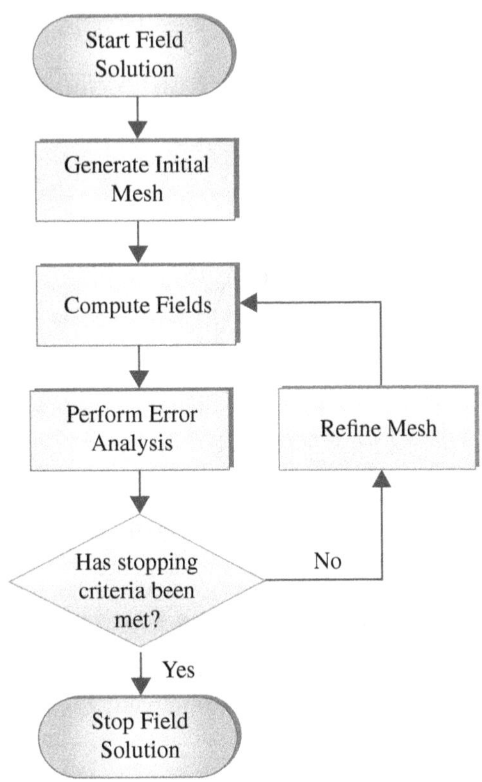

Figure 11.8 FEM Adaptive meshing procedure [4–6].

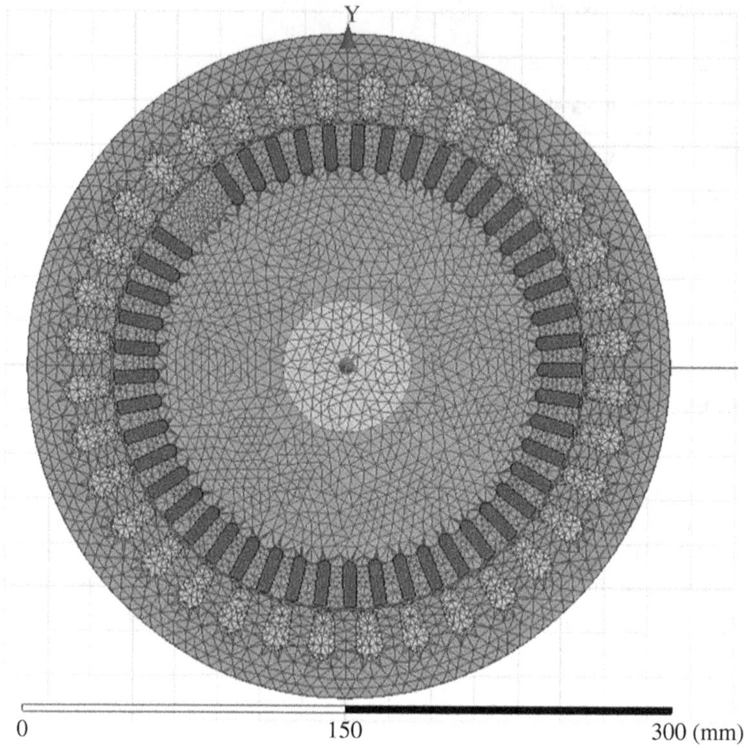

Figure 11.9 Mesh plot of the electrical machine.

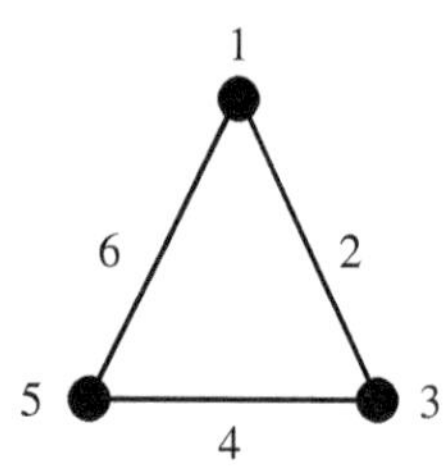

Figure 11.10 Field quantities of the 2D triangular element.

11.3.3 FEM Variation Principle

In Ansys Maxwell, the field of each finite element will approximate by using the second-order quadratic equation, i.e.

$$A_z(x, y) = a_0 + a_1x + a_2y + a_3x_2 + a_4xy + a_5y_2 \tag{11.64}$$

Field quantities of each finite element (triangular element) are calculated at six points, i.e. three corners and three midpoints of each finite element, as shown in Figure 11.10. Field quantities inside each finite are determined wrt the second-order quadratic interpolation scheme.

The FEM Variation Principle wrt the Poisson's equation is [5, 6]

$$\nabla^2 A = -\mu J$$

$$F(A) = \frac{1}{2}\int \left(\frac{\nabla A \bullet \nabla A}{\mu} + A \bullet J\right) dV \tag{11.65}$$

This function can be minimized by regulating the value of A for each finite element and respective nodes.

11.4 Analysis and Implementation of FEM Machine Models

In general, analysing electrical machines or other electrical engineering studies can be done by using FEM simulation software like Ansys Maxwell, J-Mag and CAD. The general steps involved in the analysis of electrical machines are

- The FEM model physical design: In this step geometry of the machine, material characteristics, design data details such as slots, conductors, current density and flux density and other details are considered.
- Finite element or meshing of the designed FEM model: In this step, the machine surface area is divided into small triangular or finite elements (called meshing) according to the geometry of the machine. For example, the smaller size of triangular elements is considered in the air gap of a machine as compared to the stator core to attain accurate results. In FEM simulation software, meshing can be done either manually or by auto-generation.
- Analysis of FEM model: In this step, validation check of the designed model is performed to evaluate the accuracy as well design of the FEM model and meshing.
- Optimization and interpretation of FEM results: In this step, the results are analysed and optimization techniques are implemented to attain the efficient design of the machine. Moreover, flux lines, magnetic flux density, torque, speed and loss analysis, etc., can be analysed.

In this section, the basic implementation of electrical machine FEM models with the Ansys Maxwell 2D is explained. For example, the basic configuration of electric drive implemented in Ansys Maxwell 2D is presented in Figure 11.11, where the three-phase induction motor fed with three-phase voltages (V_{as}, V_{bs}, V_{cs}, i.e. either three-phase inverter output or sinusoidal voltages). The analysis of machine in Ansys Maxwell involves three steps

1) RMxprt design to implement the Maxwell model of machine
2) Power Converter design in Simplorer
3) Integration of power converter with the Maxwell model for testing drive.

11.4.1 RMxprt Design to Implement a Maxwell Model of Machine

The Maxwell model of any machine can be done with two methods, i.e. RMxprt and Manual design. In the RMxprt method, the Maxwell model of the machine is implemented by modifying the geometrical characteristics of in-built example models or available machine project modules of various machines. In manual design of Maxwell, geometry

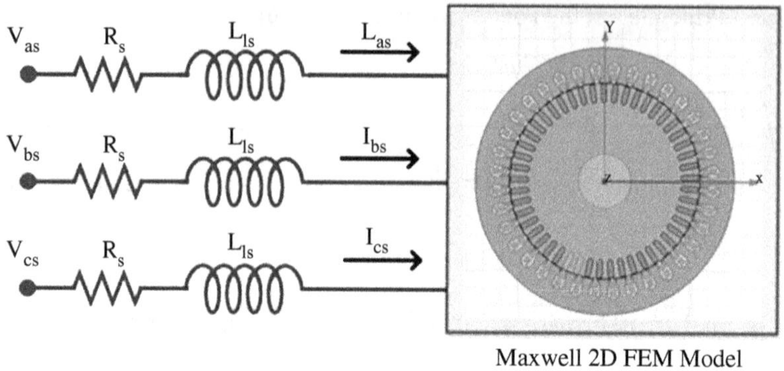

Figure 11.11 Basic configuration of the electric drive implemented in Ansys Maxwell 2D.

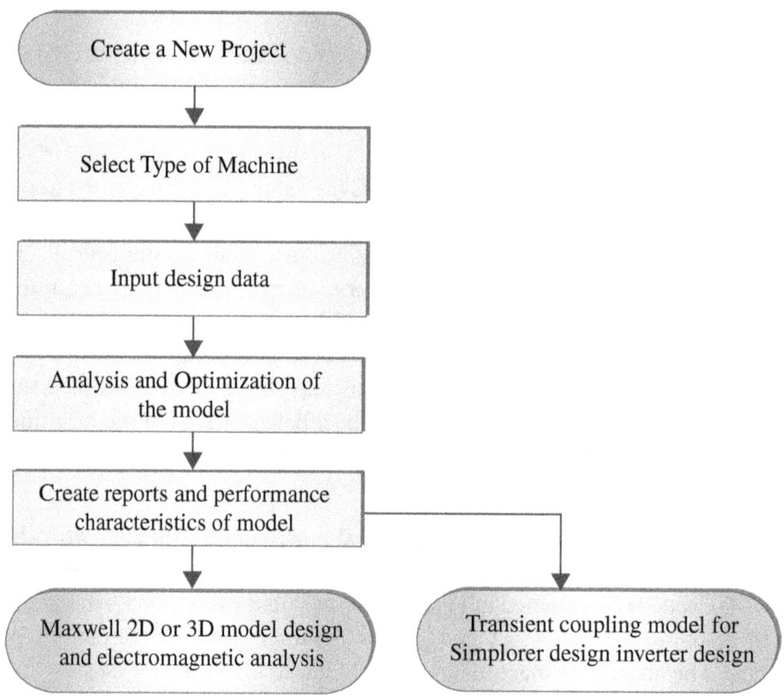

Figure 11.12 The basic flow chart of the Ansys RMxprt to design the Maxwell 2D model.

of the machine (the size, volume of slots, windings, rotor, stator etc.,) has to be drawn similar to the engineering drawing. The manual design method of a Maxwell model has limitations in terms of complexity, time-consumption, lesser the probability of attaining accurate and optimized model etc., and for 3D analysis, complexity will increase further.

The basic flow chart of the Ansys RMxprt to design the Maxwell model are shown in Figure 11.12. In RMxprt, the Maxwell machine models can be built automatically by modifying the geometry, boundary conditions, material properties or by using existing electrical machine design projects. The available models of different electrical machines

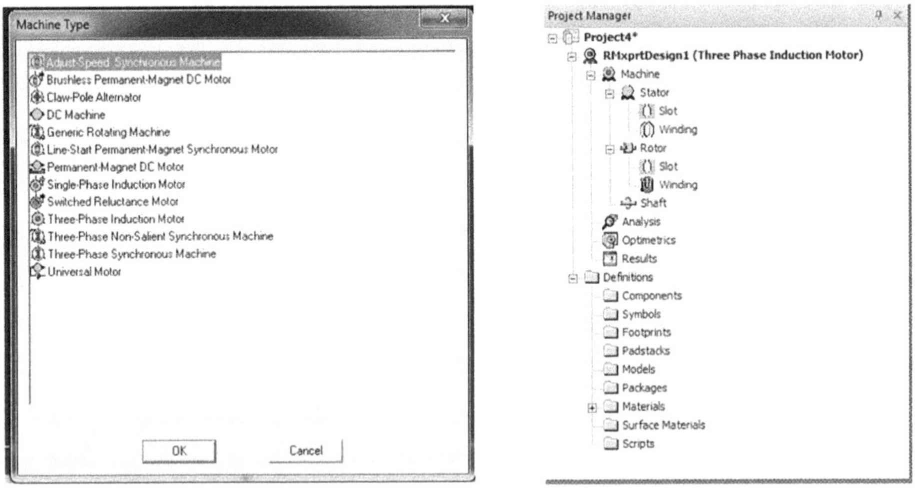

(a) (b)

Figure 11.13 (a) Available machine models in RMxprt of Ansys, (b) Project window of the RMxprt of IM.

Name	Value	Unit	Evaluated Value
Machine Type	Three Phase Inducti...		
Number of Poles	4		
Stray Loss Factor	0.005		0.005
Frictional Loss	37	W	37W
Windage Loss	0	W	0W
Reference Speed	1360	rpm	

Figure 11.14 Machine properties window in RMxprt.

(like induction, synchronous, reluctance, DC type, and other machines) in RMxprt are shown in Figure 11.13a. The RMxprt model setup includes symmetrical design parameters, excitations, non-linearities, eddy effects, coupling circuits and suitable materials for EM transient analysis. By considering all these geometrical parameters, the RMxprt generates a reduced order of machine model, which is co-simulated in the Simplorer environment where machine control, as well as performance, can be analysed with various inverter configurations.

The Project window of RMxprt in Ansys software after selecting the standard three-phase machine is shown in Figure 11.13b. From this figure, it can be observed that, the geometry of machine is predefined, i.e. stator (slots and winding), rotor (slots and winding) and shaft. In the machine properties window shown in Figure 11.14, the number of poles, reference and losses data must be defined. The stator core is made of cast iron or silicon steel with the laminations. The geometrical data wrt air-gap diameter, length of core, number of slots, material type, slot type, etc., as defined in the stator properties window, shown in Figure 11.15. Here, the stator core material can be selected based on the BH curves as well as nonlinearity behaviour. Total steel of stator area to the lamination varnish area is defined

Name	Value	Unit	Evaluated Value
Outer Diameter	210	mm	210mm
Inner Diameter	148	mm	148mm
Length	250	mm	250mm
Stacking Factor	0.92		
Steel Type	M19_24G		
Number of Slots	36		
Slot Type	2		
Lamination Sectors	0		
Press Board Thickness	0	mm	
Skew Width	0		0

Figure 11.15 Stator properties window in RMxprt.

Name	Value	Unit	Evaluated Value
Auto Design	☐		
Parallel Tooth	☐		
Hs0	0.8	mm	0.8mm
Hs1	1.05	mm	1.05mm
Hs2	15	mm	15mm
Bs0	2.8	mm	2.8mm
Bs1	5.5	mm	5.5mm
Bs2	8	mm	8mm

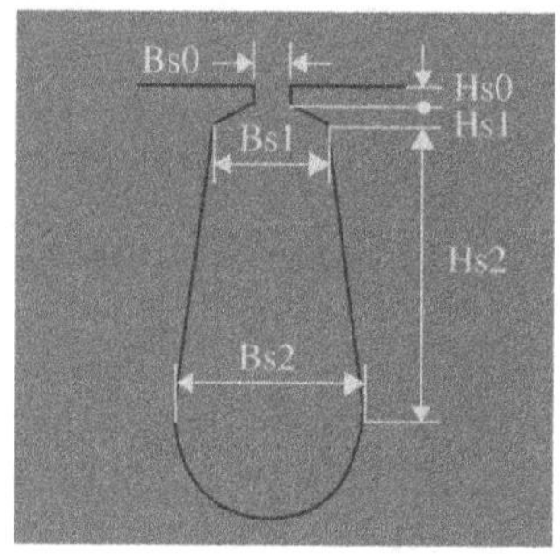

Figure 11.16 Stator slot design window in RMxprt.

by the Stacking factor. The user-defined material can be designed in the Ansys software by using the available datasheets and coefficients of the core loss component. In RMxprt, there are five different types of stator slots are available; one is shown in Figure 11.16, where the slot width, depth, height, width of slot opening and other parameters can be defined according to the user requirement. There is also the auto design option of slot design in the stator slot window, as shown in Figure 11.16. For all types of stator slots are available in the RMxprt, design procedure will be the same.

The winding design for different types of machines is given in Chapters 4 and 10. According to winding design, the data presented in the winding properties window (Figure 11.17) has to be modified, i.e. the number of layers, coil pitch, conductors per slot, wire size, type of winding (whole coiled or half coiled, full pitch or short pitch etc.,) and other details. The winding design of three-phase four-pole with 36 slots, full-pitch double-layer winding as shown in Figure 11.18. In the RMxprt, other winding types such as single layer, fractional slot, short pitch etc., can be implemented by a manual winding editor.

Similar to stator design, the rotor design wrt physical geometry in terms of diameter, core length, type of material, types of slots and other details are given according to the user requirement in the rotor properties window, as shown in Figure 11.19. The rotor slot properties window and associated data is given in Figure 11.20. Since the rotor of squirrel-cage induction motor is a cage type, the material of shorted bars, as well as end ring, width and length of each bar, has to be defined, as shown in Figure 11.21. Finally, the solution setup of RMxprt model is shown in Figure 11.22, where the output power, type of load (constant

Figure 11.17 Stator windings details of RMxprt machine model.

Name	Value	Unit	Evaluated Value
Winding Layers	2		
Winding Type	Whole-Coiled		
Parallel Branches	2		
Conductors per Slot	30		30
Coil Pitch	9		
Number of Strands	2		2
Wire Wrap	0.09	mm	
Wire Size	Diameter: 0.93mm		

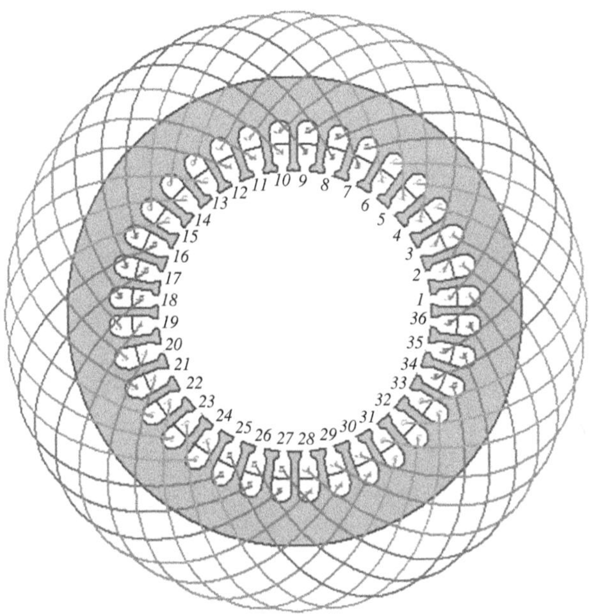

Figure 11.18 Stator winding connections in RMxprt (double layer full pitch winding of three-phase four-pole machine).

Figure 11.19 Rotor properties window in RMxprt.

Name	Value	Unit	Evaluated Value
Stacking Factor	0.92		
Number of Slots	49		
Slot Type	2		
Outer Diameter	147.3	mm	147.3mm
Inner Diameter	48	mm	48mm
Length	250	mm	250mm
Steel Type	M19_24G		
Skew Width	0		0
Cast Rotor	☐		
Half Slot	☐		
Double Cage	☐		

Name	Value	Unit	Evaluated Value
Hs0	0.5	mm	0.5mm
Hs01	0	mm	0mm
Hs1	1.2	mm	1.2mm
Hs2	14	mm	14mm
Bs0	1	mm	1mm
Bs1	5.2	mm	5.2mm
Bs2	3.5	mm	3.5mm

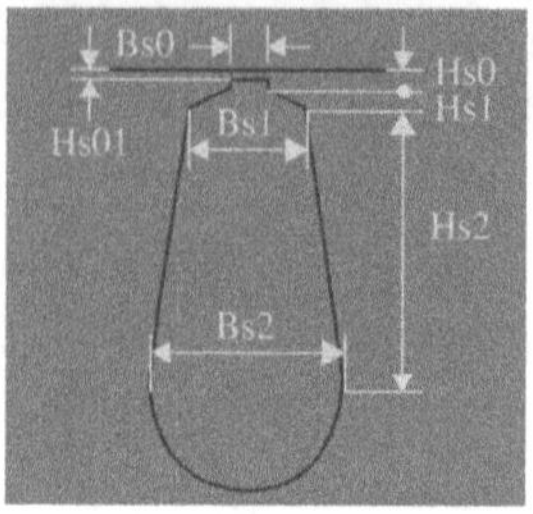

Figure 11.20 Rotor slot design window in RMxprt.

Name	Value	Unit	Evaluated Value
Bar Conductor Type	cast_aluminum_75C		
End Length	0	mm	0mm
End Ring Width	20	mm	20mm
End Ring Height	10	mm	10mm
End Ring Conductor T...	cast_aluminum_75C		

Figure 11.21 End-ring details of squirrel cage rotor in RMxprt.

Name	Value	Unit	Evaluated Value
Name	Setup1		
Enabled	☑		
Operation Type	Motor		
Load Type	Const Power		
Rated Output ...	3720	W	3720W
Rated Voltage	415	V	415V
Rated Speed	1450	rpm	1450rpm
Operating Tem...	75	cel	75cel

Figure 11.22 Solution setup of the RMxprt machine model.

power or constant torque etc.,), speed, voltage and machine mode (motor or generator) and other has to be defined. The built-in load types of Ansys RMxprt are

- Constant Speed: Speed of the machine is constant
- Constant Power: Output power of machine is constant
- Constant Torque: Torque remains constant regardless of speed
- Linear Torque: Torque increases linearly with speed
- Fan Load: The load varies non-linearly with speed.

Once the solution setup is ready, machine models are ready to analyse the performance (at full load, no-load, rated conditions, steady-state etc.,), data sheets (this is a summary of all design details of the machine, such as machine type, stator data, rotor data, rated condition, no-load conditions, full-load data, etc.,) and results or curves (like torque vs. speed, current vs. speed, efficiency vs. speed, output power vs. speed, etc.). Details of the data sheets and curves of the RMxprt are presented in Section 11.5.

After analysing the results as well as the performance of the machine in RMxprt, the final machine model has to be converted to either Maxwell 2D or 3D models. The basic view of the machine after converting to the Maxwell 2D model is shown in Figure 11.23a. In general,

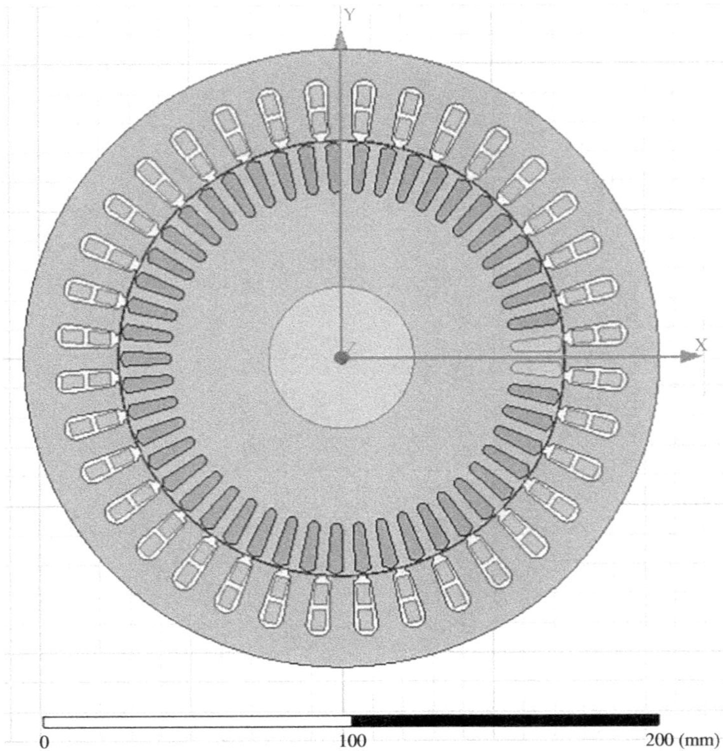

(a)

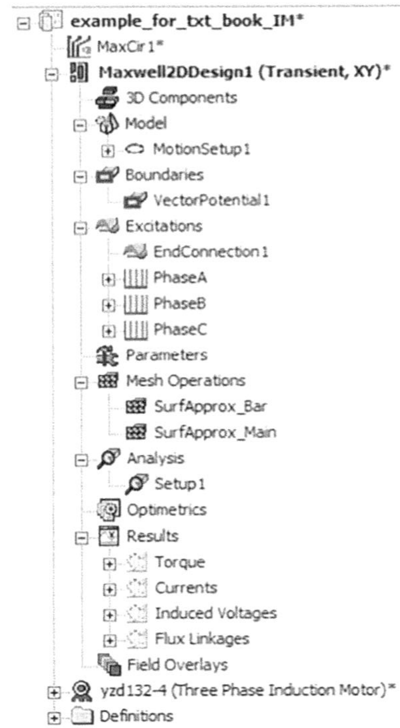

(b)

Figure 11.23 Maxwell 2D model, (a) basic view of the IM after converting RMxprt model to Maxwell 2D (b) Project window of Maxwell 2D model.

once the design of a practical machine is done, rewinding is possible by changing wire size, material, number conductors per slots, but the changing of machine core, a number of slots etc., are not possible. Similarly, once the Maxwell model is created using the RMxprt, physical design details, like number of slots, slot area, stator data, rotor data and other design parameters cannot be changed. The number of conductors per coil, the material of conductors, stator or rotor material, load torque, applied voltage and other parameters can be changed accordingly. The project window of the Maxwell model is shown in Figure 11.23b. In this figure, it can be seen that the Maxwell consists of motion setup, boundaries, excitations, mesh operations, solution setup, results, etc.

The motion step of the Maxwell model is explained in the model properties window, as shown in Figure 11.24, where the type of motion such as rotational and translational can be given. Moreover, the mechanical rotational data will be defined, for example, load torque, initial speed, a moment of inertia, damping ratio, etc., clearly shown in Figure 11.24. The boundaries, as well as initial conditions of the Maxwell model with respect to the differential equations, are defined in the boundaries window.

The powering of the three-phase machine Maxwell model can be done in three different types, i.e. voltage, current and external. The voltage excitation of a machine for phase-A is shown in Figure 11.25a, where the initial current of the machine, leakage inductance per phase, resistance per phase, phase voltage and other parameters are generated from the RMxprt model. The phase voltage can be sinusoidal voltage (like 'V_m* *sin(2*pi*f*time-2*pi/3)*') or pulse with modulation (PWM) inverter voltage (like '*pwl_periodic(ds1,time)*'). Here the PWM voltage excitation is given by using a lookup table, i.e. the inverter output voltage is imported from Simplorer or MATLAB. Similar to voltage excitation, the current excitation can be selected in the type of excitation shown in Figure 11.25a. The external type of excitation is used for integrating the power electronic converter with Maxwell machine model. In this type, the Maxwell model is exported into Simplorer and the same is powered by three-phase inverter designed in Simplorer. The detailed discussion on the external type of excitation is presented in Section 11.4.3.

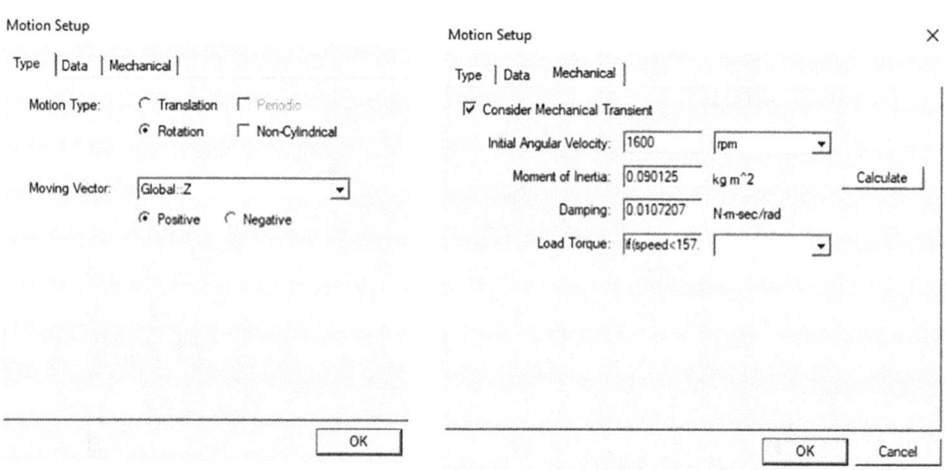

Figure 11.24 Motion setup window in the Motion properties of Maxwell 2D model.

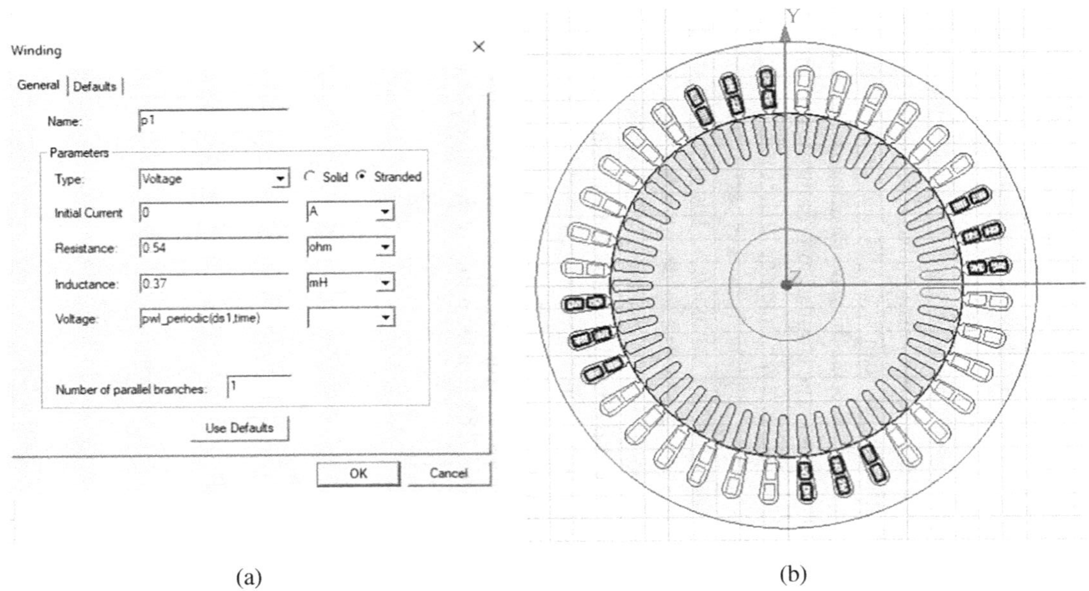

Figure 11.25 Winding details of Maxwell model (a) Phase-A Properties window (b) coils associated with Phase-A.

The coils associated with the Phase-A are shown in Figure 11.25b, where the double-layer full-pitch winding is considered and the winding connections is same as RMxprt model. The number of conductors per coil, coil pitch, the material of conductor, number of coil sides or layers per slot and others can be changed in the Maxwell model itself. However, for changing the number of slots, slot size, etc., cannot be done in the Maxwell model. The end connection of the squirrel cage rotor window is shown in Figure 11.26, where the resistance, as well as inductance between the adjacent bars, is defined.

The meshing of machine surface is done in the mesh operation window, where it can be defined that various surfaces with different size of mesh or finite elements, i.e. surface of rotor bar, surface of a stator core, rotor core, shaft, etc., The meshing can be done either manually or automatically by the software. The meshing window for a surface of rotor bar is shown in Figure 11.27 where the size of each mesh has been defined. In the same way, the meshing of other surfaces of the machine can be done. Once the meshing and excitations are done the solution setup for analysing the machine performance is presented in Figure 11.28, where the time step and stop time of the simulation is defined. The time step defines the accuracy of Maxwell model results, i.e. lesser time step in the order of microseconds (<50 μs) is preferable. After solution setup, a validation check is performed to analyse the symmetry in geometrical, excitations and other parameters. The results of the Maxwell model, like currents, voltages, induced electromotive forces (emfs), torque, speed and flux linkages, inductance variation, power and other results wrt time can be analysed in the results window. In addition to this, the flux-line distribution (A), flux density (B), magnetic-field intensity (H), and current density (J), and other magnetic characteristics can be analysed using the field overlays window, as shown in Figure 11.29. The meshing plots, magnetomotive force (mmf) calculations, user-defined plots can be done in the same field overlays window by defining the equations. The detailed results of the three-phase machine Maxwell model are presented in Section 11.5.

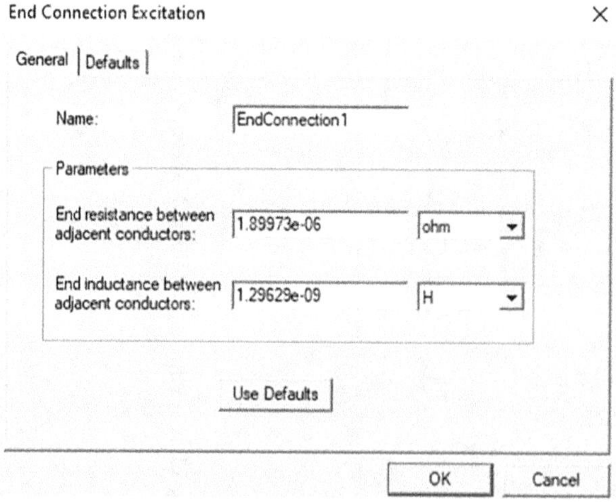

Figure 11.26 Rotor-end connection properties window of Maxwell model.

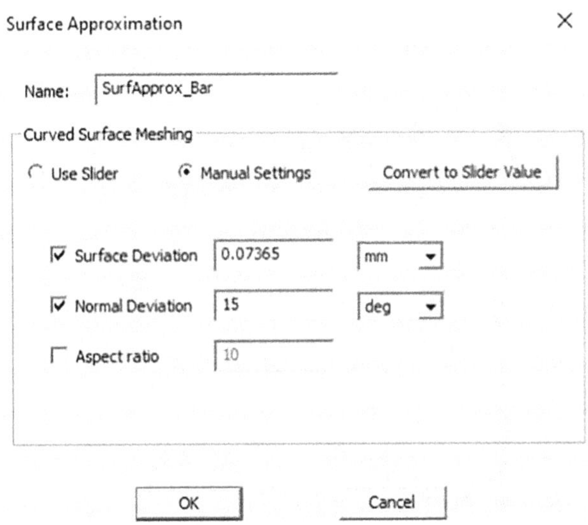

Figure 11.27 Meshing details of bar surface of Maxwell model.

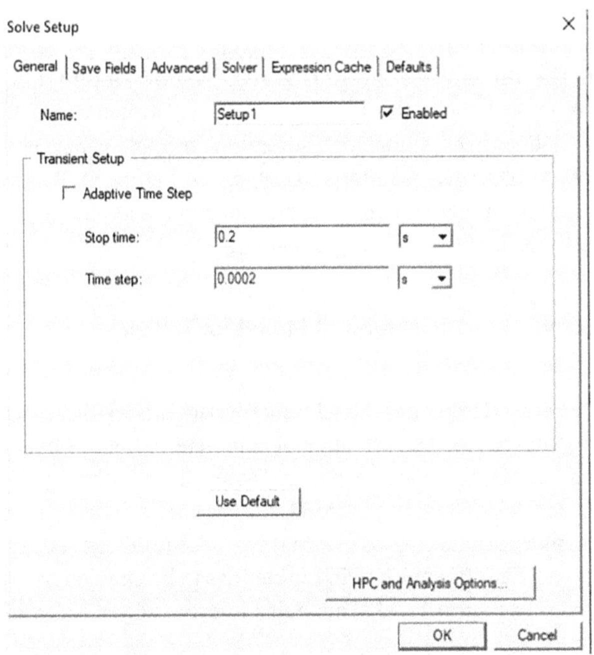

Figure 11.28 Solution setup of the Maxwell model.

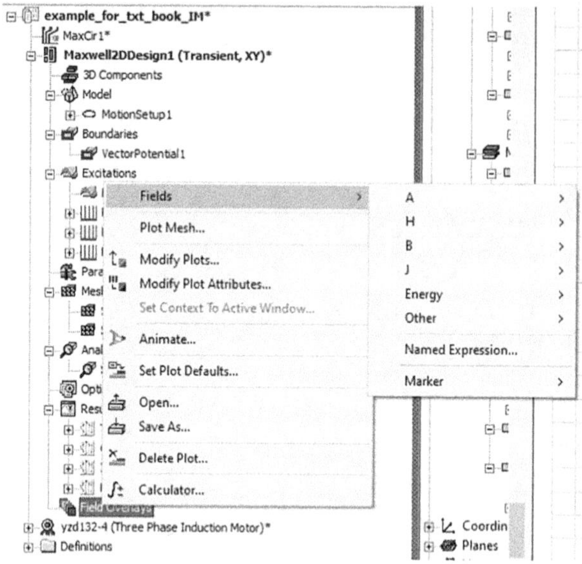

Figure 11.29 Field overlays details and quantities of Maxwell model.

11.4.2 Power Converter Design in Simplorer

Simplorer is a power electronic converter simulation tool used in Ansys Maxwell, which is similar to the MATLAB, Plexim and other power electronic simulation softwares. In the Simplorer environment, a three-phase inverter is implemented for powering the three-phase Maxwell model. The three-phase inverter project window of Simplorer environment is shown in Figure 11.30, where the inverter model is implemented by selecting the appropriate switches, DC sources, sinusoidal generators, triangle generators, comparator, etc., from the component library of Simplorer, as shown in Figure 11.31. The inverter is implemented with RL load to validate the inverter control and PWM strategy, as shown in Figure 11.32, where the sinusoidal PWM is implemented.

11.4.3 Integration of Power Converter with a Maxwell Model for Testing Drive

The Maxwell 2D machine model (discussed in Section 11.4.1) and inverter model (discussed in Section 11.4.2) are integrated to analyse the performance of complete induction motor drive with desired speed and torques. As discussed earlier in Section 11.4.1, the type of excitation in the Maxwell machine model (shown in Figure 11.25a) has to be defined as an external type. After defining all the phase excitations as external type, in the design settings of Maxwell 2D model, transient to transient link with twin builder has to be enabled in the advanced coupling section, as shown in Figure 11.33. This helps in exporting of Maxwell 2D model to the Simplorer environment.

The steps for importing the Maxwell model into the Simplorer environment are *'Twin builder => sub circuit => Maxwell component => add transient co-simulation'*. After these steps, the 'Maxwell transient-transient' coupling window will appear, as shown in

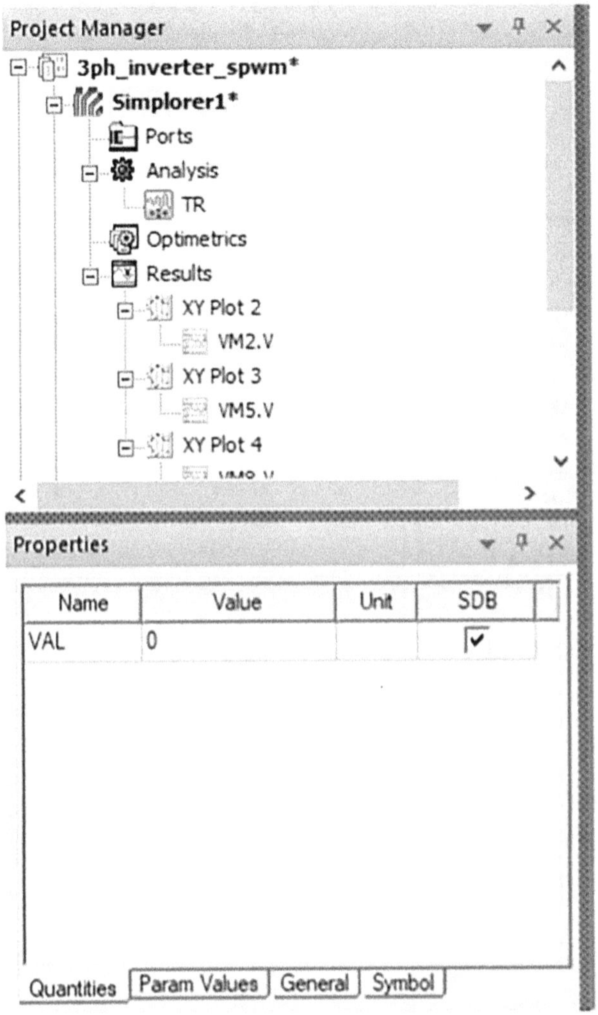

Figure 11.30 Project properties window in Simplorer environment.

Figure 11.34. In this window, details of Maxwell model in which transient link is enabled must be imported from the library to analyse the performance. The imported Maxwell model is shown in Figure 11.35, where one side of the machine terminals have to be connected to the inverter and the other sides have to be connected in either star or delta. The machine is operating as a motor, i.e. motion setup IN will be connected to mechanical ground and the motion setup OUT has to be connected to load by using the force and mass blocks, as shown in Figure 11.35. The complete setup of an induction motor drive including the inverter model and Maxwell modes is shown in Figure 11.36. By analysing this complete setup, the Maxwell model, as well as Simplorer model, will run simultaneously to visualize the system performance.

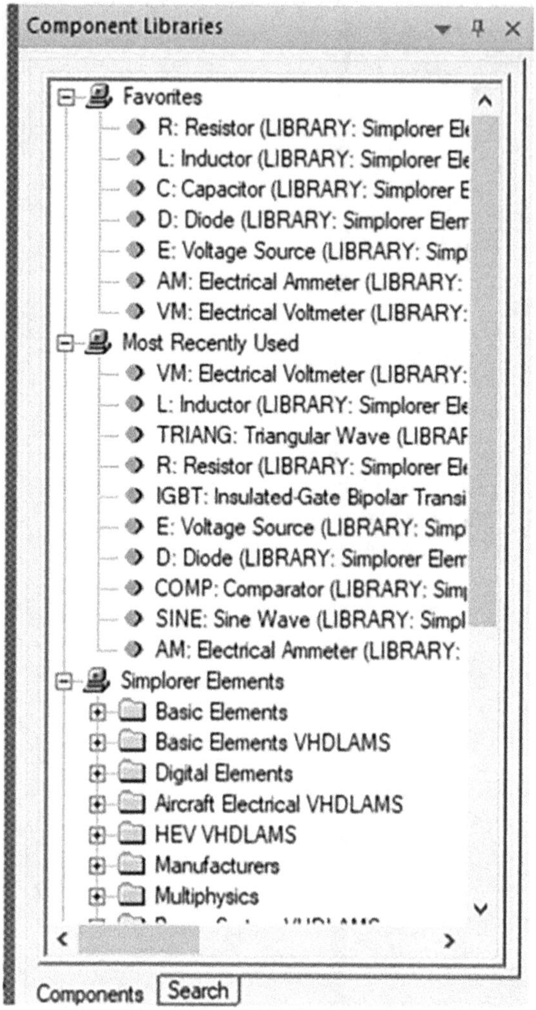

Figure 11.31 Component libraries window in Simplorer environment.

11.5 Example Model of Three-Phase IM in Ansys Maxwell 2D

The Maxwell equations involved in FEM analysis and implementation of the induction motor drive in Ansys Maxwell are discussed in the previous sections. With reference to this analysis, the three-phase induction motor drive is implemented in Ansys Maxwell 2D and results are described in this section. The RMxprt model of three-phase four-pole machine is implemented with the specifications given in Table 11.4. In this table, machine design details power, current, voltage, detailed stator as well as rotor data, machine characteristics at rated-load condition and other parameters are presented in detail. Here double-layer winding is used, since it is a full-pitch winding, the coil pitch is maintained constant, i.e. nine stator slots, as shown in Figure 11.18. From this figure, it can be observed that the top coil placed in the first slot is connected to the bottom coil placed in the 10th slot for

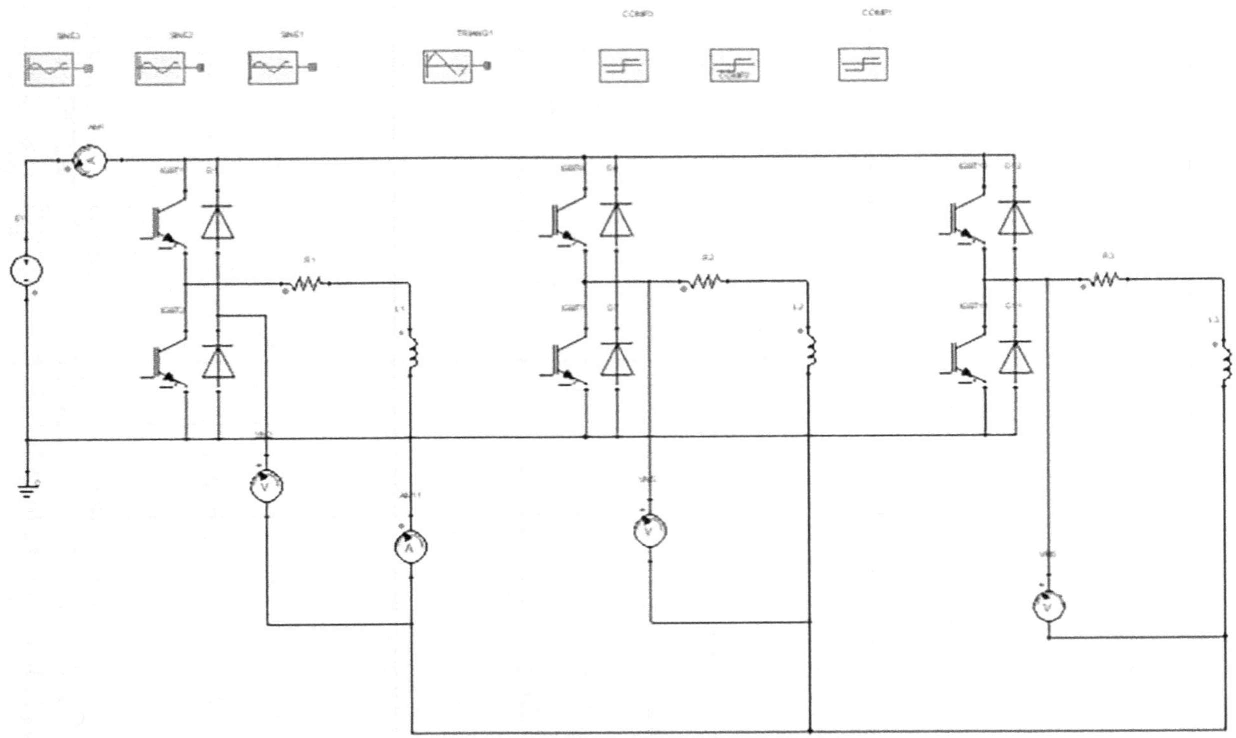

Figure 11.32 Three-phase inverter model designed Simplorer for powering the Maxwell model.

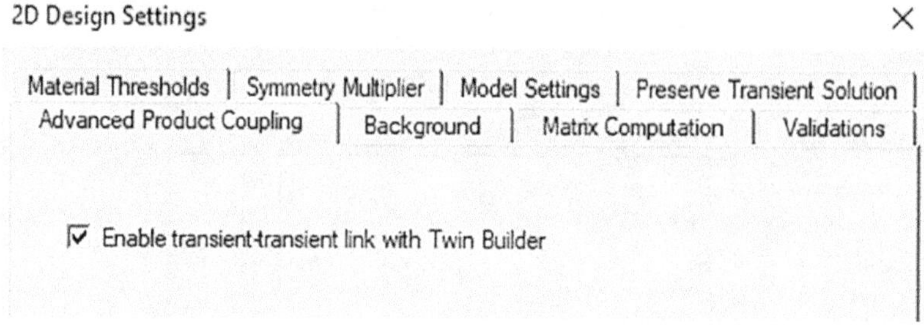

Figure 11.33 Maxwell design settings for enabling the transient-transient link.

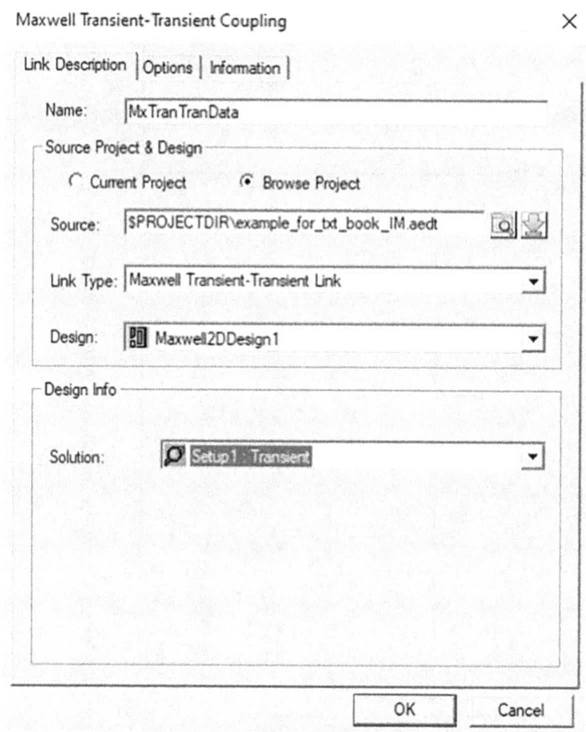

Figure 11.34 Maxwell transient-transient coupling window in Simplorer environment.

minimizing leakage flux and for getting symmetrical air-gap flux distribution. Similarly, all coils of the Phase-A are connected in series, as shown in Figure 11.18. The design details of the RMxprt model are extracted from the datasheets of a project. The performance curves of RMxprt model are shown in Figures 11.37–11.41, where the torque vs. speed, torque vs. slip, efficiency vs. speed, phase current vs. speed, slip vs. output power, torque vs. output

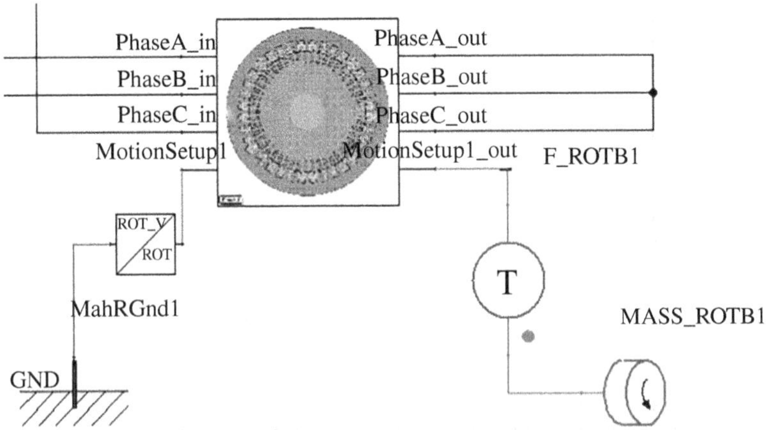

Figure 11.35 Maxwell Transient to Transient coupling model of three-phase induction machine.

power, efficiency vs. output power, power factor vs. output power, torque vs. speed in the field weakening region; voltage vs. speed in the field weakening region and other performance curves can be analysed in RMxprt. After obtaining the appropriate results according to the user specifications, the RMxprt model of a three-phase IM is converted into the FEM Maxwell model to validate the transient performance of machine with power electronic inverter configurations, i.e. shown in Figure 11.42a.

As discussed earlier, the accuracy of the FEM simulation mainly depends on a method of meshing the triangular element dimensions. The meshing near the air gap is refined by reducing the size of triangular elements. Figure 11.42b shows the triangular meshing plot of the three-phase IM for accurate flux distribution. The flux-line distribution of the three-phase IM is shown in Figure 11.43a, where it is observed that the number of poles is formed in accordance with phase-angle excitation and symmetrical-phase distribution. The magnetic-flux density plot of a three-phase four-pole machine presented in Figure 11.43b, where the maximum flux density (B) is 1.21Tesla which is within the limits of the steel material BH curve. This waveform is evidence that the machine is not saturated. From Figure 11.43, it is observed that because of the refined meshing, the flux-line distribution, as well as flux density, are smoothed.

The torque and speed response at level 2, three-phase inverted fed three-phase IM drive is shown in Figure 11.44a and b, respectively. From this figure, it can be observed that the machine is working at 1450 rpm with a load torque of 15 Nm. By using FEM analysis, torque ripple of the machine can be analysed. The phase voltages, as well as currents of the three-phase IM, are shown in Figure 11.45, where the three-phase voltages and currents are in 120° phase difference due to the symmetrical operation of machine. In addition to the speed, torque, in the Maxwell 2D, eddy currents, hysteresis losses and inductance variation and efficiency maps can be analysed in detail [14].

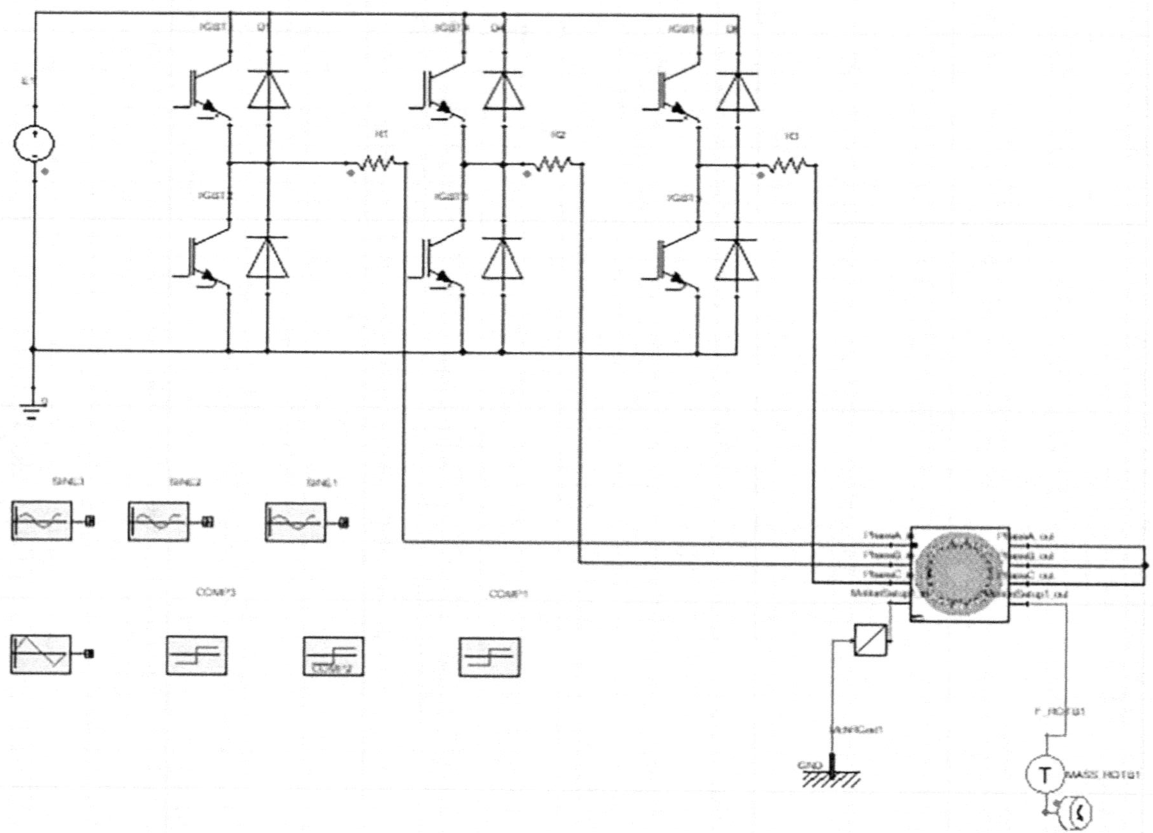

Figure 11.36 A complete system of an electrical drive with Maxwell machine as well as Simplorer inverter models.

Table 11.4 Ansys FEM model design details of the three-phase induction motor drive.

General Data	Stator Data
Given Output Power (kW): 3.72	Number of Stator Slots: 36
Rated Voltage (V): 415	Outer Diameter of Stator (mm): 210
Winding Connection: Wye	Inner Diameter of Stator (mm): 148
Number of Poles: 4	Type of Stator Slot: 2
Given Speed (rpm): 1450	Length of Stator Core (mm): 250
Frequency (Hz): 50	Stacking Factor of Stator Core: 0.92
Operation Mode: Motor	Type of Steel: M19_24G
Type of Load: Constant Power	Number of Parallel Branches: 1
Rotor Data	Coil Pitch: 9
Number of Rotor Slots: 49	Number of Conductors per Slot: 30
Air Gap (mm): 0.35	Number of Wires per Conductor: 2
Inner Diameter of Rotor (mm): 48	Wire Diameter (mm): 1.024
Type of Rotor Slot: 2	Layer Insulation (mm): 0.3
Length of Rotor (mm): 250	Slot Area (mm^2): 132.98
Stacking Factor of Rotor Core: 0.92	Net Slot Area (mm^2): 108.437
Type of Steel: M19_24G	Slot Fill Factor (%): 68.6666
Height of End Ring (mm): 10	Limited Slot Fill Factor (%): 75
Width of End Ring (mm): 20	Wire Resistivity (ohm.mm^2/m): 0.0217
Rated-Load Operation	
Stator Resistance R1 (ohm): 1.87819	Copper Loss of Stator Winding (W): 214.138
Stator Resistance at 20C (ohm): 1.54496	Copper Loss of Rotor Winding (W): 164.479
Stator Leakage Reactance X1 (ohm): 3.838	Iron-Core Loss (W): 57.9035
Slot Leakage Reactance Xs1 (ohm): 1.72229	Frictional and Windage Loss (W): 39.0982
End Leakage Reactance Xe1 (ohm): 0.393955	Stray Loss (W): 18.6
Harmonic Leakage Reactance Xd1 (ohm): 1.72176	Total Loss (W): 494.218
Rotor Resistance R2 (ohm): 1.56911	Input Power (kW): 4.21452
Rotor Leakage Reactance X2 (ohm): 2.6286	Output Power (kW): 3.7203
Resistance Corresponding to Iron-Core Loss Rfe (ohm): 2549.21	Mechanical Shaft Torque (N.m): 24.7204
Magnetizing Reactance Xm (ohm): 207.407	Efficiency (%): 88.2734
Stator Phase Current (A): 6.16476	Power Factor: 0.946897
Current Corresponding to Iron-Core Loss (A): 0.0870139	Rated Slip: 0.0419174
Magnetizing Current (A): 1.06948	Rated Shaft Speed (rpm): 1437.12
Rotor Phase Current (A): 5.91109	

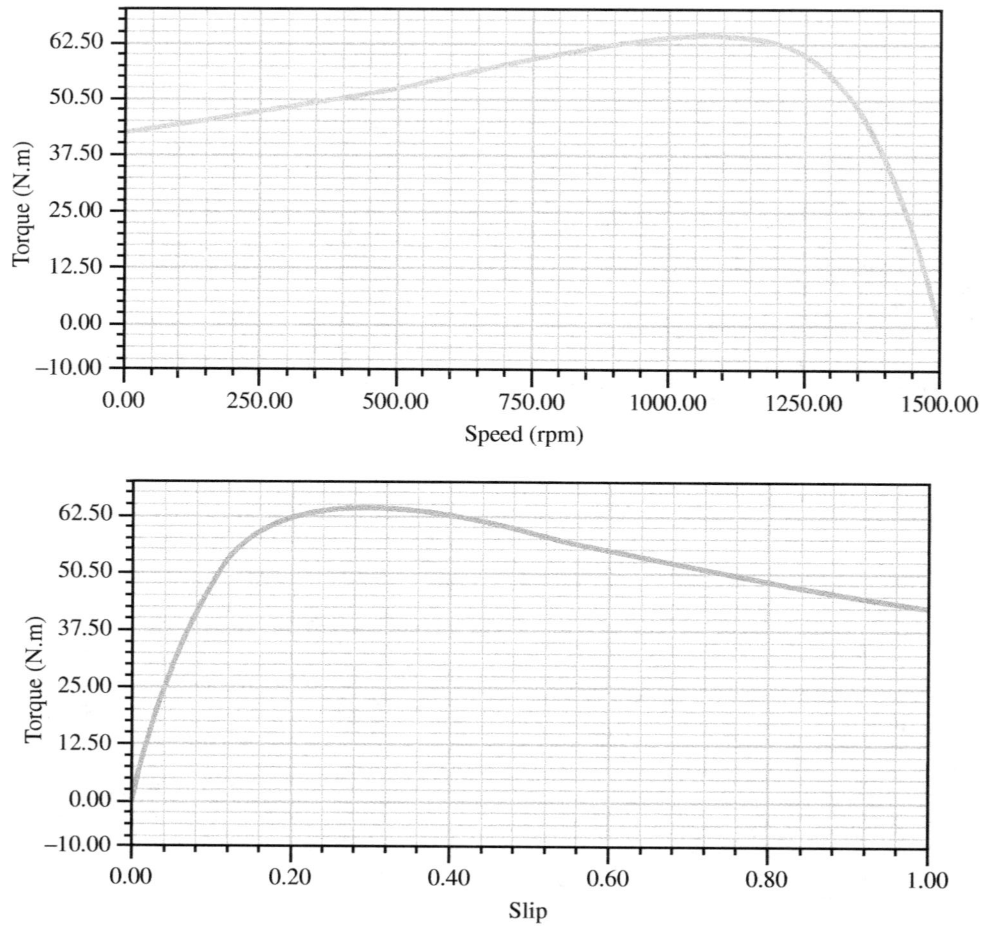

Figure 11.37 Torque vs. speed and Torque vs. slip characteristics of the three-phase induction motor in RMxprt.

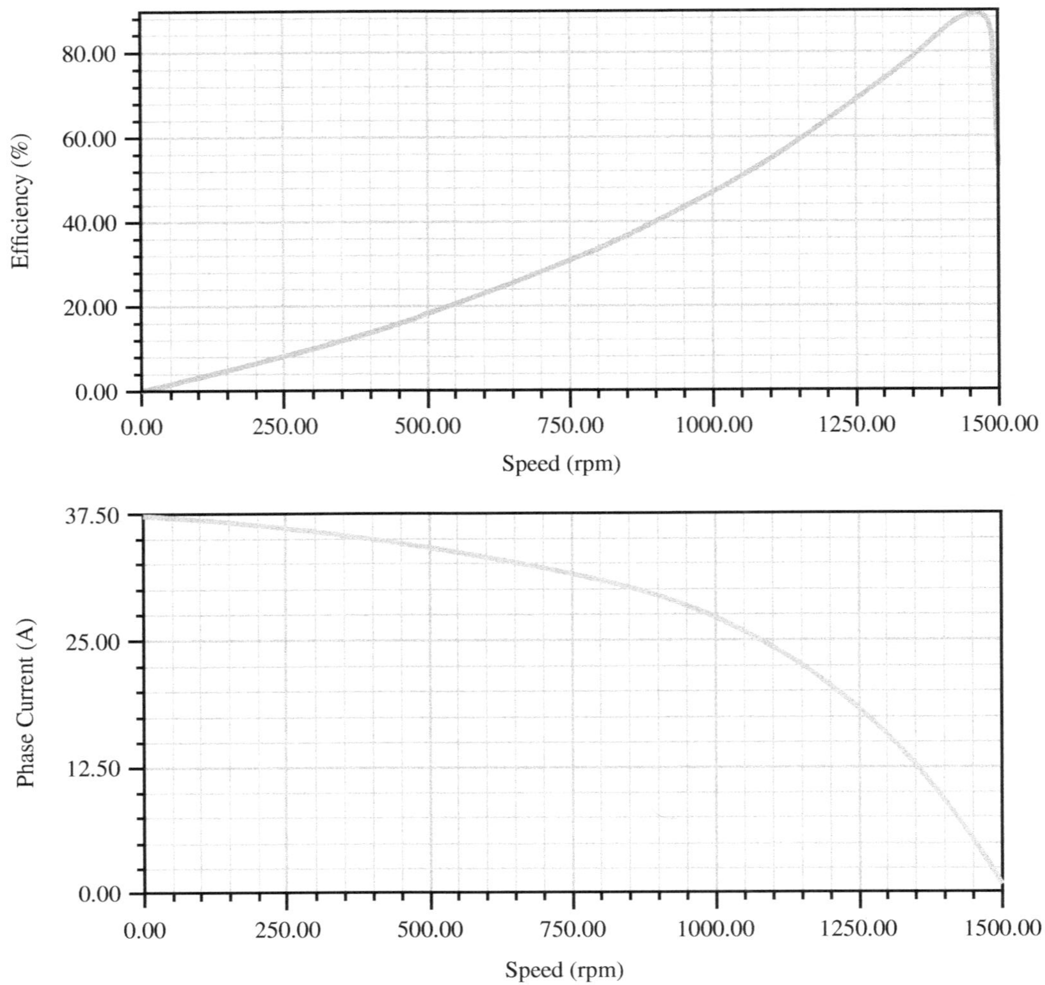

Figure 11.38 Efficiency vs. speed and Phase current vs. speed characteristics of the three-phase induction motor in RMxprt.

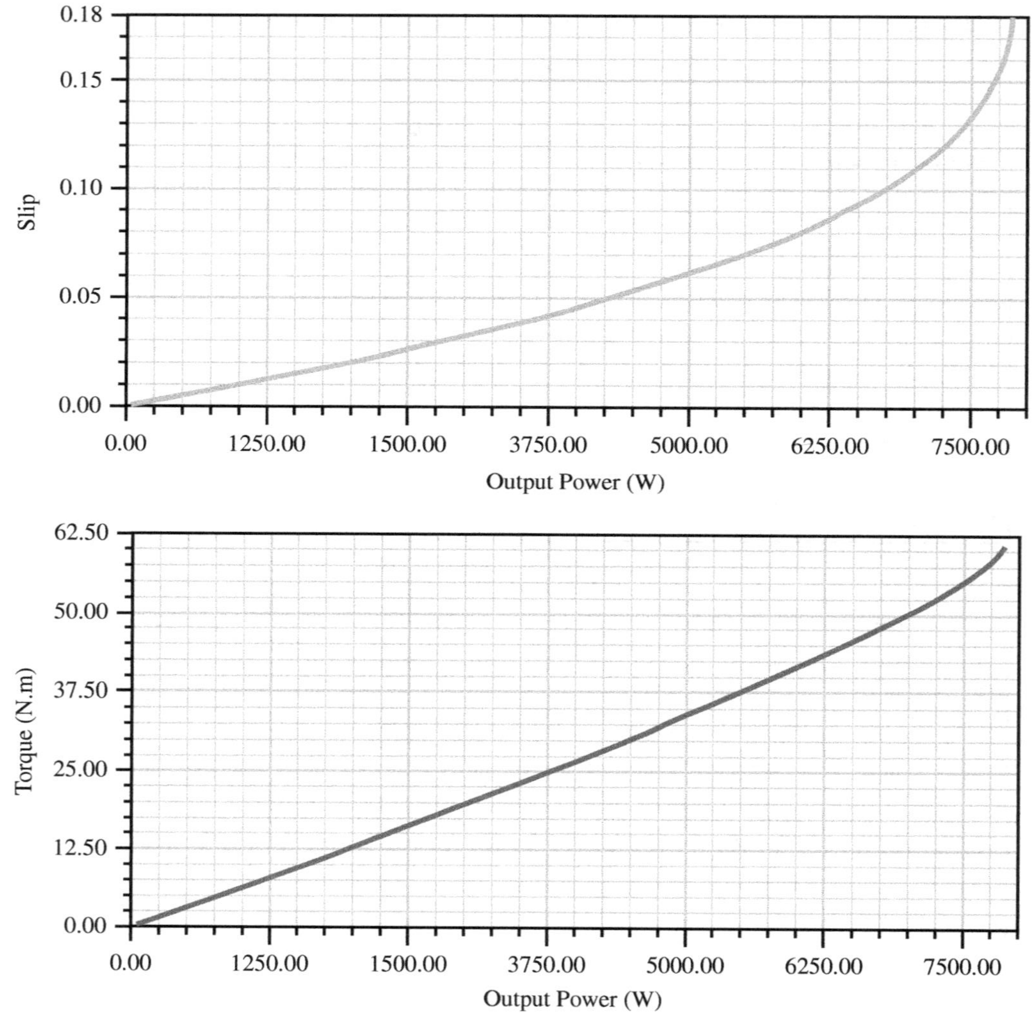

Figure 11.39 Slip vs. output power and torque vs. output power characteristics of the three-phase induction motor in RMxprt.

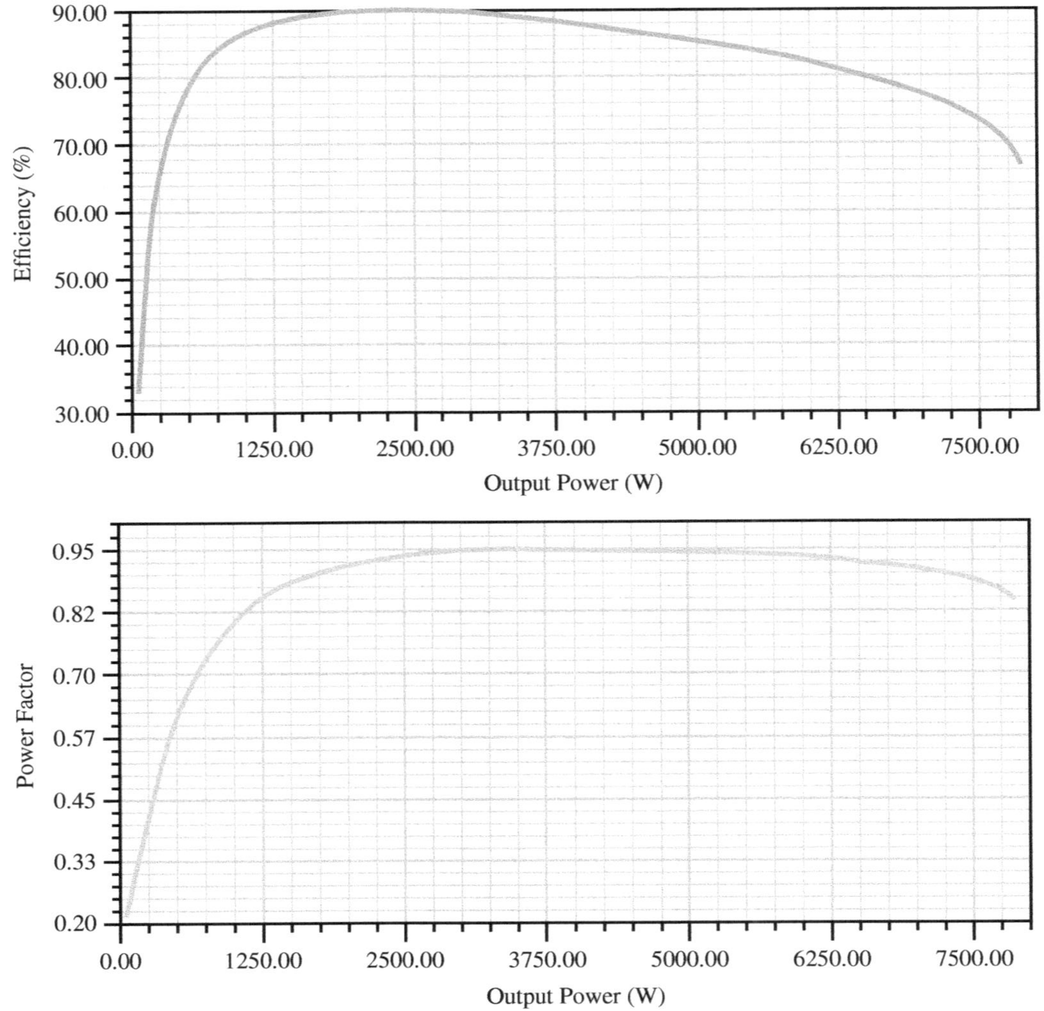

Figure 11.40 Efficiency vs. output power and power factor vs. output power characteristics of the three-phase induction motor in RMxprt.

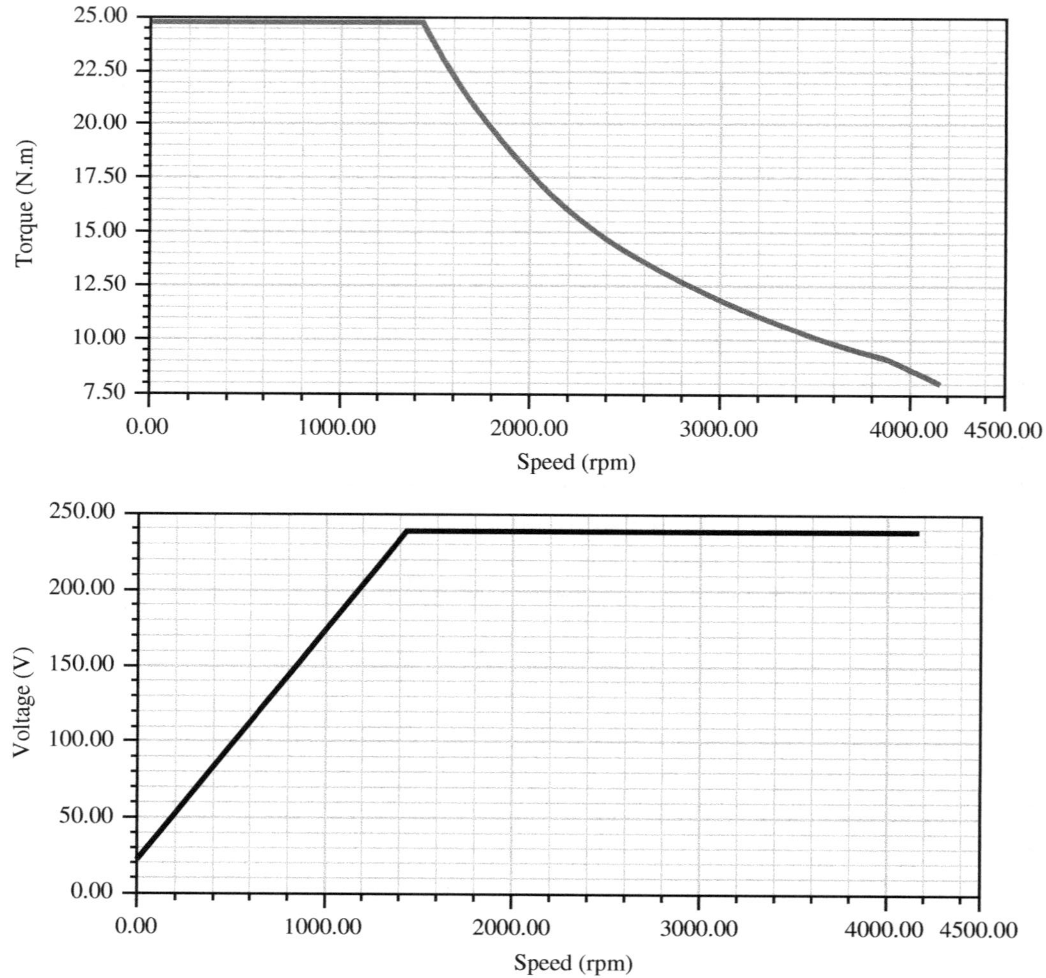

Figure 11.41 Torque vs. speed and voltage vs. speed characteristics in the field weakening region of the three-phase induction motor in RMxprt.

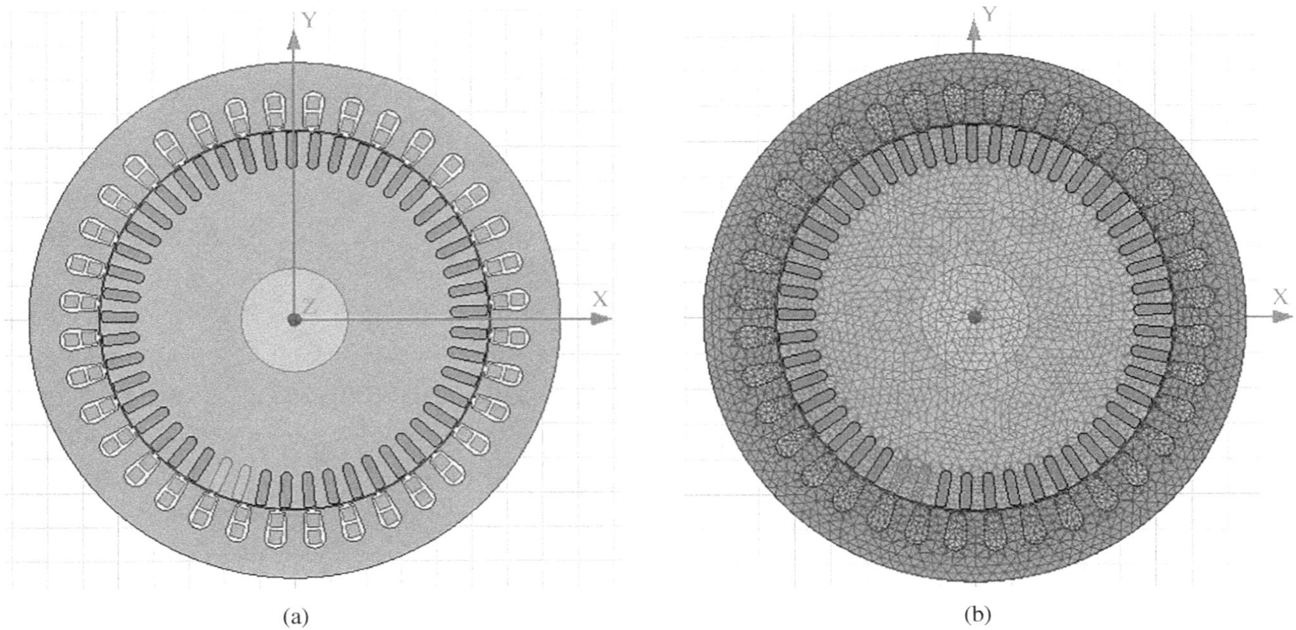

Figure 11.42 Maxwell Results: (a) Maxwell 2D model Front view, (b) Mesh plot of three-phase induction motor.

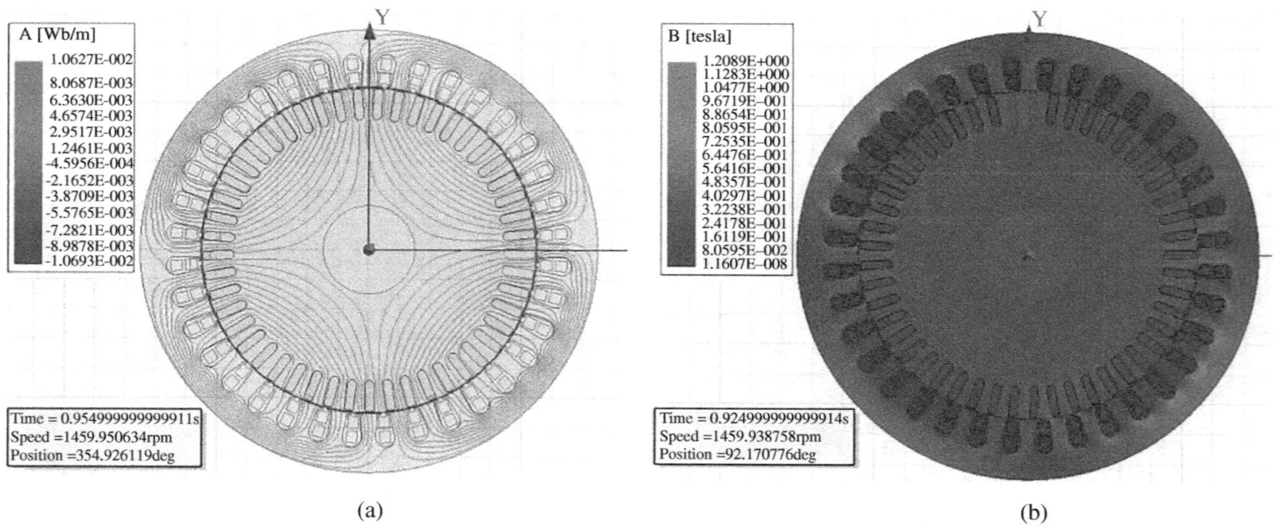

Figure 11.43 Maxwell Results: (a) Flux-line distribution, (b) Flux-density plot.

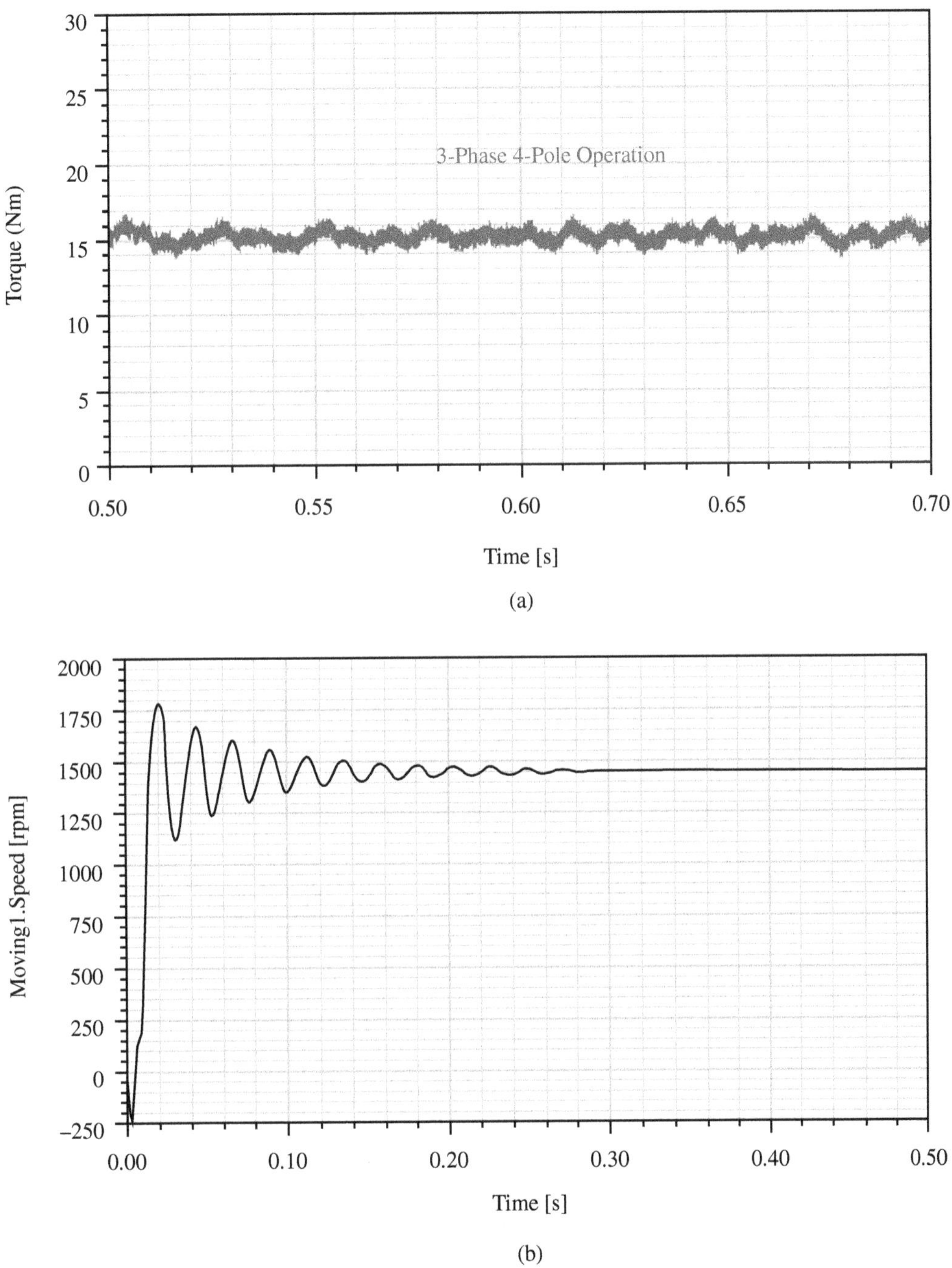

Figure 11.44 Maxwell Results: (a) Torque response, (b) Speed response of three-phase induction motor.

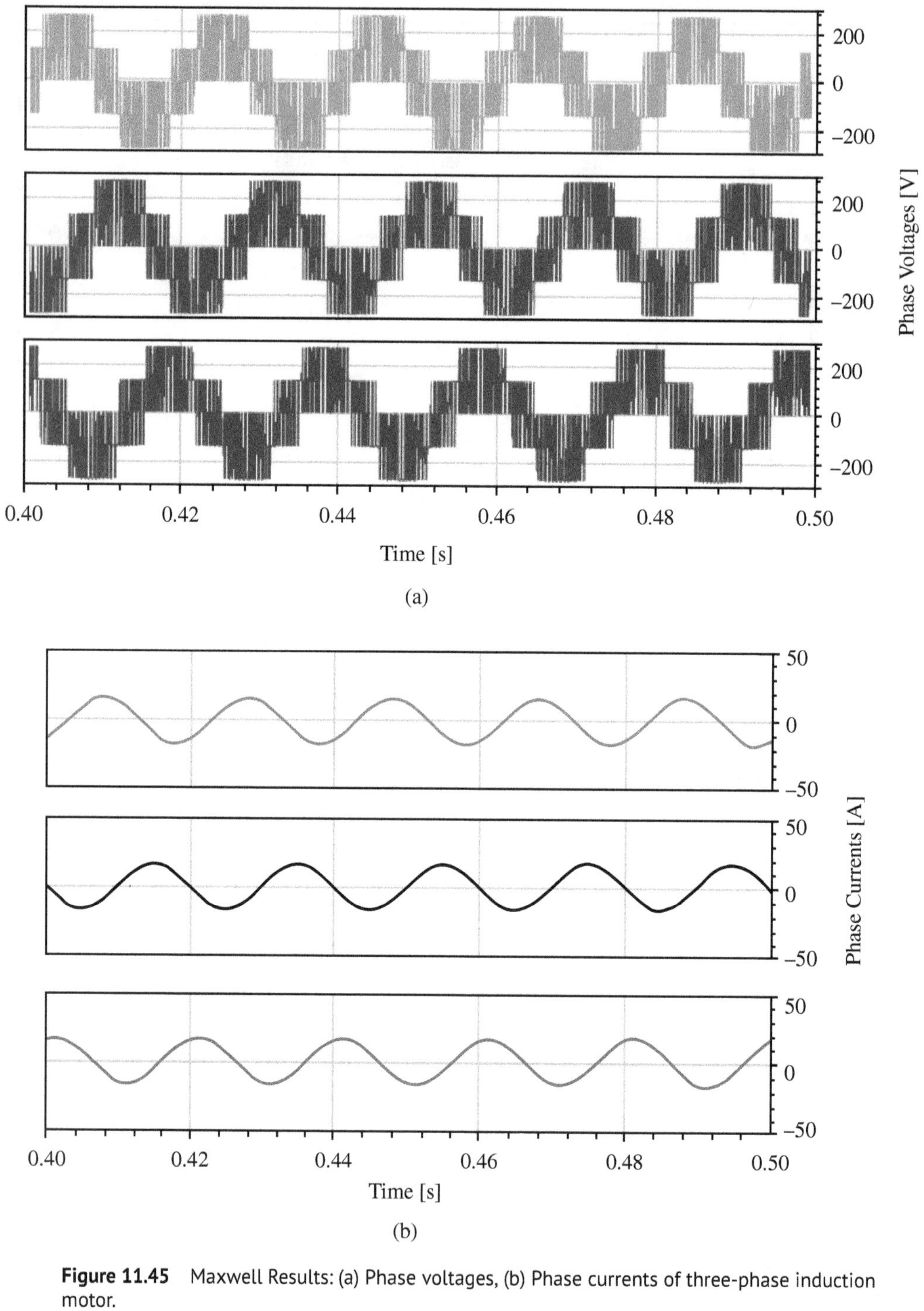

Figure 11.45 Maxwell Results: (a) Phase voltages, (b) Phase currents of three-phase induction motor.

11.6 Summary

In this chapter, the FEM analysis and Ansys Maxwell design of electrical machines are explained. The method of solving EM analysis wrt analytical as well as numerical methods are presented in detail. In the analytical techniques, the well-known separation of variables method is discussed for the heat equation. The numerical techniques for solving elliptic, hyperbolic, PPDEs are elaborated, where the finite difference and FEMs are discussed. The steps involved in the FEM, i.e. the discretization of finite elements, finite-element equations, finite-elements assembling and resulting equations of solution region are presented in detail. After that, the basic description and associated differences of 2D and 3D FEM simulations are discussed. The integral and differential form of Maxwell equations wrt Ampere, Faraday, Gauss, Lorentz laws are given. The FEM adaptive meshing and variation principles are also discussed. The analysis and implementation of FEM machine models wrt RMxprt, Maxwell and Simplorer tools of the Ansys Maxwell are presented. The three-phase induction motor is implemented in Ansys Maxwell 2D software and the respective results wrt torque, speed, magnetic flux density, flux lines, and other results are discussed for easy understanding.

References

1 Wikipedia (2020). Finite element method. https://en.wikipedia.org/wiki/Finite_element_method (accessed 30 September 2020).

2 Bargallo, R. (2006). A course manual of finite element methods for electrical engineering. Polytechnic University of Catalonia.

3 Bathe, K.-J. (2016). *Finite Element Procedures*, 2e. Watertown, MA: Massachusetts Institute of Technology.

4 Ansys (2020). Ansys RMxprt Features. https://www.ansys.com/products/electronics/ansys-rmxprt/rmxprt-features#1 (accessed 30 September 2020).

5 ANSYS Inc (2014). User's guide of 14.1 Maxwell 2D.

6 ANSYS Inc (2014). User's guide of 14.1 Maxwell 3D.

7 Sadiku, M.N.O. (2000). *Numerical Techniques in Electromagnetics*, 2e. Boca Raton, FL: CRC press.

8 Wikipedia (2020). Maxwell's equations. https://en.wikipedia.org/wiki/Maxwell%27s_equations (accessed 30 September 2020).

9 Wikipedia (2020). Partial differential equation. https://en.wikipedia.org/wiki/Partial_differential_equation (accessed 30 September 2020).

10 Bianchi, N. (2005). *Electrical Machine Analysis Using Finite Elements*. Boca Raton, FL: CRC press, Taylor & Francis.

11 Silvester, P.P. and Ferrari, R.L. (1996). *Finite Elements for Electrical Engineers*. Cambridge University Press.

12 Salon, S.J. (1995). *Finite Element Analysis of Electrical Machines*, The Springer International Series in Engineering and Computer Science. US: Springer.

13 Sadiku, M.N.O. and Nelatury, S.R. (2015). *Analytical Techniques in Electromagnetics*, 1e. Boca Raton: CRC press.

14 Reddy, B.P. (2019). Analysis and Performance Improvement of Pole Phase Modulated Induction Motor Drives for Traction Applications. PhD thesis. Indian Institute of Technology.

Index

Note: Page numbers in italics and bold to figures and tables respectively.

Electrical Machine Fundamentals with Numerical Simulation using MATLAB/SIMULINK, First Edition.
Atif Iqbal, Shaikh Moinoddin, and Bhimireddy Prathap Reddy.

Companion website: www.wiley.com/go/iqbal/electricalmachine

c

d

n

o

p

www.ingramcontent.com/pod-product-compliance
Lightning Source LLC
LaVergne TN
LVHW062202200926
848484LV00006B/1298

* 9 7 8 1 1 1 9 6 8 2 6 3 9 *